T0141879

Advances in Intelligent Systems and Computing

Volume 533

Series editor

Janusz Kacprzyk, Polish Academy of Sciences, Warsaw, Poland
e-mail: kacprzyk@ibspan.waw.pl

About this Series

The series "Advances in Intelligent Systems and Computing" contains publications on theory, applications, and design methods of Intelligent Systems and Intelligent Computing. Virtually all disciplines such as engineering, natural sciences, computer and information science, ICT, economics, business, e-commerce, environment, healthcare, life science are covered. The list of topics spans all the areas of modern intelligent systems and computing.

The publications within "Advances in Intelligent Systems and Computing" are primarily textbooks and proceedings of important conferences, symposia and congresses. They cover significant recent developments in the field, both of a foundational and applicable character. An important characteristic feature of the series is the short publication time and world-wide distribution. This permits a rapid and broad dissemination of research results.

More information about this series at http://www.springer.com/series/11156

Aboul Ella Hassanien · Khaled Shaalan
Tarek Gaber · Ahmad Taher Azar
M.F. Tolba
Editors

Proceedings of the International Conference on Advanced Intelligent Systems and Informatics 2016

 Springer

Editors
Aboul Ella Hassanien
Faculty of Computers and Information
Cairo University
Giza
Egypt

Khaled Shaalan
Dubai International Academic City
The British University in Dubai
Dubai
United Arab Emirates

Tarek Gaber
Faculty of Computers and Information
 Sciences, Department of Computer
 Science
Suez Canal University
Ismailia
Egypt

Ahmad Taher Azar
Ahmed Orabi Square
Menouf
Egypt

M.F. Tolba
Faculty of Computer and Information
 Science
Ain Shams University
Cairo
Egypt

ISSN 2194-5357   ISSN 2194-5365 (electronic)
Advances in Intelligent Systems and Computing
ISBN 978-3-319-48307-8   ISBN 978-3-319-48308-5   (eBook)
DOI 10.1007/978-3-319-48308-5

Library of Congress Control Number: 2016954538

Printed on acid-free paper

This Springer imprint is published by Springer Nature
The registered company is Springer International Publishing AG
The registered company address is: Gewerbestrasse 11, 6330 Cham, Switzerland

Preface

The 2nd International Conference on Advanced Intelligent Systems and Informatics (AISI2016), which took place in Cairo, Egypt, during October 24–26, 2016, is an international interdisciplinary conference covering research and development in the field of informatics and intelligent systems. The 2nd edition of AISI2016 is organized by the Scientific Research Group in Egypt (SRGE) and technically sponsored by the IEEE Computational Intelligence Society (Egypt chapter) and the IEEE Robotics and Automation Society (Egypt Chapter). AISI is organized to provide an international forum that brings together those who are actively involved in the areas of interest, to report on up-to-the-minute innovations and developments, to summarize the state of the art, to exchange ideas and advances in all aspects of informatics and intelligent systems, technologies and applications. The conference has four major tracks:

- Intelligent Language Processing
- Intelligent Systems
- Intelligent Robotics Systems
- Informatics

We received nearly 168 papers submissions for all research tracks, Informatics track has 28 papers out from 61 papers; Intelligent Language Processing track has 19 out of 30 papers; Intelligent systems track has 34 papers out of 51; and the Intelligent Robotics Systems has 3 papers out of 12, all submitted papers came from 28 countries, of which 98 have been accepted. All submissions were reviewed by five reviewers in average, with no distinction between papers submitted for all conference tracks. We are convinced that the quality and diversity of the topics covered will satisfy both the attendees and the readers of these conference proceedings.

We express our sincere thanks to the plenary speakers, special session chairs, workshop chairs and international program committee members for helping us to formulate a rich technical program. We would like to extend our sincere appreciation for the outstanding work contributed over many months by the Organizing Committee: Local Organization Chair, and Publicity Chair. We also wish to express our appreciation to the SRGE members for their assistance. We would like to emphasize that the success of AISI 2016 would not have been possible without the support of many committed volunteers who generously contributed their time, expertise, and resources toward making the conference an unqualified success.

Finally, thanks to Springer team for their supporting in all stages of the production of the proceedings.

We hope that you will enjoy the conference program.

Aboul Ella Hassanien
AISI 2016 General chair
Khaled Shaalan
Ahmad Taher Azar
Aboul Ella Hassanien
Tarek Gaber
AISI 2016 Track Chairs

Organization

Honarary Chair

Mohamed F. Tolba   Egypt

General Chair

Aboul Ella Hassanien   Egypt

International Advisory Board

Adel Alimi	Tunisia
Muhammad Sarfraz	Kuwait
C.Y. Suen	Canada
Mohamed Salim Bouhlel	Tunisia
Alaa Khamis	Egypt
Vaclav Snasel	Czech Republic
Beate Meffert	Germany
Dominik Slezak	Poland
Janusz Kacprzyk	Poland
Tai-hoon Kim	Korea
Qing Tan	Canada
Hiroshi Sakai	Japan
Jeng-Shyang Pan	Taiwan

Track Chairs

Intelligent Language Processing, Professor Khaled Shaalan
Intelligent Robotic System, Dr. Ahmad Taher Azar
Intelligent Systems, Professor Aboul Ella Hassanien
Informatics, Dr. Tarek Gaber, tmgaber@gmail.com

Publicity Chairs

Mohamed Abdel Fatah	Egypt
Mohammed Abdel-Megeed	Egypt
Habib Kammoun	Tunisia
Dace Ratniece	Latvia
Mohamed Tahoun	Egypt
Alaa Tharwat	Egypt
Pei-Wei Tsai	China

International Program Committee

Abdelaziz Abdelhamid	Ain Shams University, Egypt
Abdelfettah Hamdani	V University, Morocco
Abdelhak Lekhouaja Mohammed	First University, Morocco
Abdelmonaime Iachkar	ENSA USMBA, Morocco
Abdelrahman Ahmed	University of Sevilla, Spain
Ahmad Abd-alaziz	The British University in Egypt
Ahmed Anter	Benisuef University, Egypt
Ahmed Guessoum	Houari Boumediene University, Algeria
Ahmad Hossny	University of Adelaide, Australia
Ahmed Nabhan	Fayoum University, Egypt
Ahmed Sharf Eidin	Helwan University, Egypt
Ahmed Sallam	Suez Canal University, Egypt
Aladdin Ayesh	De Montfort University, UK
Almoataz B. Al-Said	Cairo University, Egypt
Amany Elshazli	Helwan University, Egypt
Amir Atiya	Cairo University, Egypt
Aya Al-Zoghby	Mansoura University, Egypt
Azza Abdel Monem	Ain Shams University, Egypt
Azzeddine Mazroui Mohammed	First University, Morocco
Azzedine Lazrek	Cadi Ayyad University, Marrakech, Morocco
Doaa Samy	Cairo University
Eman Nasr	Independent Researcher, Egypt

Farid Meziane	University of Salford, UK
Faten Khalfallah	Faculty of Economic Sciences and Management, Tunisia
Hanaa Bayomi	Cairo University, Egypt
Hanady Ahmed	Qatar University, Qatar
Hitham M. Abo Bakr	Zagazig University, Egypt
Ibrahim Bounhas	La Manouba University, Tunisia
Ibrahim Moawad	Ain Shams University, Egypt
Ibrahima Faye	Universiti Teknologi Petronas, Malaysia
Imed Zitouni	Mircosoft, USA
Ines Boujelben	Miracl-Sfax University, Tunisia
Ismaïl Biskri	Université du Québec à Trois Rivières, Canada
Kais Haddar	University of Sfax, Tunisia
Karim Bouzoubaa	Mohammed V University, Morocco
Kevin Daimi	University of Detroit Mercy, USA
Lahsen Abouenour	Mohammed V University, Morocco
Mai Oudah	Masdar Institute of Science and Technology, UAE
Majdi Sawalha	The University of Jordan, Jordan
Mervat Gheith	Cairo University, Egypt
Mahmoud El-Gayyar	Suez Canal University, Egypt
Mohamed Meselhy Eltoukhy	Suez Canal University, Egypt
Mohamed K. Hussein	Suez Canal University, Egypt
Mohamed Abd Elfattah	Mansoura University, Egypt
Mohamed Dahab	King Abdulaziz University, Saudi Arabia
Mohamed El Hannach	University of Fez, Morocco
Mohammad Abushariah	The University of Jordan
Mohammed Attia	Dublin University, Ireland
Mohammed Kayed	Beni-Suef University, Egypt
Mohsen Rashwan	Cairo University, Egypt
Mostafa Al-Emran	Al Buraimi University College, Oman
Mustafa Abdul Salam	Higher Technological Institute, Egypt
Mostafa Aref	Ain Shams University, Egypt
Mostafa Ezzat	Cairo University, Egypt
Mostafa Saleh	King Abdulaziz University, Saudi Arabia
Muhammad Shoaib	The British University in Dubai, UAE
Mustafa Abdul Salam	Modern Academy for Science and Technology, Egypt
Mustafa Jarrar	Birzeit University, Palestine
Nadim Obeid	The University of Jordan
Nagwa Badr	Ain Shams University, Egypt
Neamat El Gayar	Cairo University, Egypt
Osama Emam	IBM, Egypt
Omnia El-Barbary	Shaqraa University, Saudi Arabia
Paolo Rosso	Universitat Politècnica de València, Spain

Rana Aref	Cairo University, Egypt
Rania Al-Sabbagh	Ain shams University, Egypt
Rasha Aboul-Yazeed	Faculty of Engineering, Egypt
Salwa Hamada	Taibah University, Saudi Arabic
Sameh Alansary	Bibliotheca Alexandrina, Egypt
Samir AbdelRahman	Cairo University, Egypt
Sanjeera Siddiqui	The British University in Dubai, UAE
Santosh K. Ray	Khawarizmi International College, UAE
Sattar Izwaini	American University of Sharjah, UAE
Seham Elkareh	Alexandria University, Egypt
Sherief Abdallah	The British University in Dubai, UAE
Sherif Mahdy Abdou	Cairo University, Egypt
Sherif Khattab	Cairo University, Egypt
Sinan Al Shikh	University of Salford, UK
Slim Mesfar	RIADI, University of Manouba, Tunisia
Tarek Elghazaly	Cairo University, Egypt
Yassine Benajiba	Symanto Research, USA
Violetta Cavalli-Sforza	Al Akhawayn University, Morocco
Wajdi Zaghouani	CMU, Qatar
Yasser Hifny	Helwan University, Egypt
Yousfi Abdellah	Mohamed V University, Morocco
Sardaar Jaf	Durham University, UK
Zeinab Ibrahim	CMU, Qatar
Bappaditya Jana	Techno India University, India
Samir Elmogy	Egypt
Aarti Singh	India
Nickolas S. Sapidis	University of Western Macedonia, Greece
Ahmad Karawash	University of Quebec at Chicoutimi, Canada
Philip Moore	University College Falmouth, UK
Hannah Inbarani	India
Nizar Banu	India
Rami Belkaroui	Tunis
B.H. Shekar	India
Valentina Emilia Balas	Romania
Nizar Rokbani	Tunisia
Ram Bilas Pachori	India
Nilanjan Dey	Techno India College of Technology, India
Irene Mavrommati	Hellenic Open University, Greece
Minyar Sassi-Hidri	National Engineering School of Tunis (ENIT), Tunisia
Jin Xu	Behavior Matrix LL, USA
Anjali Awasthi	Concordia University, Canada
Evgenia Theodotou	University of East London, UK
Pavel Kromer	Czech Republic
Jan Platos	Czech Republic

Kiss Eva-Maria	Germany
Marius Balas	Romania
Muaz Niazi	Pakistan
Camelia Pintea	Romania
Mario Pavone	Italy
Rabie A. Ramadan	Egypt
N. Popescu-Bodorin	Romania
Abdelhameed Ibrahim	Egypt
Ahmed Anter	Egypt
Ahmed Elhayek	Germany
Amira S. Ashour	KSA
Edgard Marx	Germany
Eid Emary	Egypt
Fatma Helmy	Egypt
Hany Alnashar	Egypt
Hossam Zawbaa	Romania
Maryam Hazman	Egypt
Minu Kesheri	India
Mohamed Shrif	Germany
Mohammed Abdel-Megeed	Egypt
Mona Solyman	Egypt
Mona Ali	Egypt
Muhammad Saleem	Germany
Nabiha Azizi	Algeria
Namshik Han	UK
Noreen Kausar	KSA
Noura Semary	Egypt
Rania Hodhod	USA
Reham Ahmed	Egypt
Sara Abdelkader	Canada
Sayan Chakraborty	India
Shoji Tominaga	Japan
Siva Ganesh Malla	India
Soumya Banerjee	India
Sourav Samanta	India
Suvojit Acharjee	India
Swarna Kanchan	India
Takahiko Horiuchi	Japan
Tarek Gaber	Suez Canal University, Egypt
Mohamed Mostafa Eltaweel	Egypt
Tommaso Soru	Germany
Quan Zhu	Fujian University of Technology, China
Jing Zhang	Fujian University of Technology, China
Zhiming Cai	Fujian University of Technology, China

Shijian Liu   Fujian University of Technology, China
Yuan-Hsun Liao   Fujian University of Technology, China
Tien-Wen Sung   Fujian University of Technology, China
Chyun-Chyi Chen   Fujian University of Technology, China
Sheng-Hui Meng   Fujian University of Technology, China
Chia-Jung Lee   Fujian University of Technology, China
Jian-Pan Li   National Chung-Shan Institute of Science
                          and Technology, Taiwan
Chih-Wei Hsu   National Cheng Kung University, Taiwan
Jia-Shing Shih   National Cheng Kung University, Taiwan
Chih-Lun Chou   Ming Chuan University, Taiwan
Gwo-Jiun Horng   Southern Taiwan University of Science
                          and Technology, Taiwan
Yi-Chung Chen   Feng Chia University, Taiwan
Fan-Yi Hsiao   Fortune University, Taiwan
Lyuchao Liao   Fujian University of Technology, China
Xingsi Xue   Fujian University of Technology, China
Sardar Jaf   Durham University, UK

Workshop/Special Session Chair

Mohamed Abdel Aziz   Faculty of Science, Zagazig University, Egypt
                          abd_el_aziz_m@yahoo.com

Local Arrangement Committee

Mohamed Tahoun   Suez Canal University (Chair)
                          tahoun@ci.suez.edu.eg
Mohamed Mostafa Eltaweel Egypt

Plenary Speakers

The Story and Future of WiFi

Prof. Hatim Zaghloul

Calgary University, Canada

Abstract. This keynote will focus on the following points: The story behind its invention and inventors and some vision into the future and how a WiFi variant could be the basis of 6G.

Biography

Dr. Hatim Zaghloul is recognized as a visionary leader in the Canadian hi-tech community. In 1992, Dr. Zaghloul co-founded Wi-LAN Inc., and under his direction, the technology he co-invented with *Dr. Michel Fattouche* became the heart of many wireless communications standards. *Dr. Zaghloul* holds a B.Sc. in Electrical Engineering from Cairo University as well as a M.Sc. and a Ph.D. in Physics from the University of Calgary. *Dr. Zaghloul* has published extensively in technical journals and holds over ten patents. *Dr. Zaghloul* envisioned in 1991 how WOFDM technology can become the main driver for all wideband wireless communications applications including wireless local area networking, wide area networking and cellular telephony. *Dr. Zaghloul* has served as director on a variety of boards, specifically in the high-technology arena including *Wi-LAN Inc.*(www.wi-lan.com) Cell-Loc Inc (www.cell-loc.com); *DTS Inc.* (www.dtsx.com); *Imaging Dynamics Corp.*(www.imagingdynamics.com); *Wireless Inc.* (www.wire-less-inc.com); *QCC Technologies Inc.* (www.qcc.ca), *NTG Clarity Networks Inc.* (www.ntgclarity.com), *Solutrea Corp.*as well as other private entities. Dr. Zaghloul is currently the CEO and Chairman of *IPL Media Inc. for Technology*. Dr. Zaghloul co-founded *Wi-LAN Inc., Cell-Loc Inc., NTG Clarity, QCC Technologies Inc., Wireless Inc., IPL Media Inc., IPL Media Inc. for Technology, Innovative Products for Life Inc., the Solutrea* group of companies. Dr. Zaghloul has been awarded many awards: Named one of 10 Great Canadians by *MacLeans'* Magazine in July 2000; Calgarian of the Year 2000 with *Dr. Fattouche* by magazine; Entrepreneur of the Year finalist for 2 consecutive years; First Broadband Wireless Hall of Fame inductee; 2000 Pinnacle Award by Fraser Milner law firm. *Dr. Zaghloul* led *Wi-LAN Inc.* from its inception in 1993 till 2008.

A Look at the Working World of Tomorrow's Digital Society

Prof. Thea Radüntz
Federal Institute for Occupational Safety and Health, Germany

Abstract. Advanced information and communication technology has fundamentally changed the working environment. Complex and highly automated systems impose high demands on employees with respect to cognitive capacity and the ability to cope with inappropriate levels of workload. Although the main goal of automation is to simplify work, employees are complaining more frequently not only about high mental workload and stress but also about monotonous tasks. The long-term negative consequences of inappropriate workload on employee health constitute a serious problem for a digitalized society. Hence, this keynote will focus on the following points:

- Brief overview of changes in the working environment
- Opportunities and risks in the context of work and health
- Research on mental health and cognitive capacity at the Federal Institute of Occupational Safety and Health in Berlin

Biography

Thea Radüntz studied computer science with a minor in psychology. The emphasis of her education was on biosignal processing and pattern recognition. She holds a Dr. rer. nat. and a Master of Science degree from the Humboldt-Universidad zu Berlin. During her studies she worked as a research assistant at the Centre for Human-Machine Systems (ZMMS) on a project financed by Volkswagen and focused on biosignal processing for driver fatigue detection. Following an enriching time employed at the European Commission in Luxembourg, she came back to Berlin to work for an IT consulting company, where she served as project management assistant on a contract with the Bundesdruckerei (Government Printing Office). After working at the Berlin University of Technology on 3D TOF face recognition and the implementation of image processing algorithms on FPGA, she returned to the research field of biosignal processing and neuro-physiological research. Since 2011 she has been working in that field at the department of Mental Health and Cognitive Capacity of the Federal Institute of Occupational Safety and Health in Berlin.

Tutorial

Tutorial

Pattern Recognition in Intelligent Systems

Prof. Ibrahima Faye

Universiti Teknologi Petronas, Malaysia

Abstract. Pattern recognition has been playing a fundamental role in many intelligent systems. For most cases, the primary stage consists in differentiating various possible patterns of an input data. The efficacy of this primary stage highly impacts the final intelligent system. The tutorial will present various examples of implementation of pattern recognition techniques in different contexts which include industrial applications (e.g. product defect inspection), biomedical applications (e.g. computer aided diagnosis systems, protein classification), psychology (e.g. emotion recognition) etc. The tutorial builds upon a short course that has been offered from 2014. It will review fundamental steps of pattern recognition: feature extraction, dimension reduction, classification and clustering. It will highlight some recent developments, some exciting possible future applications and related challenges.

Biography

Ibrahima Faye is Associate Professor at Universiti Teknologi Petronas. His BSc, MSc and PhD degrees in Mathematics are from the University of Toulouse, France, while his MS degree in Engineering of Medical and Biotechnological data is from Ecole Centrale Paris, France. His research interests include engineering mathematics, signal and image processing, pattern recognition, and dynamical systems. He is currently leading the intelligent Medical imaging group in the Centre for Intelligent Signal and Imaging Research (CISIR) a national High Centre of Excellence (HiCoE) in Malaysia.

Contents

Intelligent Language Processing Track

Further Investigations for Documents Information Retrieval Based on DWT

Mohamed Yehia Dahab$^{(\boxtimes)}$, Mahmoud Kamel, and Sara Alnofaie

Department of Computing Science,
Faculty of Computing and Information Technology,
King Abdulaziz University, Jeddah 21589, Saudi Arabia
{mdahab,miali,salnefaie}@kau.edu.sa
http://www.kau.edu.sa/Home.aspx

Abstract. In most of the classical information retrieval models, documents are represented as bag-of words which takes into account the term frequencies (**tf**) and inverse document frequencies (**idf**) while they ignore the term proximity. Recently, term proximity among query terms has been observed to be beneficial for improving performance of document retrieval. Several applications of the retrieval have implemented tools to determine term proximity at the query formulation level. They rank documents based on the relative positions of the query terms within the documents. They must store all proximity data in the index, leading to a large index, which slows the search. Recently, many models use term signal representation to represent a query term, the query is transformed from the time domain to the frequency domain using transformation techniques such as wavelet. Discrete Wavelet Transform (**DWT**) uses multiple resolutions technique by which different frequencies are analyzed with different resolutions. The advantage of the **DWT** is to consider the spatial information of the query terms within the document rather than using only the count of terms. In this paper, in order to improve ranking score as well as improve the run-time efficiency to resolve the query, and maintain a reasonable space for the index, three different types of spectral analysis based on semantic segmentation are carried out namely: sentence-based segmentation, paragraph-based segmentation and fixed length segmentation; and also different term weighting is performed according to term position.

Keywords: Information retrieval · Term proximity · Discrete wavelet transform · Term signal

1 Introduction

Most of the classical retrieval models are based on the bag-of-words representation of documents e.g., [1–4], that take into account the **tf** and **idf** but ignore the term proximity.

© Springer International Publishing AG 2017
A.E. Hassanien et al. (eds.), *Proceedings of the International Conference on Advanced Intelligent Systems and Informatics 2016*, Advances in Intelligent Systems and Computing 533, DOI 10.1007/978-3-319-48308-5_1

Much recent work has shown that the proximity between terms is useful for improving document retrieval performance. Term proximity retrieval methods delve deeper into each document. Rather than just scratching the surface of each document by using term frequency, theses methods use the term positional information to compute the document score. Proximity represents the closeness of the query terms appearing in a document. Intuitively, document d_1 ranked higher than documents d_2 when the terms of the query occur closer together in d_1 while the query-terms appear far apart in d_2. Proximity can be seen as a kind of indirect measure of term dependence [5].

Many document retrieval techniques use term proximity information in computing document rank (e.g. [6–11]), but they are lack in time complexity as well as space complexity because they should make many comparisons and they should store large amounts of data to make these comparisons. Recently, many models use term signal representation to represent a query term and the query is transformed from time domain to frequency domain using transformation techniques such as Fourier transform [12–15] Cosine transform [16], Wavelet transform [17,18] etc. They observe the frequency spectrum of the terms through the document and determine the ranking score for each document in efficient run-time and space.

This paper presents an information retrieval framework based on **DWT**. This framework achieves the speed of the vector space methods with the benefits of the proximity methods to provide a high quality information retrieval system. The retrieval is performed by the following: (a) locating the appearances of the query terms in each document, (b) transforming the document information into the time scale, and (c) analyzing the time scale with the help of the **DWT**. By using the **DWT** we are able to combine the current vector space methods with the proximity methods.

Information retrieval systems based on **DWT** has many advantages over classical information retrieval systems but using fixed document segmentation has many disadvantages because of neglecting document semantic representation. The main research hypothesis of this paper is the necessity of using spectral analysis based on different types of semantic segmentation, namely sentence-based segmentation, paragraph-based segmentation and fixed length segmentation, to improve document score.

This paper will proceed as follows, Sect. 2 presents the related background, Sect. 3 introduces a further investigation to **DWT** method and explains how to compute the score, Sect. 4 demonstrates experiments and results using the the Text REtrieval Conference (**TREC**) collection, and finally the conclusion is shown in Sect. 5.

2   Background

In this section we survey some general aspects that are important for understanding the proposed work.

2.1 Term Signal

The notion of term signal was introduced by [14] before being reintroduced by [19] as a word signal. It is term vector representation, which defines the term number of appearance in particular bins within a document. Term signal shows how a term is spread over the document. If a term appearing in a document can be thought of as a signal, then this signal can be transformed and compared with other signals. Term-signal is an information value represented in time domain since it describes when a term occurs and its number of occurrences in certain location. Applying this scheme, for each document in the collection, a set of term signals is computed as a sequence of numbers, which display the term appearance in a specific bin of the document. In document d, the representation of term t signal is:

$$s(t, d) = \tilde{f}_{t,d} = [f_{t,1,d}, f_{t,2,d}, ..., f_{t,B,d}], \tag{1}$$

where:
$f_{t,b,d}$ is frequency of term t in segment b of document d for $1 \le b \le B$.

2.2 Weighting Scheme

Zobel and Moffat [20] provided a more general structure of the form AB-CDE-FGH, to allow for the many different scoring functions that had appeared. Their study compared 720 different weighting schemes based on the TREC document set. Their results showed that the BD-ACI-BCA method was the overall best performer. The computed segment components are weighted using a BD-ACI-BCA weighting scheme as follows:

$$\omega_{t,b,d} = \frac{1 + log_e f_{t,b,d}}{(1 - slp) + \frac{slp.W_d}{ave_{d \in D} W_d}}, \tag{2}$$

where
slp: is the slope parameter (usually equals to 0.7),
W_d: is the document vector norm,
$ave_{d \in D} W_d$: is the average document vector norm.

The term signal will be correspond to the weighted term signal as follows:

$$\tilde{\omega}_{t,d} = [\omega_{t,1,d}, \omega_{t,2,d}, ..., \omega_{t,B,d}]. \tag{3}$$

2.3 Applying Wavelet Transform

The wavelet transform has had a long journey to get to its current standing and has been used in many areas of engineering and science because of its ability to decompose natural signals into scaled and shifted versions of itself [21]. Wavelets are defined by the wavelet function $\psi(t)$ and scaling function $\varphi(t)$ in the time domain. The wavelet function is described as $\psi \in \mathbf{L}^2\mathbb{R}$ (where $\mathbf{L}^2\mathbb{R}$ is the set of functions $f(t)$ which satisfy $\int \mid f(t) \mid^2 dt < \infty$) with a zero average and norm of

1. A wavelet can be scaled and translated by adjusting the parameters s and u, respectively.

$$\psi_{u,s}(t) = \frac{1}{\sqrt{s}}\psi\left(\frac{t-u}{s}\right). \tag{4}$$

The scaling factor keeps the norm equal to one for all s and u. The wavelet transform of $f \in \mathbf{L}^2\mathbb{R}$ at time u and scale s is:

$$Wf(u,s) = \int_{-\infty}^{+\infty} f(t)\frac{1}{\sqrt{s}}\psi^*\left(\frac{t-u}{s}\right)dt, \tag{5}$$

where s and $u \in \mathbf{Z}$ and ψ^* is the complex conjugate of ψ [17].

The scaling function is described as $\varphi_{u,s}(t) \in V_n$. The scaling function satisfies the following:

$$\{0\} \leftarrow \dots \subset V_{n+1} \subset V_n \subset V_{n-1}\dots \rightarrow \mathbf{L}^2,$$

where

the set of $\varphi_{u,s}(t)$ for all u is a basis of V_n $(s = 2^n)$,
$\bigcup_{n\in\mathbb{Z}} V_n = \mathbf{L}^2(\mathbb{R})$,
$W_n = V_n \cap \overline{V_{n+1}}$

The filter bank tree-structured algorithm can use to compute the **DWT**. The signal $f(t)$ can be transformed using the filter bank tree of **DWT**:

$$f \xrightarrow{DWT^1} A^1 + D^1$$
$$\xrightarrow{DWT^2} A^2 + D^2 + D^1$$
$$\xrightarrow{DWT^3} A^3 + D^3 + D^2 + D^1$$
$$\dots\dots$$
$$\xrightarrow{DWT^s} A^s + D^s + D^{s-1} + \dots + D^1 \tag{6}$$

where A^s correspond to the approximation sub-signal and D^s corresponds to a detail sub-signal in the sth level of transforms of **DWT** of the signal $f(t)$, respectively.

One of the simplest wavelet transforms is Haar transform [22]. The Haar transform is derived from the Haar matrix. It provides the different levels of resolution of a signal. At different resolutions, the positions of the terms are showed by the transformed signal. Every possible wavelet scaled and shifted version is take in Haar transform. Given a term signal $\tilde{f}_{t,d}$ and number of segments N, the wavelet components will be $H_N\tilde{f}_{t,d}$ where H_N is the Haar matrix. When $N = 4$ we have

$$H_4 = \frac{1}{2}\begin{bmatrix} 1 & 1 & 1 & 1 \\ 1 & 1 & -1 & -1 \\ \sqrt{2} & -\sqrt{2} & 0 & 0 \\ 0 & 0 & \sqrt{2} & -\sqrt{2} \end{bmatrix}$$

The complexity time of matrix multiplication that causes this transformation is $O(N^2)$ for signals of N elements. To enhance the time of this process, one may use the wavelets scaling function and the wavelet function.

3   A Further Investigation to DWT

Park in [15–17] assumes that documents have fixed length as well as they are unformatted. Using fixed number of bins in all dataset has many disadvantages:

1. It does not distinguish between long documents and short documents. Suppose using of fixed number of bins B such that $B = 2^i$ and $i \in \mathbb{N}$. For any document d, has a number of words $|d|$, the bin b contains number of words $|b| = \frac{|d|}{B}$. That means there are different levels of proximity for different documents. Number of bins in microblogs [23] should be very small in comparison to the number of bins in scientific theses.
2. It does not consider the length of a query. If all terms of a query are found in one bin of a document, that means the document is more related to the query than a document holding the same query terms scattered in more than one bin. To consider this issue, the length of the suggested bin should be $|b| \geq |Q|$.
3. It does not consider the semantic representation of documents. A paragraph, which is a unit of thought that is focused on a single topic, may represent semantically a single bin. Also, a sentence may represent semantically a single bin.
4. Also, it does not consider the weight of matched query terms in a document according to it terms' location. The matched query terms in the first part of a document should have a greater weight more than the matched query terms in the middle and the end in the retrieved documents.

The main hypothesis in this research is to consider the necessity of using different types of segmentation and term weighting beside **DWT** to improve document score. The algorithms and techniques will be described in the following subsections.

3.1   Document Segmentation

Document may be segmented by fixed length, sentences or paragraphs. If we use fixed length the minimum distance between any occurrence of term t_i and term t_j in a document D should be considered. Suppose the document contains C words, d is the maximum distance, S represents the number of sentences and P represents the number of paragraphs. The main problem is to construct the Haar matrix in these different situations. The following algorithms show how to construct the Haar matrix.

3.2   Term Weighting

We hypothesize that the significant terms always appear in the front of a document. Formatted document may be segmented by document length, sentences, topics/subtopics or paragraphs. While unformatted documents may be segmented by document length, sentences or paragraphs. The minimum distance between any occurrence of term t_i and term t_j in a document D should be considered.

Algorithm 1. Construct The matrix

procedure CONSTRUCTMATRIX(D)

    if *UsingSentences* **then**
        $S \leftarrow D.getnumberOfSentences()$
        $N \leftarrow GetMatrixDimension(S)$
        Matrix $\leftarrow$ *GenerateHaarMatrix(N)*
    end if
    if *UsingParagraphs* **then**
        $P \leftarrow D.getnumberOfParagraphs()$
        $N \leftarrow GetMatrixDimension(P)$
        Matrix $\leftarrow$ *GenerateHaarMatrix(N)*
    end if
    if *UsingMaxDistance* **then**
        $C \leftarrow D.getnumberOfWords()$
        $Y \leftarrow \dfrac{C}{d^2}$
        $N \leftarrow GetMatrixDimension(Y)$
        Matrix $\leftarrow$ *GenerateHaarMatrix(N)*
    end if
    return *Matrix*
end procedure

4 Experiments and Results

The **TREC** dataset is used to evaluate the proposed methods. Exactly, the Associated Press disk 2 and Wall Street Journal disk 2 (**AP2WSJ2**) has been chosen. The **AP2WSJ2** set contains more than 150,000 documents. The selected query set was from 51 to 200 (in TREC 1, 2, and 3). The investigated methods will be compared with the previous high precision methods. The experiments are performed based on a variety of the following segmentation methods:

– Fixed number of segments (8 bins).
– Sentence-based segmentation.

Algorithm 2. Get Matrix Dimension

procedure GETMATRIXDIMENSION (X)

                        ▷ X is the suggested number of segments
                      ▷ the procedure returns the nearest n such that $2^n > X$
    power $\leftarrow 2$
    while (*power* $<$ *MaxPower* and $X > 2^{power}$) **do**
                     ▷ *MaxPower* is set to 13 as default
        power $\leftarrow$ *power* $+ 1$
    end while
    return *power*
end procedure

Algorithm 3. Algorithm for generating Haar matrix

procedure GENERATEHAARMATRIX(N)   ▷ to generate $N \times N$ matrix
  for $i \leftarrow 1, N$ **do**
    $H[1][i] \leftarrow 1$
  end for
  for $i \leftarrow 1, N$ **do**
    $x = \dfrac{\log i}{\log 2}$
    $y = i - 2^x$
    for $j \leftarrow 1, \dfrac{N}{2^{x+1}}$ **do**
$$H[i+1][j + y \times \frac{N}{2^x}] = 2^{x/2}$$
$$H[i+1][j + y \times \frac{N}{2^x} + \frac{N}{2^{x+1}}] = -2^{x/2}$$
    end for
  end for
  $H \leftarrow \dfrac{H}{\sqrt{N}}$
  return H
end procedure

Table 1. Comparison of Mean Average Precision (MAP), and Average Precision at K (where $K = \{5, 10, 15, 20\}$) for Different Segmentation Methods and SSR

Segmentation methods	p@5	p@10	p@15	p@20	MAP
Fixed number of segments (8 bins)	0.469	0.439	0.421	0.406	0.2322
Sentence-based segmentation	0.431	0.423	0.402	0.381	0.224
Paragraph-based segmentation	0.421	0.411	0.388	0.367	0.201
Fixed number of terms in a segment (3 terms)	0.399	0.377	0.357	0.340	0.179
Fixed number of terms in a segment (10 terms)	0.405	0.392	0.369	0.348	0.183
Sentence-based segmentation with position weight	0.421	0.409	0.387	0.370	0.219
SSR	0.3718	0.3362	0.3078	0.2856	0.163
Window based bi-gram BM25 model	0.449	0.431	0.418	0.391	0.218

- Paragraph-based segmentation.
- Fixed number of terms in a segment ($d = 3$).
- Fixed number of terms in a segment ($d = 10$).
- Sentence-based segmentation with position weight.
  These segmentation methods will be compared against another important term proximity measure namely *shortest-substring retrieval* (**SSR**) [6] and *window based bi-gram* **BM25** *model* [24].

In the experiment, the TREC official evaluation measures have been used, namely the Mean Average Precision (MAP), and Average Precision at K (where $K = \{5, 10, 15, 20\}$).

Table 1 shows that most of suggested segmentation methods did not have significant improvements. All suggested segmentation methods outperform **SSR** method. However, they did not outperform the fixed bin segmentation method and window based bi-gram **BM25** model (8 bins) while the fixed bin segmentation method outperform window based bi-gram **BM25** model.

Using Haar wavelet transform leads to zero padding which may affect the accuracy of the suggested methods.

In Sentence-based segmentation with position weight method, document has additional weight scheme for each quarter as follows [0.75, 0.50, 0.25, 0.0].

5   Conclusion

This research proposed a variety of segmentation methods to enhance document score and the accuracy of retrieving. All suggested segmentation methods are based on meaning and context. All suggested segmentation methods outperform SSR method while the sentence-based, paragraph-based and sentence-based with position weight outperform window based bi-gram BM25 model. However, they did not outperform the fixed bin segmentation method (8 bins).

Using Haar wavelet transform leads to zero padding which may affect the accuracy of the suggested methods. Another wavelet transform is proposed to be used in future work such as Daubechies.

One of the most important future work is to find the optimum bin number and weight scheme for a given dataset.

Acknowledgements. This paper was funded by the Deanship of Scientific Research (**DSR**), King Abdulaziz University, Jeddah, under grant No. (33 611-D1433). The authors, therefore, acknowledge with thanks **DSR** technical and financial support.

References

1. Salton, G., Fox, E.A., Wu, H.: Extended boolean information retrieval. Commun. ACM **26**, 1022–1036 (1983)
2. Salton, G., Wong, A., Yang, C.-S.: A vector space model for automatic indexing. Commun. ACM **18**, 613–620 (1975)
3. Kang, B., Kim, D., Kim, H.: Fuzzy information retrieval indexed by concept identification. In: International Conference on Text, Speech and Dialogue, pp. 179–186 (2005)
4. Wong, S.K.M., Ziarko, W., Wong, P.C.N.: Generalized vector spaces model in information retrieval. In: Proceedings of the 8th Annual International ACM SIGIR Conference on Research and Development in Information Retrieval, pp. 18–25 (1985)
5. Cummins, R., O'Riordan, C.: Learning in a pairwise term-term proximity framework for information retrieval. In: Proceedings of the 32nd International ACM SIGIR Conference on Research and Development in Information Retrieval, pp. 251–258 (2009)

6. Clarke, C.L.A., Cormack, G.V.: Shortest-substring retrieval and ranking. ACM Trans. Inf. Syst. (TOIS) **18**, 44–78 (2000)
7. Hawking, D., Thistlewaite, P.: Relevance weighting using distance between term occurrences (1996)
8. Bhatia, M.P.S., Kumar Khalid, A.: Contextual proximity based term-weighting for improved web information retrieval. In: Zhang, Z., Siekmann, J. (eds.) KSEM 2007. LNCS (LNAI), vol. 4798, pp. 267–278. Springer, Heidelberg (2007). doi:10.1007/978-3-540-76719-0_28
9. Aref, W.G., Barbara, D., Johnson, S., Mehrotra, S.: Efficient processing of proximity queries for large databases. In: Proceedings of the Eleventh International Conference on Data Engineering, pp. 147–154 (1995)
10. El Mahdaouy, A., Gaussier, E., El Alaoui, S.O.: Exploring term proximity statistic for Arabic information retrieval. In: 2014 Third IEEE International Colloquium on Information Science and Technology (CIST). IEEE (2014)
11. Ye, Z., He, B., Wang, L., Luo, T.: Utilizing term proximity for blog post retrieval. J. Am. Soc. Inf. Sci. Technol. **64**, 2278–2298 (2013)
12. Costa, A., Melucci, M.: An information retrieval model based on discrete fourier transform. In: Cunningham, H., Hanbury, A., Rüger, S. (eds.) IRFC 2010. LNCS, vol. 6107, pp. 84–99. Springer, Heidelberg (2010). doi:10.1007/978-3-642-13084-7_8
13. Ramamohanarao, K., Park, L.A.F.: Spectral-based document retrieval. In: Maher, M.J. (ed.) ASIAN 2004. LNCS, vol. 3321, pp. 407–417. Springer, Heidelberg (2004). doi:10.1007/978-3-540-30502-6_30
14. Park, L.A.F., Ramamohanarao, K., Palaniswami, M.: Fourier domain scoring: a novel document ranking method. IEEE Trans. Knowl. Data Eng. **16**, 529–539 (2004)
15. Park, L.A.F., Palaniswami, M., Kotagiri, R.: Internet document filtering using fourier domain scoring. In: Raedt, L., Siebes, A. (eds.) PKDD 2001. LNCS (LNAI), vol. 2168, pp. 362–373. Springer, Heidelberg (2001). doi:10.1007/3-540-44794-6_30
16. Park, L.A., Palaniswami, M., Ramamohanarao, K.: A novel document ranking method using the discrete cosine transform. IEEE Trans. Pattern Anal. Mach. Intell. **27**, 130–135 (2005)
17. Park, L.A.F., Ramamohanarao, K., Palaniswami, M.: A novel document retrieval method using the discrete wavelet transform. ACM Trans. Inf. Syst. (TOIS) **23**, 267–298 (2005)
18. Arru, G., Feltoni Gurini, D., Gasparetti, F., Micarelli, A., Sansonetti, G.: Signal-based user recommendation on twitter. In: Proceedings of the 22nd International Conference on World Wide Web Steering Committee/ACM, pp. 941–944 (2013)
19. Yang, T., Lee, D.: T3: on mapping text to time series. In: Proceedings of the 3rd Alberto Mendelzon Int'l Workshop on Foundations of Data Management, Arequipa, Peru, May 2009
20. Zobel, J., Moffat, A.: Exploring the similarity space. In: ACM SIGIR Forum, pp. 18–34 (1998)
21. Daubechies, I.: Where do wavelets come from? A personal point of view. Proc. IEEE **84**, 510–513 (1996)
22. Haar, A.: Zur theorie der orthogonalen funktionen systeme. Mathematische Annalen **69**, 331–371 (1910)
23. Diwali, A., Kamel, M., Dahab, M.: Arabic text-based chat topic classification using discrete wavelet transform. Int. J. Comput. Sci. Iss. (IJCSI) **12**, 86 (2015)
24. He, B., Huang, J.X., Zhou, X.: Modeling term proximity for probabilistic information retrieval models. Inf. Sci. **181**(14), 3017–3031 (2011)

Knowledge Representation in Intelligent Tutoring System

Alan Ramírez-Noriega[1(✉)], Reyes Juárez-Ramírez[1], Samantha Jiménez[1],
and Yobani Martínez-Ramírez[2]

[1] Universidad Autónoma de Baja California, Mexicali, Mexico
{alan.david.ramirez.noriega,reyesjua,samantha.jimenez}@uabc.edu.mx
[2] Universidad Autónoma de Sinaloa, Culiacán, Mexico
yobani@uas.edu.mx

Abstract. This study presents a compilation of techniques for Knowledge Representation (KR) in Intelligent Tutoring System (ITS). Shows pros and cons of each approach in order to use the proper technique according to the needs. Analyses literature related to ITS and KR to find the approaches. Highlights: Fuzzy Cognitive Maps, Bayesian Network, Semantic Networks, Graphs, among other methods. Each approach contributes with elements to model knowledge. We made a comparison of each model with determined factors. Each technique of KR provides his own vision of how the world should look. Besides, it shows what information is necessary to represent and what is important to ignore. Different approaches to intelligent reasoning lead to different goals and definitions of success.

Keywords: Knowledge representation · Intelligent tutoring systems · Fuzzy cognitive maps · Bayesian networks

1 Introduction

For many years technology had been involved in the educational process. This kind of technology called educational technology concerns the study and ethical practice to facilitate learning and improve the performance; made possible through the creation, use, and proper management of resources and technological processes [2]. In particular, we are focused on Intelligent Tutoring Systems (ITS) and other systems that support the teaching through the computer.

An Intelligent Tutoring System is defined as a software system that uses artificial intelligence techniques to interact with students and teach them [8,23] almost in the same way as a teacher does [4]. Carbonell [3] proposed a generalized architecture for ITS, which considers further of user interface, three core modules [4,23]: (1) Tutoring model, (2) domain model, and (3) student model.

© Springer International Publishing AG 2017
A.E. Hassanien et al. (eds.), *Proceedings of the International Conference on Advanced Intelligent Systems and Informatics 2016*, Advances in Intelligent Systems and Computing 533, DOI 10.1007/978-3-319-48308-5_2

A computer-based education system needs to know what to teach, which is called knowledge domain. When represented in a way that computer understands it. Then is called Knowledge Representation (KR). It is important to consider the KR processes and automates information management through computers. In addition, make inferences that allow decision-making as a human does in order to improve the tutoring task.

Authors such as [20,22] establish that in the human semantic memory exist a hierarchy of concepts with relations to organize the knowledge. Hence, arises the idea of representing knowledge by mean of graphs. Samples of this techniques are Concepts Maps, Bayesian Network, Cognitive Maps, Conceptual Graphs, Knowledge Maps, Semantic Network and Memory Maps. Our study analyzes the mentioned techniques, and develops a comparative schema to define the elements of knowledge considered by those techniques.

The objectives of our work are: to identify the methods used in ITS to represent knowledge, to obtain features given by each approach, to define a taxonomy about the elements that are considered to represent knowledge, to compare different approaches according to the elements identified, and finally, to highlight the use of each technique according to its characteristics.

This paper is organized as follows: Section two describes the elements obtained from each technique. Section three compares each method according to identified factors. Section four analyzes each method, and section five gives an example of the taxonomy elements in a Bayesian Network. Finally, conclusions and references are shown.

2 Features of the Knowledge Representation in ITS

In this section, we summarized the elements that were considered to represent knowledge in an educational environment. Literature review was made to find

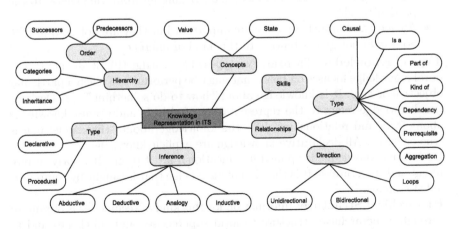

Fig. 1. Elements to KR in ITS

the previous approaches. Articles related to KR and ITS were reviewed. Figure 1 depicts the elements identified. The features of knowledge representation are:

- **Concepts:**
  The concept is an elemental piece of knowledge. According to the domain expert, it cannot be divided into smaller parts. Whereby, a concept is considered the primary unit of knowledge [13].
  The related work with computer assisted education that we analyzed is based on two essential aspects: First, the quantitative aspect, defined as the value or weight of knowledge possession [15]. It refers to a numeric value assigned to the node that represents the concept. Second, the qualitative aspect, defined as a state that refers to a discrete value where the knowledge possession can be [15].
- **Skills:** A skill is a cognitive process that interacts with one or more concepts, usually through an application. It has a particular purpose and produces an internal or external result [13,17].
- **Relationships:** Relationships are known as links. The relationships goal is to know how the concepts are related [1,6,13–15,17,24].
  Relationships can be of three types according to the link direction: (1) Unidirectional: A relation of a concept A to a concept B [10,14,18,21]; (2) Bidirectional: A relation of a concept A to a concept B and vice versa [18,21]; (3) Loops: A relation with a concept itself maybe through other concepts. [18,25].
- **Inference:** It is also called reasoning. The inference refers to obtain deductions or conclusions based on knowledge already established [5]. The main types of inference are [18]:
  - Abductive: It consists in finding the state of the world (configuration) that is the most probable given the evidence [11].
  - Deductive: It refers to predicting an effect given cause. Thus, we can get responses about the consequences of a given cause, but we cannot answer why results are produced [25]. Beginning from the general to the particular [6].
  - Analogy and inductive: They are important but they are not used in ITS. So, we do not give attention to this kind of inference.
- **Type of Knowledge:** There are two types of knowledge [10,13,26]: (1) Procedural Knowledge focuses on tasks that must be performed to reach a particular objective or goal. It is knowledge about "how to do something"; (2) Declarative knowledge refers to the representation of objects and events, knowledge about facts and relationships. It is the knowledge about "that something is true or false". All declarative knowledge are explicit knowledge.
  Declarative knowledge is applied in educational institutions. It is easy to represent and structure, so it is the kind of knowledge that is taught by computer-aided systems.
- **Hierarchical Structure:** Information for this kind of techniques is organized through a hierarchical structure to form superclasses and subclasses and to share properties [16].

Based on the hierarchical structure, some elements are derived, which are considered by different authors as important because they establish a structure in order to maintain the contents [1,15,17]. These elements are:

- Successors y predecessors: Successors concepts are knowledge elements that are considered after the current [15]. This means the current concept is the basis for learning the next idea. The predecessors concepts are considered as previous elements needed to understand the current. It is relevant prior knowledge acquired because establishes the basis for the next one [12,15,17].
- Categories: The categorization is a process through which the object is located within a class or category, involving attribution of meaning [20].
- Inheritance: It is a reasoning system that deduces properties of a concept based on the concepts properties higher in the hierarchy [27].

3   Comparison Between Knowledge Representation Approaches

This section exposes a comparative analysis of the different methods through tables, and analyzes the elements defined in the previous section. Tables referenced below consider the same symbology. The first column shows the name of the approaches discussed, while remaining columns show the elements of knowledge that are being evaluated. The x expresses those elements considered in the articles studied, while c corresponds to those elements not found in the study. However the approach could support it, if required.

Concepts and Attributes: In contrast to other techniques such as ontologies, the analyzed approaches shown in Table 1 only consider concept attributes concerning the quantifiable value or qualitative state. Regarding the first, techniques such as Bayesian Networks, Fuzzy Cognitive Maps or Maps of Knowledge consider the quantitative aspect to define a degree or probability of knowledge possession. Second, the state attribute, considers a discrete aspect about the knowledge possession, this means, knowledge can be classified in states such as excellent, very good, good, regular, undesirable, or just approved or not approved.

These attributes add extra value to the knowledge representation considering the domain uncertainty. A Bayesian Network considers the quantitative aspect to model the knowledge domain; however, the analyzed studies did not contemplate the qualitative aspect that could be easily represented. The Bayesian Causal Maps did not consider neither the quantitative value nor qualitative state, according to the analyzed articles. At last, Fuzzy Cognitive Maps and Knowledge Maps consider both quantitative and qualitative aspects.

Relationships: The Relationships contemplated in each approach vary considerably according to the author's need. Table 1 shows eight different types. The column that indicates the relation shows those structure that can define their own relations. Methods such as Concepts Maps, Conceptual Graphs, Semantic

Table 1. Relationships comparative

Approach	Concepts	attributes		Kind of relation									Relationships direction		
		Value	State	Relationships	Causal	Is a	Part of	Kind of	Dependency	Prerequisite	Aggregation	Define relation	Unidirectional	Bidirectional	Loops
Concepts Maps	x			x					x		x	c	c		
Bayesian Network	x	x	c	x	x				x				x		
Bayesian Causal Maps	x	c	c	x	x				x		x		c		x
Fuzzy Cognitive Maps	x	x	x	x	x				x				x		
Graphs	x			x		x	x					c	c		
Knowledge Maps	x	x	x	x					x				x		
Semantic Networks	x			x		x	x				x	c	x	c	c
Memory Maps	x			x								x	x		

network, and Memory Maps show great flexibility to represent the knowledge domain because they can describe their relation.

Relationships direction can be of three types such as unidirectional, bidirectional, and loops. Each approach can handle at least the unidirectional relation; these relations are the most used. A Semantic Network is the unique capable of handle connections in both directions and loops. Bayesian Causal Maps have the ability to control loops, an aspect that a simple Bayesian Network cannot do by definition.

Inference and type of knowledge: Table 2 shows information about the type of inference of each approach, and the type of knowledge that can represent. The kind of knowledge is related to the quantitative information that can be stored in the domain representation. So, Bayesian Networks, Bayesian Causal Maps and Fuzzy Cognitive Maps use the previous advantage by nature of their theories. These theories have the possibility of the abductive and deductive inference. Memory Maps are able to perform deductive inference though it was not required in the analyzed studies. Regarding to kind of knowledge, all approaches are focused in representing the declarative knowledge since this is the type of knowledge that represent the ITS.

Hierarchical Structure: All the approaches use concepts and relations; therefore, they have a similar graphic representation based in a hierarchical structure: (1) Concepts are derived from this representation such as successors and predecessors to keep an order in the learning of concepts; (2) hierarchy is used to organize knowledge in a tree structure; (3) categorization is used to group concepts with similarity; finally, (4) inheritance is used to acquire properties from the parents. All elements can be represented in the approaches, explicitly or implicitly. It depends on what the authors want to represent. Memory Maps represents explicitly each factor considered in this study (Table 2).

4 Approaches Analysis

According to Davis [5], knowledge representation approaches are a substitution of the reality, the only entirely accurate representation of an object is the object itself. All other representations are inaccurate; they inevitably contain simplifying assumptions. Each representation approach provides its own vision of how the world should be; furthermore, they define which aspects are important to represent and which to ignore. Different conceptions of intelligent reasoning nature lead to different goals and different definitions of success. A language designed to express facts declaratively is not necessarily useful to express the imperative information characteristic of a reasoning strategy.

Taking into account the previous paragraphs, the best approach to represent knowledge is given by the problem to be solved and the objectives. So, each method has advantages and disadvantages to model knowledge domains. Nevertheless, the techniques analyzed share the ability to represent knowledge through concepts and relations, forming a hierarchical structure with advantages entailed.

Concepts Maps and Conceptual Graphs have, among other advantages, the capacity to represent the type of relation between concepts as the author wants. This fact has not been as well studied as the first two; however its foundation promises great scope.

Semantic Networks can be seen as ontologies having a great capacity of knowledge representation. Ontologies have a high flexibility to represent information due to their great expressiveness to model the world. This kind of semantic network has the Web Ontology Language (OWL) standard, created by the World Wide Web Consortium (W3C). Among all the techniques mentioned in this paper, ontologies are considered the best technique when there is not need to model uncertainty. Its main disadvantage is not take the quantitative aspects of the world.

Table 2. Inference, knowledge, and structure comparative

Approach	Abductive	Deductive	Analogy	Inductive	Declarative	Procedural	Successors	Predecessors	Hierarchy	Category	Inheritance
	Inference				Knowledge		Hierarchical Structure				
Concepts Maps	c				c		c	x	c	x	c
Bayesian Network	x	x			c		c	c	c	x	c
Bayesian Causal Maps	x	x			c		c	x	c	c	c
Fuzzy Cognitive Maps	x	x			c		c	c	x	c	c
Graphs	c				c		c	c	c	c	c
Knowledge Maps	c				c		x	x	x	c	c
Semantic Networks	c				c		c	c	x	c	x
Memory Maps	c				x		x	x	x	x	x

Knowledge Maps argue good principles, even considering weights on concepts to model knowledge. However, they are not among the commonly used techniques to represent knowledge. Authors of this approach do not argue the practical use of the weights or the type of inference we can make with them.

Bayesian Networks are a technique of approximate reasoning to model the world without much semantic expressiveness, which is the main disadvantage. They were developed to resolve domains that manage uncertainty. Besides, another advantage is the abductive and deductive inference that can be achieved with its representation. Bayesian Causal Maps have arisen to give greater semantic flexibility to the original Bayesian Networks [25]. One of the major contributions of Bayesian Causal Maps are the handle of loops; not considered by the Bayesian Networks in their theory.

Finally, Fuzzy Cognitive Maps merge advantages of both, Cognitive Maps and Fuzzy Logic. This method, same as Bayesian Networks, allows to model domains with uncertainty through causal relationships. Besides, Fuzzy Logic includes abductive and deductive inference capacity. In contrast to Bayesian Causal Maps, it has been quite used in literature for different areas [7].

Table 3 displays the most significant advantage for each approach, and its main disadvantage. Even though, some techniques are emerging, they have a promising future. Other methods are well established in the literature; however, there are missing elements to represent certain domains.

Table 3. Main advantages and disadvantages of the approaches

Approach	Advantage	Disadvantage
Concepts maps	Definition of relations	Need for more efficient inference
Bayesian network	Abductive and deductive inference	Limited semantic representation
Bayesian causal maps	Abductive and deductive inference	New approach and little used
Fuzzy cognitive maps	Abductive and deductive inference	Limited availability of support tools
Graphs	Definition of relations	Need for more efficient inference
Knowledge maps	Quantitative and qualitative aspects to represent the domain	Need for more efficient inference
Semantic networks	High semantic expressivity	No handling uncertainty
Memory maps	Adequate semantic expressivity	New approach and little used

5   Representing a Bayesian Network with the Taxonomy

In order to show how to represent knowledge with a Bayesian Network, in this section we present an example of a specific domain in the field of Software Engineering. The Figure 2 displays a first chapter of the Personal Software Process (PSP) book [9], a course being taught to freshmen in computer engineering. A Bayesian Network was generated to express structured knowledge of PSP. The network includes variables (concepts), relation, and probabilistic values [19]. The network represents several aspects of the taxonomy:

1. Successors and predecessors: In relationship *Measure-Quality*, the variable *Measure* is a predecessor to *Quality* and *Quality* is a successor of *Measure*.
2. Categories and inheritance: In relation *Improvement_process* and *Measure* with *Quality*, the variable *Quality* is a category of *Improvement_process* and *Measure*. *Quality* includes and takes attributes from *Improvement_process* and *Measure* (inheritance).
3. Declarative knowledge: It is Knowledge easy to represent and structure. The knowledge represented by ITS.
4. Abductive and deductive inference: The algorithms of Bayesian Networks, combined with probabilistic values, give the possibility to do inference. For instance, does the student know the concept Quality? Taking into account that he/she knows the concept Measure. Does the student knows the concept Measure? Taking into account that he/she knows the concept Quality.
5. The direction of relationship: The relation is *unidirectional*, the arrow starts in a variable and finishes in other always with one direction.
6. Type of relationship: The relationship is *Causal*, this kind of relation gives the possibility of inference. A cause has an effect and an effect has a cause.
7. Attributes of concept value and state: The numeric values are represented by *Value* and the *State* is represented by two variables, present and absent. Present means knowledge possession and absent means the opposite.

With this representation we can manage the content of a course inside an ITS, furthermore, we can make inferences about if the student has or not a knowledge.

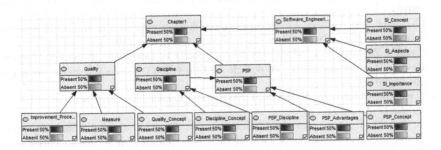

Fig. 2. Knowledge representation about PSP topic

6 Conclusions

This study collected approaches to represent knowledge in computer-aided teaching systems. All approaches are based on concepts with their relations to form a hierarchical structure. Each approach faces a particular problem according to the domain, showing its advantages and disadvantages to confront that problem. Techniques such as Concepts Maps, Conceptual Graphs, Memory Maps, and Semantic Networks are useful when semantic expressiveness is needed to model a domain, and not to deal with advanced reasoning. Ontologies stand out as Semantic Networks because they have a standard controlled by W3C. Besides, these are an important part of the development of the Semantic Web. Ontologies allow to show high semantic expressiveness to represent the domain accurately and use of ontological reasoning.

Bayesian Network, Bayesian Causal Maps, Fuzzy Cognitive Maps, and Knowledge Maps represent the best choice when do not wish to express the domain in great detail. Instead, prefers high degree of reasoning to make inferences that allow controlling uncertainty and other quantitative aspects.

For future work, a simple algorithm can make recommendations with the information obtained in this study, to adapt the user needs to represent knowledge. This can be done through the taxonomy, using the leaf nodes of the hierarchy. The taxonomy can be converted in a network, where all nodes converge in a central node. Central node gives us probabilities about what is the approach (KR) that we need to use.

Besides, it is necessary to find other knowledge representation techniques that allow to extend the options of recommendation. Finally, we will develop an ontology of the knowledge representation in ITS.

References

1. Badaracco, M., Martínez, L.: A fuzzy linguistic algorithm for adaptive test in intelligent tutoring system based on competences. Expert Syst. Appl. **40**(8), 3073–3086 (2013)
2. Campos, I.S., Mata, A.C.U.: Analysis of the debate on the impact of technological media in learning processes. Actualidades Investigativas en Educación **11**(1), 1–22 (2011)
3. Carbonell, J.R.: AI in CAI: an artificial intelligence approach to computer assisted instruction. IEEE Trans. Man Mach. Syst. **11**, 190–202 (1970)
4. Cataldi, Z., Lage, F.J.: Modelado del Estudiante en Sistemas Tutores Inteligentes. Revista Iberoamericana de Tecnologia en Educación y Educación enTecnología **5**, 29–38 (2010)
5. Davis, R., Shrobe, H., Szolovits, P.: What is a knowledge representation? AI Mag. **14**(1), 17–33 (1993)
6. de Kereki, F., Guerrero, I.: Modelo para la Creación de Entornos de Aprendizaje basados en técnicas de Gestión del Conocimiento. Ph.D. thesis, Universidad Politecnica de Madrid (2003)
7. Hossein, M., Zarandi, F., Khademian, M., Minaei-bidgoli, B.: A fuzzy expert system architecture for intelligent tutoring systems: a cognitive mapping approach. J. Intell. Learn. Syst. Appl. **4**, 29–40 (2012)

8. Huertas, C., Juárez-Ramírez, R.: Developing an intelligent tutoring system for vehicle dynamics. Procedia - Soc. Behav. Sci. **106**, 838–847 (2013)
9. Humphrey, W.S.: Introduction to the Personal Software Process. Addison-Wesley Longman Publishing Co. Inc., Boston (1997)
10. Krudysz, G.A., Sam Li, J., McClellan, J.H.: Web-based bayesian tutoring system. In: 12th Digital Signal Processing Workshop—4th Signal Processing Education Workshop, pp. 129–134 (2006)
11. Larrañaga, P., Moral, S.: Probabilistic graphical models in artificial intelligence. Appl. Soft Comput. **11**(2), 1511–1528 (2011)
12. Liu, Z., Wang, H.: A modeling method based on bayesian networks in intelligent tutoring system. Structure, 967–972 (2007)
13. Millán, E., Pérez-De-La-Cruz, J.L.: A Bayesian diagnostic algorithm for student modeling and its evaluation. User Model. User-Adapt. Interact. **12**, 281–330 (2002)
14. Millán, E., Descalço, L., Castillo, G., Oliveira, P., Diogo, S.: Using Bayesian networks to improve knowledge assessment. Comput. Edu. **60**(1), 436–447 (2013)
15. Mishra, M., Mishra, V.K., Sharma, H.R.: Intellectual ability planning for intelligent tutoring system in computer science engineering education abstract. In: Proceedings - 2012 3rd National Conference on Emerging Trends and Applications in Computer Science, NCETACS-2012, pp. 26–30 (2012)
16. Nguyen, T.A., Raspitzu, A., Aiello, M.: Ontology-based office activity recognition with applications for energy savings. J. Ambient Intell. Humanized Comput. **4**(5), 1–15 (2013)
17. Ramirez, C., Valdes, B.: A general knowledge representation model for the acquisition of skills and concepts. In: 2009 8th IEEE International Conference on Cognitive Informatics, pp. 412–417 (2009)
18. Ramirez, C., Valdes, B.: A general knowledge representation model of concepts (2012)
19. Ramírez-Noriega, A., Juárez-Ramírez, R., Huertas, C., Martínez-Ramírez, Y.: A methodology for building bayesian networks for knowledge representation in intelligent tutoring systems. In: Congreso Internacional de Investigación e Innovación en Ingeniería de Software 2015, San Luís Potosí, pp. 124–133 (2015)
20. Rivas Navarro, M.: Procesos cognitivos y aprendizaje significativo. BOCM, Madrid (2008)
21. Rodrigues, F.H., Bez, M.R., Flores, C.D.: Generating bayesian networks from medical ontologies. In: 2013 8th Computing Colombian Conference, 8CCC 2013 (2013)
22. Rodríguez, R.J.: Herramientas informáticas para la representación del conocimiento Software. Subjetividad y pocesos cognitivos **14**(1712), 217–232 (2010)
23. Santhi, R., Priya, B., Nandhini, J.: Review of intelligent tutoring systems using bayesian approach (2013). arXiv preprint arXiv:1302.7081
24. Satar, A.: Using of intelligence tutoring systems for knowledge representation in learning and teaching process. Kufa Math. Comput. **1**(5), 1–13 (2012)
25. Sedki, K., Beaufort, L.B.D.: Cognitive maps and bayesian networks for knowledge representation and reasoning. In: 24th International Conference on Tools with Artificial Intelligence, pp. 1035–1040 (2012)
26. Sharma, T., Kelkar, D.: A tour towards knowledge representation techniques. Int. J. Comput. Technol. Electr. Eng. (IJCTEE) **2**(2), 131–135 (2012)
27. Socorro, R., Simón, A., Valdés, R., Fernández, F.O., Rosete, A., Moreno, M., Leyva, E., Pina, J.: Las ontologías en la representación del conocimiento (2008)

Multimodal Graph-Based Dependency Parsing of Natural Language

Amr Rekaby Salama[✉] and Wolfgang Menzel

Department of Informatics, University of Hamburg, Hamburg, Germany
{salama,menzel}@informatik.uni-hamburg.de

Abstract. Dependency parsing is a popular approach for syntactic analysis of natural language utterances. It concerns building a dependency tree of the linguistic input relying only on a model of syntactic regularities. The cognitive process of human language processing, however, has also access to other sources of knowledge, like visual clues that can be used to improve language understanding.

In this paper, we approach integrating visual context and linguistic information to improve the reliability of dependency parsing. To achieve this goal, we modify a state-of-the-art dependency parser to make it accept visual information as extra features in addition to the original linguistic input. All these inputs (features) are considered in the learning process of the trained model. Experiments have been carried out to investigate the contribution of this additional multimodal information on ambiguity resolution and parsing quality.

Keywords: Multimodal integration · Graph-based dependency parsing · RBG parser

1 Introduction and Motivation

Natural language parsing usually suffers from the problem of ambiguity. Many ambiguities cannot be easily resolved using only the linguistic information. Trained models learn to disambiguate the syntactic structure according to the prior probability distribution found in the training data: The most frequently found interpretation will also be taken as the most plausible one, irrespective of any other factors which might contribute counter-evidence. Such behavior is highly undesirable in dynamic contexts, where the actual choice should also consider the current state-of-affairs in the world. In such a situation, including visual information into the decision process might help to find a better fitting interpretation.

One of the most noticeable sources of ambiguity in natural language is prepositional phrase (PP) attachment. As shown in Fig. 1, "I saw a girl with a telescope", the decision between high and low attachment cannot be taken based on pure syntactic information, even lexical preferences don't provide reliable clues. If the high attachment is adopted, the PP is attached to the verb which indicates its relation to the verb (I use the telescope to see the girl). In the low attachment case, PP is attached to the closest lexical item, which marks the coexistence of the PP with that item (I saw a girl who has a telescope with her). If we have available visual input in addition to the

© Springer International Publishing AG 2017
A.E. Hassanien et al. (eds.), *Proceedings of the International Conference
on Advanced Intelligent Systems and Informatics 2016*, Advances in Intelligent
Systems and Computing 533, DOI 10.1007/978-3-319-48308-5_3

Fig. 1. (A) High attachment. (B) Low attachment

linguistic one, integrating them into the learning model might help in such a situation, if the kind of knowledge provided by the visual (context) information is beneficial to disambiguate the dependencies.

In this paper, we provide a multimodal dependency parser. The parser does not only depend on the linguistic input, but also on the non-linguistic modality. Although we consider the context (visual) information as the non-linguistic modality and inject such input into the parser by providing the relations between the elements in the context, we don't work on the level of relation extraction through image processing so far. Instead we focus on trying to overcome a range of other challenges in this research such as: introducing the context knowledge into the learning model of a graph-based parser, and manipulating the scoring function to take the decision based on the linguistic and non-linguistic modalities.

We use thematic roles in the form of triples as a description language stating the situation given in the non-linguistic context. The thematic roles contain information that helps in ambiguity resolution. Both linguistic and non-linguistic information are fed into the graph based dependency parser to improve its quality.

The paper starts with a review of dependency parsing methods (transition and graph based) and previous work on the integration of context information into a rule-based parser. In Sect. 3, we present our context-integrating model. The high-level architecture of the solution is presented in Sect. 4. Then a set of experiments is discussed in Sect. 5 before we state the conclusion and make proposals for future work.

2 Previous Work

2.1 Dependency Parsing

Dependency parsing extracts a syntactic dependency tree that describes binary relationships between the words of a sentence. The nodes of the tree correspond to the word forms in the sentence while the edges represent the dependency links between them in a child-parent relationship. These links are interpreted in terms of the functions that a lexical item fulfills with respect to its governor. These functions are described using labels attached to the edges. A valid dependency tree has to be an acyclic, and connected graph with a single head of each node (Nivre 2004).

Among the machine learning approaches for dependency parsing, there are two main methods: transition-based, and graph-based parsing. The transition-based (Shift-reduce) method constructs the tree incrementally, by attaching an incoming word immediately, or delaying its attachment until a better attachment point becomes available. The decision is based on an oracle which consults the history of prior attachments decisions. MALTparser is an example of this approach (Nivre et al. 2004).

Graph-based parsers start by creating a graph where each node represents a word from the sentence (Zhang et al. 2014a). All the nodes are connected to each other. A feature vector is assigned to each edge. The cost of each edge is dynamically learned based on a function of the training dataset. The parser finds a minimum spanning tree of the graph with the optimal score (Bohnet 2010). Different algorithms are used as alternatives for the creation of the minimum spanning tree: Chu-Liu/Edmond (Chu and Liu, 1965), and Hill-climbing (Zhang et al. 2014b). RBG parser (Lei et al. 2015) claims the state-of-the-art of graph-based parsing. It uses high-order features, different spanning tree decoding algorithms (Lei et al. 2014), a passive-aggressive online learning algorithm (MIRA), and parameter averaging (Crammer et al. 2006). It outperforms other dependency parsers quality.

2.2 Context Representation

The idea of utilizing context information from the visual environment in the dependency parsing was introduced by (McCrae 2009). He injected the visual information into a constraint-based parser (Weighted Constraint Dependency Grammar WCDG), and run his research on a German language dataset. He used the Web Ontology Language (OWL) to encode high-level descriptions of the visual input. Although OWL has two main components: t-box, a-box, he considered only the a-box to describe the relations between entities in the visual context. Under a-box representation, four thematic roles (Agent, Theme, Instrument, and Owner) are used to demonstrate the conceptual relationships in the context.

In this paper we implement our ideas by means of RBG parser to develop a proof of concept model, to demonstrate that the desired fusion of multimodal information can be achieved in a learning (graph-based) parsing model. The visual information is presented in form of thematic roles. Our experiments compare the results between the new implemented context-integrating parser that combines visual and linguistic information for English sentences against the original RBG parser as a benchmark.

3 Context-Integrating Dependency Parser

RBG parser considers only the linguistic input during model learning. It hypothesizes high order scoring functions to use them in minimum spanning tree extraction. To make it sensitive to visual information, we modify RBG parser to accept additional context information as features for the learning model. Our new version of RBG keeps linguistic features on the edges between combinatorial pairs of words in addition to newly introduced visual features between the entities (words) that have a relationship in the context (visual) input. As presented in Fig. 2, a visual relation has three parts: the relation type (agent, theme, etc.), the head of the relation which is the verb (except in the "owner" relation), and the modifier of the relation.

Figures 2 and 3 represent two pictures and their (visual) context information. The linguistic input corresponding to these figures are:

- 2: "The doctor with a coat feeds at this moment the journalist with a microphone"
- 3: "The journalist with a microphone feeds at this moment the doctor with a spoon"

Fig. 2. The context information of an image (image taken from (Knoeferle 2005))

Fig. 3. The context features of an image (image taken from (Knoeferle 2005))

These sentences are confusing for a person if he/she hears them without the visual information. Using the visual information, however, one can differentiate that "microphone" in sentence 2 is an object with the journalist while the "Spoon" in sentence 3 is the tool for feeding and not an object with the doctor.

In Fig. 4, we present the adapted graph representation of the sentence in the context integrating RBG parser. We can find (t) words, $\{fv_{a=1\dots m}\}$ linguistic feature vectors (the original ones from the RBG parser). As presented in Eq. 1, each vector has n features encoding the linguistic properties of the pair of words (i,j). It has additionally p.

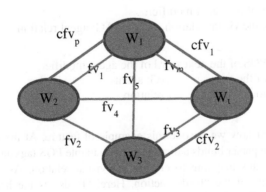

Fig. 4. Graph representation of the context-integrating dependency parsing

context feature vectors (newly introduced). These vectors consist of q visual features for the words pairs that have correlation in the context input.

As shown in Eq. 3, we build the learning model using multimodal inputs. In the testing phase, the model gets the linguistic information of the sentence in addition to the context information to find the optimal dependency tree. In equions 3, 4: x_i. is the input sentence, c_i is the context input for sentence (i), and $\tilde{y}$ the extracted dependency tree. For each sentence there is a set of possible trees $T(x_i)$, and a gold standard one $\hat{y}_i$. To represent the feature vector of the context input we use $\ddot{f}(c_i, y)$ with parameters (ω). $f(x_i, y)$ is used for the linguistic features and θ for the parameters.

$$fv_{a=1...m} = \begin{pmatrix} f_{i,j,1} \\ ... \\ f_{i,j,n} \end{pmatrix} \tag{1}$$

$$cfv_{a=1...p} = \begin{pmatrix} cf_{i,j,1} \\ ... \\ cf_{i,j,q} \end{pmatrix} \tag{2}$$

$$\tilde{y} = \max_{y \in T(x_i)} \left\{ \theta.f(x_i, y) + \omega.\ddot{f}(c_i, y) + \|y - \hat{y}_i\| \right\} \quad Train \tag{3}$$

$$\tilde{y} = \max_{y \in T(x)} \left\{ \theta.f(x, y) + \omega.\ddot{f}(c, y) \right\} \quad Test \tag{4}$$

For example, the context feature (HPp_HP_MAGP_MAGPn) describes four different aspects of a relation:

- HPp: The part of speech (POS) of the previous word of the Head.
  H: head of the visual relation.   p: previous word.   P: pos.

- HP: The head's POS.
  H: head of the visual relation.   P: pos.

- AGP: The POS of the agent modifier.
  M: modifier of the visual relation.   AG: agent relation.   P: pos.

- MAGPn: The POS of the next word of the agent modifier.
  M: modifier of the   AG: agent   P: pos.   n: next word.
  visual relation.   relation.

Now we present how we encode this example of feature. At the beginning of the parsing process, the parser builds a dictionary of available POS tags and assign an ID to each one. We use this mapping to encode the visual relation. As shown in Fig. 2, "Doctor" is "Agent" of the "feeds" action. Here, "feeds" is the head of the visual relation, "agent" is the relation type, and "Doctor" is the modifier. "Hpp" refers to the POS of the previous word of "feeds." "HP" is the POS of "feeds." "MAGP" is the POS of "Doctor." "MAGPn" is the POS of the next word to "Doctor.". Table 1 shows the encoding of this feature, and how it consists of different POS's ID referring to the mentioned dictionary. This encoding is used as a feature ID in the parser learning process. This feature is added to the visual feature vectors between the two words in the adapted graph represented above.

Table 1. Coding of the example feature

Feature code	1100	0010	0100	0010	0011
	Visual Feature ID	*POS id of "coat"*	*POS id of "feeds"*	*POS id of "Doctor"*	*POS id of "with"*

4 Solution Architecture

In this section, we present the high-level architecture of the context-integrating RBG parser and how we introduce the visual modality in it. As shown in Fig. 5, there are three types of components:

- Components of the RBG parser that are kept without modification (Old).
- Components of the RGB parser that have been changed to be compatible with multi-modal parsing (Changed).

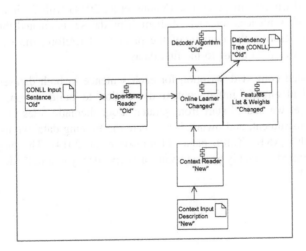

Fig. 5. The architecture of context-integrating dependency parser

• Newly introduced components (New).

The Online Learner is an existing component in RBG parser. It uses the Passive-Aggressive algorithm. We modify it to consider the additional features. Therefore, the features list and corresponding weights in RBG parser are also modified for the same purpose. On the other hand, the RBG component "Decoder Algorithm" is left unchanged. It is responsible for the minimum spanning tree decoder and implements different algorithms.

5 Experiments

5.1 Dataset Preparation

In our experiments, we developed two small corpora:

1. An extended version of Baumgärtner's dataset (Baumgärtner 2013). The original dataset condensed 24 images and 96 sentences describing these images. All sentences follow the same structure: Subject, verb, and object with adverbial modifiers. We translated this dataset from German to English, and extended it to 500 sentences that are equally distributed into the following groups:
   — The original dataset. Ex: "The Princess washes obviously the pirate."
   — A group has subject, object, and descriptions for both of them. Ex: "The Princess with long hair washes obviously the pirate with a woody leg."
   — A group has a descriptive subject, object, and a description of the action's instrument. Ex: "The Princess with long hair washes obviously the pirate with a brush."
   — A group has a subject with a description, an object with a description, and a description of the action. Ex: "The Princess with long hair washes obviously the pirate with a woody leg with a brush."
   — Sentences with subject and object in a passive form. Ex: "The pirate is washed by the Princess."
2. Part of the ILLIONS image corpus (Young et al. 2014) with 35 images and three corresponding sentences for each of them. This dataset is created through crowdsourcing to describe the content of the pictures. Therefore, the structure of the sentences varies in contrast to the first dataset.

We developed context description for each sentence in both datasets. Baumgärtner's dataset had already initial context description to build up on, but with the ILLIONS dataset we had to start from scratch. Four thematic roles have been used: agent, theme, instrument, and owner. To prepare the training data, we used the online demo of "Noah's ARK" Turbo parser (Thomson et al. 2014). The output (CONLL format) was verified manually against "Stanford dependency manual" (de Marneffe and Manning 2015).

5.2   Experiments Results

We present a set of experiments carried out to verify the effectiveness of context integration and its impact on the dependency parsing quality. We use three metrics:

- Unlabeled attached score (UAS): the percentage of correct lexical-parent attachments in the testing data.
- Labeled Attached Score (LAS): the percentage of correct labeled lexical-parent attachments in the testing data.
- Complete Attached Score (CAS): the percentage of complete sentences that have been correctly analyzed.

Experiment 1. Here we use the first dataset mentioned above. The training uses 440 sentences, and the testing data has 60 sentences. We implement two different degrees for the influence of the context features:

- "Strong Context" condition: We treated the context features during learning with an extra confidence (3 times the weight of a normal linguistic feature).
- "Normal Context" condition: The features added from the context have the normal influence on the final decision like the linguistic features.

As presented in Fig. 6, integrating the context information into the learning process slightly improves the UAS and LAS scores. This (relatively small) improvement of the attachment across all the words in the testing data, has, however, a quite big impact on the CAS score. That illustrates the benefit of using multimodal information as input for the graph based dependency parser. It helps to properly disambiguate more attachments which improves the overall CAS score by 18 %.

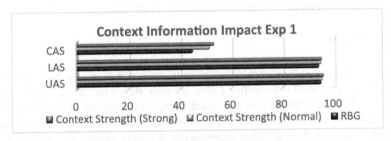

Fig. 6. Context information impact (Experiment 1)

Experiment 2. In this experiment, we used the second data set mentioned in the dataset section. Due to the small size of this dataset, and the online learning process of RBG using Passive-Aggressive (PA) algorithm, we use in this experiment the whole dataset as training data, as well as testing data. In the PA algorithm, the feature weights are updated according to the currently processed input sentence neglecting the influence of this update on the previously handled cases. Therefore, we train and test on the same dataset to check the effect of the context information. The context-integrating parser improves the overall CAS score by 8 % (Fig. 7).

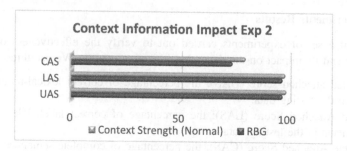

Fig. 7. Context Information Impact (Experiment 2)

Experiment 3. We trained the parser with the first dataset while testing with the second one. As shown in Fig. 8, there is no noticeable improvement in this case. We find that due to the completely different sentence structures and lexicons between the training and testing data, the context information didn't remarkably help towards better disambiguation in this case.

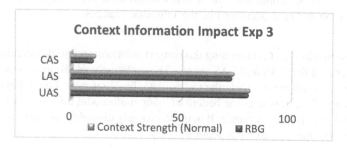

Fig. 8. Context Information Impact (Experiment 3)

6   Conclusion

In this paper, we present a context-integrating dependency parser by providing a new version of a graph-based dependency parser (namely RBG) accepting additional kind of features. These features present the visual context information of the sentence in a form of thematic roles. The experiments show an improvement of parsing quality between 8 % and 18 % in different experiments. We have shown the effectiveness of the idea on small datasets scale.

7   Future Work

While this paper is a first step in our context-integration parsing roadmap, we have identified some limitations in this work that should be tackled in future work. Currently, the system is not able to deal with a possible mismatch between the lexicons for the linguistic input and the context descriptions. So far we require them to be identical which is not a realistic assumption for richer linguistic stimuli.

In this research, we improved parsing quality using the cognitive influence of visual context information. In the future, we will work on enriching the context representations based on the linguistic input. We also need to apply the approach to a larger dataset. Additionally, we will study the behavior of the system in a situation where the context relationships contradict the linguistic content.

References

Baumgärtner, C.: On-line cross-modal context integration for natural language parsing. Ph.D. thesis Universität Hamburg, Hamburg (2013)

Bohnet, B.: Very high accuracy and fast dependency parsing is not a contradiction. In: Proceedings of the 23rd International Conference on Computational Linguistics (Coling 2010), pp. 89–97. Beijing (2010)

Chu, Y.J., Liu, T.H.: On the shortest arborescence of a directed graph. Sci. Sinica **14**, 1396–1400 (1965)

Crammer, K., Dekel, O., Shalev-Shwartz, S., Singer, Y.: Online passive-aggressive algorithms. J. Mach. Learn. Res. **7**, 551–585 (2006)

de Marneffe, M.-C., Manning, C.D.: Stanford typed dependencies manual, 1 April 2015. Retrieved from http://nlp.stanford.edu/pubs/dependencies-coling08.pdf

Knoeferle, P.: The role of visual scenes in spoken language comprehension: evidence from eye-tracking. Saarlandes: Ph.D. thesis Universität des Saarlandes (2005)

Lei, T., Xin, Y., Zhang, Y., Barzilay, R., Jaakkola, T.: Low-rank tensors for scoring dependency structures. In: Proceedings of the 52nd Annual Meeting of the Association for Computational Linguistics, ACL (2014)

Lei, T., Zhang, Y., Barzilay, R., Jaakkola, T.: RBGParser, 15 October 2015. Retrieved from github: https://github.com/taolei87/RBGParser

McCrae, P.: A model for the cross-modal influence of visual context upon language processing. In: The International Conference Recent Advances in Natural Language Processing, pp. 230–235 (2009)

Nivre, J.: Incrementality in deterministic dependency parsing. In: Proceeding Increment Parsing 2004 Proceedings of the Workshop on Incremental Parsing, pp. 50–57. Stroudsburg, PA, USA (2004)

Nivre, J., Hall, J., Nilsson, J.: Memory-based dependency parsing. In: Proceedings of the Eighth Conference on Computational Natural Language Learning (CoNLL), Boston (2004)

Thomson, S., Kong, L., Martins, A.: ARK Syntactic & Semantic Parsing Demo, 1 December 2014. Retrieved from http://demo.ark.cs.cmu.edu/parse

Young, P., Lai, A., Hodosh, M., Hockenmaier, J.: From image descriptions to visual denotations: new similarity metrics for semantic inference over event descriptions. Trans. Assoc. Comput. Linguist. **2**, 67–78 (2014)

Zhang, Y., Lei, T., Barzilay, R., Jaakkola, T., Globerson, A.: Steps to excellence: simple inference with refined scoring of dependency trees. In: Proceedings of the 52nd Annual Meeting of the Association for Computational Linguistics, ACL, pp. 197–207 (2014a)

Zhang, Y., Lei, T., Barzilay, R., Jaakkola, T.: Greedis goodif randomized: new inference for dependency parsing. In: Conference on Empirical Methods in Natural Language Processing (EMNLP), pp. 1013–1024. Doha, Qatar: Association for Computational Linguistics (2014b)

Arabic Text Classification Based on Word and Document Embeddings

Abdelkader El Mahdaouy[1,2(✉)], Eric Gaussier[1], and Saïd Ouatik El Alaoui[2]

[1] Grenoble Alpes University, CNRS-LIG/AMA, Grenoble, France
{abdelkader.elmahdaouy,eric.gaussier}@imag.fr
[2] Department of Computer Science, LIM, FSDM, USMBA, Fez, Morocco
s_ouatik@yahoo.com

Abstract. Recently, Word Embeddings have been introduced as a major breakthrough in Natural Language Processing (NLP) to learn viable representation of linguistic items based on contextual information or/and word co-occurrence. In this paper, we investigate Arabic document classification using Word and document Embeddings as representational basis rather than relying on text preprocessing and bag-of-words representation. We demonstrate that document Embeddings outperform text preprocessing techniques either by learning them using Doc2Vec or averaging word vectors using a simple method for document Embedding construction. Moreover, the results show that the classification accuracy is less sensitive to word and document vectors learning parameters.

Keywords: Arabic text classification · Arabic natural language processing · Document embeddings · Word embeddings · SKIP-Gram · Continuous Bag-of-Word · Glove · Doc2vec

1 Introduction

Text classification is the task of assigning predefined labels to natural language textual documents. It can provide conceptual views of document collections and has important applications in many information retrieval applications. With the tremendous growth of available Arabic textual information on the Web and storage supports, researchers are highly challenged to find better ways to develop effective textual information access techniques. Unfortunately, Arabic language presents serious challenges to researchers and developers of natural language processing applications. These applications have to deal with several problems related to the nature and the structure of the underlying language [10]. Hence, most researchers on Arabic NLP have focused on word stemming techniques. The aim is to reduce word variations into a concise representation while preserving most of its linguistic features. Despite the fact that these stemmers have shown a good performance for many Arabic NLP applications, most stemming

© Springer International Publishing AG 2017
A.E. Hassanien et al. (eds.), *Proceedings of the International Conference on Advanced Intelligent Systems and Informatics 2016*, Advances in Intelligent Systems and Computing 533, DOI 10.1007/978-3-319-48308-5_4

techniques introduce a large amount of noise in documents representations. In other words, root-based stemmers may group non semantically similar words into the same word index (root), while light stemming techniques may fail to group two semantically similar terms to the same stem [14,20].

Recently, NLP researchers have proposed many Word Embeddings (WEs) techniques to learn a viable representation of linguistic items based on contextual information or/and word co-occurrence [19,21]. This representation is a geometric way to capture the meaning of a word via low-dimensional dense vectors. These vectors capture dimensions of statistical and semantic information and they are used both as an end in itself (word similarity task and word analogy), and as a representational basis for NLP tasks like text classification, document clustering, part of speech tagging, named entity recognition, sentiment analysis, and Information Retrieval [9,11,27]. A more interesting model is the Document Vector model [18] that extends the WEs model (word2vec) to go beyond word level to learn phrase-level or document level representations.

In this paper, we study Arabic text classification based on Word Embeddings. Hence, documents are represented using real valued vectors rather than relying on bag-of-word representation and word stemming step. The document vectors are obtained either by averaging word vectors that are learnt using word level embedding models or simply by learning them using the document vector model (doc2vec). The Word Embeddings are learnt using the well known Continuous Bag-of-words and the Skip-Gram models of the word2vec [19], and the Glove model [21]. For document level embedding, we used the document vector model [18]. The document embeddings techniques are compared against the bag-of-words representation using word count and TFIDF weighting scheme. For representing documents within bag-of-words, we used many text preprocessing techniques such as stop word removal, normalization, and word stemming. For word stemming purpose, we used four stemming algorithms: Khoja stemmer [15], Light 10 stemmer [17], Moataz stemmer [22], and Tashaphyne stemmer[1]. The evaluations are performed using the SVM classification algorithm. The results demonstrate that the document embeddings techniques outperform all the bag-of-words representations that are obtained using the aforementioned text pre-processing techniques. Moreover, the results show that the performance is less sensitive to the learning parameters of WEs and document vectors (vectors dimensions and windows size).

2 Related Work

The automated classification of Arabic texts into predefined labels/classes has witnessed a booming interest, due to the increased availability of documents in a digital form and the ensuing need to organize them.

The large amount of Arabic text classification/categorization studies have focused on text pre-processing in order to deal with the complex morphology of the Arabic language. Said et al. [24] conducted several evaluations to

---

[1] https://sourceforge.net/projects/tashaphyne/.

study the impact of text preprocessing tools on Arabic documents classification. They compared raw text, Al-Stem stemmer, Sebawai root extractor, RDI MORPHO3 stemmer, and RDI MORPHO3 root extractor. The evaluation was performed using four feature scoring methods and different threshold values. The results showed that using light stemmer with a good-performing feature selection method such as Mutual Information or Information Gain improves the accuracy for small sized data sets and small threshold values for large data sets. In line with the previous work, Saad [22] studied the impact of pre-processing techniques (term weighting, stemming, light stemming, etc.) on several classification algorithms: KNN, Decision Tree, SVM, and Naive Bayes. The experiments were performed on several Arabic text classification data sets, including CNN, BBC, and OSAC. The results showed that light stemming and term pruning are the best feature reduction techniques. Moreover, Hmeidi et al. [12] studied the impact of raw text, light stemming and khoja root-based stemmer using standard classifiers such as Naive Bayes (NB), Support Vector Machine (SVM), K-Nearest Neighbor (KNN), Decision Tree (J48) and Decision Table classifiers. The results showed that classifying Arabic text documents via the SVM or the NB classifiers provides better accuracy, especially when used with light10 stemming. The same conclusion was drawn up by Ayedh et al. [7] and Al-Badarneh [4] using several preprocessing techniques. Furthermore, Khreisat [16] and Al-Molegi et al. [1] have proposed to classify Arabic documents by combining N-grams with several similarity measures, including Dice, Manhattan, and Euclidean distances. The overall comparison results illustrated that combining Dice measure with tri-gram gave a better performance.

In other works, researchers used dimensionality reduction methods to capture document semantics using topic models. The latter rely on word co-occurrence at documents level to model term associations. Non-probabilistic topic models [13], where documents are represented in term space of reduced dimensionality, are obtained using terms-documents matrix factorization methods. The probabilistic topic models [26] represent term associations by assuming that each topic is a probabilistic distribution over terms in the vocabulary and each document in the collection is defined as a probabilistic distribution over the set of topics. Based on the same idea, Ayadi et al. [5,6] used the LDA to extract Arabic latent topics by varying the number of topics between 25 and 100. The comparison results showed that the topics space outperforms classification in full words space as well as the LSI reduction method. Al-Anzi et al. [2] proposed a method for Arabic text classification based on LSI and clustering techniques to group similar unlabeled documents into a pre-specified number of topics. The results showed that this technique is an excellent approach to label documents without training data. In another work, Al-Anzi et al. [3] proposed a method for text classification using cosine similarity and Latent Semantic Indexing. The results demonstrated that the LSI features significantly outperform the TFIDF-based methods and that the k-Nearest Neighbors (based on cosine measure) and support vector machine achieved the best performance. Although these models have already shown a good performance, an evaluation of word embeddings against traditional word counting approaches [8] demonstrated the success of the former on a variety of NLP tasks (Semantic relatedness, Synonym detection, Concept categorization, and Selectional preferences). Moreover, experiments illustrated that the neural models are much less sensitive to parameter choices, as even the worst neural models perform relatively well, whereas counting models can fail with unsuitable parameters settings.

Although the motivation is similar, this work differ from the previous ones in that we attempt to use Word and document Embedding as a reduced semantic representation of semantic content of Arabic texts. The document vectors are obtained either by averaging word vectors that are learnt using word level embedding models or simply by learning them using the document vector model.

3   Neural Embedding Models

Unsupervised words and documents representations are very useful in many NLP tasks both as inputs to learning algorithms and as extra features in the NLP systems. Indeed, these representations reflect similarities and dissimilarities between words and documents by grouping the vectors of similar words and documents together in a low dimensional vector space.

3.1   Continuous Bag-of-Words Model (CBOW)

The CBOW model consists of representing the context by the surrounding words for a given target word [19]. The word representation is learnt by maximizing the log probability to predict the target word given its context. The CBOW model uses a simple neural architecture where the non-linear hidden layer is removed and the projection layer is shared for all words. For a given target word w_t and its context $\{w_{t-c}, ..., w_{t-1}, w_{t+1}, ..., w_{t+c}\}$, the model maximizes the following equation:

$$\frac{1}{|\mathcal{V}|} \sum_{t=1}^{|\mathcal{V}|} log[P(w_t|w_{t-c}, ..., w_{t-1}, w_{t+1}, ..., w_{t+c})] \tag{1}$$

where $|\mathcal{V}|$ is the number of words in the corpus and c is the size of the dynamic context of w_t.

3.2   Skip-Gram Model

The Skip-gram model [19] reverses the input and the output of the neural network used by the CBOW model. Hence, instead of predicting the current word using its surrounding words (context), the Skip-gram model is trained to maximize the log probability to predict the context for a given target word. Given a sequence of training words $\{w_{t-c}, ..., w_{t+c}\}$, the model maximizes the following average log probability to predict the context of the current target word:

$$\frac{1}{|\mathcal{V}|} \sum_{t=1}^{|\mathcal{V}|} \sum_{j=t-c, j\neq t}^{t+c} log[P(w_j|w_t)] \tag{2}$$

where $|\mathcal{V}|$ is the number of words in the corpus and c is the size of the dynamic context of w_t.

3.3 Glove Model

The Glove model is a global log-bilinear regression model [21] that combines the advantages of global matrix factorization and local context window methods. It is trained on the non-zero entries of the global word-word co-occurrence matrix. The model constructs a word-word co-occurrence matrix X, whose element X_{ij} represents the number of times a word j occurs in the context of word i. For each word pair, Glove defines a soft constraint: $w_i^T w_j + b_i + b_j = log(X_{ij})$ where w_i and w_j are the vectors for the main word and the context word respectively. Finally, an additional biases b_i for w_i and b_j for w_j are added to restore the symmetry. The cost function is given by:

$$J = \sum_{i=1}^{|\mathcal{V}|} \sum_{j=1}^{|\mathcal{V}|} f(X_{ij})(w_i^T w_j + b_i + b_j - \log X_{ij})^2 \tag{3}$$

where f is a weighting function to avoid weighting all co-occurrences equally:

$$f(X_{ij}) = \begin{cases} (\frac{X_{ij}}{x_{max}})^\alpha & \text{if } X_{ij} < x_{max} \\ 1 & \text{otherwise} \end{cases} \tag{4}$$

3.4 Document Embedding

The Document Embedding model [18] is a generalization of the word2vec model. It learns fixed-length feature representations from variable-length pieces of texts, such as sentences, paragraphs, and documents. The model is trained to predict words in the document. It maps every paragraph to a unique vector, represented by a column in a matrix D and every word is also mapped to a unique vector, represented by a column in a matrix W. Then the paragraph vector and the word vectors are concatenated or averaged to predict the next word in context. The model is composed of two architectures which are the document vectors distributed memory model (PVDM) and the Paragraph Vector without word ordering (PV-DBOW).

3.5 Document Embeddings Based on Word Vectors Averaging

For the document Embedding construction using Word Embeddings, we used a method similar to the ADD-SI model which has been proposed by Vulić et al. [25]. Hence, we averaged the word vectors of each document using the frequency of the target word and the document length.

Let assume that for a document $d = \{w_1, w_2, ..., w_{l_d}\}$ where w_i is a token of the document d and l_d is the document length. Then, the corresponding word vectors (trained using any Word Embedding model) for d are $\overrightarrow{W_1}, \overrightarrow{W_2}, ..., \overrightarrow{W_{l_d}}$

in k dimensional vector space. The document vector $\overrightarrow{d}$ can be obtained using addition compositional operator in k-dimensional vector space using the formula:

$$\overrightarrow{d} = \frac{x^d_{w_1}}{l_d} \cdot \overrightarrow{W_1} + \frac{x^d_{w_2}}{l_d} \cdot \overrightarrow{W_2} + ... + \frac{x^d_{w_{l_d}}}{l_d} \overrightarrow{W_{l_d}} \tag{5}$$

where x^d_w is the frequency of the word w in the document d. Thus, the word vectors are weighted based on their importance in each document.

After constructing the document features vectors representation using the aforementioned (ADD-SI) method, the data set is split using 10 folds cross-validation method. Hence, the classifier is trained and tested using the obtained features vectors rather than the bag-of-words representation of documents.

4 Experimental Results

4.1 Evaluation Method and Data

Experiments are performed on the OSAC data set [23] using Support Vector Machine (SVM). The corpus consists of 22,429 documents that belong to one of ten categories (Economics, History, Entertainments, Education & Family, Religious and Fatwas, Sports, Health, Astronomy, Law, Stories, and Cooking Recipes). The data set contains approximately 18M words. The underlying corpus has been collected from several online resources such as BBC Arabic, CNN Arabic and Aljazeera Newswire.

In order to compare the document embeddings representations of Arabic documents with their bag-of-words representations, we used stop words removal, text normalization, and stemming (Khoja stemmer, Light 10 stemmer, Moataz stemmer, and Tashaphyne stemmer). To train the word and the document embedding models, we collected a large amount of raw Arabic texts, containing about 216M tokens, including Arabic BBC, CNN and OSAC corpora[2], Arabic Newswire LDC catalog set[3], and other sentence corpora collected from WORTSHATZ[4]. For each model, we used various settings for the context windows size (5, 10, 15) and the vector dimensions (100, 200, 300, 400, 500, 600). The results are evaluated using the average weighted precision, recall, and F1 measures. The evaluation is performed using the 10 folds cross-validation method.

4.2 Evaluation Results

Firstly, we compared the documents embeddings representations of Arabic document against the bag-of-words representation that is obtained using text preprocessing techniques and stop words removal. For Word and document Embeddings, we used the default parameter setting (the dimension of vector and the windows size are fixed to 300 and 10, respectively). Table 1 presents the obtained results for the different document representations using the SVM classifier.

---

[2] https://sourceforge.net/projects/ar-text-mining/files/Arabic-Corpora/.
[3] LDC catalog number $LDC2001T55$.
[4] http://www.cls.informatik.uni-leipzig.de/langs/ara.

Table 1. The obtained results for document Embedding models and the bag-of-words representation using word count and TFIDF weighting using the SVM classifier.

	Preprocessing	Recall	Precision	F1
Word count	Text normalization	88.94	91.06	88.94
	Light 10	89.52	91.53	88.93
	Motaz	89.36	91.21	88.65
	Khoja	87.59	90.52	87.14
	Tashaphyne	89.05	91.38	89.20
TFIDF	Text normalization	89.82	92.15	89.32
	Light 10	90.39	92.11	89.86
	Motaz	90.50	92.31	89.90
	Khoja	88.67	91.49	88.56
	Tashaphyne	88.89	92.05	89.38
Document vectors	CBOW	95.05	95.49	94.85
	SKIP-gram	95.92	96.15	95.81
	Glove	96.52	96.82	96.36
	Doc2vec	94.53	94.66	94.50

The results show that document embeddings representations outperform all bag-of-words representations (obtained using text normalization and stemming algorithms) using both word count and TFIDF. Hence, the document embeddings are more suitable for representing and capturing Arabic documents content in a reduced dimension vector space. Moreover, averaging word vectors to compose document embeddings from word embeddings led to a better performance than learning document representations based on the doc2vec model. Small improvements are achieved by averaging word vectors that are learnt using the Glove model in comparison to those obtained using the word2vec models (CBOW and Skip-gram). The latter can be explained by the fact that the Glove model combines both word co-occurrence and local context methods to learn viable word vectors representation.

Lastly, we studied the accuracy sensitivity to the neural embeddings parameters as it has been mentioned in the previous subsection. Figure 1 draws up the sensitivity of the F1-measure performance to vector dimensions and context size parameters used to learn word and document embeddings.

Figure 1 shows that the performance is less sensitive to the parameters used to learn word embeddings for document embedding construction using CBOW, SKIP-gram, and Glove models. The latter finding agrees with the study of Baroni et al. [8]. For all parameters settings, the document embeddings representations (constructed using CBOW, SKIP-gram, and Glove) outperform the best stemming algorithm (light10). Moreover, the performance obtained by Doc2vec model is sensitive for small context size. For all parameter settings, constructing document embeddings from word embeddings shows better performance than

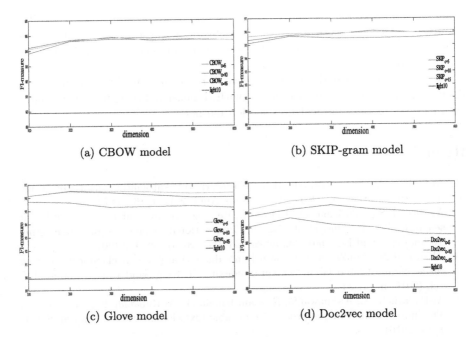

(a) CBOW model   (b) SKIP-gram model

(c) Glove model   (d) Doc2vec model

Fig. 1. Effect of vector dimensions and context size parameters for each embeddings model on F1-measure performance

learning them using the Doc2vec, especially the word vectors that are obtained using the Glove model. As a result, classifying Arabic documents based on document embeddings that are learnt directly using the Doc2vec model or simply by constructing them from word embeddings outperforms all Arabic text preprocessing techniques even in vector spaces of small dimension (100), since the latter captures implicit semantic relationships between words and documents.

5   Conclusion

In this paper, we propose distributed representation of words and documents in vector space as representational basis for Arabic text classification. The document representations are learned in an unsupervised manner to capture implicit semantic relationships between words and documents using several neural embeddings models. Hence, the document vectors are obtained either by composing them using word vectors (Compositional Distributional Semantics) or the Doc2vec model. Empirically, the document embeddings representations outperform the classical bag-of-words representations. The latter are obtained using many text preprocessing techniques, including stop word removal, text normalization, and stemming. The analysis of the document embeddings dimensions and the context size demonstrates that the classification performance is less sensitive when document embeddings are constructed from word vectors (trained using CBOW, SKIP-gram, and Glove) or simply using a large context size for training the Doc2vec model. The main conclusion is that document embeddings show better performance than text preprocessing methods though the vectors are in a lower-dimensional vector space.

A straightforward path of future research is to compare the performance of document embeddings representations with dimensionality reduction methods such as topic models and feature selection methods.

References

1. Al-Molegi, A., Izzat Alsmadi, H.N., Albashiri, H.: Automatic learning of arabic text categorization. Int. J. Digit. Contents Appl. **2**(1), 1–16 (2015)
2. Al-Anzi, F.S., AbuZeina, D.: Big data categorization for arabic text using latent semantic indexing and clustering. In: International Conference on Engineering Technologies and Big Data Analytics (ETBDA 2016), pp. 1–4 (2016)
3. Al-Anzi, F.S., AbuZeina, D.: Toward an enhanced arabic text classification using cosine similarity and latent semantic indexing. J. King Saud Univ. Comput. Inf. Sci. (2016, in press)
4. Al-Badarneh, A., Al-Shawakfa, E., Bani-Ismail, B., Al-Rababah, K., Shatnawi, S.: The impact of indexing approaches on arabic text classification. J. Inform. Sci. **1**, 1–15 (2016)
5. Ayadi, R., Maraoui, M., Zrigui, M.: Latent topic model for indexing arabic documents. IJIRR **4**(1), 29–45 (2014)
6. Ayadi, R., Maraoui, M., Zrigui, M.: LDA and LSI as a dimensionality reduction method in arabic document classification. In: Dregvaite, G., Damasevicius, R. (eds.) Information and Software Technologies. Communications in Computer and Information Science, vol. 538, pp. 491–502. Springer International Publishing, Cham (2015)
7. Ayedh, A., Tan, G., Alwesabi, K., Rajeh, H.: The effect of preprocessing on arabic document categorization. Algorithms **9**(2), 27 (2016)
8. Baroni, M., Dinu, G., Kruszewski, G.: Don't count, predict! a systematic comparison of context-counting vs. context-predicting semantic vectors. In: Proceedings of the 52nd Annual Meeting of the Association for Computational Linguistics, Baltimore, Maryland, pp. 238–247. ACL, June 2014
9. Collobert, R., Weston, J., Bottou, L., Karlen, M., Kavukcuoglu, K., Kuksa, P.: Natural language processing (almost) from scratch. J. Mach. Learn. Res. **12**, 2493–2537 (2011)
10. Farghaly, A.: Computer processing of arabic script-based languages: current state and future directions. In: Proceedings of the Workshop on Computational Approaches to Arabic Script-Based Languages, p. 1. Association for Computational Linguistics (2004)
11. Ganguly, D., Roy, D., Mitra, M., Jones, G.J.: Word embedding based generalized language model for information retrieval. In: Proceedings of the 38th International ACM SIGIR Conference on Research and Development in Information Retrieval, SIGIR 2015, New York, NY, USA, pp. 795–798. ACM (2015)
12. Hmeidi, I., Al-Ayyoub, M., Abdulla, N.A., Almodawar, A.A., Abooraig, R., Mahyoub, N.A.: Automatic arabic text categorization: a comprehensive comparative study. J. Inform. Sci. **41**(1), 114–124 (2015)

13. Hofmann, T.: Probabilistic latent semantic indexing. In: Proceedings of the 22nd Annual International ACM SIGIR Conference on Research and Development in Information Retrieval, SIGIR 1999, New York, NY, USA, pp. 50–57. ACM (1999)
14. Kadri, Y., Nie, J.Y.: Effective stemming for arabic information retrieval. In: International Conference of the Challenge of Arabic for NLP/MT at the British Computer Society (BCS), pp. 68–74 (2006)
15. Khoja, S., Garside, R.: Stemming arabic text, Computing Department, Lancaster University (1999)
16. Khreisat, L.: A machine learning approach for arabic text classification using n-gram frequency statistics. J. Informetrics 3(1), 72–77 (2009)
17. Larkey, L., Ballesteros, L., Connell, M.: Light stemming for arabic information retrieval. In: Soudi, A., Bosch, A.D., Neumann, G. (eds.) Arabic Computational Morphology, Text, Speech and Language Technology, vol. 38, pp. 221–243. Springer, Netherlands (2007)
18. Le, Q.V., Mikolov, T.: Distributed representations of sentences and documents (2014). CoRR abs/1405.4053. http://arxiv.org/abs/1405.4053
19. Mikolov, T., Chen, K., Corrado, G., Dean, J.: Efficient estimation of word representations in vector space. In: Proceedings of International Conference on Learning Representations, ICLR 2013 (2013)
20. Otair, M.A.: Comparative analysis of arabic stemming algorithms. Int. J. Managing Inf. Technol. (IJMIT) 5(2), 1–12 (2013)
21. Pennington, J., Socher, R., Manning, C.: Glove: global vectors for word representation. In: Proceedings of the 2014 Conference on Empirical Methods in Natural Language Processing (EMNLP), pp. 1532–1543. Association for Computational Linguistics, Doha, October 2014
22. Saad, M.K.: The impact of text preprocessing and term weighting on arabic text classification. Ph.D. thesis, Islamic University of Gaza, Palestine (2010)
23. Saad, M.K., Ashour, W.: Osac: open source arabic corpora. In: 6th International Conference on Electrical and Computer Systems (EECS 2010), Lefke, Cyprus, 25–26 November, pp. 118–123 (2010)
24. Said, D.A., Wanas, N.M., Darwish, N.M., Hegazy, N.: A study of text preprocessing tools for arabic text categorization, pp. 230–236 (2009)
25. Vulić, I., Moens, M.F.: Monolingual and cross-lingual information retrieval models based on (bilingual) word embeddings. In: Proceedings of the 38th International ACM SIGIR Conference on Research and Development in Information Retrieval, SIGIR 2015, New York, NY, USA, pp. 363–372. ACM (2015)
26. Wei, X., Croft, W.B.: Lda-based document models for ad-hoc retrieval. In: Proceedings of the 29th Annual International ACM SIGIR Conference on Research and Development in Information Retrieval, SIGIR 2006, New York, NY, USA, pp. 178–185. ACM (2006)
27. Zahran, M.A., Magooda, A., Mahgoub, A.Y., Raafat, H., Rashwan, M., Atyia, A.: Word representations in vector space and their applications for arabic. In: Gelbukh, A. (ed.) CICLing 2015. LNCS, vol. 9041, pp. 430–443. Springer, Heidelberg (2015). doi:10.1007/978-3-319-18111-0_32

An Enhanced Distance Based Similarity Measure for User Based Recommendations

Yasmine M. Afify$^{(\boxtimes)}$, Ibrahim F. Moawad, Nagwa L. Badr,
and Mohamed F. Tolba

Faculty of Computer and Information Sciences,
Ain Shams University, Cairo 11566, Egypt
{yasmine.afify, ibrahim_moawad,
nagwabadr}@cis.asu.edu.eg, fahmytolba@gmail.com

Abstract. Internet users are overwhelmed with a large number of choices, consequently, there is a need to filter and prioritize relevant information. Recommender System (RS) solves this problem by searching through information provided by users similar to the active user. Precise determination of similar users is the keystone to accuracy of personalized recommendation and in this regard, the contribution of this paper is two-fold. First, an enhanced Distance based similarity measure is introduced. Second, a systematic evaluation is presented of the predictive performance of the proposed similarity measure against different similarity measures in recommendations based on user based Collaborative Filtering (CF). The evaluation encompasses both numeric and non-numeric measures against the proposed measure. The performance metrics are the recommendation accuracy (statistical and decision-making) and coverage. Experimental results on three real-world datasets show that the enhanced Distance based similarity outperforms all other similarity measures for user based recommendations in respect of the recommendation accuracy and coverage.

Keywords: Recommendation · Collaborative filtering · Similarity measure · Prediction accuracy · Coverage · Precision · Recall

1 Introduction

Due to the rapid growth of Internet content in conjunction with the information overload problem, the use of Recommender Systems (RS) has proven its inevitable significance. The Collaborative Filtering (CF) is the dominant algorithm underlying the RS work and CF plays an important role in the recommendation, and is widely used due to its simplicity and effectiveness using the k Nearest Neighbors (kNN) recommendation algorithm as its reference algorithm [1]. The principle is that RS allows users to submit their ratings about the items and when it stores enough information, personalized recommendations are made for the active users. Recommendation approaches can be divided to model based and memory based approaches [1]. In the model based approach, the RS uses the rating information in order to build a model which generates the recommendations. On the other hand, the memory based approaches uses the rating information to build a user profile and identifies neighbors

© Springer International Publishing AG 2017
A.E. Hassanien et al. (eds.), *Proceedings of the International Conference
on Advanced Intelligent Systems and Informatics 2016*, Advances in Intelligent
Systems and Computing 533, DOI 10.1007/978-3-319-48308-5_5

of the active user. Afterwards, they predict the ratings of the active user on target items as an average of ratings given to them by the neighbors and recommend the top rated items. Therefore, precise identification of the neighbors is a keystone to the recommendation quality.

In order to identify the active user neighbors, RS employs basic similarity functions which determine the similarity between two users [2]. Similarity measures include: Numeric (Cosine, Pearson Correlation Coefficient (PCC) and Distance based) and non-numeric (Jaccard similarity coefficient) functions. Numeric functions consider the rates values, while the non-numeric functions consider the non-numerical information of the rates such as the proportion and structure of the rates.

Correct identification of the neighbors is significant to generate accurate recommendations. In this regard, the contributions of this paper are as follows. First, a systematic evaluation of the prediction accuracy of the basic similarity measures on user based recommendations is presented. The evaluation encompasses the most commonly used traditional numeric measures (Cosine, PCC and Distance based) [1] in addition to non-numeric measure (Jaccard similarity coefficient) and the evaluation criteria include the recommendation accuracy and coverage. Second, an enhanced Distance based similarity measure (ProDist) is introduced that rewards the increase in the number of common items between two users. Third, the performance of the proposed measure is assessed using three real-world datasets. Results show that the Pro-Dist outperforms the traditional similarity measures in respect of the recommendation accuracy and coverage.

The rest of the paper is organized as follows. Section 2 presents state-of-the-art related work. Section 3 proposes the enhanced Distance based similarity measure. The experimental evaluation is presented in Sect. 4 while Sect. 5 concludes the paper.

2 Related Work

Most popular prediction methods adopted in CF systems are memory based approaches. Memory based CF methods base rating predictions on the previously rated items by neighbors of the active user, which are users considered similar to the active user based on a similarity measure. In this section, an overview is presented on related works that introduce new similarity measures or surveys the existing ones.

Authors of [3] proposed a new similarity function in CF techniques in which the similarity between a target item and each of the co-rated items are considered to determine whether or not the two items belong to the same genre of interest. Authors of [4] proposed a content-boosted CF approach applied to movie recommendations. Their motivation is to investigate whether further success can be obtained for handling the sparsity problem by combining the previously proven methods 'local and global user similarity' and the 'effective missing data prediction'. Authors of [5] introduced a new similarity measure that mitigates new user cold start problem to provide greater precision in the results to users who cast few votes. Authors of [1] described and compared a representative group of similarity measures used in the kNN algorithm, they also provided an overview of recommender systems as well as collaborative filtering methods and algorithms. Authors of [6] proposed a new similarity function in order to select different neighbors for each different targeted item. In their work, the rating of a user on an item was weighted by the item similarity between the item and the target item.

3 Enhanced Distance Based Similarity Measure

A similarity measure determines the similarity between two users (active user and target user) by comparing the ratings of all the items rated by the two users. Most CF based RSs use one similarity measure to compute the rating based similarity of either numeric (Cosine, PCC or Distance based) or non-numeric (Jaccard similarity coefficient) [1, 6]. Nevertheless, each similarity measure computes the similarity from only one perspective.

In specific, the Cosine similarity accounts for the angle between the two rating vectors. It is a vector-space approach based on linear algebra that measures similarity between two vectors of an inner product space by measuring the cosine of the angle between them. This can be computed efficiently by taking their dot product and dividing it by the product of their Euclidean norms. The Cosine similarity cosSim between users x and y is calculated using Eq. 1:

$$cosSim(x,y) = \frac{\sum_{i \in R_x \cap R_y} R_{x,i} * R_{y,i}}{\sqrt{\sum_{i \in R_x \cap R_y} R_{x,i}^2} \sqrt{\sum_{s \in R_x \cap R_y} R_{y,i}^2}} \tag{1}$$

where $R_{x,i}$ and $R_{y,i}$ represent the rate given by users x and y to item i respectively, R_x and R_y represent the list of items rated by users x and y respectively.

On the other hand, the PCC similarity accounts for the rating patterns. It is a measure of the degree of linear dependence between two rating vectors. It computes the statistical correlation between common ratings of two users to determine their similarity. The PCC similarity pSim between users x and y is calculated using Eq. 2:

$$pSim(x,y) = \frac{\sum_{i \in R_x \cap R_y} (R_{x,i} - \bar{x})(R_{y,i} - \bar{y})}{\sqrt{\sum_{i \in R_x \cap R_y} (R_{x,i} - \bar{x})^2} \sqrt{\sum_{i \in R_x \cap R_y} (R_{y,i} - \bar{y})^2}} \tag{2}$$

where $\bar{x}$. and $\bar{y}$ represent the average rate given by users x and y respectively.

Finally, the Distance based similarity accounts for the rating magnitude. The Distance based similarity dSim between users x and y is calculated using Eq. 3:

$$dSim(x,y) = \frac{1}{1 + \sqrt{\sum_{i \in R_x \cap R_y} (R_{x,i} - R_{y,i})^2}} \tag{3}$$

On the other hand, the Jaccard similarity coefficient accounts for the arrangements of the rates. It rewards situations in which the two users have rated the same items, taking into account the proportion of the rated items to the total number of rated items by both of them. The Jaccard similarity jSim is calculated using. 4:

$$jSim(x,y) = \frac{|R_x \cap R_y|}{|R_x - R_y| + |R_y - R_x| + |R_x \cap R_y|} \tag{4}$$

During the experimental evaluation, it was noticeable that the Distance based similarity value decreases with the increasing number of common items between the two users despite the fact that the user rates are very close (difference values of 1 and 2). The spearman correlation coefficient [7], is a measure of statistical dependence between variables and was employed to validate our notice. The Spearman correlation coefficient is defined as the Pearson correlation coefficient between the ranked variables. For a dataset of size n, the n scores X_i, Y_i of two users U_1 and U_2 are converted to ranks x_i, y_i and the spearman correlation coefficient is calculated using Eq. 5. Identical values (rank ties or value duplicates) are assigned a rank equal to the average of their positions in the ascending order of the values. Table 1 depicts the calculations of the spearman correlation between the common number of items between two users X_i and their distance based similarity Y_i.

$$\rho = 1 - \frac{6\sum d_i^2}{n(n^2 - 1)} \ where \ d_i = x_i - y_i \tag{5}$$

Table 1. Pearson correlation between number of common items and distance based similarity

U_1 scores	U_2 scores	X_i	Y_i	rank x_i	rank y_i	d_i^2
2,3	3,5	2	0.309	1	4	9
1,3,2	3,4,4	3	0.25	2	3	1
3,3,4,5	1,1,3,3	4	0.217	3	2	1
1,3,4,5,2	3,5,2,4,3	5	0.21	4	1	9
$\rho = -1.$						

The spearman correlation coefficient confirmed that there is a negative tendency between the Distance based similarity and the number of common items between the two users despite their close rate values. In our opinion, this increase should be rewarded due to its relevance in identifying similar user preferences. In other words, when two users rate the same set of items, then they share the same interests and they should be perceived as similar users. Moreover, our experiments show that even in the case of two users having rated the same set of items with close rates, the Distance based similarity is low and decreases with the increasing number of common items.

Consequently, an enhanced Distance based similarity measure is proposed that acknowledges the increase in common items between the two users. The calculation of the enhanced Distance based similarity edSim is normalized using the common number of items as shown in Eq. 6.

$$edSim(x,y) = \frac{1}{1 + \frac{\sqrt{\sum_{i \in R_x \cap R_y}(R_{x,i}-R_{y,i})^2}}{|R_x \cap R_y|}} \ where \ R_x \cap R_y \neq \emptyset \tag{6}$$

Table 2. Pearson correlation between number of common items and proposed enhanced distance based similarity

X_i	Y_i	Rank x_i	Rank y_i	d_i^2
2	0.47	1	1	0
3	0.5	2	2	0
4	0.52	3	3	0
5	0.57	4	4	0
$\rho = 1$.				

In order to assess the proposed enhanced Distance based similarity measure, the spearman correlation coefficient was employed. As shown in Table 2, there is a strong positive tendency between the number of common items between the two users X_i and the enhanced Distance based similarity measure Y_i.

4 Experimental Evaluation

The experimental evaluation of the prediction performance of the basic similarity measures against the proposed enhanced Distance based similarity measure is presented. The following sub-sections present datasets, setup, metrics and experimental results.

4.1 Datasets

In order to systematically assess the prediction performance of the similarity measures in user based CF recommendations, experiments employ three real data benchmarking public datasets referenced most often in the literature [2, 8]. Datasets were selected that comprise data from different contexts. The datasets include: Book-Crossing, Movie-Lens and Armor. As shown in Table 3, the datasets differ greatly in their dimensions, rating range, size and user-matrix density.

Table 3. Properties of benchmark datasets

Dataset	Item	#Users	#Items	#Feedbacks	Rates	Format	Density
Book-Crossing [9]	Books	278,858	271,379	1,149,780	0–10	"User-ID"; "ISBN"; "Book-Rating"	0.001 %
MovieLens [10]	Movies	6,040	3,952	1,000,209	1–5	UserID::MovieID:: Rating::Timestamp	4.19 %
Armor [11]	Cloud Services	7,203	125	10,080	0–5	UserID | Service Name | Attributes | Time Stamp	1.11 %

4.2 Experiment Setup

In the real-world, sparse data is very common, as users usually use very few items compared to the number of existing items in the RS [12–15]. As shown in Table 3, the densities of the user-item matrices created from the data sets is sparse enough for the evaluation process [4].

The 10-times 10-fold cross-validation [16] is used in order to assess the prediction results. The user-item matrix is partitioned into 10 equal sized subsets. In each of the 10 folds, one subset is used for testing, while the remaining 9 subsets are used for training (10 % active users and 90 % training users) and average results were reported.

4.3 Evaluation Metrics

The evaluation metrics consist of two prediction metrics: Accuracy and coverage [1]. Two classes of prediction accuracy metrics were employed: Statistical accuracy and decision-support accuracy metrics. Statistical accuracy is measured using the Mean Absolute Error (MAE), which is the average absolute deviation of predicted values to the original values. The MAE is a linear score which means that all the individual differences are weighted equally in the average. MAE is calculated using Eq. 7:

$$MAE = \frac{\sum_{au,i} |Predicted_{au,i} - R_{au,i}|}{P} \qquad (7)$$

where au represents the active user, Predicted$_{au, i}$ represents the predicted v the active user on item i, R$_{au, i}$ represents the rate value given by the active user au to item i and P represents the total number of predictions.

The decision-support accuracy is measured using the Precision and Recall metrics. The precision represents the ratio between the number of recommended relevant items RV to the user and the number of recommended items D. The relevant items V are items with rates greater than or equal to the relevancy threshold θ. The RS precision is the average of all active users A precision values, where the precision for active user au is calculated using Eq. 8:

$$precision(au) = |RV|/|D| \qquad (8)$$

The recall represents the ratio between the number of recommended relevant items RV to the user and the number of relevant items V. RS recall is the average of all active users recall values, where recall for active user au is calculated using Eq. 9:

$$recall(au) = |RV|/|V| \qquad (9)$$

Finally, the coverage metric represents the RS prediction capacity. It calculates the percentage of cases where at least one neighbor has rated an item that is not rated by the active user. The coverage is calculated using Eq. 10:

$$Coverage = \frac{1}{|A|} \sum_{au \in A} (100 * \frac{|(I - R_{au}) \cap (Predicted_{au})|}{|I - R_{au}|}) \qquad (10)$$

where I represents set of items and R_{au} represent set of items rated by active user.

4.4   Experimental Results

In this section, the prediction performance of the different similarity measures was experimentally studied when employed as a measure for the user based similarity calculation in RS. The experimental results on the MovieLens, Book-Crossing and Armor datasets are shown in Figs. 1, 2 and 3 respectively, where the x-axis represents the similarity measures and the y-axis represents the performance metric values.

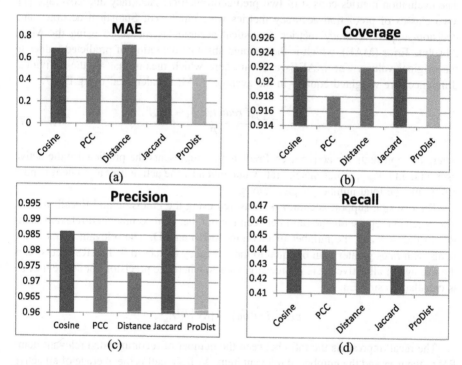

Fig. 1. Prediction performance of similarity measures on MovieLens Dataset (a) MAE, (b) Coverage, (c) Precision and (d) Recall

In these experiments, the number of recommended items is set to 20 [5, 17] and the relevance threshold θ is assigned relative to the range of rate values in each dataset. In specific, θ is assigned value of 3 for the MovieLens and Armor datasets while it is assigned value of 5 for the Book-Crossing dataset.

For the Movies dataset, Fig. 1 (a) shows that in respect of the basic numeric similarity measures, ProDist similarity surpasses all other similarity measures in respect

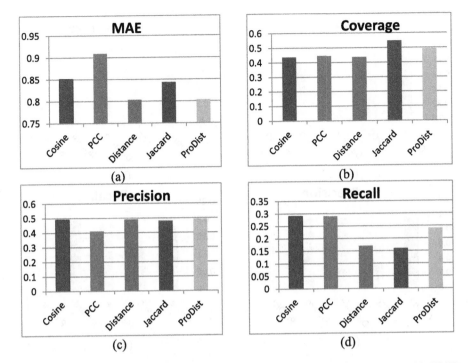

Fig. 2. Prediction performance of similarity measures on book-crossing dataset (a) MAE, (b) Coverage, (c) Precision and (d) Recall

of the recommendation accuracy and the Distance based similarities has the best performance with the lowest error rate. The Distance based similarity records the worst performance with the highest error rate.

In respect of the coverage, Fig. 1(b) shows that the PCC similarity records the lowest coverage while the ProDist similarity records the best coverage.

Regarding the precision, it is clear from Fig. 1(c) that the Jaccard similarity records the best precision values among all similarity measures while the Distance based similarity records the least precision value. The ProDist similarity stands at an excellent position (2nd) among all similarity measures in respect of the precision.

Finally, in respect of the coverage, Fig. 1(d) depicts that best coverage is reached by the Distance based similarity. The ProDist similarity and Jaccard similarities record the least coverage values among all the similarity measures.

For the Book-Crossing dataset, Fig. 2(a) shows that in respect of the numeric similarity measures, the Distance based and the ProDist similarities outperform all other similarity measures. The PCC similarity records the worst performance with the highest error rate. It is worth noting that in the Book-Crossing dataset, the performance of the enhanced Distance based similarity matches that of the Distance based similarity. This match is due to the fact that this dataset is extremely sparse (matrix density 0.001 %) and most of the users have only one or two books in common.

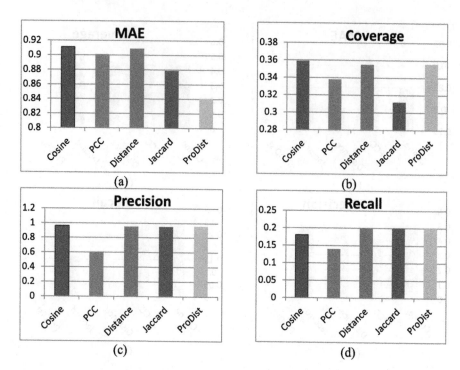

Fig. 3. Prediction performance of similarity measures on armor dataset (a) MAE, (b) Coverage, (c) Precision and (d) Recall

In respect of the coverage, Fig. 2(b) shows that the Cosine, PCC and Distance based similarities record the lowest coverage while the Jaccard similarity records the best coverage among all similarity measures. The ProDist similarity stands at an excellent position (2nd) among the similarity measures in respect of recommendation coverage.

Regarding the precision, Fig. 2(c) demonstrates that the ProDist, Cosine and Distance based similarities record the best precision with almost the same precision values followed by the Jaccard similarity while the PCC similarity comes last.

Finally, Fig. 2(d) shows that the best recall values are recorded by the Cosine and PCC similarities followed by the ProDist, Distance based and Jaccard similarities.

For the Armor dataset, Fig. 3(a) shows that the ProDist similarity records the best prediction accuracy with the lowest error rates among all similarity measures.

In respect of the coverage, Fig. 3(b) shows that the Jaccard similarity records the lowest coverage while the Cosine similarity records the highest coverage among the similarity measures. The ProDist similarity stands at an excellent position (2nd) among the similarity measures in respect of recommendation coverage.

Regarding the precision, Fig. 3(c) demonstrates that the Cosine similarity records the best precision value among all similarity measures. The ProDist stands at an excellent position (2nd) while the PCC similarity records the least precision.

Finally, in respect of the recall, Fig. 3(d) depicts that the ProDist, Jaccard and Distance based similarities record the best precision with almost the same precision values followed by the Cosine similarity and last the PCC similarity.

It is worth noting that the prediction accuracy and coverage of the MovieLens dataset surpasses other datasets with lower MAE, higher precision and coverage values. Therefore, it is concluded that the matrix density has a positive impact on the prediction performance. The more density, the better prediction performance.

5 Conclusion

Based on an extensive evaluation, it was noticeable that the Distance based similarity has a negative tendency with the number of common items between two users despite the fact that the users have submitted close rate values. Consequently, an enhanced Distance based similarity measure was proposed that addresses this problem and verified its tendency by using the Spearman correlation coefficient.

Moreover, a systematic evaluation was presented of the predictive performance of the proposed measure against different similarity measures (numeric and non-numeric) in recommendations based on the user based Collaborative Filtering (CF). The performance metrics employed were the recommendation accuracy and coverage.

Experimental results on three real-world datasets showed that in general, the prediction performance is positively affected by the increase in the dataset density. Moreover, the prediction performance of the proposed enhanced Distance based similarity measure surpasses other similarity measures in respect of the recommendation accuracy and coverage. This paper provides a comprehensive prospect for researchers and developers on the suitability of different similarity measures for user based collaborative-filtering recommender systems.

Regarding the future work, we plan to study the particle swarm optimization approach for collaborative filtering based recommender systems.

References

1. Bobadilla, J., Ortega, F., Hernando, A., Gutiérrez, A.: Recommender systems survey. Knowl. Based Syst. **46**, 109–132 (2013)
2. Ekstrand, D., Riedl, T., Konstan, A.: Collaborative Filtering recommender Systems. Found. Trends Hum. Comput. Inter. **4**(2), 81–173 (2010)
3. Yang, J.M., Li, K.F., Zhang, D.F.: Recommendation based on rational inferences in collaborative filtering. Knowl. Based Syst. **22**(1), 105–114 (2009)
4. Gözde, Ö., Hilal, K., Alpaslan, N.: A content-boosted collaborative filtering approach for movie recommendation based on local and global similarity and missing data prediction. Comput. J. **54**(9), 1535–1546 (2010)
5. Bobadilla, J., Ortega, F., Hernando, A., Bernal, J.A.: Collaborative filtering approach to mitigate the new user cold start problem. Knowl. Based Syst. **26**, 225–238 (2012). doi:10.1016/j.knosys.2011.07.021

52   Y.M. Afify et al.

6. Choi, K., Suh, Y.: A new similarity function for selecting neighbors for each target item in collaborative filtering. Knowl. Based Syst. **37**, 146–153 (2013)
7. Spearman, C.: The proof and measurement of association between two things. Int. J. Epidemiol. **39**(5), 1137–1150 (2010)
8. Yongli, R., Gang, L., Wanlei, Z.: A survey of recommendation techniques based on offline data processing. Concurrency Comput. Pract. Exper. **27**(15), 3915–3942 (2015). doi:10.1002/cpe.3370
9. Ziegler, C.N., McNee, S.M., Konstan, J.A., Lausen, G.: Improving recommendation lists through topic diversification. In: Proceedings of the 14th International Conference on World Wide Web, pp. 22–32. ACM, May 2005
10. Harper, F.M., Konstan, J.A.: The movielens datasets: history and context. ACM Trans. Inter. Intell. Syst. (TIIS) **5**(4), 19 (2015)
11. Noor, T.H., Sheng, Q.Z., Alfazi, A., Ngu, A.H., Law, J.: CSCE: a crawler engine for cloud services discovery on the world wide web. In: IEEE 20th International Conference on Web Services (ICWS), pp. 443–450. IEEE, June 2013
12. Chen, X., Zheng, Z., Liu, X., Huang, Z., Sun, H.: Personalized QoS-aware web service recommendation and visualization. IEEE Trans. Serv. Comput. **6**(1), 35–47 (2013). doi:10.1109/TSC.2011.35
13. Ma, Y., Wang, S., Hung, P.C., Hsu, C.H., Sun, Q., Yang, F.: A highly accurate prediction algorithm for unknown web service QoS values. IEEE Trans. Serv. Comput. **99**, 1–14 (2015). doi:10.1109/TSC.2015.2407877
14. Yao, L., Sheng, Q.Z., Ngu, A.H., Yu, J., Segev, A.: Unified Collaborative and Content based Web Service Recommendation. IEEE Trans. Serv. Comput. **8**, 453–466 (2014). doi:10.1109/TSC.2014.2355842
15. Afify, Y.M., Moawad, I.F., Badr, N., Tolba, M.F.: A Personalized Recommender System for SaaS Services. Concurrency Computat.: Pract. Exper. Wiley (2016). doi:10.1002/cpe.3877
16. Witten, I.H., Frank, E., Hall, M.A.: Data Mining: Practical Machine Learning Tools and Techniques, 3rd edn. Morgan Kaufmann Publishers Inc., USA (2011)
17. Symeonidis, P., Nanopoulos, A., Manolopoulos, Y.: Providing justifications in recommender systems. IEEE Trans. Syst. Hum. **38**(6), 1262–1272 (2008)

Semantic-Based Feature Reduction Approach for E-mail Classification

Eman M. Bahgat(✉) and Ibrahim F. Moawad

Faculty of Computer and Information Sciences,
Ain Shams University, Cairo, Egypt
{eman.bahgat4,ibrahim_moawad}@cis.asu.edu.eg

Abstract. E-mail is one of the most important applications for all the computer users due to its efficiency and low cost. However, some users use it in sending spam emails, which become a severe problem that has great effect on the users' performance. E-mail filtering is an important approach to identify those spam emails. In this paper, based on different machine learning algorithms, a novel semantic-based approach for email filtering is proposed. The approach analyses the content of the email and assigns a weight to each term that can help in classifying it into spam or ham email. We enhanced the traditional Email filtering approaches by applying semantic-based feature reduction model using the WordNet ontology in order to handle the high dimensionality problem of feature size. The experiments that have been conducted using Enron dataset showed great results. A comparative study has also been presented among different classifiers that prove the efficiency of the proposed approach. These classifiers are Naïve Bayes (NB), Support Vector Machine (SVM), Logistic Regression, J48 and Random Forest. The Logistic Regression classifier has the best accuracy with value of 0.96. Followed by the NB and SVM that almost have similar results of accuracy value 0.93. Finally, the Random Forest and J48 classifiers have the least accuracy values of 0.85 and 0.87 respectively.

Keywords: Email filtering · Wordnet ontology · Email classification · Spam email · Feature reduction

1 Introduction

E-mail service is one of the most important ways of communication in our life. Emails have many advantages as they are easy to use, are low cost, and can contain attached files. On the other hand, some users used email in sending computer worms and spam emails.

Spam E-mails, also known as junk emails or Unsolicited Bulk Emails (UBE), are unwanted or undesired E-mails that are sent to a group of users. According to a Cyberoam report [1], the average number of spam messages sent every day has reached 54 billion messages. The sender of spam email does not target the recipient personally, but the spam emails invade users without their assent and fill out their email inbox. Besides the time consumed in checking and deleting spam emails, they overload the network bandwidth by useless data packages. Therefore, many problems may arise:

© Springer International Publishing AG 2017
A.E. Hassanien et al. (eds.), *Proceedings of the International Conference on Advanced Intelligent Systems and Informatics 2016*, Advances in Intelligent Systems and Computing 533, DOI 10.1007/978-3-319-48308-5_6

increasing the operating cost, impacting the work productivity and privacy, and harming the network infrastructure and the recipient's machine.

To solve this problem, an Email Filtering approach is required to identify the spam emails and to dispose huge number of spam emails efficiently. The filtering approaches can be classified into two groups: based on the email origin (i.e. email source) or based on the email content (email body).

Origin-based filtering monitors the e-mail source, which is stored in the address of the sender device and the domain name. This type of filtering preserves two types of email sources; white-list and black-list. Usually, the new email source is compared with a history database to know how it is classified [2]. The problem of using Origin-based filtering is that spammers regularly change the email address, source, and IP. On the other hand, content-based filtering approaches review the email content depending on a proposed analysis technique [2]. However, the content-based filtering generates a large number of features as well as ambiguity of the meaning of the email terms. The traditional Email filtering approaches classified the emails based on the occurrence of the term in the email and neglect the syntactic and semantic properties of the email text.

In this paper, a semantic based feature reduction approach is proposed for filtering emails using the WordNet ontology [3] to handle the problem of high dimensionality of email features. We used the synonyms set of each term and group the terms that have common synonyms. Moreover, we consider the determined (meaningful) words only using the WordNet ontology as an English dictionary. In addition, we assign a weight for each term according to their occurrences in the emails. Different machine learning algorithms are applied in our approach: Naïve Bayes, Support Vector Machine, Logistic Regression, J48 and Random Forest. The results prove the efficiency of the proposed approach. The Logistic Regression recorded the best accuracy with value of 0.96. This is followed by the NB and SVM with accuracy value 0.93 (having similar results). Finally, the Random Forest and J48 classifiers have the least accuracy values of 0.85 and 0.87 respectively.

The rest of the paper is organized as follows. Section 2 outlines the related work while Sect. 3 presents the proposed system architecture of our approach. Section 4 explains how we apply semantics and how we weight the email features. Section 5 gives a brief review about the five machine learning algorithms used and Sect. 6 discusses the experiments and the results. Finally, Sect. 7 includes the conclusion and future work.

2 Related Work

As we discussed above, email filtering can be executed based on origin or content of the email. Some researchers have used origin-based filtering. For example, in [4], an email classification model has been proposed based on four machine learning algorithms: Naïve Bayes, term frequency/inverse document frequency, K-nearest neighbor, and support vector machines. The experiments have been conducted on the header part only of the email.

Most of the literature applied the email filtering based on several classification methods. Some of these classification methods are Naive Bayes (NB) [5–7], logistic regression [7], k-Nearest Neighbor (KNN) [6], C4.5 Classifier [8, 9], Artificial Neural Net-works (ANN) [8], AdaBoost [10, 11], Random Forest (RF) [11], Support Vector Machine (SVM) [5, 6, 11], and Multi-Layer Perceptron (MLP) [8]. These methods have been applied based on the content of the email.

Concerning content-based approaches, in [9], an ensemble learning and decision tree based model has been used to detect the spam emails. Four classification algorithms were applied: C4.5, NB, SVM, and KNN. Also, in [12], another filtering approach has been proposed based on two behavioral features (the URL and the time the email was dispatched) and eight keywords known as bag of spam words to classify the spam and legitimate emails. The machine learning algorithm used was RF. However, both the authors of [9, 12] didn't apply any feature reduction method.

One of the challenges for email filtering is the high dimensionality of data or feature space used. The authors in [13] have proposed a spam detection approach based on Random Forests (RF) classifier, which enables parameters optimization and feature selection approach. In [14], statistical feature selection approach based on similarity coefficients is proposed to enhance the accuracy of spam filtering and detection rate. In [15], an improvement in mutual information algorithms combined with word frequency and average word frequency to measure the relation between a feature and a class. Other feature selection methods were also presented in [16].

Some of authors conduct comparative analysis for different machine learning algorithms, in order to compare the performance different classifiers, such as in [17–19]. In [17], a spam classification approach has been proposed that uses features based on the content of the email and some readability features related to the email (e.g. document length and word length). The work has been applied using five classification algorithms: Bagging, RF, SVM, AdaBoost, and Naïve Bayes. Authors in [18] introduced other comparative analysis. Four algorithms have been used for email filtering: Logistic Regression, Neural Network, NB, and RF. In [19], they presented a comparative analysis based on four classifiers; J48, SVM, BayesNet, and LazyIBK.

A classification-based email filtering approach has been presented in [20]. This approach tries to reduce the email content features by applying stemming on the content. After that, an English dictionary was exploited in order to regard the meaningful terms only. Five classification algorithms have been used in the experiments: Naïve Bayes, SVM, Logistic Regression, J48, and Random Forest.

On the other hand, there is another way to filter emails semantically to tackle the ambiguity problems. A semantic email categorization approach based on semantic vector space model has been proposed in [21] using WordNet. The experiments have been applied using SVM, Logistic Regression, and KNN. In [22], they presented a semantic feature based model for identifying emails using general ontology to overcome the terms mismatch problem. The work has been tested using SVM classifier. Other models based on semantic text classification have been introduced in [23].

3 Proposed System Architecture

Figure 1 shows the proposed system architecture that semantically reduces the email feature dimensions to classify the emails into Ham and Spam classes. It consists of four main modules: Email Pre-processing, Semantic Based Feature Reduction, Feature Weighting, and Classification modules.

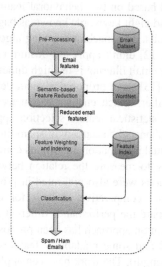

Fig. 1. The Proposed System Architecture.

In the pre-processing module, the main purpose is to remove the irrelevant terms and reduce the number of extracted features for the classifier. We extract the tokens of both subject line and the content body of the email. Then the irrelevant tokens are removed such as symbols and numbers. This is followed by eliminating the stop words. We do not apply stemming in order to keep the meaning of the words.

After the pre-processing step, the features of each email are processed semantically to reduce the size of the complete email features. We used the WordNet ontology by considering the synonymy relation among terms. This is followed by assigning a weight for each term and building the feature index.

Finally, different machine learning algorithms are applied to classify the emails into spam and ham emails. For more clearness, Fig. 2 shows the proposed system algorithm and how our semantic-based email filtering approach works.

4 Feature Reduction and Weighting

In this step, a semantic technique is used to reduce the size of extracted features in the email using the WordNet, which is a lexical database for English language. WordNet is like a supercharged dictionary/thesaurus with a graph structure. It contains a collection

> *Input:* Email Dataset
> *Output:* spam and ham Emails
> *Steps:*
> **For Each** Email **In** Dataset
>   **For Each** term **In** an Email
>   Extract tokens separately
>   Remove irrelevant tokens
>   Remove stopwords
>   Consider meaningful words only
>   Apply semantic using WordNet
>   Count frequency of each term
>   Give a weight for each term
>   **End For**
> **End For**
> Apply different classifiers
> **For Each** Classifier **In** ClassifierSet
>   Execute the classifier
>   Get the four evaluation criteria
>   Calculate Recall, Precision, F-score, and
>   accuracy
> **End For**
> Identify spam and ham Email

Fig. 2. Semantic-based Email Filtering Algorithm.

of English words (nouns, adjectives, verbs and adverbs) that are linked together by their semantic relations. English words are grouped into sets of synonyms called synsets. A synset corresponds to an abstract concept.

The purpose of using WordNet is to reduce the high dimensionality of feature using the synonyms set of each feature. We group the terms that have the same synonyms together, and scale up their weight in each email. WordNet comprises large number of English words. Hereby, it is used as an English dictionary in order to identify the meaningful words in our work, and hence the undetermined words will be discarded.

Figure 3 shows the feature reduction and weighting algorithm. After extracting the email features from pre-processing module, we used WordNet ontology as semantic resource. We tried to reduce the high dimensionality of features by considering the synonyms set of each term in an email. If one synonym in synonym set is mutual with other terms in the same email, then we can scale up the count and ignore that term. If no mutual terms, then add the new term to the feature index. In addition, we used WordNet ontology as an English dictionary to get the meaningful terms only.

After extracting the meaningful terms, each term is assigned a weigh, which is the number of occurrences of this term and its synonyms in each email. We use the term frequency/inverse document frequency Eq. (3), which is used for indexing the documents in information retrieval.

Term Frequency (TF): is defined as the number of times that term (t) occurs in document (d).

$$\text{TF}(t, d) = \frac{f_d(t)}{\max[f_d(t)]} \tag{1}$$

Where $f_d(t)$ is the frequency of term (t) in document (d).

```
Input: Email Features
Output: Feature Index
Steps:
For each term In an Email
        Get synonyms set of a term
        If (synonyms set contains any other term in email)
                Increase the count of existing term
        Else
                Add the existing term to feature index
        End If
        Check the meaningful terms
        If (term exists in WordNet)
                Add the term
        Else
                Ignore the term
        End If
        Give a weight for each term using count of term
```

Fig. 3. Feature Reduction and Weighting Algorithm.

Inverse Document Frequency (IDF): estimates the rarity of a given term in the whole document collection (If a term occurs in all documents of the collection, its IDF is zero.) and measures how important a term is within a particular document, by computing it using Eq. (2):

$$IDF(t) = \log\left(\frac{N}{df_t}\right) \tag{2}$$

Where N is the total no. of documents and df_t is the no. of documents with term (t).

Finally, The TF-IDF is the product of its TF weight in Eq. (1) and its IDF weight in Eq. (2).

$$TF - IDF = tf(t, d) \times IDF(t) \tag{3}$$

Our experiments have been conducted using TF-IDF.

5   Classifiers

In this step, we tried different classifiers to evaluate the proposed model, which are Naïve Bayes, SVM, Logistic Regression, J48 and Random Forest.

5.1   Naïve Bayes

The Naive Bayes algorithm is a simple probabilistic classifier that measures a set of probabilities by calculating the number of combinations and frequency of terms in a certain dataset. The Bayesian classifier makes a conditional independence assumption between the attributes and that significantly reduce the number of attributes, therefore it tends to proceed well and learn rapidly in different supervised classification problems [6, 7].

5.2   Support Vector Machine

Support vector machine (SVM) is a classification algorithm with strong theoretical base. SVM is a group of related supervised learning methods used in classification and regression. It separates the data into two classes by constructing a straight line (1 dimension), flat plane (2 dimensions) or an N-dimensional hyperplane. SVM can handle high dimensional feature space effectively [5, 11].

5.3   J48 Classifier

J48 is one of the most popular decision tree algorithms. J48-classifier J48 builds decision trees from a group of training data. According to the splitting node strategy, j48 selects one attribute from the training data that effectively splits its set of instances into smaller subsets. J48 classifier then visits each decision node recursively and chooses the most effective split until each leaf is pure and no more splits are available, meaning that the data has been classified as perfectly as possible [8, 9].

5.4   Logistic Regression

Logistic regression can handle the relationship between a dependent nominal variable and one or more independent variables [7]. It collects the independent variables to assess the probability that a particular event will occur.

Logistic Regression can be used when the target variable is a categorical variable with two categories – for example spam/ham emails. For a given case, logistic regression measure the probability that a case with a certain set of values of the independent variable is a member of the modeled category. If the probability is greater than 0.5, the case is classified in the modeled category. If the probability is less than 0.50, the case is classified in the other category.

5.5   Random Forest

Random Forest (RF) is an ensemble classifier that consists of a collection of decision trees. A random selected subset of training data features is used to split each tree independently and with the same distribution for all trees in the forest. A randomized selection of variables is used to divide the nodes. The main idea for using ensemble methods is that a group of weak classifiers can come together to form a strong classifier. RF runs efficiently on large datasets and its learning is fast [11].

6   Experimental Evaluation and Discussion

In our approach, we exploit the Enron-Spam, which is a large public email database collection. It contains data from about 150 employees. The corpus contains a total of about 0.5 M messages. It focuses on six Enron employees with large mail-boxes.

The Enron dataset is divided into six different subsets [24]. Our Experiments are done on a subset of the Enron Corpus containing 300 emails (32 % spam, 68 % ham). The experiments were executed on a machine with hardware specification of processor: Intel® core i7 and main memory of 8 GB.

The data set was separated randomly into two parts, the first part is used as training data set to produce the prediction model, and the other part is used for testing data set to evaluate the accuracy of our model. Testing is done by using 10-fold cross validation method. In our experimentations, we have used WEKA tool [25] to apply different machine learning algorithms. WEKA tool is an open source software that provides a collection of machine learning algorithms for data mining tasks.

6.1   Evaluation Measures

The performance of the classifiers is measured using recall, precision and accuracy.

Recall represents the percentage of correctly identified positive cases and defined as in Eq. (4):

$$Recall = \frac{TP}{TP + FN} \tag{4}$$

Precision reflects the number of real predicted examples and defined as:

$$Precision = \frac{TP}{TP + FP} \tag{5}$$

The overall accuracy has been also defined by:

$$Accuracy = \frac{TP + TN}{TP + TN + FP + FN} \tag{6}$$

Where TP is the true positive instances, TN is the true negative instances, FP is the false positive instances and FN is the false negative instances.

6.2   Results and Discussion

As explained above, the Enron dataset has been preprocessed. After applying the proposed approach, comparing to the related work in [20], the number of features was reduced from 3636 to 2309 with reduction rate equals 36.5 %. Therefore, the execution time of the classifiers was decreased due to the feature reduction.

The experiments have been conducted on our extracted feature set as mentioned above using the TF-IDF (Eq. 3) for email weighting. The classifier performance is measured using the precision, recall and accuracy.

Table 1 shows the performance of different classifiers using our proposed approach. Logistic Regression recorded the best precision with value of 0.96. Followed by the Naïve Bayes and SVM that recording almost similar results of precision value 0.93.

Table 1. Performance results for the proposed work

Evaluation Criteria	Classifiers				
	Naïve Bayesian	SVM	Logistic Regression	J48	Random Forest
Precision	0.927	0.936	0.96	0.857	0.872
Recall	0.927	0.937	0.96	0.857	0.87
Execution Time (Sec)	0.06	0.24	0.13	0.32	0.06

The J48 classifiers recorded the least precision value of 0.857. The Recall almost recorded the same performance like Precision.

In addition, the proposed work has been compared to the related work in [18, 20]. In Fig. 4, we compare the common classifiers of our proposed work with these related works, which also used the Enron dataset. As shown the proposed work has a higher accuracy than the related work in [18] with respect to Naïve Bayes, Logistic Regression, and J48 classifiers. However, for Random Forest it is slightly lower than the related work in [18]. Figure 4 also shows that the accuracy of J48 for the proposed work is 0.85, while in the related work [20] is 0.76, which is significantly better. In addition, the Naïve Bayes, Logistic Regression, and Random Forest have a slightly higher accuracy than the related work in [20]. It is clear that the proposed work has very good results.

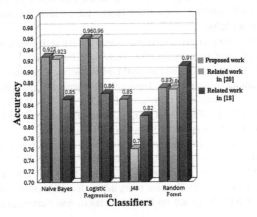

Fig. 4. Comparing accuracy value of proposed work with the related work in [20] and the related work in [18].

7   Conclusion and Future Work

In this paper, a semantic-based email filtering approach has been introduced. In this approach, we enhanced the traditional email filtering models by using WordNet ontology as a semantic resource for feature reduction. Experimental studies have been conducted using different classification algorithms. Enron dataset has been used in the experiments. A comparative study has been presented with other related work using the same dataset. The semantic-based feature reduction approach showed high performance with faster filtering execution and better accuracy. In the future work, we will enhance our proposed approach by considering some other semantic relations between term concepts for reducing the dimensionality of features.

References

1. Internet Threats Trend Report. Cyberoam® A SOPHOS Campany (2014)
2. Castillo, M.D., Serrano, J.I.: An interactive hybrid system for identifying and filtering unsolicited e-mail. In: Corchado, E., Yin, H., Botti, V., Fyfe, C. (eds.) IDEAL 2006. LNCS, vol. 4224, pp. 779–788. Springer, Heidelberg (2006). doi:10.1007/11875581_94
3. Hristea, F.T.: Semantic WordNet-based feature selection. In: Hristea, F.T. (ed.) The Naïve Bayes Model for Unsupervised Word Sense Disambiguation, pp. 17–33. Springer, Heidelberg (2013)
4. Lai, C.C., Tsai, M.C.: An empirical performance comparison of machine learning methods for spam e-mail categorization. In: Fourth International Conference on Hybrid Intelligent Systems, pp. 44–48. IEEE (2004)
5. Islam, M., Mahmud, A.A., Islam, M.: Machine Learning Approaches for Modeling Spammer Behavior. In: Kan, M.-Y., Lam, W., Nakov, P., Cheng, P.-J. (eds.) AIRS 2010. LNCS, vol. 6458, pp. 251–260. Springer, Heidelberg (2010)
6. Blanzieri, E., Bryl, A.: A survey of learning-based techniques of email spam filtering. Technical report DIT-06-056, University of Trento, Information Engineering and Computer Science Department (2008)
7. Mitchell, T.: Generative and discriminative classifiers: naive Bayes and logistic regression (2005). Manuscript http://www.cs.cm.edu/~tom/NewChapters.html
8. Renuka, D.K., Hamsapriya, T., Chakkaravarthi, M.R., Surya, P.L.: Spam classification based on supervised learning using machine learning techniques. In: International Conference on Process Automation, Control and Computing (PACC), pp. 1–7. IEEE (2011)
9. Shi, L., Wang, Q., Ma, X., Weng, M., Qiao, H.: Spam email classification using decision tree ensemble. J. Comput. Inf. Syst. 8(3), 949–956 (2012)
10. Islam, M.R., Zhou, W.: Architecture of adaptive spam filtering based on machine learning algorithms. In: Jin, H., Rana, O.F., Pan, Y., Prasanna, V.K. (eds.) ICA3PP 2007. LNCS, vol. 4494, pp. 458–469. Springer, Heidelberg (2007). doi:10.1007/978-3-540-72905-1_41
11. Islam, R., Xiang, Y.: Email classification using data reduction method. In: Proceedings of the 5th International ICST Conference on Communications and Networking in China, pp. 1–5. IEEE (2010)
12. Bhat, V.H., Malkani, V.R., Shenoy, P.D., Venugopal, K.R., Patnaik, L.M.: Classification of email using beaks: behavior and keyword stemming. In: TENCON IEEE Region 10 Conference, pp. 1139–1143. IEEE (2011)
13. Lee, S.M., Kim, D.S., Kim, J.H., Park, J.S.: Spam detection using feature selection and parameters optimization. In: Intelligent and Software Intensive Systems International Conference on Complex, pp. 883–888. IEEE (2010)
14. Abdelrahim, A.A., Elhadi, A.A.E., Ibrahim, H., Elmisbah, N.: Feature selection and similarity coefficient based method for email spam filtering. In: International Conference on Computing, Electrical and Electronics Engineering (ICCEEE). IEEE (2013)

15. Ting, L., Qingsong, Y.: Spam feature selection based on the improved mutual information algorithm. In: Fourth International Conference on Multimedia Information Networking and Security (MINES). IEEE (2012)
16. Wang, R., Youssef, A.M., Elhakeem, A.K.: On some feature selection strategies for spam filter design. In: Canadian Conference on Electrical and Computer Engineering, pp. 2186–2189, CCECE 2006. IEEE (2006)
17. Shams, R., Mercer, R.E.: Classifying spam emails using text and readability features. In: 13th International Conference on Data Mining (ICDM). IEEE (2013)
18. More, S., Kulkarni, S.: Data mining with machine learning applied for email deception. In: International Conference on Optical Imaging Sensor and Security. IEEE (2013)
19. Sharaff, A., Nagwani, N.K., Dhadse, A.: Comparative study of classification algorithms for spam email detection. In: Emerging Research in Computing, Information, Communication and Applications, pp. 237–244. Springer, India (2016)
20. Bahgat, E.M., Rady, S., Gad, W.: An e-mail filtering approach using classification techniques. In: Gaber, T., Hassanien, A.E., El-Bendary, N., Dey, N. (eds.) The 1st International Conference on Advanced Intelligent System and Informatics. AISC, vol. 407, pp. 321–331. Springer, Heidelberg (2016). doi:10.1007/978-3-319-26690-9_29
21. Lu, Z., Ding, J.: An efficient semantic VSM based email categorization method. In: International Conference on Computer Application and System Modeling, vol. 11, pp. 511–525. IEEE (2010)
22. Yoo, S., Gates, D., Levin, L., Fung, S., Agarwal, S., Freed, M.: Using semantic features to improve task identification in email messages. In: Kapetanios, E., Sugumaran, V., Spiliopoulou, M. (eds.) NLDB 2008. LNCS, vol. 5039, pp. 355–357. Springer, Heidelberg (2008)
23. Tang, H.J., Yan, D.F., Yuan, T.I.A.N.: Semantic dictionary based method for short text classification. J. China Univ. Posts Telecommun. **20**, 15–19 (2013)
24. Enron-Spam datasets: CSMINING group, http://csmining.org/index.php/enron-spam-datasets.html. Accessed 7 July 2016
25. Hall, M., Frank, E., Holmes, G., Pfahringer, B., Reutemann, P., Witten, I.H.: The WEKA data mining software: an update. ACM SIGKDD Explor. Newsl. **11**(1), 10–18 (2009)

A New Method for Interoperability Between Lexical Resources Using MDA Approach

Malek Lhioui[1], Kais Haddar[1(✉)], and Laurent Romary[2]

[1] Laboratory MIRACL, Multimedia, InfoRmation Systems and Advanced
Computing Laboratory, Sfax University, Sfax, Tunisia
ma.lhioui@gmail.com, kais.haddar@yahoo.fr
[2] Inria and Centre Marc Bloch, Berlin, Germany
laurent.romary@inria.fr

Abstract. Lexical resources are increasingly multiplatform due to the diverse
needs of linguists as well as other various user communities. Merging, comparing, making correspondences and deducing differences between these lexical
resources remain difficult tasks. Thus, interoperability between these resources is
an extremely difficult task. In this context, we define a new method based on the
MDA approach to resolve interoperability between lexical resources. The proposed method consists in building a common structure (OWL-DL ontology) for
the various resources. This common structure has the ability to communicate
different resources. Hence, we may create a complex grid between these
resources allowing transformation from one format to another. We experiment
our new built method on the basis of an LMF(LMF: Lexical Markup Framework) compliant lexical resource.

Keywords: Lexical resources · Interoperability · MDA approach · OWL
ontology

1 Introduction

Interoperability across NLP applications is a major issue since they usually have to share
the same or similar linguistic resources. Exchanging information between lexical
resources that are based upon different representation formalisms is a difficult task. This
often results in the need to transform legacy resources from one format to another,
although the discrepancies in the underlying modeling framework did not always make it
possible to achieve. The method presented here allows constructing a pivot format for
lexical resources with no prior restriction. A challenge that NLP communities is con-
fronted with is the obsolescence of older representation formalisms after long periods of
development. Our method will easily protect several resources from disappearing. So,
several formalisms will continue to persist. Thus projects, which require merging several
formalisms in the same application, will prefer to use our method. In fact, it allows using
the number of formalisms one wants. We use MDA (Model Driven Architecture)
transformation approach because of its great interest in areas handling heterogeneous
knowledge. In fact, if even a current version of a used standard (LMF for example) is not
yet stable or a new version is born, MDA approach ensures enrichment and not
destruction of the current version.

© Springer International Publishing AG 2017
A.E. Hassanien et al. (eds.), *Proceedings of the International Conference
on Advanced Intelligent Systems and Informatics 2016*, Advances in Intelligent
Systems and Computing 533, DOI 10.1007/978-3-319-48308-5_7

Building a new method for interoperability between lexical resources can stumble upon several difficulties. The first one resides in the choice of an optimal strategy for interoperability between resources: algebraic specifications, alignment ontology techniques, meta modeling, etc. In addition to this, the choice of the ontology representation language (RDF(S), OWL-Lite and OWL-DL) is also a crucial dilemma. Besides, the construction of meta-models and models in MDA approach requires a precise knowledge of the lexical resources at hand. Moreover, transformation rules have to be so definite.

The paper presents a new method strictly founded to resolve interoperability between lexical resources (LRs) using the well-known MDA Transformation approach. Indeed, we attempt to find compulsory techniques in order to establish a method for interoperability between lexical resources whatever their formats (LMF, TEI[1], HPSG[2]...). The method consists in building a pivot format by making an automatic mapping process between involved lexical resources. The target building format plays the role of the pivot. In order to build this pivot format, we have to succeed in fulfilling a set of steps. We construct OWL-DL ontology for lexical resources: construction of meta-model associated to lexical resource, construction of the ATL transformation and deduction of OWL-DL model. Thus, applying these steps, we build a new format for involving lexical resources.

The originality of this method is that there are no previous works aiming at making interoperability between lexical resources operable. Moreover, the use of MDA approach for resolving interoperability between lexical resources is in itself an innovation. Projects and NLP applications today must rely on interoperability; otherwise they are out of progress. In this context, we can read in (TAUS 2011) that: "The lack of interoperability costs the translation industry a fortune". As a matter of fact, this fortune is compensated mostly in order to adjust data formats. In addition, our method is operable whatever the language.

In the following sections, we introduce a brief state of the art in order to give a global idea about existing works related to our topic. Then, we explain precisely our method for resolving the interoperability issue between lexical resources. We apply, in the next section, the proposed method to LMF compliant lexica. Finally, we conclude with a small discussion for the obtained results.

2   State of the Art

The state of the art provides an important idea about existing works regardless of the language. There have been several works dealing with the use of MDA transformation approach for the processing of several applications. However, there is no use of the MDA approach for the processing of interoperability between lexical resources. Yet, this approach ensures interoperability according to the OMG "Portability and interoperability are built into the architecture". Since there are many related topics, we

---

[1] Text Encoding Initiative.

[2] Head-driven Phrase Structure Grammar.

classify the state of the art into three main parts: lexical resources, MDA Transformation and interoperability issue. The first part gives an idea about existing lexical resources regardless the language. We give examples of lexical resources in several languages such as Arabic. In the second part, we talk about MDA as a great method for transformation models. The last part deals with interoperability issue, and since there are no serious attempts to resolve this notion in NLP area, we will discuss the bidirectional mapping from one format to another.

2.1 Lexical Resources

Lexical resources vary in accordance with the need of linguists and this requirement varies with the NLP community development progress. This process makes the resources more complex and heterogeneous. In the literature, existing lexical resources are innumerable. We can concentrate on some of them. In the 1980ies SGML markup language was created as the first formalism representation of linguistic data. Early in the following century, several markup languages have been invented by the TEI (Wörner et al. 2006). After years and exactly in 2003, a new standard named LMF was born due to efforts provided by the community of NLP (Francopoulo 2013). Speech is one of the several areas of NLP domain. This area includes several representation formalisms as well as the other areas. For example, EXMARaLDA is one of these formalisms. It represents spoken interaction with an annotation graph (Bird and Liberman 1999). Other formalisms in this context was born such as ELAN, TASX, Praat and ANVIL. They are efficient for multimodal annotation. In the same context, there are formalisms that include several heterogeneous resource structures. The well-known example for that is Tusnelda. It is inspired typically from the work of TEI (Wagner and Zeisler 2004). There are other formalisms that take care of various linguistic levels (phonology, morphology, syntax, semantic, etc.) (Ide and Romary 2001). Thus, from an historical point of view, there is a large number of heterogeneous resources which inducing the question of transformation. This notion is the subject of the next subsection.

2.2 MDA Transformation

MDA Transformation is an approach proposed by OMG (Poole 2001) in 2001. It is increasingly used in several applications and projects whatever their kind. It consists on using different models phases. It allows interoperability between applications by connecting their models (Accord 2002). It supported evaluation and decreased manually implementation of hundred of codes for a specific domain by separating conception from implementation (Miller and Mukerji 2001). The implementation of MDA requires three main levels: MOF (Meta-Object Facility) defines the platform for implanting all models (OMG/MOF 1997). PIM (Plateform Independant Model) which serves as a basis for the business part specification of an application, PSM (Plateform Specific Model) which participates in the specification model creation of the application after projection on a platform. The major advantage of this approach apart from time saving is preoccupations separation and the transformation process. This transformation allows mapping from PIMs to PSMs using modules described in specific languages such as ATL. ATL (Atlas Transformation Language) is a language providing rules allowing transformation from source to target models. Since this approach allows interoperability between applications, it leads us to think about making evident interoperability between models. Thus, we introduce in the following subsection interoperability notion in general.

2.3 Interoperability Issue

Interoperability is the substitution, merging and sharing knowledge between different entities whatever their kind. NLP community replaces these terms by only one term, which is communication. Thus, interoperability allows communication between involved entities. Interoperability is a general notion that can be projected to many domains. In this paper, we interested to interoperability between lexical resources. Lexical resources are more and more multiplatform, multi-providers... and these characteristics are increased by the time, so that, interoperability becomes hard even impractical to achieve between lexical resources. These last suffer from several interoperability issues. For example the definition of procedures to implement a set of services in NLP applications (machine translation, named entity recognition, part of speech tagging) shall be made through LMF by ISO, TEI by TEI Consortium and HPSG by linguistics... This leads to interoperability problems when experts have to collaborate. Thus, information technology professionals consider that interoperability is an important criterion as well as security and reliability in their applications.

From an historical point of view, there are no significant efforts resolving interoperability between lexical resources. Yet, there are several challenges consisting on mapping from one format to another. The first mapping attempt is done by (Wilcock 2007) consisting on converting HPSG lexicons to an OWL ontology. In 2010, Loukil has expanded these processes by inventing a rule-based system opting to translate LMF syntactic lexicon into TDL within the LKB platform (Loukil et al. 2010). (Haddar et al. 2012) have developed a prototype for projection HPSG syntactic lexica towards LMF. In the same context, there is a mapping process already done by (Lhioui et al. 2015) aiming to convert LMF lexicons to ontologies described on OWL-DL language. Bidirectional processes are usually limited to involved formats. Whatever we desire to involve more than two formalisms, processing became hard and impossible to achieve even if we use several properties such as transitivity. For these reasons and in order to attenuate task complexity of mapping process, several organizations such as ISO give a quick solution but not efficient for interoperability using normalization. In fact, LMF is one of these solutions proposed by the ISO in 2003 (Francpolou 2013). It involves several packages aiming to cover the maximum of the large domains: phonology, morphology, syntactic, semantic, pragmatic, etc. Other researchers have used another tool for resolving interoperability which is ontologies. A famous example of these works is the General Ontology for Linguistic Description (GOLD)[3]. GOLD is an OWL ontology having specific knowledge related to linguistic domain. The GOLD ontology contains the basis linguistic knowledge of any theoretical framework. According to (Farrar and Lewis 2005), GOLD defines linguistic knowledge as axioms, for example "a verb is a part of speech", and uses at the same time language neutral, for example "parts of speech are subclasses of gold: GrammaticalUnit". The classes are presented in the protégée editor and then expressed as concepts in the GOLD ontology (Farrar and Langendoen 2003). Thus, GOLD is an abstract model and representation formalisms such as HPSG are the instantiation of this abstract model. (Farrar and Lewis 2005) consider these instantiations as sub-communities of practice noted Communities Of Practice Extension (COPEs). COPEs, sub-communities or sub-ontologies designed the same nomenclature and extend the overall GOLD ontology (Wilcock 2007). The integration of these COPEs in the GOLD ontology is a hard process and necessitates different mechanisms of ontology alignment. In the next subsection, we try to give an idea for techniques of ontologies alignment. All these notions will be strongly correlated to introduce our approach. In the following section, we define a new approach for interoperability between lexical resources using MDA Transformation.

3 Proposed Method

The new build method is based on MDA Transformation approach. This approach is well-known and has proved its importance in guaranteeing reusability. This characteristic is crucial since it makes projects up to date. The proposed method is characterized by the ability to allow involved lexical resources to operate together. The new introduced method has as input a set of lexical resources. Lexical resources are composed of a set of lexicons such as LMF lexicon. It consists of three main steps. The first one is the achievement of the two independent models PIM (source and target) and the source PSM of each LR. The second is the achievement of the transformation module in ATL and finally, the generation of the specific model PSM (OWL-DL in our case). The output of the proposed method is a set of ontologies which can operate together using several algorithms or free tools of alignment. In fact, the use of ontologies as an output is the keystone of our method. Ontology structures allow merging, comparing, making correspondences and deducing differences between lexical resources due to the tools of ontology alignment. Figure 1 describes the whole process of the proposed method.

The full schema of the proposed method will be explained carefully by examination of each step separately. In fact, MDA Transformation of the LRs to OWL ontologies is a crucial step in our method. The main idea of this step is to distinguish functional specifications from specifications of implementation related to a given platform in order to prepare structures able to operate together (in our case ontologies able to be aligned and then interacted). Thus, using MDA as an approach will make us able to elaborate independent specifications from the implementation in a specific platform using models. The first model to build is the PIM. The PIM is the model conceived to specify

---

[3] Gold is accessible and free downloadable.

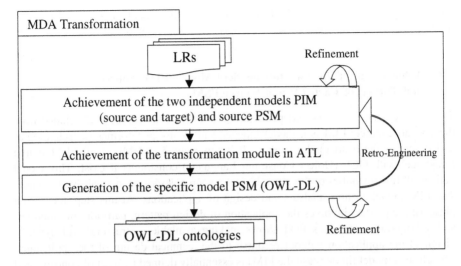

Fig. 1. Steps of the proposed method

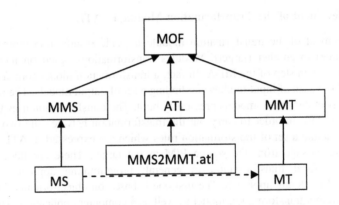

Fig. 2. MDA transformation of the LRs to OWL ontologies

involved structures independently from any specific platform. This characteristic allows us abstracting functionalities of the involved lexical resource and to compare it to other resources. If the lexical resource is updated, the associated PIM will never be destroyed, but, it will be refined as many times as possible; this makes one of the most advantages of the MDA approach when resolving interoperability issue. Figure 2 summarized the MDA Transformation in general:

Figure 2 describes the ATL transformation in the MDA approach. MOF is the meta-meta model. MMS and MMT designate respectively the meta-model source and the meta-model target. MS and MT denote respectively model source and model target. MMS2MMT.atl includes the set of transformation rules. This method is composed of three sub-steps as we have mentioned below: Achievement of the two independent models PIM (source and target) of each LR, the achievement of the transformation module in ATL and the generation of the specific model PSM (OWL-DL in our case).

3.1 Achievement of the Two Independent Models PIM (Source and Target) of Each LR and Source PSM

The achievement of the first independent model PIM of the source is concluded from the lexical resource. PIM is a model independent to any plateformes or technologies and describes the heart of the method. It is represented in UML (Unified Modeling Language) with OCL (Object Constraint Language) constraints if exist. This model defines all functionalities of the given lexical resource described in an abstract manner. The PIM model ensures analysis and design of applications. At this step, the design phase of the process involves the application of design pattern, partition into modules and sub-modules, etc. This PIM allows making available a structural and dynamic vision of the application without recourse to the technical design of the application. Therefore, a model (in our case the PIM) is essentially defined by a set of concepts and their relationships presented in a class diagram.

3.2 Achievement of the Transformation Module in ATL

The achievement of the transformation module in ATL ensures transition from one model (source) to another (target). Modules transformations based on meta-models constitute the main step of the MDA. In fact, a transformation model corresponds to a function taking a set of input models and finding a set of output models. The models, in and out, respect their meta-models previously built. The transformation uses the model manipulation API. In order to carry out the transformation between the two involved models, we define a set of transformation rules which are expressed in ATL language allowing the passage from the source PIM to the target. There are three different manners to model transformation in general: programming approach, template approach and modeling approach. The first one is based on object-oriented languages. It is to program a transformation model as well as a computer application. The second consists to define templates models and then replace them with their equivalent values in source models. The last one models transformation rules using MDA approach.

3.3 Generation of the Specific Model PSM (OWL-DL)

After achievement of the PIM model (source and target) and elaborating rules allowing the passage from the source PIM to the target, we project the source PIM to a specific model PSM (Platform Specific Model). In order to generate the target PSM, we execute the ATL rules, then, we obtain automatically the target PSM. In fact, PSM is closest to the final code. It is related to a particular platform.

4   Implantation: Transformation of LMF Lexicon to OWL-DL Ontology Using MDA Approach

In this section, we present the steps of the cited method applied to LMF lexicon: the two PIMs (source: LMF, target: OWL), the transformation rules and the two PSMs. Figure 3 represents the source PIM of the core model of the following extract of LMF lexicon developed using Eclipse Galileo:

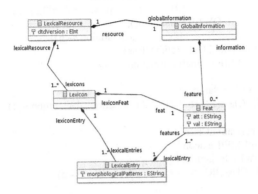

Fig. 3. The associated source PIM of the below extract of LMF lexicon

The build PIM can be refined as well as possible if the lexical resource (LMF lexicon) is updated. After building the source model, we have now obliged to build the target PIM of this lexical resource. Since we need to construct OWL-DL ontologies, we build a PIM for OWL-DL ontologies. Figure 4 represents the target PIM for the previous lexical resource (LMF lexicon):

Figure 4 defines the independent model of OWL ontologies which is related to the given lexical resource. It represents the class "Ontology" which presents the root, the prefixes which are used to abbreviate and minimize scripture of the namespaces in the entire ontology. Then, we define the set of transformation rules:

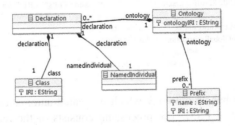

Fig. 4. The associated target PIM of the below extract of LMF lexicon

```
module LMF2OWL;
create OUT : OWL from IN : LMF;
--------------------------Ontology--------------------------
rule LexicalResource2Ontology{
        from s:LMF!LexicalResource
        to
        t:OWL!Ontology(ontologyIRI <-),d:OWL!Prefix(name<-'rdf',
                        IRI<-'http://www.w3.org/1999/02/22-rdf-syntax-ns#'),
                        u:OWL!Prefix(name<-'rdfs',
                        IRI<-'http://www.w3.org/2000/01/rdf-schema#'),
                h:OWL!Declaration(),
    g:OWL!Class(IRI <- '#LexicalResource',declaration <- h)}
    rule GlobalInformation2DeclarationClass{
        from s:LMF!GlobalInformation

        to t:OWL!Declaration(),
            g:OWL!Class(IRI <- '#GlobalInformation', declaration <- t)
}
rule Lexicon2DeclarationClass{
        from o:LMF!Lexicon
        to p:OWL!Declaration(),
            i:OWL!Class(IRI <- '#Lexicon', declaration <- p)
}
rule LexicalEntry2DeclarationClass{
        from k:LMF!LexicalEntry
        to n:OWL!Declaration(),
            j:OWL!Class(IRI <- '#LexicalEntry', declaration <- n)
}
```

These transformation rules create an OWL PSM for the LMF lexicon of Fig. 2 which is an ontology described in OWL language. These rules are stored in an ATL file. Finally, Fig. 4 represents this target PSM created automatically when executing the ATL file of the LMF Lexicon presented in Fig. 2:

```
<?xml version=      encoding=          ?>
<xmi:XMI    xmi:version=        xmlns:xmi=                          xmlns=     >        <Ontology
ontologyIRI="http://www.semanticweb.org/asus/ontologies/2016/2/"/>          <Prefix      name="rdf"
IRI="http://www.w3.org/1999/02/22-rdf-syntax-ns#"/>       <Prefix   name="rdfs"   IRI="http://www.w3.org/2000/01/rdf-
schema#"/>
    <Declaration><class IRI="#LexicalResource"/> </Declaration>
    <Declaration> <class IRI="#GlobalInformation"/></Declaration>
    <Declaration><class IRI=            /></Declaration>
    <Declaration><class IRI=                /></Declaration></xmi:XMI>
```

This output describes an ontology which is created automatically by a quick processing of the output of PSM. The processing consists of the removal of the "xmi" prefix since it is an automatic output of the ATL transformation.

5 Discussions

The new built method has proved its interest in handling heterogeneous resources. The evaluation process has been successfully led by fixing three criterions: sustainability of expertise, productivity gains and inclusion of execution platforms. The first criterion (sustainability of expertise) affects two characteristics. The first one supervises lifetime of the built models (PIM and PSM). The models must have a lifetime greater than the code. This is guaranteed by the unrestricted refinement of models. The second characteristic provides modeling languages supporting different levels of abstraction. This point is guaranteed by the fact that UML and OCL support abstraction. The second criterion concerns productivity gains. In fact, the automation operations of models guarantee the productivity gain. Moreover, the built method facilitates the creation of operations of production on the models. The last criterion concerns the taking into account of the execution platforms. This stage is explicit in the life cycle of applications. MDA approach guarantees this characteristic as platforms are related to models. These aspects make the method very robust. The other important aspect in the built method is that the transformation process is done automatically.

6 Conclusion

In this paper, we have proposed a new method for interoperability between interoperability using MDA approach. This new method allows merging, comparing, finding correspondences and deducing differences between lexical resources. Then, we implement the method by projection on LMF. Our method is reusable and generic, and operable on all lexical resources whatever the language. Our method is generated automatically. In future works, we have to extend our method using the alignment of the building ontologies. In fact, if we combine MDA Transformation and ontology alignment, interoperability appears to be quite suitable. Therefore, combining these two approach MDA Transformation and ontology alignment for this study seems to have promising results.

References

ACCORD (2002) CNAM, EDF R&D, ENST, ENST-Bretagne, France Telecom R&D, INRIA, LIFL et Softeam, Projet ACCORD (Assemblage de composants par contrats), Livrable 1.1–5, Date: Mai 2002

Bird, S., Liberman, M.: Annotation graphs as a framework for multidimensional linguistic data analysis. In: Towards Standards and Tools for Discourse Tagging, Proceedings of the Workshop. Association for Computational Linguistics (1999)

Farrar, S., Langendoen, T.: A linguistic ontology for the semantic web. GLOT Int. 7(3), 97–100 (2003)

Farrar, S., Lewis, W.D.: The GOLD community of practice: an infrastructure for linguistic data on the web (2005). http://www.u.arizona.edu/~farrar/

Francopoulo, G.: Lexical Markup Framework. ISTE Ltd and Wiley, US, Great Britain and the United States (2013)

Haddar, K., Fehri, H., Romary, L.: A prototype for projecting HPSG syntactic lexica towards LMF, JLCL (2012)

Ide, N., Romary, L.: Standards for language resources. In: Proceedings of the IRCS Workshop on Linguistic Database, pp. 141–149 (2001)

Lhioui, M., Haddar, K., Romary, L.: A prototype for projecting LMF lexica towards OWL (2015)

Loukil, N., Ktari, R., Haddar, K., Benhamadou, A.: A normalized syntactic lexicon for arabic verbs and its evaluation within the LKB platform. ACSE, Egypt (2010)

Miller, J., Mukerji, J.: Model Driven Architecture (MDA), July 2001. http://cgi.omg.org/docs/ormsc/01-07-01.pdf. Architecture Board ORMSC

OMG/MOF Meta Object Facility (MOF) Specification, OMG Document AD/97-08-14, Septembre 1997. (www.omg.org)

Poole, J.D.: Model-driven architecture: vision, standards and emerging technologies. In: ECOOP 2001, Workshop on Metamodeling and Adaptive Object Models (2001)

TAUS: Report on a TAUS research about translation interoperability, 25 February 2011

Wagner, A., Zeisler, B.: A syntactically annotated corpus of Tibetan. In: Proceeding of LREC, pp. 1141–1144, Lisboa (2004)

Wilcock, G.: An OWL Ontology for HPSG. ACL, Finland (2007)

Wörner, K., Witt, A., Rehm, G., Dipper, S. (eds.): Modelling Linguistic Data Structures. Extreme Markup Languages, Montréal (2006)

Statistical Machine Translation Context Modelling with Recurrent Neural Network and LDA

Shrooq Alsenan[✉] and Mourad Ykhlef

College of Computer and Information Sciences,
King Saud University, Riyadh, Saudi Arabia
shrooq.alsenan@hotmail.com, ykhlef@ksu.edu.sa

Abstract. Machine Translation of text is a fundamental problem in machine learning that resists solutions that do not take into account the dependencies between words and sentences. Recurrent Neural Networks have recently delivered outstanding results in learning about sequential dependencies in many languages. Arabic language as a target language has not received enough attention in the recent language model experiments due to its, structural and semantic difficulties. In this paper, we present a Statistical Machine Translation (SMT) Context Modelling using Recurrent Neural Networks (RNNs) and Latent Dirichlet Allocation (LDA). This research is based on the state-of-the-art RNN language model by Mikolov. Our preliminary contribution is in integrating and presenting a new hybridization to utilize Recurrent Neural Network sequential word learning dependencies as well as Latent Dirichlet Allocation context and topic classification ability to produce the most accurate language scoring.

Keywords: RNN · SMT · Language model · NLP · Arabic · LDA

1 Introduction

In the past few years, Natural Language Processing (NLP) has received a lot of attention in many areas such as statistical machine translation, image captioning, hand writing recognition and speech recognition. Statistical Machine Translation (SMT) can be defined as the mapping of sentences from one human language to another [1]. Machine Translation has always suffered from significant problems in producing acceptable translation results over the years due to the lack of context understanding of the sentences ignoring the natural dependencies between words that essentially allow a cohesive generating of sentences [2]. However, there has been great progress in delivering technologies in Statistical Machine Translation where translation advanced from word-based models [3] towards more sophisticated models that take contextual aspects of the words into account [2]. A powerful approach to learn about sequential dependencies between words in Machine translation that recently delivered outstanding results is Recurrent Neural Network. Recurrent Neural Network (RNN) is a connectionist model with the ability to selectively pass information across sequence steps, while processing

© Springer International Publishing AG 2017
A.E. Hassanien et al. (eds.), *Proceedings of the International Conference on Advanced Intelligent Systems and Informatics 2016*, Advances in Intelligent Systems and Computing 533, DOI 10.1007/978-3-319-48308-5_8

sequential data one element at a time [4]. In the past few years, RNN has produced some ground breaking performance improvements on Statistical Machine Translation.

In this research we investigate the capabilities of Recurrent Neural Networks (RNN) and Latent Dirichlet Allocation (LDA) in tackling English - Arabic Machine Translation. Although this topic received a lot of attention recently, it is yet to produce convincing Arabic language translation with high precision. Our preliminary finding in this approach is a new hybridization that is yet to be explored. Although LDA has been used vastly in SMT, our approach is the first to ever combine LDA topic classification ability and RNN ability to learn words dependency in building an SMT model. Such hybridization is promising to solve many statistical and structural difficulties associated with Arabic language. To accomplish our goals, we present our integrated model highlighting each component. Then we take the first steps to putting this model to test as we build a test English language model, followed by an Arabic language model on corpora of 56.000 words to comprise our bilingual corpus. We then perform corpus preprocessing on Arabic Corpora and perform successive 40 testing rounds on our data set.

The reminder of this paper is organized as follows: first we present related work in the area of study presented in this paper, then we propose our model global architecture, RNN and LDA models that we integrate in this study. Finally, we present the results of a prototyping test on our data.

2 Related Work

Neural networks are learning models that achieved state-of-the-art results in both supervised and unsupervised machine learning tasks. Most of machine perception tasks are performed using Neural Networks due to their hierarchical representations capabilities. In contrast, traditional methods rely on hand-engineered features for every specific domain [5]. Many models delivered useful results without modelling time. But despite the usefulness of those models especially Feedforward Networks, they assume independence, hence fail to model long range of dependencies or take into account an unbounded history of previous observations [2, 4, 6]. Many tasks require modelling data with a sequential structure, such as speech recognition, forecasting, image processing, and Machine Translation (MT).

RNNs are capable of dealing with sequential tasks and have recently experienced unprecedented attention in research on statistical machine translation [7]. SMT is the mapping of one sentence in one language to another. There are many approaches to accomplish this task depending on how translation is modelled. One of the first researches on recurrent networks goes back to the 1980 s when Hopfield introduced a recurrent network with pattern recognition capabilities [8]. Another early architecture on sequences with one hidden layer was introduced by Jordan on 1986 [9]. Elman [10] introduced a simpler architecture of RNN in which each hidden node has a single recurrent edge connected to it.

Recently RNN has been applied in language modelling fields and shown to perform considerably well since the state-of-the-art RRN Language model and Context dependent RNN by Mikolov et al [11, 12]. These models outperformed the previous state-of-the-art multilayer feedforward networks in perplexity and in error rate and since then this model has been adopted by many research applications in Machine Translation [7, 13–15]. Mikolov presented a Recurrent Neural Network Language Model (RNNLM) [11]. His language model using RNN obtained 50 % perplexity reduction compared to the prior models. In his experiment on speech recognition he obtained 18 % word error reduction.

Since then, Mikolov Model had many applications in Machine Translation field [2, 6, 7, 15–17]. Kalchbrenner and Blunsom [13] introduce a class of probabilistic continuous translation models that are based on continuous representations for words to overcome sparsity problem. Sparsity problems refer to the extent of having null values in a measure which affects both time and storage on the training process. These continuous representations represented no improvements on restoring results over the state of the art. Ali et al [15] represented source sentence built by Latent Semantic Analysis. However, this approach did not add any gain over systems using recurrent neural networks. Schwenk [18] proposed a feedforward network that predicts phrases of maximum length, in which all phrase words are predicted at once. This approach however, doesn't go beyond phrase boundaries to unbounded history like the model proposed in this research.

Based on Mikolov [11, 12] language model, the research goal is to present a new model of Statistical Machine Translation with Recurrent Neural Network and LDA that aims to tackle many of the difficulties faced in learning Arabic language long range dependencies, cover unbounded history of context, and improve performance precision.

3 Proposed Approach

In this section, the first step of the experiments of this paper are mentioned starting with the proposed model (SMT Context Modelling with RNN and LDA) along with its global architecture. Then an introduction and discussion of Latent Dirichlet Allocation for Context Modeling is provided.

3.1 Data Sets and Experimental Setup

Experiments of the proposed approach will be shown on English corpora first of 10,000 English sentences to test the correct performance of the system, then another test is performed on Arabic Corpora. A corpus of Arabic language of 56,000 words from Gigaword and Arabic Wikipedia was used [20]. Indeed, a corpus is different in size for different languages. For language pairs (English-Arabic) there are over 10 million sentence pairs. Most other languages though have smaller corpora available [1].

3.2 Proposed Model

The main advantage of RNN is its ability to handle sequential dependencies and to span previous time steps [10]. In the context of SMT, The translation process using RNNs can be described as a sequence input word vector $(w(1), w(2), w(3) ..., w(t))$.

As for the target sequence of translation words, it can be denoted as (y(1), y(2), y(3) ..., y(t)). For translation task, using this temporal terminology, input sequence of source language words w(t) arrive in a sequence of time steps and output sequence consist of target language words y(t).

The global architecture of this model is illustrated in Fig. 1. A Bilingual Parallel Corpus for training the SMT model is connected to the RNNs model. This corpus represents English and Arabic pairs of translated words where the source language is in English and the target language in Arabic. Context Modelling (Classification) is done using LDA in the feature layer which will be explained further in the next section.

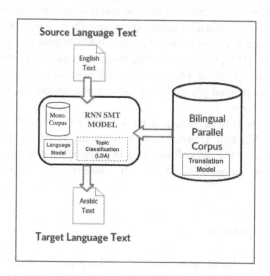

Fig. 1. Block diagram of (RNN SMT Model)

The question remains of how this model captures the history dependencies amongst words in a sentence. The architecture of RNNs has an effective representation of history learned during the data training. This characteristic that defines RNNs lies in the hidden layer where a representation of all history can be found to represent a context understanding pattern, not just a history of previous "n number of words" [4, 19].

The architecture of RNN is shown in Fig. 2. It consists of input layer, hidden layer and output layer. The input layer consists of a vector w(t) and vector s(t−1); vector w(t) represents current word w(t) encoded as 1 of V, where V is the size of the vocabulary. Vector s(t−1) is the output values in the hidden layer in the previous time steps. After the network is trained, the output layer y(t) represents: P (wt+1|wt, s(t−1)).

The input layer, hidden layer and output layer all have corresponding weight matrices. Matrices U and W is between the input and the hidden layer and matrices V between hidden and output layer. The output of computing the hidden layer and output layer are as follows, respectively:

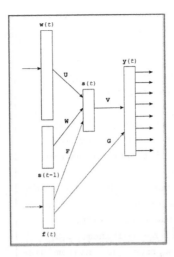

Fig. 2. Recurrent Neural Network Architecture based on [12]

$$\mathbf{s}(t) = f(\mathbf{U}\mathbf{w}(t) + \mathbf{W}\mathbf{s}(t+1) + \mathbf{F}\mathbf{f}(t)) \tag{1}$$

$$\mathbf{y}(t) = g(\mathbf{V}\mathbf{s}(t) + \mathbf{G}\mathbf{f}(t)), \tag{2}$$

$$f(z) = \frac{1}{1+e^{-z}}, \quad g(z_m) = \frac{e^{z_m}}{\sum_k e^{z_k}} \tag{3}$$

Where f(z) is a sigmoid activation function and g(z) is softmax activation functions in the output layer which is mainly used to guarantee that the output is represented in a valid probability distribution with all outputs greater than 0 and their sum is 1.

To train a neural network language model, we find the weight matrices U, V, W, F and G. Training RNNs originally happens by stochastic gradient descent is done using Back Propagation (BP) algorithm. With simple BP training, the recurrent network performs poorly in some cases. However, an improvement of this algorithm is called Back Propagation Through Time (BPTT) which propagate gradients of errors in the network back in time through the recurrent weights [12, 19]. If a network training sequence starts at time t0 and ends at time t1, the total cost function is simply the sum over time of the standard error function Esse/ce(t) at each time-step:

$$E_{total}(t_0, t_1) = \sum_{t=t_0}^{t_1} E_{sse/ce} \tag{4}$$

3.3 Latent Dirichlet Allocation for Context Modeling

To model the context, we use Latent Dirichlet Allocation (LDA) which help models the topics of a given document [24]. The word "topics" imply the variable relation (Distribution) that links words and their occurrence in documents. Topic modelling can solve identifying the topic in which a word belong to or the context identification problem in SMT by generating specific context and improve the performance of the SMT system [25].

> 1- Decide on the document length N by sampling from a Poisson distribution:
> N ~ Poisson(ξ).
> 2 - Decide on a multinomial distribution over topics for the document by sampling from a Dirichlet distribution parameterized by α: Θ~Dir(α).
> 3- For each of the N words to be generated, first decide on a topic to draw it from, and then on the word itself:
> •Choose the topic z_n~ Multinomial(Θ).
> •Choose a word w_n from the unigram distribution associated with the topic:
> p(w_n~|z_n, β).

A key parameter in LDA is a parameter named α controlling the prior distribution shape over topics for individual documents. The result of LDA is a learned value for α, and the set of topic distributions β.

4 Experiment

Following the concept introduced by Tomas Mikolov et al in [11, 12] for generating language model using RNN, we build the Arabic language model. For this experiment a corpora of Arabic language of 56,000 words from (Gigaword and Arabic Wikipedia) was used. Preprocessing of Arabic language is required as some letters in Arabic are used interchangeably by people. Therefore, normalization was needed and was carried out on three Arabic letters.

Preprocessing:
Arabic language preprocessing aims to transforming the Arabic text so that the text is processed according to the following:

Normalization of "aleph" letter:
We have normalized "aleph" letter in Arabic such as:
aleph with Hamza on top: (أ)
has been normalized to aleph with no Hamza on top: (ا)
The reason for this normalization is that all forms of Hamza are represented in the dictionary as one, and people normally misspell different forms of aleph.

Normalization of "ya'a" and "ta'a" letter:
We have normalized " ya'a" and ' ta'a' letter in Arabic such as:
the letter (ي) is normalized to (ى)
the letter (ة) is normalized to (ه)

The reason behind this normalization is that there is not a single convention for spelling those letters when they appear at the end of a word.

5   Results

For this experiment a corpora of Arabic language of 8000 sentences from (Gigaword and Arabic Wikipedia) (about 56,000 words) were used. Training and validation sets were used as follow; the training set consisted of 6500 Arabic sentences while the validation data set consists of 1500 Arabic sentences.

In the training phase, to construct the first language model using RNNs, 40 neurons were used in the hidden layer of the system. At this stage we set Back Propagation Through Time (BPTT) for 3 steps. As for the testing we used 40 testing sets each set

Table 1. Testing Scores on Arabic Sentences

ID	Sentence	Score
1	اكتساب الخبرة في الحياة	-14.60
1	اكتساب إدراك في الحياة	-15.44
1	اكتساب تجربة في الحياة	-14.96
2	الفطنة الثقافي لديه	-14.71
2	الادراك الثقافي لديه	-13.54
2	التنبه الثقافي لديه	-13.82
3	أفراد ذلك المجتمع	-14.13
3	أجزاء ذلك المجتمع	-14.86
3	أشخاص ذلك المجتمع	-13.83
4	تصنيف ضحايات الدين	-15.44
4	تصنيف ضحايا الدين	-11.47
4	تصنيفات ضحايات الدين	-14.79
5	مجموعة من الاشياء محكم	-13.90
5	مجموعة من الاشياء المرتبطة	-12.87
5	مجموعة من الاشياء متقن	-16.40
6	تصبح قيما تتوارثها الاجيال	-22.32
6	تصبح ثمناً تتوارثها الاجيال	-19.96
6	تصبح نفيسة تتوارثها الاجيال	-19.83

contains 3 candidates, that we acquired from Wikipedia [23] to make sure that the language model prefers meaningful sentences over meaning less ones.

In Table 1 we show sample of the sentences scores on our testing data. In 70 % of the times, RNN Arabic language model assigned highest scores to the correct translation of a given source sentence, while producing relatively lower scores to other meaningless sentences. However, 30 % of the sentences were given false scoring, so the error rate is relatively high due to the small training data size. Testing sentences with (ID: 6) have not been scored correctly and the most accurate translation was not chosen.

We conclude that this is due to the lack of training on larger corpora. as according to [12] a training on about 2 billion tokens need to be performed in order for the RNN to learn the language model and lead to accurate scoring results. However, training on such large corpora will require more time for each iteration on the training phase.

6   Conclusion and Future Work

RNNs produced state-of-the-art performances in recently proposed language models. In this paper, we presented a new hybridization model to Statistical Machine Translation (SMT) using Recurrent Neural Networks (RNNs) and LDA for context modelling. This research was based on the state-of-the-art RNN language model by Mikolov. Our preliminary contribution in this paper is that this hybridization model has never been proposed before. To accomplish our goal, we took the first steps to building this model. We proposed the global architecture of the model and shed light on the difficulties faced when using conventional neural network. In the implementation phase, we started building a language model using the "RNNLM" toolkit by Mikolov [21]. We built the target language "language model" in Arabic with 8000 sentences (about 56,000) words. Scoring results showed accuracy on 70 % of our testing data. We conclude that the lack of training on a larger corpora caused such lack of accuracy.

Future work involves building the complete (SMT Context Modelling with RNN and LDA). The training needs a large English and Arabic corpus (bilingual Parallel Corpora) with over 1 billion tokens. About 80 % of data will be used for training, 10 % for validating the precision and perplexity, and 10 % for testing. In addition, A big Arabic corpus will be constructed to build a dependable Arabic language model through combining number of large corpus like (Arabic Gigaword and Wikipedia).

Future work also includes comparing the model with previous state-of-the-art to showcase its performance. At the end, given the time and resources, we claim that our model can outperform previous models in learning about sequential dependencies between words and sentences in Arabic Language.

References

1. Callison-Burch, C., Talbot, D., Osborne, M.: Statistical machine translation with word-and sentence-aligned parallel corpora. In: Proceedings of the 42nd Annual Meeting on Association for Computational Linguistics, p. 175 (2004)
2. Durrani, N., Fraser, A., Schmid, H.: Model with minimal translation units, but decode with phrases. In: Proceedings of NAACL-HLT, 9–14 June 2013, Atlanta, Georgia (2013)
3. Brown, P., de Souza, P., Mercer, R., Pietra, V., Lai, J.: Class-based n-gram models of natural language. Computational Linguistics. Comput. Linguist. 18(4), 467–479 (1992)
4. Lipton, Z., Berkowitz, J., Elkan, C.: A Critical review of recurrent neural networks for sequence learning', arXiv preprint arXiv:1506.00019 (2015)
5. Farabet, C., Couprie, C., Najman, L., LeCun, Y.: Learning hierarchical features for scene labeling. IEEE Trans. Pattern Anal. Mach. Intell. 35(8), 1915–1929 (2013)
6. Zhao, B., Tam, Y.: Bilingual recurrent neural networks for improved statistical machine translation. In: Spoken Language Technology Workshop (SLT), 7–10 December 2014, South Lake Tahoe, NV. IEEE (2014)
7. Sundermeyer, M., Alkhouli, T., Wuebker, J., Ney, H.: Translation modeling with bidirectional recurrent neural networks. In: Proceedings of the 2014 Conference on Empirical Methods in Natural Language Processing (EMNLP), 14–25 October 2014, Doha, Qatar (2014)
8. Hopfield, J.J.: Neural networks and physical systems with emergent collective computational abilities. Proc. Nat. Acad. Sci. 79(8), 2554–2558 (1982)
9. Jordan, M.: Serial order: a parallel distributed processing approach. Technical report 8604, Institute for Cognitive Science, University of California, San Diego (1986)
10. Elman, J.: Finding structure in time. Cogn. Sci. 14, 179–211 (1990)
11. Mikolov, T., Zweig, G.: Context dependent recurrent neural network language model. In: 2012 workshop on Spoken Language Technology, pp. 234–239 (2012)
12. Kombrink, S., Mikolov, T., Karafiat, M., Burget, L.: Recurrent neural network based language model. In: INTERSPEECH 2010, 11th Annual Conference of the International Speech Communication Association, Makuhari, Chiba, Japan, 26–30 September 2010, pp. 1045–1048 (2010)
13. Kalchbrenner, N., Blunsom, P.: Recurrent continuous translation models. In: Proceedings of the 2013 Conference on Empirical Methods in Natural Language Processing, October 2013, Seattle, Washington, USA, pp. 1700–1709 (2013)
14. Liu, S., Yang, N., Li, Zhou, M.: A recursive recurrent neural network for statistical machine translation. In: Proceedings of ACL, pp. 1491–1550 (2014)
15. Hu, Y., Auli, M., Gao, Q., Gao, J.: Minimum translation modeling with recurrent neural networks. In: Proceedings of the 14th Conference of the European Chapter of the Association for Computational Linguistics, April 2014
16. Schuster, M., Paliwal, K.: Bidirectional recurrent neural networks. IEEE Trans. Signal Process. 45(11), 2673–2681 (1997)
17. Hochreiter, S., Schmidhuber, J.: Long short-term memory. Neural Comput. 9(8), 1735–1780 (1997)
18. Schwenk, H.: Continuous space translation models for phrase-based statistical machine translation. In: 25th International Conference on Computational Linguistics (COLING), December, Mumbai, India, pp. 1071–1080 (2012)
19. Mikolov, T.: Statistical Language Models Based on Neural Networks. Ph.D., Brno University of Technology (2012)
20. Arabic Gigaword Corpus. https://catalog.ldc.upenn.edu/LDC2011T11

21. Mikolov, T.: RNNLM Toolkit (2012). http://www.rnnlm.org/. Accessed: 28 Nov 2015
22. Guessabi, F.: The cultural problems in translating a novel from arabic to english language. AWEJ Special Issue on Translation (2), 224–232 (2013)
23. https://ar.wikipedia.org/wiki/. Accessed: 20 Dec 2015
24. Ponweiser, M.: Latent dirichlet allocation in R. Diploma Thesis, Institute for Statistics and Mathematics 2 May 2012
25. Zhengxian, G., Guodong, Z.: Employing topic modeling for statistical machine translation. In: 2011 IEEE International Conference on Computer Science and Automation Engineering, CSAE (2011)

Optimizing Fuzzy Inference Systems for Improving Speech Emotion Recognition

Reda Elbarougy[1,2(✉)] and Masato Akagi[1]

[1] Japan Advanced Institute of Science and Technology (JAIST), Nomi, Japan
elbarougy@du.edu.eg, akagi@jaist.ac.jp
[2] Department of Mathematics, Faculty of Science,
Damietta University, Damietta, Egypt

Abstract. Fuzzy Inference System (FIS) is used for pattern recognition and classification purposes in many fields such as emotion recognition. However, the performance of FIS is highly dependent on the radius of clusters which has a very important role for its recognition accuracy. Although many researcher initialize this parameter randomly which does not grantee the best performance of their systems. The purpose of this paper is to optimize FIS parameters in order to construct a high efficient system for speech emotion recognition. Therefore, an optimization algorithm based on particle swarm optimization technique is proposed for finding the best parameters of FIS classifier. In order to evaluate the proposed system it was tested using two emotional speech databases; Fujitsu and Berlin database. The simulation results show that the optimized system has high recognition accuracy for both languages with 97 % recognition accuracy for Japanese and 80 % for German database.

Keywords: Fuzzy Inference System (FIS) · Particle swarm optimization · Speech emotion recognition · Optimum clusters radius

1 Introduction

Recently speech emotion recognition is an important research topic with a huge number of potential applications. One of the most important application is the development of the human-machine interface, since human communication contains a huge amount of emotional messages which should be recognized by machines such as robot assistants in the household, computer tutors, or automatic speech recognition units in call-centers [1]. Fuzzy Inference System (FIS) is used for pattern recognition and classification purposes such as emotion classification [2]. FIS can be constructed using expert knowledge by human or from data. FIS based on expert knowledge only for complex systems such as emotion recognition system may suffer from a loss of accuracy. The relationship between the multivariable input acoustic features and output emotional state in emotion recognition system is fuzzy and complex [3]. Generated partitioning

© Springer International Publishing AG 2017
A.E. Hassanien et al. (eds.), *Proceedings of the International Conference on Advanced Intelligent Systems and Informatics 2016*, Advances in Intelligent Systems and Computing 533, DOI 10.1007/978-3-319-48308-5_9

is more difficult and meaningless for human experts. Therefore the main approach for constructing FIS is by using fuzzy rules inferred from data, such as the data-driven FIS constructed in [4].

In order to construct a FIS from data can be implemented using two main stages: automatic rule generation or initial system and the other is fine-tuning of the initial system parameters. In the first stage rule generation leads to an initial system with a given space partitioning and the corresponding set of rules and specific structure. The Adaptive Neuro-Fuzzy Inference System (ANFIS) is used as a tool for fine-tuning the parameters of initial FIS [4].

Generating the initial FIS is a very important step in constructing a classification system using ANFIS [5]. In the initialization step the structure of the FIS is determined i.e. the number of nodes, the number of membership functions for each input and output, and the number of rules [6]. This structure will be fixed through the training classification process. Therefore, this paper investigates how to optimize the structure of initial FIS for automatic rule generation. The function 'genfis2' in Fuzzy Logic Toolbox generates initial FIS model from the training data using subtractive clustering algorithm, and it requires the user to specify a parameter called 'cluster radius'. The cluster radius indicates the range of influence of a cluster when the data space is considered as a unit hypercube. Specifying a small cluster radius will usually yield many small clusters in the data, resulting in many rules and vice versa [7].

Although, the estimation results of FIS are sensitive to the initialized FIS which is determined by selecting cluster radii. However, most of the previous studies neglect this effect by randomly selecting values such as 0.5 as in [8] or 0.95 as in [9] without any investigation of the effect of these values on the final results of their systems. In order to avoid this problem it is very important to prevent subjectivity of choosing this parameter. This can be done by optimizing this parameter. The purpose of this paper is to improve the estimation accuracy of the FIS in order to accurately estimate the emotional state from the speech signal, by finding the best cluster radii which determine the influence of effect of input and output data.

This paper proposes a method for optimizing the cluster radii for the following reasons: (1) the vector of radius has very important role for the recognition accuracy. (2) The radii determine the structure of the FIS: the number of membership functions (MFs) for each input and output variable and consequently the number of fuzzy rules (3). To reduce the computation complexity by selecting the appropriate radii (small radii leads to large number of MFs very large number of parameters which will need too much time to be estimated). (4) To avoid subjectivity of choosing this parameter. Therefore, a method for selecting the optimal radii is required to improve the performance of FIS as well as avoid the above drawbacks.

This paper investigates the design of a high efficient system for emotion recognition from the speech signal. In the literature the emotional states can be represented by the categorical approach such as happy, anger or can be represented as in a two-dimensional space spanned by the two basic dimensions valence (negative-positive axis), and activation (calm-excited axis) [1]. The emotional state in this paper is represented by a hybrid model which can estimate emotion dimensions valence-activation as well as emotion category. Three FIS were used to detect the emotional state: two of them to estimate the two basic dimensions valence and activation and the third FIS was used to map the estimated emotion dimensions into emotion categories. Therefore, anl optimization algorithm is proposed for finding the best cluster radii parameter for the used FIS classifier based on Particle Swarm Optimization (PSO) technique.

2   Speech Material

In order to validate the proposed method, two emotional speech databases were used, one in the Japanese language and the other in the German language. The Japanese database is the multi-emotion single speaker Fujitsu database produced and recorded by Fujitsu Laboratory. It contains five emotional states: neutral, joy, cold anger, sad, and hot anger as described in [10].

The German database is the Berlin database. It comprises seven emotional states: anger, boredom, disgust, anxiety, happiness, sadness, and neutral speech. An equal distribution of the four similar emotional states (neutral, happy, angry, and sad) as follows: 50 happy, 50 angry, 50 sad, and 50 neutral; in total, 200 utterances were selected from the Berlin database.

For constructing a speech emotion recognition system based on the proposed method using the dimensional approach, many acoustic features must be extracted and the two emotion dimensions must be evaluated using human subjects for each utterance in the two databases. Therefore, a listening test was used to evaluate valence and activation as explained in [10], for the two databases. Then, an initial set of 21 acoustic features were extracted for each database. Moreover, the feature selection method proposed by Elbarougy and Akagi based on a three-layer model of a human perception model was used to select the most related acoustic features for each emotion dimensions [10]. Finally, 11 and 10 acoustic features were selected for Japanese and German databases, respectively.

3   The Proposed Optimization Method

Previous studies show that the cluster radii have a great effect on the accuracy of the FIS. Therefore, this study tries to find the optimal cluster radii. Our proposed radii selection method is based on the following hypothesis: the smallest root mean squared error (RMSE) for initial FIS has a greater potential for achieving a lower RMSE when applying the training using FIS. Thus, we assume that the cluster radii for the initial FIS which correspond to the minimum RMSE is the optimal radii. Therefore, initializing FIS using this optimal radii will have a large impact for predicting values of emotion dimensions. The next subsection introduces the proposed method for finding the optimal radii. Finally, a FIS will be constructed using the obtained optimal radii in order to accurately estimate valence and activation form the extracted acoustic features.

3.1   Particle Swarm Optimization Method for Cluster Radii

Particle swarm optimization is a stochastic, population based swarm intelligence algorithm developed by Kennedy and Eberhart [11]. The PSO algorithm is similar to evolutionary computation in producing a random population initially and generating the next population based on current cost. Thus, PSO is faster in finding solutions compared to other evolutionary computation techniques. In this paper we proposed PSO for optimizing the radii. This method widely used for parameter optimization in many fields for example, it has been applied to optimize premise and consequent parameters in order to make the ANFIS output fit the training data [12,13]. However, in this paper this method will be used to optimize radii in order to improve the accuracy of FIS.

In order to implement the PSO method for optimizing radii, the following steps will be done. (1) Finding the suitable range for searching the optimal radii. (2) Applying the PSO in the determined range to find the optimal radii for valence and activation. In this study, only the effect of radii parameter has been investigated. Since this parameter has the highest effect in changing the resulting clusters number and consequently the structure of FIS. The radius of each cluster specifies the range of influence of the cluster center. Specifying a smaller cluster radius will yield smaller clusters in the data, consequently more rules. Therefore, for find the suitable range of radii, the relationship between the radii and the number of rules is investigated. Figure 1 shows the relationship between the numbers of rules for different radii in the range from 0.1 to 2.

From this figure we can easily notice that a smaller value of the cluster radius results in a large number of rules, and vice versa. Larger numbers of rules usually take longer for calculations for example, using radii = 0.5 for calculating valence form the selected 10 acoustic features leads to 46 fuzzy rules as shown in Fig. 2. This figure shows the structure of the initial FIS generated for estimating valence by subtractive cluster algorithm using Matlab function 'genfis2' at radii = 0.5.

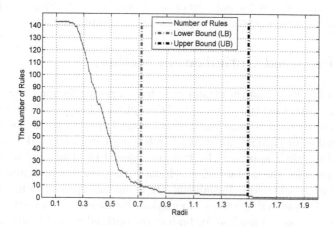

Fig. 1. Radius of clusters versus number of rules for initial FIS generated by *'genfis2'*.

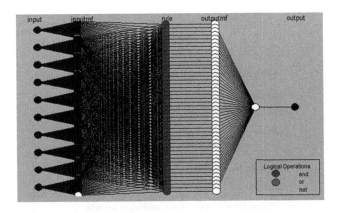

Fig. 2. The structure of the initial FIS generated by subtractive cluster at Radii = 0.5.

The number of membership functions will be 46 for each input-output which need too much time in order to calculate the parameters of each membership function

Therefore, it is better to use the suitable range which correspond to at least 11 fuzzy rules which correspond to the left vertical line in Fig. 2 at radii = 0.72 which was considered as the Lower Bound (LB) for the searching range. To construct FIS it needs at least two rules to start training using ANFIS. Therefore, the radii which correspond to less than 2 rules must be removed for searching range. Hence, the radii correspond to 3 fuzzy rules is selected as the Upper Bound (UB) for the searching range of radii.

The second step is to apply the PSO algorithm to find the optimal radii in the selected range. The mathematical representation of PSO is given as

$$x_{id_{new}} = x_{id} + v_{id_{new}} \tag{1}$$

$$v_{id_{new}} = w * v_{id} + c_1 * rand_1 * (P_{id} - x_{id}) + c_2 * rand_2 * (G_d - x_{id}) \tag{2}$$

where x_{id}, v_{id} represent the position vector and the velocity vector of the ith particle in the d-dimensional search space, respectively. In addition, c_1, c_2 are the acceleration constants, $rand_1$, $rand_2$ are uniformly generated random numbers between 0 and 1; w is the inertia weight which decreased linearly from 0.9 to 0.4 during the run [12]. The first part of Eq. 2 represents the inertia of the previous velocity. The second part is the cognition part and it provides the best own position for the particles, where P_{id} is the best previous position. The third part is the social component which represent the collaborative effect of all participles for finding the global optimal solution, where G_d is the global best position. The third component pulls the particles towards the global best position.

In this study, the PSO algorithm is used to find the optimum radii where $radii = (r_1, r_2, ..., r_d)$, where d represents the number of input-output. For example, in order to estimate valence as shown in Fig. 2 the number of input is 10 and the number of output is one i.e., $d = 11$ in this case. The flowchart shown in

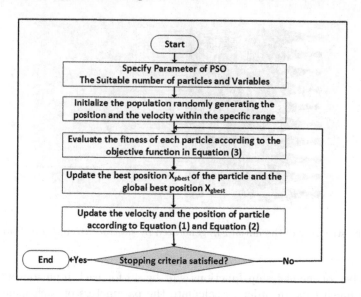

Fig. 3. The flowchart of PSO algorithm for optimizing radii.

Fig. 3 is used to optimize the radii where the fitness function here is the RMSE between the actual output and the desired output which can be described by:

$$fitness = RMSE = \sqrt{\frac{1}{N}\sum_{i=1}^{N}(\widehat{y}_i - y_i)^2} \tag{3}$$

where y_i is the actual output evaluated by human subjects and $\widehat{y}_i$ is the desired output using the system, N represents the total number of data examples in the training dataset. In multidimensional data, different radii may be specified for each input-output dimension. If the same value is applied to all data dimensions, each cluster center will have a spherical neighborhood of influence with the given radius. For the propose of comparison we use the spherical neighborhood of influence i.e., the same value is applied for all the dimensions $radii = (r_1 = r, r_2 = r, ..., r_d = r)$.

4   Speech Emotion Recognition System

In this section we evaluate the performance of the proposed PSO-FIS for estimating valence and activation. Two databases were used to train and test the proposed system. The proposed speech emotion recognition system is shown in Fig. 4. This system is consisted of two stages; in the first stage the extracted acoustic features were used as inputs to FIS to estimate emotion dimensions valence and activation. In the second stage the estimated emotion dimensions

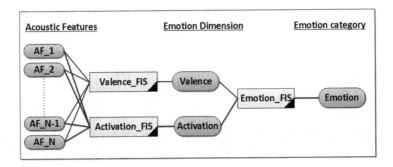

Fig. 4. The proposed speech emotion recognition system.

were used as inputs to FIS to detect the emotion categories. FIS is multiple input, single output system therefore 3 FIS were required as follows: two FISs were used to estimate emotion dimension valence and activation from acoustic features, and one FIS was used to map the estimated emotion dimensions into emotion category neutral, happy, anger, and sad. The two stages of the proposed system will be explained in details in the next sections.

4.1 Emotion Dimension Estimation

Fujitsu and Berlin database were used to evaluate the proposed system performances. The PSO algorithm is implemented to find the optimal radii for each emotion dimension individually. Then, the optimal radii is used to generate the initial FIS for the investigated dimension. In addition, the initial FIS is trained using the two databases individually. The mean absolute error (MAE) between the estimated values of emotion dimensions and the corresponding average value given by human subjects is used as a metric of the discrimination associated with each case. The MAE is calculated according to the following equation:

$$MAE = \frac{1}{N}\sum_{i=1}^{N}|\widehat{y}_i - y_i| \tag{4}$$

where y_i is the actual output evaluated by human subjects and $\widehat{y}_i$ is the desired output using FIS, N is number of utterances in the used database.

The result of PSO-FIS for each emotion dimension generated using the optimal radii can be compared with the values of radii in the range from 0 to 1. However, the optimal radii for the two databases located between 0.5 and 1. Therefore, the optimal radii was compared with that generated using six different radius of cluster (0.55 ,0.60, 0.65, 0.70, 0.75, and 0.80). Figure 5(a) and (b) show the MAEs for emotion dimensions for Japanese and German database, respectively.

For Japanese database the optimal radii for valence and activation were 0.89 and 0.78, respectively. In addition, for German database the optimal radii were

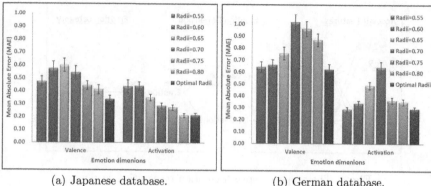

(a) Japanese database.   (b) German database.

Fig. 5. The estimation results for emotion dimensions using different radii.

0.96 and 0.91, respectively. From these figures, we can see that the best results for each emotion dimensions were obtained form the optimal radii. As we can easily see that the values of MAE is very sensitive to the values of the radii. This result confirm that the parameter radii is a very important and must be considered for initialized FIS. The optimal MAE for valence and activation for Japanese language are 0.34 and 0.21, respectively. Moreover, the results for German database for valence and activation were 0.62 and 0.29, respectively.

Table 1(a) and (b) also reveals the more detailed view of different results with various radii for FIS. These tables depicting the results of valence and activation dimension for different radii, for both databases. The results for the optimal radii is listed in the last row each table. Increasing radius causes decreasing of the number of the clusters and therefore causes the decreasing of the number of the rules. FIS has gotten 27 rules for radius 0.55 and 8 rules for radius 0.80. Decreasing the cluster radius increases complexity, however our results show that increasing radius causes performance improvement in terms of small MAE and high correlations between the estimated values and the human evaluation. Larger cluster radius assures well defined rules that helps to reduce the redundancy of data and determine the rules and membership functions, which is one of the benefits of the clustering method. The size of the cluster centers determines the number of rules associated with the fuzzy inference system.

These results for both languages revel that our proposed PSO-FIS has the ability to accurately estimate emotion dimensions using the optimal radii, with a small MAE.

4.2 Emotional Classification

The categorical and dimensional approaches are closely related, i.e. by detecting the emotional content using one of these two schemes, we can infer its equivalents in the other scheme. Findings in most of the previous studies demonstrate that the dimensional approach can actually help to improve the automatic

Table 1. Classification results using FIS classifier.

(a) Japanese database

Radii	Valence		Activation		Emotion classification
	No. of rules	MAE	No. of rules	MAE	Recognotion rate %
0.55	27	0.47	22	0.43	84 %
0.60	20	0.57	12	0.44	79 %
0.65	13	0.60	9	0.35	83 %
0.70	11	0.54	9	0.29	85 %
0.75	9	0.44	7	0.27	92 %
0.80	8	0.41	5	0.21	91 %
Optimal	**5**	**0.34**	**5**	**0.21**	**97 %**

(b) German database

Radii	Valence		Activation		Emotion classification
	No. of rules	MAE	No. of rules	MAE	Recognotion rate %
0.55	57	0.64	48	0.29	75 %
0.60	32	0.66	26	0.34	76 %
0.65	19	0.76	17	0.49	71 %
0.70	12	1.02	11	0.64	60 %
0.75	9	0.96	7	0.36	69 %
0.80	7	0.87	6	0.35	67 %
Optimal	**4**	**0.62**	**4**	**0.29**	**80 %**

emotion classification. Therefore our proposed system is hybrid system which have this advantage i.e. the proposed system estimate emotion dimensions as well as detecting the emotional category. So, the estimated values of emotion dimensions (valence and activation) were used as inputs for the FIS to predict the corresponding emotional category.

The classifications results using the estimated values of emotion dimensions as shown in Table 1(a) and (b) for the Japanese and German databases, respectively. These results for both languages revel that our proposed PSO-FIS for emotion classification has the ability to accurately predict the emotion category using the optimal radii, with a very high emotion classification rate for Japanese database with 97 % and high classification rate for German database with 80 %. The difference of the recognition rate between the two languages is due to Japanese database is a single speaker database however German database is multi-speaker database. The reason for the high classification accuracy is due the best estimation for emotion dimensions. The Simulation results show that our proposed optimized system can accurately estimate emotion dimensions valence and activation compared with randomly selecting radii, using two different emotional databases. Finally the estimated values of valence and activation were used to predict the emotional state accurately.

5   Conclusion

The aim of this paper is to improve the emotion recognition accuracy using FIS by improving estimation accuracy for emotion dimensions; valence, and activation from acoustic features. In order to accomplish this task, first, a PSO method is used to find the optimal radii for each emotion dimensions, then, the PSO-FIS system was constructed based on an initial FIS generated using the optimal radii. For estimating emotion dimensions, the proposed PSO-FIS emotion recognition system was trained and testing using two different languages. The results for both language revel that our PSO-FIS emotion recognition system initialized using the optimal radii has the ability to accurately estimate the emotion dimensions, with a small errors as well as to improve the final recognition rate. The most important contribution of this study is that the proposed method can automatically select the optimal parameter and prevent subjectivity selection which may lead to different non-optimal results.

Acknowledgments. This study was supported by the Grant-in-Aid for Scientific Research (A) (No. 25240026).

References

1. Grimm, M., Kroschel, K.: Emotion estimation in speech using a 3D emotion space concept. In: Grimm, M., Kroschel, K. (eds.) Robust Speech Recognition and Understanding, June 2007
2. Zhang, Q., Jeong, S., Lee, M.: Autonomous emotion development using incremental modified adaptive neuro-fuzzy inference system. Neurocomputing **86**, 33–44 (2012)
3. Huang, C., Akagi, M.: A three-layered model for expressive speech perception. Speech Commun. **50**(10), 810–828 (2008)
4. Lee, C., Narayanan, S.: Emotion recognition using a data-driven fuzzy inference system. In: Proceedings of 8th European Conference, EUROSPEECH, pp. 157–160 (2003)
5. Wei, M., Bai, B., Sung, A.H., Liu, Q., Wang, J., Cather, M.E.: Predicting injection profiles using ANFIS. Inf. Sci. **177**, 4445–4461 (2007)
6. Guillaume, S.: Designing fuzzy inference systems from data: an interpretability-oriented review. IEEE Trans. Fuzzy Syst. **9**(3), 426–443 (2001)
7. Zoveidavianpoor, M., Samsuri, A., Shadizadeh, S.R.: Adaptive neuro fuzzy inference system for compressional wave velocity prediction in a carbonate reservoir. J. Appl. Geophys. **89**, 96–107 (2013)
8. Entchev, E., Yang, L.: Application of adaptive neuro-fuzzy inference system techniques and artificial neural networks to predict solid oxide fuel cell performance in residential microgeneration installation. J. Power Sources **170**, 122–129 (2007)
9. Deregeh, F., Karimian, M., Nezmabadi-Pour, H.: A new method of earlier kick assessment using ANFIS. Iran. J. Oil Gas Sci. Technol. **2**(1), 33–41 (2013)
10. Elbarougy, R., Akagi, M.: Speech emotion recognition system based on a dimensional approach using a three-layered model. In: Proceedings of International Conference APSIPA, USA (2012)
11. Kennedy, J., Eberhart, R.: Particle swarm optimization. In: Proceedings of IEEE International Conferenc on Neural Network, Perth, Australia, pp. 1942–1948 (1995)

12. Toha, S.F., Tokhi, M.O.: ANFIS modelling of a twin rotor system using particle swarm optimisation and RLS. In: Proceedings of IEEE International Conferenc CIS, United Kingdom (2009)
13. Liu, P., Leng, W., Fang, W.: Training anfis model with an improved quantum-behaved particle swarm optimization algorithm. Math. Prob. Eng. **2003**, 1–10 (2013)

Tree-Based HMM State Tying for Arabic Continuous Speech Recognition

Mona A. Azim$^{(\boxtimes)}$, A. Aziz A. Hamid, Nagwa L. Badr,
and M.F. Tolba

Faculty of Computer and Information Sciences,
Ain Shams University, Cairo, Egypt
{monaabdelazim, abdelaziz, nagwabadr}@cis.asu.edu.eg,
fahmytolba@gmail.com

Abstract. One of the major challenges in building Hidden Markov Models (HMMs) for continuous speech recognition systems is the balance between the available training set and the recognition performance. For large vocabulary recognition systems, context dependent models are usually required to obtain higher recognition accuracy. This is crucial as most of the language contexts may not occur in the training set. This paper proposes an Arabic phonetic decision tree necessary to build tied state tri-phone HMMs. Experimental results based on the proposed decision tree show a promising recognition accuracy when compared with the traditional context independent models using the same training and testing sets. The maximum recognition accuracy achieved by the proposed approach was 92.8 % whereas it reached 61.5 % when tested using context independent HMMs.

Keywords: Arabic phonemes · Tri-phones hmms · Speech recognition

1 Introduction

The use of Hidden Markov Models (HMMs) has been proved to be a remarkable progress in speech recognition problems. These models can be built as a context dependent or context independent. Although the context dependent HMMs (tri-phones models) achieve higher recognition accuracy, they are not commonly used, due to difficulty of obtaining a sufficient training dataset that could cover all language contexts (tri-phones). The recognition system therefore will not be able to recognize the unseen tri-phones. In addition, training a single model for each of the possible language contexts is not an efficient solution due to the huge number of the possible tri-phones. For example, the Arabic phoneme set contains 39 phonemes with 59,319 tri-phones. In order to efficiently train this huge number of tri-phones as well as reducing the number of estimated HMM parameters, we built and trained HMMs for the available tri-phones seen in the training set then state tying algorithms were employed to expand the trained tri-phones models to synthesize the unseen tri-phones.

The most common technique used in state tying is based on phonetic decision trees to categorize the tri-phone states through utilizing a set of linguistic questions.

© Springer International Publishing AG 2017
A.E. Hassanien et al. (eds.), *Proceedings of the International Conference on Advanced Intelligent Systems and Informatics 2016*, Advances in Intelligent Systems and Computing 533, DOI 10.1007/978-3-319-48308-5_10

In this paper, we introduced an Arabic phonetic decision tree that can be used in creating a tied states tri-phones based HMMs as a part of a large vocabulary Arabic speech recognition system.

The rest of the paper is organized as follows. Section 2 presents the recent work in the Arabic speech recognition field. Section 3 discusses the proposed approach in detail. Experimental results are then described in Sect. 4. Finally, Sect. 5 includes the conclusion and future directions.

2 Related Work

State tying refers to representing the states that share the same set of parameters with only single state by tying them together. This is also can be defined as state clustering as we need to find which states are similar enough to be tied. The most common techniques used in state clustering includes data driven and phonetic decision tree clustering algorithms.

The data driven state clustering approach relies on measuring the distance among all the states of HMMs then combine them into clusters based on the distance matrix obtained. Although the data driven based approaches are suitable for multilingual speech recognition systems, they are not capable of dealing with the unseen contexts [1, 2].

The phonetic tree based state tying was introduced in [3], in which the developed recognition system proved that using phonetic decision tree in state tying is as effective as the data driven technique and has a key advantage of providing a mapping for unseen tri-phones.

Arabic speech recognition research field has gained much interest recently and has been addressed by many researchers. In [4], a phoneme based recognizer was built for the holy Quran to assist the Quran learners. The system was built using the hidden Markov models toolkit (HTK) [5] based on phoneme and allophone level transcriptions.

A novel approach was introduced in [6] for Arabic phonemes recognition. This approach introduced the usage of different number of states for each of the Arabic phoneme achieving a recognition accuracy of 56 % without using a language model [7]. When the language model was employed, the recognition accuracy was increased to 96 % [8].

A syllable based models were introduced in [9, 10] to recognize the isolated Arabic Digits. The proposed models were compared to monophonic, tri-phones and word based models and showed a better performance. A generalized set of tri-phone models was built from the generated tri-phones using a distance measure algorithm.

In [11], spoken Arabic alphabets were investigated from the speech recognition perspective. The HTK toolkit was used to implement an isolated word recognition system for Arabic Alphabets. Phoneme based models were used in the recognition system. The system achieved an overall recognition accuracy of 64.04 %.

Although most of the previously presented related work have achieved a comparable result, none of them adapted a tree based state tying for Arabic tri-phones based HMMs using an Arabic phonetic decision tree.

3   The Proposed Approach

This section discusses the process of building tied-states tri-phone HMMs using an Arabic phonetic decision tree. The process of building the tied-state tri-phones HMMs consists of two main stages:

1. Context independent (CI) stage: Includes building and training the monophonic based HMMs.
2. Context dependent (CD) stage: Aims at building the final tied-states tri-phones HMMs and this is achieved through the following steps:
   (a) Cloning the monophonic HMMs to generate context dependent tri-phones HMMs.
   (b) Using a decision tree-based tri-phones clustering algorithm to generate the final tied-states tri-phones HMMs.

3.1   Context Independent Stage

The context independent stage targets generating the monophonic-based HMMs as a preceding step towards building the context dependent tri-phones HMMs [5]. As shown in Fig. 1, this stage consists of the following steps:

1. Parameterizing the speech signal into a sequence of features vectors.
2. Create the flat start Monophonic HMMs.

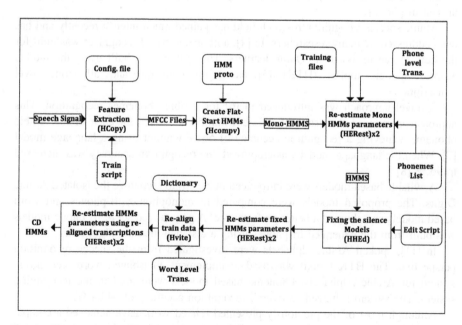

Fig. 1. The process of building context independent HMMs. HCopy, HCompv, HERest, HHEd, and HVite are HTK commands [5].

3. Re-estimate the flat start monophonic HMMs parameters.
4. Fixing the silence models.
5. Re-estimate the fixed Monophonic HMMs parameters.
6. Re-aligning the audio data.
7. Re-estimate the parameters of the monophonic based HMMs using realigned training data set.

The models obtained from the last re-estimation step are the final context independent HMMs that can be used directly in context independent recognition tasks.

3.2 Context Dependent Stage

This stage aims to create the tied states Arabic tri-phone HMMs from the Arabic monophonic HMMs. Figure 2 illustrates the steps followed to achieve this aim. The monophonic based transcriptions of the training data are firstly converted to their equivalent tri-phones transcriptions. This is realized using the HTK Label Editor (HLEd) tool. The HLEd tool can convert label files using a script file with certain commands that direct the tool to generate the tri-phones transcriptions using monophonic transcriptions. It generates the tri-phones transcriptions of the training data and a list of all the tri-phones appear in the training data. Secondly, the set of HMMs is cloned and tied so that they share the same set of parameters and this is done using the HTK HMM editor tool (HHEd). The HHEd clones the models using clone command

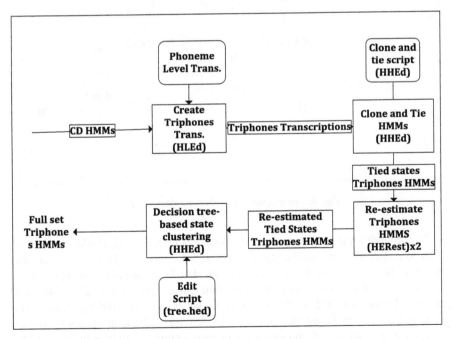

Fig. 2. The process of building context dependent HMMs.

(CL) that takes as an input the tri-phones list generated from the previous step and a set of tie commands (TI). The TI command takes as an input the macro name and the set of parameters to be shared.

The generated tri-phones based HMMs can only recognize the tri-phones that appeared in the training dataset. In order to make the proposed approach more robust, the trained models have to be extended so that they can synthesize the unseen tri-phones.

3.3   HMM State Tying Using Arabic Phonetic Decision Tree

A phonetic decision tree is a binary tree with yes/no linguistic questions attached to each node and each of these questions represents a certain phonetic context. As long as we are developing a tri-phones based model, both the left and right context of the phone were considered.

An example of the proposed phonetic tree is shown in Fig. 3. In this figure, the question "Is the phone on the left of the current phone a nasal?" is linked to the root node of the phonetic tree.

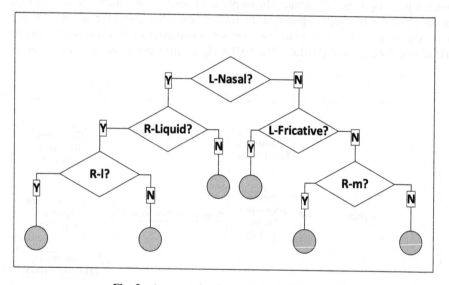

Fig. 3. An example of a phonetic decision tree

To accomplish the state tying process, one tree is built for each HMM state of each phone to cluster all of the similar states to the related tri-phones. All the tri-phones that end up at the leaf node of the tree for a certain state have their states tied [11]. HTK provides a useful tree-based clustering algorithm that accepts a list of all the allowable phonetic classes, each of these classes is represented as a question (QS) command. This command is used to create the questions needed for each of the phonetic context classes. For example, Qs Command used in representing the right context of Arabic nasal phoneme classes is QS "R_Nasal" {*+m, *+n}, where '*' represents any of the Arabic phonemes that could proceed the nasal phonemes (m, n) and '+' denotes right-hand-side part. The full set of the Arabic phonetic classes used in this research can be found in [13]. This set is used to construct the proposed decision tree through a set of QS commands.

4 Experiments and Results

A subset of the Standard Arabic Single Speaker Corpus (SASSC) [14] is used in building the HMMs of the Arabic tri-phones. This corpus contains 7 h of Arabic audio recordings represented as 4372 wave files along with their text transcriptions and phoneme level label files. In this research, a set of 14,351 words and a set of 3,284 words were employed in the training and testing experiments, respectively. The features extracted from the wave files were 13 static MFFCs along with their first and second order derivative resulting in a total of 39 features. Using these features, a set of monophonic and tri-phone based HMMs were trained to verify the proposed Arabic phonetic decision tree.

In order to choose the best configuration for the generated models in both context dependent and independent stages, different experiments using different numbers of Gaussian mixtures were conducted on the training dataset.

Firstly, a set of 41 monophonic-based models were built using a single Gaussian mixture. These models were upmixed gradually from single to 25 mixtures until the maximum recognition accuracy was obtained. Secondly, a set of 9245 tri-phones based models were built using a single mixture then upmixed the same way as monophonic based models. These models were trained then tied resulting in a set of 2678 tied tri-phone models.

A sample of the recognized text is displayed in Table 1. In this table, most of the utterances were correctly recognized although the training set employed in this research is limited. On the other hand, it is shown in this table that the article "ال" in the word "الحجز" is sometimes misrecognized due to the dominance of the phone "ل" in the article. However, when providing sufficient training data, this article is expected to be correctly recognized.

The recognition accuracy obtained from the conducted experiments are shown in Table 2. The results showed that the tied HMM tri-phone states achieved a maximum recognition accuracy of 92.8 % using 20 Gaussian Mixtures.

Table 1. Sample of the system results using tied states tri-phones

Original text	English translation	Recognized text
السلام عليكم و رحمة الله و بركاته	Peace be upon you	السلام عليكم و رحمة الله و بركاته
من فضلك أذكر أسم العائلة	Please, mention the family name	من فضلك أذكر أسم العائلة
تم الحجز	Reservation confirmed	تم لحجز
أذكر سنة صنع السيارة	What is the car manufacturing year?	أذكر سنة صنع سيارة

Table 2. Experiments results

No. Gauss. Mix.	Accuracy using context independent models	Accuracy using context dependent models
1-GM	57 %	87 %
3-GM	57.06 %	87 %
5-GM	59.67 %	86.31 %
9-GM	59.41 %	91 %
16-GM	61.51 %	92.33 %
20-GM	61.31 %	92.8 %
25-GM	60.84 %	92.6 %

However, the monophonic HMMs achieved a maximum recognition accuracy of 61.51 % using 16 Gaussian Mixtures.

These results emphasize that the context dependent models outperforming the context independent models as they take into consideration the co-articulation effect between the adjacent phonemes.

5 Conclusion and Future Work

This paper introduced building tied states HMMs for Arabic tri-phones based a proposed phonetic decision tree. That was achieved through building a monophonic based HMMs followed by creating tied state HMM tri-phones. The proposed Arabic phonetic decision tree was employed to enhance the developed tri-phones models so that they could be able to provide alternatives to the unseen tri-phones. To assess the performance of the developed approach, the number of the Gaussian mixtures was increased gradually and the recognition accuracy was recorded at each step. Experimental results showed that the tied states HMM tri-phones achieved a maximum recognition accuracy of 92.8 % using 20 Gaussian Mixtures, whereas the monophonic based HMMs achieved a maximum recognition accuracy of 61.51 % using 16 Gaussian Mixtures when trained and tested on the same data sets. Further improvements to this research may include using larger and speaker independent dataset in training and testing the generated models.

References

1. Imperl, B., Kačič, Z., Horvat, B., Žgank, A.: Clustering of tri-phones using phoneme similarity estimation for the definition of a multilingual set of tri-phones. Speech Commun. **39**(3), 353–366 (2003)
2. Žgank, A., Horvat, B., Kačič, Z.: Data-driven generation of phonetic broad classes, based on phoneme confusion matrix similarity. Speech Commun. **47**(3), 379–393 (2005)
3. Young, S.J., Odell, J.J., Woodland, P.C.: Tree-based state tying for high accuracy acoustic modelling. In: Proceedings of the workshop on Human Language Technology, pp. 307–312. Association for Computational Linguistics (1994)

4. Elhadj, Y.O.M., Alghamdi, M., Alkanhal, M.: Phoneme-based recognizer to assist reading the Holy Quran. In: Thampi, S.M., Abraham, A., Pal, S.K., Rodriguez, J.M.C. (eds.) Recent Advances in Intelligent Informatics, vol. 235, pp. 141–152. Springer International Publishing, Switzerland (2014)

5. Young, S., et al.: The HTK Book (version 3.5). University of Cambridge, December 2015

6. Nahar, K.M., Al-Khatib, W.G., Elshafei, M., Al-Muhtaseb, H., Alghamdi, M.M.: Data-driven Arabic phoneme recognition using varying number of HMM states. In: 2013 1st International Conference on Communications, Signal Processing, and Their Applications (ICCSPA), pp. 1–6. IEEE (2013)

7. Nahar, K., Al-Muhtaseb, H., Al-Khatib, W., Elshafei, M., Alghamdi, M.: Arabic phonemes transcription using data driven approach. Int. Arab J. Inf. Technol. (IAJIT) 12(3), 237–245 (2015)

8. Muhammad, G., AlMalki, K., Mesallam, T., Farahat, M., Alsulaiman, M.: Automatic Arabic digit speech recognition and formant analysis for voicing disordered people. In: 2011 IEEE Symposium on Computers & Informatics (ISCI), pp. 699–702. IEEE (2011)

9. Azmi, M., Tolba, H., Mahdy, S., Fashal, M.: Syllable-based automatic arabic speech recognition in noisy-telephone channel. WSEAS Trans. Signal Process. Proc. World Sci. Eng. Acad. Soc. (WSEAS) 4(4), 211–220 (2008)

10. Alotaibi, Y.A., Alghamdi, M., Alotaiby, F.: Speech recognition system of Arabic alphabet based on a telephony Arabic corpus. In: Elmoataz, A., Lezoray, O., Nouboud, F., Mammass, D., Meunier, J. (eds.) Image and Signal Processing, vol. 6134, pp. 122–129. Springer, Berlin (2010)

11. Odell, J.J., Woodland, P.C., Young, S.J.: Tree-based state clustering for large vocabulary speech recognition. In: Proceedings of 1994 International Symposium on Speech, Image Processing and Neural Networks, ISSIPNN 1994, pp. 690–693. IEEE (1994)

12. Ridle, J.S., Brown, M.D.: An experimental automatic word recognition system. JSRU Report 1003, p. 5 (1974)

13. Habash, N.Y.: Introduction to Arabic Natural Language Processing. Morgan & Claypool, San Rafael (2010)

14. Almosallam, I., AlKhalifa, A., Alghamdi, M., Alkanhal, M., Alkhairy, A.: SASSC: a standard arabic single speaker corpus. In: 8th ISCA Synthesis Workshop, Barcelona, Spain, pp. 249–253, September 2013

Arabic Fine-Grained Opinion Categorization Using Discriminative Machine Learning Technique

Imen Touati[(⊠)], Marwa Graja, Mariem Ellouze,
and Lamia Hadrich Belguith

MIRACL Laboratory, Arabic Natural Language Processing Research Group
(ANLP-RG), University of Sfax, Sfax, Tunisia
ismi_touati@yahoo.fr,
{marwa.graja,l.belguith}@fsegs.rnu.tn,
mariem.ellouze@planet.tn

Abstract. This paper presents an approach of fine-grained opinion categorization in Arabic news articles. This approach is based on lexical semantic analysis. We propose to categorize every opinion expression using a proposed typology of four top-level semantic categories: reporting, judgment, advice and sentiment. Each word or opinion expression will be annotated with a semantic representation which takes in consideration specificities of Arabic language. To the best of our knowledge, there is no annotated Arabic opinion corpus with the proposed semantic representation. The task of categorization is considered as a classification problem. So, we use a Conditional Random Fields (CRF) as a discriminative model that we consider as a good contribution, because of the lack of similar fine-grained opinion categorization performed with CRF. The obtained results show that the integration of CRF models is important for opinion classification of the Arabic language.

Keywords: Opinion · Semantic annotation · Fine-grained categorization · CRF

1   Introduction

Among several tasks that are related to opinion mining such as opinion holder identification, opinion target identification, opinion Summarization, we find subjectivity and polarity classification which are considered as a core task [13]. Many surveys were performed about sentiment analysis and opinion mining [11, 15, 16, 21].

Many works have dealt with sentiment classification using two main approaches: lexicon-based approach and machine learning approach [19]. Different techniques are used to resolve the problem consisting in determining the subjectivity and the polarity of multiple kinds of texts (movie reviews, product reviews, social media, news articles, blogs, web forums, etc.). The opinion mining or sentiment classification can be performed at different levels from the coarse-grain at document level, then at sentence level to a more fine-grained at word or expression level [20].

The subjectivity classification consists in determining if the word is subjective which means the presence of an opinion, private state, sentiment, speculation..., or

© Springer International Publishing AG 2017
A.E. Hassanien et al. (eds.), *Proceedings of the International Conference on Advanced Intelligent Systems and Informatics 2016*, Advances in Intelligent Systems and Computing 533, DOI 10.1007/978-3-319-48308-5_11

objective which means the presence of a simply and factual information. While the polarity classification consists in determining if a subjective word has a positive or negative polarity. [23] indicate that fine-grained Sentiment Analysis is harder than coarse-grained analysis. But, fine-grained analysis is also the more interesting and challenging problem and it has become increasingly popular.

In this work, we try to have a contribution in Arabic opinion mining by providing a fine-grained classification that does not stop at categorizing the word as objective/subjective or is positive/negative. We propose to categorize an opinion expression into semantic categories specific for Arabic based on lexical semantic analysis similarly used for English and French languages [4, 9]. Considering opinion categorization as sequence labeling problem, we propose to evaluate the conditional random fields (CRF), as a discriminative model, for the task of opinion classification in Arabic language.

The rest of the paper is organized as follows. In Sect. 2, we introduce a review or the-state-of-the-art about the opinion categorization task for English by presenting the most prominent researches performed. Then, we review the opinion mining for Arabic language with a focus on the opinion categorization task. In Sect. 3, we present our method of Arabic opinion categorization with description of the annotation of the dataset and the obtained opinion lexicon OPINAR. In Sect. 4, we present a set of experiments fulfilled on our dataset. Finally, we conclude in the Sect. 5.

2 Related Work

2.1 English Fine-Grained Categorization

The first emergence of fine-grained sentiment analysis was in [12] who performs a polarity classification of adjectives by computing the semantic orientation (positive or negative) of adjectives with consideration of context's specificity and by employing linguistic features. [7] used the Pointwise Mutual Information (PMI) technique to compute the subjectivity score of adjectives.

Others studies perform a subjectivity classification and a polarity classification. [6] proposed a lexical resource SentiWordNet based on WordNet where they associated to each synset two proprieties (subjectivity and polarity). However, Wordnet-Affect [24] provides a categorization following a set of affect labels. WordNet-Affect is an extension of WordNet Domains, including a subset of synsets suitable to represent affective concepts correlated with affective words.

[25] has developed the MPQA (Multi-Perspective Question Answering) Subjectivity Lexicon. It's a famous work based on the private states concept. A lot of methods for sentiment analysis are based on various psychological frameworks. They propose various cognitive theories of emotions and affect such as [26] who adopted a taxonomy based on appraisal theory [18]. They performed a movie review classification based on appraisal extracted groups. [4] proposed a different taxonomy of opinion annotation consisting in four categories: reporting, judgment, advice and sentiment.

2.2 Arabic Fine Grained Categorization

The most researches on opinion mining and sentiment analysis are done mainly for English and other languages like French, Chinese, etc. However, the Arabic language remains among the less studied and under-resourced languages in the domain of sentiment analysis. Little works have dealt with sentiment analysis of the Arabic language. As an instance, [22] proposed a system for subjectivity and sentiment analysis of Arabic social media SAMAR (chat, twitter…) at the sentence level. Also, [27] developed the Aara' system for polarity classification over informal colloquial Arabic comments. [10] performed many experiments of sentiment classification which were conducted on multidomain dataset.

Others studies focused on creating sentiment lexicons. [1] created a polarity lexicon (called Sifaat) using Google'API translation to expand a manually prepared list of seed terms (adjectives) tagged with semantic orientation, from other English lexica. [2] extended the former lexicon to create SANA using Point-wise mutual information. [3] introduced arabic SentiWordNet based on arabic WordNet and English SentiWordNet.

As mentioned above, and up to date, the majority of researches has focused on semantic orientation of words in the sentiment classification task. Our contribution consists in providing a fine-grained annotation of semantic categorization for Arabic and performing an opinion classification based on CRF models.

3 Our Proposed Approach

3.1 Dataset

The corpus used to perform experiments is a set of news articles from the Arabic TreeBank (ATB part3 v3.2) [17]. It contains texts from Lebanese newspaper "AnNahar" and from others sites for news channels like "Aljazeera.Net", "BBC Arabic", and "France 24 Arabic". This corpus has been manually annotated by two Arabic native speakers' experts. Each article presents a set of opinions expressed by different holders about a topic from a political domain. A holder may be a person, an organization, a country, political party, etc. The topic may be a political event, a political person, etc. We provide to annotators an annotation guideline prepared in advance for the task that explains the main annotation principles.

Figure 1   presents   a   sample   of   annotation   of   the   sentence (بســــوريا النـــار إطلاق قـف و بفـــرص تشـــكك روســيا) / Russia questioned the chances of a cease-fire in Syria/rwsyA t$kk bfrS w qf ITlAq AlnAr bswryA).

3.2 Arabic Opinion Taxonomy

The proposed analysis method aims to go deeper than binary classification. It allows a local analysis of opinions and sentiments expressed about a given topic. Our method is not following any psycholinguistic theory but it is based on a lexical semantic categorization proposed by [4] to categorize opinion expression for English and French coupled with discursive Analysis based on the discourse theory SDRT [5]. This

```
<seg texte="روسيا تشكك بلرص وقف إطلاق النار بسوريا" id="1">
    <token id="1" texte="روسيا" role="Holder" />
    <token category="judgement" type="contest" valence="-1" holderid="1" topicid="1"
    lemma="ذئب" texte="ئنكك" role="ExpOp" />
    <token type="main_topic" id="1" texte="بلرص وقف إطلاق النار بسوريا" role="Topic" />
</seg>
```

Fig. 1. A sample from our annotated corpus

semantic categorization was taken over by [9] to analyze opinions in web reviews on films, books, video games and in reactions on press articles of journal "Le Monde". We categorize Arabic opinion expressions using the same typology of four top-level categories: REPORTING, JUDGEMENT, ADVISE and SENTIMENT.

In our work, a set of semantic categories was submitted to annotators for validation on the corpus of Arabic news articles. Thus, we adopted this opinion categorization for Arabic with some modifications by requirements of sense in Arabic language. We add two semantic groups (Refuse and Contest that represent an opinion) to the category "Judgement".

Asher [4] defines the categories as follows: the REPORTING category, in which opinions are often expressed as the objects of verbs used to report the speech and opinions of others. These verbs convey the degree of the holders' commitment to the opinion being presented.

This category contains three subgroups according to the degree of commitment (Table 1). In the first subgroup, we find verbs that introduce information that (a) the author takes as established (INFORM group) or that (b) the holder is strongly committed to (ASSERT group). The second subgroup contains (c) the TELL group and (d) REMARK group. Unlike ASSERT verbs, TELL verbs do not convey strong commitments of the subject to the embedded content. It contains (e) the THINK group verbs which express the fact that the subject has a strong commitment to the complement of the verb and (f) the GUESS group verbs which express a weaker commitment on the part of the agent.

The JUDGMENT group involves words that express a positive or negative assessment of something or someone. We consider five subgroups: (g) the BLAME group, (h) the PRAISE group, (i) the APPRECIATION group, (j) The REFUSE group (k) and the CONTEST group. The two later subgroups were added in our Arabic categorization. ADVICE expressions urge the reader to adopt a certain course of action or opinion. We find here (l) the RECOMMEND group, (m) the SUGGEST group and (n) the HOPE group. Finally, words in the SENTIMENT category express an attitude toward something usually based on feeling or emotion rather than reasoning. They have a polarity as well as strength.

3.3   OPINAR Lexicon

During the opinion expression annotation process, we generate a lexical resource of opinion words: OPINAR Lexicon. The lexicon contains 381 entries. For each entry, we provide each lemma and POS tag (Part of speech) provided by the morphological

Table 1. Arabic opinion categories

Categories		Groups	Words
Reporting	a)	Inform	أعلم, افاد
	b)		أكد, أقر
	c)	Tell	قال, أعتبر
	d)	Remark	لاحظ, علق
	e)	Think	اعتقد
	f)		خمن, ظن
Judgment	g)	Blame	لام, اتهم
	h)	Praise	مدح, أشاد
	i)	Appreciation	قدر, اعجب
	j)	Refuse	امتنع, عارض
	k)	Contest	استنكر, احتج
Advice	l)	Recommend	وصى, زكى
	m)	Suggest	اقترح, لمّح
	n)	Hope	يأمل, يرجو
Sentiment	o)	Anger/Calmdown	غضب, سخط
	p)	Astonishment	انبهار, ذهل
	q)	Love, fascinate	أحب, فتن
	r)	Hate, disappoint	كره, أحبط
	s)	Fear	خوف, خشية
	t)	Offense	أذى, صدم
	u)	Sadness /Joy	حزن, سعادة, تشاؤم
	v)	Bore /Entertain	مملة, مسلية

analyzer BAMA [8], the semantic category and its group, the intensity, the commitment for opinion words that belong to REPORTING category, and the polarity for others. Figure 2 gives an instance (احتج/protest/AHtj ~) of the OPINAR lexicon.

Table 2 gives an overview about the distribution of verbs, adjectives, nouns, and expression in the OPINAR lexicon. We notice that verbs are so useful to give an opinion more than other categories.

```
<entry id="V_احتج">
<lemma>احتج </lemma>
<pos>VERB_PERFECT</pos>
<sense intensity="1" group ="contest" category="judgment" polarity="-1"/>
</entry>
```

Fig. 2. An example from the OPINAR lexicon

4 Experiments on Arabic Opinion Categorization Using CRF

4.1 CRF (Conditional Random Fields)

Conditional random fields (CRF) are undirected graphical models trained to maximize a conditional probability [14]. [14] defines the conditional probability of a label sequence $y = y_1 \ldots y_N$ given an observation sequence $x = x_1 \ldots x_N$ as:

Table 2. Distribution of opinion word by semantic category and by POS

	Verb	Adj	Noun	Expression	Total
Reporting	83	13	33	07	136
Judgment	55	10	37	02	104
Advice	32	02	08	0	42
Sentiment	39	08	44	08	99
Total	209	33	122	17	381

$$p(y/x) = \frac{1}{Z(x)} \exp\left(\sum_j \lambda_j t_j(y_{i-1}, y_i, x, i) + \sum_k \mu_k s_k(y_i, x, i)\right) \tag{1}$$

With:

$$Z(x) = \sum_y \exp\left(\sum_j \lambda_j t_j(y_{i-1}, y_i, x, i) + \sum_k \mu_k s_k(y_i, x, i)\right) \tag{2}$$

$Z(x)$ is a factor of normalization. $t_j(y_{i-1}, y_i, x, i)$ represents transition feature function of the entire sequence of words and the opinion categories (or groups) at positions i and $i-1$. $s_k(y_i, x, i)$ represents state feature function of the opinion category (or groups) at position i in the sequence of words. λ_j and μ_k are parameters which are estimated from the training set.

Given the model as defined in Eq. (1), the most probable opinion categories (or groups) sequence y^* for a sequence of words x, is:

$$y^* = arg\ max_y P(y/x) \tag{3}$$

Our results are obtained using The CRF ++ tool[1]. It is used for sequence labeling classification by [14].

All experiments reported in this work are performing using a simple validation: by keeping 80 % of the dataset as training set and the other 20 % as test set.

$$\text{Precision (P)} = \frac{Correct\ labels}{Correct\ labels + Incorrect\ labels} \tag{4}$$

$$\text{Recall (R)} = \frac{Correct\ labels}{Reference\ labels} \tag{5}$$

$$F - measure = \frac{2 \times Recall \times Precision}{Recall + Precision} \tag{6}$$

For evaluation we use recall, precision measures and the well-known F-measure, defined as the harmonic mean of recall and precision.

---

[1] (https://taku910.github.io/crfpp/).

4.2 Tasks

We have performed many experiments using CRF models. So, we have proposed many models introduced to define the CRF models. Analyzing automatically an opinion text, according to the proposed opinion scheme presented in Sect. 3.3, is a complex task. That is why we divided our classification problem to five tasks. Also, we performed many experiments to determine opinion proprieties like: semantic category, group, polarity, intensity, commitment.

In the following, we present the considered features in this work and the obtained results in different proposed tasks based on CRF models

Features.

We use a set of different features:

Tok: is the considered word.

Prec: is the word preceding the considered word.

Next: is the following word to the considered one.

Bi-gram: two successive words.

Tri-gram: three successive words.

Catg: is the opinion semantic category of the word.

Group: is the group under the opinion semantic category of the word.

Task 1: Determining the semantic category for a given opinion word

Template	Recall	Precision	F-measure
Tok	83.33 %	83.33 %	83.33 %
Tok + Prec	58.33 %	95.45 %	72.41 %
Tok + Next	69.44 %	94.34 %	80 %
Prec + Tok + Next	54.17 %	95.12 %	69.03 %
Tok + Next + Next	60.42 %	**96.67 %**	74.36 %
Prec + Prec + Tok + Next	43.75 %	88.73 %	58.60 %
Tok + Next + Next + bi-gram + tri-gram	56.25 %	96.43 %	71.05 %
Prec + Tok + Next + bi-gram	68.05 %	94.23 %	79.03 %

Task 2: Determining the group for a given opinion word

Template	Recall	Precision	F-measure
Tok	80,55 %	80,55 %	80,55 %
Tok + Catg	79,86 %	79,86 %	79,86 %
Tok + Prec	56,94 %	96,47 %	71,61 %
Tok + Next	69,44 %	95,24 %	80,32 %
Prec + Tok + Next	52,78 %	97,43 %	68,47 %
Prec + Prec + Tok + Next	40,97 %	92,19 %	56,73 %
Prec + Tok + Catg	80,55 %	80,55 %	80,55 %
Tok + Catg + Next	83,33 %	83,33 %	83,33 %
Prec + Prec + Tok + Next + Next	39.58 %	93,44 %	55,61 %
Prec + Tok + Next + bi-gram	53,47 %	**97,47 %**	69,06 %
Tok + Catg + Next + bi-gram	86.11 %	86.11 %	**86.11 %**

Task 3: Polarity classification Task

Template	Recall	Precision	F-measure
Tok	83.11 %	83.11 %	83.11 %
Tok + Catg + Group	100 %	100 %	100 %

Task 4: Commitment classification Task

Template	Recall	Precision	F-measure
Tok	95.52 %	95.52 %	95.52 %
Tok + Catg + Group	100 %	100 %	100 %

Task 5: Intensity classification Task

Template	Recall	Precision	F-measure
Tok	75.69 %	75.69 %	75.69 %
Tok + Catg + Group	86.11 %	86.11 %	86.11 %

4.3   Interpretations

Task 1 and Task 2 consist in determining respectively the semantic category and the group of an opinion word according to the presented taxonomy in Table 1. Our accomplished experiments based on CRF model show that word order and bi-gram plays a key role in expressing opinion in Arabic language. For the first task (Task 1), the CRF models are able to detect the semantic category using a simple template (Tok), since we used a lexical semantic approach. However, for the second Task (Task 2), we obtained the best result (F-measure 86.11 %) by using the most of features and the integration of the correlation between words in the same opinion expression. This is due to a linguistic specificity of Arabic language in expressing opinion: an opinion word may be followed by another word that can be: preposition, adj, noun,etc. for example: (tmnY On /he hoped that /تمنّى أن), (AbdY qlqh/he expressed his worry/ ...أبدى قلقه).

Task 3 and Task 4 consist in determining respectively polarity and commitment proprieties of an opinion word. These proprieties are not common ones for all opinion word: the commitment is considered only for opinion words which belong to the "reporting" category and according to the group we have three value of commitment.

Therefore, our experiments show that we can reach 100 % as precision measure, if we consider besides of the token, its semantic category and its group. Similarly, for determining the polarity propriety which takes into account the other categories (judgment, advice, sentiment).

Task 5 consists in determining the intensity propriety of an opinion word. Obtained results show that combining the token with its semantic category and group gives good precision (86.11 %).

5 Conclusion

In this work, we have proposed a supervised categorization approach for fine-grained analysis of opinion in Arabic news articles. The proposed approach is based on lexical semantic analysis and performed using CRF models. A semantic scheme is proposed to categorize opinion expressions of Arabic language according to taxonomy of semantic labels.

We have fulfilled many experiments to determine opinion word proprieties like: semantic category, semantic group, polarity, intensity, commitment. Obtained results show that CRF coupled with contextual, semantic information and bi-gram have achieved the best result with a F-measure of 86.11 % at the task of determining the semantic group for an opinion word. The F-measure may attend 100 % for determining the commitment propriety for an opinion word given its semantic category and its group.

As perspective, we intend in future works to determine the topic and the holder of the opinion word by investigating a set of features.

References

1. Abdul-Mageed, M., Diab, M.: Toward building a large-scale arabic sentiment lexicon. In: Proceedings of the Sixth International Global Word-Net Conference, Matsue, Japan (2012)
2. Abdul-Mageed, M., Diab, M.: SANA: a large scale multi-genre, multi-dialect lexicon for arabic subjectivity and sentiment analysis. In: Proceedings of the Ninth International Conference on Language Resources and Evaluation (2014)
3. Alhazmi, S., Black, W., McNaught, J.: Arabic SentiWordNet in relation to Sentiwordnet 3.0. Int. J. Comput. Linguist. **4**, 1–11 (2013)
4. Asher, N., BenAmara, F., Mathieu, Y.: Appraisal of opinion expressions in discourse. Lingvisticæ Investigationes **31**(2), 279–292 (2009)
5. Asher, N., Lascarides, A.: Logics of Conversation. University Press, Cambridge (2003)
6. Baccianella, S., Esuli, A., Sebastiani, F.: SentiWordNet 3.0: an enhanced lexical resource for sentiment analysis and opinion mining. In: Proceedings of the Seventh International Conference on Language Resources and Evaluation, 2200–2204. European Language Resources Association (2010)
7. Baroni, M., Vegnaduzzo, S.: Identifying subjective adjectives through web-based mutual information. In: Proceedings of the 7th Konferenz zur Verarbeitung Natrlicher Sprache (German Conference on Natural Language Processing) KONVENS04, pp. 613–619 (2004)
8. Buckwalter, T.: Buckwalter Arabic Morphological Analyzer Version 2.0 LDC2004L02. Web Download, Linguistic Data Consortium, Philadelphia (2004)
9. Chardon, B.: Chaine de traitement pour une approche discursive de l'analyse d'opinion. Ph.D. dissertation, UPS, France (2013)
10. ElSahar, H., El-Beltagy, S.R.: Building large arabic multi-domain resources for sentiment analysis. In: Gelbukh, A. (ed.) Computational Linguistics and Intelligent Text Processing. LNCS, vol. 9042, pp. 23–34. Springer, Heidelberg (2015)
11. Feldman, R.: Techniques and applications for sentiment analysis. Commun. ACM **56**, 82–89 (2013)

12. Hatzivassiloglou, V., McKeown, K.: Predicting the semantic orientation of adjectives. In: Proceedings of Annual Meeting of the Association for Computational Linguistics (ACL 1997) (1997)
13. Khan, K., Bahrudin, B., Khan, A., Ullah, A.: Mining opinion components from unstructured reviews: a review. J. King Saud Univ. Comput. Inf. Sci. **26**, 258–275 (2014)
14. Lafferty, J., McCallum, A., Pereira F.: Conditional random fields: probabilistic models for segmenting and labeling sequence data. In: Proceedings of ICML, pp. 282–289 (2001)
15. Liu, B.: Sentiment Analysis and Opinion Mining. Morgan & Claypool, San Rafael (2012)
16. Liu, B.: Sentiment Analysis: Mining Opinions, Sentiments, and Emotions. Cambridge University Press, Cambridge (2015)
17. Maamouri, M., Bies, A., Kulick, S., Krouna, S., Gaddeche, F., Zaghouani, W.: Arabic TreeBank (ATB): Part 3 Version 3.2. Linguistic Data Consortium. Catalog No: LDC2010T08 (2010)
18. Martin, J.R., White, P.R.R.: Language of Evaluation: Appraisal in English. Palgrave Macmillan, Basingstoke (2005)
19. Medhat, W., Hassan, A., Korashy, H.: Sentiment analysis algorithms and applications: a survey. Ain Shams Eng. J. **5**(4), 1093–1113 (2014)
20. Missen, M.M., Boughanem, M., Cabanac, G.: Opinion mining: reviewed from word to document level. Soc. Netw. Anal. Min. **3**(1), 107–125 (2013)
21. Pang, B., Lee, L.: Opinion mining and sentiment analysis. Found. Trends Inf. Retrieval **2**, 1–135 (2008)
22. Abdul-Mageed, M., Diab, M., Kübler, S.: Samar: subjectivity and sentiment analysis for arabic social media. Comput. Speech Lang. **28**(1), 20–37 (2014)
23. Sonntag, J., Stede, M.: Sentiment analysis: what's your opinion? In: Biemann, C., Mehler, A. (eds.) Text Mining: From Ontology Learning to Automated Text Processing Applications, pp. 177–199. Springer, Berlin (2014)
24. Strapparava, C., Valitutti, A.: WordNet-Affect: an affective extension of WordNet. In: Proceedings of the 4th International Conference on Language Resources and Evaluation (LREC 2004), Lisbon, pp. 1083–1086 (2004)
25. Wiebe, J., Wilson, T., Cardie, C.: Annotating expressions of opinions and emotions in language. Lang. Resour. Eval. **39**, 165–210 (2005)
26. Whitelaw, C., Navendu, G., Shlomo, A.: Using appraisal taxonomies for sentiment analysis, Bremen, Germany (2005)
27. Azmi, A., Alzanin, S.: Aara' – a system for mining the polarity of Saudi public opinion through e-newspaper comments. J. Inf. Sci. **40**, 398–410 (2014)

Towards Improving Sentiment Analysis in Arabic

Sanjeera Siddiqui[1(✉)], Azza Abdel Monem[2], and Khaled Shaalan[1,3]

[1] British University in Dubai, Block 11, 1st and 2nd Floor,
Dubai International Academic City, Dubai, UAE
faizan.sanjeera@gmail.com, khaled.shaalan@buid.ac.ae
[2] Faculty of Computer and Information Sciences,
Ain Shams University, Abbassia, Cairo 11566, Egypt
azza_monem@hotmail.com
[3] School of Informatics, University of Edinburgh, Edinburgh, UK

Abstract. This paper presents an innovative approach that explores the role of lexicalization for Arabic sentiment analysis. Sentiment Analysis in Arabic is hindered due to lack of resources, language in use with sentiment lexicons, pre-processing of dataset as a must and major concern is repeatedly following same approaches. One of the key solution found to resolve these problems include applying the extension of lexicon to include more words not restricted to Modern Standard Arabic. Secondly, avoiding pre-processing of dataset. Third, and the most important one, is investigating the development of an Arabic Sentiment Analysis system using a novel rule-based approach. This approach uses heuristics rules in a manner that accurately classifies tweets as positive or negative. The manner in which a series of abstraction occurs resulting in an end-to-end mechanism with rule-based chaining approach. For each lexicon, this chain specifically follows a chaining of rules, with appropriate positioning and prioritization of rules. Expensive rules in terms of time and effort thus resulted in outstanding results. The results with end-to-end rule chaining approach achieved 93.9 % accuracy when tested on baseline dataset and 85.6 % accuracy on OCA, the second dataset. A further comparison with the baseline showed huge increase in accuracy by 23.85 %.

Keywords: Sentiment analysis · Opinion mining · Rule-based approach · Arabic natural language processing

1 Introduction

The mission of Sentiment Analysis is to perceive the content with suppositions and mastermind them in a way complying with the extremity, which incorporates: negative, positive or nonpartisan. Organization's are taken to huge prestige by living up to the opinions from different people [9, 17]. Subjectivity and Sentiment Analysis characterization are prepared in four measurements: (1) subjectivity arrangement, to estimate on Subjective or Objective, (2) Sentiment Analysis, to anticipate on the extremity that could be negative, positive, or impartial, (3) the level in view of record, sentence, word

© Springer International Publishing AG 2017
A.E. Hassanien et al. (eds.), *Proceedings of the International Conference on Advanced Intelligent Systems and Informatics 2016*, Advances in Intelligent Systems and Computing 533, DOI 10.1007/978-3-319-48308-5_12

or expression order, and (4) the methodology that is tailed; it could be standard based, machine learning, or half breed [10].

Arabic is a Semitic dialect, which is distinctive as far as its history, structure, diglossic nature and unpredictability Farghaly and Shaalan [8]. Arabic is broadly talked by more than 300 million individuals. Arabic Natural Language Processing (NLP) is testing and Arabic Sentiment Analysis is not a special case. Arabic is exceptionally inflectional [4, 11] because of the fastens which incorporates relational words and pronouns. Arabic morphology is intricate because of things and verbs bringing about 10,000 root [6]. Arabic morphology has 120 examples. Beesley [6] highlighted the hugeness of 5000 roots for Arabic morphology. No capitalization makes Arabic named substance acknowledgment a troublesome mission [14]. Free order of Arabic Language brings in additional challenges with regards to Sentiment Analysis, as the words in the sentence can be swapped without changing the structure and the meaning. Arabic Sentiment Analysis has been a gigantic center for scientists [2].

The target of this study is to examine procedures that decide negative and positive extremity of the information content. One of the critical result would be to recognize the proposed end-to-end principle binding way to deal with other dictionary based and machine learning-construct approaches in light of the chosen dataset.

The rest of this paper is organized as follows. Related work is covered in Sect. 2, Data collection is covered in Sect. 3 followed by system implementation in Sect. 4. Section 5 covers evaluation and results, Sect. 6 covers error analysis and lastly Sect. 7 depicts conclusion.

2 Related Work

To take a shot at Sentiment Analysis, the key parameter is the dataset. Late endeavors by Farra et al. [13] outlined the significance of crowdsourcing as an extremely fruitful technique for commenting on dataset. Bolster Vector Machine classifier accomplished 72.6 % precision on twitter dataset of 1000 tweets [15].

The record level assessment investigation utilizing a joined methodology comprising of a dictionary and Machine Learning approach with K-Nearest Neighbors and Maximum Entropy on a blended area corpus involving training, legislative issues and games achieved an F-measure of 80.29 % [7]. Shoukry and Refea [15] achieved a precision of 72.6 % with the corpus based method. The vocabulary and estimation examination device with an exactness of 70.05 % on tweeter dataset also, 63.75 % on Yahoo Maktoob dataset [3].

With a mixed approach that is Lexical and Support Vector Machine classifier created 84.01 % exactness [1]. 79.90 % exactness was shown with Hybrid methodology which involved lexical, entropy and K-closest neighbor [5]. Shoukry and Rafea [16] independently sent two methodologies, one being Support Vector Machine accomplished a precision of 78.80 % and other one being Lexical with an exactness of 75.50 %.

3 Data Collection

The dataset utilized as a part of this paper is Twitter Tweets and film surveys. The dataset is taken from [3, 12]. These datasets are utilized as these are accessible, rich and enough to reached a conclusion, also electronic assets, for example, vocabulary is given.

Abdullah et al. [3]'s dataset contains 1000 positive tweets and 1000 negative tweets with length ranging from one word to sentences. 7189 words in positive tweets and 9769 words in negative tweets. The tweets were manually collected belong to Modern Standard Arabic and Jordanian Dialect, which covers Levantine language family. The months-long segregation procedure of the tweets was physically led by two human specialists (local speakers of Arabic).

OCA corpus, termed as Opinion Corpus for Arabic, was presented by [12]. This corpus contains total 500 opinions, of which there are 250 positive opinions and 250 negative opinions. The procedure followed by Rushdi et al. [12] included gathering surveys from a few Arabic online journal locales and site pages utilizing a straightforward bash script for slithering. At that point, they expelled HTML labels and unique characters, and spelling mix-ups were adjusted physically. Next, a preparing of each survey was done, which included tokenizing, evacuating Arabic stop words, and stemming and sifting those tokens whose length was under two characters. In their trials, they have utilized the Arabic stemmer of RapidMiner and the Arabic stop word list. At last, three distinctive N-gram plans were created (unigrams, bigrams, and trigrams) and cross validation was used to assess the corpus for which they have achieved 90.6 % accuracy.

The vocabulary utilized as a part of this paper contains the dictionary used by [3], which included opinions, named substances and a few haphazardly set words. In view of the circumspection of reiteration of the words found in negative or positive reviews and their arrangement, they were incorporated into both the rundown that is positive and negative lists.

4 Implementation of Arabic Sentiment Analysis

This paper is a significant extension of [13]. Siddiqui et al. [13] research introduced a system, which contains only two type of rules - "equal to" and "within the text rules", so as to examine whether the tweet is either negative or positive. The rules include a 360-degree coverage with an improvised segment that is end to end rule chaining principle: (1) in the middle, we termed as "within the text", (2) at the boundary we termed as either "ending with the text" or "beginning with the text", and (3) full coverage, we termed as "equal to the text". Figure 1 delineates the 360-degree rules coverage. The End-to-End mechanism with rule chaining approach introduced in this paper, includes the chaining of rules based on the positioning of the polarities in the tweets. The key underlying base ground factors which helped us formulate appropriate rules includes analysis of the tweets and the extension of positive and negative lexicons. The analysis of tweets resulted in identifying relations pertaining to words which were either disjoint, intersected or coexisted.

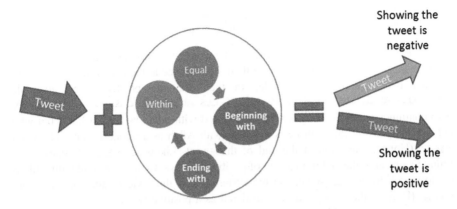

Fig. 1. A 360-degree coverage of rules to the input Tweet

The words which were disjoint that is completely indicating either positive or negative polarity were included in their respective lexicons. The words which intersected that is the ones which were found to be common in both negative and positive reviews were included in positive as well as negative lexicons. The words which coexisted at the same place in the negative and positive reviews, that is the ones which appeared at the beginning or ending were placed in either positive or negative lexicon, based on the highest frequency of the word in the respective reviews.

Rules handling intersection with the end-to-end chaining mechanism: As an example consider the following positive tweet "اوقف القرار للحفاظ على الوطن" (the decision was suspended for protecting the motherland), the word "اوقف" (suspended) appeared in the beginning of this tweet and the same word was found in the negative tweets. Hence, this set of situations was handle with the very use of positioning and chaining of rules. The steps involved to achieve the positioning and chaining of rules includes:

Rules Formation: With the logical discretion of the word "اوقف" (suspended) being seen at the "beginning of" positive tweets and "within the text" for the negative tweets, the rules thus formed were "beginning with" for positive tweets and "within the text" for negative tweets. With the correct positioning and chaining of rules this problem was resolved. In the current example "beginning with" rule needs to be positioned and chained in an orderly fashioned with the rule "within the text" so as to satisfy both positive and negative reviews. So, the rule "beginning with" was chained with the rule "within the text" for the word "اوقف" (suspended) by positioning the rule "beginning with" first followed by "within the text" rule. Hence, the rules are chained and positioned for the words which are found to repeat themselves in both positive and negative tweets.

Rules handling coexistence: The lexicons which are not seen to repeat themselves in either positive or negative tweets where handled with the rule – "within the text". For example, "الحرامية" (Thieves) is set for search "within the text" "الحرامية" (Thieves), is majorly found within the text in negative review rather than in positive ones. Hence based on the frequency of "الحرامية" (Thieves) the rule is set. Example tweet:

"انا مش مسؤول عن الحرامية يتفنجرو بمصاري البلد و انا اشحد" (I am not responsible for that thieves spend lavishly from the funds of the country and I beg.)

Rules handling disjoint: The cases wherein the words were not repeated in positive and negative cases and were found to have their significance at the end of the tweet, were handled using "ending with" rule. In the following positive tweet example "تذكر" صادقا "من، يكتب يكتب عند الله كذابا ـ اما من، يتحد ى، الصدة، يكتب عند الله صادقا" – (Remember who continues to speak falsehood he is recorded with Allah as a great liar – but who persists in speaking the truth he is recorded with Allah as an honest man). The word "صادقا" (Honest) appeared at the end of this tweet. Likewise, "صادقا" (Honest) was found to appear at the end in majority of positive reviews. Hence, the "end with" rule is set to search for "صادقا" indicating the system that if the review ends with the word "صادقا" (Honest) then it should be considered as a positive tweet.

Derivation of rules: Table 1 depicts the skeletal examples of derived rules. Conditional search includes two key phases, one includes a condition which checks on the entered tweet mapping with the rules and the second phase is the color coded output which changes its font color to - "Green fill with Green fill text" for negative tweets and "Light red fill with light red fill text" for positive tweets.

With reference to rule 1.A in Table 1 in the primary column, if the word showed up inside of the content in a positive tweet. With reference to manage 2.B or 3.B in this table, if the same word showed up toward the end or toward the starting in the negative tweet, then situating and anchoring of these two rules are finished. For this case the

Table 1. Derivation of lexicalized rule

Rule type		Derivation of rules
(1) Search within the text	A	If the Entered Tweet/Review contains the word "＿" then Fill the text with-"Green fill text" for negative tweets
	B	If the Entered Tweet/Review contains the word "＿" then Fill the text with- "Light fill text" for positive tweets
(2) Ending with the text	A	If the Entered Tweet/Review ends with the word "＿" then Fill the text with- "Green fill text" for negative tweets
	B	If the Entered Tweet ends with the word "＿" then Fill the text with- "Light red fill text" for positive tweets
(3) Beginning with the text	A	If the Entered Tweet begins with the word "＿" then Fill the text with- "Green fill text" for negative tweets
	B	If the Entered Tweet begins with the word "＿" then Fill the text with-"Light red fill text" for positive tweets
(4) Equal to the text	A	If the Entered Tweet equals to the word "＿" Then Fill the text with- "Green fill text" for positive tweets
	B	If the Entered Tweet equals to the word "＿" then Fill the text with-"Light red fill text" for negative tweets

positive principle was situated and fastened underneath the negative guideline. The principle "starting with" or "finishing with" not at all like "inside of the content" search for that word in the first place or end which will tag the tweet as negative. In the event that the situation of guideline is turned around then "inside of the content" as situated and affixed before the tenets "starting" or "consummation" with, when a negative tweet is gone through this, the tweet will be labeled positive. As this is a chained approach for a specific word, this checks its importance at the ending or starting rule, on the off chance that it doesn't have a place with that then it goes through within the text.

End-to-end mechanism with rule chaining approach.: After precisely executing the rules taking into account the fulfillment of disjoint word(s) which existed together or basic words in negative and positive surveys, the guideline anchoring was dealt with. End-to-end component with rule chaining approach fulfills a word which has a place with a positive and negative tweet regardless of its real extremity. Consequently, a negative word in a positive audit and a positive word in a negative tweet is fulfilled through apt chaining as examined beneath with the guide of case from negative and positive tweets. Example 1 Rules in use – Rule A and Rule B.

Rule   A   in   Table 1:   In   the   negative   tweet "الأردنيين ليست هناك حاجة لحماية وطننا من أيدي المنافقين" (There is no need to protect our motherland from hands of hypocrites Jordanian), contains the word "المنافقين" (hypocrites) within the text. Rule B in Table 1: In the positive tweet "اللهم لا تجعلنا من المنافقين" (O Allah! Place us not with the people who are hypocrites) contains the word "المنافقين" (hypocrites) at the last. If Rule A is not chained with Rule B, then only one of the rule will be satisfied. To satisfy both the rules, that is to correctly identify negative and positive polarity, Rule A is chained with Rule B by positioning Rule A Below Rule B, so as to allow the search to first visit the end rule first, then the within the text rule.

Example 2 with explanation on how the system works: For example, the word "تشائم" (pessimism) was found to be part of both positive and negative reviews with a slight variation. In the positive reviews appears only in the middle whereas in the negative review "تشائم" (pessimism) was found to appear at the beginning. Hence the rules thus were created covering "within the text" (refer rule 1.A in Table 1) and "beginning with" (refer rule 2.B in Table 1), but the positioning was varied. By positioning the rule "beginning with" first and then the rule "within the text" helped in satisfying both positive and negative reviews. As the system checked for the word beginning with "تشائم" (pessimism) in the entered review and if found then the review was tagged as negative. Likewise, the system proceeded with the entered review containing the word "تشائم" (pessimism), if the word "تشائم" (pessimism) was not found at the beginning then the search proceeded further and identified it as positive.

5 Evaluation and Results

To quantify the change also the nature of being trusted and had faith in, assessment plays an essential part. Cross-Validation and accuracy information are regularly used to assess the outcomes in estimation investigation. The exactness measures – Precision, Recall and Accuracy, which are generally being used was conveyed to measure the execution of the instruments utilized as a part of both the analyses. Precision, Recall and Accuracy were utilized to look at the outcomes by [3, 12, 16]. The condition is as per the following:

Precision = TP/(TP + FP)
Recall = = TP/(TP + FN)
Accuracy = (TP + TN)/(TP + TN + FP + FN)
Where:
TP – True Positive, all the tweets which were characterized accurately as positive
TN – True Negative, all the tweets which were accurately named negative
FP – False Positive, all the tweets which were mistakenly named positive
FN – False Negative, all the tweets which were mistakenly delegated negative

Results: The results fuse the relationship of the impressive number of tests coordinated in this paper. To do the connection, the accuracy of the extensive number of tests are used. Siddiqui et al. [13] system and System 2 (introduced in this paper) were attempted on [3, 12] dataset. Table 2 obviously takes after the outperformance of rules made in Siddiqui et al. [13] with enormous accuracy for Abdullah et al. [3] dataset when contrasted with the results on [12] OCA dataset.

Table 2. Comparison of System1 [13] Versus System 2

	System 1		System 2	
	Abdullah et al. [3] dataset	Rushdi et al. [12] OCA dataset	Abdullah et al. [3] dataset	Rushdi et al. [12] OCA dataset
Precision	87.4	50.4	92.5	77.4
Recall	93.3	97.6	95.8	100
Accuracy	89.6	50.1	93.9	85.6

System 1 Vs System 2 Results Comparison: Clear importance in expansion in accuracy in System 2 is seen for Abdullah et al. [3] and OCA dataset. The examination of Siddiqui et al. [13]'s System 1 and our System 2 doubtlessly answers that the end to end rule chaining improved the performance of sentiment analysis. Siddiqui et al. [13]'s-system 1 bound to limits with two norms sort with no attaching and System 2 variation of System 1 with authoritative and reasonable arranging of the standards. System 2 ended up being remarkable for both the datasets. 93.9 % accuracy for [3] dataset wherein for OCA dataset 85.6 % accuracy was measured. Still the accuracy for our system2 when tested on Abdullah et al. [3] was high with 8.3 % more exactness than OCA dataset. Recall is high for both the datasets with 3.3 % for Abdullah et al. [3] and 22.6 % for Rushdi et al. [12] as very less number of tweets are mistakenly delegated negative.

Table 3. Unhandled exceptions – Case 1

Word	Review	Comment
صحيح-true Is a positive word which appears within the text in the negative review	اذا الخبر صحيح انا اول شخص بطالب بالهجرة لأي دولة خليجية او اوروبية او حتى اسرائيل If the news is true, I am the first person claiming to migration to any GCC country, European or even Israel.	Due to the significance of the positive word – "صحيح" (true) within the text in both positive and negative reviews, restricted the rule to be created for either positive or negative. Chain was not able to form in this case.
صحيح-true Is a positive word which also appears within the text in most of the positive reviews	كلام صحيح من شان هيك الدول اللي ما فيها بطالة والمجتمعات المفتوحة بتقل فيها المشاكل النفسية Words properly stated, this is the reason that countries that do not have unemployment, and those with open societies, have less psychological problems	
	اي و الله صحيح أهم شيء الانسان كرامته I swear god that this is true, the most important thing is the human dignity	

Table 4. Unhandled exceptions – Case 1

Word	Review	Comment
حرام-Haram Is a negative word which appears within the text in a positive review	اللهم ازرقنا رزق حلالا تغننا به عن حرامك , سبحان القادر المقتدر بيده ملكوت كل شيء God grant us Muslim livelihood that enough for us to avoid Haram, glory to the capable Al-Muqtadir who has in hand the kingdom of everything.	Due to the significance of the negative word – "حرام" (Haram) within the text in both positive and negative reviews, restricted the rule to be only created for either positive or negative. Chain was not able to form in this case.
حرام-Haram Is a negative word which also appears within the text the negative reviews/sarcastic reviews.	و الله حرام و الله موتوه لشعب الاردني من وين وين بدنا نجيب الكو من وين يا الله ارحموا من في الارض يرحمكم من في السماء و الله حرام Oh my god! This is Haram, Oh god! You are killing Jordanians, from where we can afford it, from where Oh god!, Be merciful to those who are on the earth, The one who is in the sky will be merciful to you all, Oh god! This is Haram.	
	و الله حرام الناس مش لاقيه توكل حتى ترفعو البنزين Oh my god! This is Haram. The people cannot afford food such that you raise the fees of benzene (fuel).	

6 Error Analysis

Case 1: In the first case, a negative review included a positive word within the text and the same positive word was mostly found within the text in positive reviews affecting positioning of rules. Table 3 illustrates an example. Hence, the rule can either satisfy a positive review or negative review. Chain was not feasible to form in this case.

Case 2: Likewise, in the second case, a positive review included a negative word within the text and the same negative word was mostly found within the text in the negative reviews affecting positioning of rules. Rule again restricted to either positive or negative review. Table 4 illustrates an example.

7 Conclusion

This paper beats the vocabulary building process through the fitting position of words too not barring the basic words found in both the tweets for the vocabularies. The outperformance of rule chaining approach that is System 2 brought about 23.85 % in results when contrasted with Abdullah et al. [3]'s vocabulary based methodology. The incorporation of normal words in light of the examination of tweets in the negative and positive vocabulary list upgraded the general result when contrasted with the gauge dataset. Last yet not the minimum, the situating of principles has a gigantic effect to the rule based methodology as proper situating brought about fulfilling words which were observed to be regular in both negative and positive dictionaries. By and by, the end-to-end rule chaining methodology was the most difficult and costly regarding time and exertion, yet adds to the headways in the cutting edge for Arabic Sentiment Analysis, through the organized set standards and through the right utilization of various principles including "contains content", "equivalent to", "starting with" and "finishing with". In reality, displayed the recently created assessment investigation framework- System2, which beat in both arrangements of examinations when contrasted with [3]. System 2 with guidelines reached out to cover all territories was demonstrated to expand the exactness of OCA corpus by 39.8 % and 4.3 % accuracy for Abdullah et al. [3] when contrasted with System 1's principles of Siddiqui et al. [13]. Thus, starting a wakeup require every one of the scientists to redirect their enthusiasm to lead based methodology. The unmistakable hugeness in results along these lines acquired through the tenets made makes the principle based methodology the most alluring methodology.

References

1. Aldayel, H.K., Azmi, A.M.: Arabic tweets sentiment analysis–a hybrid scheme. J. Inf. Sci. (2015). doi:10.1177/0165551515610513
2. Muhammad, A.-M., Kübler, S., Diab, M.: Samar: a system for subjectivity and sentiment analysis of Arabic social media. In: Proceedings of the 3rd Workshop in Computational Approaches to Subjectivity and Sentiment Analysis, pp. 19–28. Association for Computational Linguistics (2012)

3. Abdulla, N.A., Ahmed, N.A., Shehab, M.A., Al-Ayyoub, M., Al-Kabi, M.N., Al-rifai, S.: Towards improving the lexicon-based approach for Arabic sentiment analysis. Int. J. Inf. Technol. Web Eng. (IJITWE) **9**(3), 55–71 (2014)

4. Hammo, B., Abu-Salem, H., Lytinen, S., Evens, M.: QARAB: a question answering system to support the Arabic language. In: The Proceedings of Workshop on Computational Approaches to Semitic Languages, ACL 2002, Philadelphia, PA, pp. 55–65, July 2002

5. Beesley, K.R.: Arabic finite-state morphological analysis and generation. In: Proceedings of the 16th Conference on Computational Linguistics, vol. 1, pp. 89–94. Association for Computational Linguistics, August 1996

6. Darwish, K.: Building a shallow Arabic morphological analyzer in one day. In: Proceedings of the ACL-2002 Workshop on Computational Approaches to Semitic languages, pp. 1–8. Association for Computational Linguistics, July 2002

7. Alaa, E.-H.: Arabic opinion mining using combined classification approach (2011)

8. Farghaly, A., Shaalan, K.: Arabic natural language processing: challenges and solutions. ACM Trans. Asian Lang. Inf. Process. (TALIP) **8**(4), 1–22 (2009). The Association for Computing Machinery (ACM)

9. Feldman, R.: Techniques and applications for sentiment analysis. Commun. ACM **56**(4), 82–89 (2013)

10. Korayem, M., Crandall, D., Abdul-Mageed, M.: Subjectivity and sentiment analysis of arabic: a survey. In: Hassanien, A.E., Salem, A.-B.M., Ramadan, R., Kim, T.-h. (eds.) AMLTA 2012. CCIS, vol. 322, pp. 128–139. Springer, Heidelberg (2012)

11. Syiam, M., Fayed, Z., Habib, M.: An intelligent system for Arabic text categorization. Int. J. Intell. Comput. Inf. **6**(1) (2006)

12. Rushdi-Saleh, M., Martín-Valdivia, M.T., Ureña-López, L.A., Perea-Ortega, J.M.: OCA: opinion corpus for Arabic. J. Am. Soc. Inf. Sci. Technol. **62**(10), 2045–2054 (2011)

13. Siddiqui, S., Monem, A.A., Shaalan, K.: Sentiment analysis in Arabic. In: Métais, E., Meziane, F., Saraee, M., Sugumaran, V., Vadera, S. (eds.) Proceedings of the 21st International Conference on the Application of Natural Language to Information Systems (NLDB 2016). LNCS. Springer, Heidelberg (2016)

14. Khaled, S.: A survey of Arabic named entity recognition and classification. Comput. Linguist. **40**(2), 469–510 (2014)

15. Shoukry, A., Rafea, A.: Sentence-level Arabic sentiment analysis. In: 2012 International Conference on Collaboration Technologies and Systems (CTS), pp. 546–550. IEEE (2012)

16. Shoukry, A., Rafea, A.: Preprocessing Egyptian dialect tweets for sentiment mining. In: The 4th Workshop on Computational Approaches to Arabic Script-Based Languages, pp. 47–56 (2012)

17. Taboada, M., Brooke, J., Tofiloski, M., Voll, K., Stede, M.: Lexicon-based methods for sentiment analysis. Comput. Linguist. **37**(2), 9–27 (2011)

A Spell Correction Model for OCR Errors for Arabic Text

Mariam Muhammad[✉], Tarek ELGhazaly, Mostafa Ezzat,
and Mervat Gheith

Computer Sciences Department, Institute of Statistical Studies and Research,
Cairo University, Giza, Egypt
eng_maryamadel@yahoo.com, mervat_gheith@yahoo.com,
{tarek.elghazaly,mostafa.ezzat}@cu.edu.eg

Abstract. This paper presents a new correction model for Arabic OCR errors. The proposed model is mainly based on the character segmentation and the character alignment on a single character or multi-characters. Results show that the multi-character model is better than the single character model in that it is trained on 502,167 words and can find the correct word within the top 10 proposed corrections for 94 % of the words. This model considers the effect of increasing the size of training set that perfectly leads to better results; the correction rate will approach 53 % upon using 6000 words, 80 % upon using 64,225 words, and 94 % upon using 502,167 words.

Keywords: Arabic OCR · Misspelled words · Error correction · Character model

1 Introduction

Optical Character recognition (OCR) is an automated process that converts the document images into editable text. In 1930's, the research on OCR started and focused mainly on Latin, Chinese and Japanese languages. In 1975, the research on Arabic character recognition began [1] and then it has become a primary area of interest since 1980's. Currently, many OCR systems are being developed for almost all major languages including Arabic. Arabic, as a language, has orthographic properties that reflect more complexities involved in OCR.

1.1 Arabic Orthographic Properties

Arabic language is a cursive script in which the letters are written from right-to-left and most of them are connected. It uses 28 letters, of which 15 letters have dots for distinguishing them. These letters may vary in shape according to their positions in the word. There are optional uses of diacritic and word ligatures such as "لا", which are special forms of certain letter sequences. Finally, Arabic language has a morphological complexity because its words are structured from a finite set of about 10,000 root forms and mostly contain prefix and suffix. Having the connected script and letters with dots

© Springer International Publishing AG 2017
A.E. Hassanien et al. (eds.), *Proceedings of the International Conference on Advanced Intelligent Systems and Informatics 2016*, Advances in Intelligent Systems and Computing 533, DOI 10.1007/978-3-319-48308-5_13

and using the diacritics and ligatures can make the OCR system more sensitive to dirt and speckle. Figure 1 shows some examples of ligature, dots in different letters, diacritics and letters taking the beginning, middle, or end positions.

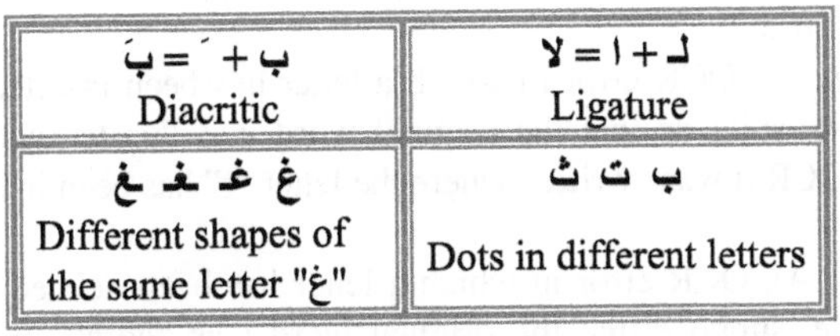

بٔ = ٔ + ب Diacritic	لا = ١ + ل Ligature
غ غـ ـغـ ـغ Different shapes of the same letter "غ"	ب ت ث Dots in different letters

Fig. 1. Orthographic properties of Arabic language

Because of Arabic orthographic properties, only a few systems of Arabic OCR have been developed. They are improved by performing post-processing process that attempts to evaluate whether each word in the OCR output is valid or not. When an invalid word is discovered (error detection), these systems usually attempt to generate a correct alternative (error correction) [2].

In the next section, we shall present the types of OCR errors provided by an example for each one.

1.2   OCR Errors

OCR errors occur when the OCR system misrecognizes the original text. These errors would be resulted from missing, adding, misspelling, or disordering a letter. There are six categories of OCR errors [3] as follows:

- *Replacement:* An OCR error in which a single character is recognized as another character.
  - For example, suppose that the original word was "Class" and when it's misspelled by OCR it was "elass" where the single character "c" is recognized as another character "e".
- *Multi replacement:* An OCR error in which a single character is recognized as multiple characters or multiple characters are recognized as one character.
  - For example, suppose that the original word was "Ram" and when it's misspelled by OCR it was "Rarn" where the single character "m" is recognized as multiple characters "r" and "n".
- *Space insertion:* An OCR error in which space has been inserted between letters causing the word to be divided into two terms.
  - For example, suppose that the original word was "Egypt" and when it's misspelled by OCR it was "Egy pt" where a space has been inserted between two letters "y" and "p" causing the word "Egypt" to be divided into two terms "Egy" and "pt".

- *Space deletion*: An OCR error in which space has been deleted between two terms causing the word to be recognized as one term.
  - For example, suppose that the original terms was "The car" and when they are misspelled by OCR they were "Thecar" where a space has been deleted between two terms "The" and "car" causing the two terms" The car" to be recognized as one term" thecar".
- *Letter insertion*: An OCR error in which a letter has been inserted in the word.
  - For example, suppose that the original word was "write" and when it's misspelled by OCR it was "writte" where the letter "t" has been inserted in the word "write".
- *Letter deletion:* An OCR error in which a letter has been deleted from the word.
  - For example, suppose that the original word was "corn" and when it's misspelled by OCR it was "crn" where the letter "o" has been deleted from the word "corn".

Most researchers classify the spelling errors into two main groups: non-word errors and real word errors; the definition of each group is as follows:

The Non-word Errors. These are words that have no meaning and don't exist in the lexicon or in the dictionary. They can be defined as a sequence of letters that doesn't form a valid word in any context of language such as "make" which is recognized by OCR as "malee".

The Real-word Errors. These are words that exist in the dictionary, but with a mistake in the context of the original word such as "make" which is recognized by OCR as "male".

The real-word errors' detection needs a higher level of knowledge compared to the non-word errors' detection. Such detection needs NLP tools to be solved. Karen Kukich shows that about 25–40 % of the spelling errors are in fact real words and this percentage is altered from a language to another [4].

In this paper, we focus on correcting the OCR errors for Arabic language using a model based on character segmentation. The effect of changing the size of training sets on the performance of the model is examined. The remainder of this paper is organized as follows: Sect. 2 presents a review of previous work in this area. The steps involved in implementing the character model are described and illustrated with an example given in Sect. 3. Section 4 shows the mechanism by which the character model is used in correcting OCR errors. Section 5 presents the results obtained by applying the proposed model and then testing the effect of changing the size of training sets on the performance of this model. Finally, Sect. 6 draws a conclusion and provides considerable insights into potential future research directions on this topic.

2   The Previous Work

There are different approaches driven by many researchers to deal with correcting words in a text in general [16–19], and with correcting OCR output in particular [4, 5].

In the early 1960s, working on the OCR error correction started until the present. During that time, many different techniques have been devised — such as minimum edit distance [6, 7, 11], similarity key techniques [20], n-gram based techniques [14, 21, 22], probabilistic techniques [10, 23] and rule-based techniques [24, 25].

Working on OCR error correction for Arabic text started in 1975 [1]. A few of these research would be considered for Arabic text because the Arabic language represents challenges as it is a language characterized by rich morphology [22, 26, 27]. Shaalan et al. [25] developed a system for Arabic spelling correction based on the rule-based approach. No experiments are reported in that research. Hassan et al. [28] used the finite state automata with the language model to develop a method to select the best correction in a given context. This method was tested on a list of 556 misspelled words where the best accuracy reported was 89 %. Haddad and Yaseen [27] introduced a hybrid model for spell checking and correcting Arabic words based on semi-isolated word recognition and correction techniques. Magdy and Darwish [7] presented a character-level model for correcting Arabic OCR degraded text that is based on character segmentation, edit distance algorithm, and language model. This model was subjected to training on 4,000 OCR corrupted words. Single character-based and character segment-based models for correcting OCR errors for Arabic text were examined then compared. The results revealed the superiority of of the character segment-based model compared to the single character based model. The main limitation considered in their work was using a small training set. Effective progress, in their future work, has shown that increasing the process of training will improve the correction.

Our proposed model relies on the work of Magdy and Darwish [7]. We tested our model on different sizes of training sets to prove the effect of increasing the size of training on the accuracy of correction. The difference between their previous work and our corresponding work is that they normalized the different forms of alif (hamza, alif maad, alif with hamza on top, hamza on waw, alif with hamza on the bottom, and hamza on ya) to alif, and ya and alif maqsoura to ya upon developing their model but we left these forms unchanged. Also, we tested our model assuming both single error and multi-error in each misspelled word and then we showed the results in each case.

3 The Proposed Model Description

Our proposed model for correcting OCR errors in Arabic text is mainly based on character segmentation; the first step involved in implementing this model is to set up an error correction model "character model" for modeling the OCR degradation text. The second step is to use this model in correcting each misspelled word in the test set to generate the candidates and choose the correct ones. It is worth noting that developing the character model for modeling the OCR degradation text is an important point because there are many challenges that OCR may encounter in order to give accurate outputs, especially for a challenging language like Arabic [8].

The next sub-sections describe the steps involved in developing the character model. There are four steps; the first step is to construct the "wrong shape" table by aligning the clear text and the OCR text, and then conduct the normalization by deleting all symbols or spaces, and then align each clear word with its corresponding

OCR word to store each character/s and its OCR shapes, and the final step is to calculate the probability.

3.1 Constructing the "Wrong Shape" Table

Developing a character model depends on training sets. Therefore, the first step involved in developing such a model is to construct the "Wrong Shape" table inside database, or DB for short, from the training sets in order to develop the character model. The main idea of constructing such table is the alignment of OCR-degraded words with clean text words. We used a software application called "Aligner" that is mentioned in [9]. The "Aligner" application takes both clear and OCR text files and tokenize them into words. Then, based on the edit distance between them, the "Aligner" updates the database that stores the clear word shape and all shapes of the words appeared in the OCR version.

3.2 Normalization

The next step involved in developing a character model is normalization. In this step, all symbols such as ":", ".", "&", "$", "#", "!" ..... etc., numbers, and any spaces will be deleted from each word either it was a clear or an OCR word existed in the "Wrong Shapes" table which has been constructed in the first step. Normalization is considered as a significant step because these symbols could be the reason for the presence of mis-spelled words, therefore, they have to be deleted so that the word may be perfectly correct.

3.3 Alignment

There are two different types of character alignment mentioned in [7] called *(one-to-one) 1:1 character alignment* where each single character is mapped to only one character [10] and *(Many-to-many) m: n character alignment* where a character segment of length m is mapped to a character segment of length n [11]. Figures 2 and 3 illustrate these types presented above. Suppose that the original word was "علم": "Aelm". This word exists in Arabic and it means "Science", but when it's misspelled by OCR it was "غلحم": "Ghelhm" that doesn't exist in Arabic.

Levenstein edit distance algorithm [15] was used to produce character model alignments. The algorithm computes the minimum number of edit operations, including insertion, deletion or replacement that are required to transform one string into another

Original	م	ع	ل	ع
Misspelled	م	ح	ل	غ

Fig. 2. one : one character alignment, where "ε"means null character.

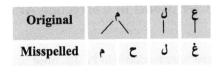

Fig. 3. m:n character alignment

in order to detect the difference between them and generate all possible conversions of the original word into OCR word.

For example, after applying the edit distance alignment on the two words as shown in Fig. 2 and using 1:1 character model, two operations will be produced and stored in the 1:1 character model DB as follows: (1) Replacement operation where the character 'ع' is replaced by 'غ' and (2) Insertion operation where the character 'ح' is inserted. But in case of m:n character model the null characters "ε" are combined with misaligned adjacent characters as shown in Fig. 3. Three replacement operations will be produced in this case and stored in m:n character model DB as follows: (1) The character 'ع' is replaced by 'غ', (2) The character 'م' is replaced by 'حم', and (3) The character 'ل' is replaced by 'حل'.

The m:n character alignment is more accurate because the character can be confused with three or four characters, especially in Arabic language.

3.4   Calculating the Probability

This step is implemented as a probabilistic string edit process. Each of 1:1 character model DB and m:n character model DB is developed by using alignment pairs in training data taking into account the type of OCR error (Replacement, Insertion or Deletion). Then, under the assumption that the clean word is $\chi = C_1..C_k.. C_l..C_n$, the resulting OCR degraded word will be $\delta = D_1..D_x.. D_y..D_m$, where $D_x..D_y$ resulted from $C_k..C_l$ and ε representing the null character. The probability estimates for each edit operation for the models are as the following formula described by Magdy and Darwish [7]:

$$P_{Replacement}\left(C_k..C_l \rightarrow D_x..D_y\right) = \frac{count(C_k..C_l \rightarrow D_x..D_y)}{count(C_k..C_l)} \tag{1}$$

$$P_{Deletion}\left(C_k..C_l \rightarrow \varepsilon\right) = \frac{count(C_k..C_l \rightarrow \varepsilon)}{count(C_k..C_l)} \tag{2}$$

$$P_{Insertion}\left(\varepsilon \rightarrow D_x..D_y\right) = \frac{count(\varepsilon \rightarrow D_x..D_y)}{count(C)} \tag{3}$$

In the next section, we will know how to use the character models in correcting the OCR errors in Arabic text.

4   Using the Character Model in Correcting Arabic OCR Errors

The steps for using the character models will be described in the next sections.

4.1   Segmentation

The first step of using the character model is to segment the misspelled word. If the 1:1 character model is used; we will need to segment the misspelled word into single characters, but if the used model was m:n character model; we will need to segment the misspelled word into all possible segmentations such that there are 2^{n-1} segments of a word with n letters. For example, given a misspelled word (غلحم) "Ghelhm", all possible segmentations can be 8 segments as follows:

غ ,ل , حم	غ , لح , م	غل , ح , م	غ , ح , ل , م
غلحم	غ , لحم	غلح , م	غل , حم

4.2   Replacement Characters and Generate Candidates

Each segment sequence of the misspelled word segments that contains a misspelled character/s will be replaced with all corresponding correct characters from the character model DB. This process will generate all possible candidates.

During candidates' generation, the character characteristics are taken into account. The character characteristics are intended to consider these two questions: what are the characters that can never have been existed so close to another letter? and what are the characters that can never have been in the first, middle or last space of the word? In order to study the character characteristic, a list of Arabic words by "Mohammed Attia" containing 9.3 million unique Arabic words[1] [12] is used. It was found that some characters can never be next to another letter. For example, the { 'ح','ص', 'ض', 'ط', 'ظ','س','ش', ' } letters cannot be next to "Thaثاء" letter. Also, the {'!','خ', 'غ', 'ء'} letters cannot be next to "Qafقاف" letter. Additionally, it was found that four characters, {'ؤ', 'ى', 'ة', 'ئ'}, can never be at the beginning of an Arabic word, and two characters, {'ى', 'ة'}, can never be at the middle of an Arabic word, while all characters can be at the end of an Arabic word.

This study will be useful during candidates' generation to eliminate all invalid candidates according to their characters. The output of the candidates' generation step is a number of candidates' words with a probability $P(\delta|\chi)$ for each word, where δ is a corrupted string composed of the characters $D_1..D_x..D_y..D_m$, χ is its correct string composed of the characters $C_1..C_k..C_1..C_n$, and $P(\delta|\chi)$ is the probability of producing δ from χ that is calculated by the following formula described by Magdy and Darwish [7]:

---

[1] Available at: https://sourceforge.net/projects/arabic-wordlist/.

$$P(\delta|\chi) = \prod_{all:Dx...Dy} P(D_x...D_y|C_k...D_l) \tag{4}$$

The candidates' words may be valid or invalid. To detect the validation of each candidate word, we used a web mined collection, which is collected from religious text by "Ibn Al-Quim" and "Ibn Tymia" books. The collection contained 7,818,052 words, ith 215,927 unique surface forms.

If a candidate word doesn't exist in the web collection, it will be considered as invalid word that will be discarded. But, if it exists in the web collection, it will be considered as valid word χ with a probability computed as follows:

$$P(\chi) = \frac{count \ of \ \chi \ in \ the \ collection}{count \ of \ all \ words \ in \ the \ collection} \tag{5}$$

4.3 Calculate the Bayes' Rule

Then, after generating the candidates and detecting the correct ones out of them. The Bayes' rule [13] will be computed for each valid candidate word as a probabilistic problem formulated as:

$$argmax_x P(\delta|\chi).P(\chi) \tag{6}$$

Where $P(\delta|\chi)$ is calculated as in Eq. (4), $P(\chi)$ is calculated as in Eq. (5), and $argmax_x$ is the scoring mechanism that computes all values of the correct word and maximizes its probability.

Then, all valid candidate words will be ranked and the top 10 of them will be chosen to generate the results.

5 Experimental Results

We evaluated the correction performance for our model on real OCR data with different sizes to study the effect of changing the size of training dataset for the correction.

The experiments were run on Intel(R) Core(TM) i3 CPU M350 @2.27 GHz with 4 GB of RAM running windows 7 and along with 64-bit architecture. Experiments execution was parallel execution. An example of the time spent in executing the two types of the proposed model on the training set of 502,167 words and the testing set of 2,000 words can be observed as follows: (i) when we used the 1:1 character model, under the assumption of a single error in each misspelled word, the time spent in its execution was 25 s, and (ii) when we used the m:n character model, under the assumption of a single error in each misspelled word, the time spent in its execution was 59 s. The reason for such an increase in time is that we need to segment the misspelled word into all possible segmentations if we used the m:n character model, as we mentioned before.

5.1   Training and Test Data

In this paper, we are interested in enhancing the results of OCR on the degraded Arabic text.

In order to train the character models and study the effect of changing the size of training dataset on the results, three training databases are used (Table 1):

First training DB contains about 6,000 words of "Zad al ma'ad" book (Arabic: زاد المعاد), or "Zad" for short. The book's name was translated into English to be "Provisions of the Hereafter", and it was written by the Islamic scholar "Ibn Al-Qayim". The 6,000 words, of which 5,740 words are correctly read by OCR, and 260 words are wrongly read by OCR.

Second training DB is the third volume of "Abgad El Aloum" book (Arabic: أبجد العلوم), or "Abgad" for short. The book's name was translated into English to be "The Alphabet of Science", and it was written by Mohammed Sedek Khan. This volume contains 64,225 words, of which 55,522 words are correctly read by OCR, and 8,441 words are wrongly read by OCR.

Third training DB is the two first volumes of "Abgad" book. These volumes contains 502,167 words, of which 182,647 words are correctly read by OCR, and 23,049 words are wrongly read by OCR.

The steps involved in constructing each training database are as follows:

1. Printing the used pages.
2. Scanning and OCR these pages using Sakhr Automatic reader version 10 [29].
3. Using the "Aligner" between both the clean text and the OCR generated text to generate the training database.
4. Detecting the wrong words that don't match with the correct words, since only wrong words are used in training the character model.

Table 1. A summary of the training dataset words

Training data set	All words	Correct words By OCR	Misspelled words by OCR
First training DB	6,000	5,740	260
Second Training DB	64,225	55,522	8,441
Third Training DB	502,167	182,647	23,049

Testing Data. We used 2,000 words of "Zad" book to test the proposed model. First, the "Aligner" software [9] are used to align the clear text and the OCR deformed text in order to detect all misspelled words that OCR can read wrongly. There are 2,000 words, of which 1,810 words are correctly read, and 190 words are wrongly read.

After detecting the misspelled words, the normalization step is performed by removing all symbols, numbers, and spaces from the clear text and the deformed text. This step corrects 107 words while 3 words are still invalid (Table 2).

Table 2. Percentage of misspelled errors after removing the symbols, numbers and spaces

	No. of misspelled errors	% From all words
All words	2000	
All misspelled words	190	9.5 %
Errors by symbols or numbers	107	5.3 %
Other errors	83	4.2 %

The results for each training set will be reported on 83 invalid words to test the character model and to know the effect of the size of training datasets on the results. The experiments were conducted on the (1:1) character model and the (m:n) character model under the assumption of one single error in each misspelled word and the assumption of multi-errors in each misspelled word.

Our evaluation metric for the proposed model performance is the correction rate, which is calculated as follows:

$$CR = \frac{No.\ of\ words\ corrected\ by\ the\ model}{No.\ of\ all\ misspelled\ words} \tag{7}$$

Table 3 reports the results of using the 1:1 character model and the m:n character model concerning the first training set that contains 6,000 words from "Zad" book, of which 260 words are wrong under the assumptions of single error and multi-errors in each misspelled word. Table 4 shows the results concerning the second training set that contains 64,225 words from the third volume of "Abgad" book, of which 8,703 words are wrong under the assumptions of single error and multi-errors in each misspelled word. Table 5 reports the results concerning the third training set that contains 502,167 words from the two volumes of "Abgad" book, of which 23,049 words are wrong under the assumptions of single error and multi-errors in each misspelled word.

Table 3. The correction rate of the top 10 generated corrections upon using the first training set

Top n generated corrections	Top 1 choice	Top 2 choices	Top 3 choices	Top 5 choices	Top 10 choices
1:1 assuming single error	36.41 %	37.34 %	37.34 %	37.34 %	37.34 %
1:1 assuming multi-error	28.9 %	30.12 %	30.12 %	30.12 %	30.12 %
m:n assuming single error	32.53 %	37.35 %	37.35 %	37.35 %	37.35 %
m:n assuming multi-error	48.19 %	52.80 %	52.80 %	52.80 %	53.01 %

Table 4. The correction rate of the top 10 generated corrections upon using the second training set

Top n generated corrections	Top 1 choice	Top 2 choices	Top 3 choices	Top 5 choices	Top 10 choices
1:1 assuming single error	71.08 %	72.28 %	72.28 %	72.28 %	73.49 %
1:1 assuming multi-error	69.87 %	69.87 %	69.87 %	71.08 %	72.28 %
m:n assuming single error	78.31 %	78.31 %	78.31 %	78.31 %	78.31 %
m:n assuming multi-error	79.52 %	79.52 %	79.52 %	79.52 %	79.52 %

Table 5. The correction rate of the top 10 generated corrections upon using the third training set

Top n generated corrections	Top 1 Choice	Top 2 Choices	Top 3 Choices	Top 5 Choices	Top 10 Choices
1:1 assuming single error	69.87 %	72.29 %	72.29 %	74.69 %	75.90 %
1:1 assuming multi-error	77.10 %	78.31 %	78.31 %	78.31 %	81.93 %
m:n assuming single error	85.54 %	89.15 %	89.15 %	90.36 %	90.36 %
m:n assuming multi-error	69.87 %	77.10 %	80.72 %	89.15 %	93.97 %

Results showed that the performance of the m:n character model is better than that of 1:1 character model, and the assumption of multi-errors in each misspelled word is better than that of single error in each misspelled word. Results also illustrated that the proposed model was contributed to the enhancement of correction for Arabic OCR degraded texts since the correction rate of m:n character model, under the assumption of multi-errors in each misspelled word, approached 94 % for the top 10 generated corrections. And after changing the sizes of the training set have been investigated, the experiments showed that the large training set leads to better results.

6 Conclusion and Future Work

In our research, we addressed the OCR error correction problem in Arabic texts focusing on the non-word errors. We described the proposed character model, the steps involved in developing the character model, and the mechanism by which it can be used for correcting OCR errors for Arabic text taking into account two types of character models (1:1) and (m:n). Then, we subjected the model to training using three different training sets where it was carried on 6,000 words, 64,225 words, and on 502,167 words. Finally, we subjected the model to testing in which the effect of changing the size of training set including 2000 words from "zad" book was investigated. Results showed that the new proposed model has contributed to the enhancement of OCR error correction in Arabic text. Results proved that if the size of the training set increased, the results will be better.

For future work, there is a need to develop an algorithm for correcting punctuations and numbers to enhance error correction. Using detection and correction techniques, a complete system of detecting and correcting errors can be developed.

Acknowledgements. The authors are grateful to the referees for their careful reading, insightful comments and helpful suggestions which have led to improvement in the paper.

References

1. Nazif, A.: A System for the Recognition of the Printed Arabic Characters. M.Sc. Thesis. Cairo University, Faculty of Engineering, Egypt (1975)
2. Habash, N., Roth, R.M.: Using deep morphology to improve automatic error detection in arabic handwriting recognition. In: Proceedings of the 49th Annual Meeting of the Association for Computational Linguistics, Human Language Technologies, vol. 1, pp. 875–884 (2011)
3. Mahdi, A.: Spell Checking and Correction for Arabic Text Recognition. M.Sc. Thesis. King Fahd University of Petroleum And Minerals, Saudi Arabia (2012)
4. Kukich, K.: Techniques for automatically correcting words in text. ACM Comput. Surv. **24**, 377–439 (1992)
5. Muhammad, M., ElGhazaly, T.: Handling OCR-degraded arabic text: a comprehensive survey. In: Proceedings of the 48th annual ISSR conference, Institute of Statistical Studies and Research, Cairo University, Egypt (2013)
6. Magdy, W., Darwish, K.: Omni font OCR error correction with effect on retrieval. In: Intelligent Systems Design and Applications (ISDA), pp. 415–420 (2010)
7. Magdy, W., Darwish, K.: Arabic OCR error correction using character segment correction, language modeling, and shallow morphology. In: Proceedings of 2006 Conference on Empirical Methods in Natural Language Processing (EMNLP 2006), Sydney, Australia, pp. 408–414 (2006)
8. Kanungo, T., Marton, G.A., Bulbul, O.: OmniPage vs. Sakhr: paired model evaluation of two arabic OCR products. In: Proceedings of SPIE Conference on Document Recognition and Retrieval (1999)
9. Ezzat, M., ElGhazaly, T., Gheith, M.: An enhanced arabic OCR degraded text retrieval model. In: Castro, F., Gelbukh, A., González, M. (eds.) MICAI 2013. LNCS (LNAI), vol. 8265, pp. 380–393. Springer, Heidelberg (2013). doi:10.1007/978-3-642-45114-0_31
10. Church, K., Gale, W.: Probability scoring for spelling correction. Stat. Comput. **1**, 93–103 (1991)
11. Brill, E., Moore, R.: An improved error model for noisy channel spelling correction. In: Proceedings of the 38th Annual Meeting of Association for Computational Linguistics, pp. 286–293 (2000)
12. Attia, M., Pavel, P., Younes, S., Shaalan, K., Josef, v., G.: Improved spelling error detection and correction for arabic. In: COLING 2012, Mumbai, India (2012)

13. Olshausen, B.A.: Bayesian Probability Theory. The Redwood Center for Theoretical Neuroscience, Helen Wills Neuroscience Institute at the University of California at Berkeley, Berkeley, CA (2004)
14. Elghazaly, T.: Improving OCR-degraded arabic text retrieval through an enhanced orthographic query expansion model. In: Tan, Y., Shi, Y., Buarque, F., Gelbukh, A., Das, S., Engelbrecht, A. (eds.) ICSI 2015. LNCS, vol. 9141, pp. 117–124. Springer, Heidelberg (2015). doi:10.1007/978-3-319-20472-7_13
15. Levenshtein, V.: Binary codes capable of correcting deletions, insertions, and reversals. Sov. Phys. Dokl. **10**, 707 (1966)
16. Mohit, B., Rozovskaya, A., Habash, N., Zaghouani, W., Obeid, O.: The first QALB shared task on automatic text correction for arabic. In: Proceedings of EMNLP 2014 Workshop on Arabic Natural Language (2014)
17. Muaidi, H., Al-Tarawneh, R.: Towards arabic spell-checker based on N-grams scores. Int. J. Comput. Appl. **53**(3), 12–16 (2012)
18. Ng, H.T., Wu, S.M., Wu, Y.,: Hadiwinoto, C., Tetreault, J.: The CoNLL-2013 shared task on grammatical error correction. In: Proceedings of CoNLL-2013 Shared Task (2013)
19. Rozovskaya, A., Habash, N., Eskander, R., Farra, N., Salloum, W.: The columbia system in the QALB-2014 shared task on arabic error correction. In: Proceedings of EMNLP 2014 Workshop on Arabic Natural Language (2014)
20. Zobel, J., Box, G., Dart, P.: Phonetic string matching: lessons from information retrieval. In: Proceedings of SIGIR-96, the 19th Annual International ACM SIGIR Conference on Research and Development in Information Retrieval (1996)
21. Riseman, E., Hanson, A.: A contextual post processing system for error correction using binary n-grams. IEEE Trans. Comput. **23**(5), 480–493 (1974)
22. Islam, A., Inkpen, D.: Real-word spelling correction using google web IT 3-grams. In: Proceedings of the 2009 Conference on Empirical Methods in Natural Language Processing, 3(August), pp. 1241–1249 (2009)
23. Kemighan, M., Church, K., Gale, W.: A spelling correction program based on a noisy channel model. In: Proceedings of the 13th conference on Computational linguistics, vol. 2, pp. 205–210 (1990)
24. Yannakoudakis, E., Fawthro, D.: The rules of spelling errors. Inf. Process. Manage. **19**(2), 87–99 (1983)
25. Shaalan, K., Allam, A., Gohah, A.: Towards automatic spell checking for arabic. In: Proceedings of the Conference on Language Engineering (ELSE), pp. 240–247 (2003)
26. Alkanhal, M.I., Al-Badrashiny, M.A., Alghamdi, M.M., Al-Qabbany, A.O.: Automatic stochastic arabic spelling correction with emphasis on space insertions and deletions. In: Proceeding of IEEE Transactions on Audio, Speech, and Language Processing, vol. 20, no. 7 (2012)
27. Haddad, B., Mustafa, Y.: Detection and correction of non-words in arabic: a hybrid approach. Int. J. Comput. Process. Orient. Lang. **20**(4), 237–257 (2007)
28. Hassan, Y., Aly, M., Atiya, A.: Arabic spelling correction using supervised learning. In: Proceedings of EMNLP 2014 Workshop on Arabic Natural Language (2014)
29. Kanungo, T., Marton, G.A., Bulbul, O.: OmniPage vs. Sakhr: paired model evaluation of two arabic OCR products. In: International Society for Optics and Photonics, pp. 109–120 (1999)

AKEA: An Arabic Keyphrase Extraction Algorithm

Eslam Amer[1(✉)] and Khaled Foad[2]

[1] Faculty of Computers and Information, Computer Science Department,
Banha University, Banha, Egypt
`eslam.amer@fci.bu.edu.eg`
[2] Faculty of Computers and Information, Information System Department,
Banha University, Banha, Egypt
`kmfi@fci.bu.edu.eg`

Abstract. Keyphrase extraction is a critical step in many natural language processing and Information retrieval applications. In this paper, we introduce AKEA, a keyphrase extraction algorithm for single Arabic documents. AKEA is an unsupervised algorithm as it does not need any type of training in order to achieve its task. We rely on heuristics that collaborate linguistic patterns based on Part-Of-Speech (POS) tags, statistical knowledge, and the internal structural pattern of terms (i.e. word-occurrence). We employ the usage of Arabic Wikipedia to improve the ranking (or significance) of candidate keyphrases by adding a confidence score if the candidate exist as an indexed Wikipedia concept. Experimental results show that on average AKEA has the highest precision value, the highest F-measure value which indicates it presents more accurate results compared to its other algorithms

Keywords: Keyphrase extraction · Natural language processing

1 Introduction

Keyphrase extraction aims to get the most informative and important terminologies from specific documents that represents the main subjects of these documents [1, 2]. Nowadays, due to the rapid increasing rate in the volume of indexed information over the World Wide Web, great challenges arises to many number of Natural Language Processing (NLP) applications that require retrieving relevant documents [2, 3], summarizing documents [4], clustering documents [5], and text mining [2]. Automatic Keyphrase extraction becomes a core task to mine effectively this enormous amount of data. It provides a promising solution to the aforementioned challenge resulted from the abundance of information which can be used to support such applications.

Keyphrase extraction can be categorized into either supervised or unsupervised techniques [7]. In supervised techniques, keyphrase extraction is mostly handled as either a classifier [8] that requires annotated training data to build models. In contrast, unsupervised techniques [9] handle keyphrase extraction as the process of training examples contain only the input patterns and no explicit target output is associated with each input.

© Springer International Publishing AG 2017
A.E. Hassanien et al. (eds.), *Proceedings of the International Conference
on Advanced Intelligent Systems and Informatics 2016*, Advances in Intelligent
Systems and Computing 533, DOI 10.1007/978-3-319-48308-5_14

In this paper, we propose a novel keyphrase extraction algorithm named AKEA that uses a heuristic that collaborate NLP techniques, statistical knowledge, and the internal structural pattern of terms (i.e. word-occurrence). AKEA make use of Arabic Wikipedia to improve the ranking (or significance) of candidate keyphrases by adding a confidence score if the candidate exist as a Wikipedia concept.

In our evaluation, we compare AKEA with other unsupervised-based keyphrase extraction algorithms like KPE [10] and KP-Miner [11]. The evaluation results show that AKEA significantly outperforms KP-Miner and showing slight improvement to KPE in terms of widely used evaluation metrics in Information Retrieval, which are F-measure and average precision [12].

This article is organized as follows. Section 2 presents a detailed review of related studies on keyphrase extraction algorithms. Section 3 provides the details of the AKEA approach and its main steps. Section 4 describes the comparative evaluation of AKEA in terms of F-measure and average precision. Section 5 summarizes the findings and results and concludes the paper.

2 Related Work

You, Fontaine and Barthès [13] have developed an automatic keyphrase extraction system for scientific documents then they evaluated their proposed system by comparing it with a previous work. They proposed their system to concentrate on two important factors. The first factor was the more precise position for proposed keyphrases. They proposed a candidate phrase generation method which is based on the core word expansion algorithm. This algorithm reduced the size of the candidate set without increasing the computational complexity. The second factor was overlap elimination for the output list. This is occurred when a phrase and its sub-phrases coexist as candidates so an inverse document frequency feature is introduced for selecting the proper granularity.

Hong and Zhen [14] have described a keyword extraction methodology. Each word in a text is endowed with seven properties is based on the linguistic properties of key words analysis. These seven properties are the word frequency, part of speech, syntactical function of words; location appeared, as well as word's morphology. Weighting methods were given to the different features based on the characteristics of each feature. The Classification model using SVM model was used on the results of keyword extraction for further optimization.

El-Ghannam and El-Shishtawy [15, 16] have proposed two keyphrase techniques aims at extracting summarization of multi-document. These techniques are Sen-Rich and Doc-Rich. Each technique involves four processes. The processes are extracting local keyphrases, constructing cluster topics, scoring sentences and documents and extracting summary sentences. The fourth process is different for both techniques. A cluster topic is constructed by finding the union of the extracted keyphrases. In the fourth step, Sen-Rich prefer rich sentences which include more important cluster topics. Doc-Rich favors sentences located in more important (centroid) documents. The document of centroid score is computed by counting the number of matched keyphrases with other documents.

An approach which aims at extracting a key-phrase from a domain has been proposed by Huang and Ciou [17]. The proposed approach generated a list of key-phrases based on a domain which involves a collection of web pages for a specific topic. The system firstly extracts top candidate key-phrases for each web page, then collects all the key-phrases of the same domain from all web pages, and finally generates a set of terms using union operators. The system secondly builds the relationships between terms in the term set by using the Free Dictionary, and then generates a sorted list of terms based on its weights. Finally, by using the top terms specified by users, a semantic graph can be constructed which shows the relationships between terms in the same domain.

KP-Miner system which aims at extracting keyphrases from both English and Arabic documents was proposed by El-Beltagy and Rafea [11]. The KP-Miner aims to extract simple and compound words from English and Arabic general and specific documents. The proposed system was configurable as the rules and heuristics involved in the system are related to the general nature of documents and keyphrase. Non-specialists can tune the system to their specific documents. System developers have capability to easily change the introduced constants to fit their requirements.

The automated methodology for taxonomy learning from a set of text documents which each explain a concept was proposed by Paukkeri, Garca-Plaza, Fresno, Unanue and Honkela [18]. The taxonomy learning was carried out by a hierarchical approach to the Self-Organizing Map clustering of documents. The document feature vectors were clustered with a hierarchical approach to the Self-Organizing Map (SOM) but any hierarchical clustering method could be used. The proposed methodology used three different methods to represent the documents. The first method was a combination of rule-based stemming and fuzzy logic-based feature weighting and selection. The second method was an automatic keyphrase extraction with statistical stemming. The third method was the traditional tf-idf measure with the traditional rule-based stemming.

Chen, Yin, Zhu and Qiu [19] have proposed a word features method to address the problem of automatic keyword extraction. They proposed some word features for keyword extraction. Using background knowledge of the input document, these word features were derived. They used a first querying patent data to acquire the background knowledge, and then mining the hidden knowledge by exploring the inventor, assignees, and citation information of the patent data search result file set. The method of word features are derived from this background knowledge. Experimental results have proved that using these word features, superior performance in keyword extraction can be achieved to other state-of-the-art approaches.

Rodas [20] has proposed a semantic metadata extraction engine which is built for text documents written in natural language. The semantic structure of a training corpus was learned and then dynamically generates a metadata structure. A metadata structure was generated based on the hierarchical structure of the thematic roles in the sentences. A syntactic analysis is required for both the learning process and the analysis of metadata to determine phrases and their elements, and use dependency grammar to create relationships between words leading to a hierarchal organization.

A method for extracting domain specific important key terms/phrases from a set of Web pages (in this research the Web pages crawled from University Web sites) have proposed in [21]. In the proposed method, they first created indexes of the titles of Wikipedia articles and of the titles of the crawled Web pages, and then they took intersection indexes between the two indexes. Finally, the valid n-grams to single terms are filtered in order to determine single significant keywords.

An algorithm to extract keyword and keyphrase from Islamic document is proposed in [22]. The algorithm used the lexico-syntactic method (focusing on noun phrase) and statistical method to acquire an object. This is an initial step which can be used in developing an Islamic ontology.

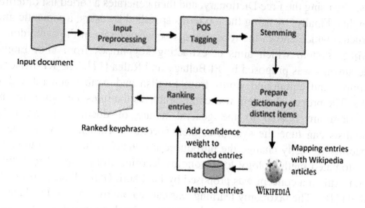

Fig. 1. AKEA flow diagram

3   Proposed Arabic Keyphrase Extractor Algorithm (AKEA)

We propose an unsupervised model to extract keyphrases from unstructured Arabic text. The model relies solely on statistical and linguistic features of words inside the document.

The flow process of Arabic Keyphrase Extraction Algorithm (AKEA) is outlined in Fig. 1. AKEA is divided into five steps namely; input preprocessing, Part-of-speech (POS) tagging, stemming, preparing dictionary of distinct entries, and finally ranking entries. A description of each step will be discussed in the following subsections.

A. Input Preprocessing.

The goal of the preprocessing is to partition sentences into separate lines and adjusting phrase boundary. To partition input document into line of sentences, we rely on two categories of splitting characters in Arabic:

1. ***Punctuation characters***: like ".", "!", "،" Enable the reader to either full stop, or moderate stop, or short stop respectively. The main goal is to acquire comfort and necessary breath in order to continue reading process.
2. ***Signs of vocal tones***: like "!", ":","؟" They are also stop signs, however – in addition to the moratorium, it has certain special psychological and emotions sound style while reading. All splitting characters, numbers, and stop words[1] are removed and replaced by "*".

---

[1] http://www.ranks.nl/stopwords/arabic.

B. Part of Speech (POS) tagging.

Experimental observations showed that most keyphrases annotated by human are considered noun phrases [23, 24]. In our proposed work words are tagged using Stanford Log-linear Part-Of-Speech Tagger[2]. We are concerned to extract all nouns and adjectives or patterns like adjective + noun and prune the rest.

C. Stemming

The objective is to find the root form of a word by pruning affixes associated with it. Affixes attached to word can be: antefixes, prefixes, suffixes and postfixes. Arabic word considered sophisticated if all types of affixes are attached to its root. Table 1 shows an example of Arabic word (ليفاوضونهم) which contains all forms of affixes.

Table 1. Attached forms of affixes for Arabic word (ليفاوضونهم) meaning "to negotiate with them"

Antefix	Prefix	Core	Suffix	Postfix
ل Preposition means "to"	ي A letter referring to the present tense and the person of conjugation	فاوض Negotiate	ون End of conjugation	هم A pronoun refer to meaning "them"

In our work we rely on work done in [25] to stem nouns and adjectives resulted from part B.

Although severe stemming may incorrectly group different words in semantics into the same index, however, we rely on the word co-occurrence and word associations in third parity information resource like Wikipedia to decrease such types of errors.

D. Prepare dictionary of distinct items

The purpose of this step is to build a hierarchical n-grams of distinct word and/or co-occurrence with other words in the processed document along with their frequencies. We utilize LZ78 compression technique [26] to deal with words instead of characters. To get the idea behind this step, let us consider the following example:

Given the following two sentences that are taken from the processed input document:

Sentence #1: "*t1 t2 t3*t3 t4* t1 t2 t3* t4 t5* t6*"
Sentence #2: "* t4 t5 * t1 t2 t4 * t3 t6 *"

Where *t1, t2, t3, t4, t5,* and *t6* represents different words, and "*" is the phrase boundary in sentence. The demonstration of how distinct items dictionary is prepared is fully illustrated in Table 2.

In Table 2 sentences are split into sequences based on "*". Each sequence is divided initially into patterns of bigram. If the pattern hasn't indexed before in

---

[2] http://nlp.stanford.edu/software/tagger.shtml.

Dictionary, it is added; and initially given a frequency value of "one", otherwise the frequency of pattern is incremented by "one".

Dictionary entries are assigned two different weight values, the first value which is the entry frequency weight which depicts the occurrence time in the processed document. The second value is the entry influence weight which is frequency times calculated according to a grammatical rule [27].

Table 2. Dictionary preparation of distinct words and co-occurrence of words

Input lines:
Sentence #1: "*t1 t2 t3*t3 t4* t1 t2 t3* t4 t5* t6*"
Sentence #2: "* t4 t5 * t1 t2 t4 * t3 t6 *"

Sequence	Pattern	Status	Action	Dictionary
t1 t2 t3	t1 t2	New	Add to dictionary	t1 t2
t1 t2 t3	t2 t3	New	Add to dictionary	t2 t3
t3 t4	t3 t4	New	Add to dictionary	t3 t4
t1 t2 t3	t1 t2	Exist	Increment frequency	–
t1 t2 t3	t1 t2 t3	New	Add to dictionary	t1 t2 t3
t4 t5	t4 t5	New	Add to dictionary	t4 t5
t6	t6	New	Add to dictionary	t6
t4 t5	t4 t5	Exist	Increment frequency	–
t1 t2 t4	t1 t2	Exist	Increment frequency	–
t1 t2 t4	t2 t4	New	Add to dictionary	t2 t4
t3 t6	t3 t6	New	Add to dictionary	t3 t6
–	–	Null	Quit	

The rule at [27] favor noun phrases that appear earlier or at the end of sentences. The later weight is calculated according to the following equation:

Where N_i is the number of words in sentence i, p_0 is the index of first word in phrase p in the sentence.

Mapping Dictionary entries with Wikipedia titles.

All indexed entries in the dictionary are mapped to Arabic Wikipedia titles[3]. If a match occurs between an entry and Wikipedia title(s), the entry will receive confidence value equal to one that means that the dictionary entry is a Wikipedia concept otherwise it will receive value of zero.

E. Ranking Entries

Eateries are ranked according to the following equation,

$$Rank(i) = \log\left(p_i x \frac{TF_i + TI_i}{L} + CF_i \right)$$

---

[3] https://ar.wikipedia.org/.

Where p_i is the position of dictionary entry i. The position is calculated as $(L - Ls)$ where L is total lines in document, Ls is the first sentence where dictionary entry i occurs. TF_i, TI_i and CF_i denotes the term frequency, influence weight, and Wikipedia confidence factor for dictionary entry i respectively.

4 Evaluation

4.1 Data Set

The corpus used in this paper is Arabic news datasets which are related to different news' domains. We rely on El-Watan[4] daily News, and Aljazeera[5] news to collect our dataset. Selected articles are related to the following domains: economy (الإقتصاد),, sport (الرياضة), culture (ثقافة), and international news (أخبار عالمية). Articles are varies in length. The lengths range from (140) to (1541) words.

4.2 Evaluation Metrics

AKEA will be compared against KPE [10], and KP-minor. The performance evaluation will be according to precision, recall, and F-measure for top-10 entries returned from each algorithm for an article.

PRECISION is the ratio of the number of relevant entries retrieved to the total number of irrelevant and relevant entries retrieved [12], it is calculated as:

$$precision = \frac{Number\ of\ relevant\ documents\ retrieved}{Total\ number\ of\ documents\ retrieved}$$

RECALL is the ratio of the number of correct or relevant entries retrieved to the total number of relevant entries indexed in a database [12], it is calculated as:

$$recall = \frac{Number\ of\ relevant\ documents\ retrieved}{Total\ number\ of\ relevant\ documents}$$

F-MEASURE is the weighted harmonic mean of precision and recall [12], it is calculated as:

$$F - measure = \frac{2.precision.recall}{precision + recall}$$

We add Arabic Part-of-speech tagging as a component to KPE to allow it to deal with articles in Arabic. All other components in KPE left without change.

---

[4] http://www.elwatannews.com/.

[5] http://www.aljazeera.net/portal.

Every news article contains annotated keywords' tags (كلمات مفتاحية) that will be considered as the golden human annotated keywords of an article. Such golden keywords will be used to evaluate the performance of AKEA, KPE, and KP-minor.

4.3   Evaluation Results

A sample output is presented in Table 3 which show the top-10 keyphrases returned by our proposed algorithm AKEA, compared with KPE algorithm, and KP-miner algorithm. The table is organized as follows: the article title along with its URL is placed in the top cell of table, then a list of assigned keyphrases as appeared in article, top-ten results output of each algorithm.

Overall performance shown in Table 4. AKEA has been evaluated against KPE and KP-miner using Arabic news dataset that are related to different news' domains. Each domain has its own data size that is indicated in the second column (# files) in Table 4. Experimental results indicates that AKEA outperforms KP-miner in most domain,

Table 3. Sample case for Top-10 results returned by AKEA, KPE, and KP- miner to a sample news article

Document name: 50 قتيلا من الجيش والحشد الشعبى بمحيط الفلوجة		
Document URL: http://www.aljazeera.net/news/arabic/2016/6/11/قتلى-بالعشرات-من-الجيش-والحشد-بمحيط-الفلوجة		
Assigned keyphrases:		
التحالف الدولى – الحشد الشعبى – تنظيم الدولة – بغداد – الفلوجة - العراق		
AKEA algorithm 12- matches found, presented by bold font	KPE algorithm 12- matches found presented by bold font	KP-miner algorithm 8- matches found presented by bold font
الحشد الشعبى **شرق الفلوجة** مصادر أمنية عراقية **القوات العراقية** القوات الأمنية وزارة الدفاع العراقية **تنظيم الدولة الإسلامية** **الجيش** **التحالف الدولى** **هجوم للتنظيم**	**القوات العراقية** **الحشد الشعبى** **شرق الفلوجة** **تنظيم الدولة الإسلامية** **وزارة الدفاع العراقية** **التحالف الدولى** **هجوم للتنظيم** الطائرات المروحية مصادر أمنية عراقية القوات الأمنية	**الحشد الشعبى** **الفلوجة** **تنظيم الدولة** الطائرات المروحية **الجيش** وزارة الدفاع **القوات العراقية** قتلى مصادر أمنية القوات الأمنية

Table 4. Performance comparison between AKEA, KPE, and KP-miner for top–10 extracted keyprases

Domain	# files	AKEA			KPE			KP-miner		
		Average precision	Average recall	Average F-measure	Average precision	Average recall	Average F-measure	Average precision	Average recall	Average F-measure
Sport	4550	0.248	0.291	0.268	**0.260**	**0.295**	**0.276**	0.191	0.262	0.221
Economics	3468	**0.296**	0.301	0.298	0.283	**0.311**	0.296	0.281	0.275	0.278
Culture	2782	**0.273**	**0.286**	**0.279**	0.260	0.278	0.269	0.252	0.266	0.259
International news	2035	**0.263**	0.288	**0.275**	0.251	**0.290**	0.269	0.249	**0.290**	0.268

however, KP-miner showed an equal Recall performance in *International news* domain. Experimental indicate that, for long articles KPE showed a comparative recall performance compared to AKEA, and KP-miner. On average AKEA has the highest precision value, the highest F-measure value which indicates it presents more accurate results compared to its other algorithms.

5 Conclusion

This paper introduced AKEA system, an Arabic keyphrase extraction mechanism. AKEA incorporate many heuristics that collaborate linguistic patterns based on Part-Of-Speech (POS) tags, statistical knowledge, and the internal structural pattern of terms (i.e. word-occurrence). Arabic Wikipedia has been used as a third parity to enhance the ranking (or significance) of candidate keyphrases by adding a confidence score if the candidate keyphrase is matched with an indexed Wikipedia concept. As shown through the evaluation, AKEA improve the effectiveness of keyphrase extraction. AKEA is unsupervised which means it neither require any training documents to build the model nor corpus resources to compare results. The work presented can be easily applied to deal with any other language.

Experimental results showed that the performance of AKEA outperforms other unsupervised algorithms as it has the highest precision value, the highest F-measure value which indicates it presents more accurate results compared to its other algorithms.

References

1. Jean-Louis, L., Zouaq, A., Gagnon, M., Ensan, F.: An assessment of online semantic annotators for the keyword extraction task. In: Pham, D.-N., Park, S.-B. (eds.) PRICAI 2014. LNCS, vol. 8862, pp. 548–560. Springer, Heidelberg (2014)
2. Harb, H., Fouad, K., Nagdy, N.: Semantic retrieval approach for web documents. Int. J. Adv. Comput. Sci. Appl. (IJACSA) **2**(9), 4673–4681 (2011)
3. Fouad, K., Khalifa, A., Nagdy, N., Harb, H.: Web-based semantic and personalized information retrieval. Int. J. Comput. Sci. Iss. (IJCSI) **9**(3), 3 (2012)
4. Babekr, S., Fouad, K. Arshad, N.: Personalized semantic retrieval and summarization of web based documents. Int. J. Adv. Comput. Sci. Appl. (IJACSA) **4**(1) (2013)
5. Fouad, K., Hassan, M.: Agent for documents clustering using semantic-based model and fuzzy. Int. J. Comput. Appl. (0975–8887) **62**(3), 10–16 (2013)
6. Gupta, V., Lehal, G.: A survey of text mining techniques and applications. J. Emerg. Technol. Web Intell. **1**(1), 60–76 (2009)
7. Wang, R., Liu, W., McDonald, C.: How preprocessing affects unsupervised keyphrase extraction. In: Gelbukh, A. (ed.) CICLing 2014, Part I. LNCS, vol. 8403, pp. 163–176. Springer, Heidelberg (2014)
8. Deng, Z., Zhu, X., Cheng, D., Zong, M., Zhang, S.: Efficient KNN classification algorithm for big data. Neurocomputing **195**, 143–148 (2016)

9. Hastie, T., Tibshirani, R., Friedman, J.: The Elements of Statistical Learning. In: Hastie, T., Tibshirani, R., Friedman, J. (eds.) Unsupervised Learning, 2nd edn. Springer, New York (2008)

10. Aliaa, A., Ghalwash, Y., Amer, E.: KPE: an automatic keyphrase extraction algorithm. In: IEEE Proceeding of International Conference on Information Systems and Computational Intelligence (ICISCI 2011), pp. 103–107 (2011)

11. El-Beltagy, S., Rafea, A.: KP-Miner: a keyphrase extraction system for English and Arabic documents. Inf. Syst. **34**, 132–144. Elsevier B.V. (2009)

12. Wong, W., Liu, W., Bennamoun, M.: Ontology learning from text: a look back and into the future. ACM Comput. Surv. **44**, 20:1–20:36 (2012)

13. You, W., Fontaine, D., Barthès, J.-P.: An automatic keyphrase extraction system for scientific documents. Knowl. Inf. Syst. **34**, 691–724. Springer (2013)

14. Hong, B., Zhen, D.: An extended keyword extraction method. In: 2012 International Conference on Applied Physics and Industrial Engineering. Physics Procedia, vol. 24, pp. 1120–1127. Elsevier B.V. (2012)

15. El-Ghannam, F., El-Shishtawy, T.: Multi-topic multi-document summarizer. Int. J. Comput. Sci. Inf. Technol. (IJCSIT) **5**(6), 77–90 (2013)

16. Al-Saleh, A., Menai, M.: Automatic Arabic text summarization: a survey. Artif. Intell. Rev. **45**, 203–234 (2016)

17. Huang, Y.-F., Ciou, C.-S.: Constructing personal knowledge base: automatic key-phrase extraction from multiple-domain web pages. In: Cao, L., Huang, J.Z., Bailey, J., Koh, Y.S., Luo, J. (eds.) PAKDD Workshops 2011. LNCS, vol. 7104, pp. 65–76. Springer, Heidelberg (2012)

18. Paukkeria, M., Garca-Plazab, A., Fresnob, V., Unanueb, R., Honkelaa, T.: Learning a taxonomy from a set of text documents. Appl. Soft Comput. **12**, 1138–1148. Elsevier B.V. (2012)

19. Chen, Y., Yin, J., Zhu, W., Qiu, S.: Novel word features for keyword extraction. In: Dong, X.L., Yu, X., Sun, Y., Dong, X.L., Li, J., Sun, Y. (eds.) WAIM 2015. LNCS, vol. 9098, pp. 148–160. Springer, Heidelberg (2015). doi:10.1007/978-3-319-21042-1_12

20. Rodas, A.: Semantic metadata extraction from open domain texts in natural language. Master of Science in Computer Engineering University Of Puerto Rico Mayaguez Campus. ProQuest LLC (2013)

21. Qureshi, M., O'Riordan, C., Pasi, G.: Short-text domain specific key terms/phrases extraction using an n-gram model with wikipedia. ACM (2012). 978-1-4503-1156-4/12/10

22. Saad, S., Salim, N., Omar, N.: Keyphrase extraction for Islamic knowledge ontology. IEEE (2008). 978-1-4244-2328-6/08

23. Hulth, A.: Improved automatic keyword extraction given more linguistic knowledge. In: Collins, M., Steedman, M. (eds.) Proceedings of the 2003 Conference on Empirical Methods in Natural Language Processing, pp. 216–223 (2003)

24. Wan, X., Xiao, J.: Single document keyphrase extraction using neighborhood knowledge. In: Proceedings of the 23rd National Conference on Artificial intelligence, AAAI 2008, vol. 2, pp. 855–860. AAAI Press (2008)

25. Khoja, S., Garside, R., Knowles, G.: An Arabic tagset for the morphosyntactic tagging of Arabic (2001)

26. Pu, M.: Fundamental data Compression, 1st edn. Elsevier, UK (2006)

27. Kumar, N., Srinathan, K.: Automatic keyphrase extraction from scientific documents using N-gram filtration technique. In: Proceeding of the Eighth ACM Symposium on Document Engineering, pp. 199–208 (2008)

Lexicon Free Arabic Speech Recognition Recipe

Abdelrahman Ahmed[1(✉)], Yasser Hifny[2], Khaled Shaalan[3,4], and Sergio Toral[1]

[1] Electronic Engineering Department, University of Seville, Seville, Spain
abdahm@alum.us.es, storal@us.es
[2] Department of Information Technology, University of Helwan, Cairo, Egypt
yhifny@fci.helwan.edu.eg
[3] School of Informatics, Edinburgh, UK
[4] The British University in Dubai, Dubai, UAE
Khaled.shaalan@buid.ac.ae

Abstract. We present the first end-to-end recipe of Arabic speech recognition using lexicon free Connection Temporal Classification (CTC) and Recurrent Neural Networks (RNN). The study describes in details the decisions made, step by step, in building Arabic system including transcription method, feature extraction, training process and decoding optimization. The results are compared with Hidden Markov Models (HMM), Gaussian mixture models (GMM), and tandem baseline in Arabic using the same data set. The corpus is Aljazeera broadcast and language model extracted from the Aljazeera corpus, web and twitter crawling using different n-grams. We measure both word error rate (WER) and character error rate (CER) for each n-gram order. The results achieved are very close to the baseline with some recommendations.

Keywords: Speech recognition · Connectionist Temporal Classification · Recurrent neural network · Language model

1 Introduction

Arabic language is a challenging language and it is considered as one of the most morphologically complex languages (Ali et al. 2014a; Farghaly and Shaalan 2009; Othman et al. 2004; Shaalan et al. 2015). Automatic Speech Recognition (ASR) for Arabic has been a research concern over the past decade (Diehl et al. 2012; Farghaly and Shaalan 2009). Arabic language has limited resources in speech recognition because it requires high experience in speech technology using HMM/GMM models and linguistic experts as well (Shaalan 2014). Most of the speech recognition systems are developed for Indo-European language (Diehl et al. 2012). The language is structured from right to left with different pattern of vowelization (Shaalan et al. 2009). There were many attempts for presenting Arabic speech recognition recipes (Ali et al. 2014a), however, the methods used are still challenging to learn and to get competing results compared to other

© Springer International Publishing AG 2017
A.E. Hassanien et al. (eds.), *Proceedings of the International Conference on Advanced Intelligent Systems and Informatics 2016*, Advances in Intelligent Systems and Computing 533, DOI 10.1007/978-3-319-48308-5_15

languages (Radha and Vimala 2012). The motivation of this paper is to present smooth and easy procedures in development Arabic ASR system using a neural network method (NN). This paper proposes Arabic ASR using recurrent neural network (RNN). It discards HMM/GMM methods complications, decision trees, and dictionary. The study presents a recipe of Arabic transcription and feature extractions based on Stanford CTC source code. Next sections briefly introduce ASR methods of HMM/GMM, DNN and RNN. The experiment, Sect. 5, highlights building robust Arabic ASR using CTC and estimates the parameters of training process. Furthermore, it explores the effect of Out Of Vocabulary (OOV) on results in CTC compared to other methods. We offer analysis to the HMM/GMM, tandem and RNN for different types of language models using Aljazeera corpus 8 h. The results, Sect. 6, discusses the WER/CER and some important observations regarding the training and decoding process. Then conclusion, Sect. 7, with proposed future work of technical considerations for next experiments.

2   Related Work

The HMM/GMM acoustic models have proved significant improvement in speech recognition in different languages including Arabic in terms of accuracy and word error rate (WER) (Motlicek et al. 2015). The word alignment is performed by HMM state modeling and each state is presented by more than one Gaussian model. The HMM training is based on maximum likelihood. The lexicon and language model are prepared before decoding process so that the decoder matches the best score for each word defined in the dictionary (Attia et al. 2012). HMM/GMM modeling approach still has some drawbacks of (1) it requires a deep knowledge in HMM (2) it is developed under assumption that observations are independent which does not comply with vocal tract.

In speech recognition, the neural network achieved outstanding jumps in this field and others i.e. handwriting, visual recognition etc. (Raschka 2015). The deep neural network (DNN) is a conventional multilayer perceptron (MLP) with many hidden layers for data processing (Yu and Deng 2012). By mixing both of neural network (NN) and HMM/GMM, the speech recognition outperformed the HMM/GMM method which is called tandem. The tandem method (Hermansky et al. 2000) is a feature extraction approach using deep neural network (DNN) in order to obtain complementary features to MFCC (Hinton et al. 2012). This setup outperforms the HMM model. The hybrid approach, which is not included in this study, is another approach of combining the HMM and artificial neural networks (ANN). In hybrid approach, HMMs are used for sequential modeling and ANN models are used as flexible discriminant classifiers to estimate a scaled likelihood (Bourlard and Morgan 1998). This approach also combined with DNN and DBRNN which gain prominence with performance improvement (Graves et al. 2013). The hybrid method can be generalized using deep conditional random fields methods (Hifny 2015). Yet, the process of data preparation and integration between HMM/GMM tandem or hybrid are very complicated and requires much time processing. There were attempts for presenting full recipe of Arabic speech recognition using Kaldi[1] (Ali et al. 2014a), HTK[2] and Sphinx[3] (Radha and Vimala 2012). They have trained acoustic models using HMM or HMM/GMM or combined them with DNN. The acoustic model was based on morphological or grapheme approaches. The decoding process was depending on lexicon and based on word error rate (WER). Connectionist Temporal Classification (CTC) is a novel technique (Graves et al. 2006) for automatic speech recognition and other applications (i.e. handwriting) using recurrent neural network (RNN). This approach uses only neural network to map acoustic input to the transcribed text (label by label). This technique avoids legacy complicated approaches of Hidden Markov Models (HMM) and Gaussian Mixture Model (GMM).

3   Arabic Speech Recognition System

The conceptual framework of the speech recognition system in this study is described in Fig. 1.

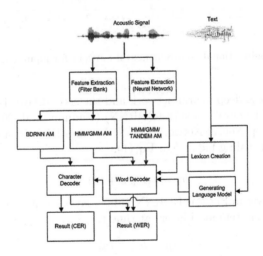

Fig. 1. The study framework

The automatic speech recognition system consists of three main components: acoustic model, language model and decoder. Next subsections discuss the acoustic model in more details using CTC/BDRNN, the language model using n-gram and the decoder based on characters.

---

[1] http://www.kaldi-asr.org/.
[2] http://htk.eng.cam.ac.uk/.
[3] http://cmusphinx.sourceforge.net/.

3.1   Acoustic Model

In acoustic model, we are searching for parameters that maximize the objective function. CTC is an objective function for temporal classification. BDRNN estimates the parameters for best alignment of label to input audio.

3.1.1   Bidirectional Recurrent Neural Networks (BDRNN)

RNN presents the dynamic speech features by feeding the low-level acoustic features into the hidden layer together with the recurrent hidden feature from the past history (Yu and Deng 2012). The model is initially built on previous state as nothing to do with the future (Graves et al. 2013), hence a separate hidden layers are built for forward and backward training to overcome this limitation. Each layer is computed separately and summed together as the following Fig. 2:

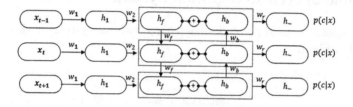

Fig. 2. Stanford bidirectional recurrent neural network for input x and output $p(c|x)$

The first layer is deep neural network and the second layer is temporal bidirectional recurrent neural network (BDRNN) (Hannun et al. 2014). BDRNN is different than Deep Neural Network (DNN) because DNN is one layer of input audio frames and scaling factor b in Eq. (1).

$$h_1 = f(w_1 x_t + b_1) \tag{1}$$

The second layer in Fig. 2 is BDRNN hidden layer j is partial sum of forward and backward layer (temporal layer) at time t.

$$h_{jt} = h_{ft} + h_{bt} \tag{2}$$

The forward and backward hidden layers are computed independently with weight matrix w_f and w_b. The partial hidden layer takes an input from previous hidden layer $h_{(j-1)t}$. Therefore, the hidden layer h_f and h_b at time t are in Eq. (3).

$$h_{ft} = f(w_j h_{(j-1)t} + w_f h_{(t+1)f} + b_j)$$

$$h_{bt} = f(w_j h_{(j-1)t} + w_b h_{(t-1)b} + b_j) \tag{3}$$

where $f(z) = min(max(z, 0), \mu)$ is a rectified linear activation function clipped to a maximum possible activation of μ to prevent overflow (Glorot et al. 2011).

Rectified linear hidden units have been show to work well in general for deep neural networks, as well as for acoustic modeling of speech data (Glorot et al. 2011). The softmax function as an output distribution of $p(c|x)$ for expected characters π at c_k.

$$p(c_k|x_t) = \frac{e^{-(w_k h_\varsigma + b_k)}}{\sum_{j=1}^{\pi} e^{-(w_j h_\varsigma + b_j)}} \tag{4}$$

where h_ς is the final hidden layer following the temporal layer (Maas et al. 2015) and b_k is the scalar bias term. The gradient is the derivative of loss function which is calculated for all weights for DBRNN.

3.1.2   Connectionist Temporal Classification (CTC)

In BDRNN, the objective function is defined separately for each point in the training sequence (Maas et al. 2015). Therefore, the output is fragmented and in the form of independent labels, which requires pre-segmentation to the training data and post-processing to the output labels. CTC is a way to compute the probability of a sequence of labels and the corresponding audio frames. It calculates the probability of all possible alignments. It interprets the network output as a probability distribution over all possible label sequences, conditioned on a given input sequence (Graves et al. 2006). Let y be the sequence of network outputs, and denoted by y_t^k the activation of output unit k at time t. Then y_t^k is interpreted as the probability of observing label k at time t, which defines a distribution over the set L of length T sequences over the alphabet L. π is the predicted sequence of labels in Eq. (5) (Graves et al. 2006).

$$p(\pi|x) = \prod_{t=1}^{T} y_\pi^t, \forall \pi \in L^T \tag{5}$$

Our study is based on foundation of Stanford (Maas et al. 2015) using character based approach. The training and decoding process will be performed on characters one by one, which are our labels in this case. The words are constructed according to minimization of CTC lose function (Graves et al. 2006) with no need for lexicon in decoding process compared to word level. Despite we decode on character level, the evaluation for the test set will be performed on both of word and character level, to guarantee the validity of the proposed approach. Furthermore, the study becomes reliable by comparing the results with previous method HMM/GMM using the same data set for training and testing. The probability of a character sequence C given the input sequence X.

$$P(C|X) = \prod_{t=1}^{T} p(c_t|X) \tag{6}$$

3.2   Language Model

Language model is the probability distribution of hypothesized words by scoring probability of the words correlation to previous word sequence (Yu and

Deng 2012). The scores are estimated prior to training process when the domain is defined (Yu and Deng 2012). Language model is used to resolve ambiguous utterances (Chen and Goodman 1996; Othman et al. 2004). For example, two sentences *it takes two* or *it takes too* could not be decoded by acoustic model but combined with language model. The n-gram language model is the number of words count (frequency) proceeding the n-words in sequence. As we used the characters instead of words, Eq. (7) is the n-gram character based prior probability.

$$p(c_1 \ldots c_m) = \prod_{i=1}^{m} p(c_i | c_1 \ldots c_{i-1}) \qquad (7)$$

where c_i is the character order in stream of characters. Most of the speech recognition systems uses 3-grams, 4-grams, up to 7-grams. The higher gram order, the higher certainty (low entropy) we get. However, as we are targeting characters, we need to extend the n-grams to the highest possible order to increase the certainty per word as well as preceding words.

3.3   Decoding

The beam search decoder is performed on a character level. This method gives more advantages than word level decoding because of two reasons: Firstly, the decoding speed in character level is much higher compared to word level because of lexicon. The word decoding consumes much time every space reached between labels to match the word from predefined dictionary (Graves and Jaitly 2014). Secondly, it overcomes the out of vocabulary (OOV) problem compared to word decoding (Maas et al. 2015). Algorithm 1 illustrate the decoding pseudo code developed by Stanford (Maas et al. 2015).

The decoder ignores the blank symbols due to spaces or non-defined characters in the character set, i.e. noise. It ignores the non-blank symbols due time shift of character alignment which produces the same character again. This is called collapse function which controls the hypothesized characters to be repeated or the characters reside between two blanks. The parameter of α is the scaling factor of the language model. We sum the logs of acoustic model and language model probabilities To avoid underflow and increase the speed of the algorithm. The resulted values have different values because to different type of models. The different values make biasing for some values than others which is adjusted by the scaling factor (α).

The following equation presents the relationship between acoustic and language model as following:

$$c^{\psi} = argmax[\log p(x|c) + \alpha \log p(c)] \qquad (8)$$

Where c^{ψ} is the hypothesized character.

The β is the insertion bonus which is a scaling factor of the final insertion of the characters string. It is the exponent value of the length of the generated

Algorithm 1. The Decoder input is the CTC likelihood of the characters $p_{ctc}(c|x_t)$ and language model $p_{clm}(c|s)$. For each time step t and for each string s in current and previous hypothesized set Z_{t-1}, we extend s with a new character excluding blanks and repeated character with no separation. We apply the language model for computing probability of s. Z_0 initial value is empty (ϕ). Notation: π is the character set with no blanks (_). Output string is concatenated $s := s + c$. $|s|$ is the string length of s. $p_b(c|x_{1:t})$ is the probability of s ending by blank conditioned on input x up to time t. $p_{nb}(c|x_{1:t})$ is the same as p_b but not ending with blank. $p_{tot}(c|x_{1:t}) = p_b(c|x_{1:t}) + p_{nb}(c|x_{1:t})$.

Inputs: $p(c|x_t)$ and $p_{clm}(c|s)$.
Parameters: language model weight $Î±$, insertion bonus $Î²$ and beam width k.
$Z_0 \leftarrow \phi, p_b(\phi|x_{1:0}) \leftarrow 1, p_{nb}(\phi|x_{1:0}) \leftarrow 0$
for t =1 ...T do
   $Z_t \leftarrow \{\}$
   for s in Z_{t-1} do
      $p_b(s|x_{1:t}) \leftarrow p_{ctc}(\_|x_t)p_{tot}(s|x_{1:t-1})$
      $p_{nb}(s|x_{1:t}) \leftarrow p_{ctc}(c|x_t)p_{nb}(s|x_{1:t-1})$
      $Z_t \leftarrow Z_{t+s}$
      for c in π do
         $S \leftarrow s + c$
         if $c \neq s_{t-1}$ then
            $p_{nb}(S|x_{1:t}) \leftarrow p_{ctc}(c|x_t)p_{clm}(c|s)^\alpha p_{tot}(c|x_{1:t-1})$
         else
            $p_{nb}(S|x_{1:t}) \leftarrow p_{ctc}(c|x_t)p_{clm}(c|s)^\alpha p_b(c|x_{1:t-1})$
         end if
         $Z_t \leftarrow Z_{t+s}$
      end for
   end for
   $Z_t \leftarrow k$ most probable s by $p_{tot}(s|x_{1:T})|s|^\beta$
end for
$Return argmax_{s \in Z_t} p_{tot}(s|x_{1:T})|s|^\beta$

---

string multiplied by hypothesized string probability. In case of $\beta < 1$, a reduction factor is applied which reduces the opportunity of the decoder to insert the hypothesized string (conservative decoder) and vice versa. The beam length is the length of the hypothesized characters sequence length to be processed through probability calculation.

4 Front-End Preparation

We mean by front-end preparation is the input data transcription from Arabic to Latin as well as preparing all input features to be ready for training. The Latin characters are transcribed into numerical values and ready for the neural network processing. Then the audio feature extraction is applied to build the input data matrix. The following subsections detail each step.

4.1   Convert the Arabic Text to Latin (Transcription Process)

The Arabic transcription is converted into Latin characters so that we can make processing easier. A sample of character set is shown in Table 1.

Table 1. Sample of Arabic letters, corresponding characters transcription and its English equivalent.

Arabic Letter	أ	د	ش	ل	ك
Transcription	ga	d	sh	l	k
Equivalent English	A	D	SH	L	K

The transcription process maps each letter from Arabic to the corresponding Latin character. We added spaces between each character because the characters length are different (some characters are transcribed in one character, two characters and three). We used special characters as a hash (#) for the sentence start, star (*) for the end of the statement and separator (|) to mark the spaces between the words. The special characters help the decoder to detect the sentence and word boundaries. The next example shows a transcription of a statement:

حيث تقوم الحياة السياسية على التعددية الحزبية

hh y th | t q w m | a l hh y a t | a l s y a s y t | aa l a | a l t aa d d y t | a l hh z b y t *

The transcription process transforms the statement from right to left (Arabic writing direction) to left to right (Latin). The special characters inform the decoder of the statement start and end, as well as the start and end of each word indicated by " | ". Buckwalter[4] is a powerful open source tool to transcribe the Arabic to Latin, and it is used in many Arabic transcription purposes. However, we built our own transcription lookup table for editing the character set much easier as we did many experiments using different character set. Each character set eliminates some similar spoken characters to study the CTC response in each case. We started with character set of 33 character, then 35, then finally 39 characters for best results.

4.2   Convert the Transcription to Alias

The character set and the special characters are listed to have a number for each character. The next step is to convert the transcription to the corresponding number mapped in the list as shown in Table 2:

Table 2. Numerical transformation from Latin to numbers.

Transcription	#	ga			sh	k	*
Alias	0	2	39	22	31	20

---

[4] http://www.qamus.org/transliteration.htm.

For example:

$$\text{\# hh y th} - \text{t q w m} - \text{a l hh y a t}^*$$
$$1\ 14\ 37\ 12\ 39\ 11\ 30\ 36\ 33\ 39\ 7\ 32\ 14\ 37\ 7\ 11\ 20$$

Example 2. Numerical transformation

4.3   Voice Feature Extraction

The feature extraction is based on a filter bank (FB) instead of Mel frequency cepstral coefficient (MFCC). The empirical results show that FB outperformed the MFCC as a feature extraction technique in speech and speaker recognition technologies (Alfred 1999; Nadeu et al. 1995). FB acts as a bandpass filter for the audio signal in the frequency domain (Alfred 1999). FB processing consists of projecting the features to a higher dimensional space in which classification can be more easily (Yu and Deng 2012). Our current representation is based on using a Log-Fourier-transform-based filterbank with 40 coefficients (plus energy) distributed on a mel-scale, together with their first and second temporal derivatives resulting in a 123 element feature vector. The features are pre-processed having zero mean, unit variance and acoustic context information. The context window is 10 of the frames before and after the current frame (21 frames). Hence the feature dimensions are 123×21.

5   The Experiment

The corpus consists of 8 h Aljazeera broadcast, which was collected and transcribed by QCRI[5] using advanced transcription system (Ali et al. 2014b). The training set is 4270 files and the test set is only 23 files, with each audio file duration from 10–20 s. Stanford recipe trains the audio files which are packed into batches, each batch produces a three feature files: Alias#, Feat# and key#. The input files contain the aliases (transcription in numbers), number of frames extracted and the features extracted. In this study, we have two types of language models, pseudo language model (PLM) and real language model (RLM). The PLM consists of training and test corpus in order to adjust the parameters of α, β for optimum value. The parameters estimated are $\alpha = 5$, $\beta = 3.8$, beam length =150. Afterward, we replaced the PLM with real language model (RLM) of 980k unique words and 6.9 million words in the same domain of Aljazeera broadcast prepared by QCRI. Furthermore, we have added conversational text by crawling twitter for optimum compliance with the domain as the training corpus consists of news reports and talk shows ($OOV < .005$). We built 7, 9, 14 and 15 g orders for both the PLM and RLM with modified Kneser-Ney smoothing using the KenLM toolkit[6]. The KenLM source code has compiled to accept more than 9 g. The 15 g LM is about 64 GB ARPA format and 32 GB binary format. The binary format has an advantage of getting faster decoding process

---

[5] http://www.qcri.com/.

[6] https://kheafield.com/code/kenlm/.

and less memory compared to ARPA. 15th gram is the highest gram we could reach because the memory limitation of the OS. The experiment setup is 24 core, 32 GB RAM, NVidia Tesla 80 K graphical processing unit (GPU). Tesla k80 is the state of art GPU technology, at the time of writing this paper, of 24 GB memory cards and 2900 processing core for training process. Decoding and language model is performed over the CPUs parallel processing. The training parameters were adjusted to 50 epoch, 5 hidden layers each layer size is 1840 layers. The step is 1e-5, and maximum sentence length 6000 frames.

6   Results

The experiment baseline of HMM/GMM and HMM/GMM/Tandem compared to CTC are shown in Table 3.

Table 3. HMM/GMM/Tandem, CTC results using PLM/RLM

Model	CER-PLM	WER-PLM	CER-RLM	WER-RLM
HMM/GMM	NA	29 %	NA	40 %
HMM/GMM/Tandem	NA	10.99	NA	37 %
CTC-7 g	18 %	34 %	29 %	55 %
CTC-9 g	16 %	29 %	27 %	55 %
CTC-14 g	4.3 %	12 %	24 %	47 %
CTC-15 g	1.3 %	3.9 %	22 %	40.8 %

The HMM/GMM and tandem are built using the Hidden Markov Model Toolkit (HTK)[7] version 3.5. This version has improvement for language model to accept more than 64k words and supports DNN model. The results in Table 3 based on $\alpha = 5$, $\beta = 3.8$, beam length 150. We used the HRESULT command of HTK and built two output files (Young et al. 2013), the first file contains the words sequence that is compared to the test set. The second file is generated as a sequence of labels (characters) and it is compared to test set prepared in the same manner (label sequence). The results started to be steady from epoch 30 because the small size of training set (50 epoch per week). The increasing of network hidden layer size 2048 and beam length 150 improved less than 1 %. The CTC results of CER at 15 g is very close to the WER of HMM/GMM baseline (40.8 %, 40 %, 37 %). Although the decoder processes the utterances with no lexicon, the CTC acoustic and language model succeeded to distinguish the different way of writing characters that have the same pronunciation but different ways of writing. For example, letter ت, which is equivalent to letter t in English, has different ways of writing in the middle of word like above and the end of word like ة. When it comes at the end of the word, it can be spoken or silent which has slightly lower probability than pronounced character (Maas et al. 2015). CTC proves ability to overcome this challenge side by side with

---

[7] http://htk.eng.cam.ac.uk/.

the language model for distinguishing the different letters. Figure 3 is sketchy figure to illustrate the DBRNN character probability over time s and collapsed output $k(s)$. k function ignores the blank symbols due to spaces or non-defined characters in the character set, i.e. noise. It ignores the non-blank symbols due time shift of character alignment which produces the same character again. For example, *m nn hathaa* may be collapsed to *mn hatha*.

BDRNN has sometimes the ability to transcribe OOV. Table 4 illustrates a phrase from the test set and the corresponding transcription output by different methods. The word between double brackets لحرمانها is an OOV. The HMM/GMM failed to deal with this word and find the most matching labels which is three instead of one. The CTC recognizes it without altering the meaning of the whole statement. This helps in natural language understanding systems applications (SLU) for constructing meaning from the text.

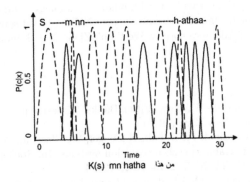

Fig. 3. Figure 3 sketchy figure illustrates collapse function output for blank (_ _ _) and non-blank ( ) symbols.

Table 4. Illustrates a phrase from the test set and the corresponding transcription output by different methods.

Method	Transcription
Test Set	من جانب أكد مرشد الجمهورية الإيرانية آية الله علي خامنئي إن بلاده لن تستسلم أبدا للضغوط الدولية التي ترمي ((لحرمانها)) من حقها في التكنولوجيا النووية
GMM/HMM	من جانب أكد مرشد الجمهورية الإيرانية آية الله علي خامنئي إن بلاده لن تستسلم أبدا للضغوط الدولية التي طب ((إلى حد ما نها)) قادرة تقول انا وليام
GMM/HMM- Tandem	من جانبه أكد مرشد الجمهورية الإيرانية آية الله علي خامنئي إن بلاده لن تستسلم أبدا للضغوط الدولية التي ترمي ((لحرمانها)) من حقيقة تتوج انا وياه
CTC	من جانبه أكد مدير دورية الإيرانية آية الله علي خامنئي إن بلاده لن تستسلم أبدا من الدولة الدولة التي ترمي ((لحرمانها)) من حق في التطوير الجنوبية

7 Conclusion

We presented end-to-end CTC LVCSR Arabic speech recognition using Aljazeera 8 h corpus and language model. We compared the results with HMM/GMM and tandem. The study use CTC source code developed by Stanford University with additional development effort, around 500 line of code for transcription, feature extraction, twitter crawling and results preparation. We intend to have CTC training for 1200 h QCRI corpus Aljazeera broadcast and having different training models of the language model using RNN. We intend develop Arabic transcription for EESEN[8] project setup, which is word base approach using weighted finite state transducer (WFST), to compare the results to CTC lexicon free. Furthermore, the long short term memory (LSTM) may improve the results in Arabic language compared to RNN, which needs more investigation.

Acknowledgment. Many thanks for Luminous technology center (info@luminous-technologies.com) for having full access to server NVidia based setup. Special thanks for QCRI for proving Aljazeera corpus. We thank Ziang Xie, Standard University.

References

Alfred, M.: Signal Analysis Wavelets, Filter Banks, Time-Frequency Transforms and Applications. Wiley, New York (1999)

Ali, A., Zhang, Y., Cardinal, P., Dahak, N., Vogel, S., Glass, J.: A complete KALDI recipe for building Arabic speech recognition systems. In: Spoken Language Technology Workshop (SLT), IEEE 2014. IEEE (2014a)

Ali, A., Zhang, Y., Vogel, S.: QCRI advanced transcription system (QATS). In: Spoken Language Technology Workshop (SLT) (2014b)

Attia, M., Samih, Y., Shaalan, K.F., van Genabith, J.: The floating Arabic dictionary: an automatic method for updating a lexical database through the detection and lemmatization of unknown words. In: COLING (2012)

Bourlard, H., Morgan, N.: Hybrid HMM/ANN systems for speech recognition: overview and new research directions. In: Giles, C.L., Gori, M. (eds.) NN 1997. LNCS, vol. 1387, pp. 389–417. Springer, Heidelberg (1998). doi:10.1007/BFb0054006

Chen, S.F., Goodman, J.: An empirical study of smoothing techniques for language modeling. In: Proceedings of the 34th Annual Meeting on Association for Computational Linguistics. Association for Computational Linguistics (1996)

Diehl, F., Gales, M.J., Tomalin, M., Woodland, P.C.: Morphological decomposition in Arabic ASR systems. Comput. Speech Lang. **26**(4), 229–243 (2012)

Farghaly, A., Shaalan, K.: Arabic natural language processing: challenges and solutions. ACM Trans. Asian Lang. Inf. Process. (TALIP) **8**(4), 14 (2009)

Glorot, X., Bordes, A., Bengio, Y.: Deep sparse rectifier neural networks. In: International Conference on Artificial Intelligence and Statistics(2011)

Graves, A., Fernnndez, S., Gomez, F., Schmidhuber, J.: Connectionist temporal classification: labelling unsegmented sequence data with recurrent neural networks. In: Proceedings of the 23rd International Conference on Machine Learning. ACM (2006)

---

[8] http://arxiv.org/abs/1507.08240.

Graves, A., Jaitly, N.: Towards end-to-end speech recognition with recurrent neural networks. In: Proceedings of the 31st International Conference on Machine Learning (ICML-2014) (2014)

Graves, A., Mohamed, A.-R., Hinton, G.: Speech recognition with deep recurrent neural networks. In: IEEE International Conference on Acoustics, Speech and Signal Processing (ICASSP). IEEE (2013)

Hannun, A., Case, C., Casper, J., Catanzaro, B., Diamos, G., Elsen, E., Prenger, R., Satheesh, S., Sengupta, S., Coates, A.: Deep speech: scaling up end-to-end speech recognition. arXiv preprint. arXiv:1412.5567

Hermansky, H., Ellis, D. W., Sharma, S.: Tandem connectionist feature extraction for conventional HMM systems. In: Proceedings of IEEE International Conference on Acoustics, Speech, and Signal Processing, ICASSP 2000. IEEE (2000)

Hifny, Y.: Unified acoustic modeling using deep conditional random fields. Trans. Mach. Learn. Artif. Intell. 3(2), 65 (2015)

Hinton, G., Deng, L., Yu, D., Dahl, G.E., Mohamed, A.-R., Jaitly, N., Senior, A., Vanhoucke, V., Nguyen, P., Sainath, T.N.: Deep neural networks for acoustic modeling in speech recognition: the shared views of four research groups. IEEE Sig. Process. Mag. 29(6), 82–97 (2012)

Maas, A. L., Xie, Z., Jurafsky, D., Ng, A.Y.: Lexicon-free conversational speech recognition with neural networks. In: Proceedings of the 2015 Conference of the North American Chapter of the Association for Computational Linguistics: Human Language Technologies (2015)

Motlicek, P., Imseng, D., Potard, B., Garner, P.N., Himawan, I.: Exploiting foreign resources for DNN-based ASR. EURASIP J. Audio Speech Music Process. 2015(1), 1–10 (2015)

Nadeu, C., Hernando, J., Gorricho, M.: On the decorrelation of filter-bank energies in speech recognition. In: Eurospeech. Citeseer (1995)

Othman, E., Shaalan, K., Rafea, A.: Towards resolving ambiguity in understanding arabic sentence, In: International Conference on Arabic Language Resources and Tools, NEMLAR. Citeseer (2004)

Radha, V., Vimala, C.: A review on speech recognition challenges and approaches. World Comput. Sci. Inf. Technol. J. 2(1), 1–7 (2012)

Raschka, S.: Python Machine Learning. Packt Publishing, Birmingham (2015)

Shaalan, K.: A survey of arabic named entity recognition and classification. Comput. Linguist. 40(2), 469–510 (2014)

Shaalan, K., Bakr, H.M.A., Ziedan, I.: A hybrid approach for building Arabic diacritizer. In: Proceedings of the EACL 2009 Workshop on Computational Approaches to Semitic Languages. Association for Computational Linguistics (2009)

Shaalan, K., Magdy, M., Fahmy, A.: Analysis and feedback of erroneous Arabic verbs. Nat. Lang. Eng. 21(02), 271–323 (2015)

Young, S., Evermann, G., Gales, M., Hain, T., Kershaw, D., Liu, X., Moore, G., Odell, J., Ollason, D., Povey, D.: The HTK Book (for HTK Version 3.5). Cambridge University Engineering Department, Cambridge (2015)

Young, S., Evermann, G., Gales, M., Kershaw, D., Moore, G., Odell, J., Ollason, D., Povey, D., Valtchev, D., Woodland, P.: The HTK Book. Cambridge University Engineering Department, Cambridge (2013)

Yu, D., Deng, L.: Automatic Speech Recognition. Springer, London (2012)

Agent Productivity Measurement in Call Center Using Machine Learning

Abdelrahman Ahmed[1(✉)], Sergio Toral[1], and Khaled Shaalan[1,2,3]

[1] Electronic Engineering Department, University of Seville, Seville, Spain
abdahm@alum.us.es, storal@us.es
[2] School of Informatics, Edinburgh, UK
[3] The British University in Dubai, Dubai, UAE
Khaled.shaalan@buid.ac.ae

Abstract. We present an application of sentiment analysis using natural language toolkit (NLTK) for measuring customer service representative (CSR) productivity in real estate call centers. The study describes in details the decisions made, step by step, in building an Arabic system for evaluation and measuring. The system includes transcription method, feature extraction, training process and analysis. The results are analyzed subjectively based on the original test set. The corpus consists of 7 h real estate corpus collected from three different call centers located in Egypt. We draw the baseline of productivity measurement in real estate sector.

Keywords: Sentiment analysis · Agent productivity · Natural language toolkit · productivity measurement

1 Introduction

The call centers are the front door of any organization where crucial interactions with the customer are handled (Reynolds 2010). The effective and efficient operations are a key ingredient to the overall success of insource and outsource call center profitability and reputation. It is very difficult to measure productivity objectivity because the agent output, as a firm worker, is the spoken words delivered to the customer over the phone. The team of quality assurance is responsible for listening and evaluating the customer service representatives (CSR). However, the evaluation is handled in a subjective way. In other words, they listen to the recorded calls to evaluate the performance and productivity of the call center agent according to their previous experience. The quality team is responsible for reviewing the calls and measuring the level of service quality in term of conversation evaluation with the customer. In the best case, the quality team may evaluate a small number of the total number of agents calls because of limited number of quality team or huge number of calls or both. The mission

© Springer International Publishing AG 2017
A.E. Hassanien et al. (eds.), *Proceedings of the International Conference on Advanced Intelligent Systems and Informatics 2016*, Advances in Intelligent Systems and Computing 533, DOI 10.1007/978-3-319-48308-5_16

of the quality team is to listen to the agents recorded calls and to use predefined evaluation forms (evaluation check list) (Reynolds 2010) as an objective evaluation. Evaluation forms state the main aspects of the productive call, i.e. saying the greeting, mentioning his/her name, or well describing the product or the service when answering customer inquiries. The evaluation process has many drawbacks because of the following reasons: (1) although the evaluation is performed through predefined check lists, the quality team marks the list according to evaluator perception and previous experience (Judkins et al. 2003). The subjectivity of this evaluation restrains generalizing the evaluation results for the rest of the agents calls. (2) The limited number of quality members may not be able to cover all the agents consistently per time. Accordingly, some agents are evaluated in rush time under very high work pressure while others are evaluated in normal operation. This leads to inconsistent evaluation from agent to another. These reasons and others have drastic impact on agent performance and lead to high turnover which negatively affects call center business. This paper proposes a method of objectively measuring agent productivity through machine learning. The rest of the paper is structured as follows: Sect. 2 discussed the conceptual framework and overview about each framework component. Section 3 details the front end preparation for feature extraction and Sect. 4 explains the experiment carried out. Obtained results are shown in Sect. 5 and finally, Sect. 6 concludes the paper.

2   Proposed Framework

In this section, we review previous literature about productivity measurement as well as general concepts and methods of agent evaluation in call centers environment.

2.1   Productivity Measurement Definition

A productivity measure is commonly understood as a ratio of outputs produced to resources consumed (Steemann Nielsen 1963). However, the observer has many different choices with respect to the scope and nature of both the outputs and resources considered (Card 2006). For example, outputs can be measured in terms of delivered product or services, while resources can be measured in terms of effort or monetary cost (Card 2006). An effective productivity measurement enables the establishment of a baseline against which performance improvements can be measured (Thomas and Zavrki 1999). This is the crucial part in productivity measurement because each call center has its own constructed reality, which differs from one domain to another. Therefore, it requires a dynamic approach of grasping eminent productivity characteristics for each call center, which helps organizations make better decisions about investments in processes, methods, tools, and outsourcing (Card 2006). The productivity measurement can be formulated using the following equation:

$$Productivity = \frac{\text{Agent Output}}{\text{Input Effort}} \tag{1}$$

Most of the call centers focus on the agents working time and schedule analysis as a baseline of performance evaluation. The call centers key performance indicators (KPIs) are widely used to measure agent performance (Reynolds 2010). However, they mainly measure agent discipline and the time duration sitting front the desk. Although the monitoring and reporting call center systems have a wide improvement for better evaluating the agent productivity, the agent output performance is gauged and only measured through call recording systems (Reynolds 2010). The agent output (spoken words) is our objective in this study for analysis and study. The conceptual framework of the study is described in Fig. 1, where the evaluation process is automated from the beginning to the end. The block diagram includes a speaker diarization process, speech recognition, sentiment modeling and classification process. The next section discusses each building block in the framework.

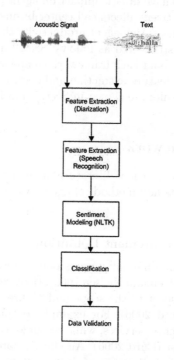

Fig. 1. The study Frame work

2.2   Speech Recognition and Speaker Diarization

Speech recognition systems started in the 80 s and obtained a significant improvement by the era of neural network for machine learning (Yu and Deng 2012). By transcribing the calls into text, the content analysis has become a powerful tool for features prediction and interpretation (Carmel 2005; Othman et al. 2004).

The speech in Arabic language achieved high accuracy in terms of word error rate (WER) (Ali et al. 2014). The word error rate is the main indicator of the speech recognition accuracy (Young et al. 2013). The lower WER, the higher performance of speech recognition (Woodland et al. 1994). The inbound or outbound call is divided into agent talk part, customer part, silences, music on hold and noise. As the agent part is the target of the analysis, it requires a diarization process. The diarization process is some sort of acoustic models that can make sophisticated signal and speech processing to split the one channel mono recorded voices into different speakers (Tranter and Reynolds 2006). It removes silences, music as well as giving clear one speaker voice (Tranter and Reynolds 2006).

2.3   Sentiment Analysis

Once speech is transformed to text, words (features) are the input data like raw material for sentiment analysis. Sentiment analysis refers to detecting and classifying the sentiment expressed by an opinion holder, and considers the word frequency and word probability following specific words (context) (Chen and Goodman 1996). Sentiment analysis, sometimes also called opinion mining, is the way to classify the text based on opinions and emotions regarding a particular topic (Richert et al. 2013). This technique classifies the text in polarity way (on/off/yes/no/good/bad), and it is used for assessing people opinion in books, movies etc. It deals with billions of words over the web and classifies the positive and negative opinions according to the most informative features (word) extracted from the text. Agent productivity is different than opinion mining or emotions classification because productivity is assessment of the output of the agent as mentioned in Eq. (1) regardless emotional effect. One of the most important quality standards of call centers is emotional control of the agent while speaking to the customer (Wegge et al. 2006). This means less emotions or opinion expressions are expressed through the call. Hence, the study proposes sentiment analysis to extract a hidden features that are different than opinion mining and help classifying productivity on stochastic basis. This can be achieved by high accuracy of success instances compared to training set of the input calls.

Sentiment feature is the probability resulted of the class type (productive/non-productive) given probability of class features p(Cx) being x is the features (words). A Naïve Bayes classifier was selected to train the model and classify the results. The Naïve Bayes classifier is built on Bayes theorem by considering that features are independent from each other. Naïve Bayes satisfies the following equation:

$$p(c|x) = \frac{p(x|c)p(c)}{p(x)} \tag{2}$$

where c is the class type (productive/non-productive), and $x_1, x_2, , x_n$ are the text features. The $p(x|c)$ is the likelihood of the features given the class in the training process. Naïve Bayes is a generative model so that the model is trained to generate the data x from class c in Eq. (2). The computation of the term $p(x|c)$ means features prediction to be seen in input x given the class c (Martin and Jurafsky 2000). We ignore $p(x)$ in the equation denominator as it is a constant value that never change. Accordingly, we are looking for maximum value of both of the likelihood value $p(x|c)$ and prior probability $p(c)$ to predict the class probability given input features. Hence the predictive class is as follows:

$$p(c|x) = argmax[p(x|c)p(c)] \tag{3}$$

We calculate the joint probability by multiplying the probability of words given class $p(x|c)$ with class probability $p(c)$ to get the highest features for each class (highest probability) as follows (Murphy 2006):

$$p(C_k|x_1, x_2, , x_n) = p(C_k) \prod_{i=1}^{n} p(x_i|C_k) \tag{4}$$

The next sections will illustrate the training of the classifier and how to satisfy the Eq. (4).

2.4   Data Validity

The data classification accuracy is one of the most important parts in this study. The human accuracy shows that the level of agreement regarding sentiment is about 80 %[1]. However, we cannot consider this as a baseline for two reasons. The first reason, the accuracy is dependent on the domain of the text collected, and varies from one domain to another. For example, the productivity features are perceived in different way than other domains like spam emails or movies review. The second reason, machine learning approach is incomparable to human perception as we could not find similar research in the same domain. Hence, we draw the first baseline for agent call center in real estate in Arabic. For the data validation of the study, the naïve classifier should be able to classify the test set accurately as intended. The accuracy calculation is given by Eq. (5).

$$Accuracy = \frac{F_{cor}^c}{F_{tot}} \tag{5}$$

where F_{cor}^c is the correctly classified features per class and F_{tot} is the total features extracted.

3   Front-End Preparation

Front-end preparation refers to the input data diarization, the transcription of voice into text, and the transcription of the Arabic text into Latin to be ready for training. The following subsections detail each step.

---

[1] http://www.webmetricsguru.com/archives/2010/04/sentiment-analysis-best-done-by-humans.

3.1 Audio Diarization and Speech Recognition

The diarization process was intended to be performed using LIUM diarization toolkit (Meignier and Merlin 2010). It is a java based open source toolkit specialized in diarization using speech recognition models. It required Gaussian Mixture Model (GMM) for training voice and corresponding labels using two clustering states or more according to the number of speakers (number of states equal to the number of speakers). It uses GMM mono-phone to present the local probability for each speaker (Meignier and Merlin 2010). For speech recognition, we uses both of GMM and Hidden Markov Model (HMM) methods but in different configurations. In Arabic, we use 3 HMM states for each phone (Arabic proposed phones are 40 phones), each state is presented by 16 Gaussian models. There are some challenges we have faced when applying this process in our study. The first challenge that the corpus is around 7 h (small corpus) to build mature acoustic model for diarization and speech recognition. The second challenge that the corpus is a telephone conversations (8 k bit). LIUM requires at least 16 k bit to work. Hence, we manually transcribed and split the agent talk.

3.2 Converting the Arabic Text into Latin (Transcription Process)

The Arabic transcription is converted into Latin characters so that we can make processing easier. The character set are 36 character as shown in Table 1.

Table 1. Sample of Arabic letters, corresponding characters transcription and its English equivalent.

Arabic Letter	أ	د	ش	ل	ك
Transcription	ga	d	sh	l	k
Equivalent English	A	D	SH	L	K

The transcription process maps each letter from Arabic to the corresponding Latin character. The next example shows a transcription of a statement:

عليكم السلام ورحمة الله وبركاته

Îykm aslam w r@mt all* w brkat*

Example 1. Sample of Arabic statement transcribed in Buckwalter.

The transcription shown above transforms the statement from right to left (Arabic writing direction) to left to right (Latin). Buckwalter[2] is a powerful open source tool to transcribe the Arabic to Latin, and it is used in many Arabic transcription purposes.

---

[2] www.qamus.org/transliteration.htm.

3.3   Training the Naïve Bayes Classifier

We have manually transcribed and classified the data (training set) into productive/non-productive. For training the classifier, we have to find the probability of maximum likelihood of $p(x|c)$ and $p(c)$. The $p(c)$ is simply calculated as following:

$$p(c) = \frac{N_c}{N_{tot}} \tag{6}$$

where N_c is the number of text files annotated per class divided by total number of files N_{tot}. To calculate the maximum likelihood, we count the frequency of word per class and divide it to overall words counted per the same class (Martin and Jurafsky 2000) as following:

$$p(x|c) = \frac{count(x_i|c)}{\sum count(x|c)} \tag{7}$$

As some words may not exist in a class, the result of $count(x_i, c)$ will be zero. The total multiplied probabilities will be zero as well. A Laplace smoothing (Martin and Jurafsky 2000) is used to avoid this problem by adding one:

$$p(x|c) = \frac{count(x_i|c) + 1}{\sum count(x|c) + 1} \tag{8}$$

To avoid underflow and to increase the speed of the code processing, we use $log(p)$ in Naïve Bayes calculations (Yu and Deng 2012).

4   The Experiment

The corpus consists of 7 h real estate call centers hosted in Egypt, which is collected and transcribed by Luminous Technologies center (info@luminous-technologies.com). We have listened to the calls carefully (30 calls) and categorized them to productive and non-productive. The criterion used in text files annotation is the ability of the agent or CSR to respond to the customer inquiries with right answers and appropriate words (features) being used. The 30 calls are split into smaller audio chunks each audio file duration is around 10–20 s. This process is mandatory for diarization and speech recognition for files decoding process because longer files are failed to be automatically recognized (Meignier and Merlin 2010). Therefore, total files are 500 files divided to training set of 400 files and test set of 10 % of the files (100 files). While the solution proposed is based on machine learning, we avoid the problem of productivity definition and left the quality team in each call center decide the criteria to categorize the calls according to their definition. At the end, a modeling of the training set is built for a unique features according to each call center criteria. The number of productive files was 400 files versus only 100 files for non-productive set. This unbalanced annotation biases the results of probability of each class $p(c)$ in Eq. (3) which gives (scaling) for one class than the other. It should be for

Table 2. The most informative features

Word Index	Feature in BuckWalter	Feature in Arabic	Translation into English	Classification
1	trw@	تروح	To go	Non Productive
2	ambar@	امبارح	Yesterday	Non Productive
3	M∧rf$	معرفش	I have no idea	Non Productive
4	Fwq	فوق	Upper	Productive
5	Alsaylz	السايلز	Sales	Productive
6	Nmrty	نمرتي	My mobile number	Non Productive
7	Aktwbr	أكتوبر	October (City in Egypt)	Productive
8	m%*r	مظهر	The view	Productive
9	alsT@	السطح	The roof	Productive
10	Ktyr	كتير	Expensive	Non Productive

example 400 productive versus 400 non-productive. In this case, we have to use a scaling factor i.e. α to adjust the probability per class $p(c)$. The equation will be as follows:

$$C^{\psi} = argmax[logp(x|c) + \alpha logp(c)] \tag{9}$$

where C^{ψ} is the predicted class $\in C$

We convert the text from Arabic to Latin using Buckwalter script. Then we return back the result into Arabic. The code is developed in python using NLTK naïve classifier (Raschka 2015). The code uses bag-of-words which is an unordered set of words regardless to its position but its frequency (Martin and Jurafsky 2000).

5   Results

The accuracy result of this experiment is 67 % success of the calls classified either productive or non-productive. The corpus (30 calls - 7 h) is small as expected to get higher accuracy for bigger corpus. The test set has been classified as per resulted features extracted from the corpus which is called the most informative features (MIF). MIF is feature or set of features that have very high frequency presenting productive/non-productive words that the decision of classification is built on them. The features have been converted from Buckwalter to Arabic, then to English as following:

The classifier provides the frequency of the whole words in the test set according to each class (productive/non-productive). As shown in Table 2, we selected only the 10 most informative features out of more than 100 features extracted and classified. We try subjectivity to explore the meaning behind classification for better understanding the definition of productivity. For feature number 3 - معرفش or in English saying (**I have no idea**) is non-productive according to lack of awareness of the product or the service. In feature number 6, the agent dictates his/her mobile number over the phone, and this is considered no productive as it consumes much time. The prices for the feature number 10 - (**expensive**) is classified non-productive feature because it drags the CSR in debate and consumes much time with no benefit. Furthermore, it might be an unjustified answer by agent which may indicate less awareness to the market updates and prices changes. This feature may be categorized under the same feature number 3. For productive agents, they mentioned features of the apartments or villas such as (**the view**), (**the roof**) and the city (**October**) can be considered as product awareness. Nevertheless, the feature in itself works perfectly in evaluating the agent for mentioning some selling points through the call. There are other features meaningless, for example, (**to go**), (**yesterday**), (**Upper**), and (**Sales**). We think these may relevant to the error occurred of 33 % of sentiment analysis which is expected from the beginning. The accuracy can be improved by increasing the corpus and getting balanced classification for the training set for productive and non-productive.

6   Conclusion

We have proposed a method of evaluating agents performance using a machine learning approach to evaluate call center CSR objectively. The sentiment analysis is proposed using NLTK naïve classifier to classify the agents as productive/non-productive. The study also proposes end-to-end evaluation system using diarization and speech recognition. There are research gaps for finding another approach of sentiment analysis using logit regression. Comparing the results of this study (generative model) with Logit regression as a discriminative model may give significant explanation in productivity measurement. The statement fragmentation techniques, i.e. fragment significance estimation, are statement context analysis for better evaluating the text contents and context. For Agent evaluation and in order to be more informative, the productivity classification can also be extended to a range of scales rather than binary mode (productive/non-productive). Supervised neural network can also help to obtain better results for problem classifications with less time consumption.

Acknowledgment. Many thanks for Luminous technology center (info@luminous-technologies.com) for the corpus and giving full access to experiment server. Special Thanks for Dr. Kyoko Fukukawa, Bradford University, Bradford, UK and Dr. Yasser Hifny, Helwan University, Cairo, Egypt for their outstanding effort in this paper.

References

Ali, A., Zhang, Y., Cardinal, P., Dahak, N., Vogel, S., Glass, J.: A complete KALDI recipe for building Arabic speech recognition systems. In: 2014 IEEE Spoken Language Technology Workshop (SLT). IEEE (2014)

Card, D.N.: The challenge of productivity measurement. In: Proceedings of the Pacific Northwest Software Quality Conference (2006)

Carmel, D.: Automatic analysis of call-center conversations. In: Ron Hoory, A.R. (ed.) (2005). https://www.researchgate.net/publication/221614459

Chen, S.F., Goodman, J.: An empirical study of smoothing techniques for language modeling. In: Proceedings of the 34th Annual Meeting on Association for Computational Linguistics. Association for Computational Linguistics (1996)

Judkins, J.A., Shelton, M., Peterson, D.: System and method for evaluating agents in call center. Google Patents (2003)

Martin, J.H., Jurafsky, D.: Speech and Language Processing. Prentice Hall PTR, Upper Saddle River (2000). International Edition

Meignier, S., Merlin, T.: LIUM SpkDiarization: an open source toolkit for diarization In: CMU SPUD Workshop (2010)

Murphy, K.P.: Naive Bayes Classifiers. University of British Columbia, Vancouver (2006)

Othman, E., Shaalan, K., Rafea, A.: Towards resolving ambiguity in understanding arabic sentence. In: International Conference on Arabic Language Resources and Tools, NEMLAR. Citeseer (2004)

Raschka, S.: Python Machine Learning. Packt Publishing, Birmingham (2015)

Reynolds, P.: Best practices in performance measurement and management to maximize quitline efficiency and quality. North American Quitline Consortium (2010)

Richert, W., Chaffer, J., Swedberg, K., Coelho, L.: Building Machine Learning Systems with Python. Packt Publishing, Birmingham (2013). 1. GB

Steemann Nielsen, E.: Productivity, definition and measurement. In: Hill, M.N. (ed.) The Sea, vol. 2, pp. 129–164. Wiley, New York (1963)

Thomas, H.R., Zavrki, I.: Construction baseline productivity: theory and practice. J. Constr. Eng. Manag. **125**(5), 295–303 (1999)

Tranter, S.E., Reynolds, D.A.: An overview of automatic speaker diarization systems. IEEE Trans. Audio Speech Lang. Process. **14**(5), 1557–1565 (2006)

Wegge, J., Van Dick, R., Fisher, G.K., Wecking, C., Moltzen, K.: Work motivation, organisational identification, and well-being in call centre work. Work Stress **20**(1), 60–83 (2006)

Woodland, P.C., Odell, J.J., Valtchev, V., Young, S.J.: Large vocabulary continuous speech recognition using HTK. In: 1994 IEEE International Conference on Acoustics, Speech, and Signal Processing, ICASSP-94. IEEE (1994)

Young, S., Evermann, G., Gales, M., Kershaw, D., Moore, G., Odell, J., Ollason, D., Povey, D., Valtchev, D., Woodland, P.: The hkt book (2013)

Yu, D., Deng, L.: Automatic Speech Recognition. Springer, London (2012)

Natural Language Processing for Arabic Metaphors: A Conceptual Approach

Manar Alkhatib[1(✉)] and Khaled Shaalan[1,2]

[1] The British University in Dubai, Dubai, UAE
2015246033@student.buid.ac.ae,
Khaled.shaalan@buid.ac.ae
[2] School of Informatics, Edinburgh, UK

Abstract. Metaphor is a literary device that allows us to express a concept in terms of another. In other words, it is based on similarity between concepts. Metaphorical expressions represent a great variety and they are used in conventional metaphors, which we reproduce and comprehend every day, poetic, novel, and Holy Qur'an. The use of metaphor is ubiquitous in natural language text and it is a serious bottleneck in automatic text understanding, and developing methods to identify and deal with metaphors is an open problem in Arabic natural language processing, especially Machine Translation. Due to the complexities involved in metaphor, it semantically influenced the meaning of machine-translated text. This makes metaphor an important research area for computational and cognitive linguistics, and its automatic identification and interpretation is indispensable for any semantics-oriented Arabic natural language processing. In this paper, we present the challenges of Arabic NLP of metaphors, which is very important in developing a computational NLP-based system for in Classical Arabic, Modern Standard Arabic and Dialect Arabic. We also highlight main problems that arises when translating an Arabic metaphor to another language.

Keywords: Arabic · Natural language processing · Metaphors · Holy qur'an · Modern standard arabic · Dialect arabic

1 Introduction

Arabic is the language of a large part of our planet. It is the main language in 22 countries, spoken by more than 250 million people (Shaalan 2014). It is also the second language in many Islamic countries because it is considered the spiritual language of Islam-one of the world's major religions. It is one of the official languages in the United Nations. However, separating it from another components of Arabic such as grammar, orthography, morphology, literature, writing, reading and conversation is necessary in order to facilitate a focus on its teaching and learning. The different NLP approaches for metaphor interpretation mainly depend on how the relation between the source and the target is viewed as a(n): analogy, novelty, or anomaly. Metaphorical expressions represent a great variety, ranging from conventional metaphors, which we reproduce and comprehend every day such as "السيارة دي لهلوبة بنزين" (This car consumes a lot of

A.E. Hassanien et al. (eds.), *Proceedings of the International Conference on Advanced Intelligent Systems and Informatics 2016*, Advances in Intelligent Systems and Computing 533, DOI 10.1007/978-3-319-48308-5_17

petrol), to poetic, and novel such as "وَلَيْلٍ كَمَوْجِ البّحِر أَرْخَى سُدُولَهُ" (Like heavy waves, long nights 'pon me descend), and to Holy Qur'an (HQ) "وَجَعَلْنَا سِرَاجًا وَهَّاجًا" (and we placed a radiant lamp). In this paper, we present the Arabic metaphors that are in use in Holy Qur'an, Modern Standard Arabic (MSA) and Dialect Arabic (DA), which should be addressed by scholars interested in Arabic computational linguistics.

1.1 Definition of Metaphors

Metaphor is an expression used in everyday life in languages to compare between two dissimilar things. It signifies a situation in which the unfamiliar is expressed in terms of the familiar. In addition, it is a central concept in literary studies. A metaphor is sometimes confused with a simile especially for translators who may translate metaphor into simile or the vice versa. However, it is not too difficult to decide the case of simile because of the correlative existence of the simile markers like "as, similar to and like" which are not found in the metaphor (Eldin 2014).

Metaphors are figurative expressions that have specific cognitive and cultural significances. It has been generally contemplated and analyzed inside the schema of verbal discourse, scholarly works and humanistic studies (Eldin 2014). Lakoff (2008) indicates that no one can imagine any language without metaphor, so it has an inherent value in the use of any language. It is pervasive in everyday life, not only in language but also in thought and action. Our ordinary conceptual system, in terms of which we both think and act, is fundamentally metaphorical in nature.

It is agreed that a metaphor involves two concepts: a source concept, which is related to the words used metaphorically, also called the vehicle of the metaphor, and a target concept, which is what the metaphor is used for and tries to describe, also called the tenor of the metaphor (BEUST 2003).

All languages contain metaphors (so-called in Arabic الاستعارة, Al-Isti'ara or المجاز, Al-Majaz(. A lot of them are used in their own language. Ahmed Yasen (2013) claims that "the metaphorical statement and the corresponding similarity statement cannot be equivalent in meaning because they have different truth conditions". There is a difference between a simile التشبيه, and metaphors (الاستعارة); a simile is where two things are directly compared because they share a common feature. The use of the prefix letter (ك) as or the word like (مثل) compares two words, but a metaphor compares two things directly without using them. It is more eloquent than simile. In translation, it exists in a source language (SL) such as Arabic and in a target language (TL) such as English. However, each language is created from the culture of their people in which the metaphor is differnt (Muttalib 2014).

In the linguistic view of metaphor, there are three components that form a metaphor. They have been referred to as topic (tenor), vehicle and ground. The topic is an entity referred to, and the vehicle is the notion to which this entity being compared. The base in which this comparison is being made from is called the ground. Knowles and Moon (2005) also identify these three components, they stated that metaphor consists of the metaphor (a word, phrase, or longer stretch of language); its meaning (what it refers to metaphorically); and the similarity or connection between the two.

In linguistics, the terms vehicle and tenor are replaced by 'source' and 'target', respectively. The commonality between the target (or tenor) and source (or vehicle) may be referred to as a 'ground' (Eldin 2014).

1.2   Types of Metaphors

Metaphor can function as a means of formatting language in order to describe a certain concept, an action or an object to make it more comprehensive and accurate.

Ahmed Yasen (2013) classifies metaphors, i.e. isti'ara (الاستعارة), into two groups:

1. Declarative metaphors (تصريحية, Tasrihiyya) in which, only the vehicle is mentioned and the tenor is deleted. In this type of isti'ara, the vehicle is explicitly stated and used to make a comparison between two different concepts that share a feature or a property in order to reveal the senses. Declarative Metaphor is also considered as a decorative addition to the ordinary plain speech. It is also used at certain times to achieve aesthetic effects (ibid). For example, in Arabic language one might say (غزالة, gazalleh, a deer) instead of saying (a beautiful woman), which is the vehicle in a metaphor based on the similarity between this animal and the person in terms of beauty and elegance.
2. Cognitive Metaphor (مكنية, Makniya) in which, only the tenor is mentioned and the vehicle is deleted. In this type of isti'ara, the vehicle is only implied by mentioning a verb or a noun that is always accompanies it. The Cognitive Metaphor is used as a means of formatting language in order to describe a certain concept, an action or an object to make it more comprehensive and accurate. In this case, it focuses on the denotation rather than the connotation of the metaphor that addresses the receptor in order to highlight its cognitive function.

Another classification of metaphor is that of Newmark (1988:105-113) which includes six types of metaphors: dead, cliche, stock, adapted, recent and original.

2   Arabic Language

2.1   Classical Arabic

Classical Arabic is the language of the Qur'an. The Quran is held by Muslims to be a single-authored text, the direct words of God (Allah), conveyed by the angel Gabriel to Mohammed 1355–1378 years ago, and later transcribed verbatim to be used as the sole authoritative source of knowledge, wisdom and law. "أَنْزَلْنَاهُ قُرْآنًا عَرَبِيًّا لَعَلَّكُمْ تَعْقِلُونَ إِنَّا", "We have sent it down as an Arabic Qur'an, in order that you may learn wisdom".

2.2   Modern Standard Arabic

Modern Standard Arabic (MSA) is the official Arabic language nowadays. It is either written or spoken without any different in the form. MSA is the language of literature and the media; books, newspapers, magazines, official documents, private and business correspondence, street signs and shop signs – all are written in Modern Standard Arabic. MSA has been developed out of Classical Arabic, the language of the Quran. During the era of the caliphate, Classical Arabic was the language used for all religious, cultural, administrative and scholarly purposes. The linguistic features for this holy book provided unique aspects to MSA from literary, structural and stylistic points of view. MSA omits some classical grammatical constructs, has a stricter word order, uses a simpler numeral system, and obviously includes some more recently coined or borrowed words (Diab and Habash 2014).

2.3 Dialect Arabic

Tongues are the essential type of Arabic utilized as a part of all unscripted talked classifications: conversational, television shows, interviews, and so on. Dialects are progressively being used in digital media like newsgroups, weblogs, discussions and the like. Different countries such as Syria, Lebanon, Jordan, Palestine, Gulf and Egypt though uses Arabic, but in reality they all are differ in dialects. Researchers need to consider this as a major fact and should not assume if a system is designed for Arabic dialect in Syria, then the same could benefit Morocco. Dissimilitude dialect is seen in terms of the variations from one another, which could be phonologically, lexically, morphologically, and linguistically; many sets of variations are commonly muddled. In unscripted circumstances where spoken MSA would typically be required (e.g. television shows on TV), the users more often than not depend on rehashed code-exchanging between their tongue and MSA, as almost all local speakers of Arabic cannot create supported unconstrained talk in MSA (Habash and Rambow 2006), (Diab and Habash 2014). For Example the sentence "how are you?" in different dialect: Egypt dialect "ازايك" (azyk), in Gulf "شو اخبارج" (shw akhbarij), in Syria and Lebanon "كيفكن" (kifakun) and in Jordan and Palestine "كيفكم" (kayfakum) or "شيفكم" (shayfkum).

The following is only one of many that covers the main Arabic dialects (Habash and Rambow 2006):

- **Gulf Arabic (GLF)** includes the dialects of Bahrain, Kuwait, Qatar, United Arab Emirates, Saudi Arabia, and Oman. It is the closest of the regional dialect to MSA, perhaps because the current form of MSA evolved from an Arabic variety originating in the Gulf region
- **Iraqi Arabic (IRQ)** is the dialect of Iraq. In some dialect classifications, Iraqi Arabic is considered a sub-dialect of Gulf Arabic; though it has distinctive features of its own in terms of prepositions, verb conjugation, and pronunciation.
- **Levantine Arabic (LEV)** includes the dialects of Lebanon, Syria, Jordan, and Palestine. It differs somewhat in pronunciation and intonation, but are largely equivalent in written form; closely related to Aramaic.
- **Egyptian Arabic (EGY)** covers the dialects of the Nile valley: Egypt and Sudan. It is the most widely understood dialect, due to a thriving Egyptian television and movie industry, and Egypt's highly influential role in the region for much of 20th century.

- **Maghrebi** Arabic covers the dialects of Morocco, Algeria, Tunisia, Mauritania, and Libya. It is a large region with more variation than is seen in other regions such as the Levant and the Gulf, and could be subdivided further, even though it is heavily influenced by the French and Berber languages.

Socially, it is common to distinguish three sub dialects within each dialect region: city dwellers, peasants/farmers and Bedouins. The three degrees are often associated with a class hierarchy from rich, settled city-dwellers down to Bedouins. Different social associations exist as is common in many other languages around the world.

3 Metaphors in Arabic

3.1 Metaphors in MSA

Metaphors are of great importance for all aspects of life, newspapers, TV, Magazines, Internet news…etc. This applies, in particular, to politics, since politics and political discourse are domains of high abstraction and complexity, and metaphors can provide ways of simplifying complexities and making abstractions accessible, "Politics without metaphor is like a fish without water" (Torlakova 2014). Metaphor (and figurative language) may heavily influence not only people's general perceptions of reality but also impact or manipulate their attitudes, ideas and value systems. News is a representation of the world in language; because language is a semiotic code, it imposes a structure of values, social and economic in origin, on whatever is represented; and so inevitably news, like every discourse constructively patterns that of which it speaks. News is a representation in this sense of construction; it is not a value-free reflection of facts (Torlakova 2014).

For instance, "طار الخبر في المدينة" (the news flew in the city) there is (isti'ara makniya) in which the news which is the tenor is likened to a bird flying in the city which is the vehicle and absent in the sentence. It means that the news spread quickly in the city as a bird flying in the sky very fast. Another examples in the poem: the face of the moon وجه القمر and the hand of the fate "يد القدر". Where the metaphors are the expressions temporarily created by authors or speakers, which can inspire readers or audiences rich imagination; rained pearls and watered flower "أمطرت لؤلؤا وسقت وردا" that means tears as a pearls and cheeks as red flowers.

3.2 Metaphors in Holy Qura'n

The Quran uses a lot of metaphors and figurative language. Throughout the history, different critics have rarely defined this word alike (تشبيه ، tashbeh). The first who is known to have used the term Al-majaz is Abu Ubayda in his book, "Majazal-Quran". However, he did not mean by that the counterpart of haqiqa and figurative language. He mostly uses the word in the formula: "A, its majaz is B", where A denotes the Qur'anic word or phrase and B its "natural" equivalent. In fact, Ubayda was concerned with the first meaning of the term "majaz" which means 'explanatory re-writing' in 'natural' language, of idiomatic passages in the Scripture, while the second sense of "majaz" is figurative language which was developed later. In his Majaz alQur'an, Ubayda does not define majaz, but at the beginning of his work he does give a list of thirty nine cases of deviation from the 'natural' language that can be found in the Qur'an (Alshehab 2015). The following is an instance of the word "آية" Ayah from Qur'an which is interpreted as a metaphor; a device for presenting a concept.

$$\text{(وَآيَةٌ لَهُمُ اللَّيْلُ نَسْلَخُ مِنْهُ النَّهَارَ فَإِذَا هُم مُّظْلِمُونَ)}$$

"The night also is a sign unto them: we withdraw the day from the same, and behold, they are covered with darkness" (Qur'an: Surah Yasin, Verse: 37). Here the tenor (المستعار له) is the appearance of the day after the darkness of the night; the vehicle (المستعار منه) is the appearance of something or an animal after its skin has been pulled off. To strip is "to remove a layer or layers of coverings, clothes, … from sth" (Hornby 1995). The word (يسلخ) use the word withdraws where its meaning is to move away something. Despite the words strip & draw forth are more compatible to the Arabic word but using withdraw gives a metaphorical sense of meaning, i.e. when the day comes, an army of darkness will withdraw (Ahmed Yasen 2013).

Example

$$\text{(وَاخْفِضْ لَهُمَا جَنَاحَ الذُّلِّ مِنَ الرَّحْمَةِ وَقُل رَّبِّ ارْحَمْهُمَا كَمَا رَبَّيَانِي صَغِيرًا)}$$

"HE sends down out of heaven water, and the wadis flow each in its measure, and the torrent carries a swelling scum." (Qur'an: Surah Al-Ra'd, Verse: 17). One of the beautiful metaphors in Qur'an, in this ayah, the expression زَبَدًا رَّابِيًا is the vehicle being the main figure of speech utilized. The tenor, as clarified in the ayah itself, and confirmed by common interpretation (tafaseer such Ma'ariful Quran, is Falsehood (actually, false beliefs) (Alshehab 2015).

Another metaphor in the verse:

$$\text{(لَا تَعْمَى الْأَبْصَارُ وَلَكِن تَعْمَى الْقُلُوبُ الَّتِي فِي الصُّدُورِ)}$$

IT IS not the eyes that are blind, but it is the hearts in the bosoms, that are blind. (Qur'an: Surah Al-Hajj, vese 46). It shows two absolute metaphors. Heart is a well-known idiomatic reference to 'sense', 'affect', and 'feeling'. Blindness is also a rather common representation of the state of senselessness, lack of insight, and affective insensitivity (Al-Ali et al. 2016).

Another metaphor in the verse:

$$\text{(وَاخْفِضْ لَهُمَا جَنَاحَ الذُّلِّ مِنَ الرَّحْمَةِ وَقُل رَّبِّ ارْحَمْهُمَا كَمَا رَبَّيَانِي صَغِيرًا)}$$

"And, out of kindness, lower to them the wing of humility, and say: My Lord! bestow on them they Mercy even as they cherished me in childhood." (Qur'an: Surah Al-Isra'a, Verse 24). The Arabic metaphor here is embedded in the wing of the bird (جناح الطائر) . The wing is used in Arabic as a metaphor in many expressions just as in English language. The far meaning here is evidence in obeying our parents and to be humble with them, saying nothing to annoy them. Their rendering meets the Arabic interpretation in (Al-Jalalain) ألن لهما جناح الذليل من الرحمة لرقتك عليهما "leniency to them the wing of humility and forbearance upon them".

The Qur'anic text is a linguistic miracle and was intended to challenge Arabs who are fluent in classic Arabic analogy, and what makes the Qur'an a miracle, is that it is impossible for a human being to compose something like it, as it lies outside the productive capacity of the nature of the Arabic language. The productive capacity of nature, concerning the Arabic language, is that any grammatically sound expression of the Arabic language will always fall within the known Arabic literary forms of prose and poetry. All of the possible combinations of Arabic words, letters and grammatical rules have been exhausted and yet its literary form with metaphors have not been matched linguistically. The Arabs, who were known to have been Arabic linguists par excellence, failed to successfully challenge the Qur'an (Mohaghegh and Dabaghi 2013).

3.3　Metaphors in Dialect Arabic

Arabic dialects, collectively henceforth Dialectal Arabic (DA), are the day to day vernaculars spoken in the Arab world. The pervasiveness of metaphor expressions in day-to-day speech. The Arabic language is a collection of historically related variants, that live side by side with Modern Standard Arabic (MSA). As spoken varieties of Arabic, they differ from MSA on all levels of linguistic representation, from phonology, morphology and lexicon to syntax, semantics, and pragmatic language use. The most extreme differences are at the phonological and morphological levels. We can see the difference in meaning with the use of the word white in metaphor expressions. For example, the expression سامي قلبو أبيض "Sami's heart [is] white" expresses that Sami is a good person, whereas the expression "كدبة بيضة" "a white lie" means a lie which is "honest and harmless", respectively. Another example is the praise with the word "donkey" in the expression سامي حمار شغل "Sami is a donkey at work" which means "Sami is a very patient and hard worker". The same as the donkey in his patience and hardworking. Describing a person as donkey in Arabic is very offensive which has meaning such as foolish or stupid. In dialect metaphors, we usually use the bad words (bad expressions) to express a good adjective and the vice versa.

　Dialect Metaphors expressions are day-to-day speech, that we are using all the time (Biadsy et al. 2009):

- **Arguments** "مافيك تدافع عن موقفك" (You cannot **defend** your position), the word "تدافع" (defend), it must be for something like country, building. We consider the person in the argument with us as an opponent and we attack his position. Another example, "حكيو صدمني", His speech shocked/blew me
- **Ideas and Speech are Food and Commodities:** "أفكاره لذيذة وحلوة" (His ideas [are] tasty and sweet), it means his ideas are good, while "أفكاره بلا طعمه" (his ideas [are] without taste) it means not useful. Another example "حط الموضوع على نار هادية" (He put the subject on a low fire), it means work on it slowly without a rush. Also "أكل الكتب أكل" (He ate the books), it means he studied the books thoroughly.
- **Time** "إجا وقت الجد" (the time of seriousness has come), it means it is the time to work hard and be serious. Another example "راح آذار" (March went away), it means March ended. Also "رمضان صار على" "الابواب" (Ramadan has reached the doorsteps) it means Ramadan month will start soon.

- **Times are location** "طلع فوق التسعين" (He stepped over ninety), it means his age more than ninety, and "العمر الو حدود" (Age has limits), it means each age period has capabilities.

The dialect metaphors are difficult to understand and comprehend, unless we are familiar with them and we are from the same culture with the same dialect, as each country has its own metaphor dialect.

4 Arabic Metaphors in Natural Language Processing

In general, there is an agreement that a metaphor involves two concepts: a source concept, related to the words used metaphorically, also called the vehicle of the metaphor, and a target concept, which is what the metaphor is used for and tries to describe, also called the tenor of the metaphor. This is best illustrated by examples. In Example (1):

"طار الخبر بالمدينة" (the news flew in the city).

The source of the metaphor is the action of flying, and the target might be the description of the news. Arabic NLP approaches for metaphor interpretation are mainly different in the views of the relation between the source and the target concepts: as an analogy, as a novelty, or as an anomaly. This relation is mostly viewed as an analogy. Thus, interpreting a metaphor requires deeply structured knowledge representations in order to trace back and describe the analogy between concepts. The relation between the source and the target is viewed as a novelty: it is not a pre-existing similarity but one created by the existence of the metaphor. Thus, interpreting it requires the dynamic selection and transfer of knowledge from the source domain to the target domain. Metaphor may also be viewed as a semantic anomaly. In Example(1), there is an anomaly if one considers that "flying" does not normally apply to unphysical thing such as the news, but on physical objects such as a bird. Metaphors are not always anomalies, and the vice versa. In Example (2):

"كلامو قتلني" (his talk killed me)

There is no anomaly, nonetheless "قتلني" (killed me) may be viewed as metaphoric. Only contextual information can help for disambiguating the whole sentence. It is not necessary to focus on relation between the source and the target to interpret metaphors. The meaning of a metaphor can be interpreted and represented by a multi-dimensional vector, exactly like other meanings in the Latent Semantic Analysis approach (BEUST 2003).

Automatic processing of metaphor can be clearly divided into two subtasks: metaphor recognition, and metaphor interpretation. The Metaphor recognition distinguishing between literal and metaphorical language in text. Whereas, the Metaphor interpretation (identifying the intended literal meaning of a metaphorical expression. Both of them have been repeatedly addressed in NLP (Shutova 2010).

The Arabic NLP community has been especially interested in analyzing text-based inputs and outputs, primarily because computers readily accept text inputs in standard orthographies, not inputs in a phonetic alphabet (without special provision). Using text inputs is also standard practice in linguistics among those who study syntax, semantics, pragmatics, and discourse theory. NLP is complementary to and has much to contribute to the success of speech recognition, speech synthesis, and handwriting recognition technologies, but, from the NLP point of view, these are extended capabilities. Consider the following questions: What do we mean by understanding when we talk of language understanding? Can Arabic NLP understand the metaphors? What would it take to convince us that our computer understands Metaphors language? It is hard to answer precisely, since there is no exact formulation of what we mean by ordinary human understanding of language (Hoard 1998).

4.1 Machine Translation for Arabic Metaphors

Machine translation (MT) is simply defined as the translation of human languages, such as English and Arabic, using the area of information technology and applied linguistics. In Arabic language, machine translation is bound to face many problems in producing meaningful coherent translations between Arabic and English language. When evaluating the output of MT, the transferred meaning is the most significant point of focus. Semantics is a very important aspect in translation both as a theory and application; thus, it requires our utmost attention. Besides, Arabic Language is also challenging in that it is a derivational or constructional language rather than a concatenative one. Arabic has relatively free world order, mainly, nominal Sentence Subject-Verb-Object (SVO) and Verbal Sentence Verb-Subject-Object (VSO) (Shaalan 2010). As an example of a sentence that has the same meaning but differ in word order, consider the verbal sentence "طار الخبر في المدينة" (flew the news in the city) versus the nominal "المدينة الخبر طار في" (the news flew in the city).

Metaphor is common in the Qur'an, but its use in scripture takes on a special meaning because of the creedal presumption that the entire Qur'an is the direct articulation of God. The classical scholar Ibn Qutaibah also thought that no one could accurately translate the Holy Quran (HQ) into another language because he felt that there is no other language that can capture as many metaphorical patterns as Arabic. Holy Quran is considered the most rhetorical holy book; it is a challenge for the whole world. Allah sent His HQ to His Messenger Profit Mohammed with an Arabic tongue, it is full of rhetorical styles that cannot be compared to any holy scripture as Allah said in (Qur'an: Surah AL Baqarah, Verse: 23):

"وَإِن كُنتُمْ فِي رَيْبٍ مِّمَّا نَزَّلْنَا عَلَى عَبْدِنَا فَأْتُواْ بِسُورَةٍ مِّن مِّثْلِهِ وَادْعُواْ شُهَدَاءكُم مِّن دُونِ اللّهِ إِنْ كُنْتُمْ صَادِقِينَ"

that might be translated as: produce a surah of the like thereof and call your witnesses besides Allah, if you are truthful. Translation of the Qur'an has always been a problematic and difficult issue in Islamic theology. Since Muslims revere the Qur'an as miraculous and inimitable (i'jaz al-Qur'an), they argue that the Qur'anic text should not be isolated from its true form to another language or form, at least not without keeping the Arabic text along with (Fatani 2006). To build a machine translation for the Holy Quran we have to adherent, to one of the four Sunni Schools Shafi'i, Hanafi, Maliki, and Hanbali, and the rule based approaches will be based on them.

MSA has a wealth of resources in terms of morphological analyzers, disambiguation systems, annotated data, and parallel corpora. In comparison, research on dialectal Arabic (DA), the unstandardized spoken varieties of Arabic, is still lacking in NLP in general and metaphor Machine Translation in particular. Translating MSA is much easier than DA. MSA has a standard orthography and is used in formal settings; while DA does not have standard orthographies, and at the same time it has an increasing presence on the web. For example the translation of the sentence قلبه توقف using google translator produces "His heart stopped", but when we try this sentence in its Dialect Arabic form: ألبو ونف it gives "Albu refraining", or an another form "كلبه وكف" it gives "My dog and keeping". The machine translation system either changes the meaning of the sentence completely or produces an output that has no meaning. The translation of metaphor is not an exception and likewise, we can get the same results. For example, the metaphor "احمر وجه الفتاة خجلاً" which means "the girl blushed away in shyness" can be expressed Dialect Arabic form as "ألب وش البت طماطمايه". When give both metaphorical sentences to google translator as input it produces the following English translation as output:

- Red girl's face in shame   احمر وجه الفتاة خجلاً

- No translation   ألب وش البت طماطمايه

The first sentence is translated literally by google translate and; hence, is not correct. However, for the second one it does not produce any output.

To build an Arabic machine translation we have to follow the three different approaches of rule-based translation systems: direct, transfer, and interlingual (Abdel-Monem et al. 2009). Which one is preferred depends on the availability of the linguists with the required knowledge and background and/or a large tagged datasets for training and testing.

Metaphor holds a number of connotative meanings, and conveying these meanings is not an easy task for MT. The system should match the translation close to the source text's connotative meaning. The problem of translating an Arabic metaphor to another language, is basically derived from finding a target language equivalent that reflects the meaning, effect, and image of the original. For example, in rule-based MT, we have to put rules of translation that match between words, with their indicates of simile "التشبيه" like in.

- Colours
  - red to blush in shyness
  - white to kindness and something good
  - black to darkness, night, and bad things.
  - Yellow to illness and weaknesses

- Vegetables
  - Tomato to shyness
  - Limon to illness
  - Watermelon to bigness

- Objects
  - Sun to shine
  - Moon to shine and beauty
  - Palm to tallness

5 Conclusion

The present paper has tackled the challenges of metaphor in the Qur'an, Modern Standard Arabic (MSA) and Dialect Arabic (DA) within the theoretical framework. Metaphor is an expression in everyday life in languages to compare between two dissimilar things. It signifies a situation in which the unfamiliar is expressed in terms of the familiar. All languages contain metaphors, and a lot of them are used in their own language. The use of a metaphor makes the reader clarify and define the relationship between object and image. Meanwhile, this process serves two purposes: first, it forces the reader to participate actively in the Qur'an, i.e. consider its message and follow its teachings. Second, it gives him knowledge about something he did not know or only partly knew by making it analogous to something he can imagine. Moreover, the concept of the metaphor as a means of transferring meaning continues to be its principal function in current linguistic theories. Therefore, if this is not done, we would not understand them. Thus, as we have seen, the linguistic creativity of the Qur'an is extraordinary. Dialectal Arabic (DA), are the day to day vernaculars spoken in the Arab world. Dialect Arabic metaphors show the pervasiveness of metaphorical expression in our daily speech to the extent it is sometime not easy to recognize them.

In the Arabic NLP of metaphors, we do not have to focus on the relationship between the source and the target concepts, we have to use a symbolic representation in order to provide an understandable Arabic metaphors, as the meaning of a metaphor can be interpreted and represented by a multi-dimensional vector, exactly like other meanings representation in the semantic analysis approach. Furthermore, the Arabic NLP of metaphors requires the same interpretation process as other meanings. It is especially important for translators and researchers in the field of translation, with particular emphasis on the translation of metaphor and simile in Arabic. This might need certain rules in both linguistic and cultural phenomenon especially that Arabic metaphor is not given special consideration and the study of metaphor in the context of machine translation studies has not, unfortunately, kept pace with the discoveries about the nature and role of metaphor in Arabic language.

References

Abdel-Monem, A., Shaalan, K., Rafea, A., Baraka, H.: Generating arabic text in multilingual speech-to-speech machine translation framework. Mach. Transl. 22(4), 205–258 (2009). Springer, Netherlands

Yasen, A.: The commonest types of metaphor in English (2013)

Al-Ali, A., El-Sharif, A., Alzyoud, M.: The Functions And Linguistic Analysis of Metaphor in the Holy Qur'an European Scientific Journal May 2016 edition 12(14) ISSN: 1857 – 7881 (Print) e - ISSN 1857– 7431 (2016)

Alshehab, M.: Two english translations of arabic metaphors in the Holy Qura'n. Arab World English Journal (AWEJ) Special Issue on Translation (4) May, 2015

Biadsy, F., Hirschberg, J., Habash, N.: Spoken arabic dialect identification using phonotactic modeling. In: Proceedings of the EACL 2009 Workshop on Computational Approaches to Semitic Languages, pp. 53–61, Athens, Greece, 31 March 2009. Association for Computational Linguistics (2009)

Diab, M., Habash, N.: Natural Language Processing of Arabic and its Dialects, EMNLP 2014, Doha, Qatar Tutorial (2014)

Eldin, A.: A cognitive metaphorical analysis of selected verses in the Holy Quran. Int. J. Engl. Linguist. **4**(6) (2014). 2014ISSN 1923-869X E-ISSN 1923-8703

Fatani, A.: Translation and the Qur'an. In: Leaman, Oliver. In: The Qur'an: An Encyclopaedia, pp. 657–669. Routeledge, Great Britain (2006)

Habash, N., Rambow, O.: MAGEAD: a morphological analyzer and generator for the Arabic dialects. In: Proceedings of the 21st International Conference on Computational Linguistics and the 44th annual meeting of the Association for Computational Linguistics, pp. 681–688. Association for Computational Linguistics (2006)

Hoard, J.E.: Language understanding and the emerging alignment of linguistics and natural language processing. In: Lawler, J., Dry, H. (eds.) Using Computers in Linguistics, pp. 197–230. Routledge, New York (1998). A chapter overview http://www.routledge.com/linguistics/introduction.html#chapter.7

Hornby, A.S.: Oxford Advanced Learner's Dictionary of Current English, 5th edn. Oxford University Press, Oxford (1995)

Knowles, M., Moon, R.: Introducing Metaphor. Routledge, New York (2005)

Lakoff, G.: The neural theory of metaphor. In: Gbbs, J.R. (ed.) The Cambridge Handbook of Metaphor and Thought, pp. 17–38. Cambridge University Press, Cambridge (2008). http://dx.doi.org/10.1017/CBO9780511816802.003

Mohaghegh, A., Dabaghi, A.: A comparative study of figurative language and metaphor in English, Arabic, and Persian with a focus on the role of context in translation of Qur'anic metaphors. J. Basic Appl. Sci. Res. **3**(4), 275–282 (2013). 2013TextRoa, d Publication

Muttalib, N.: Rendering of Metaphor in Yasin Surat (Chap. 36) of the Holy Quran, No. 210 vol. 2, 2014 AD, 1435 AH

Raii, J.: Metaphor in day-to-day Arabic speech: a conceptual approach. Tishreen University Journal for Research and Scientific Studies - Arts and Humanities Series **31**(1) (2009)

Shaalan, K.: Rule-based approach in Arabic natural language processing. Int. J. Inf. Commun. Technol. (IJICT) **3**(3), 11–19 (2010)

Shaalan, K.: A survey of arabic named entity recognition and classification. Comput. Linguist. **40** (2), 469–510 (2014)

Shutova, E.: Models of metaphor in NLP. In: Proceedings of the 48th Annual Meeting of the Association for Computational Linguistics, pp. 688–697, Uppsala, Sweden, 11–16 July 2010. c 2010 Association for Computational Linguistics (2010)

Torlakova, L.: Metaphors of the Arab spring: figurative construals of the uprisings and revolutions. J. Arabic Islamic Stud. **14**, 1–25 (2014). © Ludmila Torlakova, University of Bergen, Norway

Alserag: An Automatic Diacritization System for Arabic

Sameh Alansary[1,2]([⊠])

[1] Bibliotheca Alexandrina, Alexandria, Egypt
sameh.alansary@bibalex.org
[2] Phonetics and Linguistics Department, Faculty of Arts,
Alexandria University, Alexandria, Egypt

Abstract. Diacritization of written text has a significant impact on Arabic NLP applications. We present an approach to Arabic automatic diacritization that integrates morphological analysis with shallow syntactic analysis. The developed system (Alserag) is a rule based system. The system depends on three modules in order to provide fully diacritized Arabic words namely, morphological analysis module, syntactic analysis module and morph-phonological processing module. The results of the system were evaluated for accuracy against the reference using two metrics; diacritization error rate (DER) and word error rate (WER). The DER measurement was 8.68 % while WER measurement was 18.63 %. The system is benchmarked against three known diacritization systems; Harakat, Mishkal, and Aldoaly.

1 Introduction

Diacritizing Arabic written text is crucial for many NLP tasks, translation can be enumerated among a longer list of applications that vitally benefit from automatic diacritization [1–3]. Arabic diacritics are superscript and subscript diacritical marks (vocalization or voweling), defined as the full or partial representation of short vowels, shadda (gemination), nunation, and hamza [4]. Diacritization helps the reader in disambiguating the text or simply in articulating it correctly. As Arabic is a language where the intended pronunciation of a written word cannot be completely determined by its standard orthographic representation; it rather depends on a set of special diacritics. The absence of these diacritics in Arabic text increases lexical and morphological ambiguity, because one written form can have several vocalizations, each vocalization may have different meaning(s) [5, 6]. However, these diacritics are generally left out in most genres of written Arabic which results in widespread ambiguities in vocalizations and meaning.

Although native speakers are able to disambiguate the intended meaning and pronunciation from the surrounding context with minimal difficulty, it is not the case with automatic processing of Arabic which is often hampered by the lack of diacritics. Several applications can radically benefit from automatic diacritization, such as Text-to-speech (TTS), Part-Of-Speech (POS) tagging, Word Sense Disambiguation (WSD), and Machine Translation [6].

© Springer International Publishing AG 2017
A.E. Hassanien et al. (eds.), *Proceedings of the International Conference on Advanced Intelligent Systems and Informatics 2016*, Advances in Intelligent Systems and Computing 533, DOI 10.1007/978-3-319-48308-5_18

Much work has been done on Arabic diacritization. The actually implemented systems can be divided into two categories [7]: Systems implemented by individuals as part of their academic activities and systems implemented by commercial organizations for realizing market applications. One of the advantages of the first type is that they present some good ideas as well as some formalization. The weak point about these systems is that they are mostly partial demo systems [7]. The following are examples of these systems: Vergyri and [8–11]. For the second category, the most representative commercial Arabic morphological processors are Sakhr, Xerox, and RDI [7]. There are also other available systems as Mishkal Arabic diacritizer[1], and Harakat Arabic diacritizer[2]; they are free Arabic diacritizers which are available online. Finally, on March Google has launched an innovative new Google Labs Arabic tool called Tashkeel, a tool that adds the missing diacritics to Arabic text. Unfortunately, the tool is not available now.

There is another system [12] that has integrated three different proposed techniques, each of which has its own strengths and weaknesses. They are lexicon retrieval, diacritized bigram and SVM statistical-based diacritizer. Most of the previous approaches cited above utilize different sequence modeling techniques that use varying degrees of knowledge from shallow letter and word forms to deeper morphological information. None of the previous systems make use of syntax with the exception of [13] which have integrated syntactic analysis; however, they are not rule based. In this paper, Alserag; an Arabic diacritizer, is proposed. Alserag is based on different steps: retrieval of unambiguous lexicon entries, disambiguating between the different stored possible solutions of the words to realize their internal diacritization through the morphological analysis step (the system tokenizes a text and provides a solution for each token and restore the appropriate internal diacritics from the dictionary), the syntactic processing step that is responsible for the case ending detection is based on shallow parsing and finally the morpho-phonological step that is developed to fulfill the requirements of vowel harmony and assimilation. Section 2 demonstrates the system architecture. Section 3 explains the different applied modules to fully diacritize texts. Section 4 evaluates the output and discusses the results and benchmarking process. Finally, Sect. 5 concludes the paper.

2 System Architecture

In this system, a rule-based approach was adopted. In this section, the different processes that took place in order to convert a plain text into a fully diacritized text will be described. Figure 1 presents the system's overall architecture, where the diacritization is achieved through 7 main phases: (i) Preprocessing which is responsible for auto-correcting the raw text and segmenting the Arabic text into sentences. (ii) Tokenization which is the process of splitting the natural language input into lexical items. (iii) Disambiguation which is a process of choosing the right internal diacritization for

---

[1] http://tahadz.com/mishkal.

[2] http://harakat.ae/.

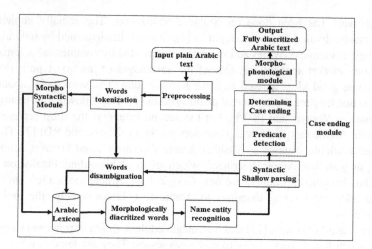

Fig. 1. Architecture of alserag system.

the word from the dictionary. (iv) Name entity recognition (stored in the dictionary and have been obtained from the UNLarium[3] [14]). (v) Syntactic shallow parsing which is an analysis of a sentence by identifying its constituents (NPs, JPs—etc.). (vi) Case ending module which is responsible for predicting the arguments of the predicate and assigning the diacritical marks that are attached to the ends of words to indicate their grammatical function. (vii) Morph-phonological module which is a series of rules that focus on the sound changes that take place in morphemes (minimal meaningful units) when they are combined to form words.

There are two engines that are used in Alserag, the first is Interactive ANalyzer (IAN)[4] which is used in the analysis process, it includes a grammar for natural language analysis. The syntactic processing is done automatically through the natural language analysis grammar, the second is dEep-to-sUrface natural language GENErator engine (EUGENE)[5] which is used in the generation process, It receives the analyzed input and provides a diacritized output without any human intervention, for more details see [15].

3 Development of the System Resources

Alserag depends on two resources; the Arabic diacritized dictionary and a set of linguistics rules. Each one will be described in details in the following subsections.

---

[3] http://www.unlweb.net/unlarium/index.php?unlarium=dictionary.

[4] it is a web application developed in Java and available at http://dev.undlfoundation.org/index.jsp.

[5] it is a web application developed in Java and available at http://dev.undlfoundation.org/index.

3.1 Dictionary

The Arabic diacritized dictionary is a dictionary where Arabic natural language words exist with their diacritics, along with the corresponding linguistic features which describe the Arabic word morphologically, syntactically and semantically. For example, the Arabic word "كتب" 'ktb' 'write' its diacritics "كَتَبَ" 'kataba' and a list of linguistic features such as part of speech, tense, transitivity, person, gender, number, etc. are included in the dictionary. The words in the Arabic diacritized dictionary are extracted from the Arabic dictionary in UNLarium. The process of diacritizing the entries mainly depends on two resources: BAMA and Alkhalil Arabic Morphological Analyzer.

The diacritizing process begins with Buckwalter's analysis. Some words have only one solution, other words have more than one solution and some words couldn't be analyzed in Buckwalter. These are analyzed by Alkhalil which also suggests different solutions to some words. Then, these words are verified manually to select their correct diacritization. Not all of the Arabic diacritized dictionary entries are fully diacritized. Nouns, adjectives, subjunctive and indicative verbs are partially diacritized, since their case endings depend on the context. By default, a present tense verb is marked by a short /o/ (الضمة), in this case it is called indicative (المرفوع المضارع). However, if a present verb is preceded by certain particles, the verb will be marked by a short /a/ (الفتحة), and if the verb ends by one of the three suffixes (ين ون، ان)، the final (ن) will be deleted, in this case it is called subjunctive (المضارع المنصوب). Nevertheless, imperative verb forms and past verb forms are fully diacritized, because their case endings are not affected by the context. Some enhancements have to be made in the Buckwalter solutions. For example, some solutions have a missing vocalization "ْ" before "ا" as in "عالِم" 'EAlim' 'scientist', "مكتبات" 'makotabAt' 'libraries'. So, these missing vocalizations have been added manually.

3.2 The Linguistic Rules

Alserag depends on three modules in order to provide fully diacritized Arabic words namely, morphological analysis module, syntactic analysis module and morphphonological processing module.

Morphological analysis: is responsible for the morphological analysis of Arabic words and assigning the correct POS and the internal diacritization of words which is achieved through two processes; tokenization process and disambiguation process. However, before the tokenization process began, a preprocessing phase should take place over the string stream to fix the most common spelling mistakes, if needed. First, the tokenization algorithm is based on the entries of the dictionary. It starts from left to right trying to match the longest possible string with dictionary entries. The process starts with preventing joined lexical items. Then, it identifies the different suffixes and prefixes that could be attached to each lexical category. Disambiguation rules apply over the natural language list structure to constrain word selection and to correctly disambiguate the POS. They have the following format:(node 1) (node 2) (...)(node n) = P; Where (node 1), (node 2) and (node n) are nodes, and P is an integer expressing the possibility of occurrence. The engine is able to tokenize automatically some words correctly based on the dictionary and assign the correct POS to words. On the other hand, the larger the number of entries in the dictionary, the more the ambiguity during tokenization increases. For example, the sequence "بالفيضانات" 'by the floods' would be automatically segmented as "بال" 'bAl' (worn) + "فيضانات" 'fayaDAnAt' (floods), according to the longest match algorithm, given the fact that the dictionary includes [ART ال] 'Al' (the), [ADJ بال] 'bAl' (worn), [ب bi] 'bi' (by) and [N فيضانات] 'fayaDAnAt' (floods).

Rule in (1) states that adjectives can only be followed by a blank space (BLK), suffix (SFX) or to occur at the end of the sentence (STAIL), where (^) means not. So, [فيضانات] + [ال] + [ب] will be chosen as the appropriate combination.

1. (ADJ)(^SFX,^BLK,^STAIL) = 0;

If words have spelling mistakes or undergo morpho-syntactic changes, rules will investigate the morphological form of those words. For example, the most common mistake in Arabic writing is /Hamza/in the initial position as in "انتزع" ">anotaziE". Rules will investigate the morphological pattern of the wrongly spelled word by the regular expression techniques. For example, if a five-letters word begins with the sequence "ت.(إ||أ|آ)./", the wrong written /Hamza/("إ", "أ" or "آ") will be modified to the correct "إ", according to the Arabic grammar, by the rule in (2), as in the pattern "افتعل" //?ifta?ala//. Then the correct diacritized form will be retrieved from the dictionary "إِنْتَزَعَ" ">inotazaE".

2. ([/..ت.(إ||أ|آ)/],^Hamza_modified)(%y,PUT): = ("1 >"إ, Hamza_modified)(%y);

Second, disambiguation is concerned with preventing the wrong automatic lexical choices and obtaining the right internally diacritized words. Some linguistic indicators can help in solving the lexical ambiguity which are morphological and adjacency indicators.

Morphological indicators: affixation has an important role as the first level of part of speech disambiguation, as prefixes and suffixes are the smallest processing units rules can begin with. Prefixes can help as indicators in determining correct lexical choices. For example, in the word "للدفع" 'liDafoE', the noun "دفع" 'DafoE' (push) is chosen instead of the verb (V) "دَفَعَ" 'DafaE' (to push), since it is preceded by the preposition "ل" 'li' (to) by the rule in (3).

3. (P)(V) = 0;

Adjacency indicators: After disambiguating the POS on the word level, the role of the adjacent word will take its effect as the second level of disambiguation. In this level, disambiguating the part of speech could be controlled by many qualifiers.

Number and Gender qualifiers: as in "وهم يسمون" (and they call). According to the longest match algorithm, the engine will automatically choose the noun "وَهُم" 'wahom' (illusion). But, because it is followed by a plural verb "يُسَمُّونَ" 'yusam ~ uwna' (they call) and subject and verb should agree in number and gender in Arabic, this tokenization will be rejected and will be retokenized as "وَ" 'wa' (and) + "هُم" 'hum' (they) by the rule in (4).

4. (SHEAD) ([وهم],%x) (BLK) (V, ^NUM = %x) = 0;

Functional word qualifier: Particles could be used as indicators for disambiguating the part of speech, as there are particles for verbs and others for nouns. For example, the particle "أي" '>ay ~' (any) is a noun particle. Therefore, in the combination "أي شرط" (any condition), rule in (5) will reject the word "شرط" if it is chosen as a verb "شَرَطَ" '$ar ~ aTa' (slit), since it is preceded by the particle (PTC) "أي" '>ay ~' (any). Then, it will backtrack it to the noun "شَرْط" '$aroT' (condition).

5. (PTC, "أيّ") (BLK) (V) = 0;

The co-occurrence of specific words with words with specific semantic features is used as an indicator. The word "تقلع" has different internal diacritizations that depend on the different meanings, such as "تُقلِع" 'tuqoliE' 'take off' with the semantic feature motion (MOT) and "تَقلَع" 'taqolaE' 'strip' which has the semantic feature contact (CTC). If the verb "تَقلَع" 'taqolaE' 'strip' is followed by a noun such as "طائرة" 'TA} irap' 'airplane' which has the semantic feature artifact (ARF) (Nouns denoting man-made objects). Rule in (6) will reject "تَقلَع" 'strip' "تُقلِع" 'tuqoliE' 'take off'.

6. (V, SEM = CTC) (BLK)(ART)(N, SEM = ARF) = 0;

Syntactic analysis: Shallow parsing is considered necessary for case ending assignment. Transformation rules have been developed to group words under the different phrasal categories. The rules follow the very general formalism α: = β; where the left side α is a condition statement, and the right side β is an action to be performed over α. Phrasal grouping is necessary for identifying the sentence components and linking them by predicate. Then, the different functions of the sentence components can be identified and assigned the suitable case ending. This process will be illustrated in the following. Rules were developed to syntactically mark the phrasal units of the partially diacritized sentence in (7).

7. وَلِذَلِكَ لَمْ تَبْعَثْ اَلدِّرَاسَةَ اَلْمَدْرَسِيَّة لِتَارِيخ اَلْفَرَاعِنَة أَيّ شَوْق بَيْن اَلطَّلَبَة أَوْ اَلْخِرِّيجِين لِلِاِسْتِزَادَة

'wali*`lika lam taboEavo Ald ~ irAsap Almadorasiy ~ ap litAriyx AlfarAEinap ayo $awoq bay ~ na AlT ~ alobap > aw Alxir ~ iyjiyna lilAisotizAdap'
'Therefore, the school study for the Pharaohs history did not provoke any urge between the students or the graduates to increase.'

In sentence (7), different NPs structures are established. The first is established by rule in (8a); it combines the definite article "ال" 'the' and the following noun to project a noun phrase (NP) "اَلدِّرَاسَة" 'Ald ~ irAsap' 'the school study', "اَلْفَرَاعِنَة" 'Alfar-AEinap' 'the Pharaohs', "اَلطَّلَبَة" 'AlT ~ alobap' 'the students' and "اَلْخِرِّيجِين" 'Alxir ~ iyjiyna" the graduates'. The second NP structure is formed by rule in (8b); it combines the indefinite noun "تَارِيخ" 'tAriyx' 'history' and the NP "اَلْفَرَاعِنَة" 'the Pharaohs', the composed NP "تَارِيخ اَلْفَرَاعِنَة" is automatically assigned with the features of its head "تَارِيخ" such as gender, number, animacy and semantic class that are necessary to describe the NP. The third NP structure consists of two coordinated elements and a conjunction; "اَلطَّلَبَة أَوْ اَلْخِرِّيجِين" 'AlT ~ alobap > aw Alxir ~ iyjiyna 'the students or the graduates' by rule in (8c). Moreover, adverbial phrase (AP) consists of the adverb "بين" 'bay ~ na' 'between' and the coordination NP; "الطَّلَبَة أو الخرِّيجين" 'AlT ~ alobap > aw Alxir ~ iyjiyna' is established by rule in (8d). However, the AP "الخرِيجين بين الطلبة أو" 'between the students or the graduates' is considered as an optional argument in the sentence in (7). Next, prepositional phrases (PPs) will be established; the two previously composed NPs "الإستِزاَدَة" '>alisotizAdap' and "تاَريخ الفرَاعِنَة" 'tAriyx AlfarAEinap' '' will be combined with the preceding preposition "لِ" 'li' (to) to form the prepositional phrases (PPs) "للاستزادة" 'to increase' and "للتاريخ الفراعنة" 'for the Pharaohs history' by rule in (8e).

8. (a) (ART, %a)(%y, N): = ((%a)(%y), NP, ANI = %y, GEN = %y, NUM = %y, SEM = %y);
   (b)(^ART, %a)(%y, N)(NP, %x): = (%a)((%y, np)(%x), NP, ANI = %y, GEN = %y, NUM = %y, SEM = %y);
   (c)(NP, %a)(%y, COO)(NP, %x): = ((%a, np)(%y)(%x), NP, ANI = %a, GEN = %a, NUM = %a);
   (d) (ADV, %a)(NP, %x)(%j): = ((%a)(%x), AP)(%j);
   (e) (%x, P, ^pp)(%n, NP): = ((%x, pp)(%n), PP);

Different syntactic functions of the predicate arguments should be identified in order to assign the case ending after the shallow parsing stage. In (7), the arguments of the verb should be identified which will be illustrated in the following.

Verbs and their Arguments Diacritization: The sentence in (7) contains a verb, it is considered as the core of the sentence, since it is the verb that answers the three most important elements of any message - the what, who and when. In terms of the importance of the verb in the diacritization process, verb decides the case ending of the sentence elements. In sentence in (7), the verb "تَبْعَث" 'taboEavo' 'provoke' is a transitive verb that requires two arguments, one to function as a subject "الدِّرَاسَة" 'Ald ~ irAsap' 'study' and another as an object "أيّ" 'ayo' 'any'. After identifying the phrasal constructions, grammar rules have been developed to assign the function and the case ending of the composed verb arguments by rule in (9a). The rule states that, if a verb is followed by two noun phrases and there is gender agreement between the verb and the following noun phrase (NP, GEN = %v), this noun phrase will be considered as the subject of the verb (SBJ). The second will be considered as the object (OBJ). Once the functions of the arguments have been determined, the case ending will be assigned to each noun phrase; the nominative case (NOM) will be assigned to the subject and the accusative case (ACC) will be assigned to the object.

9. (a) (V, TSTD, %v)(NP, GEN = %v, ^CAS, %n)(PP, %a)(NP, ^CAS, %n2): = (%v) (SBJ, CAS = NOM, %n)(%a)(OBJ, CAS = ACC, %n2);
   (b) (لم, %a)(%x, PRS, ^MOO): = (%a)(MOO = JUS, %x);

In the rule in (9a), the nominative and the accusative cases have been assigned to the heads of the two composed NPs; the words "الدِّرَاسَة" (CAS = NOM) and "أيّ" (CAS = ACC). Rule in (9b) assigns the mode of the verb "تَبْعَث" as jussive (JUS) "تَبْعَث", because it is preceded by a jussive particle, as illustrated in (10).

10. وَلِذَلِكَ لَمْ تَبْعَثِ اَلدِّرَاسَةُ اَلْمَدْرَسِيَّة لِتَّارِيخ اَلْفَرَاعِنَة أَيَّ شَوْق بَيْن اَلطَّلَبَة أَوِ اَلْخِرِّيجِين لِلِاسْتِزَادَة

The modifiers are diacritized accordingly. The genitives such as "muDAf > ilayhi" 'مضاف إليه' and the constituents after prepositions do not depend on the case ending of the preceding elements. In sentence in (7), genitive case (GNT) is assigned to genitives, as "تَّارِيخ" and "إِسْتِزَادَةً", "اَلطَّلَبَة", "شَوْق" and "اَلْفَرَاعِنَة".

Adjectives, coordinated elements and nouns in apposition are assigned the same case ending of the preceding element. The adjective "اَلْمَدْرَسِيَّة" 'Almadorasiy ~ ap' is assigned with the same case ending of the preceding noun "اَلدِّرَاسَةُ" 'Ald ~ irAsap', so it is assigned nominative case "اَلْمَدْرَسِيَّةُ". As for the coordinated elements, as in "اَلْخِرِّيجِين اَلطَّلَبَة أَو", the case of the NP "اَلطَّلَبَة" which is genitive is assigned to the NP "اَلْخِرِّيجِين". However, in Arabic, masculine plural noun ending with "ين" suffix, does not permit the genitive case marker (kasra), its genitive case is marked by fatha "ó" 'a'. The final diacritization for the sentence in (7) is as in (11).

11. وَلِذَلِكَ لَمْ تَبْعَثِ اَلدِّرَاسَةُ اَلْمَدْرَسِيَّةُ لِتَّارِيخِ اَلْفَرَاعِنَةِ أَيِّ شَوْقٍ بَيْنَ اَلطَّلَبَةِ أَوِ اَلْخِرِّيجِينَ لِلِاسْتِزَادَةِ

Nominal sentences Diacritization: Nominative case is directly assigned to the topic of the sentence (noun or noun phrase in the beginning of sentences), because it is considered as "مبتدأ" 'mobtadaa'. Since Arabic is a free word-order language, comment may precede topic in nominal sentences such as the sentence in (12).

12. في اَلْحَدِيقَة بَيْت 'A house in the garden'

The case of the topic "بَيْت" 'A house' can be detected in the system in the case of the prepositional phrase "في اَلْحَدِيقَة" "خبر شبه الجملة" 'in the garden' precedes it by rule (13).

13. (SHEAD,   %x)(%c,   PP)(NP,   ^CAS,   %y): = (%x)(%c)(%y,   mobtadaa, CAS = NOM);

Rule in (13) states that if a prepositional phrase "في اَلْحَدِيقَة" comes in the beginning of the sentence is followed by a noun phrase "بَيْت", where (SHEAD) means beginning of the sentence. This noun phrase "بَيْت" is assigned with nominal case (NOM).

However, these nominal phrases cases change if Anna and her sisters precede them. In the example, "إِنَّ في اَلْحَدِيقَة بَيْتًا" '<in ~ a fiy AlHadiyqap bayotAF', the NP "بَيْت" 'bayot' 'house' became accusative.

Morpho-phonological process: Many morpho-phonological alternations occur in Arabic due to the concatenative nature of Arabic morphology, the interaction between morphological and phonological processes is usual. There are two cases where morpho-phonological change is necessary; vowel harmony and assimilation necessity. Vowel harmony takes place in the diacritization process (i.e. phonological). For example, a morpho-phonological rule is necessary for "لهُ" 'lahu' 'for him' that consists of two morphemes "لِ" 'li' + "هُ" 'hu', to change the vowel "ó" 'i' in "لِ" to "ʼʼ" 'a' to be more harmonious with the vowel "ó" 'u' on the suffix "هُ". Moreover, the phonological Arabic system doesn't permit the "moon letters" '(ا-ب-غ-ح-ج-ك-و-خ-ف-ع-ق-ي-م-ه)' to be assimilated with the /l/ of the definite article "الـ" 'Al' 'the', as they are not near in the place of articulation, but they can assimilate with the other Arabic alphabets which are called "sun letters". When the definite article is followed by a sun letter, the /l/ of the Arabic definite article al- assimilates to the initial consonant of the following noun, resulting in a doubled consonant (phonologically) which is orthographically expressed by putting a shaddah 'ـّ' on the consonant after اﻟ. For example, the word "الصباح" 'the morning' before applying the morpho-phonological rule, is diacritized as "الصّبَاح" 'AlsabAH'. Another rule adds a shaddah before the vowel ("ـَ", "ـُ" or "ـِ"), if the diacritical mark is on a sun letter, it would be diacritized "الصّبَاح" 'Als ~ abAH'.

4 Evaluation and Benchmarking

The corpus has been selected from the International Corpus of Arabic (ICA). The selected corpus size is 400,000 Modern Standard Arabic words; they are divided into 300,000 words as tuning data and 100,000 words as testing data. The selected texts are from different sources; Newspapers, Net Articles and Books representing the following genres; politics: 148,211, miscellaneous: 100,253, child stories: 57,174, economy: 34,930, society: 32,955 and sports: 26,477.

The results were evaluated automatically for accuracy against the reference which is a fully diacritized texts by Arabic linguist using the following two metrics; diacritization error rate (DER) which is the proportion of characters with incorrectly restored diacritics. Word error rate (WER) which is the percentage of incorrectly diacritized white-space delimited words: in order to be counted as incorrect, at least one letter in the word must have a diacritization error.

These two metrics were calculated as: (1) all words are counted excluding numbers and punctuators, (2) each letter in a word is a potential host for a set of diacritics, and (3) all diacritics on a single letter are counted as a single binary (True or False) choice. Moreover, the target letter that is not diacritized is taken into consideration, as the output is compared to the reference.

In addition to calculating DER and WER, the evaluation system calculates internal diacritics and case ending separately. Alserag results were compared with the output of other three known diacritization systems; Harakat, Mishkal, and Aldoaly as they are the only available systems. The outputs of these three systems were evaluated using the same data. Table 1 shows benchmarking of the whole data of Alserag among the other three systems.

According to the results obtained by the benchmarking process, our system scored the least error rate followed by Harakat and Mishkal and finally Aldoaly which scored over 80 % error rate. Future plan is associated with improving parsing phase as it is the main source of problems that raised DER and WER. In addition, it is planned to perform the evaluation and benchmarking of Alserag by using the same dataset of LDC (Arabic Treebank) used by more robust systems such as Sakhr, RDI and Microsoft system in the evaluation process so at least we can compare results published by such systems.

Table 1. Benchmarking of the whole data of alserag among the other three systems.

	Alserag	Harakat	Mishkal	Aldoaly
Int.	5.77%	43.30%	32.53%	80.92%
C.E	14.77%	16.23%	31.15%	89.72%
DER	8.68%	37.63%	32.24%	82.76%
WER	18.63%	43.49%	65.00%	97.87%

5   Conclusion

The paper presents an automatic diacritization system Alserag that is developed based on the rule- based approach which is considered as our contribution to the subject of automatic diacritization. All of the other available systems that were mentioned are statistical based. The results of the system were evaluated against the reference. The DER was 8.68 % while WER measurement was 18.63 %.

References

1. Smr, O.: Yet Another Intro to Arabic NLP (2005). http://ufal.mff.cuni.cz/~smrz/ANLP/anlp-lecture-notes.pdf
2. Rashwan, M., Abdou, S., Rafea, A.: Stochastic arabic hybrid diacritizer. In: IEEE Transactions on Natural Language Processing and Knowledge Engineering, pp. 1–8 (2009)
3. Attia, M., Rashwan, M.A.A., Al-Badrashiny, M.A.S.A.A.: Fassieh®, a semi-automatic visual interactive tool for morphological, PoS-tags, phonetic, and semantic annotation of arabic text corpora. IEEE Trans. Audio Speech Lang. Process. 17(5), 916–925 (2009)
4. Maamouri, M., Bies, A., Kulick, S.: Diacritization: A Challenge to Arabic Treebank Annotation and Parsing. Linguistic Data Consortium, University of Pennsylvania, USA (2006)
5. Bouamor, H., Zaghouani, W., Diab, M., Obeid, O., Oflazer, K., Ghoneim, M., Hawwari, A.: A pilot study on arabic multi-genre corpus diacritization annotation. In: Proceedings of the Second Workshop on Arabic Natural Language Processing, pp. 80–88. c2014 Association for Computational Linguistics, Beijing, China (2015)
6. EL-Desoky, A., Fayz, M., Samir, D.: A smart dictionary for the arabic full-form words. IJSCE 2(5) (2012). ISSN: 2231-2307
7. Al Badrashiny, M.: Automatic Diacritizer for Arabic Text. A Thesis Submitted to the Faculty of Engineering. Cairo University in Partial Fulfillment of the Requirements for the Degree of master of science in electronics and electrical communication (2009)
8. Vergyri, D., Kirchhoff, K.: Automatic diacritization of arabic for acoustic modeling in speech recognition. In: COLING Workshop, Geneva, Switzerland (2004)
9. Ananthakrishnan, S., Narayanan, S., Bangalore, S.: Automatic diacritization of arabic transcripts for asr. In: Proceedings of ICON-2005, Kanpur, India (2005)
10. Zitouni, I., Sorensen, J.S., Sarikaya. R.: Maximum entropy based restoration of arabic diacritics. In: Proceedings of the 21st International Conference on Computational Linguistics and 44th Annual Meeting of the Association for Computational Linguistics (ACL), Workshop on Computational Approaches to Semitic Languages, Sydney-Australia (2006)

11. Habash, N., Rambow, O.: Arabic diacritization through full morphological tagging. In: Proceedings of the 8th Meeting of the North American Chapter of the Association for Computational Linguistics (ACL), (HLT-NAACL) (2007)
12. Shaalan, K., Abo Bakr, H.M., Ziedan, I.: A hybrid approach for building Arabic diacritizer. In: Semitic 2009 Proceedings of the EACL Workshop on Computational Approaches to Semitic Languages (2009)
13. Shahrour, A., Khalifa, S., Habash, N.: Improving arabic diacritization through syntactic analysis. In: Proceedings of EMNLP, Lisbon (2015)
14. Alansary, S.: MUHIT: A multilingual harmonized dictionary. In: The 9th Edition of the Language Resources and Evaluation Conference, Reykjavik, Iceland, 26–31 May 2014
15. Alansary, S.: A Suite of Tools for Arabic Natural Language Processing: A UNL Approach, the special session on Arabic Natural Language Processing: Algorithms, Resources, Tools, Techniques and Applications, (ICCSPA 2013), Sharjah, UAE (2013)

Using Text Mining to Analyze Real Estate Classifieds

Sherief Abdallah$^{(\boxtimes)}$ and Deena Abu Khashan

The British University in Dubai, Dubai, United Arab Emirates
sharic@ieee.org

Abstract. There has been an explosion of websites that manage classifieds in general, and real estate listings in particular. Many brokers have adapted their operation to exploit the potential of the web. Despite the importance of the real estate classifieds, there has been little work in analyzing such data. In fact, we are not aware of any work that attempted to analyze the textual data in real estate classifieds using data mining techniques.

In this paper we propose a data mining process that exploits the textual data in real estate classifieds. We conduct the analysis on a large data set, which we gathered from three different property websites. We show that our process exploits the unstructured and ungrammatical textual features to significantly improve the prediction accuracy of a real estate unit price. We also illustrate how text mining combined with linear regression can be used to identify keywords that affect the price negatively or positively.

1 Introduction

There has been an explosion of websites that manage classifieds.[1] We focus in this paper on real estate classifieds, which describe a real estate unit that is available either for rent or sale. Although at some point it was feared that such a new trend in advertising properties may threaten the profitability of traditional brokerage companies, many brokers have adapted their operation to exploit what the web has to offer; and they integrated the web technology into their process [5]. Nowadays, many large brokerage companies developed their own websites to list their properties, in addition to listing their properties on 3rd party web portals. In Dubai, where the real estate market constitutes 12.5 % of the Gross Domestic Product (GDP),[2] several major website portals targeted real estate

---

[1] Criagslist (www.craigslist.com) is a clear example.

[2] "Dubai's GDP climbs 4.4 %", Khaleej Times, 12th June 2013.

© Springer International Publishing AG 2017
A.E. Hassanien et al. (eds.), *Proceedings of the International Conference on Advanced Intelligent Systems and Informatics 2016*, Advances in Intelligent Systems and Computing 533, DOI 10.1007/978-3-319-48308-5_19

classifieds.[3] Furthermore, major real estate brokering companies list classifieds on their own websites.[4]

Despite the importance of the real estate classifieds, there has been little work in analyzing such data. In particular, most of the previous work that applied data mining to real estate data focused on structured attributes such as number of bedrooms, area, location, etc. [7] (a broader view of related work is given in Sect. 5). However, for classifieds, the unstructured and ungrammatical[5] textual attributes (such as the classified title and description) are important components of a classified that should not be ignored.

In this work we apply text mining, along with regular data mining, to analyze real estate classifieds. Our aim is to answer two important research questions:

- Can the use of text mining improve the accuracy of predicting the price of a real estate classified?
- Can we identify which keywords affect the price of a real estate unit either positively or negatively?

Answering these questions will be very valuable for real estate agents, and even home owners who directly post classifieds. Predicting a more accurate price, for a real estate unit, that takes into account the textual unstructured data (not just the structured data) will prevent the stakeholder from overestimating or underestimating the price. Further more, by identifying the important keywords, the stakeholder can refine the unit description and title to better reflect the price being asked.

We conduct the analysis on a large data set (+50 K records) of real estate classifieds, which we gathered from three different websites that post real estate classifieds. We show that our proposed approach significantly improves the prediction of a property price. We also illustrate how text mining combined with a linear regression model can be used to identify keywords that affect the price negatively or positively.

The rest of the paper is organized as follows. The following section describes how the data was collected and prepared. We then explain our proposed data mining process for predicting the price of real estate units using text mining and linear regression model. This is followed by the evaluation and analysis of our proposed approach using the collected data. We then discuss the related work and conclude our paper.

2   Data Preparation

We extracted our data from three major websites that post on-line residential real estate classifieds in the city of Dubai, United Arab Emirates. The data

---

[3] Examples include Gulf News Ads (www.gnads4u.com), Dubizzle (www.dubbizle. com), Bayut (www.bayut.com), and Property Finder (www.propertyfinder.ae).

[4] Such as such as Better Homes (www.bhomes.com) and Hamptons (www.hamp tons.ae).

[5] Ungrammatical means the text does not strictly conform to grammatical rules.

Table 1. The features of the data set.

Name	Type	Description	Number of values
Type	Nominal	Type of classified (flat/villa, for sale/rent)	4
Beds	Integer	Number of bedrooms	14
Location	Nominal	Neighborhoods	161
Title	Text	Title of the classified	-
Description	Text	Description of the property	-
Price	Integer	The renting or selling price of the property	-

was collected in the period from February 2011 to April 2011, using our own web crawler.[6] The collected data contained both apartments (flats) and villas (houses) that are offered for either sale or rent. A total of 66,388 records were extracted. Table 1 illustrates the extracted features.

The extracted data was then cleaned as follows. Records with unreasonably low price/rent were removed (less than AED 10,000).[7] Some unwanted texts were removed from the "Description" feature and replaced with white spaces, including HTML tags, email addresses, website addresses, and few irrelevant phrases related to contacting agents (please contact, for more information, for further information, for international call please dial).

In our analysis we partitioned the data in to 6 different subsets depending on the type: one data set for each of the four types (renting apartments, renting villas, selling apartments, and selling villas), one data set for all the ads for rent (apartments or villas), and one data set for all the ads for sale (apartments or villas).

3   Using Text Mining to Analyze Classifieds

Figure 1 illustrates the process we propose to analyze real estate classifieds data. The process was implemented using RapidMiner.[8] The following sub-sections describe the different components of our proposed approach.

3.1   Text Mining

The text mining stage converts the text attributes to numerical features. As we mentioned earlier, The purpose of applying text-mining is to discover the effect of the hidden information involved in "Title" and "Description" features that

---

[6] While more sophisticated crawlers do exist, [13], for our purposes we preferred a simple and efficient crawler.

[7] The rental price in Dubai is annual, and the currency is Arab Emirates Dirham (AED). US Dollar = 3.67 AED.

[8] RapidMiner is an open-source system for data mining that allows building a rich data-flow process of data-mining operators (http://rapid-i.com).

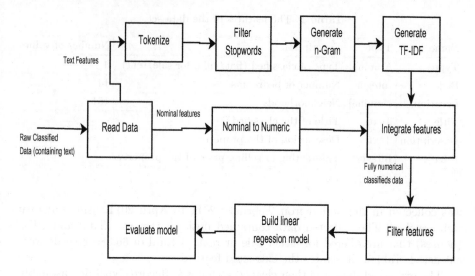

Fig. 1. The data mining process we used to analyze real estate classifieds.

might enhance the accuracy of predicting a property's price. First, the text is tokenized by splitting the text into sequence of tokens, words. The characters of every single word (token) is converted to lowercase. Then, stop-word tokens are removed using a predefined stop-word list. Also any word that is shorter than two characters is removed. This is followed by the generation of term n-Grams of tokens. A term n-Grams is a series of successive tokens of length n. The last step of the text mining process is generating the Term Frequency-Inverse Document Frequency (TF-IDF) for each token. TF-IDF counts how many times a particular token (n-Gram) appears in text, which is then inversely weighted by how common the token is across different classifieds. The output of the text-mining model is thousands of new numerical features. Each feature corresponds to a token, where the feature value is the TF-IDF of the token.

3.2   Feature Selection and Preprocessing

In the second phase, the features are filtered to reduce their numbers. Here we use the correlation between each feature and the target attribute (the property price) as the feature weight. Only the features with minimum weight threshold are then selected (we have experimented with different threshold values as we show later). Also to facilitate linear regression, nominal features (such as the "Type" and "Location") are converted to binary features, where each feature corresponds to a value of the original nominal attribute. For example, "Dubai Marina" is a possible value of the location attribute. After pre-processing we have a binary feature "loc_dubai_marina", which equals 1 if and only if the location attribute equals "Dubai Marina".

3.3   Linear Regression

Finally the linear regression model learning algorithm is applied to the pre-processed dataset. The linear Regression (LR) model assumes the price of a property is a linear combination of the property features.[9] The learning algorithm finds the best weight for each feature based on the dataset. The weight intuitively reflects the feature's effect on the price. For example, when a feature has a positive weight value, it means that the feature works toward increasing the price. Similarly, having a feature with negative weight has the effect of decreasing the price. We show in the following section how this intuition helps in identifying important keywords.

4   Analysis

After building the LR model and using it for prediction, the evaluation metrics are calculated. Table 2 displays the Root Mean Squared Error (RMSE), and the Correlation Coefficient (CC) for the different datasets without using the textual features (i.e. relying primarily on structured features).

Table 2. Performance measurements of linear regression using only structured features.

Dataset	RMSE	CC
Type 0	110,263.85	0.479
Type 1	204,650.55	0.543
Type 2	8,676,287.33	0.296
Type 3	4,756,663.73	0.802
Renting	144,483.34	0.586
Selling	7,738,338.21	0.501

As shown in Table 2, the RMSE for Type0 and Type1 are much lower than the RMSE for Type2 and Type3. This is expected because Type0 and Type1 subsets represent renting price of real estate property and Type 2 and Type3 subsets represent selling price. On the other hand, Type2 dataset has the greatest RMSE and the lowest linear correlation between regular features and the price feature. Also, although the dataset of Type3 has the highest correlation, its RMSE is relatively high. Having high correlation means that there is a strong linear relationship between the predicted price and the other regular (structured) features so that when values of the regular features increase (decrease) the price value increases (decreases) as well.

Looking at the LR model for each data subset and analyzing it, we found (not surprisingly) that the features related to locations have the biggest effect on

---

[9] A common assumption in automatic real estate valuation.

Table 3. Results of linear regression experiments with and without text mining

Type0 Renting apartments dataset	LR	Text Mining + LR	
		Uni-gram	Bi-gram
Selected weight	-	$w >= 0.2$	$w >= 0.3$
No. of features	122	321	343
RMSE	110,263.85	82,044.01	61,715.12
CC	0.479	0.757	0.871

Type1 Renting villas dataset	LR	Text Mining + LR	
		Uni-gram	Bi-gram
Selected weight	-	$w >= 0.7$	$w >= 0.16$
No. of features	92	558	548
RMSE	204,650.55	99,021.76	89,710.59
CC	0.543	0.914	0.93

Type2 Selling apartments dataset	LR	Text Mining + LR	
		Uni-gram	Bi-gram
Selected weight	-	$w >= 0.06$	$w >= 0.25$
No. of features	87	503	407
RMSE	8,676,287.33	5,490,825.94	3,853,490.15
CC	0.296	0.797	0.906

Type3 Selling villas dataset	LR	Text Mining + LR	
		Uni-gram	Bi-gram
Selected weight	-	$w >= 0.15$	$w >= 0.2$
No. of features	76	331	508
RMSE	4,756,663.73	3,778,524.95	3,750,734.50
CC	0.802	0.88	0.882

Renting apartments and villas dataset	LR	Text Mining + LR	
		Uni-gram	Bi-gram
Selected weight	-	$>= 0.2$	$>= 0.2$
No. of features	143	237	361
RMSE	144,483.34	105,307.09	96,334.73
CC	0.586	0.807	0.841

Selling apartments and villas dataset	LR	Text Mining + LR	
		Uni-gram	Bi-gram
Selected weight	-	$>= 0.1$	$>= 0.1$
No. of features	102	239	283
RMSE	7,738,338.21	5,788,073.98	4,440,108.38
CC	0.501	0.762	0.868

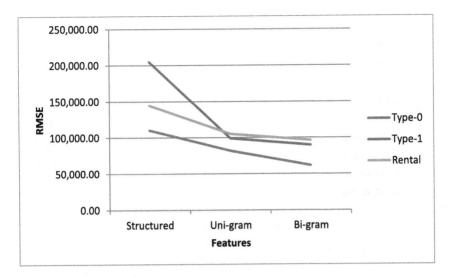

Fig. 2. The RMSE, for the three datasets related to renting property, declines as more sophisticated textual features are added.

increasing or decreasing the price. In the second part of our experiments we apply the linear regression model after the text mining process. The experiments were repeated twice for each dataset. First experiment considered uni-grams tokens and the second experiment considered bi-grams tokens. Table 3 shows the results.

For better illustration, Figs. 2 and 3 visualize the decline in RMSE for the different datasets. As more textual features are incorporated, from only structured features to uni-grams and finally to bi-grams, the RMSE is consistently declining across the different datasets. This clearly confirms that exploiting textual features in real estate classifieds can greatly improve the accuracy of price prediction.

Few points are worth noting. The RMSE (even after the decline) is relatively high. This is due to the wide variety of offered units. For example, the price of a 2-bedroom apartment for sale in Dubai Marina (a neighborhood in Dubai) ranges from just above 1 million, to over 25 million AED (one of the 25+million unit is a luxurious furnished penthouse). Given the variety, an RMSE of less than 5 million AED (as shown in Fig. 3 with bi-gram textual features) is actually very good.

To understand why text mining reduced the RMSE, we investigated the linear regression model to identify which words affected the price positively or negatively. Some of the words that affected the price are related to the location. While the location attribute in the original dataset[10] did specify the location of the unit in a structured manner, the words that were discovered through text-mining were at a finer grain. For example, there is a location code reserved for

---

[10] Recall that the location attribute was converted to binary attributes corresponding to each location value, as we explained in Sect. 3.2.

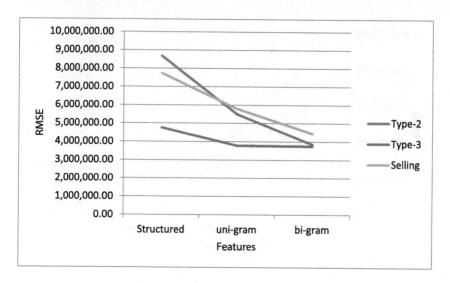

Fig. 3. The RMSE, for the three datasets related to buying property, declines as more sophisticated textual features are added.

Table 4. Sample of words (uni-gram and bi-gram) that affect the price of a real estate classified (either positively or negatively)

Dataset	Sample of words affecting positively	Sample of words affecting negatively
Type-0	investment, balcony	bus, unfurnished
Type-1	wardrobes, driving_mall	fronds_palm, truly_luxurious
Renting	shores_palm, signature_villas	european_miditranian,building
Type-2	hotel_barsha, penthouse_car	road_plot, ground_floor
Type-3	views_beachfront, billiard	palm_offer,school_springs
Buying	full_building, mixed_jumeirah	apartment

Palm Jumeirah. Through text mining and linear regression, the textual feature "fronds_palm" (which refers to a neighborhood within Palm Jumeirah) was discovered to affect the price negatively. Table 4 lists a sample of the discovered important words that affect the price of a real estate unit either positively or negatively.

5   Related Work

Due to the importance of valuating a real estate property, there has been extensive research on automatic valuation [2,3,9,11,14,15]. Most of that work used hedonic models, which assumed the price can be predicted from a combination (usually linear) of the property (structured) features. Traditional hedonic models are based on human expertise, where the model parameters are usually hand-coded by experts, unlike our proposed method here, which uses data mining. There has been growing literature on the use of data mining techniques to analyze real estate data [4,6–8,10], however, most of the previous work focused only on structured features and ignored textual features. We review sample of these works in the remainder of this section.

One of the early works [10] used decision tree and neural network techniques to predict the sale price of a house. The analysis used data with 15 numerical features that represent the houses' characteristics plus a categorical feature that corresponds to the address. The dataset consisted of 1000 records that were collected from the houses' sales transactions in Miami, US. Unlike our work, the analysis focused only on properties for sale (did not include rentals), used only structured features (no text mining) and relied on a much smaller dataset (compared to our +50,000 records). A broader analysis was conducted in [17], covering 295,787 transactions from four cities in the US. Again, only numerical features were used (although more extensive features were used, almost 200) and no textual features were used (also despite attempting to predict the price, no performance criterion was reported). A more recent work [7] proposed Adaptive Neuro Fuzzy Inference System (ANFIS) and tested the system over 360 records of past sales properties in Midwest, US. The dataset had 14 numerical features and again no textual feature was used.

Another research paper [6] focused on studying the prediction of prices of apartments in city in Macedonia. Among the three data mining techniques that were applied on a dataset of 1200 sales transactions, the logistic regression (very similar to linear regression) was found to be the superior in prediction accuracy over decision tree and neural network techniques. Like the other earlier mentioned papers, there was no use of textual data. Some attempted to add structure to unstructured and ungrammatical data. However, this requires domain knowledge to build a reference structure (model) which can be used to extract the corresponding features [12]. Our proposed approach does not require deep domain knowledge (aside from simple data cleansing, the whole process is mostly automated).

The most related work to ours was concurrently and independently developed for analyzing real estate classifieds in the United States [16].[11]

6 Conclusion and Future Work

We propose in this paper a data mining process that uses text mining along with linear regression to improve the prediction of the price of real estate classifieds. We show that using text mining significantly reduces the RMSE of prediction. We also show how our proposed approach can identify keywords that affect the price positively or negatively.

---

[11] Our work was actually published as an MSc dissertation a couple of months earlier than Dick Stevens' dissertation. However, we can not provide further evidence due to the double-blind review process.

One of the direction we want to pursue is extending our analysis to the Arabic language (which is commonly used in our region). We are also considering the integration of our system with a named-entity recognition component [1], particularly for identify locations in ungrammatical text, to improve accuracy.

References

1. Abdallah, S., Shaalan, K.F., Shoaib, M.: Integrating rule-based system with classification for arabic named entity recognition. In: Gelbukh, A. (ed.) Computational Linguistics and Intelligent Text Processing. LNCS, vol. 7181, pp. 311–322. Springer, Heidelberg (2012)
2. Bourassa, S.C., Cantoni, E., Hoesli, M.: Predicting house prices with spatial dependence: a comparison of alternative methods. J. Real Estate Res. **32**(2), 139–159 (2010)
3. Case, B., Clapp, J., Dubin, R., Rodriguez, M.: Modeling spatial and temporal house price patterns: a comparison of four models. J. Real Estate Finan. Econ. **29**(2), 167–191 (2004)
4. Chen, T.H., Chen, C.W.: Application of data mining to the spatial heterogeneity of foreclosed mortgages. Expert Syst. Appl. **37**(2), 993–997 (2010)
5. Crowston, K., Wigand, R.T.: Real estate war in cyberspace: an emerging electronic market? Int. J. Electron. Markets **9**(1–2), 1–8 (1999)
6. Gacovski, Z., Kolic, J., Dukova, R., Markovski, M.: Data mining application for real estate valuation in the city of skopje. In: ICT Innovations 2012, Web Proceedings, pp. 537–538 (2012). ISSN 1857-7288
7. Guan, J., Zurada, J., Levitan, A.S.: An adaptive neuro-fuzzy inference system based approach to real estate property assessment. J. Real Estate Res. **30**(4), 395–422 (2008)
8. Helbich, M., Brunauer, W., Hagenauer, J., Leitner, M.: Data-driven regionalization of housing markets. Ann. Assoc. Am. Geog. **103**(4), 871–889 (2013)
9. Helbich, M., Jochem, A., Mcke, W., Hfle, B.: Boosting the predictive accuracy of urban hedonic house price models through airborne laser scanning. Comput. Environ. Urban Syst. **39**, 81–92 (2013)
10. Jaen, R.D.: Data mining: an empirical application in real estate valuation. In: Haller, S.M., Simmons, G. (eds.) FLAIRS Conference, pp. 314–317. AAAI Press (2002)
11. McGreal, S., de La Paz, P.T.: An analysis of factors influencing accuracy in the valuation of residential properties in spain. J. Property Res. **29**(1), 1–24 (2012)
12. Michelson, M., Knoblock, C.A.: Creating relational data from unstructured and ungrammatical data sources. J. Artif. Intell. Res. (JAIR) **31**, 543–590 (2008)
13. Pera, M.S., Qumsiyeh, R., Ng, Y.K.: Web-based closed-domain data extraction on online advertisements. Inf. Syst. **38**(2), 183–197 (2013)
14. Rossini, P.: Accuracy issues for automated and artificial intelligent residential valuation systems. In: International Real Estate Society Conference (1999)
15. Rossini, P., et al.: Using expert systems and artificial intelligence for real estate forecasting. In: Sixth Annual Pacific-Rim Real Estate Society Conference, Sydney, Australia, pp. 24–27. Citeseer (2000)
16. Stevens, D.: Predicting real estate price using text mining. Master's thesis, Tilburg University School of Humanities, The Netherlands (2014)
17. Wedyawati, W., Lu, M.: Mining real estate listings using ORACLE data warehousing and predictive regression. In: Proceedings of the 2004 IEEE International Conference on Information Reuse and Integration, IRI, pp. 296–301 (2004)

Intelligent Systems Track

Mitigating Malware Attacks via Secure Routing in Intelligent Device-to-Device Communications

Hadeer Elsemary[(⊠)]

Faculty of Mathematics and Computer Science,
Georg-August-University, Göttingen, Germany
hadeer.el-semary@informatik.uni-goettingen.de

Abstract. Device-to-Device (D2D) communications have received significant attention nowadays due to the excess number of applications and services. D2D communication promises a higher data rate, lower communication delays, and better power efficiency. Therefore, D2D is expected to be a vital technical component in Internet of Things (IoT) and play an important role with the next generation 5G. Moreover, the rapid growth in mobile capabilities opens the door to the cyber criminals that explore new avenues for malware attacks. Although the literature is proposed security schemes for malware attacks. However, the research field is still immature and unexplored in depth due to the fast evolution of malware at a rate far exceeding the evolution of security techniques. In this paper, the problem of detecting malware attacks is considered in D2D network and a secure energy-efficient routing protocol is proposed. The protocol aims at detecting malware attached to message before it infects the targeted device through optimal secure energy-efficient routes. Moreover, the protocol takes into account the attacker's behavior, computation of players' strategies including different attack cases and consideration of the dynamic scheme in terms of calculating malware detection capabilities and malware types due to the fast evolution of the malware. Through simulations, the proposed routing protocol is evaluated in terms of the detecting rate of the malicious messages and overall expected payoff of the defender compared with other non-strategic routing protocols. Results show that the game achieves Nash equilibrium, and leads to an optimal defense strategy for the network.

Keywords: Device-to-Device communication · Game theory · Malware · Security · Energy efficient routing

1 Introduction

Due to the recent rapid growth in demand of the mobile communication network, new technologies are proposed to improve throughput, communication delay and

© Springer International Publishing AG 2017
A.E. Hassanien et al. (eds.), *Proceedings of the International Conference on Advanced Intelligent Systems and Informatics 2016*, Advances in Intelligent Systems and Computing 533, DOI 10.1007/978-3-319-48308-5_20

computational offloading. Due to this fact, Device-to-Device (D2D) communications are recognized as promising technologies for communications nowadays and in the near future [1]. D2D communication are proposed as a means of gathering the proximity, hop gains and reuse [2]. In addition, D2D enables the device to communicate directly without the involvement of Base Station (BS) or any central entity such that the communication occurs on either licensed or unlicensed spectrum [3].

Due to the recent market demand for new services such as context-aware, proximity services, the industry is exploiting new use cases and new business models based on D2D communication e.g., pervasive health-care monitoring, social networking, public safety and rescue and location based services. Therefore, D2D is expected to be a vital technical component in Internet of Things (IoT) and play an important role with the next generation 5G.

1.1 Problem Definition and Related Work

Due to impressive demand and benefits of D2D communication in different and large areas, new severe security threats are expected on D2D network. Furthermore, the direct connections between devices via short range technologies (i.e., WiFi, Bluetooth) are more vulnerable to security threats. The authors in [9] identify the important security requirements in D2D communication as well as survey and evaluate the existed security schemes. However, the academia and industry have not yet investigated the security issues of the D2D communication seriously.

Due to these capabilities, mobile devices are considered an attractive launching pad for malware attacks. However, this research field is still immature and unexplored in depth [5]. As a result, the mobile malware threats are considered as a hot topic in the next future.

In this paper, we review briefly some of the proposed secure routing protocols based on game theory that optimizing the intrusion detection in mobile wireless network (i.e., multi-hop D2D fashion) [13,14]. Apart from the secure routing protocols, another set of work based on game theory aims at optimizing the intrusion detection [7,8]. The presented stochastic routing protocol in [13] aims at mitigating the eavesdropping from the insider attacker and improving the fault tolerance. They select randomly among paths to forward the packets. In [10–12], the secure routing and packet forwarding game is proposed and they used game theory to study the interaction between the good nodes and malicious nodes under noise and imperfect monitoring. They derived the optimal defense strategies with extensive evaluation of the effectiveness of these strategies. The aforementioned work considered only the insider attackers. In [14], the authors proposed a zero-sum complete information game between network and attacker. They derived the defense strategy for network to detect malicious attacks based on complete information taking into consideration the energy consumption and the quality of service.

The proposed static Bayesian game [6] is one-shot game, where the defender does not take into consideration the evolution of the game. Thus, the defender maximizes his payoffs based on fixed prior beliefs about the types of his opponent.

1.2   Summary of Contributions

Due to the fact that this aforementioned research field is not fully developed, with this paper, a game theoretic routing protocol for D2D communications is proposed to the malware detection problem. It is worth mentioning that the game theory provides an extensive set of mathematical tools to plan for the real life security problems. In particular, adversaries are attempting to infect targeted devices residing in the D2D network through the compromised gateway. On the worst case, we assume that the adversaries exploit the vulnerabilities of the gateways to inject messages attached with malware as well as exploit the operating system vulnerabilities of the smart devices to mount sophisticated malware attacks. Thus, the adversaries need to be considered as rational players and their expected behaviors need to be taken into consideration. Our protocol RMSR models the malware detection problem as two players non-cooperative zero-sum repeated game between D2D network and the adversaries. The main objective of this paper is to propose an optimal secure and energy-efficient routing protocol taking into account the attacker strategies and actions as well as the fast evolution of the malware. We formulate the zero-sum repeated game case where the defender routes the traffic such that the value of the game is maximized. On other words, the defender selects the most secure routes that have the maximum capability of malware detection and enough residual energy for routing process. On the other hand, the attacker tries to be unpredictable to the defender by enriching his actions with different malware types.

This paper extended the work in [6] and the main contribution of this paper summarized as the following: first, presentation of realistic pervasive health monitoring scenario in case of outsider attacker as a case study. Second, consideration of a D2D network contains heterogeneous devices in terms of operating systems. Third, investigation of the repeated game for malware detection problem taking into account the attacker's behavior, computation of players' strategies including different attack cases and consideration of the dynamic scheme in terms of calculating malware detection probabilities and malware types due to the fast evolution of the malware. Finally, conduction of simulations using Omnet++ where the optimality of the defender strategies is verified. The rest of the paper is organized as follows: In Sect. 2, we present the system model describing the environment, the attack model and network setup. In Sect. 3, we introduce the mathematical notations used and formulate the game theoretic framework. In Sect. 4, we describe in details the RMSR routing protocol. In Sect. 5, we conduct simulation and compare RMSR with other protocols. Finally, Sect. 6 contains conclusion and remarks as well as an outlook on further extensions of the paper.

2 System Model

2.1 Environment

As a motivating example, a real world pervasive health monitoring scenario is considered as depicted in Fig. 1. In fact, real world mobile cloud-based health monitoring faces excessive networking latency and longer response time. As a result, there is a growing demand to improve human health and well-being in health care systems [15]. As shown in Fig. 1, smart gateway is exploited to offer several higher-level and low latency local services such as local storage and local data processing. In order to benefit from network-controlled D2D communications [2], the mobile devices in the immediate physical proximity are connected to other devices and form a local D2D network. The mobile devices communicate with one another in D2D multi-hop fashion by using short-range wireless connections such as WiFi in ad-hoc mode. The smart gateway act as a hub between mobile devices in the local D2D network and the remote health data center. The smart gateway provides the management of health records for the patients by storing, updating and retrieving all the medical information about the patients.

We denote any source device by S, which acts as a coordinator which manages the queries of patient health care record locally or remote data center in the Internet if necessary. Since the wireless channel is open and accessible to both legitimate network users and malicious attackers, adversaries can build their malicious access points known as Evil Twin attack which provides deceptive SSID and replace a legitimate gateway. As a result, the adversary is able to eavesdrop on network traffic, inject the infectious message and infect any device residing in a given local D2D network.

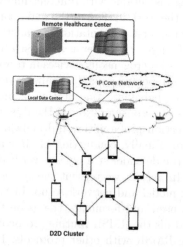

Fig. 1. Example of pervasive healthcare monitoring mobile network

2.2 Network Set-Up

In this paper, Consider a local D2D network of $\mathcal{N}$ heterogeneous mobile devices denoted by $[\mathcal{N}]$ such that the mobile devices are heterogeneous in terms of operating systems. For each message, there is a set of all routes $[R]$ from the S to targeted device. The S selects $r_j \in [R]$ to deliver message, where $[N_j]$ is the set of devices along the route r_j. We assume software-based malware detection systems with sophisticated detection capabilities to be deployed on each device. Every device is running anti-malware control and it can also carry out the real time network traffic monitoring. We have Ω different mobile operating systems [6], expressed by the finite set $[\Omega]$. We denote by $[\mathcal{M}_\omega]$ the set of $\mathcal{M}_\omega$ as a different malware available to the attacker to infect devices that run the mobile operating system ω. For each $\omega \in [\Omega]$, we assume $\mathcal{C}_\omega$ anti-malware software (i.e., Resources) expressed by the finite set $[\mathcal{C}_\omega]$. Anti-malware detection software is residing on each mobile device n_i and each anti-malware software has its detection rate to detect successfully certain malware type. Since the routing is a cooperative process where the messages are relayed among devices. Any device along the route detects the intrusion with strong evidence of anomalies, it is responsible on responding quickly to the intrusion. as a result every device is responsible on inspecting the received message using its detection capability.

We denote by $D(c_k^i, \mathcal{M}_m)$ is the disability of the device n_i to detect the malware $\mathcal{M}_m$ (i.e., False negative [17]). As a result, for the fixed route r_j, the disability of r_j to detect $\mathcal{M}_m \in [\mathcal{M}_\omega]$ is given by:

$$D(r_j, \mathcal{M}_m) := \prod_{n_i \in N_j} D(c_k^i, \mathcal{M}_m) \tag{1}$$

then the route detection capability of r_j to successfully detect the $\mathcal{M}_m \in [\mathcal{M}_\omega]$ before it reaches the targeted device that runs ω (i.e., True Positive [17]) is given by:

$$\psi(r_j, \mathcal{M}_m) := 1 - D(r_j, \mathcal{M}_m) \tag{2}$$

In addition, the multi-hop D2D communication and malware detection process will necessitate cooperation between devices. Some devices may not collaborate to relay other device's traffic because of their limited available energy. Therefore, our protocol ensures the route availability during the routing process and considers the battery-level of the devices in the routing decision. It chooses the routes with highest energy devices on the basis of residual energy of the device. Formally the battery-level of device n_i, $n_i \in [\mathcal{N}]$ is given by: $\mathcal{E}(n_i) = \frac{\mathcal{E}_r}{\mathcal{E}_{max}}$, such that $\mathcal{E}_r$ is the remaining energy and $\mathcal{E}_{max}$ is the maximum energy available for the device. Therefore the route battery-level on r_j is derived by multiplying the battery-level of all the devices along the route as follows:

$$\mathcal{E}(r_j) := \prod_{n_i \in N_j} \mathcal{E}(n_i) \tag{3}$$

3   Repeated Malware-Defense Secure Routing Games (RMSR)

To investigate the interactions between the defender and the adversary, a non-cooperative two-players game is considered that played by the D2D devices (defender) and the adversary (attacker) in order to derive the optimal defense strategic decisions for the defender. We assume that the attacker exploits zero-day vulnerability of the mobile operating system. Therefore, he selects the malware that targets the vulnerability of certain platform running on the targeted device. On the other hand, the defender has the statistics about different existing malware types for each mobile platform ω. Furthermore, the mobile devices learns more about the attacker actions from the IDS during the subsequent repeated game. Continuing with the notations mentioned in [6]:

- **Strategy Set**: The strategy set of a player refers to all available moves the player is able to take. We consider that the defender's pure strategies is a set of all possible routes $r_j \in [R]$ from the S to the targeted device. The attacker's pure strategy is a set of different malware types $\mathcal{M}_m \in [\mathcal{M}_\omega]$ from which the attacker selects to send to the targeted device aiming its infection.
- **Payoff**: we define the $\mathcal{U}_\Theta$ as the payoff of the defender. The payoff of the defender depends on the route detection capability and the route availability. We define $\mathcal{U}_\Psi$ as the payoff of the attacker of type $\omega \in \Omega$, where the attacker's payoff is opposite to defender's payoff (i.e. zero sum game).

We consider the defender is the row player in the payoff matrix and the attacker are the column player. For a given pure strategy profile $(r_j, \mathcal{M}_m), r_j \in [R], \mathcal{M}_m \in [\mathcal{M}_\omega]$ where $\omega \in \Omega$, the payoff of the defender is given by

$$U_\Theta(r_j, \mathcal{M}_m; \omega) = [\psi(r_j, \mathcal{M}_m)\mathcal{V} + \mathcal{E}(r_j)] \tag{4}$$

We assume the $\mathcal{V}$ is the defender's security gain value (monetary), where $\mathcal{V} > 0$. The defender's payoff is the expected gain of detecting the malware before infecting the targeted device depends on the route detection rate in Eq. (1) summed up the route energy level in Eq. (2).

The defender's mixed strategy $X = [x_1, x_2, \ldots, x_\xi]$ is the probability distribution over different routes in $[R]$ (i.e. Pure strategies) from the source device S to the targeted device. Where x_j is probability that the defender will choose its j-th route to send the message.

For each $\omega \in \Omega$, the attacker's mixed strategies $Y^\omega = [y_1^\omega, y_2^\omega, \ldots, y_\eta^\omega]$ is the probability distribution over different malware (i.e. Pure strategies) against targeted devices that run ω. Where y_l^ω is probability that the attacker will choose its l-th malware to infect device that runs ω. In two players zero-sum game with finite number of actions for both players, there is at least a Nash equilibrium in mixed strategy [16]. The RMSR game consists of mixed strategy profile (X, Y^ω), therefore the expected payoff of the defender is denoted by:

$$\mathcal{U}_\Theta \equiv U_\Theta(X, Y^\omega) = \sum_{x=1}^{\xi} \sum_{l=1}^{\eta} x_j y_l^\omega U_\Theta(r_j, \mathcal{M}_m; \omega) \tag{5}$$

For zero sum game, $\mathcal{U}_\Psi = -\mathcal{U}_\Theta$, This means that the defender's gain is considered the attacker's loss.

4   Repeated Malware-Defense Secure Routing Protocol

Our proposed protocol RMSR stages as follows:

Route Discovery Stage: First the S node broadcasts a Route Request message, the devices that receive this message should broadcast it to their neighbors. If the receiving device is the targeted device, it sends back the Route Reply message containing the full reverse source route. We have appended two new fields in the Route Reply message (route detection capability, route battery-level). Each intermediate device on receiving the Route Reply, updates the route detection rate field by multiplying its detection capabilities using Eq. (1) and updates the route battery-level field using Eq. (3). When the Route Reply reaches the S, it calculates overall route detection rate using Eq. (2)

Route Selection Stage: After the S node receives several routes, then stores its routing table. It uses its routing table to solve the RMSR game and deriving the optimal defense strategy X^*. The S node selects the best route probabilistically according to X^* to send the message.

5   Performance Evaluation

In this section, we have conducted the simulations to evaluate the effectiveness of the optimal defense strategies of the defender and identify when these strategies can work well. In these simulations, mobile devices are randomly deployed inside a rectangular area of 800×800 m and are conducted in Omnet++ network simulator, INET framework and fixed parameters are shown in Table 1. The simulation time changes (10, 20, 40, 60 min) with total number of 7000 messages and 1000 from total messages is malicious messages. We consider one attacker who sends a sequence of malicious messages to a certain target aiming at infecting the targeted device. We plot the detecting rate of malicious messages for three routing protocols (RMSR, DSR, AODV) in case of 3 different attack cases. We evaluate the performance of the optimal defense strategy for RMSR in terms of the detected rate of the malicious messages comparing with other nonstrategic protocols DSR and AODV. We investigate the properties of the Nash equilibrium in RMSR game through the simulations and conduct the results of our study. In the Fig. 2 shows the results of the detecting rate of the malicious messages, varying the pause times in case of different attack cases. In case of Optimal attack, it can be observed that the detecting rate of the malicious messages are small for AODV and DSR protocols, while the RMSR protocol still keeps the detecting rate high even in a hostile environment. Whilst in the case of Uniform attack, it can be noticed that RMSR protocol outperforms the other two ad-hoc protocols and still has high detecting rate of malicious messages. On the other hand, in case of the worst weighted attack, RMSR protocol still outperforms

Table 1. Simulation parameter values

Parameter	Value
Number of nodes	20
Mobility model	Linear Mobility
Mobility Speed	10 mps
Mobility Update Interval	0.1 s
Packet size	512 bytes
Packet generation rate	2 packets/s

the other two protocols and has highest detecting rate of malicious messages. We can notice the average values of the detecting rate of malicious messages in case of a Uniform attack as non-strategic attack within the range [83 %, 78 %]. Whilst in case of Optimal attack the average values of the detecting rate of the malicious messages within the range [75 %, 71 %] as well as in the worst case of the weighted attack the average values of the detecting rate of the malicious messages are approximately 70 %. Likewise in Fig. 3 shows the results of the expected payoff of the defender in case of different attack cases. It is observed that RMSR protocol outperforms the other non-strategic protocols and achieved the highest expected payoff for the defender in all cases. Therefore, we can notice that the RMSR protocol have steadily the best performance even in the worst case.

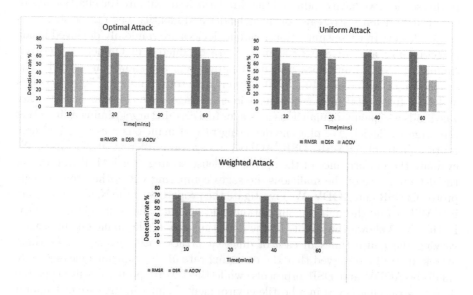

Fig. 2. Detecting rate of malicious packets for 3 different attack cases

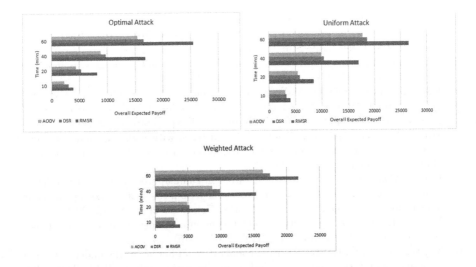

Fig. 3. Expected payoff of the defender for 3 different attack cases

6   Conclusion and Future Work

In this paper, the problem of detecting the malicious messages is considered in the D2D network in a game-theoretic framework in the presence of outsider attacker. A Repeated security game between the D2D network and the attacker is modeled. We show that the game has the Nash equilibrium leading to optimal defense strategy. The defender must design an effective defense scheme to detect malicious messages injected into the network by the attacker. We considered the scenario where the defender has the information about the optimal routes obtained from the RMSR game and can pick the routes to maximize the chances of the detection. Results show that RMSR protocol based on strategic plan outperforms the other non-strategic protocols. Finally, we have estimated the performance of the proposed RMSR protocol on the D2D network. In future work, we plan to consider that the insider attacker case that has control on the IDS, then there is no trust guarantee among the devices and each device will inspect message independently and will try to find the next hop to forward the message.

References

1. Mumtaz, S., Rodriguez, J. (eds.): Smart Device to Smart Device Communication. Springer International Publishing, Switzerland (2014)
2. Fodor, G., Parkvall, S., Sorrentino, S., Wallentin, P., Lu, Q., Brahmi, N.: Device-to-device communications for national security and public safety. IEEE Access **2**, 1510–1520 (2015)
3. Asadi, A., Wang, Q., Mancuso, V.: A survey on device-to-device communication in cellular networks. IEEE Commun. Surv. Tutorials **16**, 1801–1819 (2014)

4. Khouzani, M.H.R., Sarkar, S., Altman, E.: Maximum damage malware attack in mobile wireless networks. IEEE/ACM Trans. Netw. **20**(5), 1347–1360 (2012). Dreese Laboratories, Ohio State University, Columbus, OH, USA

5. Khouzani, M.H.R., Sarkar, S., Altman, E.: Saddle-point strategies in malware attack. IEEE J. Sel. Areas Commun. **30**, 31–43 (2012)

6. Elsemary, H., Hogrefe, D.: Malware-defense secure routing in intelligent device-to-device communications. In: 1st International Conference on Advanced Intelligent System and Informatics (AISI2015), Beni Suef, Egypt. Springer (2015)

7. Wang, W., Chatterjee, M., Kwiat, K.: Coexistence with malicious nodes: a game theoretic approach. In: International Conference on Game Theory for Networks, GameNets 2009, pp. 277–286 (2009)

8. Liu, Y., Comaniciou, C., Man, H.: A Bayesian game approach for intrusion detection in wireless ad hoc networks. In: Proceedings from the 2006 Workshop on Game Theory for Communications and Networks, GameNets 2006, p. 112 (2006)

9. Wang, M., Yan, Z.: Security in D2D communications: a review. In: Trustcom/BigDataSE/ISPA, vol. 1, pp. 1199–1204 (2015)

10. Yu, W., Ji, Z., Liu, K.J.R.: Securing cooperative ad-hoc networks under noise and imperfect monitoring: strategies and game theoretic analysis. IEEE Trans. Inf. Forensics Secur. **2**, 240–253 (2007)

11. Yu, W., Liu, K.J.R.: Game theoretic analysis of cooperation stimulation and security in autonomous mobile ad hoc networks. IEEE Trans. Mob. Comput. **6**, 507–521 (2007)

12. Yu, W., Liu, K.J.R.: Secure cooperation in autonomous mobile ad-hoc networks under noise, imperfect monitoring: a game-theoretic approach. IEEE Trans. Inf. Forensics Secur. **3**, 317–330 (2008). Department of Electrical & Computer Engineering, University of Maryland, College Park, MD, USA

13. Bohacek, S., Hespanha, J.P., Lee, J., Lim, C., Obraczka, K.: Game theoretic stochastic routing for fault tolerance and security in computer networks. IEEE Trans. Parallel Distrib. Syst. **18**, 1227–1240 (2007)

14. Panaousis, E., Alpcan, T., Fereidooni, H., Conti, M.: Secure message delivery games for device-to-device communications. In: Poovendran, R., Saad, W. (eds.) GameSec 2014. LNCS, vol. 8840, pp. 195–215. Springer, Heidelberg (2014). doi:10.1007/978-3-319-12601-2_11

15. Rahmani, A.-M., Kumar, N., Gia, T.N., Granados, J., Negash, B., Liljeberg, P., Tenhunen, H.: Smart e-Health gateway: bringing intelligence to internet-of-things based ubiquitous healthcare systems. In: 12th Annual IEEE Consumer Communications and Networking Conference (CCNC) (2015)

16. Nash, J.F.: Equilibrium points in N-Person games. Proc. Natl. Acad. Sci. **36**, 48–49 (1950)

17. Fawcett, T.: An introduction to ROC analysis. Pattern Recogn. Lett. **27**, 861–874 (2006)

An Efficient System for Finding Functional Motifs in Genomic DNA Sequences by Using Nature-Inspired Algorithms

Ebtehal S. Elewa[1(✉)], Mohamed B. Abdelhalim[1],
and Mai S. Mabrouk[2]

[1] College of Computing and Information Technology, Arab Academy
for Science, Technology and Maritime Transport Cairo, Cairo, Egypt
ebtehale@hotmail.com
[2] Biomedical Engineering Department, Misr University for Science
and Technology, 6 October, Giza, Egypt

Abstract. Motifs are short patterns in Deoxyribonucleic Acid (DNA) that indicate the presence of certain biological characteristics. Motifs finding is the process of successfully finding meaningful motifs in large DNA sequences. Nature-inspired algorithms have been recently gaining much popularity in solving complex and large real-world optimization problems similar to the motif finding problem. This work aims on investigating the application of nature-inspired algorithms in motif finding problem. The investigation methodology is divided into three main approaches; the first is to apply well-known nature-inspired algorithms in solving the problem, then the enhancement of an algorithm is investigated, and finally the hybridization between two algorithms is investigated. Experiments are performed on synthetic as well as real data sets. The results show that the combination provides the best results, however, individual and modified algorithms provide also good results compared to some state-of-the-art tools.

Keywords: Cuckoo search · Gravitational search algorithm · Particle swarm optimization · Motif finding · DNA · Nature-inspired algorithms

1 Introduction

The basic unit of living cells is the Deoxyribonucleic Acid (DNA). DNA sequences consist of Adenine (A), Cytosine (C), Guanine (G), and Thymine (T) [1], and contains the genetic information required for the proper functioning of all living organisms [2] Motifs are short, recurring patterns in DNA that are presumed to have a biological function. Motif finding is the process of finding these short and meaningful patterns. The purpose of motif finding is to discover repeated patterns in DNA sequences in order to understand the structure of certain cells, find the triggers of some diseases and create new treatments. So far, nature-inspired algorithms have been successfully able to solve many complex problems in almost all areas [3–6]. The success of these algorithms is due to their ability to mimic nature in solving dynamic and complex problems in a reasonable amount of time and optimal cost.

© Springer International Publishing AG 2017
A.E. Hassanien et al. (eds.), *Proceedings of the International Conference on Advanced Intelligent Systems and Informatics 2016*, Advances in Intelligent Systems and Computing 533, DOI 10.1007/978-3-319-48308-5_21

This work introduces enhanced use of nature-inspired algorithms to solve the motif finding problem. We followed the methodology of applying individual algorithms, then we modified one of them while hybridizing the other two to explore the ability to enhance individual algorithms. The first individual algorithm used is the cuckoo search algorithm, and then the modified adaptive cuckoo search that uses new strategies to improve the original algorithm is implemented. The gravitational search algorithm and the famous Particle Swarm Algorithm (PSO) are the other algorithms used and finally the hybrid of them is implemented.

The rest of the article is organized as follows: Sect. 2 provides a brief explanation of the motif finding problem and related work. Section 3 describes the implemented algorithms. Section 4 contains the problem formulation and the overall process. Section 5 shows the experimental results and discussion. Section 6 concludes the article and introduces possible directions for future work.

2 Problem Definition and Related Work

2.1 The Motif Finding Problem

Given a set of DNA Sequences defined by the alphabets {A, C, G, T}, motifs are defined as short repeated segments in the DNA. The motif finding problem has been studied since the early years of bioinformatics, a number of methods, algorithms and tools have been developed in the recent years to solve this problem However, recent show that their prediction accuracy is still low. Also most of the algorithms suffer with local optima [7].

2.2 Current Motif Finding Algorithms and Related Work

Most motif finding algorithms have their own strengths and weaknesses and fall into two major groups based on the combinatorial approach used: The first group is Word-based methods. The advantage of the word-based method is that it exhaustively searches the whole search space and therefore guarantees finding a global optimum. However, this also means that they are only suitable for short motifs. The main drawback of word-based algorithms is that they are usually computationally expensive. Some popular tools that use word-based methods are MITRA [8], and Weeder [9]. The second group is probabilistic sequence methods. The most popular methods MEME [10], AlignACE [11]. However, these algorithms are not guaranteed to find globally optimal solutions. In literature, evolutionary algorithms have been also used to find DNA motifs, however most of them depend on Genetic Algorithm (GA) [12, 13] and Particle swarm optimization (PSO) [14, 15]. So far, surveys have been made to evaluate motif finding tool [16]. The results show that combining more than one approach in finding motifs provides better results than using a single approach, and that no specific tool provides the best performance for all datasets.

3   Methods and Materials

The proposed system is comprised of the following three fundamental building phases: (1) The first phase 'Read Input' includes reading the file that contains the DNA Sequence, (2) The second phase' Execute optimization algorithm' is to execute the optimization algorithm, and (3) The last step 'Calculate evaluation criteria' is to compare the predicted motif and compare it with the actual motif and calculating the evaluation criteria. These three phases of the introduced approach is described in Fig. 1.

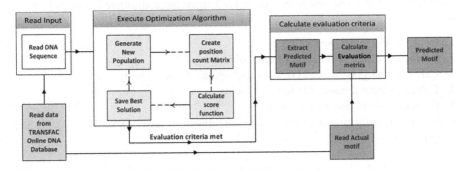

Fig. 1. Overall system block diagram

3.1   Cuckoo Search Algorithm (CS)

Cuckoo search was introduced by Yang and Deb in 2009 [17], it is inspired by the cuckoo birds and their parasitism reproduction behavior in nature.

Cuckoo search has both global search and exhaustive local search capabilities, which explains why it is so efficient in finding almost optimal solutions. These capabilities are due to the idea of Lévy flights used by the algorithm. The Lévy flight process, is a random walk that is characterized by a series of instantaneous jumps chosen from a probability density function in which the step lengths are distributed according to a heavy-tailed probability distribution [25].

3.2   Modified Adaptive Cuckoo Search (MACS)

Despite the success of the original cuckoo search algorithm, many variants and hybridizations have been suggested by many researches to improve its efficiency. One important variation is the modified adaptive cuckoo search suggested by Zhang [18]. It introduces new strategies such as grouping, parallel, information sharing and adaptive strategies.

3.3   Hybridizing GSA and PSO

GSA has been used to solve many optimization problems, however the main drawback of GSA is that it can easily fall into local optima and it also suffers from slow exploitation. These drawbacks have been addressed by researches by using hybrid algorithms to overcome the slow exploitation of GSA [19, 20]. GSA and PSO have been used alone to solve the motif finding problem in [14, 15, 21, 25]. However, combining the features of the two algorithms has never been used to find motifs. PSO is widely used in hybrid algorithms due to its ability to find global optimal, it's speed of convergence and simplicity.

In this work, PSO and GSA are combined to produce a hybrid GSA-PSO algorithm to find motifs. The hybrid algorithm combines between the social behavior of PSO in updating the particles locations and velocities according to gbest and the local search capabilities of GSA The new hybrid GSA-PSO algorithm is illustrated in the following steps:

1. Generate the initial population.
2. Evaluate the fitness function of each object.
3. Calculate the gravitational constant and update the local best.
4. Calculate inertial mass (M) for each object.
5. Calculate forces and acceleration.
6. Update the velocity and position using the following classical PSO Eqs. (1), (2) and (3):

$$V_i(t+1) = Rand \times V_i(t) + C_1 \times ac_i(t) + C_2 \times (gbest - X_i(t)) \tag{1}$$

$$X_i(t+1) = X_i(t) + V_i(t+1) \tag{2}$$

$$ac_i(t) = f_i(t)/M_i(t) \tag{3}$$

Where $f_i(t)$ is the total forces that act on the agent object i and $M_i(t)$ is the mass of the object i.

7. Repeat steps from 2 to 6 until stop criteria is reached.

4 Problem Formulation

4.1 Motifs Representation

An individual is represented by a vector of integers (X); each integer in the vector represents the starting positions of the potential motif in each DNA sequence. Accordingly, the length of the vector equals to the number of input sequences, so each vector (X) contains (d) items and (d) is the number of DNA sequences. Hence a candidate vector can be represented in the following form given in Eq. (4):

$$X_i = (x_{i1}, x_{i2}, x_{i3}, \ldots\ldots, x_{id}) \tag{4}$$

4.2   Objective Function

In this work, the score function described by Reddy and Arock [14] is used. Equation (5) describes the objective function where p(s) is the profile matrix corresponding to starting positions and $M_{p(s)}(J)$ is the largest count in column j of P(s).

$$Score = \sum_{j=1}^{l} M_{p(s)}(J) \tag{5}$$

5   Experimental Results

The proposed algorithms have been tested on synthetic as well as real data and the average results are calculated. MATLAB Bioinformatics Toolbox is used to implement the algorithms. Synthetic data used in testing the algorithms was implemented by the model provided by Pevsner and Size [22]. The goal of any motif finding algorithm is to find and discover these motifs without previously knowing their locations in the DNA sequence regardless of the type of DNA sequence. TRANSFAC database (available at http://www.gene-regulation.com/pub/databases.html#transfac) is used in this work to evaluate the algorithms on real datasets. The reason for that choice is that the real sequence data from the biological database TRANSFAC are a part of the freely accessible benchmark data ensemble constructed by Tompa et al. [23], also this real data belongs to different kinds of species mainly human, mouse, rat, yeast, and plants to ensure that the algorithms work on different types of species.

In order to analyze the performance of these algorithms; recall, precision and F-score metrics are used. Recall and precision are defined in Eqs. (6) and (7):

$$Recall = n_c/n_t \tag{6}$$

$$Precision = n_c/n_p \tag{7}$$

Where n_c is the number of motifs that are correctly predicted, n_p is the total number of predicted motifs and n_t is the total number of actual true motifs? F-score value that combines both recall and precession terms is defined in Eq. (8).

$$F - Score = \frac{2 \cdot Precision \cdot Recall}{Precision + Recall} \tag{8}$$

6   Results and Discussions

For synthetic data, different motifs are used by using different (l, d) pairs to test the implemented algorithms where l is the length of the motif and d in the maximum number of mutations. The famous challenging instances of (13,4), (15,5), (17,6) [22] are used in this work. The algorithms have also been tested against the real data described in the previous section and compared with well-known algorithms that have been used in finding motifs, such as AlignACE [11] that uses Gibbs sampling to find sequence elements conserved in a set of DNA sequences, Multiple EM for Motif Elicitation or MEME [10] which is a tool for discovering motifs in a group of related DNA or protein sequences. MEME represents motifs as position-dependent letter-probability matrices which describe the probability of each possible letter at each position in the pattern and finally compared with GALF [24] which is a tool based on Genetic Algorithm with Local Filtering. Figure 2 shows the average precision of motifs instances of (13,4), (15,5), (17,6), Fig. 3 shows the average recall and finally Fig. 4 provides the average F-score on the synthetic data with different motif lengths. The results on synthetic data show that GSA-PSO provides the best recall and precision values and accordingly provide the best F-score values. In addition, GSA-PSO is able to provide the best results for short and long motifs as well. The results also show that MACS is the second best in precision, recall and F-score values.

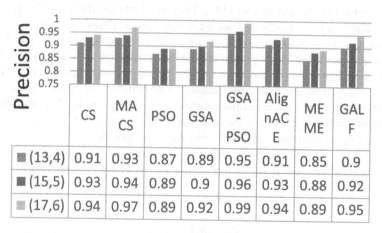

	CS	MACS	PSO	GSA	GSA-PSO	AlignACE	MEME	GALF
(13,4)	0.91	0.93	0.87	0.89	0.95	0.91	0.85	0.9
(15,5)	0.93	0.94	0.89	0.9	0.96	0.93	0.88	0.92
(17,6)	0.94	0.97	0.89	0.92	0.99	0.94	0.89	0.95

Fig. 2. Precision of instances (13,4), (15,5), (17,6) (Color figure online)

The F-score values show that GSA-PSO correctly identifies motifs and provides better results than GALF which is based on Genetic Algorithm. The enhancement in precession recall and F-score values is up to 0.05.

Table 1 shows the average precision, recall and F-score respectively on real DNA datasets. The results are the average of 20 runs on all the species.

7 Discussion

The results show that implemented algorithms perform better than well-known algorithms on the benchmark data set. Except the basic PSO algorithm that performs less than GALF but almost the same as MEME & AlginACE, that have the highest precision.

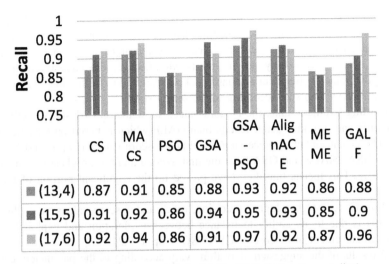

Fig. 3. Recall of instances (13,4), (15,5), (17,6) (Color figure online)

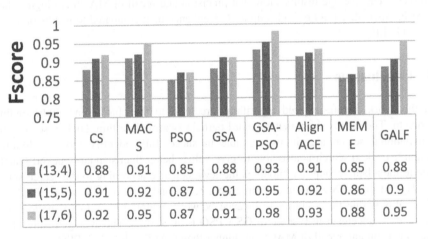

Fig. 4. F-score of instances (13,4), (15,5), (17,6) (Color figure online)

Table 1. Average precision, recall and f-score values on real datasets

Algorithm	Precision	Recall	f-score
CS	0.59	0.64	0.61
MACS	0.73	0.81	0.76
PSO	0.53	0.57	0.55
GSA	0.81	0.76	0.78
GSA-PSO	**0.85**	**0.81**	**0.83**
AlignACE (10)	0.64	0.57	0.59
MEME (11)	0.67	0.53	0.58
GALF (24)	0.80	0.74	0.76

MACS, GSA and PSO-GSA algorithms have higher recall values than MEME & AlginACE. This means that these algorithms have less false positives. The results show that standard particle swarm algorithm, cuckoo search and gravitational search algorithms are able to successfully identify motifs. Modified adaptive cuckoo search is also able to give values better than standard cuckoo search, due to the adaptive step that modifies the lévy step size according to the search stage.

Regarding Synthetic data, the F-score values show that GSA-PSO correctly identifies motifs and provides better results than GALF which is based on Genetic Algorithm. The enhancement in precession recall and F-score values is up to 0.05.

PSO-GSA, MACS and GSA have the first, second and third highest F-score value respectively. This means that they have picked up the significant portion of the real motifs.

Hybrid GSA-PSO has higher precision, recall and F-score values than PSO and GSA, this is due to its ability to balance between exploration and exploitation the average enhancement in F-score of GSA-PSO is up to 0.24.

The results of the suggested algorithms vary according to the parameters used in each algorithm. The parameters were tuned to achieve the best result in this specific problem. The average results show that precision and recall of MACS are higher than GALF. Also GSA-PSO provides up to 3 % enhancement compared to results obtained from GALF

8   Conclusions

In this work, nature-inspired algorithms are used to the challenging motif finding problem. The results show that standard particle swarm algorithm, cuckoo search and gravitational search algorithms are able to successfully identify motifs. Modified adaptive cuckoo search is also able to give values better than standard cuckoo search, due to the adaptive step that modifies the lévy step size according to the search stage.

Hybrid GSA-PSO has higher precision, recall and F-score values than PSO and GSA, this is due to its ability to balance between exploration and exploitation. The average enhancement in F-score of GSA-PSO is up to 0.24. The average results show that precision and recall of MACS are higher than GALF. Also GSA-PSO provides up to 3 % enhancement compared to results obtained from GALF. The results obtained show that these nature-inspired algorithms are generally able to correctly identify meaningful motifs instances in synthetic as well as real datasets and the quality of the results is better than the other tools considered for comparison. In addition, hybrid PSO-GSA provides the best results. Moreover, the enhanced CS and GSA algorithms provide good results that are better than the other considered tools.

References

1. D'haeseleer, P.: What are DNA sequence motifs? Nat. Biotechnol. **24**(4), 423–425 (2006)
2. Marbrouk, M., Hamdy, M., Mamdouh, M., Aboelfotoh, M., Kadah, Y.M.: BIOINFTool: bioinformatics and sequence data analysis in molecular biology using Mat Lab. In: Proceedings of Cairo International Biomedical Engineering Conference, 01–09 October 2006
3. Zelinka, I.: A survey on evolutionary algorithms dynamics and its complexity–Mutual relations, past, present and future. Swarm Evol. Comput. **25**, 2–14 (2015)
4. Smolinski, T.G., Milanova, M.M., Hassanien, A.E.: Applications of Computational Intelligence in Biology: Current Trends and Open Problems. Studies in Computational Intelligence Springer, Heidelberg (2008)
5. Smolinski, T.G., Milanova, M.M., Hassanien, A.E.: Applications of Computational Intelligence in Bioinformatics and Biomedicine: Current Trends and Open Problems. Springer, Heidelberg (2008)
6. Abdelhalim, M.B., Habib, S.E.D.: Particle swarm optimization for HW/SW partitioning. In: Lazinica, A. (ed.) Particle Swarm Optimization, pp. 49–76. Tech Education and Publishing, New York (2009)
7. Wei, W., Xiao-Dan, Yu.: Comparative analysis of regulatory motif discovery tools for transcription factor binding sites. Genom. Proteomics Bioinf. **5**(2), 131–142 (2007)
8. Eskin, E., Pevzner, P.A.: Finding composite regulatory patterns in DNA sequences. Bioinformatics **18**(suppl 1), S354–S363 (2002)
9. Pavesi, G., et al.: Weeder web: discovery of transcription factor binding sites in a set of sequences from co-regulated genes. Nucl. Acids Res. **32**, 199–203 (2004)
10. Bailley, T., Elkan, C.: Unsupervised learning of multiple motifs in biopolymers using expectation maximization. Mach. Learn. **21**(1–2), 51–80 (1995)
11. Roth, P., et al.: Finding DNA regulatory motifs within unaligned noncoding sequences clustered by whole-genome mRNA quantitation. Nat. Biotechnol. **16**(10), 939–945 (1998)
12. Vijayvargiya, S., Shukla, P.: Identification of transcription factor binding sites using genetic algorithm. Int. J. Res. Rev. Comput. Sci. **2**(2), 100–107 (2011)
13. Basha Gutierrez, J., Frith, M., Nakai, K.: A genetic algorithm for motif finding based on statistical significance. In: Ortuño, F., Rojas, I. (eds.) IWBBIO 2015, Part I. LNCS, vol. 9043, pp. 438–449. Springer, Heidelberg (2015)
14. Reddy, U., et al.: A particle swarm solution for planted(l, d)-Motif problem. In: IEEE Symposium in Bioinformatics and Computational Biology (CIBCB), pp. 222–229 (2013)
15. Lei, C., Ruan, J.: A particle swarm optimization-based algorithm for finding gapped motifs. BioData Min. **3**(1), 3–9 (2010)
16. Das, M.K., Dai, H.K.: A survey of DNA motif finding algorithms. BMC Bioinf. **8**(7), 1 (2007)
17. Yang, X., Deb, S.: Cuckoo search via Lévy flights. In: World Congress on Nature and Biologically Inspired Computing (NaBIC 2009), pp. 210–214. IEEE Publications (2009)
18. Zhang, Y., Wang, L., Wu, Q.: Modifed adaptive cuckoo search (MACS) algorithm and formal description for global optimisation. Int. J. Comput. Appl. Technol. **44**(2), 73–79 (2012)
19. Sinaie, S.: Solving shortest path problem using gravitational search algorithm and neural networks (Doctoral dissertation, Universiti Teknologi Malaysia) (2010)
20. Zhang, Yu., Wu, L., Zhang, Y., Wang, J.: Immune gravitation inspired optimization algorithm. In: Huang, D.-S., Gan, Y., Bevilacqua, V., Figueroa, J.C. (eds.) ICIC 2011. LNCS, vol. 6838, pp. 178–185. Springer, Heidelberg (2011)

21. González-Álvarez, D.L., Vega-Rodríguez, M.A., Gómez-Pulido, J.A., Sánchez-Pérez, J.M.: Applying a multiobjective gravitational search algorithm (MO-GSA) to discover motifs. In: International Work-Conference on Artificial Neural Networks, pp. 372–379 (2011)
22. Pevzner, P.A., Sze, S.H.: Combinatorial approaches to finding subtle signals in DNA sequences. In: Proceedings of the Eighth International Conference on Intelligent Systems for Molecular Biology, California USA, pp. 269–278 (2000)
23. Tompa, M., et al.: Assessing computational tools for the discovery of transcription factor binding sites. Nat. Biotechnol. **23**(1), 137–144 (2005)
24. Chan, T.M. et al.: TFBS identification by position and consensus-led genetic algorithm with local filtering. In: GECCO 2007: Proceedings of the 2007 Conference on Genetic and Evolutionary Computation, pp. 377–384. ACM, London, England (2007)
25. Hassanien, A.E., Alamry, E.: Swarm Intelligence: Principles, Advances, and Applications. CRC – Taylor & Francis Group (2015). ISBN 9781498741064 - CAT# K26721

Security as a Service Model for Cloud Storage

Alshaimaa Abo-alian[⊠], Nagwa L. Badr, and M.F. Tolba

Faculty of Computer and Information Sciences,
Ain Shams University, Cairo, Egypt
{a_alian,nagwabadr}@cis.asu.edu.eg,
fahmytolba@gmail.com

Abstract. Along with the widespread interest in cloud computing, there are still concerns that hinder the proliferation and the adoption of cloud services. One of the main concerns is data security in cloud storage environments. Numerous research problems belonging to cloud storage security have been studied intensively before. However, addressing the three dimensions of outsourced data security (i.e., confidentiality, availability, and integrity) as a cloud service is still a challenge in cloud storage. As there is always a tradeoff between maintaining security and obtaining efficiency, it is difficult but nevertheless essential to explore how to efficiently address security challenges over dynamic cloud data. This paper proposes an integrated security model for data storage in the cloud that provides authentication, access control, auditing and data management services. A strong authentication sub-system that is based on the traditional password and the user keystroke dynamics is presented. A dynamic access control system is proposed to ensure data confidentiality in cloud computing. A public auditing system is proposed to delegate the integrity verification of outsourced data in the cloud storage to a third party auditor while maintaining the data privacy. Experimental results demonstrate the effectiveness and efficiency of the proposed model.

Keywords: Access control · Auditing · Authentication · Cloud computing · Data security

1 Introduction

Cloud computing is the long dreamed vision of computing as a utility, where cloud customers can remotely store their data into the cloud so as to enjoy high-quality networks, servers, applications and services from a shared pool of configurable computing resources [1, 2]. Although outsourcing data to the cloud is economically attractive for long-term large-scale storage, the data security in cloud storage systems is a prominent problem because of the lack of physical possession and control of the data for data owners [3]. Additionally, the shared cloud infrastructure between organizations is still facing a broad range of both internal and external threats to the outsourced data [4]. Cloud storage systems do not immediately offer any guarantee of data integrity, confidentiality, and availability. As a result, cloud service providers should adopt data security solutions to ensure that the data of their customers is available, correct and safe from unauthorized access and disclosure [5].

© Springer International Publishing AG 2017
A.E. Hassanien et al. (eds.), *Proceedings of the International Conference on Advanced Intelligent Systems and Informatics 2016*, Advances in Intelligent Systems and Computing 533, DOI 10.1007/978-3-319-48308-5_22

Consequently, this paper aims to identify potential security problems of outsourced data in the cloud and to propose a Security-as-a-Service model that addresses the security requirements for cloud storage as identified from the literature. The main contribution of this paper is to propose a Security-as-a-Service model that enables cloud customers to share their data over any cloud storage platform while data confidentiality, privacy, integrity and availability are preserved. The proposed model includes: (1) A new strong (multi-factor) user authentication sub-system that is based on the traditional passwords and the behavioral keystroke dynamics. The proposed authentication sub-system eliminates the tradeoff between the authentication accuracy and the elapsed time of the verification process by utilizing different feature extraction methods and clustering the user profile templates in the keystroke dataset, (2) A new access control sub-system that integrates hierarchical role based access control with attribute based encryption, in order to support automatic user role assignments and relieve the data owner from the online and computational burdens of user role/permission assignment processes and (3) A new auditing sub-system that supports batch auditing where multiple auditing tasks with different data files can be performed simultaneously. Moreover, the proposed auditing sub-system supports efficient data recovery to repair any corrupted data.

The rest of the paper is organized as follows. Section 2 overviews related work. Section 3 provides the detailed description of the proposed security as a service model. Then, Sect. 4 further discusses the evaluation criteria and experimental results and finally, the conclusion is presented in Sect. 5.

2 Related Work

Cloud security is an active research area; numerous research problems belonging to this area have been studied intensively before, [6–8]. Recently, there have been considerable research studies [9–11, 24] regarding user authentication in cloud environments which can be categorized as: (1) Knowledge based authentication methods that rely on factors the user knows such as; a password or a PIN code, (2) Possession based authentication systems that require the user to physically possess an object such as a token, certificate and ID card, (3) Biometric based authentication systems that identify users according to certain physiological attributes or behavioral characteristics that are uniquely associated with a person such as; fingerprint, iris, voice and keystroke dynamics, and (4) Multi-factor or strong authentication systems are referred to as a combination of two or three of these authentication methods. In addition, various access control models have emerged to assure confidentiality, including; Discretionary Access Control (DAC) [12], Mandatory Access Control (MAC) [13], Role-Based Access Control (RBAC) [14], identity-based encryption [15] and attribute based encryption [16, 17]. Moreover, many data auditing systems have been proposed to ensure the integrity of remote data in the cloud classified into several categories according to: (1) The type of the auditor: Public auditing [18] or private auditing [19], (2) The distribution of data to be audited: Single-copy [20] or replicated data [21] and (3) The data persistence: Static [22] or dynamic data [23].

However, there is always a tradeoff between maintaining security and obtaining efficiency. Thus, it is difficult but nevertheless essential to explore how to efficiently address security challenges over dynamic cloud data.

3   The Proposed Security as a Service Model

As illustrated in Fig. 1, the proposed model involves four main entities; the cloud service provider (CSP), a set of data owners, a set of users and the third party auditor (TPA). In order to ensure the three main security dimensions (i.e., confidentiality, integrity and availability) for outsourced data in the cloud, the proposed security model consists of four principal sub-systems; authentication, access control, auditing and data management sub-systems.

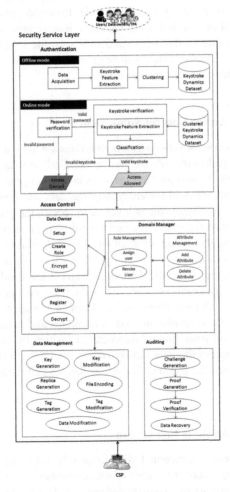

Fig. 1. The proposed architecture of the security service

3.1   The Proposed Authentication Sub-system

The proposed authentication sub-system supports keystroke authentication for cloud environments that is applicable to fixed text and free text typed on traditional and touchscreen keyboards. The proposed authentication sub-system operates in two modes: Offline/registration and online/login. During registering a new user, the new user firstly gives some enrollment information such as ID and password and inputs a certain number of typing samples. Secondly, the sub-system collects the raw keystroke timing measurements for that user during the data acquisition process. Then, the feature extraction process removes outliers and selects the most relevant feature set by combining different feature selection methods via the weighted sum fusion rule. Subsequently, a profile template for that user is built and finally, the user profile template can be included in an existing cluster if it satisfies the predefined cluster rules or establishes a new cluster. During login, the user provides an ID and a password and the login information will be firstly verified. If the user provides the wrong ID and/or password, access will be denied. However, if the user provides the correct password, the extracted keystroke features will be verified using the classification process. If the user's keystroke features are similar to that of a valid user's profile template in the clustered keystroke dataset, access to the system will be granted; otherwise, access will be denied even if the user provides the correct password.

3.2   The Proposed Access Control Sub-system

In the proposed access control sub-system, permissions are assigned to roles rather than users. The users who qualify the role are granted permissions to access outsourced data files. The data owner firstly runs the Setup process in order to generate a master key and a group public key that are used for encrypting the data and generating the keys for users and roles. Then, the Create role is invoked to specify the proper access structure for each role and generate the secret key and the public parameters for each role. Data may be encrypted using the public information of the role that is allowed to access it.

Any user can register to the system and provide the essential information such as user name, occupation, etc. Then, the domain manager is responsible for verifying and assigning attributes such as years of experience, education, etc., generating a secret key for the user associated with their attributes. The Domain Manager may also execute the Revoke attribute process to revoke any user's attributes and consequently update the user's secret key and all roles' public parameters in which the user is involved. The user may request to access the data outsourced in the CSP via the Decrypt process. Then, the process outputs the plaintext if the user has the permission to access the data and fails if otherwise.

3.3   The Proposed Auditing Sub-system

The auditing sub-system is concerned with frequently verifying the integrity of the outsourced data and data recovery in case of data corruption. First, the TPA challenges the CSP to verify the integrity of all outsourced replicas. This process randomly specifies the positions of the blocks to be checked. Upon receiving the challenge request, the CSP sends a tag proof and a data proof as well, without revealing any information about the challenged data blocks to the TPA. Upon receiving the proofs from the CSP, the TPA checks whether the proofs are valid or not. If there is any corruption, the TPA should notify the data owner and run the data recovery process. In the case of multiple replica data files, the CSP could recover the original file from a healthy replica and restore the corrupted replica. In the case of single copy encoded data files, the corrupted data can be recovered by the parity data in the erasure correction codes.

3.4   The Proposed Data Management Sub-system

The data management sub-system is responsible for generating the keys of the users, preparing the data before being moved to the cloud storage and maintaining data dynamics. The data owner may choose to outsource single copy or multiple replica data files. In the case of a single copy data file, the file needs to be first encoded using any erasure correction codes before being outsourced. Encoded files can mitigate arbitrary amounts of data corruption. In the case of a multiple replica data file, unique differentiable replicas of the data file are generated in order to prevent the CSP from storing fewer replicas than what have been paid for. Unique copies of each file block are created by encrypting it using a homomorphic probabilistic encryption scheme. Then, the tag generation process pre-computes a set of integrity tags for each data block for further data integrity auditing. These tags will be used by the TPA to verify the integrity of the outsourced data. When any data owner modifies the data, the data modification process whether the CSP has performed the modification on all the replicas' blocks and their corresponding tags as well. When a new user is authorized to modify the data file, the Key Modification process updates the system keys. When revoking a user, the Tag Modification process updates the system keys and regenerates the tags that are previously computed by the revoked user.

4   Experiments and Discussion

The experiments are conducted using MATLAB R2013a (version 8.1) and Java JDK 1.7.60 on a system with an Intel Core i5 processor running at 2.2 GHz and 4 GB RAM running Windows 7. All files used in the experiments are downloaded from the Human Genome Project at NCBI [25]. For assessing the proposed authentication sub-system, we use the Clarkson University's keystroke dataset [26] that is based on a mixed data of passwords, fixed text, and free text. All results are the averages of 20 trials.

The objective of the following experiments is proving the efficiency and scalability of the proposed security model. Figure 2 illustrates the effect of the number of users on the computational times of the proposed authentication, access control, and auditing sub-systems, using a 128 MB file. As illustrated in Fig. 2, the computational times of access control and auditing sub-systems remain constant, irrespective of the number of existing users which maintain the model scalability, especially with large scale systems of huge numbers of users. Although the computational time of the authentication

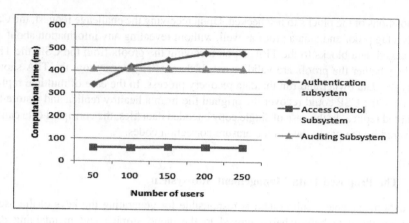

Fig. 2. The effect of the number of users on the computational time of the proposed security service

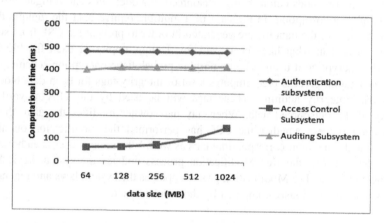

Fig. 3. The effect of data size on the computational time of the proposed security service

sub-system grows linearly with the number of users, it turns out to be constant when the number of users exceeds 200.

For a system with 200 users, Fig. 3 demonstrates that the computational times of the proposed authentication and auditing sub-systems remain constant, irrespective of the outsourced data size, which proves the efficiency of the proposed service. However, the large file sizes generally increase the computational time of the access control sub-system, especially when the data size exceeds 256 MB. Additionally, the storage overheads of the access control sub-system (i.e., the cipher text size and the user decryption key size) and the auditing sub-system (i.e., integrity tags) are independent of the number of users in the system. However, the storage overhead of the authentication sub-system (i.e., keystroke samples and features) grows linearly with the number of users (Fig. 4).

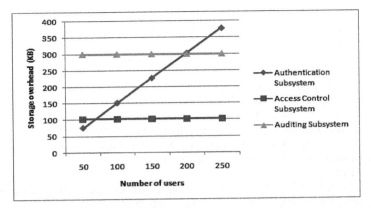

Fig. 4. The effect of the number of users on the storage overhead of the proposed security service

5 Conclusion

Data outsourcing raises serious security and privacy issues that slow the rate of cloud migration. This paper introduces security as a service model that offers a baseline security to data owners to protect their own outsourced data on the cloud. A new multi-factor authentication sub-system is proposed based on the traditional password and user key-stroke dynamics. The proposed authentication sub-system utilizes different feature extraction methods, as a pre-processing step, to minimize the feature space dimensionality. Additionally, a dynamic access control system is introduced in order to support automatic user role assignments by integrating hierarchical role based access control with attribute based encryption. Moreover, a multi-owner dynamic public auditing sub-system is proposed to ensure the integrity of the outsourced data without downloading the entire data files or revealing any private information about the verified data. So the proposed model does not only prevent the deception and forgery of cloud storage providers but also prevents the leakage of outsourced data during the process of verification.

References

1. Cao, N., Wang, C., Li, M., Ren, K., Lou, W.: Privacy-preserving multi-keyword ranked search over encrypted cloud data. IEEE Trans. Parallel Distrib. Syst. **25**(1), 222–233 (2014)
2. Abo-alian, A., Badr, N.L., Tolba, M.F.: Auditing-as-a-Service for cloud storage. In: Intelligent Systems 2014, pp. 559–568. Springer International Publishing (2015)
3. Wang, C., Chow, S.S., Wang, Q., Ren, K., Lou, W.: Privacy-preserving public auditing for secure cloud storage. IEEE Trans. Comput. **62**(2), 362–375 (2013)
4. Abo-alian, A., Badr, N.L., Tolba, M.F.: Keystroke dynamics-based user authentication service for cloud computing. Concurrency Comput. Pract. Experience **28**, 2567–2585 (2015)

5. Abo-alian, A., Badr, N.L., Tolba, M.F.: Hierarchical attribute-role based access control for cloud computing. In: The 1st International Conference on Advanced Intelligent System and Informatics (AISI2015), pp. 381–389 (2016)
6. Abo-alian, A., Badr, N.L., Tolba, M.F.: Integrity as a service for replicated data on the cloud. Concurrency Comput. Pract. Experience (2016)
7. Abo-Alian, A., Badr, N.L., Tolba, M.F.: Integrity verification for dynamic multi-replica data in cloud storage. Asian J. Inf. Technol. **15**(6), 1056–1072 (2016)
8. Ramachandran, M., Chang, V.: Towards performance evaluation of cloud service providers for cloud data security. Int. J. Inf. Manage. **36**(4), 618–625 (2016)
9. Liu, H., Ning, H., Xiong, Q., Yang, L.: Shared authority based privacy-preserving authentication protocol in cloud computing. IEEE Trans. Parallel Distrib. Syst. **26**(1), 241–251 (2015)
10. Liu, Z., Yan, H., Li, Z.: Server-aided anonymous attribute-based authentication in cloud computing. Future Gener. Comput. Syst. **52**, 61–66 (2015)
11. Abo-alian, A., Badr, N.L., Tolba, M.F.: Authentication as a service for cloud computing. In: The International Conference on Internet of Things and Cloud Computing, pp. 36–42. ACM (2016)
12. Younis, Y., Kifayat, K., Merabti, M.: An access control model for cloud computing. J. Inf. Secur. Appl. **19**(1), 45–60 (2014)
13. Zhu, Y., Zhu, H.: Formal verification of mandatory access control for privacy cloud. In: 3rd International Conference on Computer Science and Network Technology, pp. 297–300. IEEE (2013)
14. Luo, J., Wang, H., Gong, X., Li, T.: A novel role-based access control model in cloud environments. Int. J. Comput. Intell. Syst. **9**(1), 1–9 (2016)
15. Shen, J., Liu, D., Liu, Q., Wang, B., Fu, Z.: An authorized identity authentication-based data access control scheme in cloud. In: 18th International Conference on Advanced Communication Technology, pp. 56–60. IEEE (2016)
16. Liang, K., Au, M., Liu, J., Susilo, W., Wong, D., Yang, G., et al.: A secure and efficient ciphertext-policy attribute-based proxy re-encryption for cloud data sharing. Future Gener. Comput. Syst. **52**, 95–108 (2015)
17. Li, M., Yu, S., Zheng, Y., Ren, K., Lou, W.: Scalable and secure sharing of personal health records in cloud computing using attribute-based encryption. IEEE Trans. Parallel Distrib. Syst. **24**(1), 131–143 (2013)
18. Wang, B., Li, B., Li, H.: Panda: Public auditing for shared data with efficient user revocation in the cloud. IEEE Trans. Serv. Comput. **8**(1), 92–106 (2015)
19. Ren, Y., Shen, J., Zheng, Y., Wang, J., Chao, H.C.: Efficient data integrity auditing for storage security in mobile health cloud. Peer-to-Peer Netw. Appl. **9**, 1–10 (2015)
20. Shuang, T., Lin, T., Xiaoling, L., Yan, J.: An efficient method for checking the integrity of data in the cloud. China Commun. **11**(9), 68–81 (2014)
21. Zhang, Y., Ni, J., Tao, X., Wang, Y., Yu, Y.: Provable multiple replication data possession with full dynamics for secure cloud storage. Concurrency Comput. Pract. Experience **28**, 1161–1173 (2015)
22. Yuan, J., Yu, S.: Proofs of retrievability with public verifiability and constant communication cost in cloud. In: The 2013 International Workshop on Security in Cloud Computing, pp. 19–26. ACM (2013)
23. Liu, C., Ranjan, R., Yang, C., Zhang, X., Wang, L., Chen, J.: MUR-DPA: top-down levelled multi-replica merkle hash tree based secure public auditing for dynamic big data storage on cloud. IEEE Trans. Comput. **64**(9), 2609–2622 (2014)

24. Hassanien, A.E.: Hiding iris data for authentication of digital images using wavelet theory. Int. J. Pattern Recogn. Image Anal. **16**(4), 637–643 (2006)
25. National Center for Biotechnology Information (2014). http://www.ncbi.nlm.nih.gov
26. Vural, E., Huang, J., Hou, D., Schuckers, S.: Shared research dataset to support development of keystroke authentication. In: The 2014 IEEE International Joint Conference on Biometrics, pp. 1–8. IEEE (2014)

Fuzzy Rough Set Based Image Watermarking Approach

Musab Ghadi[1(✉)], Lamri Laouamer[1,2], Laurent Nana[1], and Anca Pascu[1]

[1] Lab-STICC (UMR CNRS 6285), University of Bretagne Occidentale, Brest, France
e21409716@etudiant.univ-brest.fr, {Laurent.Nana,Anca.Pascu}@univ-brest.fr
[2] Department of Management Information Systems, Qassim University,
Buraydah, Kingdom of Saudi Arabia
laoamr@qu.edu.sa

Abstract. Computational intelligent techniques can be useful in developing efficient watermarking approaches that are able to maintain and reduce risks to integrity, confidentiality, and availability of information and resources in computer and network systems. This paper aims to develop a new spatial domain-based watermarking approach that uses the fuzzy rough set to select well thought out blocks to embed secret data with acceptable rate of imperceptibility and robustness against different scenarios of attacks. The proposed model focuses on analyzing the host image to discover specified features in some blocks that in turn will be considered in the watermarking process. These features include the characteristics of the Human Visual System (HVS) regarding the color sensitivity and the textured/semi-smooth regions, where embedding the watermark in low color sensitivity to the human eye and more textured regions gains high imperceptibility and robustness. The experiment results show that the proposed approach gives interesting and remarkable results to preserve the image authentication.

Keywords: Intelligent techniques · Watermarking · Fuzzy rough set · Imperceptibility · Robustness · Attacks

1   Introduction

Digital watermarking is one of the appropriate solutions, which can contribute significantly to the authentication of the transmitted text, images, and videos on the Internet. Many watermarking approaches are proposed both in spatial and frequency domains [1–5]. taking into account the complexity of modern systems and the diversity of attacks are needed. Computational intelligent techniques exhibit many capabilities to adapt and provide multimodal solutions for these complex systems. The scope of computational intelligent methods involves fuzzy logic, genetic algorithm, artificial neural network, and rough set. Some frequency domain-based watermarking approaches use these techniques with a goal to develop an efficient watermarking approach using Discrete Cosine Transform

© Springer International Publishing AG 2017
A.E. Hassanien et al. (eds.), *Proceedings of the International Conference on Advanced Intelligent Systems and Informatics 2016*, Advances in Intelligent Systems and Computing 533, DOI 10.1007/978-3-319-48308-5_23

(DCT), Discreet Wavelet Transform (DWT), and Singular Value Decomposition (SVD) [4,5]. Intelligent techniques in spatial domain based watermarking approaches have not been widely used. Few works such as those proposed in [6–8] explore the artificial neural network and rough set principles. This paper aims to develop a new spatial domain watermarking approach based on the fuzzy rough set technique by selecting particular blocks to embed secret data with an acceptable rate of imperceptibility by achieving a remarkable robustness against different kind of attacks and maintaining a low process complexity. The rest of this paper is organized as follows: In Sect. 2, we present a literature review. Section 3, illustrates the rough set principle. Then, the system model is illustrated in Sect. 4. The experiments result is presented in Sect. 5 and a comparative study is conducted in Sect. 6. Finally, we conclude the presented approach in Sect. 7.

2 Literature Review

This section presents a literature review on spatial and frequency domain-based digital image watermarking approaches using computational intelligent techniques in both watermark embedding/extraction processes.

In the case of spatial domain, authors in [6] proposed an efficient watermarking approach that embeds an encoded watermark image in the image blue components. The different intensities of the original blue components are used as a feature to train an Artificial Neural Network (ANN) and extract the attacked watermark image. The experiments result showed that the Bit Correct Rate (BCR) ratio ranged between [57.23–100]% and the Peak Signal-to-Noise Ratio (PSNR) ranged between [10.03–39.6] dB. As well, authors in [8] proposed a fragile watermarking scheme that operates in spatial domain based on the K-means method. The proposed scheme computes the distance between each 2×2 block and its centroid to find the feature sequence that XOR-ed with random sequence number to generate a 8 authenticated bits that replace the Least Significant Bits (LSBs) in each 2×2 block of the original image. This model was tested in tamperproofing under a variety of attacks like cut-paste and collage attacks. The experiments result showed that the PSNR ratio reached the 46.2 dB. For the case of frequency domain, a probabilistic neural network technique was proposed in [9] based on a DWT watermarking scheme. As well, a genetic algorithm was applied to develop an SVD-based watermarking scheme in [10] and a DCT-based scheme in [11]. In addition, [12] utilized a fuzzy rough set to design a DWT-based watermarking scheme.

3 Rough Set Principle

Rough set is one of the computational intelligent tools that deals with the induction of concept approximation to process information in a database. It is concerned with classification and analysis of imprecise and incomplete information

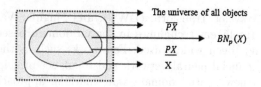

Fig. 1. The three approximation regions in rough set theory

or knowledge to facilitate the tasks of feature selection and knowledge discovery [13,14]. The imprecision and uncertainty in rough set theory is expressed by boundary region of a set. Rough set technique is based on sets of objects (U) and attributes (A) that represent an information system I such as $I = (U,A)$, where $X \subseteq U$ and $P \subseteq A$. The system table is built such as its rows correspond to objects U and columns correspond to attributes A. Based on the defined attributes P, the rough set can approximate X using only the information contained in P by defining three regions, as illustrated in Fig. 1; (1) The B-upper region $\overline{PX}$, which is the set of all of objects that can be possibly classified as member of X with respect to P. It can be denoted by $\overline{PX} = \{x | p(x) \cap X \neq \phi \}$. (2) The B-lower region $\underline{PX}$, which is the set of all objects that can be classified with certainly as member of X regarding P, and certainly belong to the subset of interest. It can be denoted by $\underline{PX} = \{x | p(x) \subseteq X\}$. (3) The boundary region BNp(X), represents the set of all of objects, which cannot be classified neither in $\overline{PX}$ nor $\underline{PX}$. If the BNp(X) is empty, then the set is called 'Crisp set'. Otherwise, if the BNp(X) is non-empty, then the set X is called 'Rough set' in other terms $BNp(X) = \overline{PX} \text{-} \underline{PX}$ [14].

4   System Model

The proposed system considers two types of colored image: semi-smooth and textured images to construct a system table, where any colored image I consists of three components R, G, and B that can be converted to YCbCr components to display each of luminance component(Y), blue-difference (Cb), and red-difference (Cr). Then the proposed model exploits the low HVS sensitivity to the blue component (Cb) [6] and the highly textured regions in the original image to be concerned in the embedding process. Selecting those attributes is justifited by the availability of imperceptibility and robustness against attacks [6,11].

4.1   Construct an Information System for both Semi-smooth and Textured Images

Table 1 below, illustrates the information systems for both semi-smooth and textured images by taking into account that the average value of 8×8 block pixels in Cb matrix corresponds to Cb attribute and the range of the DC coefficient in each block corresponds to DC coefficient attribute. The decision of the information system is based on two thresholds: T1 that corresponds to the average value

Table 1. Information systems of semi-smooth and textured images

Information system of semi-smooth images				Information system of textured images			
Class (C)	C_b	DC coefficient	Decision (D)	Class (C)	C_b	DC coefficient	Decision (D)
1	$X \leq 127$	0	Yes	1	$X \leq 127$	0	No
2	$X > 127$	0	Yes	2	$X > 127$	0	No
3	$X \leq 127$	[1-3]	Yes	3	$X \leq 127$	[1-3]	No
4	$X > 127$	[1-3]	Yes	4	$X > 127$	[1-3]	Yes
5	$X \leq 127$	[4-5]	Yes	5	$X \leq 127$	[4-5]	Yes
6	$X > 127$	[4-5]	No	6	$X > 127$	[4-5]	Yes
7	$X \leq 127$	[6-7]	No	7	$X \leq 127$	[6-7]	Yes
8	$X > 127$	[6-7]	No	8	$X > 127$	[6-7]	Yes
9	$X \leq 127$	[8-9]	No	9	$X \leq 127$	[8-9]	Yes
10	$X > 127$	[8-9]	No	10	$X > 127$	[8-9]	Yes
11	$X \leq 127$	[10-11]	No	11	$X \leq 127$	[10-11]	Yes
12	$X > 127$	[10-11]	No	12	$X > 127$	[10-11]	Yes

of Cb component and T2 corresponds to the range of DC coefficient in any 8×8 block. By demonstrating the information systems in Table 1 for semi-smooth and textured images respectively, we can find that the decision for semi-smooth images depends on T1 ≤ 127 and T2 $= [4\text{-}5]$. This can be theoretically justified by noting that the probability to see an increasing value in the average Cb in each block is very low, particularly when the average value did not exceeded 127, compared to when the average value of Cb is greater than 127. As well as regard, the value of DC coefficient where the probability to see a change in DC coefficient may be increased if the DC coefficient becomes greater than or equal to [4,5]. On the other hand, the decision regarding the textured images depends on T1 > 127 and T2 $= [1\text{-}3]$. This can be justified by the fact that embedding watermark in more textured regions is imperceptible and highly robust against attacks. So the probability to see an increasing value in the average of Cb for each block in the textured images is very low, particularly when the average value exceeds 127 and the value of DC coefficient is equal to or greater than [1–3].

4.2   The Employment of Rough Set Technique

From the information systems illustrated in Table 1, we can build a unified information system as illustrated in Table 2 to be manipulated in one of computational intelligent technique to deal with a variety of images regardless of its nature. Rough set is our choice to deduce a concepts approximation of existing information in our model. By rough set, we can extract the CD-upper approximation $\overline{CD}$, the CD-lower approximation $\underline{CD}$, and the boundary BNc(D) regions as following:

$$Yes \rightarrow \{1, 3, 5, 7, 8, 9, 10, 12, 14, 16, 18, 20, 22, 24\}$$

$$NO \rightarrow \{2, 4, 6, 11, 13, 15, 17, 19, 21, 23\}$$

$$Yes \setminus No \rightarrow \{1, 2, 3, 4, 5, 6, 11, 12, 13, 14, 15, 16, 17, 18, 19, 20, 21, 22, 23, 24\}$$

$$\overline{CD} \rightarrow \{1, 2, 3, 4, 5, 6, 11, 12, 13, 14, 15, 16, 17, 18, 19, 20, 21, 22, 23, 24, 7, 8, 9, 10\}$$

Table 2. Information systems of semi-smooth and textured images

Class (C)	C_b	DC coefficient	Decision (D)	Class (C)	C_b	DC coefficient	Decision (D)
1	$X \leq 127$	0	Yes	13	$X \leq 127$	[6-7]	No
2	$X \leq 127$	0	No	14	$X \leq 127$	[6-7]	Yes
3	$X > 127$	0	Yes	15	$X > 127$	[6-7]	No
4	$X > 127$	0	No	16	$X > 127$	[6-7]	Yes
5	$X \leq 127$	[1-3]	Yes	17	$X \leq 127$	[8-9]	No
6	$X \leq 127$	[1-3]	No	18	$X \leq 127$	[8-9]	Yes
7	$X > 127$	[1-3]	Yes	19	$X > 127$	[8-9]	No
8	$X > 127$	[1-3]	Yes	20	$X > 127$	[8-9]	Yes
9	$X \leq 127$	[4-5]	Yes	21	$X \leq 127$	[10-11]	No
10	$X \leq 127$	[4-5]	Yes	22	$X \leq 127$	[10-11]	Yes
11	$X > 127$	[4-5]	No	23	$X > 127$	[10-11]	No
12	$X > 127$	[4-5]	Yes	24	$X > 127$	[10-11]	Yes

$$\underline{CD} \rightarrow \{7, 8, 9, 10\}$$

$$BNc(D) \rightarrow \{1, 2, 3, 4, 5, 6, 11, 12, 13, 14, 15, 16, 17, 18, 19, 20, 21, 22, 23, 24\}$$

Then our model concerned with those blocks that are matching the condition for any rough set element (i.e. inside the BNc(D) set).

4.3 Parsing JPEG Bitstream

As proposed in [15], the JPEG file structure is constructed by combining many segments that represent the bitstream file for any image. Each 8×8 DCT block was compressed by 1 DC and possibly 63 AC coefficients. In our model, we parse the encoded data to find the DC coefficient category based on Huffman table that is illustrated in [16].

4.4 Model Initialization

The proposed model has to read the Cb component from the RGB image to build an avg_matrix and category_matrix that define the attributes of the pre-arranged information systems to deduce the decision in the proposed model. The initialization process is illustrated in Fig. 2.

4.5 Watermark Embedding Process

After defining the avg_matrix and category_matrix from Cb matrix of RGB image, we can check whether the given block of Cb matrix matches the rough set rules by means of BNc(D) set or not. Subsequently, any satisfied block will be in-queued in rough set RS queue, and will be embedded with a watermark w by a linear interpolation equation. This equation gives us the ability to control the visibility/invisibility of embedded watermark by factor t. The obtained

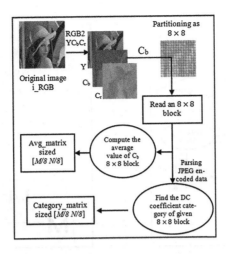

Fig. 2. Model initialization structure

watermarked image iw_cb is illustrated in Eq. (1), and the embedding process is illustrated in Fig. 3.

$$i_w = (1 - t)w + t * i, \ \ 0 < t < 1 \tag{1}$$

where i is a squared size Cb component of the RGB image, w is the watermark image sized 8×8, and t is the linear interpolation factor. The value of the linear interpolation factor t reflects the degree of visibility/invisibility of the embedded watermark in the original image within three cases.

Case 1 (visible case): if t closes to 0, then $i_w = \underbrace{(1 - t) * w}_{\text{goes to 1}} + \underbrace{(t * i)}_{\text{goes to zero}}$. The result will be as $i_w = $ w, which means that w is the dominant in i_w.

Case 2 (invisible case): if t closes to 1, then $i_w = \underbrace{(1 - t) * w}_{\text{goes to zero}} + \underbrace{(t * i)}_{\text{goes to 1}}$. The result will be as $i_w = $ i, which means that i is the dominant in i_w.

Case 3 (semi-visibility case): if $t \in [0.4 - 0.6]$, then w will be clear with some visibility degrees. So, in our experiments we used $t = 0.98$, since the invisibility is achieved if t is closes to 1. After embedding the watermark, the resulting iw_cb matrix and the RS queue would be sent to the receiver via public network, which requires to concatenate the Y and Cr of the original image with iw_cb. The Receiver will receive the attacked iwa_RGB image and the attacked RS queue, because the RS may be also prone to different attacks, we need to secure it. One-Time Pad (OTP) algorithm can be a best choice to encrypt it due to its unbreakability.

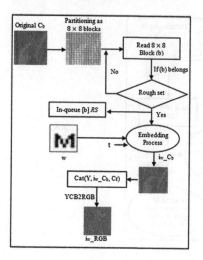

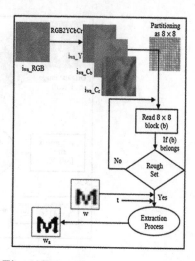

Fig. 3. Embedding process framework   **Fig. 4.** Extraction process framework

4.6   Watermark Extraction Process

The watermark extraction will be applied on the attacked watermarked image iwa_RGB in order to extract the attacked watermark w_a also via linear interpolation as illustrated in Eq. (2).

$$w_a = (1/t)w - (1-t)/t * i_w a, \ 0 < t < 1 \tag{2}$$

where w_a is the extracted attacked watermark image, i_wa is the attacked watermarked image, and $t = 0.98$. The extraction process is illustrated in Fig. 4.

5   Experiments Result

The proposed model processed a square color images sized 128×128 as an original image and 8×8 watermark image. The resulted watermarked images are exposed to a variety of attacks including: geometric attack (i.e. such as rotation) and non-geometric attacks (i.e. like JPEG compression, Gaussian noise, and median filtering) by StirMark Benchmark v.4 [17]. Figure 5 illustrates a sample of the processed images. The tests are conducted on these seven images that are in turn partitioned as 8×8 blocks and a watermark image w sized 8×8 is embedded in particular blocks of the original images that satisfy the rough set rules.

To evaluate the performance of our model, we applied three metrics (i.e. PSNR, Bit Error Rate (BER), and Correlation Coefficient (CC) [6,11]), where the PSNR and CC measured the robustness and the similarity between the original watermark w and all extracted attacked watermark w_a, while the BER metric measured the stabilization (i.e. the number of correct bits that can be extracted from the attacked watermarked image). Due to the large number of

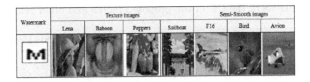

Fig. 5. Samples of original images and the watermark images

Table 3. Average results for PSNR, BER, and CC for processed images

Images name	Robustness against attacks					
	Attacks	JPEG_80	Medain_9	Noise_10	Rot_10	Rot_45
Lena	PSNR	45.5	45.54	44.8	45.17	44.75
	BER	0.18	0.18	0.19	0.18	0.18
	cc	1	1	1	1	1
Baboon	PSNR	44.5	44.45	43.8	44.2	44.0
	BER	0.19	0.19	0.19	0.19	0.19
	cc	1	1	0.99	1	1
Peppers	PSNR	45.2	45.17	44.6	44.8	44.5
	BER	0.18	0.18	0.19	0.18	0.19
	cc	1	1	0.99	1	1
Sailboat	PSNR	44.8	44.9	44.1	44.5	44.2
	BER	0.19	0.19	0.19	0.19	0.19
	cc	1	1	1	1	1
F16	PSNR	43.4	43.4	43.4	43.5	43.6
	BER	0.20	0.20	0.20	0.20	0.19
	cc	1	1	1	1	1
Bird	PSNR	45.2	45.2	44.1	44.8	44.5
	BER	0.19	0.19	0.19	0.19	0.19
	cc	1	1	1	1	1
Avion	PSNR	43.5	43.5	43.5	43.6	43.7
	BER	0.20	0.20	0.20	0.19	0.19
	cc	1	1	0.99	1	1

blocks that satisfy the rough set rules, our experiment's result are displayed in average as showed in Table 3.

The results of PSNR, CC, and BER are very interesting, where the PSNR ratio ranges between (43.4–45.5) dB and the CC ratio is equal to 1 in most cases, while the BER ratio ranges between (18–20) %. By comparing the results with the different kind of attacks, we can note that the result involved with Noise_10 and Rot_45 are the least ones in terms of PSNR, while JPEG_80 and Median_9 present the same results in PSNR, CC, and BER. From the displayed

Table 4. Comparisons of the CC of the proposed model and other related approaches

Image's name	Attacks	Metric	Han et al. [11]	Lu et al. [18]	Proposed model
Lena image	JPEG_80	cc	NA	0.98	1
	Noise_10	cc	0.92	0.91	1
	Median_9	cc	0.87	0.95	1
Baboon image	JPEG_80	cc	NA	0.99	1
	Noise_10	cc	0.91	0.91	1
	Median_9	cc	0.82	0.93	1
F16 image	JPEG_80	cc	NA	0.99	1
	Noise 10	cc	0.91	0.94	1
	Median 9	cc	0.82	0.91	1

Table 5. Comparisons of PSNR of the proposed model and the model in [11]

Attacks	Metric	Han et al. [11]	Proposed model
Lena	PSNR	42.54	43.9
Baboon	PSNR	41.5	43.5
F16	PSNR	42.7	42.6

Table 6. Comparisons of BER of the proposed model and the model in [6]

Image		Lena image		Baboon image		F16 image	
Attacks	Metric	Findik et al. [6]	Proposed model	Findik et al. [6]	Proposed model	Findik et al. [6]	Proposed model
JPEG_80	BER	0.40	0.156	0.42	0.159	0.29	0.16
Noise_10	BER	0.07	0.156	0.13	0.159	0.07	0.16
Rot_10	BER	0.08	0.155	0.12	0.159	0.08	0.16
Rot_45	BER	0.15	0.156	0.19	0.157	0.14	0.158

results we can also conclude that our proposed model is compatible to deal with both textured and semi-smooth images. This will give us a sense that the rough set-based watermarking technique is very useful to achieve the robustness and imperceptibility in terms of stabilization and similarity.

6   Comparative Study

In this section we compare the results achieved by the proposed model with those of other interesting related approaches. An adaptive work has been made on our model to be compatible with other related work, in terms of the host image and watermark sizes. All of these approaches involved a particular computational intelligent technique. Our comparisons held on three images (i.e. Lena, Baboon

Table 7. Clarification of the Weakness/strength of the obtained approaches comparing with our model

The model	Is our model outperforming the obtained approach? And in which terms?	The evidence of the weakness/strength of the obtained approaches comparing with our model
Findik et al. [6] (2011)	Our model outperform [6] in terms of BER in case of high scale factor of attacks.	The weakness of this technique appears in embedding a watermark image in blue components of the host image, without consideration to the nature of embedding region (i.e. textured or smooth).
Han et al. [11] (2016)	Our model outperform [11] in terms of CC and PSNR	The weakness of this technique appears in embedding watermark in the DC coefficient of all DCT blocks, without consideration to the HVS sensitivity properties.
Lu et al. [18] (2006)	Our model outperform [18] in terms of CC.	The weakness of this technique appear in selecting the textured regions by considering the green and red components of the host image, which is not suitable for availability of imperceptibility and robustness against attacks, since the HVS sensitivity is rematkable in cases of choosing red and green components comparatively to the blue component.

and F16) with different attacks scenario. Firstly, our results are compared with [11,18] in terms of CC. Secondly, our results are compared with [11] in terms of PSNR and finally, our results are compared with [6] in terms of BER. Tables 4, 5 and 6 present these results, followed by a discussion in Table 7. Through Tables 4, 5 and 6, we can note that our model outperformed [11,18] models, in terms of

CC under Noise_10 and Median_9 attacks, whereas in the case of JPEG_80 the CC in all approaches are convergent. Our model also enhanced a robustness ratio by (1–2) dB in all attacks scenario comparing with [11]. Similarly, the BER was enhanced by 25 % with JPEG_80 attack, and it is approximately the same in Rot_45 attack. In terms of Noise_10 and Rot_10, the results of BER ratio in [6] outperformed our model by 10 %. This is due the low noise and rotation used factors. However, we think that our model is more practical than model in [6], especially with the high used factors in noise, rotation, or JPEG attacks, which appear clearly in the case of Rot_45 attack.

7 Conclusion

This paper aims to develop a new spatial domain-based watermarking approach that utilizes the fuzzy rough set technique by selecting many blocks to embed secret data with acceptable rate of imperceptibility and robustness against different kinds of attacks. Based on the proposed model structure, we can conclude that the rough set-based watermarking model in spatial domain is capable to achieve a significant rate of robustness and imperceptibility with means of similarity and stabilization. The experiment's result showed that the results of PSNR, CC, and BER are very interesting. This gives a sense that the rough set-based watermarking can serve significantly in achieving both watermark imperceptibility and robustness.

References

1. Sun, Q., et al.: A blind color image watermarking based on DC component in the spatial domain. Optics **124**(23), 6255–6260 (2013). Elsevier
2. Laouamer, L., et al.: Improving authenticity and robustness of medical images watermarking schemes based on multi-resolution decomposition. In: International Conference on Imaging Systems and Techniques, pp. 331–336. IEEE (2015)
3. Ouyang, L., et al.: A blind robust color image watermarking method using quaternion fourier transform. In: Congress on Image and Signal Processing, vol. 1, pp. 485–489. IEEE (2013)
4. Laouamer, L., Tayan, O.: A semi-blind robust DCT watermarking approach for sensitive text images. Arabian J. Sci. Eng. **40**(4), 1097–1109 (2015)
5. Benhocine, A., et al.: New images watermarking scheme based on singular value decomposition. J. Inf. Hiding Multimed. Sig. Process. **4**(1), 9–18 (2013)
6. Findik, O., et al.: Implementation of BCH coding on artificial neural network-based color image watermarking. Int. J. Innov. Comput. Inf. Control **7**(8), 4905–4914 (2011)
7. Bhattacharya, S.: Watermarking digital image using fuzzy matrix compositions and rough set. Int. J. Adv. Comput. Sci. Appl. **5**(6), 135–140 (2014)
8. Albakrawy, L., et al.:A rough k-means fragile water-marking approach for image authentication. In: Proceedings of the Federated Conference on Computer Science and Information Systems, pp. 19–23. IEEE (2011)

9. Alnabhani, Y.: Robust watermarking algorithm for digital images using discrete wavelet and probabilistic neural network. J. King Saud Univ. Comput. Inf. Sci. **27**(4), 1–9 (2015). Elsevier
10. Aslantas, V.: A singular-value decomposition-based image watermarking using genetic algorithm. J. Electron. Commun. **62**(5), 386–394 (2008). Elsevier
11. Han, J.: A digital image watermarking method based on host image analysis and genetic algorithm. J. Ambient Intell. Humanized Comput. **7**(1), 37–45 (2016)
12. Cong, J., et al.: Robust digital image watermark scheme on wavelet domain using fuzzy rough sets. J. Int. Fuzzy Sys. **30**(1), 1–12 (2015). IOS
13. Swiniarski, R., Skowron, A.: Rough set methods in feature selection and recognition. Pattern Recogn. Lett. **24**, 833–849 (2003)
14. Rissino, S., Lambert-Torres, G.: Rough set theory-fundamental concepts, principals, data extraction, and applications, data mining and knowledge discovery in real life applications. In: Data Mining and Knowledge Discovery in Real Life Applications, pp. 35–58, Chap. 3. InTech (2009)
15. Ghadi, M., et al.: JPEG bitstream based integrity with lightweight complexity of medical image in WMSNS environment. MEDES, pp. 53–58. ACM (2015)
16. Ghadi, M., et al.: Enhancing digital image integrity by exploiting JPEG bitstream attributes. JIDES **2**(1–2), 20–31 (2015). Elsevier
17. Petitcolas, F.A.P., Anderson, R.J., Kuhn, M.G.: Attacks on copyright marking systems. In: Aucsmith, D. (ed.) IH 1998. LNCS, vol. 1525, pp. 218–238. Springer, Heidelberg (1998). doi:10.1007/3-540-49380-8_16
18. Lu, Y., et al.: A novel color image watermarking method based on genetic algorithm. In: International Conference on Intelligent Computing, vol. 345, pp. 72–80 (2006)

EEG Signal Classification Using Neural Network and Support Vector Machine in Brain Computer Interface

M.M. El Bahy, M. Hosny, Wael A. Mohamed[(✉)], and Shawky Ibrahim

Faculty of Engineering (at Benha), Benha University, Cairo, Egypt
{mmbahy, mohamed.hosny, wael.ahmed,
shawky.Ibrahim}@bhit.bu.edu.eg

Abstract. Classification of EEG signals is one of the biggest problems in Brain Computer Interface (BCI) systems. This paper presents a BCI system based on using the EEG signals associated with five mental tasks (baseline, math, mental letter composing, geometric figure rotation and visual counting). EEG data for these five cognitive tasks from one subject were taken from the Colorado University database. Wavelet Transform (WT), Fast Fourier Transform (FFT) and Principal Component Analysis (PCA) were used for features extraction. Artificial Neural Network (ANN) trained by a standard back propagation algorithm and Support Vector Machines (SVMs) were used for classifying different combinations mental tasks. Experimental results show the classification accuracies achieved with the three used feature extraction techniques and the two classification techniques.

Keywords: Brain computer interface (BCI) · Artificial neural network (ANN) · Support vector machine (SVM)

1 Introduction

Today, a lot of communication systems between disabled peoples and external devices have been suggested and developed. Among them, brain computer interface (BCI) and brain machine interface (BMI) has been very attractive recently [19]. One mental task can be visualized by a person then the brain wave signals are measured, processed and evaluated to identify this mental task. Moreover, external devices such as computers or machines can be controlled [1]. AT first BCIs were used for medical reasons, but now BCI systems are also being developed for general people purposes mostly for entertainment. The main technology used in BCI systems for recording brain activity is electroencephalography (EEG). Although it is an imperfect indication of brain activity, compared to other technologies (MEG, FMRI and FNIR), but EEG has the most advantages for BCI systems. The main advantages are high temporal resolution, portability, doesn't expose patients to high-intensity magnetic fields and low cost of EEG hardware [2]. We can divide BCI systems in to the following classes: invasive BCI in which sensors have surgically implanted into the brain and non-invasive BCI

© Springer International Publishing AG 2017
A.E. Hassanien et al. (eds.), *Proceedings of the International Conference on Advanced Intelligent Systems and Informatics 2016*, Advances in Intelligent Systems and Computing 533, DOI 10.1007/978-3-319-48308-5_24

that uses sensors located on the scalp [3]. The use of additional motor movements is required in dependent BCI, while independent BCI doesn't require any muscle activity. A synchronous BCI where user interacts with the system only in specific time frames, however asynchronous can be used for any time frames. BCI systems methodology consist of signal acquisition, preprocessing, feature extraction and classification. So far the accuracy of classification has been one of the main drawbacks of the current BCI systems. Enhancing the accuracy may be achieved through enhancements in the three main phases of BCI.

This paper presented a non-invasive offline system for classifying different combination of mental tasks using EEG signals from the publicly available dataset of Kein and Aunon's and it achieved a high classification rates. In this research three different techniques of features extraction were used which are: Wavelet Transform, Fast Fourier Transform and Principal Component Analysis. Two classifications techniques were used which are: Neural Network trained by a standard back propagation algorithm and Support Vector Machines.

The rest of this paper continues as follows. Section 2 presents the previous work for brain computer interface systems. Section 3 describes the system methodology and explains the used techniques for feature extraction and data classification. Section 4 illustrates the experimental results of the proposed methodology for classifying the mental tasks. Conclusion and future work are illustrated in Sect. 5.

2   Related Work

N. Saadat, and H. Pourghassem [4] acquired EEG signals from three normal subjects during three tasks: Imagination of the left, right hand movements and generating some words. Band pass filter between 0–30 Hz was used to remove noise from EEG signals and the transition matrix of EEG signals was scaled between [0, 1] using non-linear normalization technique. Discreet Fourier Transform (DFT) was used to extract the spectral and spatial features from EEG signals as a feature extraction method. Classification was done using multi-layer perceptron neural network trained by back propagation algorithm. Classification rates between 73 % and 81 % were achieved for all subjects. Kenji Nakayama et al. [5] presented efficient pre-processing techniques in order to achieve high classification accuracy of mental tasks. The preprocessing techniques like segmentation along time axis, amplitude of FFT of EEG signals and reduction of samples by averaging and nonlinear normalization. Classification accuracy of 78 % was achieved for the recognition of five tasks. Anderson et al. [6] proposed a system by which EEG features were extracted through the short time principal component analysis (STPCA) and the EEG data was classified by the linear discriminant analysis (LDA). Classification accuracy for the recognition of five tasks was 77.9 %. Yuji Mizuno et al. [7] employs the maximum entropy method (MEM) for frequency analyses and investigates an alpha frequency band and beta frequency band in which features are more apparent In addition, learning vector quantization (LVQ) is used for clustering the EEG data with features extracted and classification accuracy for the recognition of five tasks was 81 %. Hosni et al. [8] used three of the five mental tasks from Keim and Aunon's dataset. These tasks were baseline task, letter task and math task. Eye blinks were identified and removed using Independent Component Analysis (ICA). Three different feature extraction techniques were used in this paper which are Parametric Auto Regressive (AR) modeling, AR spectral analysis and band power differences. Classification was done using Radial Basis Function (RBF) and Support Vector Machines (SVM). Best classification accuracy achieved was 70 %. Martina Tolić and Franjo Jović [9] extracted the features of EEG signals using Discrete Wavelet Transform and Neural Network was used as a classifier. Mean classification accuracy for the recognition of all five tasks was 90.75 %.

3 Proposed System Design

The aim of this research is to compare between three different features extraction techniques with two classifiers and classifying different combinations of three, four and five mental tasks. The system's methodology comprises of four main stages as illustrated in Fig. 1. The first step was Signal acquisition. The second step was signal preprocessing, to remove noises, artifacts and unwanted data. The third step was features extraction from the EEG signals. The fourth and the final step was classification of the signals to different classes that corresponding to the different mental tasks.

3.1 Signal Acquisition

The EEG data used in this study were collected by Keirn and Aunon [10]. This dataset can be described as follows: Several trials of five mental tasks were recorded and the number of times that each mental task was repeated is different from one subject to another. The number of trials for each subject as shown in Table 1 [11]. Each channel from each trial produced 2500 sample points for the 10 s recording because the amplified EEG signals were sampled at 250 samples per second. The selected subject was 6. EEG signals were recorded in two different days so, there were two sessions of

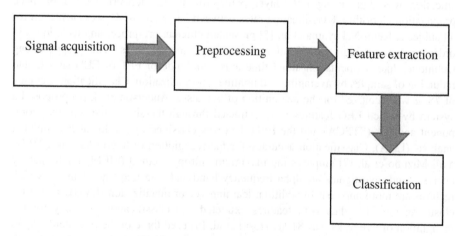

Fig. 1. Proposed system methodology.

Table 1. No of trials for each subject

Subject	Number of trials
1	10
2	5
3	10
4	10
5	15
6	10
7	5

recordings. First session differs from second session, this is possible since the recording sessions were separated by two weeks and it is known that the statistics of the brain waves are non-stationary over extended periods of time. If the subject is losing concentration throughout the task then mixing early and late portions of the EEG may degrade classification performance. In our research EEG signals of both sessions were used together and this is one of the main challenges in our research.

Data for seven subjects were recorded, that every subject was seated in an industrial acoustics company sound controlled booth with dim lighting and a noiseless fan. EEG signals were recorded from positions C3, C4, P3, P4, O1 and O2 (shown in Fig. 2) using an electro-cap (elastic electrode cap) defined by the 10–20 system of electrode placement [12]. The impedances of all electrodes were retained below 5 kilo ohm. They made measurements with reference to electrically linked mastoids, A1 and A2. A bank of amplifiers (Grass7P511) were connected through the electrodes whose band-pass filter were set at 0.1 to 100 Hz to preprocess the data. The EEG signals were sampled at 250 Hz with a Lab Master 12-bit A/D converter mounted on a computer.

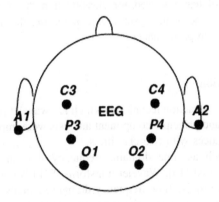

Fig. 2. Electrodes placement.

In this paper, EEG signals from subject 6 performing five different mental tasks have been used. These mental tasks are:

Baseline task. Every subject was asked to relax and think of nothing in particular. This task can be used as a control and as a baseline measure of the EEG signals.

Math task. The subjects were given none trivial multiplication problems, such as 45 times 18 and they were supposed to solve them without vocalizing or making any movements. The task was none repeating and designed so that an immediate answer was not apparent. At the end of the task the subjects verified whether or not he/she attained the solution and no subject finished the task before the end of the 10 s recording session.

Mental letter-composing task. Every subject was asked to mentally compose a letter to a close friend without vocalizing. This task was done several times and every subject was asked to continue with the letter from where they left off the previous time rather than starting again each time.

Geometric figure rotation *task*. Every subject was given 30 s to study a particular three-dimensional object, after which the drawing was removed and the subjects were asked to visualize that object being rotated about an axis.

Visual counting task. Every subject was asked to imagine a blackboard and to visualize numbers being written on the board consecutively, with the previous number being erased before the next number was written. They were also told to resume counting from where they left off in the previous time rather than starting again each time.

3.2 Preprocessing

In preprocessing stage noise and artifacts should be removed to enhance classification accuracy so, band pass filter between 1 and 45 Hz was used to filter the EEG signals and a 5th order Butterworth filter was used to remove the unwanted artifacts.

Band pass filter helps to select the frequency band containing useful information, reducing the number of features used for classification, have a direct influence on reducing the execution time of the system, and increasing the utilization of Memory which improve the system performance.

3.3 Feature Extraction

Three different features extraction techniques which are wavelet packet decomposition, fast Fourier transform and principal component analysis were implemented to compare between their performances with two classifiers.

Many methods such as time domain, frequency domain, and time-frequency domain methods were used [13]. Wavelet transform (WT) is considered to be one of the most suitable choice to use time-frequency domain methods for feature extraction so it was the first used feature extraction technique [14]. The output of the Wavelet packet decomposition can be computed by the following equation:

$$Wp_t = wpdec(x, Level, 'haar')$$ (1)

Where wpdec is a one-dimensional wavelet packet analysis function and Level split the data vector x into tree nodes for making the computation in each node. The EEG data were decomposed into (Haar) mother wavelet with five level wavelet packet decomposition in our system.

The second used feature extraction technique was the fast Fourier transformation, to extract the frequency components of the signal, select the required components and calculate the power for these components which were considered to be the input vector of each classifier. FFT computes the DFT where:

$$X_k = \sum_{n=0}^{N-1} x_n e^{-i2\pi k_n/N}$$ (2)

Where k = 0, 1 ... N-1, Xn is the sampled values and N is the total number of samples in the vector [15].

The last used method was the Principal Component Analysis technique that generally used to dimensionally reduce the original data to first n Eigen values [16]. Principal component analysis is a statistical procedure that uses an orthogonal transformation to convert a set of observations of possibly correlated variables into a set of values of linearly uncorrelated variables called principal components. Principal components number is less than or equal to original variables number.

In this transformation the first principal component has the largest possible variance value. The PCA transformation matrix W = [e.g. w_1, w_2, ...,w_n] can be obtained by performing a general eigenvalue decomposition of the covariance matrix R = XX^t where X is the input signal(s) and w_1, ...,w_n are n normalized orthogonal eigenvectors of X^tX corresponding to n different eigenvalues λ_1, λ_2, ..., λ_n in descending order.

The PCA transformation (Y) of X is then given by: Y = W^tX where the rows of Y are uncorrelated to each other.

3.4   Classification

Neural Network. Several researchers have been used neural network to classify the EEG signal. In this research artificial neural network trained by a standard back propagation algorithm was used for classification. Data were recorded from seven subjects during performing five mental tasks which were (baseline, math, mental letter composing, geometric figure rotation and visual counting). The selected subject was 6. The five mental tasks were measured 10 times each one. The length of EEG signals for each trial was 10 s. The recorded data were divided into training and testing sets Therefore, 10 trials are available for each task. Among them, 9 trials are used for training and one trial is used for testing.

Ten trials are selected for testing and classification accuracy was evaluated based on the average over these 10 trials. Extracted features were considered as input neurons to the neural network algorithm. The output layer should contain 5 neurons for the five classes that represent the five mental tasks that we want to classify. The number of neurons in the input layer varied according to the length of the input features vector.

Support vector machine (SVM). Support vector machine is a supervised learning method to analyze data and distinguish patterns, frequently used for classification and regression analysis. SVM constructs a discriminant hyperplane that maximizes the margins to identify classes, compared with other classifiers; SVM has a good generalization property, insensitive to overtraining and has a good performance with limited data. The five mental tasks were measured 10 times each one. The length of EEG signals for each trial was 10 s. The recorded data were divided into training and testing sets Therefore, 10 trials are available for each task. Among them, 9 trials are used for training and one trial is used for testing. Ten trials are selected for testing and classification accuracy was evaluated based on the average over these 10 trials.

4 Experimental Results

EEG data of subject 6, which can be obtained from the web site of Department of computer science, Colorado state university, were used. Three different features extraction methods were used as follows.

4.1 Wavelet Transform (WT)

Wavelet Packet Decomposition was applied on the EEG signals. The EEG data were decomposed into Haar mother wavelet with five level wavelet packet decomposition. Coefficients from nodes (5 0), (5 1), (5 2), (5 3), (5 4) and (5 5), which represents frequencies from 1 Hz to 45 Hz were extracted. The mean, $\mu(x)$, standard deviation, $\sigma(x)$, and entropy, $\varepsilon(x)$ were calculated for these coefficients by the following equations respectively:

$$\mu = \frac{1}{N} \sum_{i=1}^{N} x_i \tag{3}$$

$$\sigma = \sqrt{\frac{1}{N} \sum_{i=1}^{N} (x_i - \mu)^2} \tag{4}$$

$$\varepsilon = -\sum_{i=1}^{N} P(x_i) \log_2(x_i) \tag{5}$$

Where N is the total number Coefficients in the vector, P is the probability of x_i Values of each coefficient vector were calculated and used as features. Thus we have 6*3 = 18 features for each channel and a total of 108 features for each task.

4.2 Fast Fourier Transform (FFT)

Spectrum of EEG signals was calculated and the top hundred Fast Fourier Transform power values were taken. Thus we have 100 features for each channel and a total of 600 features for the 6 channels used.

4.3 Principal Component Analysis (PCA)

PCA technique reduced the original data to first n Eigen values. The highest variance value was taken as a feature so we have one feature for each channel and a total of 6 features for the 6 channels.

Neural Network trained by a standard back propagation algorithm and support vector machine were used for classification.

4.4 Multi-layer Perceptron Neural Network

Neural Network trained by a standard back propagation algorithm was used in our research. The number of neurons in the input layer varied according to the length of the input features vectors. Many tests were done to find the best configuration for the neural network in terms of: number of neurons in the hidden layer and the maximum number of iterations (epochs) in the learning process.

For each features set, the configuration that produced optimal weights (which lead to maximum correct classification rate in the testing) for I/O mapping was used which were:

Number of neurons in the hidden layer = 100.

Maximum number of iterations (epochs) in the learning process = 1000.

The activation function used was the sigmoid function, the learning rate was 0.1 and the training stopped when either the maximum number of epochs reached 1000 or the mean square error reached to a small value such as 0.001.

4.5 Support Vector Machine

The SVM classifier in this paper was based on LIBSVM implementation from [17]. Many tests were done to find the optimal parameters for SVM in terms of: type of the kernel, the Coefficient in kernel function, Degree in kernel function. Parameters that lead to maximum correct classification rate in the testing for I/O mapping were used which were: Polynomial kernel was used, Degree in kernel function = 3, and Coefficient in kernel function = 0.

Data were analyzed using MATLAB 2013 and a computer (Intel Core i7 CPU 2.20 GHz, 8 GB DDR RAM, Windows 7). Total classification accuracies for classifying different combination of three mental tasks using the three feature extraction techniques and two classifiers as shown in Tables 2 and 3 shows the effect of increasing the frequency band from [1 45] to [1 100] on it. Total classification accuracies for classifying different combination of four mental tasks as shown in

Table 2. Classification accuracies of different three mental tasks, frequency band [1 45].

	B,m,l	B,m,r	B,m,c	B,l,r	B,l,c	B,r,c	M,l,r	M,l,c	L,r,c	M,r,c
Svm fft	70	90	70	80	56.67	83.34	80	53.33	63.333	83.335
Svmwav	86.668	96.67	86.67	90	76.67	93.33	96.67	73.33	76.668	86.67
Svm pca	73.335	80	83.335	86.668	70	90	66.667	56.666	73.335	73.336
Nn fft	66	82.33	66	67.67	46	78.33	71	50.33	64.67	76.33
Nn wav	90	96.67	77.67	90	77.67	86.67	93.33	80.67	83.33	83.33
Nn pca	75	76.67	80	86.67	78.67	86.67	73.33	72.67	79	70

B ....Base line task.
M....Math task.
L.....Letter composing task.
R.....Geometric figure rotation task.
C.... Visual counting task.

Table 3. Classification accuracies of different three mental tasks, frequency band [1 100].

	B,m,l	B,m,r	B,m,c	B,l,r	B,l,c	B,r,c	M,l,r	M,l,c	L,r,c	M,r,c
Svm fft	70	90	70	76.67	56.67	83.33	80	50	60	83.33
Svmwav	86.67	96.67	90	90	80	93.33	96.67	83.33	83.33	90
Svm pca	76.67	80	83.33	90	70	90	70	63.33	76.67	76.67
Nn fft	64.67	82.33	66	66.67	43.33	81.67	73.33	54.33	64.33	75
Nn wav	90	96.67	83.33	90	79	87.33	93.33	86	83.33	80
Nn pca	78	73.33	86.67	86.67	80.67	90	73.67	81.67	90	83.33

Table 4. Classification accuracies of different four mental tasks, frequency band [1 45].

	B,m,l,r	B,m,l,c	B,m,r,c	B,l,r,c	M,l,r,c
Svm fft	77.5	57.5	77.5	65	60
Svmwav	90	77.5	87.5	80	77.5
Svm pca	70	65	75	75	57.5
Nn fft	59	46.5	65.75	51.5	56.5
Nn wav	92.5	80	85.25	81.25	82.5
Nn pca	74	77.75	74.25	80	68.5

Table 5. Classification accuracies of different four mental tasks, frequency band [1 100].

	B,m,l,r	B,m,l,c	B,m,r,c	B,l,r,c	M,l,r,c
Svm fft	75	55	77.5	62.5	57.5
Svmwav	90	82.5	90	82.5	85
Svm pca	72.5	65	77.5	75	62.5
Nn fft	59.25	44.25	64.25	56	59.25
Nn wav	92.5	82.75	85	82.25	82.5
Nn pca	73.75	78.25	75	86	70.25

Table 6. Classification accuracies of all five mental tasks, frequency band [1 100].

Svm fft	60
Svmwav	84
Svm pca	64
Nn fft	45.2
Nn wav	82.6
Nn pca	74.2

Tables 4 and 5 shows the effect of increasing the frequency band from [1 45] to [1 100] on it. Table 6 shows classification accuracies for classifying all five mental tasks using frequency band [1 100].

As shown in the above tables the performance of wavelet transform is better than fast Fourier transform and principal component analysis whether with neural network or support vector machine. Performance of principal component analysis is better than fast Fourier transform with the two classifiers. Increasing the frequency band from [1 45] to [1 100] improves the classification accuracies. Best classification accuracy for classifying five mental tasks was 84 % and it was obtained for wavelet packet decomposition with support vector machine.

A tri-state Morse code scheme could be used as an application of our system to help disabled peoples having problems in speech as it could translate different combination of three mental tasks into English words like food, water and TV. The basic alphabets in the conventional Morse code scheme are dot and dash so two mental tasks will be sufficient to be used but, the use of an additional mental task was proposed to represent space between dot and dash. This space will represent the end of either a dot or dash and starting of a new dot or dash, which help users to concentrate on the sequence of mental tasks not the time duration of each mental task. Therefore, to use this tri-state Morse code, we need three different combinations of mental tasks where each task will correspond to either a dot, a dash or a space. Using this tri-state Morse code, English letters, Arabic numerals and punctuation marks to form words and complete sentences could be constructed [18].

5 Conclusion

This paper presented a non-invasive system for classifying different combinations of mental tasks using Brain EEG signal processing. In the proposed model EEG data of subject 6 was obtained from the web site of Department of computer science, Colorado state university. The EEG signals were extracted by six electrodes (C3, C4, P3, P4, O1 and O2) for five mental tasks. Wavelet Transform (WT), Fast Fourier Transform (FFT) and Principal Component Analysis (PCA) techniques were used for features extraction. Data were classified using the Back Propagation neural network and support vector machine. Experimental results show the classification accuracies achieved with the three used feature extraction techniques and the two classification techniques.

References

1. Dornhege, G., Millan, J.D.R., Hinterberger, T., Mc-Fanland, D.J., Muller, K.R.: Toward Brain-Computer Interfacing. The MIT Press, Cambrigde (2007)
2. Berger, T.W., Chapin, J.K., Gerhardt, G.A., McFarland, D.J., Principe, J.C., Soussou, W.V., Taylor, D.M., Tresco, P.A.: Brain-Computer Interfaces: An International Assessment of Research and Development Trends. Springer, New York (2008)
3. Hortal, E.: Online classification of two mental tasks using a SVM based BCI System. IEEE Neural Engineering (2013)
4. Saadat, N., Pourghassem, H.: Mental task classification based on HMM and BPNN. In: International Conference on Communication Systems and Network Technologies (CSNT), pp. 210–214. IEEE (2013)
5. Nakayama, K., Inagaki,K.: A brain computer interface based on neural network with efficient pre-processing. In: International Symposium on Intelligent Signal Processing and Communication Systems, Yonago Convention Center Tottori Japan December, pp. 673–676 (2006)
6. Anderson, C.W., Bratman, J.A.: Translating thoughts in to actions by finding patterns in brain wave. In: Proceedings of the Fourteenth Yale Workshop on Adaptive and Learning Systems, Yale University New Haven CT June, pp. 1–6 (2008)
7. Mizuno, Y., Mabuchi, H.: Clustering of EEG data using maximum entropy method and LVQ. Int. J. Comput. 4(4), 193–200 (2010)
8. Hosni, K., Gadallah, M., Bahgat, S., AbdelWahab, M.: Classification of EEG signals using different feature extraction techniques for mental-task bci. In: Computer Engineering & Systems. ICCES 2007 International Conference on IEEE, pp. 220–226 (2007)
9. Tolić, M., Jović, F.: Classification of wavelet transformed EEG signals with neural network for Imagined mental and motor tasks. Kinesiology, pp. 130–138 (2013)
10. Keirn, Z.A., Aunon, J.I.: a new mode of communication between Man and His surroundings. IEEE Trans. Biomed. Eng. 37, 1209–1214 (1990)
11. Huan, N.-J., Palaniappan, R.: Neural network classification of autoregressive features from electroencephalogram signals for brain–computer interface design. J. Neural Eng. 1, 142–150 (2004)
12. Jasper, H.: The ten twenty electrode system of the international federation electroen-cephalographic clin. Neurophysiol. 10, 371–375 (1958)
13. Shedeed, H. et al.: Brain EEG signal processing for controlling a robotic arm. In: IEEE, pp. 152–157 (2013)
14. Yang, G., Nakayama, K., Hirano, A.: A dual-class voting mechanism for brain computer interface based on wavelet packet and support vector machine. In: IEEE (2013)
15. Shedeed, H.A., Issa, M.F.: Brain-EEG signal classification based on data normalization for controlling a robotic arm. Int. J. Tomogr. Simul. 29, 72–85 (2016)
16. Kottaimalai, R. et al.: EEG Signal classification using principal component analysis with neural network in brain computer interface applications. In: IEEE International Conference on Emerging Trends in Computing, Communication and Nanotechnology (ICECCN) (2013)
17. Chang, C.-C., Lin, C.-J.: LIBSVM: A library for support vector machines. ACM Trans. Intell. Syst. Technol. 2, 1–27 (2011)
18. Palaniappan, R. et al.: A new brain–computer interface design using fuzzy ARTMAP. In: IEEE Transactions on Neural Systems and Rehabilitation Engineering, vol. 10 (2002)
19. Ella Hassanien, A.: Ahmad taher azar brain computer interface: trends and applications. Intelligent Systems Reference Library, vol. 74. Springer (2015)

Content Based Image Retrieval with Hadoop

Heba Gaber[✉], Mohammed Marey, Safaa E. Amin,
and Mohamed F. Tolba

Computer and Information Sciences, Ain Shams University, Cairo, Egypt
eng.heba.gaber@gmail.com, mohammedmarey@hotmail.com,
safaa_amin007@htomail.com, fahmytolba@gmail.com

Abstract. Hadoop has become a widely used open source framework for large scale data processing. MapReduce is the core component of Hadoop. It is this programming paradigm that allows for massive scalability across hundreds or thousands of servers in a Hadoop cluster. It allows processing of extremely large video files or image files on data nodes. This can be used for implementing Content Based Image Retrieval (CBIR) algorithms on Hadoop to compare and match query images to the previously stored terabytes of an image descriptors databases. This work presents the implementation for one of the well-known CBIR algorithms called Scale Invariant Feature Transformation (SIFT) for image features extraction and matching using Hadoop platform. It gives focus on utilizing the parallelization capabilities of Hadoop MapReduce to enhance the CBIR performance and decrease data input\output operations through leveraging Partitioners and Combiners. Additionally, image processing and computer vision tools such as Hadoop Image Processing (HIPI) and Open Computer Vision (OpenCV) are integration is shown.

Keywords: Hadoop · HIPI · OpenCV · Big data · Crowd sourcing · MapReduce · Combiners · Partitioners

1 Introduction

CBIR Systems are used for searching and retrieving query images from large databases based on the image content, which is derived from images themselves by using computer vision techniques. SIFT is one of the most important CBIR techniques that depends on extracting distinctive invariant features from images that can be used to perform reliable matching between different views of an object or scene. For large scale images or large numbers of images stored in the database, SIFT feature extraction is a computationally intensive problem, as it takes a long time to extract and match them to the extracted SIFT features. Therefore, there exists a need for an efficient platform to handle image descriptor features extraction and processing that may reach Terabytes of data in size.

In this work Hadoop is used as a reliable, scalable, distributed computing platform. It is an open source project from the Apache Software Foundation [14, 15] and its core consists of MapReduce programming model implementation. Hadoop enables the execution of applications on thousands of nodes and petabytes of unstructured or semi-structured data which have been impossible to process efficiently (cost and time) so far.

© Springer International Publishing AG 2017
A.E. Hassanien et al. (eds.), *Proceedings of the International Conference on Advanced Intelligent Systems and Informatics 2016*, Advances in Intelligent Systems and Computing 533, DOI 10.1007/978-3-319-48308-5_25

This paper presents the implementation of the SIFT descriptor extraction and matching using Hadoop MapReduce, integrated with HIPI and OpenCV libraries for SIFT features extraction and matching. Moreover, we show how to enhance the matching performance by leveraging parallelization mechanisms for MapReduce Partitioners and Combiners.

The paper is organized as follows: Sect. 2 discusses the related work, Sect. 3 presents CBIR with SIFT algorithm, Sect. 4 presents Hadoop MapReduce, Combiners and Partitioners. Section 5 discusses the integration of HIPI and OpenCV computer vision libraries utilized in the proposed system. Section 6 explains implementing CBIR with SIFT for image processing on Hadoop and how to enhance its performance, Sect. 7 concludes the presented work and shows our future research direction.

2   Related Work

Recently, using Hadoop for image processing has been receiving attention, there exists some recent implementations for many domains like biomedical, social media, surveillance, satellite and geographical images processing applications summarized in Table 1.

In [1] a machine learning tool "HaarFilter" is implemented detecting human faces by using the Haar Cascading technique. It was implemented using HIPI and OpenCV on Hadoop platform. Compressing social media images where HIPI is used as image processing tool as shown in [2]. Providing image processing as a service published to the public and implemented on Hadoop platform using integration for OpenCV library is presented in [3]. In [4] feature extraction using Hadoop and SIFT is shown.

Image processing with Hadoop for health care in India using HIPI that allows is presented in [5]. Using HIPI and Avro for processing surveillance images is shown in [6]. In [7], analyzing terabytes of microscopic medical images on Hadoop using Tera-soft library is demonstrated. Processing satellite and geospatial huge images databases on Hadoop is shown in [8, 9].

Table 1. Image processing with Hadoop platform

Ref	Tools	Application
[1]	OpenCV, HIPI	Machine learning for face detection using Hadoop
[2]	HIPI	Compress social media images
[3]	OpenCV	Provide image processing tools as a service
[4]	Custom Development	Feature extraction (target tracking, traffic management and accident discovery) using Hadoop and SIFT
[5]	HIPI	Image processing with Hadoop for application of heath care applications in India
[6]	HIPI, Avro	HIPI and Avro for processing surveillance images
[7]	Tera-soft	Analyzing terabytes of microscopic medical images
[8]	Custom	Satellite image processing
[9]	Custom	Processing 100 M Geographical images on Hadoop Geographical files processing

3   CBIR with SIFT

CBIR with SIFT can be implemented as illustrated in Fig. 1 using four main components (1) Image collection, (2) Image features extraction, which mainly affects the quality of searching, (3) Features Indexing and (4) Searching using Query images, which mainly affects the efficiency.

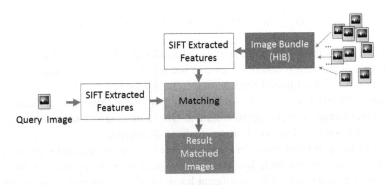

Fig. 1. CBIR with SIFT Architecture

SIFT algorithm can be implemented effectively for content based image retrieval based on Region of Interest (ROI). The SIFT feature is invariant to rotation, image scaling and transformation and partially invariant to illumination changes and affine transformation. The main advantage of using SIFT is that it describes the same features on different spatial scales.

SIFT is based on convolving an image with a Gaussian kernel for several values of the variance σ and then taking the difference between adjacent convolved images. If $D(x, y, \sigma)$ denotes such differences, then the method is based on finding the extrema of this function with respect to all the three variables. Such maxima are candidates for key points that, after additional analysis, are used to characterize objects in an image. SIFT calculate result descriptor is shown in Fig. 2 [10]. DoG $D(x, y, \sigma)$ provides a good approximation for the Laplacian-of-Gaussian. It can efficiently be computed by subtracting adjacent scale levels of a Gaussian pyramid.

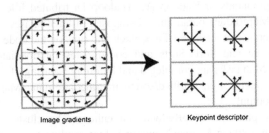

Fig. 2. SIFT Descriptor Generation [10]

SIFT feature extraction is composed of four important phases:

- Phase 1- Scale-space Extrema Detection: potential interest points are identified by searching the overall scales and image locations in the image matrix. The scale space of an image is defined as a function $L(x, y, k\sigma)$ and produced from the convolution of a variable scale Gaussian $G(x, y, k\sigma)$, with an input image, $I(x, y)$

$$L(x, y, k\sigma) = G(x, y, k\sigma) * I(x, y) \tag{1}$$

$$D(x, y, \sigma) = (L(x, y, k\sigma) - L(x, y, \sigma)) * I(x, y) \tag{2}$$

- Phase 2 Keypoint localization: A detailed model is created to determine location and scale of all interest points detected in phase1. After that, key points are selected based on their stability and their resistance to distortion.
- Phase 3- Orientation assignment: For each key point identified in phase 2, the direction of gradient is computed around it. One or more orientations are assigned to each key point based on local image gradient directions.
- Phase 4: Keypoint descriptor: In this phase, the local image gradients are measured in the region around each key point. This gradient is then transformed into a representation that allows for significant levels of local shape distortion and change in illumination.

One of the most featured implementations for CBIR is vision based Navigation, based on a Visual Information System (NAVVIS) project for indoor navigation for the Technical University of Munich (TumIndoor). The navigation application uses images that are captured by a smart phone as visual fingerprints of the environment. The captured images are matched to the previously recorded geotagged reference database with CBIR techniques. TumIndoor introduces an extensive benchmark dataset for visual indoor localization, available to the research community for downloading, thus any experimental results can be compared with the published results [10, 11]. Some other feature extraction and matching techniques exist such as; Speed-Up Robust Features (SURF) [12] and Binary Robust Independent Elementary Features (BRIEF) [13]. SIFT's accuracy holds up against more modern algorithms yet it's computationally expensive.

4 Hadoop, Map-Reduce, Combiners and Partitioners

The two main components of Hadoop are: Hadoop Distributed File System (HDFS) and MapReduce. MapReduce paradigm is composed of three phases. (1) The Mapper: Each Map task operates on a single HDFS block and runs on the node where the block is stored. (2) Shuffle and Sort: Sorts and consolidates intermediate data from all mappers, after the Map tasks are completed and before the starting of the Reduce tasks. (3) The Reducer: Operates on shuffled/sorted intermediate data (Mapping output) to produce the final output.

The Combiner phase in MapReduce can enhance the cluster performance by reducing results on a single Mapper's output before sending the data to the reducer.

A Combiner, also known as a semi-reducer, which is an optional class that operates by (1) accepting the inputs from the Map class as key value pairs, (2) summarizing the map output records with the same key, (3) passing its output as new key-value pairs to the original Reducer class.

The Cluster performance can also be enhanced through partitioning (indexing) the keys using custom or default Partitioner, where each partition is passed to a single reducer. A Partitioner partitions the key value pairs of intermediate Map outputs. It partitions the data using a user defined condition, which works like a hash function. The total number of partitions is the same as the number of Reducer tasks for the job.

In Fig. 3 using Hadoop MapReduce Paradigm by integrating HIPI and OpenCV libraries for extracting SIFT features is presented, the first step is collecting image data-base into a HIB bundle, afterwards OpenCV is used for extracting image data, pre-processing (grayscale conversion) and feature extraction afterwards. In this work we show how processing data in combiner and Partitioner phase and changing the number of processing units can effectively enhance the overall performance and reduce the input/output operations of the system.

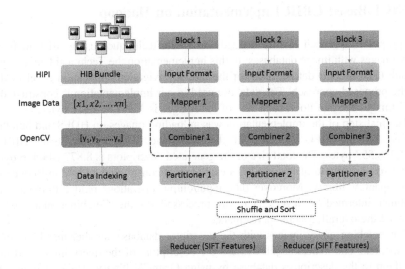

Fig. 3. Map-Reduce Paradigm

5   Integrating Computer Vision Libraries

For image processing on a Hadoop platform, HIPI was introduced. HIPI [15] is an image processing library designed to be used with the Apache Hadoop Map-Reduce parallel programming framework. HIPI facilitates efficient and high throughput image processing with Map-Reduce style parallel programs typically executed on a cluster. It provides a solution for how to store a large collection of images on the HDFS and make them available for an efficient distributed processing. HIPI is developed and maintained by a growing number of developers around the world.

The primary input object to a HIPI program is a HIPI Image Bundle (HIB) which is a collection of images represented as a single file on the HDFS [16], the first processing stage of a HIPI program is a culling step that allows filtering the images in a HIB, based on a variety of user defined conditions like spatial resolution or criteria related to the image metadata.

Further processing can be done using OpenCV [17] for feature extraction. We selected the widely used OpenCV library as the base image processing library integrated with the proposed platform. OpenCV is an open source library written in C++ and it also has Java and Python interfaces supporting Windows, Linux, Mac OS, iOS and Android. OpenCV is optimized and parallelized for multi-cores and accelerators using the Open Computer Library (OpenCL). OpenCV was installed on the Hadoop cluster to enable image processing capability with Map-Reduce parallel programming model.

In the presented work integrating HIPI and OpenCV was introduced, the images folder will be uploaded to the HDFS as a HIB bundle using HIPI, afterwards in the feature extraction job, OpenCV library is used to get SIFT features from the HIB. The result of the feature extraction job will be saved on Hadoop. Another job for matching query image to the descriptors database was implemented to find the matched image.

6   SIFT Based CBIR Implementation on Hadoop

The proposed approach mainly depends on CBIR techniques using SIFT to match query images with image databases. In this implementation the Technical University of Munich (TumIndoor) dataset used for the Indoor navigation application is selected to test the proposed approach. The database used in our implementation is constructed of 9,437 images for one floor, total size of the images is 1,395,864,371 bytes.

The first step of the implementation was uploading images on HDFS and bundling images utilizing HIPI. The size of the constructed HIB bundle using HIBI is 1,503,238,553 bytes. The total size of the generated descriptors is 8,877,244,416 bytes. Afterwards a database of image SIFT descriptors is constructed with Map-Reduce job using OpenCV library. Moreover data input/output operations reduce through using Combiner intermediate layer, data was processed on the Combiner phase which enhanced the overall systems performance.

For matching query images with the descriptor database, another matching job is implemented. The matching job extracts the descriptors of the query image and matches them to the descriptors database by using OpenCV library. The matching job's runtime performance was tested with different numbers of reducers and was enhanced through grouping the descriptors based on the descriptors' number of rows in-stead of using the image name as the mapper output key.

The implementation performed on Cloudera virtual machine [18] where Java Eclipse and Hadoop were already installed and HIPI and OpenCV is configured. The current machine consists of 4 processors and 10 GB RAM. By adding more memory, processors and additional physical nodes the run-time performance of the system will increase and the processing time is expected to decrease. The algorithm is tested with different parameters on the same environment by changing the number of reducers and adding Combiners and Partitioners on different phases of computation. We noticed that grouping the data under 100 reducers distributed based on the number of the rows in each descriptor

Table 2. HDFS and File execution KPIs

Environment Configuration/Counter	FILE: Number of (GB) read	FILE: Number of (GB) Written	HDFS: Number of bytes read	HDFS: Number of read operations
1000 reducer	16.56	24.46	28.4	12,650
100 reducers	16.14	24.36	28.4	9,950
50 reducer	16.12	24.35	28.4	9,800
100 reducer and classifier num-rows	16.14	24.36	12.8	2,631
Partitioner and 100 reducer with classifier num-rows %100	16.14	24.36	12.8	2,631
Partitioner, combiner and 100 reducers classifier num-rows %100	0.0036	0.018	19.17	5,618
Partitioner, combiner and 100 reducers classified by num-rows	0.0036	0.018	21.05	6,499

matrix enhanced the performance more than using very large number of reducers (1000) or small number of reducers (50). This can't be taken for granted as the optimal number of reducers and the partitioning parameter shall depend on the input data size and the variation of the size of the generated descriptors. We also show the impact of adding Combiners and the Partitioners after grouping the data under a 100 reducer.

Hadoop offers Job Tracker, an UI tool to determine the status and statistics of all jobs. Using the job tracker UI, we can view the Counters that have been created and investigate the execution results. Table 2 shows the results of the comparison between different execution parameters in terms of mapping time, reduced shuffling processing time, and numbers of read and write operations and overall performance of processing the same dataset. Examples for those Counters are:

1. *File-number of bytes read:* is the number of bytes read by local file systems and it denotes the total bytes read by reducers. They also occur during the shuffle phase when the reducers spill intermediate results to their local disks while sorting.
2. *File-number of bytes written:* consists of two parts. The first part comes from mappers. The second part comes from reducers. In the shuffle phase, all the reducers will fetch intermediate data from mappers and merge and spill to reducer-side disks. All the bytes that reducers write to disk will also be included in this counter.
3. *HDFS-number of bytes read:* Denotes the bytes read by mappers from HDFS when the job starts. This data includes not only the content of source file but also metadata about splits.
4. *HDFS-number of read operations:* Denotes the bytes written to HDFS. It's the number of bytes of the final output. Since HDFS and local file systems are different file systems, so the data from the two file systems will never overlap. From this

implementation, the least file system access was achieved through using Partitioners and Combiners so that data is accessed after data grouping in the Map Phase.

The other performance KPIs are related to the time and the memory consumed in each phase in the Map-Reduce. Total time spent by all maps (5), All reducers (6), All map tasks (7), All reduce tasks (8) in milliseconds, total megabyte seconds taken by all map tasks (9), All reduce tasks (10). It was noticed the least mapping time was achieved through partitioning the data with descriptor metadata element ex: "Number of rows in the SIFT descriptor matrix" and without using Partitioners or Combiner layers, this can be explained as the data won't be re-copied after the mapping phase. On the opposite side the least reduce time was achieved through adding Combiners and Partitioners and getting reduced input/output through the intermediate layer before Shuffle and Sort process.

Comparing the two approaches the second approach is more efficient since mapping time can be enhanced through adding more slots to the system and distribute processing on them, yet the reduce phase is considered a critical phase as it will be responsible to process all the Mappers output to get the final results and sometimes considered a bottleneck phase.

Also, we noticed enhanced performance counters by using Partitioners and Combiners for example (11) GC (Global Cash) time elapsed, (12) Processing time (CPU time spent), (13) Data snapshot physical and (14) Virtual memory and the (15) Total heap usage.

7   Conclusion and Future Steps

In this work we implemented CBIR based on SIFT for features extraction and matching of query images to huge image descriptors on Hadoop. SIFT is very intensive and requires very high processing and storage power. In this work Hadoop MapReduce is used as a Big Data platform for processing and storage of the image descriptors data. HIPI is integrated in the system and used to upload and filter data to HDFS as HIB bundle. This work also shows integrating HIPI and OpenCV computer vision libraries with Hadoop to accelerate features extraction and matching. The Feature extraction job was implemented to generate image descriptors database based on SIFT features. Another job was implemented to match query images to the database. We have shown how to use Map-Reduce Combiners and Partitioners for enhancing performance and reduce the input/output operations. The proposed approach seeks to maximize the full potential of cloud computing as well as ubiquitous sensing, a viable way to achieve this was to have a combined framework with a cloud at the center.

We have demonstrated the practicality of our approach by extracting SIFT features from TumIndoor images, confirming that the Map-Reduce based approach can accelerate the feature extraction process effectively through a Map-Reduce paradigm for a database of about 8 GB in size. Our next step is to enhance the matching with the SIFT descriptors through categorizing the data as visual words and involving machine learning techniques to classify the bag of visual words.

References

1. Nausheen, K.M., Ram, M.S. Haarfilter: a machine learning tool for image processing in Hadoop. Int. J. Technol. Res. Eng. 3 (2015)
2. Barapatre, M.H., Nirgun, M.V., Jagtap, M.H., Ginde, M.S.: Image processing using mapreduce with performance analysis. Int. J. Emerg. Technol. Innovative Eng. I(4) (2015)
3. Yan, Y., Huang, L.: Large scale image processing research cloud. In: Cloud Computing, pp. 88–93 (2014)
4. Cheng, E.: Efficient feature extraction from a wide area motion imagery by MapReduce in Hadoop. In: SPIE Defense + Security. International Society for Optics and Photonics (2014)
5. Augustine, D.P.: Leveraging big data analytics and hadoop in developing India's healthcare services. Int. J. Comput. Appl. **89**(16), 44–50 (2014)
6. Gawde, A.U., Shah, M., Ukaye, I., Nanavati, M.: Object detection in hadoop using HIPI. Int. J. Adv. Res. Eng. Technol. (2013)
7. Bajcsy, P.: Terabyte-sized image computations on Hadoop cluster platforms. In: IEEE International Conference on Big Data, pp. 729–737. IEEE (2013)
8. Han, W., Kang, Y., Chen, Y., Zhang, X.: A MapReduce approach for SIFT feature extraction. In: International Conference on Cloud Computing and Big Data, pp. 465–469 (2013)
9. Moise, D., Shestakov, D., Thor, G., Amsaleg, L.: Indexing and searching 100 M images with Map-Reduce. In: ACM International Conference on Multimedia Retrieval, pp. 17–24 (2013)
10. Huitl, R., Schroth, G., Hilsenbeck, S., Schweiger, F., Steinbach, E.: TUMindoor: An extensive image and point cloud dataset for visual indoor localization and mapping. In: 19th IEEE International Conference on Image Processing (ICIP) Orlando, FL, pp. 1773–1776. IEEE (2012)
11. Schroth, G.: Mobile Visual Location Recognition. Ph.D. Thesis. Munich: Technische Universität München, July 2013
12. Panchal, P.M., Panchal, S.R., Shah, S.K.: A Comparison of SIFT and SURF. Int. J. Innovative Res. Comput. Commun. Eng. **1**(2), 323–327 (2013). ISSN: 2320–9798
13. Calonder, M., Lepetit, V., Strecha, C., Fua, P.: BRIEF: binary robust independent elementary features. In: Daniilidis, K., Maragos, P., Paragios, N. (eds.) ECCV 2010. LNCS, vol. 6314, pp. 778–792. Springer, Heidelberg (2010). doi:10.1007/978-3-642-15561-1_56
14. http://hadoop.apache.org
15. White, T.: Hadoop: The Definitive Guide, 2nd edn. O'Reilly Media, Sebastopol (2011)
16. http://hipi.cs.virginia.edu/
17. http://opencv.org/
18. http://www.cloudera.com/

A Comparison Between Optimization Algorithms Applied to Offshore Crane Design Using an Online Crane Prototyping Tool

Ibrahim A. Hameed$^{(\boxtimes)}$, Robin T. Bye, and Ottar L. Osen

Software and Intelligent Control Engineering Laboratory,
Faculty of Engineering and Natural Sciences,
Norwegian University of Science and Technology,
NTNU in Ålesund, Postboks 1517, 6025 Ålesund, Norway
{ibib,robin.t.bye,ottar.l.osen}@ntnu.no
http://blog.hials.no/softice/

Abstract. Offshore crane design requires the configuration of a large set of design parameters in a manner that meets customers' demands and operational requirements, which makes it a very tedious, time-consuming and expensive process if it is done manually. The need to reduce the time and cost involved in the design process encourages companies to adopt virtual prototyping in the design and manufacturing process. In this paper, we introduce a server-side crane prototyping tool able to calculate a number of key performance indicators of a specified crane design based on a set of about 120 design parameters. We also present an artificial intelligence client for product optimisation that adopts various optimization algorithms such as the *genetic algorithm*, *particle swarm optimization*, and *simulated annealing* for optimising various design parameters in a manner that achieves the crane's desired design criteria (e.g., performance and cost specifications). The goal of this paper is to compare the performance of the aforementioned algorithms for offshore crane design in terms of convergence time, accuracy, and their suitability to the problem domain.

Keywords: Virtual prototyping · Product optimisation · Artificial intelligence · Genetic algorithm (GA) · Particle swarm optimization (PSO) · Simulated annealing (SA) · Automation

1 Introduction

The need to reduce the time and cost involved in taking a product from conceptualisation to production and the desire to meet customers' demands have encouraged producers to adopt new technologies in manufacturing such as virtual prototyping (VP) [1]. VP refers to the process of simulating the user, the

© Springer International Publishing AG 2017
A.E. Hassanien et al. (eds.), *Proceedings of the International Conference on Advanced Intelligent Systems and Informatics 2016*, Advances in Intelligent Systems and Computing 533, DOI 10.1007/978-3-319-48308-5_26

product, and their combined (physical) interaction in software through the different stages of product design, and the quantitative performance analysis of the product [2]. Being a relatively new technology, VP typically involves the use of virtual reality (VR), virtual environments (VE), computer-automated design (CautoD) solutions, computer-aided design (CAD) tools, and other computer technologies to create digital prototypes [3].

Members of the Software and Intelligent Control Engineering (SoftICE) Laboratory at NTNU in Ålesund have previously presented a CautoD solution for product design optimization, initially focused on the design of offshore cranes [4–6]. Specifically, Bye et al. [4] introduced a server-side crane prototyping tool (CPT) whose key component is a crane calculator able to calculate a number of key performance indicators (KPIs) of a specified crane design based on a set of about 120 design parameters. In the same paper, the authors also introduced a client-side web graphical user interface (GUI) that facilitates the process of manually selecting the design parameters in the CPT to obtain a simple visualisation of the designed crane and its 2D safe working load (SWL) chart. Subsequently, we have presented a client-side artificial intelligence for product optimisation (AIPO) module that uses a genetic algorithm (GA) for optimising the design parameters in a manner that achieves the crane's desired design criteria (e.g., KPIs related to performance and cost specifications) [5]. Hameed et al. [6] replaced the AIPO module and its GA library with a new Matlab software module, the Matlab crane optimisation client (MCOC), that also used a GA for optimization. Additionally, and contrary to the AIPO module, the MCOC also included the possibility for multiobjective optimization (MOO).

The purpose of this paper is to further extend the MCOC with particle swarm optimization (PSO) and simulated annealing (SA), apply these various optimization algorithms to offshore crane design using the previously developed CPT, and compare the performance of these algorithms in terms of convergence time and accuracy. Offering a number of different objective functions and algorithms for optimizing these objectives, users will have several optimization options they can choose from with an ultimate goal to reduce cost and time of the design process. The remainder of the paper is organized as follows: A review of our offshore crane design optimization framework is presented in Sect. 2. An introduction to the various optimization algorithms and various objective functions are presented in Sect. 3. In Sect. 4, results of the optimization algorithms applied to the offshore crane design problem are presented. Finally, a comparison between different algorithms in terms of convergence time and accuracy, concluding remarks, and future work is presented in Sect. 5.

2   Sofware Framework for Offshore Crane Design

This section outlines the software architecture and describes the main components of the software framework.

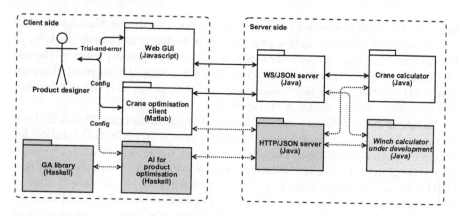

Fig. 1. Generic and modular software architecture for intelligent CautoD of offshore cranes, winches, or other products. The modules in white (grey) and their solid (dashed) interconnections are inside (outside) the scope of this paper [5,6].

2.1   Client-Server Architecture

The diagram in Fig. 1 shows the client-server architecture of the software framework [5].

On the server-side, the CPT is able to calculate a number of KPIs of a specified crane design based on a set of about 120 design parameters. On the client-side, the web GUI facilities the process of manually selecting the design parameters of the designed CPT and providing a simple visualisation of the designed crane and its 2D workspace safe working load (SWL) chart [4]. Additionally, the AIPO module uses a GA for optimising the design parameters in a manner that achieves the crane's desired design criteria. Each KPI is typically related to overall performance, weight and cost of the designed crane. Both WebSocket (WS) and the hypertext transfer protocol (HTTP) have been implemented as communication interfaces (see Fig. 1). Whereas Bye et al. [5] tested and used their AIPO client using HTTP, [6] tested and used WS for their MCOC. Both approaches in [5,6] used the JavaScript Object Notation (JSON), a lightweight human-readable data-interchange format, for data messages. In this study, we adopt the same approach as in [6] but extend the MCOC module with the PSO and SA optimization algorithms. We use WS with JSON for communication, and keep the web GUI used previously in order to obtain visualizations of the 2D load charts.

2.2   Online Crane Prototyping Tool (CPT)

The CPT server consists of a crane calculator and two modules for handling WS/JSJON and HTTP/JSON connections (see Fig. 1). Here, the MCOC connect to the CPT via WS/JSON. Messages are sent as JSON objects in a standardised format that the CPT accepts, consisting of three parts (subobjects) [5]: (i) a **base** object with a complete set of default design parameter values;

(ii) a `mods` object with a subset of design parameter values that modifies the corresponding default values; and (iii) a `kpis` object with the desired KPIs to be calculated from `base` and `mods` and returned to the MCOC by the CPT.

2.3   Crane Calculator

The components of an offshore crane may consist of several thousand parameters, however, with the help from crane designers we have been able to isolate the most important ones and reduce this number to a set of about 120 design parameters. Based on the values of these parameters, which can be set manually or by a CautoD tool such as MCOC, the crane calculator is able to calculate a fully specified crane design and its associated KPIs [4]. The goal of the designer is therefore to determine appropriate design parameter values that achieve some desired design criteria based on KPIs, or to try to improve an existing design. The design must simultaneously meet requirements by laws, regulations, codes and standards, as well as other constraints, such as a maximum total delivery price and total crane weight.

The accuracy of our crane calculator has been verified against other crane calculators and spreadsheets that are commonly used in the industry. As a proof of its trustworthiness, Seaonics AS[1] has already adopted the CPT server and web GUI client for manual crane design [4].

2.4   Web Graphical User Interface (GUI)

For the sake of simplicity and practicality, the previously developed web GUI is used to interact with the crane calculator via WS/JSON communication. The web GUI can be used to manually adjust the 120 design parameters in the crane calculator by a trial-and-error approach. The effect of the chosen parameter values on a number of KPIs and other design criteria can then be investigated numerically, with the possibility of exporting the resulting crane design data to text files. The GUI also provides a simple visualization of the crane's main components and its 2D SWL load chart, as shown in Fig. 2 [4]. The load chart is updated in real-time when the user modifies either of the design parameters.

2.5   Matlab Crane Optimisation Client (MCOC)

A manual trial-and-error design process using the web GUI together with the CPT is time-consuming and cost-inefficient, since there are more than 120 parameters that must be specified by the crane designer [5]. This large number of parameters makes the search space (the space of all possible combinations of parameter values) very large and a manual trial-and-error approach will necessarily be cumbersome and lead to suboptimal designs.

---

[1] http://www.seaonics.com/.

Bye et al. proposed an AIPO software module to replace the human crane designer in order to automate and optimise the design process [5]. Hameed et al. used Matlab to implement such a crane optimisation client, the MCOC module [6]. For the WS/JSON interface, they used two libraries freely available from the MathWorks File Exchange[2]. The first was MatlabWebSocket, which is a simple library consisting of a websocket server and client for Matlab, whilst the other was JSONlab, which is a toolbox to encode/decode JSON files in Matlab. Extending the work of [6], who used the single-objective and multi-objective GA solvers in the Matlab Global Optimization Toolbox [7], we also employ the PSO and SA algorithms from the same toolbox. As objective functions, the same cost and fitness functions defined in [6] are adopted.

3   Optimization Algorithms

This section briefly describes the optimization algorithms used for off-shore crane design in this paper.

3.1   GA

GA is a search method based on principles of natural selection and genetics [8]. GAs encode the decision variables of a search problem into finite-length strings of alphabets of certain cardinality. The strings, which are candidate solutions to the search problem, are referred to as *chromosomes*, the alphabets are referred to as *genes*, and the values of the genes are called *alleles*. Once the problem is encoded in a chromosomal manner and a fitness or cost measure for discriminating good solutions from bad ones has been chosen, a GA can start to evolve a *population* of candidate solutions to the search problem using the following standard steps in loop: initialization, evaluation, selection, recombination, mutation, and replacement. For more technical details about GA, please refer to [9].

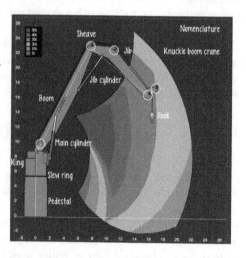

Fig. 2. Illustration of the main components of an offshore knuckleboom crane and its 2D load chart. Image courtesy of ICD Software AS.

3.2   PSO Algorithm

PSO is a population-based stochastic optimization technique, developed by Eberhart and Kennedy in 1995 [10,11]. A simple PSO is initialized with a group

---

[2] http://www.mathworks.com/matlabcentral/fileexchange.

of random particles (candidate solutions) moving in a search space with the same velocity, searching for an optimum solution by updating each particle's velocity at each iteration based on a weighted sum of three factors: (i) its inertia (previous velocity); (ii) its own personal best value found; and (iii) and the global best found among all or a subset of all particles. By choosing appropriate weights, the algorithm is able to explore and exploit the search space by moving the particles to better solutions. PSO has been successfully applied in many areas, for example function optimization, artificial neural network training, and fuzzy system control [13].

3.3   SA Algorithm

SA is an optimization algorithm that is emulating the cooling and crystallizing behaviour of chemical substances. Since annealing in nature results in *low-energy* configurations of crystals, it can be simulated in an algorithm to minimize cost functions. SA was first presented by Kirkpatrick, Gelett and Vecchi in 1983 for the optimal solution of problems related to computer design, such as component placement and wire routing [14]. SA was independently derived by Vlado Cerny in 1985, who used it to solve the travelling salesman problem [15]. The literature often distinguishes SA from evolutionary algorithms because SA does not involve a population of candidate solutions and therefore SA is considered a single-individual stochastic algorithm [16].

3.4   Objective Functions

In this paper, a set of crane design objective functions adopted from [6] are used for testing the optimization algorithms. All of the objectives are functions of two crane KPIs, namely SWL_{max}, which is the maximum safe working load, and W, which is the total crane weight.

- f_1 is given as the ratio of SWL_{max} to W,

$$f_1 = \frac{SWL_{max}}{W}.$$ (1)

We note that f_1 will increase when SWL_{max} increases and/or W decreases, and vice versa.
- $f_2(w_1, w_2)$ is given as the weighted sum of both SWL_{max} and W,

$$f_2(w_1, w_2) = w_1 SWL_{max} + w_2 \frac{1}{W}.$$ (2)

where w_1 amd w_2 are weight values used to reflect the importance or amount of contribution of SWL_{max} and W. We note that the total fitness will increase as SWL_{max} increases and/or W decreases, and vice versa.

– $f_3(w_1, w_2)$ and $f_4(w_1, w_2)$ are used to design a crane where we aim for respectively W and SWL$_{\max}$ to be as close as possible to the parameter values of the nominal benchmark crane, W$_{\text{target}}$ and SWL$_{\text{target}}$, respectively, while we optimise the remaining KPI:

$$f_3(w_1, w_2) = w_1 \frac{1}{\text{SWL}_{\max}} + w_2 \left| \text{W}_{\text{target}} - \text{W} \right| \tag{3}$$

and

$$f_4(w_1, w_2) = w_1 \left| \text{SWL}_{\text{target}} - \text{SWL}_{\max} \right| + w_2 \text{W}, \tag{4}$$

where w_1 and w_2 are weight values defined as before. The weight values w_1 and w_2 in this paper are set to one.

3.5   Case Study: Optimization of a Real Knuckleboom Crane

We adopt the same case study as we have presented previously in [5,6], where a real-world knuckleboom crane designed and delivered to a company in Baku, Azerbaijan, is used as a nominal benchmark against optimised crane designs determined by the optimization algorithms. As mentioned previously, such offshore cranes have about 120 different design parameters and a number of KPIs that can be derived from the chosen design. In an attempt to reduce design time, cost and satisfy customers' needs, a CautoD solution in which the MCOC module uses various optimization algorithms is used to automate the process and optimise the design. Here, GA, PSO and SA algorithms are applied to the offshore crane design problem for the four different objective functions described in previous section.

4   Results

Table 1 provides a summary of the results. For each optimization algorithm, the table compares the boom length (mm), L_{boom}, the jib length (mm), L_{jib}, the maximum boom pressure (bar), $P_{\text{max,boom}}$, and the maximum jib pressure (bar), $P_{\text{max,jib}}$, as well as the resulting maximum safe working load (ton), SWL$_{\max}$, and the total crane weight (ton), W, for the benchmark nominal crane, f_n, with each of the crane designs optimized using objective functions f_1–f_4. The table also shows the total convergence times T (hours) and the total number of iterations for each optimization algorithm employing each of the objective functions, and the mean and median values across each configuration.

Table 1 show that the SA required orders of magnitude more iterations to converge than the GA and PSO. This is due to the fact that SA only improves a single candidate solution at any iteration, whereas the GA and PSO both employ sets with many candidate solutions. The PSO converged slightly earlier than the GA for f_1 and f_2, whilst the GA converged much earlier than the PSO for f_3 and f_4. Looking at the processing time T, the same relationship between PSO and GA holds, whereas the processing time of SA is much longer for f_1 and f_2

Table 1. Optimization results for various objective functions compared to benchmark nominal crane design f_n.

	Measure	f_n	$f_1 - f_n$	$f_2 - f_n$	$f_3 - f_n$	$f_4 - f_n$	Mean	Median
GA	L_{boom}	15800	−3763	−3533	3666	−2356	−1497	−2945
	L_{jib}	10300	−4176	−4179	−4015	−1971	−3585	−4096
	$P_{max,boom}$	315,0	68,0	25,9	77,1	−41,6	32,4	47,0
	$P_{max,jib}$	215,0	47,0	51,8	79,7	−114,0	16,1	49,4
	fitness	-	1,26	40,66	0,02	50,89	23,21	20,96
	SWL_{max}	100,0	42,2	40,7	3,0	−0,0	21,4	21,8
	W	50,7	−6,8	−6,6	−0,0	−3,8	−4,3	−5,2
	T	-	1,64	2,30	1,73	1,52	1,80	1,69
	iterations	-	26	31	23	22	25,5	24,5
PSO	L_{boom}	15800	−3800	−3800	3940	−3789	−1862	−3795
	L_{jib}	10300	−4300	−4300	−4300	−4297	−4299	−4300
	$P_{max,boom}$	315,0	45,6	45,6	66,5	−87,7	17,5	45,6
	$P_{max,jib}$	215,0	85,0	85,0	13,8	−65,4	29,6	49,4
	fitness	-	1,30	28,29	0,01	50,86	20,12	14,80
	SWL_{max}	100,0	43,4	43,4	3,2	0,0	22,5	23,3
	W	50,9	−7,0	−7,0	0,0	−7,0	−5,2	−7,0
	T	-	1,50	1,67	4,04	4,37	2,89	2,85
	iterations	-	21	21	61	66	42,25	41
SA	L_{boom}	15800	−3799	−3793	−220	−92	−1976	−2006
	L_{jib}	10300	−1445	−3364	246	−219	−1195	−832
	$P_{max,boom}$	315,0	22,5	44,1	65,7	−0,5	33,0	33,3
	$P_{max,jib}$	215,0	−41,5	61,0	2,6	−59,8	−9,5	−19,5
	fitness	-	1,00	25,92	0,01	50,86	19.45	13,46
	SWL_{max}	100,0	37,0	40,0	4,0	0,0	20,1	20,5
	W	50,9	−4,6	−6,2	0,0	−0,3	−2,8	−2,4
	T	-	8,14	8,13	4,44	1,54	5,56	6,29
	iterations	-	11708	11708	6418	2198	8008	9063

than that of the GA and PSO but only slightly worse for f_3 and almost three times faster than the PSO for f_4.

For optimization of f_1 and f_2, PSO is the best choice (biggest increase of SWL_{max} and biggest decrease of W), slightly outperforming the GA, which in turn provides a better result than SA. For optimization of f_3, where the optimal W must be close to W_{target} SA has the greatest maximisation of SWL_{max}, before PSO followed by the GA. For optimization of f_4, however, where the optimal SWL_{max} must be close to W_{target}, SA is only able to slightly improve (reduce) W, whilst PSO is the best choice followed by the GA.

The abovementioned observations are summarised in Fig. 3, which shows the median convergence time and median difference of the optimized SWL_{max} and W using GA, PSO and SA and the nominal crane. The figure shows that when considering median values, the GA converged faster compared to both PSO and SA, whereas PSO yields the highest value of SWL_{max} and the lowest value of W compared to the values of both the GA and SA.

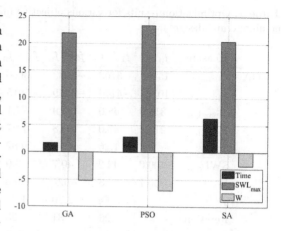

Fig. 3. Mean convergence time and mean difference between the optimized SWL_{max} and W using GA, PSO and SA and the nominal crane values.

5   Conclusions

In this paper, the GA, PSO, and SA optimization algorithms are applied to the optimization of an offshore crane design problem using four different objective functions f_1–f_4 that are all related to two KPIs, the maximum SWL, SWL_{max}, and the total crane weight, W.

The results show that all three optimization algorithms are able to determine optimized crane designs that outperforms a nominal benchmark crane, for a number of different objective functions, and in only a matter of hours. PSO has the best optimization for all the objective functions, except for f_3, where it is slightly worse than SA. Its convergence rate and processing time outperform those of the GA and SA for f_1 and f_2, whilst being 2–3 times worse than the GA for f_3 and f_4. However, one may speculate that the GA will require more iterations and a longer processing time to achieve the same optimization results as the PSO. Finally, we note that in addition to its implementation simplicity, PSO has the smallest number of tuning parameters compared to GA and SA. Given all these factors, we conclude that PSO is the best general purpose choice, but note that the SA seems particularly well suited for optimization of f_3. The GA has a higher implementation complexity than PSO but can also be a suitable general purpose choice at the expense of requiring more iterations and processing time than used here.

The CautoD solution proposed in this paper is not limited to crane design but can easily be extended and applied to the design of various other products or complex systems. The online system framework described in this paper is currently in operation at the SoftICE lab at NTNU in Ålesund and the authors are encouraging interested researchers and readers to contact them to use it in their own applications.

Acknowledgements. The SoftICE lab at NTNU in Ålesund wishes to thank ICD Software AS for their contribution towards the implementation of the simulator, and Seaonics AS for providing documentation and insight into the design and manufacturing process of offshore cranes. We are also grateful for the support provided by Regionalt Forskningsfond (RFF) Midt-Norge and the Research Council of Norway through the VRI research projects *Artificial Intelligence for Crane Design (Kunstig intelligens for krandesign (KIK))*, grant no. 241238 and *Artificial Intelligence for Winch Design (Kunstig intelligens for vinsjdesign (KIV))*, grant no. 249171.

References

1. Mujber, T.S., Szecsi, T., Hashmi, M.S.J.: Virtual reality applications in manufacturing process simulation. J. Mater. Process. Technol. **155**, 1834–1838 (2004)
2. Peng, S., Venkat, K., Vijay, V., Richard, M.: Design and virtual prototyping of humanworn manipulation devices. In: The ASME Design Technical Conference and Computers in Engineering Conference, DETC99/CIE-9029, Nevada, Las Vegas (1999)
3. Gowda, S., Jayaram, S., Jayaram, U.: Architectures for internet-based collaborative virtual prototyping. In: The ASME Design Technical Conference and Computers in Engineering Conference, DETC99/CIE-9029, Nevada, Las Vegas (1999)
4. Bye, R.T., Osen, O.L., Pedersen, B.S.: A computer-automated design tool for intelligent virtual prototyping of offshore cranes. In: Proceedings of the 29th European Conference on Modeling and Simulation (ECMS 2015), pp. 147–156, Albena, Bulgaria (2015)
5. Bye, R.T., Osen, O.L., Pedersen, B.S., Hameed, I.A., Schaathun, H.G.: A software framework for intelligent computer-automated product design. In: Proceedings of the 30th European Conference on Modeling and Simulation (ECMS 2016), pp. 534–543, Regensburg, Germany (2016)
6. Hameed, I.A., Bye, R.T., Osen, O.L., Pedersen, B.S., Schaathun, H.G.: Intelligent computer-automated crane design using an online crane prototyping tool. In: Proceedings of the 30th European Conference on Modeling and Simulation (ECMS 2016), pp. 564–573, Regensburg, Germany (2016)
7. Mathworks Inc.: MATLAB Global Optimization Toolbox. The Mathworks Inc., Natick, Massachusetts (2015). http://mathworks.com/products/global-optimization/
8. Holland, J.H.: Adaptation in Natural and Artificial Systems: An Introductory Analysis with Applications to Biology, Control, and Artificial Intelligence. University of Michigan Press, Oxford (1975)
9. Simon, D.: Evolutionary Optimization Algorithms: Biologically Inspired and Population-Based Approaches to Computer Intelligence. Wiley, Hoboken (2013)
10. Kennedy, J., Eberhart, R.C.: Particle swarm optimization. In: Proceedings of IEEE International Conference on Neural Networks, vol. IV, pp. 1942–1948. IEEE service center, Piscataway, NJ (1995)
11. Eberhart, R.C., Kennedy, J.: A new optimizer using particle swarm theory. In: Proceedings of the Sixth International Symposium on Micro Machine and Human Science, pp. 39–43. IEEE service center, Piscataway, NJ, Nagoya, Japan (1995)
12. Eberhart, R.C., Shi, Y.: Comparison between genetic algorithms and particle swarm optimization. In: Porto, V.W., Saravanan, N., Waagen, D., Eiben, A.E. (eds.) EP 1998. LNCS, vol. 1447, pp. 611–616. Springer, Heidelberg (1998). doi:10.1007/BFb0040812

13. Eberhart, R.C., Shi, Y.: Particle swarm optimization: developments, applications and resources. In: Proceedings of Congress on Evolutionary Computation 2001. IEEE service center, Piscataway, NJ., Seoul, Korea (2001)
14. Kirkpatrick, S., Gelett, C.D., Vecchi, M.P.: Optimization by simulated annealing. Science **220**, 621–630 (2001)
15. Cerny, V.: A thermodynamic appraoch to the traveling salesman problem: an efficient simulation. J. Optim. Theory Appl. **45**, 44–51 (2001)
16. Davis, L., Steenstrup, M.: Genetic algorithms and simulated annealing: An overview. In: Davis, L. (ed.) Genetic Algorithms and Simulated Annealing, Pitman Publishing, San Diego (1987)

Direct Adaptive Control of Process Control Benchmark Using Dynamic Recurrent Neural Network

Mohamed A. Hussien$^{(\boxtimes)}$, Tarek A. Mahmoud, and Mohamed I. Mahmoud

Faculty of Electronic Engineering,
Industrial Electronics and Control Engineering Department,
Menoufia University, Shibin Al Kawm, Egypt
mohamed.hussien@el-eng.menofia.edu.eg

Abstract. In this article, we develop a direct adaptive control scheme based on Dynamic Recurrent Neural Network (DRNN) for a process control benchmark. The DRNN is represented in a general nonlinear state space form for producing the control action that force the system output to a desired trajectory. The control algorithm can be implemented without a priori knowledge of the controlled system. Indeed, the weights of the DRNN controller are adjusted on-line using the truncated Back Propagation Through Time (BPTT) method. Unlike the approaches in the literature, the learning signal of the network weights is generated by a control error estimator stage in the developed controller. Finally, the developed controller is applied to a laboratory flow control system with two experimental scenarios.

Keywords: Direct adaptive control · Dynamic recurrent neural network · State space neural network · Flow control system

1 Introduction

During the past decades, Neural Networks (NNs) have been become attractive paradigms in the modeling and control of nonlinear processes. In literature, the structure of NNs is classified as feed forward and recurrent. The feed forward structure uses static discrete-time models that capture the dynamics of the real process through the use of tapped-delay lines in the model inputs and outputs. The feed forward neural networks have been developed in the direct inverse control [1,2], the internal model-based control(IMC) [3–5] and predictive control [6–11]. They have introduced improved modeling performance than linear models and as a result they are able to achieve better control performance of nonlinear systems. However, the feed forward neural networks suffer from number of drawbacks. In the identification of complex dynamic systems, the feed forward networks are unable to identify time dependent nonlinear dynamics with

© Springer International Publishing AG 2017
A.E. Hassanien et al. (eds.), *Proceedings of the International Conference on Advanced Intelligent Systems and Informatics 2016*, Advances in Intelligent Systems and Computing 533, DOI 10.1007/978-3-319-48308-5_27

high accuracy [12]. Besides, these networks suffer from large number of neurons, number of required delays and the weight updates do not use the internal neural network information.

To address these issues, the second structure of NNs, the dynamic recurrent NNs (DRNNs) have been introduced for the identification and control of non-linear systems such as [13–17]. The advantage of the dynamic recurrent neural networks (DNN) is threefold with respect to static networks [18]. First, the DRNN has self-loops and backward connections to memorize past information, hence it can capture the dynamic of the system. Second, the number of the network parameters is considerably lower. It is only necessary to identify the dimension of the state space, since the number of inputs and outputs is specified by their counterparts in the real process. Third, they can be represented in the state-space form, which is more suitable to most control schemes [19].

Motivated by the aforementioned review, this work develops a direct adaptive control scheme using the dynamic recurrent neural network for a practical system. In this scheme, DRNN is designed to represent the control action the make the system output to track a desired action. The network structure is implemented without any information about the controlled system. In the online stage, the network parameters are adjusted by using the truncated BPTT algorithm. Likewise to [20], a control error estimator is introduced to the control algorithm to estimate the learning error needed for the DRNN weights adaptation. The bounded input bounded output stability of the DRNN controller is discussed. Finally, the control algorithm is applied in the control of a process control laboratory system which is a real-life installation. In the real process, the problem of control is much more harder to achieve in the presence of noise and disturbances. The obtained results of the process control system depicted that the DRNN controller can be apply to the real life system with acceptable performance despite the external disturbance. Moreover, as the process works in the real-time, the computation complexity is also an important issue, because there are hard time constraints imposed on the control algorithm. The work shows that software implementation of the developed control method is possible, and the methods can be practically used in real industrial process with the sampling time equal to 0.22 s.

The rest of the paper is structured as follows. In Sect. 2, the description of the DRNN in the state space form is briefly provided. Section 3 develops the proposed direct adaptive control scheme based on the DRNN model. The benchmark system description and the control results are given in Sect. 4. Finally, Sect. 5 provides a concluding summary of this work.

2   Dynamic Recurrent Neural Network

In this section, we describe the structure of the DRNN which is investigated as a direct controller in this article. Let's consider a fully recurrent neural network with n neurons and m inputs as depicted in Fig. 1. Define $y(k) \in \Re$, $x(k) = [x_1(k), x_2(k), .., x_n(k)]^T \in \Re^n$ and $u(k) = [u_1(k), u_2(k), ..., u_m(k)]^T \in \Re^m$ the

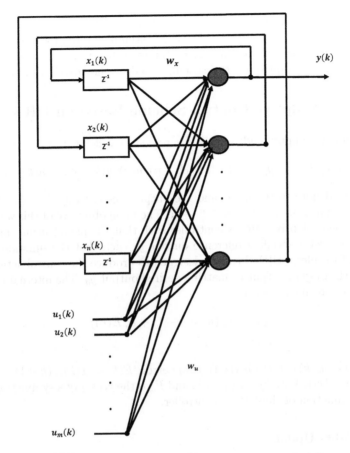

Fig. 1. The DRNN structure (where Z^{-1} denotes to the back shift operator)

network output, the hidden neuron outputs, and the external inputs at time step k respectively; $w_x \in \Re^{n \times n}$ is the matrix of the weights of the feedback connection of the hidden neurons; $w_u \in \Re^{n \times m}$ is the matrix of the weight connections between the external input and the hidden neurons. The equation of the DRNN model at time step k in the form of the state space equations is described by:

$$x(k) = F(w_x x(k-1) + w_u u(k))$$
$$y(k) = C^T x(k) \tag{1}$$

where $F(.)$ stands for the vector valued activation function of the hidden neurons and $C \in \Re^n$ is the vector of the output connection weights. Typically, the hyperbolic tangent activation function is selected giving well modeling results. In this work, the output weight vector is given as $C = [1, 0, ..., 0]^T$.

In control system applications, the NNs can be used in the controllers in either indirect or direct control schemes. In the former scheme, NN is designed to model the system dynamics. In such a way, the neural network model can be implicitly used to compute the control action that satisfies the controller's design specifications. In the direct scheme, NN is employed to produce the control action that force the system states to its target states. In the following section, a proposed direct adaptive control scheme based on the DRNN will be given.

3   Direct Adaptive Control Scheme Based on DRNN

Consider a single-input single output discrete system defined by:

$$y_p(k) = f(y_p(k-1), y_p(k-2), ..., y_p(k-n), u(k), u(k-1), ..., u(k-m)) \quad (2)$$

where u, y_p denote to the input and the output, respectively, k is the discrete time index, $n, m > 0$, and $f :\in \Re^{n+m} \to \Re$. The main objective of this work is to design a direct adaptive DRNN controller such that the plant output described in Eq. (2) tracks a specified reference output y_d such that the function $f(.)$ is unknown. Consider a DRNN can be designed to produce the control action $u(k)$ such that the system output y_p match a desired output y_d. The internal network states $x(k)$ and the output are:

$$x(k) = F_u(w_x x(k-1) + w_e E(k))$$
$$u(k) = x_1(k) \quad (3)$$

where $e(k) = y_d(k) - y_p(k)$ is the tracking error, $E(k) = [e(k), e(k-1), ..., e(k-m)]^T$, $x(k) = [x_1(k), x_2(k), ..., x_n(k)]^T$, and F_u is the vector of a symmetric tanch activation function of the DRNN controller.

3.1   Weights Update

In this subsection, we present the adaptation law of the DRNN controller weights in the proposed scheme. The learning law of the DRNN network weights w_x and w_e is derived through the minimization of a suitable cost function. Despite the training of the static neural networks, the cost function for training the DRNN is a time varying error. The back-propagation through time (BPTT) algorithm is the basic method for the dynamic recurrent neural networks learning [18,21]. In fact, the stander algorithm of the BPTT is more theoretical than practical interest because it makes use of potentially unbounded history storage. Therefore, more practical extension methods to the BPTT have been developed such as the truncated BPTT methods for the on-line training phase [18]. Define the following error for the DRNN controller:

$$e_u(k) = u_d(k) - u(k) \quad (4)$$

where $u_d(k)$ is the desired control action the make the system output $y_p(k)$ to track a desired signal $y_d(k)$. Then, the cost function of the DRNN can be given by:

$$E_u(k) = \frac{1}{2}e_u^2(k) \quad (5)$$

In order to describe the details of the on-line training of the DNNN controller using the truncated BPTT method, merge $E(k)$ and $x(k)$ to form a vector $Z(k) \in \Re^{n+m}$ as $Z(k) = [e(k), e(k-1), ..., e(k-m), x_1(k), x_2(k), ..., x_n(k)]^T$. Also, all the weights of the network w_e and w_x can be collected into a single matrix $W \in \Re^{n \times (n+m)}$. Thus, for $l = 1, 2, ..., n$ and $i = 1, 2, ..., n+m$, the network equations becomes:

$$S_l(k) = \sum_{i=1}^{n+m} w_{li} z_i(k)$$

$$x_l(k) = f_u(S_l(k))$$

$$u(k) = x_1(k) \tag{6}$$

where w_{li} represents the connection weight to the lth state unit (i.e., $x_l(k)$) from the ith unit of of the input (i.e., $z_i(k)$), and $f_u(.)$ is the tanch function. The learning law of the network weights W can be derived through the minimization of the cost function defined in (5) using the truncated BPTT method. Then, the on-line updating rule of any particular weight w_{li} of the DRNN controller can be calculated by

$$\Delta w_{li} = -\eta \sum_{k-h+1}^{t} \frac{\partial E_u(k)}{\partial w_{li}(t)} \tag{7}$$

where η is the learning rate, h is the past history considered in the calculation and must be fixed and chosen longer for better. We can calculate the term $\frac{\partial E_u(k)}{\partial w_{li}(t)}$ as:

$$\frac{\partial E_u(k)}{\partial w_{li}(t)} = \frac{\partial E_u(k)}{\partial S_l(t)} \frac{\partial S_l(k)}{\partial w_{li}(t)} = \frac{\partial E_u(k)}{\partial S_l(t)} z_i(t-1) \tag{8}$$

where

$$\frac{\partial E_u(k)}{\partial S_l(t)} = \frac{\partial E_u(k)}{\partial x_l(t)} f_u(S_l(t)) \tag{9}$$

and

$$\frac{\partial E_u(k)}{\partial x_l(t)} = \begin{cases} -e_u(t) & \text{if } t = k, \\ \sum_{j=1}^{n} \frac{\partial E_u(k)}{\partial S_l(t)} w_{li} & \text{if } t < k. \end{cases} \tag{10}$$

Define $\delta_l(t)$ as

$$\delta_l(t) = \begin{cases} -e_u(t) \acute{f}_u(S_l(t)) & \text{if } t = k, \\ \acute{f}_u(S_l(t)) \sum_{i=1}^{n} \delta_l(t+1) w_{li} & \text{if } t < k. \end{cases} \tag{11}$$

Finally, the on line adaptation rule for the DRNN controller is

$$w_{li}(new) = w_{li}(old) + \eta \sum_{k-h+1}^{t} \delta_l(t) z_i(t) \tag{12}$$

Due to the desired control action $u_d(k)$ is not available, the learning error defined in (4) cannot be obtained. Then, an estimated control error expressed by $\hat{e}_u$ can be introduced in the adaptive law of the DRNN controller (11) and (12) as in the following subsection.

3.2 Learning Error Estimator

In [20], a simple method was proposed to estimate the learning error $\hat{e}_u$ such that it can be estimated directly as:

$$\hat{e}_u(k) = k_e e(k) + k_c \Delta e(k) \tag{13}$$

where $e(k) = y_d(k) - y_p(k)$ is the tracking error, $\Delta e(k)$ is the change of this error, k_e and k_c are positive small values. Finally we can write:

$$\hat{e}_u(k) = G e_u(k) \tag{14}$$

where G is a positive scaling factor that will be included in the learning rate η. Consequently, the adaptive laws (11) and (12) of the controller's weights can be rewritten as:

$$\delta_l(t) = \begin{cases} -\hat{e}_u(t) \acute{f}_u(S_l(t)) & \text{if } t = k, \\ \acute{f}_u(S_l(t)) \sum_{i=1}^{n} \delta_l(t+1) w_{li} & \text{if } t < k. \end{cases} \tag{15}$$

$$w_{li}(new) = w_{li}(old) + \eta \sum_{k-h+1}^{t} \delta_l(t) z_i(t) \tag{16}$$

3.3 Convergence Analysis

Here, we discuss the bounded input bounded output (BIBO) stability of the DRNN controller. To investigate the BIBO stability of the DRNN controller, consider the network states has the following linear form:

$$x(k) = w_x x(k-1) + w_e E(k) \tag{17}$$

Let λ_{max} as the largest absolute eigenvalue of w_x and according to the linear system theory, if $|\lambda_{max}| < 1$, the linear system defined in (17) will be bounded. For different eigen values of w_x, the linear system (17) has different transient properties. Thus, for the DRNN defined in (3) with the tanch activation function, we have

$$\|tanch(w_x x(k-1) + w_e E(k))\| \leq \|w_x x(k-1) + w_e E(k)\| \tag{18}$$

Therefore, the DRNN defined in (17) with the tanch activation function and the condition that the largest absolute eigenvalue of w_x is smaller than 1, we can say that the DRNN output is bounded for all inputs.

To summarize, Fig. 2 describes the overall diagram of the proposed direct adaptive control scheme based on the DRNN.

4   Experimental Results

4.1   Benchmark System

The considered benchmark system is the Process Control System (PCS) developed at Research Laboratory in Industrial Electronics and Control Engineering Department, Faculty of Electronic Engineering, Menoufia University, Egypt. The system consists of two water tanks and one pump as illustrated in Fig. 3. In turn, the PCS configuration diagram is shown in Fig. 4. The pump allows to move the water from the lower tank to the upper one through the valve V1 or from the lower tank to itself through the valve V2. Under the influence of gravity, the water is automatically transfered from the upper tank to the lower one through V3 to ensure a water supply. In the upper tank, an ultrasonic sensor is installed to measure the water level. A pressure sensor is used to measure the pressure in the pipes. There is also a flow sensor to measure the water flow after pump. Different water loops can be implemented by a set of different types of valves (solenoid and proportional ones). Moreover, a heater installed in the lower tank and a temperature sensor to measure the temperature when water circulates through the pipes. The specifications of the process parts are shown in Table 1.

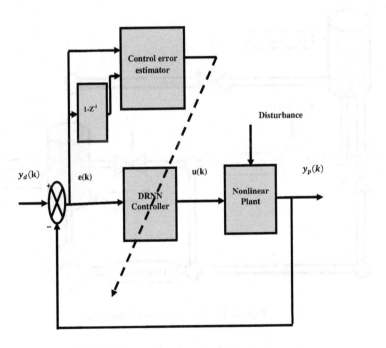

Fig. 2. The proposed control structure

Fig. 3. The laboratory installation actual view

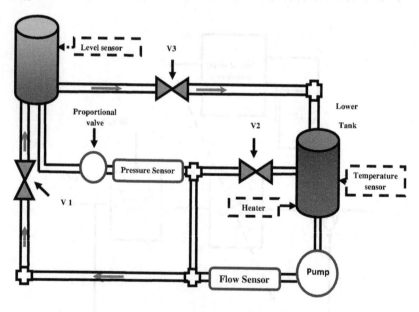

Fig. 4. PCS configuration diagram

4.2   Control Results

The benchmark system allows for many different configurations. It can be configured as different simple of SISO (Single Input Single Output) systems based on the variable to be controlled. In this work, the flow control is selected where

Table 1. The specifications of the process components

The component	Specifications
Pump	Pressure 2.8 bar : Flow 12 L/min
Flow sensor	Flow range : [2 16] L/min
Pressure sensor	Pressure range : [0 10] bar
Ultrasonic sensor	Sensing range : [60 300] mm
Solenoid valves (V1,V2,V3)	Operating pressure range is 7 bar
Proportional valve	Operating : [0 8] bar

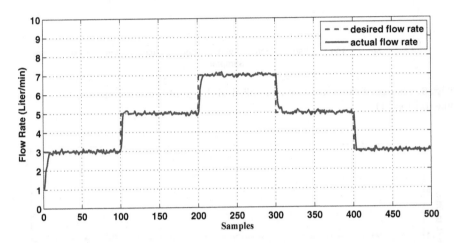

Fig. 5. The flow rate of the PCS using the proposed direct adaptive control scheme (experimental task 1)

the flow sensor signal is the input of the control system and the control effort is applied to the motor pump. When the pump is running, it realizes two tasks, speeding up the flow of water to the upper tank and also makes the water circulation in the outer loop. Herein, the controller aims to maintain a constant flow of water from the lower tank to the upper tank regardless of the water amount in the tanks. In the closed loop system, the process control system is interfaced with a personal computer (PC) through a data acquisition card (NI-CDAQ-9171) which receives the input values (0 to 9V) from the computer and transmits it to the system with sampling period of 0.22 s. An amplifier is used to convert the sampled-input signal from the data acquisition card to variable DC voltage from 0 to 24 V and fed it to the motor pump. Thus, PC (Processor Intel core i3, CPU 3.06 GHz, Ram 2.00 GB, operating System 32 bit Windows 7) that runs Lab view program code is adopted as the controller. During the experiments, the valves V1 and V3 were permanently opened to continuously create connection between the two tanks, and the valve V2 was permanently closed.

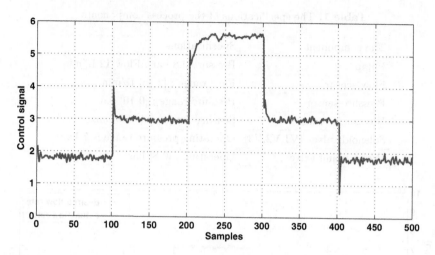

Fig. 6. Change of the control signal applied to the amplifier of the motor pump (experimental task 1)

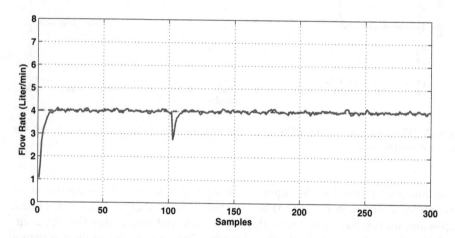

Fig. 7. The flow rate of the PCS using the proposed direct adaptive control scheme (experimental task 2)

Table 2. Comparison of RMSE index for the proposed controller, conventional PI and fuzzy PI controllers

Controller	Experimental task (1)	Experimental task (2)
The proposed adaptive DRNN controller	0.2649	0.2652
PI controller	0.6288	0.5490
Fuzzy PI controller	0.3085	0.2738

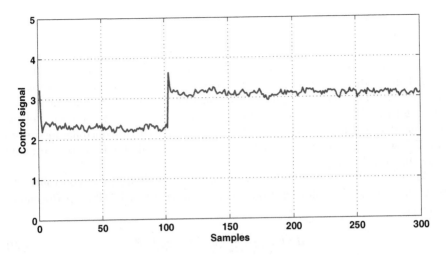

Fig. 8. Change of the control signal applied to the amplifier of the motor pump (experimental task 2)

In order to apply the proposed direct control scheme to the flow control system, several parameters should be selected priori such as the network parameter structure (i.e., n and m), the initial values of the network weights (i.e., w_e and w_x), and the parameters of the control error estimator parameters (i.e., K_e and K_c). In this study, we have employed DRNN controller with $n = 3$, and $m = 1$. When we have increased both n and m values, no further improvements in the control performance have been observed. Actually, the initial values of DRNN controller weights are of no importance as they are tuned on line and they eventually converge to their suitable values. Thus, w_e and w_x were initialized at small random values. Also, the two parameters k_e and k_c were selected as small positive values to avoid the output oscillation and overshoots (i.e., $k_e = 0.5$ and $k_c = 0.35$).

Two experimental tasks were achieved to investigate the performance of the developed controller. In the first one, we have investigated the controller performance with the set point changes which was assumed as:

$$y_d(k) = \begin{cases} 3 & \text{if } k < 100, \\ 5 & \text{if } 100 \leq < k < 200, \\ 6 & \text{if } 200 \leq < k < 300, \\ 5 & \text{if } 300 \leq < k < 400, \\ 3 & \text{if } k \geq < 400. \end{cases} \tag{19}$$

where k is the sampling instant. Figure 5 illustrates the change of flow rate according to the desired flow defined in (19). The obtained result of the proposed controller showed that it can achieve acceptable tracking performance with set-point changes. Moreover, the control effort applied to the amplifier of the motor pump is smooth as depicted in Fig. 6.

During the last experiment, the controller performance was tested when the flow process control system was corrupted by external disturbance. An external disturbance, 30% of the set-point, was added to the system and the set point was set to 4 liter/min. Figures 7 and 8 show the obtained flow rate and the control signal for the second experimental task, respectively. Despite the presence of the external disturbance, the results of this task ensure that the controller algorithm can follow the set-point accurately with a very small steady state error and smooth control effort.

Moreover, the performance of the proposed DRNN adaptive controller can be verified using the following Root Mean Squared Error (RMSE) performance index:

$$RMSE \sqrt{\frac{1}{N} \sum_{k=1}^{k=N} (y_d(k) - y_p(k))^2} \tag{20}$$

where N is the number of samples, y_d and y_p are the desired and the actual flow rate, respectively. For the comparison reason, a conventional PI and a fuzzy PI controllers are tested in the same platform for the two experimental tasks. Table 2 provides comparative results of the proposed controller and both PI and fuzzy PI controllers. These comparative results reavel that the overall performance of the proposed DRNN controller for the two experimental scenarios is better to the classical PI and the fuzzy PI controllers. The aforementioned experimental results show that the efficancy of the direct adaptive DRNN controller method in controlling a real time application.

5   Conclusion

In this paper, a direct adaptive control using a dynamic recurrent neural network for a real time application is developed. The developed control scheme design has the following features: (1) The configuration parameters of the DRNN controller used in the controller method is considerably lower. Thus, the proposed scheme is more suitable for real time implementation. (2) The DRNN controller is implemented without a priori knowledge of the controlled and the off line training of the network is not required. The experimental results illustrated that the developed DRNN controller method can perform good tracking. Furthermore, a successful control and a desired performance can be obtained in the presence of external disturbances.

References

1. Cabrera, J., Narendra, K.: Issues in the application of neural networks for tracking based on inverse control. IEEE Trans. Autom. Control **40**(11), 2007–2027 (1999)
2. Hussain, M.A., Kershenbau, L.: Implementation of neural network-basedinverse-model control strategies on an exothermic reactor. Sci. Asia **27**, 41–50 (2001)

3. Awais, M.: Application of internal model control methods to industrial combustion. Appl. Soft Comput. **5**(2), 223–233 (2005)
4. Chidrawar, S., Patre, B.: Implementation of neural network for internalmodel control and adaptive control. In: Proceedings of the International Conference on Computer and Communication Engineering, Malaysia (2008)
5. Deng, H., Xu, Z., Han-Xiong, L.: A novel neural internal model control formultiinput multi-output nonlinear discrete-time processes. J. Process Control **19**, 1392–1400 (2009)
6. Yu, D.L., Yu, D.W., Gomm, J.B.: Neural model adaptation and predictive control of a chemical process rig. IEEE Trans. Control Syst. Technol. **14**(5), 828–840 (2006)
7. Kittisupakorn, P., Thitiyasook, P., Hussain, M.A., Daosud, W.: Neural network based model predictive control for a steel pickling process. J. Process Control **19**(4), 579–580 (2009)
8. Peng, H., Wu, J., Inoussa, G., Deng, Q., Nakano, K.: Nonlinear system modelingand predictive control using the RBF nets-based quasi-linear ARX model. Control Eng. Pract. **17**(1), 59–66 (2009)
9. Tiwari, S., Naresh, R., Jha, R.: Neural network predictive control of UPFC for improving transient stability performance of power system. Appl. Soft Comput. **11**(8), 4581–4590 (2011)
10. Nikdel, N., Nikdel, P., Badamchizadeh, M.A., Hassanzadeh, I.: Using neuralnetwork model predictive control for controlling shape memory alloy-basedmanipulator. IEEE Trans. Ind. Electron. **61**(3), 1394–1401 (2014)
11. Yan, Z., Wang, J.: Robust model predictive control of nonlinear systems withunmodeled dynamics and bounded uncertainties based on neural networks. IEEE Trans. Neural Netw. Learn. Syst. **25**(3), 457–469 (2014)
12. Chun-Fei, H., Chin-Min, L., Ang-Bung, T., Chao-Ming, C.: Adaptive control for MIMO uncertain nonlinear systems using recurrent wavelet neural network. Int. J. Neural Syst. **22**(1), 37–50 (2012)
13. Ku, C.C., Lee, K.Y.: Diagonal recurrent development neural networks for dynamic systems control. IEEE Trans. Neural Netw. **6**(1), 144–156 (1995)
14. Pan, Y., Wang, J.: Model predictive control of unknown nonlinear dynamical systems based on recurrent neural networks. IEEE Trans. Ind. Electron. **58**(8), 3089–3101 (2011)
15. Fairbank, M., Li, S., Fub, X., Alonso, E., Wunsch, D.: An adaptive recurrentneural-network controller using a stabilization matrix and predictive inputs to solve a tracking problem under disturbances. Neural Netw. **49**, 74–86 (2014)
16. Xiao, L., Chao, J., De-Xin, L., Da-Wei, D.: Nonlinear adaptive control using multiple models and dynamic neural networks. Neurocomputing **136**, 190–200 (2014)
17. Hong, G., Lu, Z., Ying, H., June-Fei, Q.: Nonlinear model predictive control based on a self-organizing recurrent neural network. IEEE Trans. Neural Netw. Learn. Syst. **27**(2), 402–415 (2016)
18. Williams, R.J., Zisper, D.: An efficient gradient-based algorithm for on-line training of recurrent networks. Neural Comput. **2**(4), 490–501 (1990)
19. Zamarreno, J.M., Vega, P.: State-space neural network, properties and application. Neural Netw. **11**, 1099–1112 (1998)
20. Mahmoud, T.A., Elshenawy, L.M.: Echo state neural network based statefeedback control for SISO affine nonlinear systems. In: Proceedingsof 1st IFAC Conference on Modeling, Identification and Control of NonlinearSystems (MICNON-2015), Saint-Petersburg, Russia, 24-26 June 2015 (2015)
21. Pearlmutter, B.A.: Gradient calculations for dynamic recurrent neural networks: a survey. IEEE Trans. Neural Netw. **6**(5), 1212–1227 (1995)

An Enhanced Distributed Database Design Over the Cloud Environment

Ahmed E. Abdel Raouf$^{(\boxtimes)}$, Nagwa L. Badr, and M.F. Tolba

Faculty of Computer and Information Sciences,
Ain Shams University, Cairo, Egypt
ahmed_ezzat991@yahoo.com, nagwabadr@cis.asu.edu.eg,
fahmytolba@gmail.com

Abstract. The design of a distributed database is one of the major research issues within the distributed database system area. The main challenges facing distributed database systems (DDBS) design are fragmentation, allocation and replication. In this paper, we present an enhanced distributed database design over the cloud environment. It minimizes the execution time needed for the transactions and for noticing an error when an invalid query or data cannot found occurs. It also allows users to access the distributed database from anywhere. Moreover, it allows fragmentation, allocation and replication decisions to be taken statically at the initial stage of designing the distributed database. Experimental results show that the proposed system design, results in a significant reduction of the execution time needed for the transactions to reach the data in an appropriate site and the time taken to notice an error when an invalid query or data not found occurs.

Keywords: Distributed database system (DDBS) · Distributed database management system (DDBMS) · Fragmentation · Replication · Allocation · Cloud computing

1 Introduction

A database that consists of very large amounts of data used by different applications at different physical locations needs efficient support. Large distributed enterprise databases, telecom databases and scientific databases are examples of application areas [5]. The delay of accessing remote databases is the main problem of many of these applications. As a result, it is necessary to use a distributed database that employs fragmentation, allocation and replication [5].

Distributed database systems (DDBS) typically consist of a number of distinct database fragments and its replicas which allocate and replicate at different geographic sites. The sites of DDBSs are managed by distributed database management systems (DDBMS) and can also communicate through a network. Each site has autonomous processing capability and also participates in the execution of at least one global database application, which requires accessing data residing at several different sites [1].

The major research issue in a distributed database system area is the design of the distributed database system. The main challenges facing the DDBS design are the

© Springer International Publishing AG 2017
A.E. Hassanien et al. (eds.), *Proceedings of the International Conference on Advanced Intelligent Systems and Informatics 2016*, Advances in Intelligent Systems and Computing 533, DOI 10.1007/978-3-319-48308-5_28

following questions: Which type of fragmentation will be used to fragment the database relations and where to allocate, replicate the fragments and its replicas to the sites of the distributed database system to enhance system performance and increase availability.

The main contribution in this extended research is adding new layers with modules to our previous DDBS design in [1]. The new modules aim to minimize the execution time needed for the transactions to reach the data in an appropriate site and also minimize the required time to notice an error when an invalid query or data not found occurs. The proposed system design allows the distributed database system to be accessed from anywhere without owning any technology infrastructure. Moreover, it allows static fragmentation, allocation and replication decisions to be taken at the initial stage of designing the distributed database, without the help of empirical data about query executions. Experimental results show that the proposed system design, results in a significant reduction of the execution time needed for the transactions to reach the data in an appropriate site and the time taken to notice an error when an invalid query or data not found occurs.

The rest of this paper is organized as follows. Section 2 reviews the related work. The enhanced distributed database design over the cloud environment and its components proposed in Sect. 3. Section 4 presents the experimental results. Finally the conclusion and future work are presented in Sect. 5.

2   Related Work

A dynamic DDBS over the cloud environment was proposed by our previous work in [1]. Moreover, an enhanced allocation and replication technique was proposed, which allocates the fragments and its replicas to the sites that already requires it without taking into account the location of the sites. The enhanced allocation and replication technique aims at maximizing the number of local accesses compared to access from the remote sites, enhance the system performance and increase the availability.

An optimized scheme for vertical fragmentation, allocation and replication of a distributed database called (VFAR) was proposed by our previous work in [2]. The proposed VFAR scheme partitions the distributed database relations vertically at the initial stage of the database design by using the enhanced minimum spanning tree (MST) Prim's algorithm. Moreover, it allocates and replicates the resulted fragments and its replicas to the sites that require it.

The authors of [4] propose a new technique for horizontal fragmentation of the distributed database relations. This technique is used for partitioning the relations at the initial stage as well as in later stages of DDBS. The authors of [6] provide some algorithms to ensure generality of the technique developed in [4] by addressing some important scalability issues. However, the allocation strategy of this technique [4] doesn't meet the goal of data allocation. The goal of data allocation can be achieved by allocating the fragments to the sites that require it only. As a result the user can access data with low cost and time.

A decentralized approach for dynamic fragmentation, allocation and replication in DDBS was proposed in [5]. It performs fragmentation, allocation and replication based on recent access history. However, the main drawback of this approach is that it doesn't consider the load of the sites after the process of allocation and replication, site constraints as well as the optimal number of replicas of each fragment to enhance the system's performance and increase availability.

Chord was designed to create a network that is decentralized, reliable and scalable [3]. It uses the consistent-hashing method to distribute hash keys to nodes. The nodes in chord are organized in a ring topology, in which each node in chord contains a finger table of its neighbors and their possible assignments of keys.

A cluster based peer to peer architecture for a distributed database named Flexipeer was proposed in [3]. They try to implement the concept of chord in peer-to-peer based data management. The proposed architecture in [3] contains a local cluster administrator (LCA) which manages the sites of each cluster. It also contains a global cluster administrator (GCA) which manages the whole architecture. This work uses the fragmentation technique of previous work done [4]. In addition, a clustering approach for partitioning database sites and allocating fragments across the sites of each cluster was proposed.

However, this approach [3] wastes a lot of time between nodes, LCA, LCA Validator, Resource Checker and GCA until it reaches the required data. Moreover, the allocation technique used in this work doesn't meet the goal of data allocation by allocating fragments to the sites that need it. In addition, it doesn't address the replication phase of database design.

The authors of [7] present a synchronized horizontal fragmentation, allocation and replication model. It performs horizontal fragmentation of distributed database relations based on the attribute retrieval and update frequency.

The works in [7] and [8] perform the allocation process based on the cost of moving the data fragments from one site to the other and the fragment access pattern. A new algorithm called the Region Based Fragment Allocation (RFA) was proposed in [9]. It considers the frequency of fragments accessed by region as well as individual nodes to move the fragment from the source node to target node. It decreases the fragments migration using its knowledge of the network topology in comparison to optimal [10] and threshold [11] algorithms. It also reduces the amount of topological data required in decision making in comparison to the BGBR [12] algorithm.

A model that takes site constraints into account in the process of re-allocation was presented in [13]. However, when information queries continuously change in a faster way and when the number of fragments and sites largely increase, this model will be more complicated.

A Near Neighborhood Allocation (NNA) algorithm which moves data to a neighborhood node that is placed in the path to the node with the maximum access counter was proposed in [14]. It could be of greater use in larger networks to decrease the delay of response times.

The authors of [15] propose an algorithm that dynamically re-allocates the fragments to sites of the distributed database system at runtime. It takes into account the time constraints, volume threshold and the volume of data transmitted in successive time intervals in accordance with the changing access patterns to dynamically re-allocate the fragments to the sites. However, the number of time intervals, their duration and the volume threshold are the most important factors that determine the frequency of fragment re-allocations.

The authors of [16] perform the re-allocation process given the time and sites constraints as well as changing the data access patterns of the DDBS. It adopts the shortest path algorithm once the data movement decision is taken to reduce data transmission costs compared to the previous methods. The drawback of this algorithm is the storage required compared to some previous algorithms.

The limitations of existing literature can be summarized in these points. Firstly, the earlier approaches that suggest a distributed database design wastes a lot of time between nodes to reach outsourced data in an appropriate site and to notice an error when an invalid query or data not found occurs.

Secondly, the proposed approaches do not consider the load of the sites after the process of allocation and replication, site constraints as well as the optimal number of replicas of each fragment to enhance the system's performance and increase availability.

Finally, the allocation strategy that's used at the initial stage of distributed database design doesn't meet the goal of the data allocation. The goal of the data allocation can be achieved by allocating the fragments to the sites that require it only. As a result the user can access data with low costs and time. In addition, it doesn't address the replication phase of the distributed database at the initial stage of designing distributed databases.

3   The Enhanced Distributed Database Design Over Cloud Environment Architectures

To overcome the limitations of existing literature highlighted by the above survey, this research proposes an enhanced DDBS design over the cloud environment. In our previous work in [1] a DDBS design over the cloud was introduced. To extend this work, the main contribution in this extended research is adding new layers with modules to our previous DDBS design in [1]. The new modules aim to minimize the execution time needed for the transactions to reach the data in an appropriate site and also minimize the required time to notice an error when an invalid query or data not found occurs. The proposed design allows distributed database systems to be accessed from anywhere without the need to own any technology infrastructure. The proposed system can be accessed through: A web browser, mobile application or desktop application while the database is stored on servers at remote sites. In addition, it allows fragmentation, allocation and replication decisions to be taken statically at the initial stage of designing the distributed database, without the help of empirical data regarding query executions.

The proposed design is shown in Fig. 1 and consists of two layers: Distributed database system manager and distributed database cluster layers.

3.1   Distributed Database System Manager (DDBSM)

The distributed database system manager (DDBSM) is used to manage the distributed database systems. It has special operations to firstly, partition the distributed database

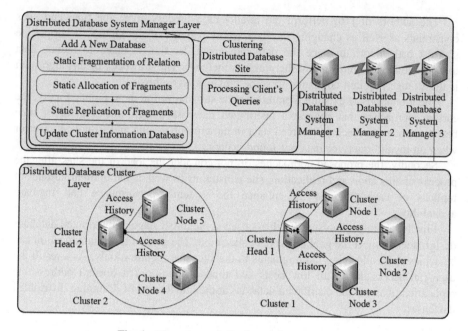

Fig. 1. The enhanced distributed database design

relation into fragments and also allocates and replicates the resulted fragments to the sites. Secondly, it clusters the distributed database sites into disjoint clusters and finally, processing the client queries. This layer consist of three modules: Adding a new database, clustering distributed database sites and processing client queries.

Add New Database. This module is used to add the database tables to DDBS. Firstly, the new tables are fragmented, allocated and replicated to the sites of the distributed database using the static fragmentation, allocation and replication technique proposed by our previous work in [1], which can be used at the initial stage of the DDBS design using the enhanced CRUD (Create, Read, Update and Delete) matrix, without the help of empirical data about query executions and also without taking into account the location of the sites.

Secondly, the cluster information database is updated to save the locations of each fragment and replica in the distributed database system. The Cluster Information Database is a database that holds complete information about each fragment or replica in which site or cluster.

Clustering Distributed Database Sites. This module is used to cluster the distributed database sites into disjoint clusters. In addition, it also allocates sites to each cluster. After that it updates the cluster information database with the recent location of each site in DDBS.

Client Queries Processing. This module is used for processing the client queries. When the client sends a query to the DDBSM, firstly the distributed database systems manager checks the syntax of the query, if the query doesn't satisfy the syntax, then the DDBSM sends a message to the client saying the query is invalid.

However, if the query satisfies the syntax, then the DDBSM uses the cluster information database to determine the location of the sites that hold the required data based on the type of query. If the locations of the data are found, then the DDBSM sends the locations of the site that holds the data to the clients.

After that, the client uses the information about the communication costs between the sites of the DDBS to determine the closest site that holds the data. Finally, the query is sent to the selected site and the results are sent from the site to the client. However, if the locations of the data aren't found in the cluster information database then, the DDBSM sends a message to the client that the data was not found. The flow diagram of the query processing is shown in Fig. 2.

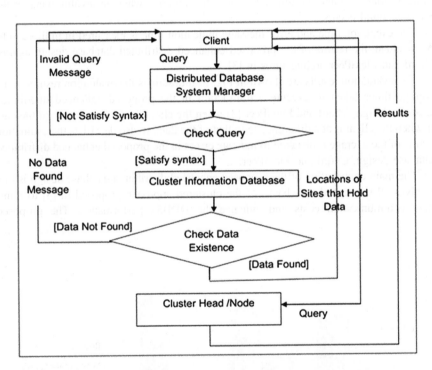

Fig. 2. Flow diagram of query processing

3.2   Distributed Database Clusters Layers

The distributed database clusters layer consists of more than one cluster. Each cluster consists of one cluster head and more than one cluster node. The cluster head is the same as a cluster node; however it has especial and additional operations to manage the other cluster nodes.

At each access to the cluster node, firstly, it checks whether it is a local access or remote. Secondly, it allows the user to access the database of the node to run the query and fetch the required data. Finally, the node access record is updated to save the information about the user's access.

4   The Experimental Results

The performance of the proposed enhanced distributed database design is studied in a simulated environment. The simulated environment consists of three HP Compaq computers with Core-two Duo 2.10 processors and 2 GB RAM using SQL Server as DBMS. This simulation is intended to evaluate the performance of the proposed enhanced distributed database design in execution of transactions reaching the data in an appropriate site and the time taken to notice an error when an invalid query or the data not found occurs.

The execution time and error message indication time is measured in milliseconds. We evaluate the results of the proposed enhanced distributed database design against Chord and FlexiPeer architectures in [3].

The evaluation results are shown in Fig. 3. It describes the evaluation results of the average times taken to execute a transaction in the proposed enhanced distributed database design, Chord and FlexiPeer [3] when the client is at the same site of the data or when the client is at a different site to that of the data. It also describes the evaluation results of the average time taken to indicate errors in the proposed enhanced distributed database design, Chord and FlexiPeer.

The main contribution of the proposed enhanced distributed database design is that it solves all mentioned drawbacks of the clustered approach proposed in [3] to mini-mize communication costs and enhance the DDBSs performance. The proposed

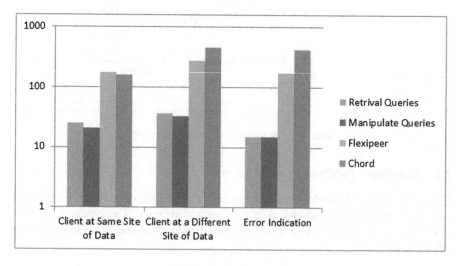

Fig. 3. Evaluation results

enhanced distributed database design firstly, uses our previous enhanced fragmentation, allocation and replication technique proposed in [1], which takes the fragmentation, allocation and replication decisions at the initial stage of designing the distributed database. It is knowledge based on the requirement analysis phase by using the enhanced CRUD (Create, Read, Update and Delete) matrix without help of empirical data regarding query executions and without taking into account the location of the sites. It also aims to maximize the number of local accesses compared to remote accesses in order to enhance the systems performance and increases the availability. Secondly, it results in a significant reduction of the execution times needed for the transactions to reach the data in an appropriate site and the time taken to notice an error when an invalid query or data not found occurs.

5 Conclusion and Future Work

Improving the performance, increasing the availability of data and access facility are the main motivation for distributed databases. The efficient distribution of fragments and its replicas to the sites of the distributed database system play the critical role of the performance and the cost of a distributed database system. In this paper, we present an enhanced distributed database design over a cloud environment. The proposed system design allows distributed database systems to be accessed from anywhere. In addition, it allows fragmentation, allocation and replication decisions to be taken statically at the initial stage of designing the distributed database, without the help of empirical data about query executions. Experimental results show that the enhanced distributed database design, results in a significant reduction of the execution time needed for the transactions to reach the data in an appropriate site and the time taken to notice an error when an invalid query or data not found occurs. As proposed future work, firstly, we plan to use an enhanced clustering technique to cluster distributed database sites into disjoint clusters. Secondly, we plan to implement the remaining parts of the proposed architecture to efficiently allow fragmentation, allocation and replication decisions to be taken automatically at run time.

References

1. Abdel Raouf, A.E., Badr, N.L., Tolba, M.F.: Dynamic distributed database over cloud environment. In: Hassanien, A.E., Tolba, M.F., Taher Azar, A. (eds.) AMLTA 2014. CCIS, vol. 488, pp. 67–76. Springer, Heidelberg (2014). doi:10.1007/978-3-319-13461-1_8
2. Abdel Raouf, A.E., Badr, N.L., Tolba, M.F.: An optimized scheme for vertical fragmentation, allocation and replication of a distributed database. In: 2015 IEEE Seventh International Conference on Intelligent Computing and Information Systems (ICICIS), pp. 506–513. IEEE, Cairo (2015)
3. Amalarethinam, D., Balakrishnan, C.: A study on performance evaluation of peer-to-peer distributed databases. IOSR J. Eng. 2, 1168–1176 (2012)
4. Khan, S., Hoque, A.: A new technique for database fragmentation in distributed systems. Int. J. Comput. Appl. 5, 20–24 (2010)

5. Hauglid, J.O., Ryeng, N.H., Nørvåg, K.: DYFRAM: dynamic fragmentation and replica management in distributed database systems. Distrib. Parallel Databases **28**, 157–185 (2010)
6. Khan, S., Hoque, A.: Scalability and performance analysis of CRUD matrix based fragmentation technique for distributed database. In: 15th International Conference on Computer and Information Technology (ICCIT), pp. 557–562. IEEE, Chittagong (2012)
7. Abdalla, H.I.: A synchronized design technique for efficient data distribution. Comput. Hum. Behav. **30**, 427–435 (2014)
8. Abdalla, H.I.: An efficient approach for data placement in distributed systems. In: 2011 Fifth FTRA International Conference Multimedia and Ubiquitous Engineering, pp. 297–301. IEEE, Loutraki (2011)
9. Varghese, P.P., Gulyani, T.: Region based fragment allocation in non-replicated distributed database system. Int. J. Adv. Comput. Theory Eng. **1**, 62–70 (2012)
10. Corcoran, L.: A genetic algorithm for fragment allocation a distributed database system. In: Proceedings 1994 ACM Symposium on Applied Computing, SAC 1994, pp. 247–250. ACM, USA (1994)
11. Ulus, T., Uysal, M.: Heuristic approach to dynamic data allocation in distributed database systems. Inf. Technol. J. **2**, 231–239 (2003)
12. Bayati, A., Ghodsnia, P.: A novel way of determining the optimal location of a fragment in a DDBS: BGBR. In: Systems Networks, pp. 64–69. IEEE Computer Society, Washington (2006)
13. Abdalla, H.: A new data re-allocation model for distributed database systems. Int. J. Database Theory **5**, 45–60 (2012)
14. Gope, D.: Dynamic data allocation methods in distributed database system. Am. Acad. Sch. Res. J. **4**, 1–8 (2012)
15. Mukherjee, N.: Synthesis of non-replicated dynamic fragment allocation algorithm in distributed database systems. Int. J. Inf. Technol. **1**, 36–41 (2011)
16. Abdallaha, H.I., Amer, A.A., Mathkour, H.: Performance optimality enhancement algorithm in DDBS (POEA). Comput. Hum. Behav. **30**, 419–426 (2014)

An Energy Management Approach in Hybrid Energy System Based on Agent's Coordination

Djamel Saba[1,2](✉), Fatima Zohra Laallam[2], Brahim Berbaoui[1],
and F.H. Abanda[3]

[1] Unité de Recherche en Energies Renouvelables en Milieu Saharien,
URER-MS, Centre de Développement des Energies Renouvelables,
CDER, 01000 Adrar, Algeria
saba_djamel@yahoo.fr, berbaoui.brahim@gmail.com
[2] Fac. des Nouvelles Technologies de l'Information et de la Communication,
Dépt. d'Informatique et Technologies de l'Information,
Univ Ouargla, 30000 Ouargla, Algeria
laallamfz@gmail.com
[3] Faculty of Technology, Design and Environment,
School of the Built Environment, Oxford Institute for Sustainable Development,
Oxford Brookes University, Oxford OX3 0BP, UK
fabanda@brookes.ac.uk

Abstract. Recently, the field of energy management is studied and most solutions are based on systems with centralized architectures; the latter is characterized by several advantages, but also disadvantages such as fault tolerance or the adaptability to changes in the SEH. In addition, these systems are often difficult to design because of the "top-down" approach used: the designer usually knows how each element must respond autonomously, but a centralized management system focuses only on the overall system response. A proposal for a solution to this problem is presented in this document; it relates to a distributed management solution based on the paradigm of multi-agent systems (MAS). This solution is based on a "bottom-up" approach to ensure better system reliability. After analyzing the previous work, a description of MAS and communication protocol between agents for energy management in a hybrid energy system (photovoltaic, wind) with energy storage is presented.

Keywords: Multi agents system · Hybrid energy system · Communications protocols · Energy management · Centralized and decentralized architectures · Top-down and bottom-up approaches · Centralized and distributed management

1 Introduction

In early twentieth century, the environmental impacts associated with human energy consumption began to appear: inadequacies of primary energy sources, climate change, pollution, acid rain, etc. [1]. To remedy this, we have to invent a way to develop sustainably. In addition to reducing energy consumption, many tracks related to the way we consume must be explored. A consumer may still draw energy at any time of the day. Until now, energy suppliers are equipped to handle all the requests of people

© Springer International Publishing AG 2017
A.E. Hassanien et al. (eds.), *Proceedings of the International Conference on Advanced Intelligent Systems and Informatics 2016*, Advances in Intelligent Systems and Computing 533, DOI 10.1007/978-3-319-48308-5_29

without considering the environmental impact and the automated management aspect of energy. The importance of these problems is growing with the demands of users that increase more and more. The energy management should improve energy efficiency by reducing losses from primary energy, that is to say energy "potential" contained in natural resources and energy that will be consumed. This management should be thinking in terms of production, distribution and end-use of energy.

We are interested in this research on energy management to make the system able to dynamically find a production and energy consumption policy while taking into account the criteria set by the user and the availability of sources energy. The development of the exploitation of clean energy has prompted many experts to study wind, photovoltaic, energy storage and control systems based on coordination between various units of the system. Different countries have worked on some projects that show improvements in the power of a controlled energy storage system [2–6].

Classically, the energy management of electrical systems (photovoltaic, wind …) is based on an intelligent and centralized algorithm. This method shows their performance but, as it is characterized by certain defects such as difficulty in the extensibility of the system, shown by [7–9]. Moreover, if an event is not taken into consideration by the system, the system is unable to respond properly and the control program must be completely redesigned [10–13]. In the same context, Roche et al. [14] offer work that includes an energy control strategy based on coordination between MAS agents to establish optimal planning methodology that reduces operating costs and Nitrogen oxides (NOx) emissions. With the same approaches to artificial intelligence, Lagorse et al. [15] adopt the bottom-up approach to design a management system for distributed energy systems based on cooperation between agents of the MAS approach and verified by simulations that this system is more effective in treating problems better than the energy management systems centralized. However, Jiang [16] uses a high energy density to design an effective and robust energy plan for a hybrid system based on coordination between the MAS agents. In the same context, Wu and Zhou [17], propose a solution for managing a hybrid system based on coordination between agents.

The hybrid system concerned with the study includes two energy sources (photovoltaic, wind) accompanied by energy storage. The main objective of this work is to create a balance between production and consumption of energy to improve the effectiveness of overall functioning and reduce costs of consumption. In the same context, Kovaltchouk et al. [18], present a comparison between centralized and decentralized control on the basis of their cost and the life cycle. The storage system optimization was also treated. The objective of this study was to determine the economic benefit that could be drawn from the centralized management of energy storage systems. They determined that the centralized control of the storage system is more reliable.

Based on the existing research results, we proposed a design for an energy control system based on multi-agent coordination. The proposed system stimulates the tendering system using communication protocols. The proposed system solves the optimal distribution plan for the global energy system and optimizes solar wind energy storage performance for large-scale power generation units based on the following steps: (1) Design an architecture of multi-agent system based on the physical structure of the hybrid system (solar, wind) and energy storage batteries, and (2) Propose appropriate control strategies based on the output characteristics of the various elements of the hybrid system.

2 The Energy Management: Centralized or Decentralized

We distinguish two approaches to energy management which are being examined in various research projects aimed to know the energy capacity of each.

2.1 The Centralized Energy Management

In this type of power management, we have a various energy generators in combination with storage where it can be controlled in a centralized manner. The disadvantage of this approach is the difficulty of controlling the system with increasing elements of the system. The complexity of the centralized optimization increases exponentially with the number of generators and loads. Therefore, technically, it is only sensible to implement centralized management for a small number of generators and loads.

2.2 The Decentralized Energy Management

This type of management is less complex and decisions are taken by the decentralized energy management systems that optimize power generation and consumption (load).

2.3 Centralized Production to Decentralized Production

Until the 1990s, the electrical energy is produced exclusively centrally and consumed in a decentralized way which requires the establishment of a network capable of transporting the energy produced in a few production plants to the millions of consumers spread over the entire territory.

For economic and ecological reasons for the growing interest in the environment, producers are encouraged to develop means of decentralized electricity production based on the use of renewable energy sources and cogeneration in order to increase the energy efficiency of production facilities.

3 General Organization of a Control System

A control system comprises software and hardware components. The main components are as follows and illustrated in Fig. 1.

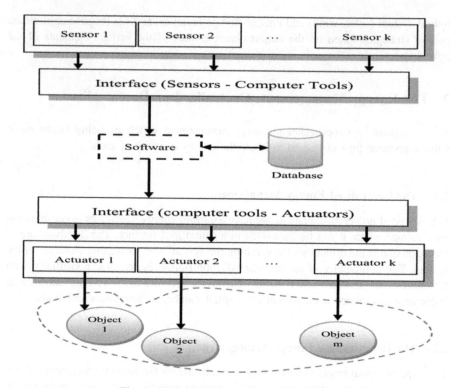

Fig. 1. The main control system components

3.1   The Sensors

These are the technological components that ensure the measurement interface; they are very different types and using the most varied physical phenomena for measuring physical parameters of the process transforming them into usable electrical quantities by digital computers.

3.2   The Actuators

They are used to convert a control signal generated by the control system in an effective action at the level of the controlled physical system.

3.3   The Software

Operation of the system is ensured by a set of software for processing the particular applications of the controlled system.

3.4   The Communications Equipment

These tools ensure the transmission and exchange of information safely and reliably to ensure the proper functioning of the system and the coordination of its various elements. Communications tend to be organized around local networks that are designed to connect in a simple way all the automated system stakeholders.

3.5   The Interfaces

They liaise between the sensors and the computer equipment on one side and between the computer equipment and actuators of another.

4   Presentation of Hybrid Energy System and Environment

The hybrid system and environment consist of four parts are: energy production, energy storage, energy consumption and the influences part (see Fig. 2).

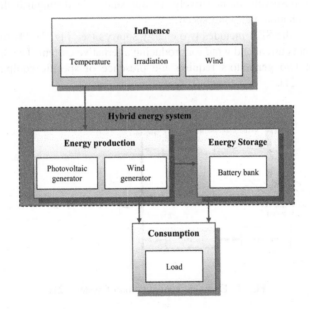

Fig. 2. The essential elements of the hybrid energy system and environment.

4.1   Energy Production

This part concerns the sources of energy; generally we select one as the main source and the other as secondary or alternative source. This classification based on a meteorological study of the installation site. In our work, we are interested in photovoltaic-wind system.

4.2 Energy Storage

The battery is used to store the energy produced by the energy sources. There is a need of storage every time of energy demand that is shifted in time a result of external energy supply. The volume of the battery bank is attached with the number of days of autonomy (the total absence of electricity generation by sources). A good study based on a good optimization of these systems.

4.3 Energy Consumption

Concerns of users need electrical energy which is measured in Watts. In the new generation of energy management designer gives importance to the economy of energy and expenses are classified according to their priority.

4.4 Influences

It concerns the elements of the external environment that influence on the production of energy, either positively or negatively, in our study we distinguish three elements temperature, radiation and wind speed.

Electrically, the SEH includes two energy sources (see Fig. 3). Photovoltaic is for producing direct current and wind for producing alternative current. The diversity in the production of two generators requires the presence of specific equipment such as converters [19–21].

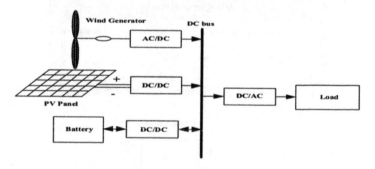

Fig. 3. Electrical configuration of system [21]

5 Agent Modeling and Characteristic Analysis

Distributed Artificial Intelligence (DAI) deals with situations where multiple systems interact to solve a common problem [22, 23]. DAI is divided into two main branches: solving distributed problems (RDP) and the simulation of complex systems (CSS). However, an agent is an encapsulated system located in a computing environment in which it is capable of performing a flexible and autonomous action consistent with the design goals [24]. The meeting of a group of agents build a multi-agent system (MAS) which includes a set of dynamic agents that interact more often, depending on modes of cooperation, competition or coexistence [25–28].

5.1   Basic Structure of Agent

From the perspective internal architecture of the agent, various types of (Multi-Agent-based Energy-Coordination Management System) MA-ECMS agents are distinguished by the nature of the internal modules that compose and by the manner in which these modules are interleaved with each other. Indeed, in our conception of SMA, we opted for the decomposition of the internal architecture of agents in several modules. Each module is dedicated to a specific function required for the fulfillment of the mission of the agent [29]. This modular decomposition of the internal architecture of agents promotes the improvement, adding and reuse of the main modules. In the case of our multi-agent platform, the internal architecture of agents includes seven modules. Figure 4, illustrates the internal structure of our agents and the imbrications of modules used. These main modules are:

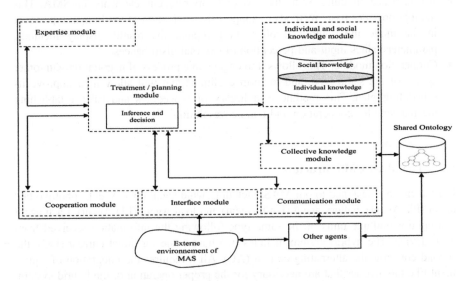

Fig. 4. Basic structure of agent

- Individual and social knowledge module: This is in two parts: individual knowledge and social knowledge. Individual knowledge reflect seen that the agent itself, including name, address, individual goals, decision protocols and different states. Social skills reflect the representation that the agent of the social environment in which it operates. This special knowledge includes agents that can contact or with whom to cooperate as well as communication protocols or interaction to use. Indeed, they allow the agent to select the most appropriate agents who will be able to help him to achieve his goal or improve its utility function. Each agent has appropriate information modeling and knowledge associated with it.

- A collective knowledge module: It allows establishing communications with a shared ontology between agents of the MA-ECMS. This ontology includes all concepts and informational details on the hybrid energy system.
- Treatment/planning module: This is based on a body of knowledge acquired by the agent and messages from other agents. At this module, the decision-making process allows the agent to choose the solution or the appropriate response. Based either on performance criteria encapsulated by a utility function or a set of goals.
- A module of expertise: This describes the basic actions that the agent can perform as well as their skills. It reflects the business capabilities and cognitive organizational entity encapsulated by the modeler agent. Indeed, this module details the behavior of the agent which will enable it to improve its utility function or achieve his goal through a series of basic actions.
- A module for the management of communications: With other agents. On the one hand, this module receives from the "processing module/planning" message transmission requests to one or more agents; on the other, he relays his messages from other agents.
- An interface module: With the external environment elements in SMA. This module provides basic methods to support the interaction of the human type/agent in the mode input/output. It offers in particular the results display methods, parameter values input and perception of external disturbances.
- Cooperation module: This allows it to engage in a process of negotiation with other agents of the same type, and to form coalitions that will allow it to improve its scheduling plan. This module is a library of procedures and rules establishing mechanisms for cooperation and conflict resolution.

5.2   The MA-ECMS Structure

Four agents are designed to our MAS, including wind agent (WT Agent), photovoltaic agent (PV Agent), battery agent (BAT Agent) and load Agent (LD Agent). Each agent can control multiple units of the same type following the information received from each device. The hybrid system includes two buses, one for direct current (DC), the second concerns the alternating current (AC). All it requires the integration of equipment like the inverter that are necessary for the proper operation of the hybrid system. The structure of the MA-ECMS (Multi-Agent-based energy-coordination management system) is shown in Fig. 5. The system adopts the multi-agent client-server mechanism of cooperation, following the changes of the load and weather data. After receiving the request of the load, the agent of the main energy source begins the process of supply of energy (To avoid conflicts caused by multiple initiators on coordination of tasks energy at the same time, a priority between energy sources (Agent) are stored in the knowledge base associated with the MA-ECMS). Other officers determine how to participate in the coordination of energy according to their own characters and control objectives.

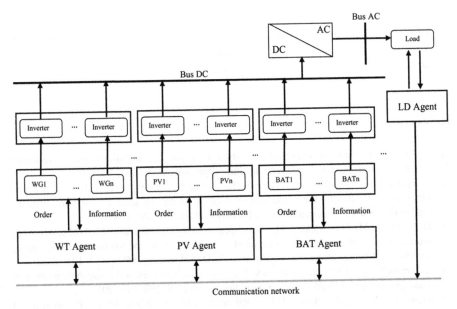

Fig. 5. The general structure of MA-ECMS

- LD Agent: The demand for energy is represented by "LD Agent". This agent will check if the load demand for each time step is met accordingly by production and energy storage systems.
- PV Agent and WT Agent: Represent respectively the control units of the PV array and wind generator. Both generators are characterized by a power curve that depends on the meteorological data of the site concerned by the installation of the hybrid system. The principle of maximizing the use of energy and the characteristics of generators are designed to participate in the bidding actively until all available equipment operates at maximum output power. At the same time, they can initiate task of coordination of energy when changing the output power.
- **BAT Agent:** their role is to monitor the state of charge (SOC) battery and manage the operation of the charge/discharge. On the basis of information coordination with other agents determines the energy value to load or unload.

6   Conclusion and Perspectives

The growth of the field of systems management energy, the need to provide designers with a quality work and traditional modeling techniques are static and can consider the hybrid energy system as an evolving system. This has led us to seek to model the energy management in these systems by multi-agent technology that led to very interesting results in many other areas. We have tried, through this initiation to present the future SMA for power management in a hybrid system and give a description of each agent (missions, internal architecture). The external environment was also presented, such as climate data influencing the hybrid system and the necessary sensors to data acquisition. The next step is the implementation of the proposed solution to visualize simulation results for energy management. In this part, we choose a model and a platform most suitable for our application.

References

1. Duy-Long, H.: Un système avancé de gestion d'énergie dans le bâtiment pour coordonner production et consommation. Thèse, Institut polytechnique de Grenoble (2007)
2. Campoccia, A., Dusonchet, L., Telaretti, E., Zizzo, G.: Comparative analysis of different supporting measures for the production of electrical energy by solar PV and wind systems: four representative European cases. Sol. Energy 83(3), 287–297 (2009)
3. Mohammadi, M., Hosseinian, S.H., Gharehpetian, G.B.: Optimization of hybrid solar energy sources/wind turbine systems integrated to utility grids as microgrid (MG) under pool/bilateral/hybrid electricity market using PSO. Sol. Energy 86(1), 112–125 (2012)
4. Askarzadeh, A.: Developing a discrete harmony search algorithm for size optimization of wind–photovoltaic hybrid energy system. Sol. Energy 98, 190–195 (2013)
5. Merei, G., Berger, C., Sauer, D.U.: Optimization of an off-grid hybrid PV-wind-diesel system with different battery technologies using genetic algorithm. Sol. Energy 97, 460–473 (2013)
6. Bayod-Rujula, A.A., Haro-Larrode, M.E., Martınez-Gracia, A.: Sizing criteria of hybrid photovoltaic–wind systems with battery storage and self-consumption considering interaction with the grid. Sol. Energy 98, 582–591 (2013)
7. Kremers, E., Gonzalez de Durana, J., Barambones, O.: Multi-agent modeling for the simulation of a simple smart microgrid. Energy Convers. Manage. 75, 643–650 (2013)
8. Da Rosa, M.A., Leite da Silva, A.M., Miranda, V.: Multi-agent systems applied to reliability assessment of power systems. Int. J. Electr. Power Energy Syst. 42(1), 367–374 (2012)
9. Pipattanasomporn, M., Feroze, H., Rahman, S.: Securing critical loads in a PV-based microgrid with a multi-agent system. Renew. Energy 39(1), 166–174 (2012)
10. El-Shater, T.F., Eskander, M.N., El-Hagry, M.T.: Energy flow and management of a hybrid wind/pv/fuel cell generation system. Int. J. Sustain. Energy 25(2), 91–106 (2006)
11. El-Shater, T.F., Eskander, M.N., El-Hagry, M.T.: Energy flow and management of a hybrid wind/pv/fuel cell generation system. Energy Convers. Manage. 47(9–10), 1264–1280 (2006)
12. Becherif, M., Paire, D., Miraoui, A.: Energy management of dolar panel and battery system with passive control. In: International Conference on ICCEP 2007 (2007)
13. Paire, D., Becherif, M., Miraoui, A.: Passivity-based control of hybrid sources applied to a traction system. In: Workshop on Hybrid and Solar Vehicles, Italy (2006)
14. Roche, R., Idoumghar, L., Suryanarayanan, S., Daggag, M., Solacolu, C.A., Miraoui, A.: A flexible and efficient multi-agent gas turbine power plant energy management system with economic and environmental constraints. Appl. Energy 101, 644–654 (2012)
15. Lagorse, J., Paire, D., Miraoui, A.: A multi-agent system for energy management of distributed power sources. Renew. Energy 35(1), 174–182 (2010)
16. Jiang, Z.: Agent-based power sharing scheme for active hybrid power sources. J. Power Sources 177(1), 231–238 (2008)
17. Wu, K., Zhou, H.: A multi-agent-based energy-coordination control system for grid-connected large-scale wind–photovoltaic energy storage power-generation units. Sol. Energy 107, 245–259 (2014)

18. Kovaltchouk, T., Blavette, A., Ben Ahmed, H., Multon, B., Aubry, J.: Energy converter farm comparison between centralized and decentralized storage energy management for direct wave. In: Tenth International Conference on Ecological Vehicles and Renewable Energies (EVER) (2015)

19. Saba, D., Laallam, F.Z., Hadidi, A.E., Berbaoui, B.: Contribution to the management of energy in the systems multi renewable sources with energy by the application of the multi agents systems "MAS". Energy Procedia **74**, 616–623 (2015)

20. Saba, D., Laallam, F.Z., Hadidi, A.E., Berbaoui, B.: Optimization of a multi-source system with renewable energy based on ontology. Energy Procedia **74**, 608–615 (2015)

21. Saba, D., Laallam, F.Z., Belmili, H., Hadidi, A.: Contribution of renewable energy hybrid system control based of multi agent system coordination. In: The Symposium on Complex Systems and Intelligent Computing (CompSIC 2015) (2015)

22. Omatu, S., Neves, J., Corchado Rodríguez, J.M., González, S.R., Paz Santana, J.F., Gonzalez, S.R. (eds.): Distributed Computing & Artificial Intelligence. AISC, vol. 217. Springer, Heidelberg (2014). doi:10.1007/978-3-319-00551-5

23. Omatu, S., Bersini, H., Corchado Rodríguez, J.M., González, S.R., Pawlewski, P., Bucciarelli, E. (eds.): Distributed Computing and Artificial Intelligence, 11th International Conference. AISC, vol. 290. Springer, Heidelberg (2014). doi:10.1007/978-3-319-07593-8

24. Håkansson, A., Hartung, R., Nguyen, N.T. (eds.): Agent and Multi-agent Technology for Internet and Enterprise Systems. SCI, vol. 289. Springer, Heidelberg (2010). ISBN 978-3-642-13525-5

25. Ferber, J.: Les Systèmes Multi Agents: vers une intelligence collective. InterEditions IIA, Paris (1995)

26. Kaczorek, T.: A new formulation and solution of the minimum energy control problem of positive 2D continuous-discrete linear systems. In: Szewczyk, R., Zieliński, C., Kaliczyńska, M. (eds.) Recent Advances in Automation, Robotics and Measuring Techniques. AISC, vol. 267, pp. 103–114. Springer, Heidelberg (2014)

27. Kociszewski, R.: Minimum energy control of fractional discrete-time linear systems with delays in state and control. In: Szewczyk, R., Zieliński, C., Kaliczyńska, M. (eds.) Recent Advances in Automation, Robotics and Measuring Techniques. AISC, vol. 267, pp. 127–136. Springer, Heidelberg (2014)

28. Charif, Y., Sabouret N.: Interaction protocol for service composition in the room. In: JFSMA2006, pp. 253–266 (2006)

29. Bryson, J., Stein, L.A.: Modularity and specialized learning in the organization of behaviour. In: French, R.M., Sougné, J.P. (eds.) Connectionist Models of Learning, Development and Evolution, pp. 53–62. Springer, Heidelberg (2000)

Improving the Predictability of GRNN Using Fruit Fly Optimization and PCA: The Nile Flood Forecasting

Mohammed E. El-Telbany[(⊠)]

Computers and Systems Department, Electronics Research Institute, Giza, Egypt
telbany@eri.sci.eg

Abstract. Generalized regression neural network (GRNN) is commonly used for function approximation. The predictability of GRNN for Nile flood forecasting is investigated through optimization of the smoothing parameter using fruit-fly (FOA) algorithm. Due to the excess number of inputs that are highly correlated, principal component analysis (PCA) is used to reduce the dimension of a set of linearly correlated variables to improve forecasting task. Our empirical experiment shows that the performance of GRNN is better than other neural network included in this paper. The results of the prediction made by the proposed model showed a close fit with the actual data.

Keywords: Generalized regression neural network · PCA · Fruit-fly optimization · Prediction · Neural networks · Nile river flood

1 Introduction

The goal of time-series prediction is to identifying the near future movements of given variable(s) by looking at their behaviour in the past. The Nile river flood time-series have a large number of specific properties that together makes the prediction task unusual (i.e., the process is time varying and volatility in the time-series change as the flood moves). There a variety of techniques that are applied for time-series forecasting including autoregressive integrated moving average(ARIMA) [1] and artificial neural networks(ANNs) [2]. There are various ANN forecasting models that are applied for Nile time-series forecasting. The most used among them are the multi-layer perceptrons (MLPs) [3]. The ANNs showing encouraging results due its flexible mathematical structure which is capable of identifying complex non-linear relationships between inputs and outputs [2]. Actually, the success of ANNs modelling mostly depends on selecting suitable structure and the learning algorithm used to fit the model and prediction accuracy. However, ANNs suffers from difficulty in trapping into local minima, over-fitting and selecting relevant input variables [4]. There are some

© Springer International Publishing AG 2017
A.E. Hassanien et al. (eds.), *Proceedings of the International Conference on Advanced Intelligent Systems and Informatics 2016*, Advances in Intelligent Systems and Computing 533, DOI 10.1007/978-3-319-48308-5_30

other types of neural networks models, such as generalized regression neural network (GRNN) [5] which are proposed for regression tasks. Since, the characteristics of different ANNs are related to their structure and learning methods. The GRNN learning algorithm is easier than other existing algorithms (one-pass learning algorithm) and required one parameter (smoothing parameter σ) to be optimized. Swarm intelligence algorithms commonly used to search and optimization domains. The most widely used algorithms in applications are, Particle swarm optimization (PSO) [6,7], ant colony optimization (ACO) [8], artificial bee colony algorithm (ABC) [9], Cuckoo Search(CS) [10], Harmony Search(HS) [11], Bacterial Colony Chemotaxis (BCC) [12], Fruit Fly Optimization algorithm (FOA) [13–15], and Bat-Inspired Algorithm(BA) [16] are some of them. FOA has been successfully applied to solve the function optimization, GRNN parameters optimization in order to forecast the annual power load [17], least squares support vector machine (LSSVM) parameters optimization for short-term Load forecasting [18]. Utilizing the GRNN for Nile flood forecasting as one of the attractive neural network used for function approximation is implemented, where the studies show that GRNN has better function approximation performance than feed-forward networks and other statistical neural networks on some datasets [19,20]. The principal advantages of the GRNN as one pass learning are its quick learning and fast convergence to an optimal regression surface as the number of samples becomes large [21]. However, its efficiency decreases with large datasets due to complexity of resulted huge learned neural network and direct influence of the smoothing parameter. Determining optimal smoothing parameter value [19,22] and decreasing pattern layer size are the major problems of GRNN. Clustering techniques reduce the number of neurons at the pattern layer by grouping data into clusters [23]. Feature extraction methods are also used for improving the performance of GRNN [24]. In this paper, we present a forecasting method by using of GRNN and utilizing both of FOA and PCA algorithms to Nile flood time-series forecasting. The goal will be twofold:

- to take into account in the GRNN dependency on smoothing parameter, to find the optimal or effective smoothing parameter;.
- to reduce, whenever possible, the dimensionality of the data through PCA technique, to improve the performance of GRNN.

The rest of the paper is organized as follows. In Sects. 2 and 3 briefly explains the theory of the GRNN and FOA algorithms. Section 4 briefly introduces the Nile river flood forecasting problem and describes the data set and processing step. Section 5 describes the proposed methodology. Section 6 describes an evaluation of Nile flood forecasting using GRNN modelling and prediction results. Finally Sect. 7 presents the findings and conclusions.

2   Generalized Regression Neural Network

The GRNN was proposed by Specht [5] as radial basis function (RBF) network with a normalized linear output. The GRNN has the capability of approximating

312   M.E. El-Telbany

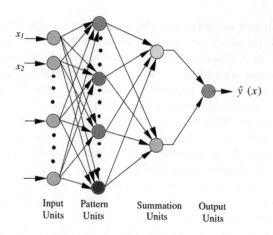

Fig. 1. Block diagram of a general regression neural network

any function between input and output vectors from past training data [25]. GRNNs feature fast training times, can model non-linear functions, and perform well against infrequent outliers and erroneous observations. The architecture of a basic GRNN, shown in Fig. 1, has four layers (1) input layer, (2) pattern layer, (3) summation layer, and (4) output layer. The input layer provides all the variables to the neurons in the pattern layer; The neurons in the pattern layer perform a non-linear transformation of the input patterns, the Gaussian kernel function is used for GRNN training and the smoothing factor σ in the kernel function is the only free parameter as shown in Eq. 1:

$$f(\mathbf{X}, y) = \hat{y}_i(\mathbf{X}) = \frac{\sum_{i=1}^{n} y^i \exp \frac{-D_i^2}{2\sigma^2}}{\sum_{i=1}^{n} \exp \frac{-D_i^2}{2\sigma^2}} \tag{1}$$

where y is the output predicted by GRNN, $\mathbf{X}$ the input vector $(x_1, x_2, \ldots, x_m)$ which consists of m predictor variables and $f(\mathbf{X}, y)$ the joint probability density function of $\mathbf{X}$ and y learned by GRNN. The $D_i^2 = (\mathbf{X} - \mathbf{X}_i)^T(\mathbf{X} - \mathbf{X}_i)$ presents the squared Euclidean distance between the input vector $\mathbf{X}$ and the ith training input vector $\mathbf{X}_i$, y_i is the output vector corresponding to the vector $\mathbf{X}_i$, $\hat{y}_i(\mathbf{X})$ is the estimate corresponding to the vector $\mathbf{X}$, n is the number of samples, and $0 < \sigma \le 1$ is a smoothing parameter that controls the size of the receptive region. The output from the pattern layer then becomes input for all neurons in the summation layer. The summation layer calculate the sum of the weighted outputs of the hidden layer. Finally, the output layer performs a normalization step, a dot product between a weight vector and a vector composed of the signals from the pattern units, to yield the predicted value of the output variable.

3   Fruit Fly Optimization Algorithm

The Fruit Fly Optimization Algorithm (FOA) is a new swarm-based method based on the food finding behaviour of the fruit fly [13,14]. Then, after it gets close to the food location, it can also use its sensitive vision to find food and the company's flocking location, and fly towards that direction too. The behaviours of the fruit flies could be demonstrated in Fig. 2. Each fruit flys position correspond a candidate value for smoothing parameter of the GRNN. When fruit fly swarm flies towards one location, it is treated as the evolution of each iterative swarm. After hundreds of iteration, the best smoothing parameter can minimize an accuracy error function.

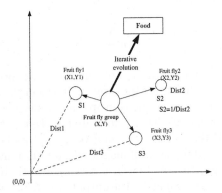

Fig. 2. Food searching iterative process of fruit fly swarm [26]

4   Nile Flood Time Series Processing

The Nile flood forecasting is an important process for optimum utilization of water resources and retaining sufficient water in reservoir for irrigation and power generation at High Dam of Aswan, Egypt. The Nile flood time-series $x(t_i)$ are discrete projections of a continuous signal $x(t)$ in time domain. The flood signal is usually measured as a chronological sequence at discrete time steps t_i with constant increment Δt. The used data is the readings of average daily flow volume $[m^3/day]$ for each ten-day period at the Dongola station, located in Northern Sudan (south of the High Dam) from 1975 to 1993 years. The data point presented the ten-day average flood value. The Nile river time-series is far from steady and can be actually involved stochastic and deterministic components and exhibits a seasonal behaviour. The flow is low during the winter months, and peaks during the months of August and September. Transforming this non-stationary time-series into a stationary one by differencing the mean transformation then normalize the values of stationary time-series in the range of 0 to 1. Finally, the time-series is processed using PCA for reduction of the dimensionality or preventing over-fitting to increase generalization ability. Using

PCA which is a dimensionality reduction technique based on feature transformation [27]. The projection of a d-dimensional multivariate time-series, represented as a set of vectors $x_i \in \mathbb{R}^d$ transformed onto a subspace of dimension $m < d$ that minimizes the reconstruction error

$$E_{PCA} = \sum_{i=1}^{m} \| x_i - \sum_{j=1}^{m} (x_i.e_j)e_j \|^2, \tag{2}$$

where the vectors $\{e_k\}, 1 \leq k \leq m$ define a partial orthonormal basis of the input space and called principal components(PCs). From Eq. 2, it show that the subspace with minimum reconstruction error is also the subspace with maximum variance. The basis vectors of this subspace are given by the top m eigenvectors of the $d \times d$ covariance matrix,

$$\mathbf{C} = \frac{1}{n} \sum_{i=1}^{n} x_i x_i^T, \tag{3}$$

Assuming that the input patterns x_i are centred on the origin. The outputs of PCA are simply the coordinates of the input patterns in the new subspace, using the directions of eigenvectors as the principal axes. The covariance matrix $\mathbf{C}$ is a symmetric matrix, meaning that it can write the eigenvector decomposition of $\mathbf{C}$ as $\mathbf{C} = \mathbf{X}^T\mathbf{X} = \mathbf{E}\Lambda\mathbf{E}^T$, with $\Lambda = diag(\lambda_1, \ldots, \lambda_p)$ a diagonal matrix of the eigenvalues λ_j in descending order of magnitude, and $\mathbf{E}$ a matrix whose columns are the corresponding orthogonal eigenvectors e_j. Identifying e_j as the jth top eigenvector of the covariance matrix, the output $z_i \in \mathbb{R}^m$ for the input data time-series pattern $x_i \in \mathbb{R}^d$ has elements $z_{ij} = x_i.e_j$. The eigenvalues of the covariance matrix in Eq. 3 measure the projected variance of the high dimensional data set along the principal axes. Thus, PCA transforms the set of delay variables to a new set of uncorrelated ones with reduced dimension.

5   Proposed Methodology

In recent years many new forecasting methods based on swarm intelligence algorithms were used for Nile flood forecasting, where the GA and PSO trained the MLPs neural networks [28,29]. In this paper, Nile flood forecasting problem is solved by building a regression model using GRNN with a number of values from the time-series, and with output the predicted value. The GRNN training is essentially an optimization process of the smoothing parameter σ using FOA algorithm. Due to non-random characteristic of Nile flood time-series, there is dependencies exist between the dimension of the inputs variables. In this situation, where dimensions reduction should be possible, leading to improved prediction performances. A linear PCA is applied to first reduce the dimensionality of the variables. The trained GRNN is used to make successive one step ahead forecasts. The block diagram of the methodology is shown in Fig. 3. The data set is divided two sections, training and test sets, comprising of 78 % and 22 % of the total data set respectively. So, the first 500 data samples are taken for training,

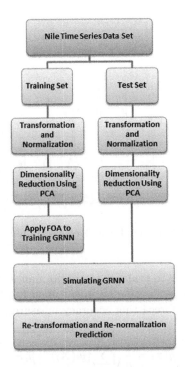

Fig. 3. Methodology flowchart for Nile River Flood forecasting

the next 140 samples for validation and test. Without using PCA as data reduction technique, each training sample comprises the flood of the previous eight samples and the prediction of the future flood for ninth sample. This means that by using the flood values of $x(t-7), x(t-6), \ldots, x(t-1), x(t)$ as inputs of the GRNN and MLPs neural network, can predict the flood value $\hat{x}(t+1)$ as an output.

6   Experimental Results

This section demonstrated the results of proposed approach with a comparison with MLPs as a well known forecasting approach. We implement the proposed algorithms using the toolboxes MATLAB toolbox. MLPs is used and different network architectures are used in order to determine the optimal solution in terms of accuracy or minimum error function. The best architecture that observed have one hidden layer with ten nodes with learning rate is 0.02. The learning algorithm used was the *conjugated gradients* one and an activation function of the hidden and output neurons, the sigmoid and linear ones, respectively. For GRNN tuning by FOA, the number of flyers are 25, the spread parameter of GRNN is searched in the range of $[0.01, 1]$, and the number of iterations is 50.

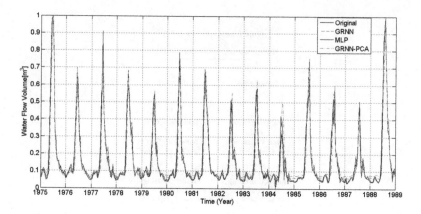

Fig. 4. The predicted Nile flood for training data set

There are various ways to evaluate the performance of predictor. The easiest way is by measuring the estimation accuracy, as in Eq. 4, the root mean square error (RMSE) is used as forecasting accuracy measures.

$$g(\sigma) = \frac{1}{n} \sqrt{\sum_{i=1}^{n} (\hat{y}_i(\mathbf{X}_i) - y_i)^2} \tag{4}$$

The GRNN performance is controlled only by the smoothing parameter during the training phase which significantly affects the accuracy of GRNN predictions. Therefore, GRNN training was mainly optimization of the smoothing parameter. The predicted results from different training algorithms are shown in Figs. 4 and 5. Due to the dimension of the inputs are highly correlated, the dimensions of the input variables are reducing using PCA by eliminating the principal components corresponding to smaller *eigen* values. The best number of principle components is three which gave the minimum error as shown in Fig. 6. The results of runs on the Nile flood data set summarized in Table 1. From the results listed in Table 1; we can conclude that the FOA algorithm was able to find the best smoothing parameter for the GRNN which approximate the Nile time-series data. Moreover, the results demonstrate the predictive strength of GRNN and its potential for solving Nile flood forecasting problem in comparison with MLPs. The results of the prediction made by the proposed model indicated a a close fit with the actual data. Through the resultsas shown in Fig. 7, the stability of learning process using FOA is obviously superior to the GRNN model's stability.

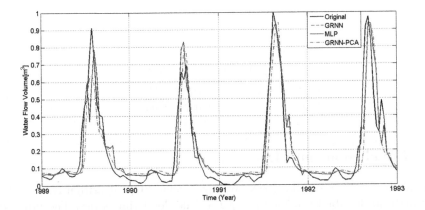

Fig. 5. The predicted Nile flood for testing data set

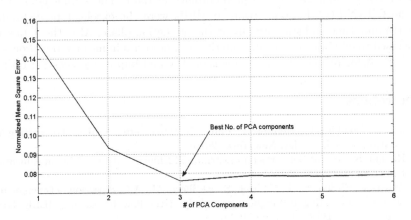

Fig. 6. The effect of number components on GRNN accuracy

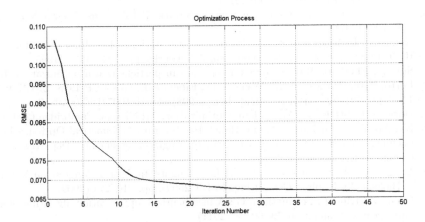

Fig. 7. The convergence curve of the optimization process

Table 1. The comparison between predicted and actual measured for proposed methodology

	MLPs RMSE	GRNN RMSE	GRNN-PCA RMSE
Training set	0.078	0.0072	0.066
Test set	0.086	0.0078	0.063

7 Conclusions and Discussions

In this paper, a proposed GRNN optimized using FOA algorithm is applied to evaluate its feasibility in forecasting the Nile flood time-series. The proposed forecasting model owes its good performance not only to the valuable properties of GRNN but also to the predictability of data. The significant features of the GRNN-PCA in comparison to MLPs were the excellent predictability, fast training time and stability during the prediction phase. Also, the data set is processed to eliminating non-stationarity of time-series in mean and is processed using PCA for reduction of the dimensionality and improve forecasting task.

References

1. Box, G., Jenkins, G., Reinsel, G., Ljung, G.: Time Series Analysis Forecasting and Control, 5th edn. Wiley, New York (2016)
2. Haykin, S.: Neural Networks and Learning Machines, 3rd edn. Pearson Education, Upper Saddle River (2009)
3. Atiya, A., El-Shoura, S., Shaheen, S., El-Sherif, M.: A Comparison between neural-network forecasting techniques case study: river flow forecasting. IEEE Trans. Neural Netw. **10**(2), 402–409 (1999)
4. Yu, X.: Can back-propagation error surface not have local minima? IEEE Trans. Neural Netw. **45**, 1019–1021 (1992)
5. Specht, D.: A general regression neural network. IEEE Trans. Neural Netw. **2**, 568–576 (1991)
6. Kennedy, J., Eberhart, R., R.: Particle swarm optimization. In: Proceedings of IEEE International Conference on Neural Networks, pp. 1942–1948 (1995)
7. Shi, Y., Eberhart, R.C.: Parameter selection in particle swarm optimization. In: Porto, V.W., Saravanan, N., Waagen, D., Eiben, A.E. (eds.) EP 1998. LNCS, vol. 1447, pp. 591–600. Springer, Heidelberg (1998). doi:10.1007/BFb0040810
8. Dorigo, M., Sttzle, T.: Ant Colony Optimization. MIT Press, Cambridge (2004)
9. Karaboga, D.: An Idea Based on Honey Bee Swarm for Numerical Optimization. Technical report-TR06, Erciyes University, Engineering Faculty, Department of Computer Engineering (2005)
10. Yang, X., Deb, S.: Cuckoo search via Levy Flights. In: World Congress on Nature & Biologically Inspired Computing (NaBIC2009), pp. 210–214 (2009)
11. Geem, Z., Kim, J., Loganathan, G.: A new heuristic optimization algorithm: harmony search. Simulation **76**, 60–68 (2011)
12. Li, W., Wang, H., Zou, Z., Qian, J.: Function optimization method based on bacterial colony chemotaxis. J. Circ. Syst. **10**, 58–63 (2005)

13. Pan, W.: A new fruit fly optimization algorithm: taking the financial distress model as an example. Knowl. Based Syst. **26**, 69–74 (2011)
14. Pan, W.: A new evolutionary computation approach: fruit fly optimization algorithm. In: Conference of Digital Technology and Innovation Management (2011)
15. Xing, B., Gao, W.: Innovative Computational Intelligence: A Rough Guide to 134 Clever Algorithms. Intelligent Systems Reference Library. Springer, Heidelberg (2014)
16. Yang, X.: A nature-Inspired metaheuristic bat-inspired algorithms. Stud. Comput. Intell. **284**, 65–74 (2010)
17. Li, H., Guo, S., Li, C., Sun, J.: A hybrid annual power load forecasting model based on generalized regression neural network with fruit fly optimization algorithm. Knowl. Based Syst. **37**, 378–387 (2013)
18. Sun, W., Ye, M.: Short-term load forecasting based on wavelet transform and least squares support vector machine optimized by fruit fly optimization algorithm. J. Electr. Comput. Eng., pp. 2–9 (2015)
19. Ren, S., Yang, D., Ji, F., Tian, X.: Application of generalized regression neural network in prediction of cement properties. In: International Conference on Computer Design and Applications (2010)
20. Wang, Z., Sheng, H.: Rainfall prediction using generalized regression neural network: case study Zhengzhou. In: International Conference on Computational and Information Sciences (2010)
21. Al-Daoud, E.: A comparison between three neural network models for classification problems. J. Artif. Intell. **2**, 56–64 (2009)
22. Ozyildirim, B., Avci, M.: Generalized classifier neural network. Neural Netw. **39**, 18–26 (2013)
23. Zhao, S., Zhang, J., Li, W., Song, W.: A generalized regression neural network based on fuzzy means clustering and its application in system identification. In: International Symposium on Information Technology Convergence (2007)
24. Erkmen, B., Yildirim, T.: Improving classification performance of sonar targets by applying general regression neural network with PCA. Expert Syst. Appl. **35**, 472–475 (2008)
25. Leung, M., Chen, A., Daouk, H.: Forecasting exchange rates using general regression neural networks. Comput. Oper. Res. **27**, 1093–1110 (2000)
26. Lin, S.: Analysis of service satisfaction in web auction logistics service using a combination of fruit fly optimization algorithm and general regression neural network. Neural Comput. Appl. **22**, 783–791 (2013)
27. Burges, C.: Dimension reduction: a guided tour. Found. Trend Mach. Learn. **2**(4), 275–365 (2009)
28. El-Telbany, M., Abdelwhab, A., Shaheen, S.: Forecasting of the nile river inflow using genetic algorithms. In: Proceedings of International Conference on ANN and GA (1997)
29. El-Telbany, M., Konsowa, H., El-Addawi, M., M.: Studying the predictability of neural network learned by particle swarm optimization. Egypt. J. Eng. Appl. Sci., pp. 377–390 (2005)

Abrupt Cut Detection in News Videos Using Dominant Colors Representation

Ibrahim A. Zedan$^{(\boxtimes)}$, Khaled M. Elsayed, and Eid Emary

Faculty of Computers and Information, Cairo University, Cairo, Egypt
{i.zedan,k.mostafa,e.emary}@fci-cu.edu.eg

Abstract. In this paper, we propose a new representation of images. We called that representation as "Dominant Colors". We defined the dissimilarity of two images as a vector contains the difference in order of each dominant color between the two images representations. Our new image representation and dissimilarity measure are utilized to detect the abrupt cuts in news videos. A neural network trained with our new dissimilarity measure to classify between two classes of news videos frames: cut frames, and non-cut frames. Our proposed system tested in real news videos from different TV channels. Experimental results show the effectiveness of our new image representation and dissimilarity measure to describe the images, and detect the abrupt cuts in news videos.

Keywords: News videos · Shot boundary detection · Abrupt cut · Dominant colors

1 Introduction

The amazing increase in video data on the internet leads to an urgent need of research work to develop efficient procedures for summarization, indexing, and retrieval of this video data. Among all video data the news videos have gained a special importance as many individuals are concerned about news video. News videos are closely related to us as it gives us the main news in the world recently. It is highly needed to extract key frames of news video in order to know the main content of news and avoid spending a lot of time to watch every detail of the video. News video should be segmented into shots in order to extract the key frames [1]. From this point of view, we can see that shot boundary detection and Key frame Extraction are essential steps for the organization of huge video data.

News video has relatively fixed structure and the shots of news video are categorized to several classes [1, 3]. The Introduction and the ending sequences are usually contain graphics. Separator is a sequence of frames inserted to separate between different news stories and its pattern is similar to the beginning and the ending sequences. The appearance of the anchor person shots is similar among a large portion of the TV channels. Interview shots are a special type from studio shots where an interview between the anchor person and the interviewed person is taking place. Weather forecast shots are studio shots show a weather map that generally contains animated symbols that represent the weather conditions. Sport shots contain a lot of motion like football matches. Report shots are recorded outside the studio.

© Springer International Publishing AG 2017
A.E. Hassanien et al. (eds.), *Proceedings of the International Conference on Advanced Intelligent Systems and Informatics 2016*, Advances in Intelligent Systems and Computing 533, DOI 10.1007/978-3-319-48308-5_31

In this paper we introduce a new image representation and dissimilarity measure. To prove the effectiveness of our new image descriptor, we applied it on the problem of shot boundary detection in news videos as SBD is normally the first and essential step for content based video summarization and retrieval. We focused on the detection of abrupt cuts as it is the most common shot boundaries.

The rest of the paper is organized as follows; an overview of SBD foundation, challenges, and methods is given in Sect. 2. A detailed description of the proposed system is given in Sect. 3. Experimental results are shown in Sect. 4. Section 5 presents our conclusion and future work.

2 Shot Boundary Detection

Shot boundary detection is the initial step of video content analysis, indexing, and retrieval. Shot Boundary Detection (SBD) or Video Segmentation is the method of automatically identifying the shots transitions inside a video [4]. Video shot is a series of consecutive interrelated frames represents continuous activities in time and space and recorded contiguously by a single camera. Two shots can be combined by a gradual transition or an abrupt cut. The abrupt cut is the simplest transition between the shots and is defined as an instantaneous transition from one shot to the next shot that occurs in a single frame and is significantly more common than gradual transitions. The gradual transition is a more complex transition that artificially combine two shots by using special effects like fade in/out, dissolve, and wipe [5, 6].

2.1 SBD Foundation and Challenges

The foundation of SBD is detecting the frames discontinuities. The similarity between adjacent frames is measured and compared with a threshold to detect the significant changes that corresponds to shot boundaries. The SBD performance relies on the suitable choice of the similarity measure that used to detect shot boundaries. In SBD algorithms, threshold selection is an important Issue. The utilization of a dynamic adaptive threshold is more favored than global thresholding as the threshold used to detect a shot boundary changes from one shot to another and should be based on the distribution of the video inter-frames differences [4, 6].

Shot boundary detection faces many challenges such as detection of gradual transitions, flash lights, and object/camera motion. Many examples of camera motion exist such as panning, tilting, and zooming [6]. Detection of gradual transitions is the most challenging problem in SBD as the editing effects used to create a gradual transition are very comparable to the patterns of object/camera motion. The vast majority of the video content representations utilize color features. Color is the essential component of video content. Abrupt illumination changes such as flashlights might be recognized as a shot boundary by a large portion of the SBD tools [2].

2.2 Shot Boundary Detection Methods

Diverse methodologies have been proposed to segment video into shots such as the compressed domain methods, and the un-compressed domain methods [5]. In the un-compressed domain, shot boundary detection methods are mainly based on computing an image-based features. Image features can be based on pixel differences, statistical-based methods, edge differences, or their combination [4]. Methods based on pixel differences are easy to implement, but it is very sensitive to motion and noise [1].

Statistical-based methods include histogram differences which can be calculated on the entire frame level or at a block level [7]. In [2] each video frame is partitioned into blocks. For each successive frame pair, the histogram matching difference between the corresponding blocks is computed. Video frame is assumed as a shot boundary if it has histogram matching difference greater than a threshold. In [4] R, G and B pixels are quantized and only one color index is computed. JND histogram is computed at the entire frame level. In JND histograms, each bin contains visually similar colors and visually different from other bins. For every two consecutive frames, histogram intersection is computed. For each video frame, a sliding window is centered. A frame is considered as a cut if its similarity degree is the window minimum and much less than the window mean and the second minimum. A frame is considered a part of gradual transition if it is less than a threshold related to the window similarity mean. In [8] histogram intersection in HSI color space is used as similarity measure between frames. A placement algorithm up to eight frames is applied to convert gradual transition to abrupt cut. Shot boundaries are extracted by thresholding. Generally spatial information is missed in the statistical based methods and away to enhance results is to work on a block level rather than the entire image level. In [9] histogram of gradient is combined with HSV color histogram to enhance results.

Edge based SBD methods identified two sorts of edge pixels which are the entering and exiting edges. Entering edges are new edges show up far from the locations of the previous shot edges. Exiting edges are the old shot edges that vanish. The main disadvantage of the edge based methods is its execution time [2].

Several SBD algorithms that depend on the scale and rotation invariant local descriptors were proposed as in [1, 7]. In [7] for every pair of frames, a similarity score is calculated as a function of the normalized histogram correlation, and the ratio of the matched SURF descriptors. Cut is detected if the similarity score is less than a threshold. A frame that corresponds to a local minimum of the moving similarity average is considered a part of gradual transition. In [1] only abrupt cuts are detected in news video. Frames are converted to HSV color space and divided into uneven blocks with different weighting coefficients. A sliding window is utilized. Window is divided into right and left sub-windows. The difference between sub-windows is calculated as weighted frame difference. If the difference is large, an adaptive binary search process is continued on the sub-window whose have larger discontinuity until obtaining a window size of 2 frames represents the candidate cut. Final cuts are assumed if the SURF points matching ratio is below a threshold. In [10] a combination of local and global descriptors is integrated to represent the video frames. Dissimilarity scores between the frames are calculated and shot boundaries are extracted utilizing adaptive thresholding. In [11] a temporal window is centered in each frame index and the cumulative mutual information is calculated. The cumulative mutual information combines the mutual information between multiple pairs inside the window. Shot boundaries are detected by identifying the local minimums of the cumulative mutual information curve.

Motion based algorithms are not favored in the uncompressed domain as the estimation of motion vectors expends huge computational power [2]. In [5] a method for abrupt cut detection based on motion estimation is introduced. Video frame is divided into blocks. Block-matching algorithm is applied to calculate the blocks motion vectors. Candidate abrupt cut is identified when both the average motion magnitude and the quantitative angle histogram entropy are greater than specified thresholds as the camera motion cannot destroy the motion regularity.

Audio features have been utilized to enhance shot boundary detection as in [3, 6]. In [3] a scene boundary detection and Scene classification for news video is combined in one process using HMM. For each video frame, a feature vector that combines several features is derived. Average color components and the luminance histogram difference are used as image features. For adjacent video frames, some features derived from the difference image are computed as motion indicators. The logarithmic energy and the cepstral vector of the audio samples of each video frame are used as audio features. In [6] a technique to segment video into shots is presented. Multiple independent features are combined within an HMM framework. The absolute gray-scale histogram difference between adjacent frames is used as an image feature. The acoustic difference in long intervals before and after the adjacent frames is used as an audio feature to reflect the audio types. Video frames are divided into blocks. A block matching algorithm is applied between adjacent frames to extract the blocks motion vectors. For the blocks motion vectors, the magnitude of the average and the average of the magnitude are used as motion features to detect camera movements. Several machine learning techniques such as Support Vector Machines (SVM) are used for video segmentation by classifying the video frames into boundary frames and non-boundary frames [7].

3 Proposed System

In our previous work [12] we introduced a method for caption detection and localization in news videos. We showed the patterns of caption insertion and removal. Actually the process of caption insertion or removal takes a range of frames. Caption insertion and removal introduce many changes and can happen during the same video shot. Also caption may be static and shared in several consecutive shots. For all this reasons it is clear that captions in news videos disrupt the shot boundary detection process. Our idea for detecting shots boundaries is based on the elimination of the candidate caption area and a new feature for describing an image we called it as "dominate colors". We represent the image by the order of its colors sorted from the highest frequency to the lowest frequency. If two images are similar so its order of dominate colors representations will be similar. We defined the dissimilarity between two images as a vector contains the difference in order for each color. Our idea for detecting the shots boundaries is based on the fact that if a shot boundary exist so

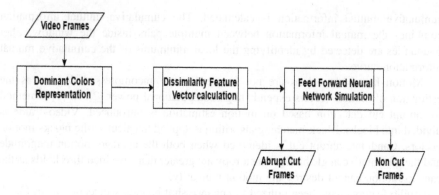

Fig. 1. The block diagram of the proposed system

a large difference of the dominate colors order will occur. We trained a feed forward neural network to classify the video frames to abrupt cut frames or non-cut frames. The proposed system is shown in Fig. 1.

3.1 Dominant Colors Representation

As shown in Fig. 2 for each video frame, we extract the R, G, and B components. At first we truncate the bottom 25 % of the three color components as it is the candidate caption area and actually the most important part of image is always at the frame center. Then we generate the dominate colors of each color component as shown in Fig. 3. The RGB representation of the video frame is the concatenation of the dominate colors of the three components.

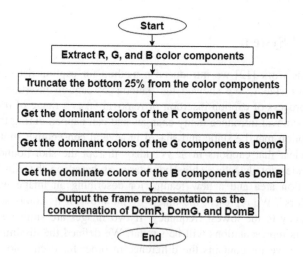

Fig. 2. RGB frame representation extraction

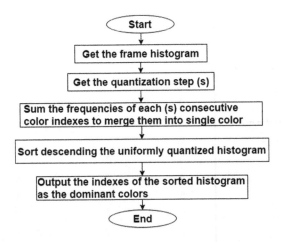

Fig. 3. Dominate color extraction.

Colors quantization is involved in the process of dominate colors extraction. The main purpose of the colors quantization is to absorb the lighting conditions that result in variation of the pixels color intensities of the same object. We carried out several experiments with different quantization steps and traced the accuracy results to conclude the colors quantization impact and select the best model. Also as shown in Fig. 4 we carried out several experiments based on the gray representation of video frames. For each video frame, we transform it from RGB to HSI. We work only on the intensity component. Dominate gray intensities are extracted with several quantization steps.

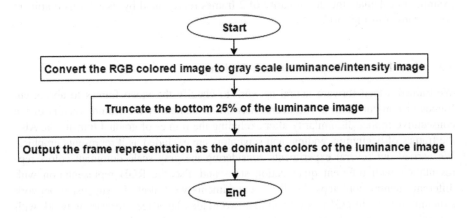

Fig. 4. Gray representation extraction of a video frame

3.2 Dissimilarity Feature Vector Calculation

We defined the dissimilarity between two video frames as a vector contains the difference in order for each color. To calculate the dissimilarity vector (dist) between

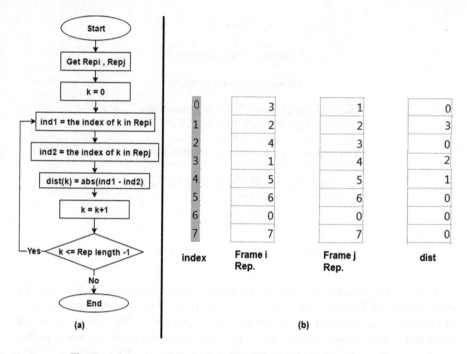

Fig. 5. (a) Dissimilarity vector calculation algorithm. (b) An example.

frame i (Fi) and frame j (Fj) first we get its dominant colors representation Repi, Repj. The algorithm of calculating the dissimilarity vector is shown in Fig. 5(a) and an example to calculate the dissimilarity of 2 frames represented by its 8 levels dominant colors shown in Fig. 5(b).

3.3 Feed Forward Neural Network Training

We trained a feed forward neural network to classify the video frames to abrupt cut frames or non-cut frames. The neural network is trained by the feature vectors of the consecutive frames dissimilarity after excluding the frames of gradual transitions. Also the frames of beginning, ending, and separators are excluded as it represents graphics. We carried out several experiments considering the gray scale dominant colors representation with different quantization steps and also the RGB representation with different quantization steps. In all our experiments we follow the suggested network structure in [13]. In [13] it is proved that two hidden layer feed forward network with the below structure significantly decreases the required number of hidden neurons while preserving the learning capability with any small error. Let M be the No of output classes = 2 and Let N be the length of dissimilarity feature vector that varied according to the used representation (gray or RGB) and the used quantization step. The number of neurons in the two hidden layers is as follows:

$$\text{No of neurons in the first hidden layer} \; = \; \text{sqrt}((M+2)*N) + 2*\text{sqrt}(N/(M+2)) \quad (1)$$

$$\text{No of neurons in the second hidden layer} \; = \; M*\text{sqrt}(N/(M+2)) \quad (2)$$

3.4   Abrupt Cut Detection Process

The process of the abrupt cut detection is carried out by simulating the neural network with the dissimilarity vectors between each two consecutive video frames. Each candidate detected cut between frame i, and i−1 will be verified by checking the neural network with the dissimilarity vector between frame i−1, and i+2. This verification step aims to eliminate the false alarms come from flashes as flashes effect are introduced in maximum in two successive frames.

4   Experimental Results

We build our data set as RGB colored videos collected form YouTube from different news TV channels as the Egyptian first channel, AlArabiya, and Al Jazeera. Shot boundaries are marked manually. Statistics about our data set is shown in Table 1. We developed a MATLAB tool to help us to manually mark the shot boundaries and decide each shot type (studio, report, etc.) and each shot boundary type(cut, fade-in, fade-out, dissolve, and wipe) as shown in Fig. 6. We used the following Measures for assessing the system results:

Table 1. Data set statistics

# of videos	53	# of abrupt cuts	1040
Total duration in minutes	85	# of gradual transitions	268

$$\text{Precision (P)} \; = \; \#\text{ Correctly Detected Cuts} \; / \; \#\text{ Detected Cuts} \quad (3)$$

$$\text{Recall (R)} \; = \; \#\text{ Correctly Detected Cuts} \; / \; \#\text{ Ground Truth Cuts} \quad (4)$$

$$\text{F Measure} \; = 2* \text{Precision} * \text{Recall} \; / \; (\text{Precision} + \text{Recall}) \quad (5)$$

We carried out several experiments with the dominate colors gray representation and RGB representation with several quantization steps as shown in Table 2. We used the F Measure alone for evaluation as it combines the precision and recall together with equal weight. From the results it is clear that increasing the quantization step will result in enhancement in the results up to limit after that will result in negative feedback and that

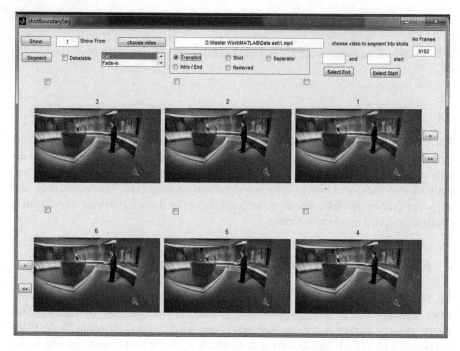

Fig. 6. Our manual shot boundary marker tool

is logical. Also RGB representation results is slightly better than the gray representation and the best accuracy is reached when using RGB representation with quantization step = 4 so we will use this network. The test data results when using RGB representation with quantization step = 4 is detailed in Table 3 utilizing flash light elimination.

By observing the reasons of false alarms that are generated from the proposed system, we will found mainly two reasons for false alarms: rapid motion and dissolves especially short dissolves that happen in one frame. As shown in Fig. 7(a), the reason of false alarm at frame 4370 is the rapid zoom on the money paper. In Fig. 7(b) false alarm is generated at frame 1499 because the video frame is changed suddenly due to firing a bomb. In Fig. 8 the trained neural network generates two successive cuts at the

Table 2. Gray and RGB representations results on test data without flash light elimination

Quantization step	Gray representation	RGB representation
	F Measure	F Measure
2	91.00 %	91.56 %
4	91.08 %	93.40 %
8	92.50 %	93.30 %
16	91.36 %	91.40 %
32	86.83 %	89.49 %
64	63.00 %	82.05 %

Table 3. Test data results using RGB dominate colors representation with quantization step = 4 utilizing flash light elimination

Video	Ground truth	Correct	Miss	False alarms	P %	R %	F %
Vid1	68	62	6	8	88.57	91.17	89.86
Vid2	27	26	1	2	92.86	96.30	94.55
Vid22	23	23	0	0	100.00	100.0	100.0
Vid30	26	25	1	2	92.59	96.15	94.34
Vid43	25	25	0	0	100.00	100.0	100.0
Vid9	29	29	0	0	100.00	100.0	100.0
Overall	198	190	8	12	94.06	95.96	95.00

Fig. 7. False alarms observation

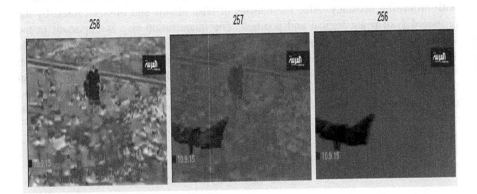

Fig. 8. Short dissolve

Fig. 9. Misses observation

frames 257,258 as this is a case of short dissolve. Also by observing the miss cases as shown in Fig. 9, the system fail to detect its cuts. In Fig. 9(a) the system misses a cut at frame 1851, in (b) misses a cut at frame 4512, in (c) misses a cut at frame 1151, and in (d) misses a cut at frame 2362. We found the main reason is the very similar background or in other words the two shots involved in the cut are recorded in the same place and corresponds to same scene.

5   Conclusion and Future Work

There are several highlights in our work. Although the training data is not big enough and the proposed system is tested on challenging data, the new image representation and our image dissimilarity measure succeeded to detect the abrupt cuts in news videos with promising accuracy. Also the reasons of false alarms and misses caused by the proposed system are logical and justified. The quantization step is important in the dominant colors extraction process as it absorbs the variation in video frame pixels values due to lighting condition but this is valid up to limit after that loss of information will occur. The RGB dominant colors representations have slightly better results than the gray scale representation. We are planning to extend our data set, publish it, and retrain our models to enhance results. Also we are planning to add gradual transition detector to our system.

References

1. Jiang, M., Huang, J., Wang, X., Tang, J., Wu, C.: Shot boundary detection method for news video. J. Comput. **8**(12), 3034–3038 (2013)
2. Dhagdi, T.S., Deshmukh, P.R.: Keyframe based video summarization using automatic threshold & edge matching rate. Int. J. Sci. Res. Publ. **2**(7), 1–12 (2012)
3. Eickeler, S., Muller, S.: Content-based video indexing of tv broadcast news using hidden markov models. In: International Conference on Acoustics, Speech, and Signal Processing, Phoenix, AZ, 15–19 March 1999, vol. 6, pp. 2997–3000 (1999)
4. Janwe, N.J., Bhoyar, K.K.: Video shot boundary detection based on JND color histogram. In: IEEE Second International Conference on Image Information Processing, Shimla, 9–11 December 2013, pp. 476–480 (2013)
5. Wang, C., Sun, Z., Jia, K.: Abrupt cut detection based on motion information. In: International Conference on Intelligent Information Hiding and Multimedia Signal Processing, Dalian, 14–16 October 2011, pp. 344–347 (2011)
6. Boreczky, J.S., Wilcox, L.D.: A hidden Markov model framework for video segmentation using audio and image features. In: IEEE International Conference on Acoustics, Speech and Signal Processing, Seattle, WA, 12–15 May 1998, vol. 6, pp. 3741–3744 (1998)
7. Apostolidis, E., Mezaris, V.: Fast shot segmentation combining global and local visual descriptors. In: IEEE International Conference on Acoustics, Speech and Signal Processing, Florence, 4–9 May 2014, pp. 6583–6587 (2014)
8. El-bendary, N., Zawbaa, H.M., Hassanien, A.E., Snasel, V.: PCA-based home videos annotation system. Int. J. Reasoning-based Intell. Syst. **3**(2), 71–79 (2011)
9. Shao, H., Qu, Y., Cui, W.: Shot boundary detection algorithm based on HSV histogram and HOG feature. In: 5th International Conference on Advanced Engineering Materials and Technology, pp. 951–957 (2015)
10. Tippaya, S., Sitjongsataporn, S., Tan, T., Chamnongthai, K., Khan, M.: Video shot boundary detection based on candidate segment selection and transition pattern analysis. In: IEEE International Conference on Digital Signal Processing, Singapore, 21–24 July 2015, pp. 1025–1029 (2015)
11. Cernekova, Z., Nikolaidis, N., Pitas, I.: Temporal video segmentation by graph partitioning. In: IEEE International Conference on Acoustics Speech and Signal Processing Proceedings, Toulouse, 14–19 May 2006, vol. 2, pp. 209–212 (2006)
12. Zedan, I. A., Elsayed, K.M., Emary, E.: Caption detection, localization and type recognition in Arabic news video. In: The 10th International Conference on Informatics and Systems Proceedings (INFOS 2016), Cairo, Egypt, 9–11 May 2016
13. Huang, G.: Learning capability and storage capacity of two-hidden-layer feed forward networks. IEEE Trans. Neural Netw. **14**(2), 274–281 (2003)

An Improved DV-Hop Localization Algorithm

Enas Mohamed[1(✉)], Hatem Zakaria[1], and M.B. Abdelhalim[2]

[1] Benha Faculty of Engineering, Electrical Engineering Department,
Benha University, Banha, Egypt
enasbhit@gmail.com, hatem.radwan@bhit.bu.edu
[2] College of Computing and Information Technology, Arab Academy of Science
and Technology and Maritime Transport, Alexandria, Egypt
mbakr@ieee.org

Abstract. This paper proposes two improved DV-hop localization algorithms named an Equal Sub-Area-Based DV-Hop (ESAB-DV-Hop) and Equal Sub-Area-Based DV-Hop with RSSI (ESAB-DV-Hop with RSSI) which achieve less error than an original algorithm. The two proposed algorithms are derived from the DV-hop method. The 1st one depends on dividing the network area into sub-areas and setting the nodes with unknown locations to calculate their positions from anchor nodes in the same sub-area and the 2nd one depends on dividing the network area into sub-areas and each node with unknown location calculates the distance between itself and other nodes using hop count number if they are not neighbor anchor nodes and RSSI if they are neighbor anchor nodes. Simulation results show that the localization accuracy of the Equal Sub-Area-Based DV-Hop (ESAB-DV-Hop) method is better than the original algorithm by 13.5 % and by 12 % compared to a recent algorithm that solves the same problem when the total number of the nodes equals 100, the communication range is 20 m and the anchors ratio is 10 %. The Equal Sub-Area-Based DV-Hop with RSSI (ESAB-DV-Hop with RSSI) method is better than the original algorithm by 17.5 % and by 16 % compared to a recent algorithm that solves the same problem when the total number of the nodes equals 100, the communication range is 20 m and the anchors ratio 10 %.

Keywords: Wireless sensor networks · Localization · DV-Hop

1 Introduction

Wireless sensor network has distributed nodes. These nodes send the data to the monitor base station. However, the monitor base station does not have correct view of the monitored area since those data are sent without any knowledge of the location where it is taken. In addition to that, GPS cannot be used in many applications as it consumes power and cannot be used in indoor applications. Many applications of wireless sensor networks are based on its position such as intrusion detection, inventory and supply chain management, and surveillance, environmental monitoring, indoor user tracking and so on. Therefore, localization is an important field of research for sensor networks that depends in its application on knowing sensors positions [1, 2].

© Springer International Publishing AG 2017
A.E. Hassanien et al. (eds.), *Proceedings of the International Conference on Advanced Intelligent Systems and Informatics 2016*, Advances in Intelligent Systems and Computing 533, DOI 10.1007/978-3-319-48308-5_32

There are two main divisions of localization algorithms: range-based and range-free. Range-based method depends on the properties of the signals transmitted between neighboring nodes. There are many ranging techniques such as Time Of Arrival (TOA) [3], Time Difference Of Arrival (TDOA), Received Signal Strength Indicator (RSSI), and Angle Of Arrival (AOA) [4]. Range-free techniques estimate nodes locations based on connectivity information. Range-free algorithms do not require additional hardware [5]. There are many algorithms of range-free techniques such as Approximate Point In Triangulation (APIT) [6], Amorphous [7], Distance Vector-Hop method (DV-Hop) [2], Centroid method [8], and MDS-MAP [9, 10].

DV-Hop localization algorithm is a range-free technique. The main error source in DV-hop is that the real path distance between nodes does not reflect the calculated path distance because of the hop size value is not the same for all hops.

In this paper, we introduce an improved DV-hop technique to reduce the error and raise the accuracy of localization by dividing the area of the network into equal sub-areas, and each node with unknown location calculates its position from anchor nodes in the same sub-area. By this way, the hop size would be calculated from the anchor nodes in the same sub-area not the whole network.

Many approaches on wireless sensor network localization have been reported. Qian et al. proposed a modified DV-hop algorithm [11], which chooses specific anchors to find the position of the nodes with unknown locations. Ji and Liu proposed a modified DV-hop algorithm [12], which gets the value of hop size depending on communication radius of the node. Sassi, et al. selected 3 anchors including the nearest one from the node with unknown location depending on the more accurate position calculation results of this node with unknown location [13]. Yu and Li, proposed a method depending on adjusting the measured distance between the node with unknown location and Anchor node to be close from the real distance [17]. The previous works concentrate on decreasing the gap between the calculated hop size and the real distances between the nodes. Therefore, they can decrease the error of the position of unknown nodes.

The remainder of the paper is organized as follows: In Sect. 2, problem statement is mathematically represented and discussed and the form of the solution is given. In Sect. 3, the background of the original DV-Hop is introduced. Our Equal Sub-Area-Based DV-Hop and Equal Sub-Area-Based DV-Hop with RSSI is discussed in Sects. 4 and 5. In Sect. 6, simulation results and comparison are presented. Finally, in Sect. 7, the conclusion is presented.

2 Problem Statement

The problem of DV-hop is that the distance calculated from counting the hops between the node with unknown location and the anchors are not equal to the real distance between them, therefore, the number of hops does not reflect the actual distance. For example, as depicted in Fig. 1, the real distance between node U with unknown location and anchor node R1 is not equal the calculated distance (i.e., the size of the 2 hops between node U with unknown location and the anchor node R1). Therefore, decreasing the error of the calculated distance is the way to decrease the error of DV-hop and improve its accuracy.

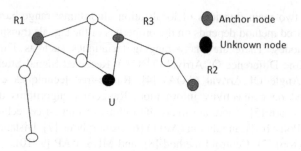

Fig. 1. Example of DV-hop algorithm

Our contribution proposes a modified DV-hop using 2 ways to decrease the error and increase the accuracy by using the 1st way which depends on dividing the area of the network into equal subareas, and each node with unknown location computes its position from its subarea's anchors and the 2nd way which also depends on dividing the area of the network into equal subareas, and each node with unknown location calculates the distance between itself and other nodes using hop count number if they are not neighbor anchor nodes and RSSI if they are neighbor anchor nodes. Using these 2 ways the hop-size will be calculated from the anchors' hop count in the same sub-area not in the whole network or will be calculated using RSSI from anchors' packets from the same sub-area. Hence, it reflects the real distance between nodes.

3 Dv-Hop Localization Algorithm

DV-Hop algorithm has been proposed by Niculescu and Nath [14, 15]. The original algorithm consists of three steps [1]:

1. Finding the shortest path between each anchor node and sensor nodes. Each anchor node broadcast message contains ID, position, hop count number. Each sensor node save the message with the minimum hop count to each anchor node and increment the hop count value by one then propagate the message. At the end of this stage, each node maintains routing table {ID, A_{xi}, A_{yi}, A_{hi}} where {A_{xi}, A_{yi}} is the location of node A_i and hop_{ij} is the minimum hop count number between sensor node and anchor node A_i.
2. Each anchor node finds the hop size value using the minimum hop count number hop_{ij} between it and the other anchors. Using this formula

$$hop\_size_i = \frac{\sum_{i \neq j} \sqrt{\left(A_{xi} - A_{xj}\right)^2 + \left(A_{yi} - A_{yj}\right)^2}}{\sum_{i \neq j} hop_{ij}} \tag{1}$$

Where (A_{xi}, A_{yi}), (A_{xj}, A_{yj}) are position coordinates of anchor A_i, and anchor A_j, hop_{ij} is the minimum number of hops between anchor A_i and anchor A_j. Then each anchor node propagates its estimated hop size value. Nodes with unknown locations receive the first one only to ensure it is the nearest anchor node. Then it calculates the distance between itself and the anchors by multiplying the hop size with the hop count.

When nodes with unknown locations have the distance of anchors, it computes its position coordinates using trilateration method which is a mathematical method used for finding the position of a node with unknown location using the distance between it and the anchors [2].

4 Proposed Equal Sub-Area Based DV-Hop

This paper proposes two methods to improve the localization accuracy. The 1st method depends on dividing the area of the network into equal sub areas, and each node with unknown location calculates its position from anchor nodes in the same sub-area. By this way, the hop size will be calculated from the anchor nodes in the same sub-area not in the whole network. The proposed method consists of the following steps (see flow chart in Fig. 2):

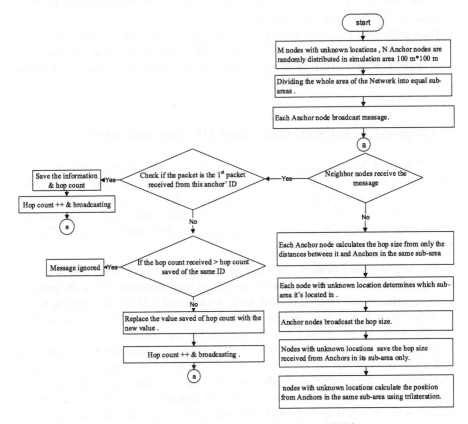

Fig. 2. Flow chart of the equal sub-area based DV-hop

1. Dividing the whole area of the network into equal sub-areas. The number of sub-areas is defined by:

$$Number\ of\ sub - areas(x) = Number\ of\ anchor\ nodes(y)/z$$

Where Z is the smallest number x divisible by it with condition that, it is equal to or more than three and it's the number of anchor nodes in each sub-area.

2. The difference in this step and step 1 in the original algorithm is that the message broadcasted by anchor nodes contains a new additional field for sub- area number to anchor packet, which represents a negligible overhead on the communications payload. Therefore, by the end of this stage, each node maintains a table $\{A_{xi}, A_{yi},$ sub-area number, $h_i\}$. Once an anchor A_i gets hop count value from other anchor nodes, it estimates the hop size from distances of anchors in the same sub-area not anchors over the entire network.

3. Each node with unknown location determines which sub-area it is located in by finding the smallest $\sum_{at\_specific\_sub-area} hop\_count$ of each anchor node to determine in which sub-area it is located.

4. Each node with unknown location uses the hop size to compute its position from anchor nodes in this sub-area using trilateration method [2]. This hop size is an average hop size received from anchors of the node with unknown location sub-area.

5 Proposed Equal Sub-Area Based DV-Hop with RSSI

The 2nd method depends on dividing the network area into sub-areas and each node with unknown location calculates the distance between itself and other anchor nodes using hop count number if they are not neighbors and RSSI if they are neighbor anchor nodes. By this way, the value of distances between node with unknown location and anchor node will be more accurate. The proposed method consists of the following steps (see flow chart in Fig. 3):

1. This step is similar to step 1 in the proposed Equal sub-area based DV-Hop.
2. The difference in this step and step 1 in the original algorithm is that the message broadcasted by anchor nodes contains a new additional field for sub- area number to anchor packet, which represents a negligible overhead on the communications payload. So, by the end of this stage, each anchor node maintains a table $\{A_{xi}, Ayi,$ sub-area number, $h_i\}$ and each node with unknown position maintains table $\{A_{xi}, A_{yi},$ sub-area number, h_i, RSSI from anchors' neighbors to itself$\}$. Once an anchor A_i gets hop count value from other anchor nodes, it estimates the hop size from distances of anchors in the same sub-area not anchors over the entire network.
3. Each node with unknown position determines which sub-area it is located in by finding the smallest of each anchor node to determine in which sub- area it is located.

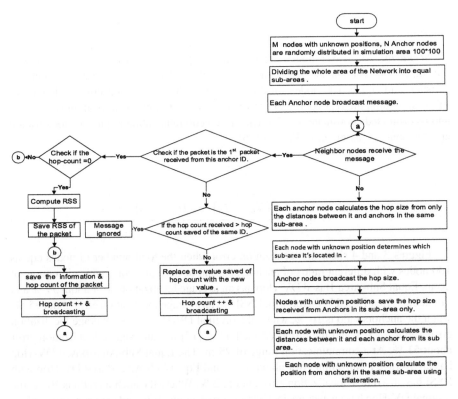

Fig. 3. Flow chart of the equal sub-area based DV-hop with RSSI

4. Nodes with unknown positions save the hop size received from Anchors in its sub-area only.
5. Each node with unknown position calculates the distances between it and each anchor from its sub- area. If the hop count between the node with unknown position and anchor node is one (i.e., its neighbor), it calculates the distance using RSSI. If the hop count between the nodes with unknown position and anchor node is more than one, it calculates the distance by multiplying the average hop size by the hop count.
6. Each node with unknown position calculates the position from anchors in the same sub-area using trilateration.

6 Performance Analysis and Comparison

In this section, the performance of the Equal Sub-Area-Based DV-Hop and Equal Sub-Area-Based DV-Hop with RSSI has been evaluated and compared with the traditional DV Hop and the recent research work of Yu and Li, [17]. Our developed algorithm has shown better localization accuracy than the improved dv-hop discussed in [17] using the same case studies.

Two *main scenarios* were foreseen for the performance analysis of the localization algorithm:

1st Scenario: All sensor nodes are randomly deployed in two dimensional area of *100* m*$*$*100* m. The anchor nodes ratio is variable, while the total number of nodes is constant and equals 100 nodes. In the simulation, the localization error is defined as follows: assume the real coordinate of the node *Ui* with unknown location (Uxj, Uyj), and its estimated coordinates (Uxi, Uyi), r is the communication radius, the localization error is obtained as given in Eq. (2) [16]:

$$error_i = \frac{\sqrt{\left(U_{xi} - U_{xj}\right)^2 + \left(U_{xi} - U_{xj}\right)^2}}{r} \tag{2}$$

Figures 3 and 4 show the localization error when the total number of nodes equals 100 nodes with different ratios of anchors. As can be seen at a communication range of 20 m, Equal Sub-Area-Based DV-Hop has an average localization error about 39 % and Equal Sub-Area-Based DV-Hop with RSSI has an average localization error about 35 %. when the anchors ratio is 10 %, where the original DV-Hop has an average localization error about 52.5 % and the proposed method in [17] has an average localization error about 51 %. At a communication range of 25 m, The Equal Sub-Area-Based DV-Hop has an average localization error about 35 % and Equal Sub-Area-Based DV-Hop with RSSI has an average localization error about 32 %. Where the anchors ratio is 10 %, the original DV-Hop has an average localization error about 40 % and the proposed method in [17] has an average localization error about 38 %.

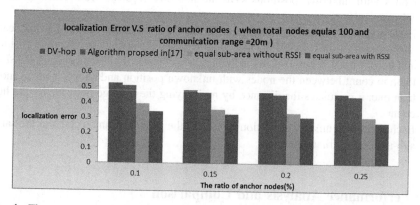

Fig. 4. The average localization error when the total nodes 100 node and the communication range 20 m.

2nd Scenario: The total number of nodes is variable while the anchor nodes ratio is constant.

Figures 5 and 6 show that the localization error when the anchors ratio is 10 % with different total number of nodes. As can be seen at a communication range of 20 m, Equal Sub-Area-Based DV-Hop has an average localization error about 29.58 % and Equal Sub-Area-Based DV-Hop with RSSI has an average localization error about 27 %. When the total number of nodes equals 200 nodes, where the original DV-Hop has an average localization error about 36 % and the proposed method in [17] has an average localization error about 33 %. At a communication range of 25 m, Equal Sub-Area-Based DV-Hop has an average localization error about 27.05 % and Equal Sub-Area-Based DV-Hop With RSSI has an average localization error about 24 %, when the total number of nodes equals 200 nodes, where the original DV-Hop has an average localization error about 33 % and the proposed method in [17] has an average localization error about 30.5 % (Fig. 7).

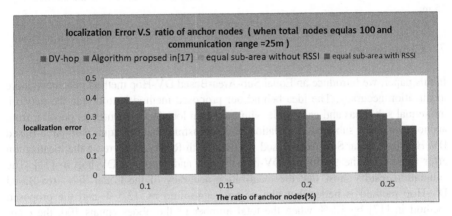

Fig. 5. The average localization error when the total nodes 100 node and the communication range 25 m.

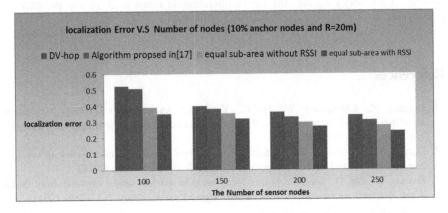

Fig. 6. The average localization error when anchors ratio is 10 % and the communication range 20 m.

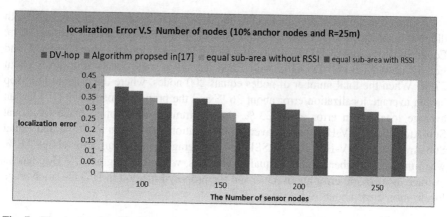

Fig. 7. The average localization error when anchors ratio is 10 % and the communication range 25 m.

7   Conclusion

In this paper, we introduce an Equal Sub-Area-Based DV-Hop method to achieve more localization accuracy. The idea behind our proposed method is to divide the network into equal sub-areas and each node with unknown location computes its position from anchors within its sub-area. Simulation results illustrate that the Equal Sub-Area-Based DV-Hop and Equal Sub-Area-Based DV-Hop with RSSI can decrease the localization error more than the traditional DV-hop method and proposed DV-hop in [17]. The simulation results show that the localization accuracy of the Equal Sub-Area-Based DV-Hop method is better than the original algorithm by 13.5 % and the proposed method in [17] by 12 % when the total number of the nodes equals 100, the communication range is 20 m and the anchors ratio 10 % and the Equal Sub-Area-Based DV-Hop with RSSI method is better than the original algorithm by 17.5 % and by 16 % compared to a recent algorithm that solves the same problem when the total number of the nodes equals 100, the communication range is 20 m and the anchors ratio 10 %.

References

1. Jungang, Z., Chengdong, W., Hao,C., Yang, X.: An improved DV-Hop localization algorithm. In: IEEE International on Progress in Informatics and Computing (PLC), Shanghi, vol. 1, pp. 469–471 (2010)
2. Dargie, W., Poellabauer, C.: Fundamental of wireless sensor Networks. Series on wireless communications and mobile computing, 1st edn. Wiley, New York (2010)
3. Enyang, X., Zhi, D., Sour, D.: Source localization in wireless sensor networks from signal Time of arrival measurements. IEEE Trans. Signal Process. 59(6), 2887–2897 (2011)

4. Rong, P., Sichitiu, M.L.: Angle of arrival localization for wireless sensor networks. In: 2006 3rd Annual IEEE Communications Society on Sensor and Ad Hoc Communications and Networks, SECON, Reston, VA, vol. 1, pp. 374–382 (2006)
5. Rakshit, S.M., Hate, S.G.: Range free localization using expected hop progress in wireless sensor network. Int. J. Innov. Res. Adv. Eng. 1(6), 172–179 (2014)
6. Ji, Z.W., Hongxu, J.: Improvement on APIT localization algorithms for wireless sensor networks. In: IEEE International on Networks Security Wireless Communications and Trusted Computing, Vol.1, pp. 719–723 (2009)
7. Nagpal, R.: Organizing a Global Coordinate System from Local Information on an Amorphous Computer, Technical report AI Memo 1666, MIT Artifical Intelligence Laboratory (1999)
8. Nirupama, B., John, H., Deborah, E.: GPS-less low cost outdoor localization for very small devices. IEEE Pers. Commun. Mag. 7, 28–34 (2000)
9. Yi, S., Wheeler, R.: Improved MDS-based localization. In: Twenty Third Annual Joint Conference of the IEEE Computer and Communications Societies, Vol. 4, pp. 2640–2651 (2004)
10. Shang, Y., Wheeler, R., Zhang, Y., et al.: Localization from mere connectivity. In: 4th ACM International Symposium on Mobile Ad Hoc Networking, Annapolis, pp. 201–212 (2003)
11. Qian, Q., Shen, X., Chen, H.: An improved node localization algorithm based on DV-Hop for wireless sensor networks. Comput. Sci. Inf. Syst. 8(4), 953–972 (2011)
12. Ji, W.W., Liu, Z.: An improvement of DV-Hop algorithm in wireless sensor networks. In: IEEE International on Wireless Communications, Networking and Mobile Computing, Wuhan, PP. 1–4 (2006)
13. Hichem, S., Tawfik, N., Noureddine, L.: A selective 3-anchor DV-Hop algorithm based on the nearest anchor for wireless sensor network. Int. J. Comput. Electr. Autom. Control Inf. Eng. 8(10) (2014)
14. Niculescu, D., Nath, B.: Ad Hoc positioning system (APS) using AOA. In: Twenty–Second Annual Joint Conference of IEEE Computer and Communications, vol. 3, pp. 1734–1743 (2003)
15. Niculescu, D., Nath, B.: Ad hoc positioning system (APS). In: IEE International on Global Telecommunication Conference, vol. 5, pp. 2926–2931 (2001)
16. Zhang, J., Guo, N., Lil, J.: An improved DV-Hop localization algorithm based on the node deployment in wireless sensor networks. Int. J. Smart Home 9(10), 197–204 (2015)
17. Yu, W., Li, H.: An improved DV-Hop localization method in wireless sensor networks. In: Computer Science and Automation Engineering (CSAE), pp. 199–202 (2012)

New Quantum Image Steganography Scheme with Hadamard Transformation

Bassem Abd-El-Atty, Ahmed A. Abd El-Latif[✉],
and Mohamed Amin

Faculty of Science, Department of Mathematics,
Menoufia University, Shibin Al Kawm 32511, Egypt
ahmed_rahiem@yahoo.com

Abstract. Based on Hadamard transformation and the novel enhanced quantum representation for quantum images (NEQR), a quantum image steganography scheme to embed a quantum text message into a quantum image is proposed. The extraction process can recover the text message with the stego image only. In the earlier works, there is no quantum image steganography algorithm to embed the quantum text message into a quantum image. The simulation results demonstrated that the proposed scheme has high-capacity, good invisibility, and high security.

Keywords: Quantum steganography · Quantum image processing · Hadamard transformation

1 Introduction

Quantum information processing has great deal of attention from both engineers and quantum scientists. The main task of quantum information is to develop new quantum algorithms for processing and storing quantum information. Quantum computer is a machine accepts input quantum states as a superposition of several inputs in another state as output quantum state [1]. Quantum steganography utilizes the effects of quantum mechanics such as quantum computation and quantum communication to achieve tasks of quantum information hiding. Quantum steganography, can be defined as the classical steganography in viewpoint of quantum mechanics, has become important researching area of quantum cryptography [2]. There are four types of quantum steganography classified as per embedding strategies [3]. The first type is information hiding of quantum data [4–10]. The second type is quantum error code [11, 12]. The third type based on the quantum cryptography protocols to hide secret messages [13, 14]. The last type, which based on quantum image processing techniques named as quantum image steganography.

The effects of quantum parallelism used in processing quantum images (PQI) to improve many processing tasks in classical image processing. The first step of PQI is to represent and store the classical images on quantum computers. There are many representations for classical images on quantum computers, such as Entangled Image [15], Real Ket [16], Multi-Channel representation of quantum image [17], log-polar representation [18], flexible representation of quantum images (FRQI) [19], which uses

© Springer International Publishing AG 2017
A.E. Hassanien et al. (eds.), *Proceedings of the International Conference on Advanced Intelligent Systems and Informatics 2016*, Advances in Intelligent Systems and Computing 533, DOI 10.1007/978-3-319-48308-5_33

number of qubits 2x + 1 to represent a gray image with size 2x × 2x and the NEQR model for represent quantum images [20]. In spite of the used qubits of NEQR increases from 2x + 1 qubits used in FRQI to 2x + q qubits, it is excellent for processing quantum image because the quantum representation is very similar to the representation of a classical image.

Recently, a number of quantum image security algorithms including encryption [21–23] and steganography based on NEQR have been proposed [24–27]. Jiang *et al.* introduced in [24] a scheme based on Moir´e Pattern, which embeds the secret binary image with size 2n × 2n into a cover gray image with size 2n × 2n. Afterward, Jiang *et al.* introduced in [25] a new quantum image steganography algorithm based on LSB, which embeds a message as binary 2n × 2n image into a cover image 2n × 2n gray image.

In the earlier works, there is no previous quantum image steganography algorithm to embed quantum secret message as a text into a quantum image. The quantum image steganography algorithms [24, 25] embed binary images or message as a binary image with maximum capacity one bit per pixel. However, quantum image steganography algorithms [24, 25] were broken to embed quantum text message into a cover image. In this paper, we will propose a quantum image steganography scheme to embed quantum text message instead of a message as an image into a quantum image. It utilizes Hadamard transformation to increase the security of embedded quantum data. Experimental results demonstrated that the maximum capacity increases from one bit per pixel to two bit per pixel in the proposed scheme and the invisibility is good [24, 25].

The rest of this paper is as follows. The Hadamard transformation and NEQR representation model are briefly reviewed in Sect. 2. Section 3 introduced the proposed quantum image steganography scheme. Section 4 gives analyses and results. The comparison with related schemes provided in Sect. 5. Finally, in Sect. 6 the conclusion is drawn.

2 Background

In this section, the NEQR model [20] for representation quantum images and Hadamard transformation are briefly reviewed, which the essentials of the proposed scheme.

2.1 NEQR Quantum Representation

By using the representation pixels idea for images in traditional computers, the NEQR representation is proposed in [20]. The NEQR model has information about the pixels color and its related position of each pixel in the image. The mathematical representation of a quantum image for an $2^n \times 2^n$ image can be expressed as follows.

$$|I\rangle = \frac{1}{2^n} \sum_{i=0}^{2^{2n}-1} |c_i\rangle \otimes |i\rangle \tag{1}$$

$$|c_i\rangle = |c_i^{q-1}....c_i^1 c_i^0\rangle, \; c_i^j \in \{0,1\}, i = 0,1,....,2^{2n}-1, j = 0,1,....,q-1$$

where the sequence $c_i^{q-1}....c_i^1 c_i^0$ encodes the color value with range 2^q, $|i\rangle$ *for* $i =$ $2^{2n} - 1,..., 1, 0$, *are* 2^{2n} dimension computational basis quantum states. There are two parts of $|I\rangle$ as follows $|c_i\rangle$, which encodes the information about the pixels color and $|i\rangle$ which encodes the information about the related position for colors in the image.

2.2   Hadamard Transformation

The basic unite of classical informationand classicalcomputation is bit. Quantum information and quantum computation are based upon an analogue idea, quantum bit (qubit). Qubits are represented by state vectors

$$|0\rangle = \begin{pmatrix} 1 \\ 0 \end{pmatrix} |1\rangle = \begin{pmatrix} 0 \\ 1 \end{pmatrix}$$

and a superposition of two states

$$|\psi\rangle = \alpha|0\rangle + \beta|1\rangle \tag{2}$$

where α and β are complex numbers and $|\alpha|^2 + |\beta|^2 = 1$

To express the operation of the classical gates, we have used a truth table. For quantum gates, we use matrix representation. Hadamard transformation is represented by the following matrix.

$$H = \frac{1}{\sqrt{2}} \begin{pmatrix} 1 & 1 \\ 1 & -1 \end{pmatrix}$$

3   Proposed Quantum Image Steganography Scheme

We introduce a quantum image steganography scheme based on NEQR representation and Hadamard gate in this section. The proposed quantum image steganography procedures are shown in Fig. 1.

3.1   Embedding Process

The proposed quantum image steganography embedding procedures consist of two phases, which are given by the following steps.

Phase 1 Preparation of quantum states

Let the cover image $|I\rangle$ is $2^n \times 2^n$ image and the secrete message encoded into two binary matrixes with size $2^n \times 2^n$.

Note the size of the secrete text message must be less than or equal to $2^{2n-2}(2 \times 2^n \times 2^n \text{bit} = 2 \times 2^n \times 2^n/8 \text{ byte})$ characters.

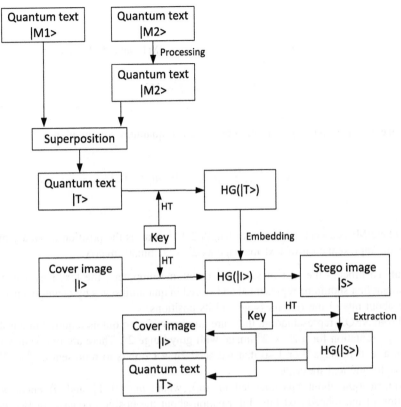

Fig. 1. Embedding and extraction procedures of the proposed quantum image steganography

The NEQR representations of $|I\rangle$, $|M1\rangle$ and $|M2\rangle$ are shown as follows:

$$|I\rangle = \frac{1}{2^n} \sum_{i=0}^{2^{2n}-1} |c_i\rangle \otimes |i\rangle = |c_i\rangle = |c_i^{q-1}....c_i^1 c_i^0\rangle, \ c_i^j \in \{0,1\} \tag{3}$$

$$|M1\rangle = \frac{1}{2^n} \sum_{i=0}^{2^{2n}-1} |m1_i\rangle \otimes |i\rangle, \ |m1_i\rangle = |m1_i^{q-1}....m1_i^1 m1_i^0\rangle, \ m1_i^j \in \{0,1\} \tag{4}$$

$$|M2\rangle = \frac{1}{2^n} \sum_{i=0}^{2^{2n}-1} |m2_i\rangle \otimes |i\rangle, \ |m2_i\rangle = |m2_i^{q-1}....m2_i^1 m2_i^0\rangle, \ m2_i^j \in \{0,1\} \tag{5}$$

Note that the representation of quantum text is represented only in m_i^0 and all others $m_i^j = 0$ where $j = 1, 2, 3, ..., q-1$, according to Theorem 1.

Then shift the coefficients of vector $|M2\rangle$ by 2^{2n+1} as follows:

$$\left|M2_{mod(i+2^{2n+1},\,2^{2n+q})}\right\rangle = |M2_i\rangle,\, i = 1,\, 2, \ldots,\, 2^{2n+q} \tag{6}$$

Then take the superposition of the two states $|M1\rangle$ and $|M2\rangle$ to state $|T\rangle$.

$$|T\rangle = \frac{1}{\sqrt{2}}(|M1\rangle + |M2\rangle) \tag{7}$$

Theorem 1 Text message can be represented in quantum computation as

$$|M\rangle = \frac{1}{2^n} \sum_{i=0}^{2^{2n}-1} |m1_i\rangle \otimes |i\rangle,\, m_i \in \{0, 1\} \tag{8}$$

where $|M\rangle$ is a vector space with length 2^{2n+1} and i is the position of binary bit m_i after the representation of text message as $2^n \times 2^n$ binary matrix.

Proof To process the text message on quantum computers, the text message information at first should be represented and stored in quantum states by quantum encoding state about bits of the text message and its positions.

From NEQR representation, there are number of 2n +q qubits required to construct the representation for a $2^n \times 2^n$ image with gray range 2^q. There are two values for a bit in a text message 0 or 1 so that we need 2n + 1 qubits to represent a $2^n \times 2^n$ bit matrix for the text message.

Information about bits encoded in $|m_i\rangle$, where $m_i \in \{0, 1\}$ and $|i\rangle$ encodes the position of the related text bits. Information about the position consists of two parts: horizontal and vertical coordinates. By taking into consideration a quantum text message is a system in 2n-qubit,

$$i\rangle = |v\rangle|h\rangle = |v_{n-1}v_{n-2}, \ldots \ldots, v_0\rangle|h_{n-1}h_{n-2}, \ldots \ldots, h_0\rangle, \tag{9}$$

$$h,\, v \in \{0,\, 1,\, \ldots \ldots,\, 2^n - 1\},$$

$$|v_k\rangle, |h_k\rangle \in \{|0\rangle,\, |1\rangle\},\, k = 0, 1, 2,\, \ldots \ldots,\, n-1,$$

here $|v_{n-1}v_{n-2}, \ldots \ldots, v_0\rangle$ encodes the first n-qubit over the vertical axis and $|h_{n-1}h_{n-2}, \ldots \ldots, h_0\rangle$ by encoding the second n-qubit over the horizontal axis. The state of quantum text message is a normalized as shown in (10),

$$\||M\rangle\| = \frac{1}{2^n} \sqrt{\sum_{i=0}^{2^{2n}-1} |m_i\rangle \otimes |i\rangle} = \frac{1}{2^n}\sqrt{2^{2n}} = \frac{1}{2^n} 2^n = 1 \tag{10}$$

Phase 2 Apply unitary Hadamard transformation

Firstly, form unitary Hadamard transformation gate (HG) controlled by secrete key K in binary form where $K = k_1k_2\ k_3\ ...k_i....\ K_{2n+q}$, $i = 2, 3, ..., 2n + q$, $k_i \in \{0,1\}$, $k_1 = 1$ is to ensure the color qubit transformation.

$$HG = H \overset{2n+q}{\underset{i=2}{\otimes}} H^{k_i}, H^{k_i} = \{ \begin{matrix} H, k_i = 1 \\ I, k_i = 0 \end{matrix} \, i = 2, 3, ..., 2n+q \qquad (11)$$

where I is identity matrix in 2-dimantional form and H is the Hadamard transformation matrix.

Then, execute HG on the quantum cover state $|I\rangle$ and quantum text state $|T\rangle$, getting vector $HG(|I\rangle)$ and vector $HG(|T\rangle)$ as follows:

$$HG(|I\rangle) = \frac{1}{2^n} \sum_{i=0}^{2^{2n}-1} HG(|c_i\rangle) \otimes |i\rangle \qquad (12)$$

$$HG(|T\rangle) = \frac{1}{2^n} \sum_{i=0}^{2^{2n}-1} HG(|m_i\rangle) \otimes |i\rangle \qquad (13)$$

Then, take the superposition of the two states $HG(|I\rangle)$ and $HG(|T\rangle)$ to the stego image state $|S\rangle$.

$$|S\rangle = \frac{1}{\sqrt{2}} (HG(|I\rangle) + HG(|T\rangle)) \qquad (14)$$

3.2   Extraction Operation

At the extraction operation, we utilize the secret key to extract embedded text message from the quantum cover image. The quantum process is totally revertible, because all used transforms are unitary matrices in quantum computation.

By executing inverse of HT on the stego-image $|S\rangle$, getting the cover image $|I\rangle$ and the text message $|M\rangle$.

$$\begin{aligned} inv(HG(|S\rangle)) &= inv(HG(\frac{1}{\sqrt{2}}(HG(|I\rangle) + HG(|T\rangle)))) \\ &= \frac{1}{\sqrt{2}} (inv(HG(HG(|I\rangle))) + inv(HG(HG(|T\rangle)))) \\ &= \frac{1}{\sqrt{2}} (|I\rangle + |T\rangle) \end{aligned}$$

4 Analyses and Results

We perform the simulation of quantum operations on the classical computer to analyses our proposed scheme. We introduce several analyzes, such as visual quality, payload capacity and security analysis.

4.1 Visual Quality

Visual quality is the amount of difference between the stego image and the original cover image in pixels values. There are many quantities to measure the difference of pixels between the stego image and the original cover image; one of the most used quantities is PSNR (peak signal to noise ratio). It is defined as the MSE (mean squared error) for two images X and Y with size i × j.

$$MSE = \frac{1}{ij}\sum_{x=0}^{i-1}\sum_{y=0}^{j-1}[X(x,y) - Y(x,y)]^2 \tag{16}$$

PSNR is defined as follows.

$$PSNR = 20\log_{10}(\frac{MAX_X}{\sqrt{MSE}}) \tag{17}$$

Where, MAX_X is the maximum value of pixels in the cover image X. In this scheme, Y related to the stego image and X related to the original cover image. To simulate the proposed scheme we use images with size 64 × 64 in these experiments. The following Fig. 2 describes experimental results of our scheme and Figs. 3 and 4 describe two different text messages that embedded in the stego image and Table 1 shows the PSNR values among the original cover image and the stego image for different two text messages.

original image   stego image

Fig. 2. The visual effects

Embedded quantum text, Embedded quantum text,Embedded quantum text, Embedded quantum text, Embedded quantum text, Embedded quantum text, Embedded quantum text, Embedded quantum text, Embedded quantum text, Embedded quantum text, Embedded quantum text, Embedded quantum text, Embedded quantum text, Embedded quantum text, Embedded quantum text, Embedded quantum text, Embedded quantum text, Embedded quantum text, Embedded quantum text, Embedded qua

Fig. 3. Text message 1 with length 1024 character

Steganography is the art and science of invisible communication, which serves to hide sensitive or secret information by embedding it in a large innocent message.

Fig. 4. Text message 2 with length 165 character

Table 1. PSNR values for different secrete text messages

Secrete message	Cover image
Message1	68.499
Message2	73.27

Obviously, from Fig. 2, human eyes can't identify the difference between the original coverimage and the stego image. From Table 1, the values of PSNR are enough. So our scheme has a good visual effect.

4.2 Payload Capacity

The steganography capacity can be stated as the ratio between the number of embedded bits of the text message and the number of pixels for cover image. The proposed scheme's capacity is given as follows:

$$C = \frac{number of message bits}{number of cover image pixels} = \frac{2 \times 2^n \times 2^n bit}{2^n \times 2^n pixel} = 2\, bit/pixel$$

The proposed algorithm has Payload capacity two bits per pixel, which is higher than of majority of the quantum image steganography algorithms. So we can conclude that our scheme has high-capacity.

4.3   Security Analysis

The embedded text message is same as the extracted text message so that there is no BER (Bit Error Rate) in extracted text message. The extract operation needs only quantum stego image and the binary key. The key utilized to control unitary Hadamard transformation, which is secret. Therefore, the effect of the proposed quantum image steganography scheme based on NEQR for quantum images and Hadamard transformation is excellent and has no error (error-free).

5   Comparison with Related Schemes

From the above section, the proposed algorithm has high PSNR compared to the schemes of [24, 25] based on the simulation results. The payload capacity of our scheme is two bit per pixel while the capacity of algorithms in [24, 25] are one bit per pixel. From Table 2, the proposed scheme has high-capacity, high security, good visual effect and embedding quantum text message instead of a message as an image.

Table 2. Comparison with related schemes

Items	Proposed scheme	Scheme in [24]	Scheme in [25]
Maximum capacity	2 bit/pixel	1 bit/pixel	1 bit/pixel
Security	Yes, using key	No	No
Embedding data	Quantum text message	Binary image	Message as binary image
Visual quality (PSNR) using Lena as cover image	73.27	29.2717	50.8426

6   Conclusion

A quantum image steganography scheme based on NEQR representation of quantum images and Hadamard transformation is proposed. The proposed scheme has good advantages such as the extracting operation does not require the original image or information about the embedded text message. Simulation results proved that the efficiency of the proposed scheme.

Acknowledgements. The authors would like to thank Dr. Li-Hua Gong For the valuable comments. This work was supported by Ministry of Higher Education and Scientific Research (Egypt-Tunisia Cooperation Program: 4-13 A1) and the National Natural Science Foundation of China (61462061 and 61561033).

References

1. Tseng, C.C., Hwang, T.M.: Quantum digital image processing algorithms. In: 16th IPPR Conference on Computer Vision. Graphics and Image Processing. Kinmen, ROC, pp. 827–834 (2003)
2. Xu, S.J., Chen, X.B., Niu, X.X., Yang, X.Y.: High-efficiency quantum steganography based on the tensor product of Bell states. Sci. China **56**, 1745–1754 (2013)
3. Jiang, X.S., Bo, C.X., Xin, N.X., Xian, Y.Y.: Steganalysis and improvement of a quantum steganography protocol via a GHZ4 state. Chin. Phys. B **22**, 060307 (2013)
4. Terhal, B.M., DiVincenzo, D.P., Leung, D.W.: Hiding bits in Bell states. Phys. Rev. Lett. **86**, 5807–5810 (2001)
5. Eggeling, T., Werner, R.F.: Hiding classical data in multipartite quantum states. Phys. Rev. Lett. **89**, 097905 (2002)
6. DiVincenzo, D.P., Leung, D.W., Terhal, B.M.: Quantum data hiding. IEEE Trans. Inf. Theory **48**, 580–598 (2002)
7. Hayden, P., Leung, D., Smith, G.: Multiparty data hiding of quantum information. Phys. Rev. A **71**, 062339 (2005)
8. Guo, G.C., Guo, G.P.: Quantum data hiding with spontaneous parameter down conversion. Phys. Rev. A **68**, 044303 (2003)
9. Chattopadhyay, I., Sarkar, D.: Local indistinguishability and possibility of hiding cbits in activable bound entangled states. Phys. Rev. A **365**, 273–277 (2007)
10. Matthews, W., Wehner, S., Winter, A.: Distinguishability of quantum states under restricted families of measurements with an application to quantum data hiding. J. Commun. Math. Phys. **291**, 813–843 (2009)
11. Banacloche, J.G.: Hiding messages in quantum data. J. Math. Phys. **43**, 4531 (2002)
12. ShawB, A., Brun, T.A.: Quantum steganography with noisy quantum channels. Phys. Rev. A **83**, 022310 (2011)
13. Zhang, D., Li, X.: A quantum information hiding scheme using orthogonal product states. WSEAS Trans. Comput. **6**, 757–762 (2007)
14. Liao, X., Wen, Q., Sun, Y., Zhang, J.: Multi-party covert communication with steganography and quantum secret sharing. J. Syst. Softw. **83**, 1801–1804 (2010)
15. Venegas-Andraca, S.E., Ball, J.L.: Processing images in entangled quantum systems. Quantum Inf. Process. **9**, 1–11 (2010)
16. Latorre, J.I.: Image compression and entanglement. arXiv: quant-ph/, 0510031 (2005)
17. Sun, B., Le, P.Q., Iliyasu, A.M., Yan, F., Garcia, J.A., Dong, F., Hirota, K.: A multi-channel representation for images on quantum computers using the RGB a color space. In: Proceedings of the IEEE 7th International Symposiumon Intelligent Signal Processing, pp. 1–6 (2011)
18. Zhang, Y., Lu, K., Gao, Y., Xu, K.: A novel quantum representation for log polar images. Quantum Inf. Process. **12**, 3103–3126 (2013)
19. Le, P.Q., Dong, F., Hirota, K.: A flexible representation of quantum images for polynomial preparation, image compression and processing operations. Quantum Inf. Process. **10**, 63–84 (2011)
20. Zhang, Y., Lu, K., Gao, Y., Wang, M.: NEQR: a novel enhanced quantum representation of digital images. Quantum Inf. Process. **12**, 2833–2860 (2013)
21. Hua, T.X., Chen, J.M., Pei, D.J., Zhang, W.Q., Zhou, N.R.: Quantum image encryption algorithm based on image correlation decomposition. Int. J. Theor. Phys. **54**, 526–537 (2015)

22. Zhou, N.R., Hua, T.X., Gong, L.H., Pei, D.J., Liao, Q.H.: Quantum image encryption based on generalized Arnold transform and double random-phase encoding. Quantum Inf. Process. **14**, 1193–1213 (2015)
23. Liang, H.R., Tao, X.Y., Zhou, N.R.: Quantum image encryption based on generalized affine transform and logistic map. Quantum Inf. Process. **15**, 1–24 (2016)
24. Jiang, N., Wang, L.: A novel strategy for quantum image steganography based on Moir´e pattern. Int. J. Theor. Phys. **54**, 1021–1032 (2014)
25. Jiang, N., Zhao, N., Wang, L.: LSB based quantum image steganography algorithm. Int. J. Theor. Phys. **55**, 107–123 (2015)
26. Jiang, N., Wang, L.: Quantum image scaling using nearest neighbor interpolation. Quantum Inf. Process. **14**, 1559–1571 (2015)
27. Wang, S., Sang, J., Song, X., Niu, X.: Least significant qubit (LSQb) information hiding algorithm for quantum image. Measurement **73**, 352–359 (2015)

Prediction of Antioxidant Status in Fish Farmed on Selenium Nanoparticles using Neural Network Regression Algorithm

Ahmed Sahlol[1,4]([⊠]), Ahmed Monem Hemdan[2,4], and Aboul Ella Hassanien[3,4]

[1] Faculty of Specific Education, Damietta University, Damietta, Egypt
atsegypt@du.edu.eg
[2] Faculty of Veterinary Medicine, Kafer-Elsheikh University, Kafr el-sheikh, Egypt
[3] Faculty of Computers and Information, Cairo University, Giza, Egypt
[4] Scientific Research Group in Egypt (SRGE), Cairo, Egypt
http://www.egyptscience.net

Abstract. Oxidative stress is the most common stress form which is responsible for the increased mortality as well as retardation of productivity in sheries. Selenium plays a vital role in combating oxidative stress. It appears as a potent antioxidant with reduced toxicity in its nanoscale form. In this paper, the effect of the different concentrations of Nano-selenium in the diet on the antioxidant status of common carp was investigated through the estimation of antioxidant enzymes activity and some biochemical blood prole. The adopted regression algorithm for prediction was Back-propagation Neural Network. The model compromised between fast analytical technologies and biological aspect through prediction the healthy status and expected hazards related to oxidative stress. The experiment was performed on four groups of common carp with measured rearing parameters and the same amount of diet at the rates of 0 (control), 0.5, 1 and 2 mg/k gm amount of Nano-selenium concentration in the ration, aiming to build preliminary prediction models to know the antioxidant status activity. The regression performance was tested by several mathematical validations including MSE (Mean squared error), RMSE (Root mean squared error), MSRE (mean squared relative error), MARE (mean absolute relative error), RMSRE (root mean squared relative error), MAE (Mean absolute error), MAPE (Mean absolute percentage error), MSPE (mean squared percentage error), RMSPE (root mean squared percentage error) which showed promising results of the regression model.

Keywords: Antioxidant status · Nanoselenium · Regression models · Back-propagation neural network · Mathematical validation

1   Introduction

Selenium nanoparticles play a vital role to prevent oxidative damages in fish which change the structure of the cell. Oxidative stress (OS) leads to the

© Springer International Publishing AG 2017
A.E. Hassanien et al. (eds.), *Proceedings of the International Conference on Advanced Intelligent Systems and Informatics 2016*, Advances in Intelligent Systems and Computing 533, DOI 10.1007/978-3-319-48308-5_34

generation of reactive oxygen species (ROS) that trigger the apoptosis in tissues. To overcome these adverse effects of ROS, the living systems activate antioxidant enzymes via selenium as they present at the catalytic site of antioxidant enzymes leading to their activation [1]. Nanoparticles of selenium are also more efficient to increase antioxidant activity than other forms [2,3]. They show more scavenging to H2O2 and lipid hydroperox-ides to give more protection cellular membrane from oxidative damage [4] Otherwise, Selenium deficiency decreases growth rate and increase mortality rate [5]. Lots of researches suggested that pollution can generate ROS that leads to cell damage and increase mortality rate [6]. The current paper used a set of parameters (including superoxide dismutase, glutathione perox- idase, Catalase, Low-density lipoprotein and high-density lipoprotein.) those affected by deviations in the selenium concentrations in breeding Common Carp species in vitro.

Successful machine learning applications have to take into account many practical considerations as they require pre-knowledge about the specific data type and the studied domain, as well machine learning has been successfully applied in several biological disciplines, such as phenotypic profiling of drug development and effects [7–11], high-content screening [12–19], proteomics [20–22], and DNA sequence analysis [23,24].

The main contribution of the paper is the application a machine learning approach to predict some antioxidant concentrations depending on nanoselenium concentrations and fish weights (initial and final weight) using a well-known regression algorithm; Back propagation-neural network. the regression algorithm parameters were tuned precisely and the prediction accuracy was also measured by some familiar mathematical validation tests.

The organization of this paper is as follows: Sect. 2 presents the materials and the methods. Results with discussions are described in Sect. 3. Finally, conclusions and future work are provided in Sect. 4.

2   Materials and Methods

2.1   Dataset Preparation

Selenium Nanoparticles were applied to four groups of fish with the different concentrations at the rates of 0 (control), 0.5, 1 and 2 mg/kg. By the end of the experiment, blood samples and parts of liver were stored at 80 C till used for enzyme assays. Glutathione peroxidase (GPx) activity was assayed according to [26]. Liver superoxide dismutase (SOD) was measured according to [25]. Liver catalase (CAT) was determined according to [27]. Triglycerides, high- density lipoprotein (HDL) and low-density lipoprotein (LDL) were estimated using commercial kits (Chemiscience Chemical Company, Cairo, Egypt). Figure 1 shows samples from the dataset.

INPUTS		control without nano particles						Outputs										
initial weight	final weight	protein	moisture	lipid	ash	liver	muscle	MDA	GPX	SOD	CAT	globulin	albumin	cholestrol	HDL	LDL	AST	ALT
15.1	21.87	22.63	76.96	3.26	0.09	0.34	0.29	0.56	1.07	54.3	0.84	2.18	1.73	151.11	43.43	72.22	12.75	29.8
15	21.82	22.89	76.95	3.26	0.09	0.34	0.29	0.57	1.06	53.06	0.83	2.19	1.72	151.08	44.64	72.19	12.73	28.26
14.99	21.72	22.52	77.21	3.26	0.09	0.35	0.28	0.62	1.05	52.35	0.82	2.02	1.85	151.6	44.11	72.87	12.6	29.11
14.95	21.42	22.22	77.41	3.37	0.09	0.36	0.26	0.64	0.97	45.75	0.8	1.84	2.07	152.4	44.37	74.54	12.31	28.75
14.98	21.71	22.44	77.24	3.28	0.09	0.35	0.28	0.62	1.04	51.43	0.81	2.04	1.89	151.66	44.12	73.1	12.5	29.1
14.96	21.54	22.32	77.29	3.32	0.09	0.35	0.27	0.63	1.02	49.43	0.81	1.92	1.99	152.05	44.16	73.13	12.43	28.83
14.95	21.42	22.26	77.38	3.37	0.09	0.36	0.26	0.64	0.98	45.76	0.8	1.86	2.06	152.31	44.38	74.32	12.34	28.76
14.96	21.51	22.3	77.3	3.31	0.09	0.36	0.27	0.63	0.98	48.22	0.81	1.9	2.02	152.11	44.18	73.17	12.41	28.82
14.97	21.65	22.13	77.46	3.41	0.09	0.36	0.26	0.65	0.97	45.12	0.8	1.82	2.11	153.5	44.62	75.28	12.27	28.74
14.95	21.44	22.28	77.35	3.38	0.09	0.36	0.27	0.63	0.98	46.76	0.8	1.88	2.04	152.27	44.34	74.19	12.35	28.79
15.1	21.84	22.61	76.98	3.26	0.09	0.34	0.29	0.57	1.06	53.14	0.83	2.16	1.8	151.26	43.75	72.49	12.71	29.32
15.1	21.74	22.56	77.19	3.24	0.09	0.35	0.28	0.58	1.05	52.65	0.82	2.07	1.84	151.53	44.05	72.58	12.64	29.18
15	21.82	22.86	76.96	3.27	0.09	0.34	0.29	0.57	1.06	53.06	0.83	2.18	1.73	151.1	44.72	71.32	12.72	28.26
14.95	21.4	22.2	77.43	3.42	0.09	0.36	0.26	0.64	0.97	45.42	0.8	1.83	2.08	152.44	44.43	75.11	12.3	28.75
14.99	21.73	22.55	77.2	3.27	0.09	0.35	0.28	0.61	1.05	52.43	0.82	2.04	1.84	151.54	44.09	72.83	12.61	29.16

Fig. 1. Samples of the dataset

2.2   Prediction Algorithms BP-NN (Back Propagation Neural Network)

Neural networks can be considered as a group of simple components working in parallel. These components are inspired by biological nervous systems. As in nature, the connections between components largely determine the network function. A neural network can be trained to perform a particular function (a prediction in this work) by adjusting the values of the connections (weights) between components, Fig. 1. In another way, it is trained so that a specific input leads to a specific output; this happened based on a comparison between the output and the target, until the network output matches the target. To achieve that, many such input/target pairs are needed to train a network.

Neural networks have been designed to perform complex problems that are difficult for conventional computers or human beings. It has been used in several fields like in manufacturing as Manufacturing process control, product design and analysis, process and machine diagnosis, also in financial as loan advising, mortgage screening, corporate bond rating, credit card activity tracking, also in medical fields as Breast cancer cell analysis, prosthesis design, optimization of transplant times, hospital expense reduction, hospital quality improvement, and many other fields that imply on recognition or regression for a very complex problems.

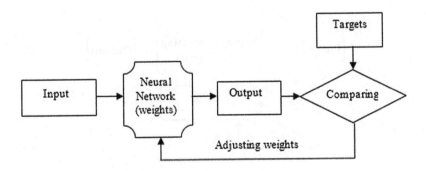

Fig. 2. Neural network working mechanism

In the training phase, inputs are presented to a perceptron, so the perceptron weights and biases changed according to the error, the perceptron will eventually find weight and bias values that solve the problem. As seen in Fig. 2, suppose we have individual element inputs $p_1, p_2, ...p_R$, multiplied by weights w_1, w_2, w_R, then the weighted values are fed to the summing junction. Their sum is simply W_p, the dot product of the single row matrix W and the vector p. The neuron has a bias b, which is summed with the weighted inputs to form the net input n. This sum, n, is the argument of the transfer function f.

$$n = w_1, 1, p_1 + w_1, 1, Rp_R + b \qquad (1)$$

This transfer function is commonly used in backpropagation networks, in part because it is differentiable. The sigmoid transfer function takes the input, and scale the output into the range of 0 to 1, Fig. 3. Neural networks are good at fitting functions and recognizing patterns. In practice, there is proof that a fairly simple neural network can fit most complex functions.

For achieving more accurate results, then the following approaches have to be justified:

1. Reset the initial network weights and biases to new values and achieve the training again.
2. Increase the number neurons of hidden layer.
3. Increase the number of training samples.
4. Try a different training algorithm [28].

2.3   Mathematical Validation

In order to evaluate the performance of the prediction models, external test samples were used to validate the models developed by the testing set. The acknowledged statistical parameters are shown in Table 1: From Table 1, it is important to be noted that MSE (Mean Square Error) is a measure of how close a fitted line is to the data points. For every data point, we take the distance vertically from the point of the corresponding value on the curve t (the error)

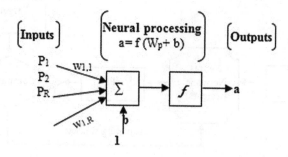

Fig. 3. Neural network structure

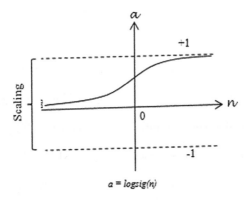

Fig. 4. Log-sigmoid transfer function

and square the value. Then we add up all those values for all data points and divide by the number of points minus two. The smaller the Mean Squared Error, the closer the t is to the data. The MSE has the units squared of whatever is plotted on the vertical axis, whereas, RMSE (root mean square error) gives the same weight to all errors, it penalizes variance as it gives errors with larger absolute values more weight than errors with smaller absolute values. RMSE is used as a standard statistical metric to measure model performance. Also, MAE is used to measure how close predictions are to the eventual outcomes, it is less sensitive to the occasional very large error because it does not square the errors in the calculation.

If RMSE is close to MAE, the model makes many relatively small errors while if RMSE is close to MAE2, the model makes few but large errors. However, MAPE is used for purposes of reporting, it is expressed in generic percentage terms which will make some kind of sense for validating the used predictor. Some other mathematical validation methods were used to validate how and which prediction algorithm is better than others like msre (mean squared relative error), mare (mean absolute relative error), rmsre (root mean squared relative error), mspe (mean squared percentage error), rmspe (root mean squared percentage error). Relative error methods give indications of how incorrect a quantity is from the true value, unlike absolute error methods which give indications about how much the approximately measured value varies from the true value.

2.4   Modelling and Implementation

Simulation experiments were implemented by MATLAB (2014b version) on a 1.80 GHz Core(TM) i7-4500U CPU personal computer with 8.0 G memory under Microsoft Windows 10 professional.

Table 1. Statistical parameters

Evaluation index	Formula
Mean squared error	$\text{MSE} = \frac{1}{n}\sum_{t=1}^{n} e_t^2$
Root mean squared error	$\text{RMSE} = \sqrt{\frac{1}{n}\sum_{t=1}^{n} e_t^2}$
MSRE (mean squared relative error)	$\text{MSRE} = \frac{1}{n}\sum_{1}^{n}\left\{\frac{Q_i - Q_j}{Q_j}\right\}^2$
MARE (mean absolute relative error)	$\text{MARE} = \frac{1}{n}\sum_{t=1}^{n}\frac{|w_o - w_p|}{w_o}$
RMSRE (root mean squared relative error)	$\text{RMSRE} = \sqrt{\frac{1}{n}\sum_{1}^{n}\left\{\frac{Q_i - Q_j}{Q_j}\right\}^2}$
MAE (Mean absolute error)	$\text{MAE} = \frac{1}{n}\sum_{t=1}^{n}|e_t|$
MAPE (Mean absolute percentage error)	$\text{MAPE} = \frac{100\%}{n}\sum_{t=1}^{n},\left|\frac{A_t - F_t}{A_t}\right|$
MSPE (mean squared percentage error)	$\text{RMSPE} = \frac{\sum_{j=1}^{V}\left(\frac{E_j}{A_j}\right)^2}{V}$
RMSPE (root mean squared percentage error)	$\text{RMSPE} = \sqrt{\frac{\sum_{j=1}^{V}\left(\frac{E_j}{A_j}\right)^2}{V}}$

2.5   Regression Parameters

This step is considered one of the most important in any regression model. In any machine learning model, there are usually one or two parameters that affect the performance of the model significantly; those parameters have to be justified carefully. The parameters in Table 2 were chosen carefully, so as to achieve the best regression performance. Tuning parameters can be such as the performance function in which stop the training of the network once it achieves the lowest value of it (MSE), it can be also the number of hidden layers which can also improve the training performance. The training function applies the perceptron learning rule in its pure form, in that individual inputs are applied individually, in sequence, and corrections to the weights and bias are made after each presentation of the inputs, the used training function in this work is Levenburg Marquardt algorithm. It was chosen because it is the fastest learning algorithms, however, it uses large amounts of memory.

3   Results and Discussions

Selenium plays an important role in activating antioxidant defence system as it forms selenocysteine, which is part of the active centre of the glutathione peroxidase [29]. Our model (based on BP-NN) was built to predict basic antioxidant enzymes activity and their dynamics in serum and liver tissue (SOD, CAT, GPX, HDL and LDL) based on variable Nano-Selenium concentrations in the diets.

Table 2. Backpropagation neural network parameters

Regression model	Parameters	value
Backpropagation neural network	Performance function	MSE (mean squared error)
	Number of neurons in the hidden layers	20
	Training function	Levenburg Marquardt algorithm

The initial weight (the weight of the fish before performing the experiment) and the final weight (the weight of the fish after performing the experiment) were incorporated with selenium Nanoparticles concentrations as variable features. Otherwise, the estimated water parameters (PH-chemical oxygen demand, biological oxygen demand and temperature) were expressed in the experimental data as xed features. In the training phase, the prediction models adapt their parameters based on matching between the variable inputs (Nano-Selenium concentrations, the initial and the final weight) and outputs (SOD, CAT, GPX, HDL and LDL concentrations). In the testing phase, we test our models efficiency by exposing them to new data. In order to build our prediction model, 80 % of our data was chosen randomly for training the prediction model and the rest (20 %) was used for testing the model performance.

3.1 Prediction of SOD (Superoxide dismutase) and GBX (Glutathione peroxidase enzyme) Antioxidant Activity

Superoxide dismutases (SODs) are enzymes that break down the superoxide anion to oxygen and hydrogen peroxide. These enzymes are synthesized not only in aerobic cells but also in extracellular fluids. Their activity depends on cofactors that depend on selenium particles.

Also, Nano selenium is an important Co-factor for selenoproteins. So, selenium intake lead to increase the selenoprotein expression, including GPx activity, in different tissues including liver due to Nano-selenium availability.

The main purpose of our model is to achieve better prediction accuracy as much as possible, this can be measured by comparing the output of our model with the actual values.

From Fig. 5, it can be seen that BP-NN achieved also most accurate prediction of GPX and SOD concentrations in most samples as the output of the model (the predicted values) are almost the same as the actual values. This gives an indication of how BP-network is a reliable predictor, by adjusting the number of neurons (to a specific value) in the hidden layer of the network so that the prediction accuracy could be improved, however, increasing the neurons than a specific limit can cause overfitting. The figure shows that the more selenium nanoparticles concentration the more enhance the activity of the enzyme, as selenium enter in the activation pathway of the enzyme. The activity of this

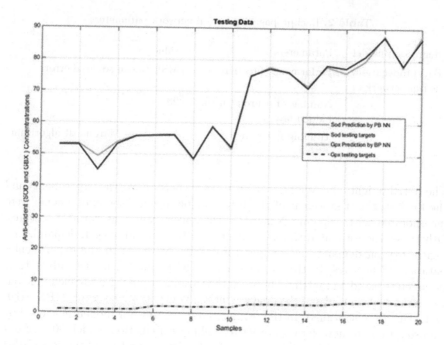

Fig. 5. The testing targets and the prediction of SOD and GBX antioxidants by BP-NN

enzyme was predicted [30]. Our prediction model is able to predict of GPXs and SODs concentrations depending on the dose of nano-selenite with fish weights (initial and final), Fig. 4. This study has attempted to describe quantitatively the nature of the relationship between Nano-selenium supplementation and GPx activity in liver tissue [12].

3.2 Prediction of CAT (Catalase) Antioxidant, HDL (High Density Lipoprotein) and LDL (Low Density Lipoprotein) Activity

Selenium nanoparticles enhance Catalases enzyme activity that catalyses the conversion of hydrogen peroxide to water and oxygen, using cofactors. This is found in peroxisomes in most eukaryotic cells. Its only substrate is hydrogen peroxide. Here, its cofactor is oxidised by one molecule of hydrogen peroxide and then regenerated by transferring the bound oxygen to the second molecule of substrate [31]. Also, Selenium nanoparticles enhance HDL conc. which has the ability to promote cholesterol efflux and protect against stressors. In addition, the anti-inflammatory properties of HDL are important as well. High concentration of Nano-Selenium enhances Apolipoprotein-AI, which is considered as the main antioxidant factor in HDL [32]. Increasing the Nano-selenium concentration inhibits Lipid peroxidation in LDL and help in lowering the LDL level (Bad Cholesterol). So, its concentration is considered a biomarker for antioxidant status in the fish [32].

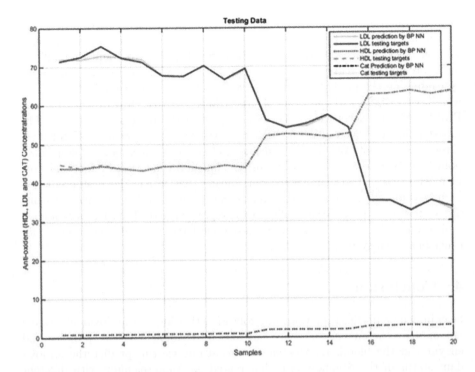

Fig. 6. The testing targets and the prediction of HDL, LDL and CAT antioxidants by BP-NN

Table 3. Comparison of anti-oxidants prediction errors of BP-Network by well-known statistical methods for evaluation.

Anti-oxidant	MSE	MAE	MAPE	MSRE	RMSE	MARE	RMSRE	MSPE	RMSPE
SOD	0.86	0.54	0.95	3.61	0.93	0.009	0.019	3.617	1.90
GBX	1.034	0.666	1.14	0.0003	1.017	0.011	0.019	3.71	1.92
CAT	**2.36 e05**	**0.003**	**0.300**	**1.54**	**0.004**	**0.003**	**0.003**	**0.154**	**0.392**
HDL	0.057	0.130	0.276	2.899	0.240	0.002	0.053	0.289	0.538
LDL	0.297	0.408	0.683	7.27	0.544	0.006	0.008	0.727	0.852

In this experiment, we aim at minimizing the error rate resulting from the predicted values (by our model) and the real values.

Figure 6 shows the proposed prediction model and its ability to predict of CATs, LDL and HDL concentrations depending on the dose of nano-selenite with fish weights (initial and nal).

From Fig. 6, when comparing the prediction line with the target line for HDL, LDL and CAT concentrations, it is obvious that BP-NN has the most accurate prediction of them in most samples, especially, for HDL and CAT, the two lines look identical. This gives an indication of how the proposed regression model is accurate and how the purpose of this work in achieved. Several experiments were achieved until we optimize the BP-network parameters and that reflects on improving the regression accuracy. Table 3 shows the prediction accuracy depending on the prediction error. As seen, some familiar statistical validation were used (MSE, MAE, MAPE, MSRE, RMSE, MARE, RMSRE, MSPE and RMSPE), each one of them reflects an aspect of the regression accuracy. The lowest number of each of them the highest the regression accuracy. Error measurements calculate of how a fitted line is close to the data points (every data point). As seen from Table 3, CAT anti-oxidant has the lowest prediction errors in most statistical measurements; this means that it has the best prediction accuracy among the other anti-oxidants. So, in most statistical measurements are a good indication of the quality of the regression model, it gives also a good indication of the reliability of the regression model and that it is appropriate for other biological systems.

4   Conclusion

Selenium nanoparticles play a vital role in activating the antioxidant system not only by the enzymes activity but also by forming bioactive proteins and enzymes in the blood. By the end of this study, we can predict the antioxidant status of the common carp that reared on Nano-selenium with different concentration through antioxidant enzymes, GPx SOD, HDL and LDL prediction. Clinically, supplementation of Nano-selenium in the diet at the rate of 1 mg/kg could enhance growth performance of common carp. Two mg Nano-selenium/kg diet showed the highest muscle and liver selenium concentration and consequently, the best antioxidant status in fish. So, the supplementation of selenium nanoparticles at 2 mg/kg concentration showed positive effects on the immune status of the carp. Choosing a suitable regression technique (BP-NN) as well as tuning its parameters accurately can lead to a good prediction model. The chosen regression algorithm shows accurate results in the testing phase, the achieved results were validated by well-known mathematical validators (MSE, RMSE, MAE, MAPE, MSRE, MARE, RMSRE, MSPE, RMSPE) which show numerically how reliability are the used predictors.

References

1. Rayman, M.P.: The importance of selenium to human health. Lancet **356**, 233–241 (2000)
2. Wilhelm-Filho, D., Torres, M.A., Tribbes, T.B., Pedrosa, R.C., Soares, C.H.L.: nfluence of season and pollution on the antioxidant defenses of the cichlid fish acara (Geophagus brasiliensis). Braz. J. Med. Biol. Res. **34**, 719–726 (2001)
3. Peng, D., Zhang, J., Liu, Q., Taylor, E.W.: Size effect of elemental selenium nanoparticles (Nano-Se) at supra nutritional levels on selenium accumulation and glutathione S-transferase activity. J. Inorg. Biochem. **101**, 1457–1463 (2007)

4. Wang, K.Y., Peng, C.Z., Huang, J.L.: The pathology of selenium deciency in Cyprinus carpio. J. Fish Dis. **36**, 609–615 (2013)
5. Rumelhart, D.E., McClelland, J. (eds.): Parallel Distributed Processing, vol. 1. Massachusetts Institute of Technology Press, Cambridge (1986). Explorations in the Microstruciure of Cognition, vol. I and II. M.I.T. Press
6. Watanabe, T., Kiron, V., Datoh, S.: Trace minerals in fish nutrition. Aquaculture **151**, 185–207 (1997)
7. Perlman, Z.E., Slack, M.D., Feng, Y., Mitchison, T.J., Wu, L.F., Altschuler, S.J.: Multidimensional drug profiling by automated microscopy. Science **306**, 1194–1198 (2004)
8. Slack, M.D., Martinez, E.D., Wu, L.F., Altschuler, S.J.: Characterizing heterogeneous cellular responses to perturbations. Proc. Natl. Acad. Sci. USA **105**, 19306–19311 (2008)
9. Loo, L.H., Lin, H.J., Singh, D.K., Lyons, K.M., Altschuler, S.J., Wu, L.F.: Heterogeneity in the physiological states and pharmacological responses of differentiating 3T3-L1 preadipocytes. J. Cell Biol. **187**, 375–384 (2009)
10. Castoreno, A.B., Smurnyy, Y., Torres, A.D., Vokes, M.S., Jones, T.R., Carpenter, A.E., Eggert, U.S.: Small molecules discovered in a pathway screen target the Rho pathway in cytokinesis. Nat. Chem. Biol. **6**, 457–463 (2010)
11. Murphy, R.F.: An active role for machine learning in drug development. Nat. Chem. Biol. **7**, 327–330 (2011)
12. Kittler, R., Putz, G., Pelletier, L., Poser, I., Heninger, A.K., Drechsel, D., Fischer, S., Konstantinova, I., Habermann, B., Grabner, H., et al.: An endoribonuclease-prepared siRNA screen in human cells identifies genes essential for cell division. Nature **432**, 1036–1040 (2004)
13. Lansing Taylor, D., Haskins, J.R., Giuliano, K.A.: High Content Screening. Humana Press, Totowa (2007)
14. Doil, C., Mailand, N., Bekker-Jensen, S., Menard, P., Larsen, D.H., Pepperkok, R., Ellenberg, J., Panier, S., Durocher, D., Bartek, J., et al.: RNF168 binds and amplifies ubiquitin conjugates on damaged chromosomes to allow accumulation of repair proteins. Cell **136**, 435–446 (2009)
15. Collinet, C., Stter, M., Bradshaw, C.R., Samusik, N., Rink, J.C., Kenski, D., Habermann, B., Buchholz, F., Henschel, R., Mueller, M.S., et al.: Systems survey of endocytosis by multiparametric image analysis. Nature **464**, 243–249 (2010)
16. Fuchs, F., Pau, G., Kranz, D., Sklyar, O., Budjan, C., Steinbrink, S., Horn, T., Pedal, A., Huber, W., Boutros, M.: Clustering phenotype populations by genome-wide RNAi and multiparametric imaging. Mol. Syst. Biol. **6**, 370 (2010)
17. Neumann, B., Walter, T., Hrich, J.K., Bulkescher, J., Erfle, H., Conrad, C., Rogers, P., Poser, I., Held, M., Liebel, U., et al.: Phenotypic profiling of the human genome by time-lapse microscopy reveals cell division genes. Nature **464**, 721–727 (2010)
18. Schmitz, M.H.A., Held, M., Janssens, V., Hutchins, J.R.A., Hudecz, O., Ivanova, E., Goris, J., Trinkle-Mulcahy, L., Lamond, A.I., Poser, I., et al.: Live-cell imaging RNAi screen identifies PP2A-B55alpha and importin-beta1 as key mitotic exit regulators in human cells. Nat. Cell Biol. **12**, 886–893 (2010)
19. Mercer, J., Snijder, B., Sacher, R., Burkard, C., Bleck, C.K., Stahlberg, H., Pelkmans, L., Helenius, A.: RNAi screening reveals proteasome- and Cullin3-dependent stages in vaccinia virus infection. Cell Reports **2**, 1036–1047 (2012)
20. Yang, Z.R., Chou, K.C.: Bio-support vector machines for computational proteomics. Bioinformatics **20**, 735–741 (2004)
21. Datta, S., Pihur, V.: Feature selection and machine learning with mass spectrometry data. Methods Mol. Biol. **593**, 205–229 (2010)

22. Reiter, L., Rinner, O., Picotti, P., Httenhain, R., Beck, M., Brusniak, M.Y., Hengartner, M.O., Aebersold, R.: mProphet: automated data processing and statistical validation for large-scale SRM experiments. Nat. Methods **8**, 430–435 (2011)

23. Castelo, R., Guig, R.: Splice site identification by idlBNs. Bioinformatics **20**(Suppl), 69–76 (2004)

24. Ben-Hur, A., Ong, C.S., Sonnenburg, S., Schlkopf, B., Rtsch, G.: Support vector machines and kernels for computational biology. PLOS Comput. Biol. **4**, e1000173 (2008)

25. McCord, J.M., Fridovich, I.: Superoxide dismutase an enzymatic function for erythrocuprein (hemocuprein). J. Biol. Chem. **244**(22), 6049–6055 (1969)

26. Noguchi, T., Cantor, A.H., Scott, M.L.: Mode of action of selenium and vitamin E in prevention of exudative diathesis in chicks. J. Nutr. **103**, 1502–1511 (1973)

27. Abei, H.: Catalase in vitro. Methods Enzymol. **272**, 121–126 (1984)

28. Beale, M.H., Hagan, T., Demuth, B.: Neural Network Toolbox 7, Users Guide. The MathWorks Inc., Natick (2010)

29. Kohrle, J., Brigelius-Floh, R., Bick, A., Grtner, R., Meyer, O., Floh, L.: Selenium in biology: facts and medical perspectives. J. Biol. Chem. **381**, 849–864 (2000)

30. Hao, X., Ling, Q., Hong, F.: Effects of dietary selenium on the pathological changes and oxidative stress in loach (Paramisgurnus dabryanus). Fish Physiol. Biochem. **40**, 1313–1323 (2014)

31. Atencio, L., Moreno, I., Prieto, A.I., Moyano, R., Blanco, A., Camen, A.M.: Effects of dietary selenium on the oxidative stress and pathological changes in tilapia (Oreochromis niloticus) exposed to a microcystin-producing cyanobacterial water bloom. Toxicon **53**, 269–282 (2009)

32. Podrez, E.A.: Antioxidant properties of highdensity lipoproetein and atherosclerosis. Clin. Exp. Pharmacol. Physiol **37**, 719–725 (2010)

A Hybrid Krill-ANFIS Model for Wind Speed Forecasting

Khaled Ahmed[1,7], Ahmed A. Ewees[2,7($\boxtimes$)], Mohamed Abd El Aziz[1,3,7],
Aboul Ella Hassanien[4,7], Tarek Gaber[5,7], Pei-Wei Tsai[6], and Jeng-Shyang Pan[6]

[1] Faculty of Computers and Information, Cairo University, Giza, Egypt
[2] Department of Computer, Damietta University, Damietta, Egypt
ahmed.abdelghany@egyptscience.net
[3] Faculty of Science, Zagazig University, Zagazig, Egypt
[4] Faculty of Computer Science, Nahda University, Beni-suef, Egypt
[5] Faculty of Computers and Informatics, Suez Canal University, Ismailia, Egypt
[6] Fujian Provincial Key Laboratory of Big Data Mining and Applications,
Fujian University of Technology, Fuzhou, Taiwan
[7] Scientific Research Group in Egypt (SRGE), Cairo, Egypt
http://www.egyptscience.net

Abstract. Finding an alternative renewable energy source instead of using traditional energy such as electricity or gas is an important research trend and challenge. This paper presents a new hybrid algorithm that uses Krill Herd (KH) optimization algorithm and Adaptive Neuro-Fuzzy Inference System (ANFIS) to be able to fit for wind speed forecasting, which is an essential step to generate wind power. ANFIS's parameters are optimized using KH. The proposed model called (Krill-ANFIS). This model is compared with three models basic ANFIS, PSO-ANFIS, and GA-ANFIS. Krill-ANFIS proved that it can be used as an efficient predictor for the wind speed as well as it can achieve high results and performance measures of root mean square error (RMSE), Coefficient of determination R^2 and average absolute percent relative error (AAPRE).

Keywords: Renewable energy · Adaptive Neuro-Fuzzy Inference System · Wind speed forecasting · Krill herd optimization · Hybrid model

1 Introduction

The repaid increase in human activities presents an urgent need for clean, green, and cheap energy source, so searching for an alternative solution instead of a transitional solution is an important task [1]. Renewable energy is generated based on nature resource such as wind, biochar, biomass, solar, and rain [2]. Wind power is used as renewable energy source in many countries such as Portugal [3], Egypt [4] and USA [5] and proved its quality as an alternative energy source in wind farms to be used in water quality projects [6] or on generating electricity [7].

© Springer International Publishing AG 2017
A.E. Hassanien et al. (eds.), *Proceedings of the International Conference on Advanced Intelligent Systems and Informatics 2016*, Advances in Intelligent Systems and Computing 533, DOI 10.1007/978-3-319-48308-5_35

Forecasting wind speed is an essential step to generate wind power which can lead to solve many gaps and challenges such as predicting the amount of generated electricity, the required number of turbine or number of pumps for pushing water for grains and the connection between these turbines or pumps and managing failure points [8]. Traditional models which used for wind forecasting were a physical model or statistical model. Recently there are a lot of studies which deal with handling wind speed forecasting based on traditional data Mining models based on fuzzy logic, Adaptive Neuro-Fuzzy Inference System (ANFIS) or hybrid swarms model for this task.

ANFIS is a hybrid approach based on two concepts, the first one is neural network and the second is fuzzy logic. There a lot of studies which used ANFIS for renewable energy prediction as in solar [9], wind [8], and biochar [10]. Swarms optimization [11,12] are algorithms which inspired from intelligence swarms in nature that based on objective quality function and search for food or facing an enemy this can handle many research gaps such as renewable energy prediction tasks.

Krill Herd (KH) [13] is a swarm algorithm which basically has three phases induced motion, foraging motion, and random diffusion which simulates the krill herd in nature while searching for food. KH algorithm can be used for optimizing ANFIS as a hybrid model to be used for the forecasting task.

This paper presents a wind speed forecasting model. It based on hybrid KH optimization algorithm with ANFIS (called Krill-ANFIS) and used to forecast Egyptian wind speed as a case study.

The rest of this paper is arranged as follows, Sect. 2 presents basic ANFIS. Section 3 presents Krill herd Optimization algorithm. Section 4 introduced the proposed model. Section 5 discusses experimental results; and finally, conclusions and future work are shown in Sect. 6.

2   Basic ANFIS Model

Adaptive Neuro Fuzzy Inference (ANFIS) [14] is based on Takagi-Sugeno-Kang model which is using a set of IF-Then to generate fuzzy rules nonlinear mapping from its input space to the output space. The first order Sugeno type fuzzy model can be given by:
Rule(1) IF x is

$$A_1 \ AND \ y \ is \ B_1, \ THEN \ f_1 = p_1 x + q_1 y + r_1 \tag{1}$$

Rule(2) IF x is

$$A_2 \ AND \ y \ is \ B_2, \ THEN \ f_2 = p_2 x + q_2 y + r_2 \tag{2}$$

where A_1, A_2 and B_2, B_2 are the member functions for x and y respectively. p_1, q_1, r_1 and p_2, q_2, r_2 are the associated parameters of the output functions that are defined in the training phase.

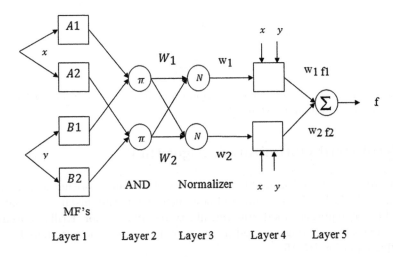

Fig. 1. The structure of ANFIS model.

The ANFIS model consists of five layers as illustrated in Fig. 1. In the first layer: The output of each node is calculated by:

$$O_{1i} = \mu_{1i}(X), i = 1, 2, O_{2i} = \mu_{B_{1-2}}(Y), i = 1, 2 \tag{3}$$

where x, y are the crisp input to node i, and A_i, B_i are the membership values of the membership functions μ_A and μ_B, respectively. $\mu_{A_i}(x)$ and $\mu_{B_{i-2}}(y)$ are the generalized Gaussian membership function that defined by:

$$O_{1i} = \mu_{1i}(X), i = 1, 2, O_{2i} = \mu_{B_{1-2}}(Y), i = 1, 2 \tag{4}$$

where X, Y are the crisp input to node i, and A_i, B_i are the membership values of the membership functions μ_A and μ_B, respectively. $\mu_{A_i}(x)$ and $\mu_{B_{i-2}}(y)$ are the generalized Gaussian membership function that defined by:

$$\mu(x) = e^{-(x - P^i / \sigma_i)^2} \tag{5}$$

where σ_i and P^i are the premise parameters set, which represent the standard deviation and mean of data. In the second layer, each node computes the firing strength of a rule as:

$$O_{2i} = \mu_{Ai}(X) * \mu_{B_{1-2}}(Y) \tag{6}$$

In layer three, each node computes the ratio of the ith rule's firing strength to the sum of all rules' firing strengths (i.e. normalized firing strength) as:

$$O_{3i} = \varpi_i = \omega_i / \Sigma \omega_i \tag{7}$$

The node in layer four is an adaptive node and its output is computed as:

$$O_{4,i} = \varpi_i f_i = \varpi_i (p_i x + q_i y + r_i) \tag{8}$$

where p_i, q_i and r_i is the parameter set (the consequent parameters) of the node. w_i referred to as the normalized firing strengths.

$$O_5 = \sum \varpi_i F_i \tag{9}$$

In the last layer 5, there is a single node and it is output is defined as (the summation of all incoming signals).

3   Krill Herd Optimization Algorithm

Krill Herd (KH) optimization algorithm is a new swarm algorithm which represents how the krills life in nature and how their own nature inspired intelligence is used for searching for food. Figure 2 illustrates its behavior. Krill's movement Dx in time Dt searching for food in the search space can be formulated as the following equation [13,15]:

$$\frac{Dx}{Dt} = Ni + Fi + Di \tag{10}$$

where Ni is induced motion, Fi is foraging motion and DI is random diffusion.

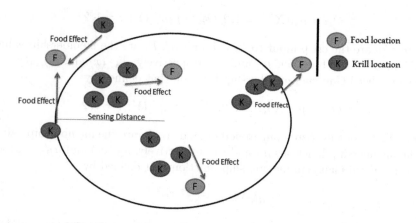

Fig. 2. Krill herd behavior.

Induced motion (Ni) is defined as:

$$N_i^{New} = N^{Max}\alpha_i + \omega N_i^{New} \tag{11}$$

where N^{Max} maximum induced speed, ω the weight of individual krill, and α_i value is equal to the summation of results Eq. 12 and the results of Eq. 13.

$$\alpha_{local} = \Sigma_{j=1}^{N} K_{ij} X_{ij} \tag{12}$$

where X is krill's related position and K is objective fitness value.

$$\alpha_{target} = C^{Best} K_{ibest} X_{ibest} \tag{13}$$

where i iteration number and Best is related $Nbest$ position or related best fitness value.

$$C^{Best} = 2(Rand + \frac{I}{Imax}) \tag{14}$$

where $Rand$ is a value from 0 to 1 randomly generated, I is the initial number of iteration and $Imax$ represent maximum iteration number.

Foraging behavior (Fi):

$$F_i = V_f B_i + \omega F_{old} \tag{15}$$

where F_i is the foraging behavior V_f is the speed of krill ω is weight of krill and F_{old} is last foraging motion.

$$B_i = B_{food} + B_{best} \tag{16}$$

where B_{food} is food effect on the krill and B_{best} is the best fitness.

Random diffusion (Di):

$$D_i = D_{max}(1 - \frac{I}{Imax})\delta \tag{17}$$

where D_i is random diffusion and D_{max} is maximum speed, I is the initial number of iterations and $Imax$ represent maximum number of iterations.

4   The Proposed Model

In this section, the proposed model is introduced, in which it is based on the ANFIS and the KH algorithm (called Krill-ANFIS). The overall model is illustrated in Fig. 3. The first step in the proposed model is to divide data into training set and testing set, then the second step is to learn the ANFIS based on a training set. In this step the parameters of ANFIS are updated by using the KH algorithm, that starts by generating the positions of the population. For each krill the fitness function is computed and then the three Krill's movements are used. The position of each krill is then updated after using genetic operators. The krill still repeating these steps until the stop condition is stratified. The position of each krill is then passed again to ANFIS as parameters. The next step in the proposed model is to input the testing set and forecast the wind speed, finally, the performance of the proposed model is computed.

5   Experiments and Results

The experimental data consists of 1500 wind instances which are collected from Egypt, each instant has five parameters (low temperature, out temperature,

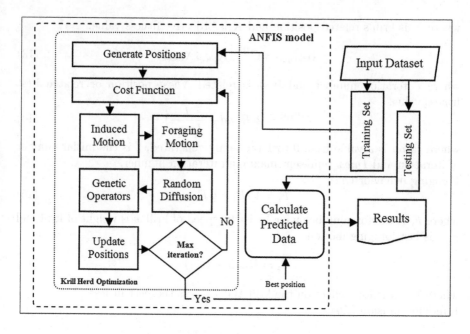

Fig. 3. Krill-ANFIS proposed model.

humidity, rain index, and wind speed). These parameters are divided into input data (the first four parameters) and output data (the wind speed parameter).

In the experiments, the data was divided randomly into 70 % samples for training set and 30 % for testing set. The parameters values which used for this experiment are set to 25 for the size of the population and the maximum number of iteration is 100, while the experiment where executed using "Matlab" over "Windows 10".

The performance of Krill-ANFIS is calculated by Root Mean Square Error (RMSE), Average Absolute Percent Relative Error (AAPRE), and Coefficient of Determination (R^2) as in Eqs. (18), (19), and (20) respectively.

$$RMSE = \sqrt{\frac{1}{n} \Sigma_1^n (y_i - Y_i)^2)} \tag{18}$$

$$AAPRE = \frac{100}{N} (\sum_{i=1}^{N} |\frac{(x_i - y_i)}{y_i}|) \tag{19}$$

$$R^2 = 1 - \sum_{i=1}^{n} \frac{(y_i - x_i)}{(y_i - \bar{y}_i)} \tag{20}$$

The results of Krill-ANFIS were compared against three models, namely, basic ANFIS [16], PSO-ANFIS [8] and GA-ANFIS [17], and proved good results. So, using KH optimization algorithm to learn the parameters of ANFIS lead to

Table 1. The quality measures average of 19 runs.

Model/measure	RMSE	AAPRE	R2
Krill-ANFIS	0.3617	1.6294	0.99
ANFIS	0.5442	1.6147	0.98
PSO-ANFIS	0.3723	1.6262	0.99
GA-ANFIS	0.3736	1.5843	0.99

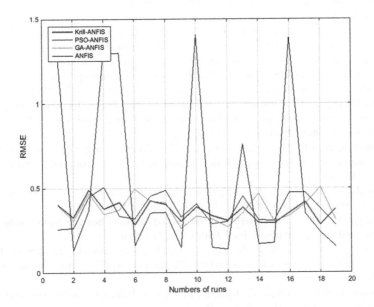

Fig. 4. The quality measures of Krill-ANFIS and other models in RMSE.

improve the performance of it. Table 1 shows the average of the results of 19 runs.

The quality measures of Krill-ANFIS in term of RMSE is illustrated in Fig. 4, and it shows best results in different runs.

6   Conclusion and Future Work

This paper introduces an alternative efficient renewable energy predictor (Krill-Anfis). Krill-Anfis is a hybrid of Adaptive Neuro-Fuzzy Inference System (ANFIS) and Krill Herd (KH) optimization algorithm. KH optimization algorithm is used for optimizing the ANFIS. This model was applied to forecast the wind speed of Egypt. Krill-ANFIS achieved higher results than basic ANFIS, PSO-ANFIS, and GA-ANFIS. In Future, we will update the proposed model using chaotic krill instead of krill herd, and present comparative study about hybrid ANFIS and swarms in renewable energy prediction.

References

1. Everett, R., Boyle, G., Peake, S., Ramage, J.: Energy Systems and Sustainability: Power for a Sustainable Future. Oxford Univerity Press, Oxford (2012)
2. Twidell, J., Weir, T.: Renewable Energy Resources. Routledge, London (2015)
3. Söder, L., Abildgaard, H., Estanqueiro, A., Hamon, C., Holttinen, H., Lannoye, E., Gómez-Lázaro, E., O'Malley, M., Zimmermann, U.: Experience and challenges with short-term balancing in european systems with large share of wind power. IEEE Trans. Sustain. Energy **3**(4), 853–861 (2012)
4. Ibrahim, A.: Renewable energy sources in the egyptian electricity market: a review. Renew. Sustain. Energy Rev. **16**(1), 216–230 (2012)
5. Foley, A.M., Leahy, P.G., Marvuglia, A., McKeogh, E.J.: Current methods and advances in forecasting of wind power generation. Renew. Energy **37**(1), 1–8 (2012)
6. Peñate, B., García-Rodríguez, L.: Current trends and future prospects in the design of seawater reverse osmosis desalination technology. Desalination **284**, 1–8 (2012)
7. Díaz-González, F., Sumper, A., Gomis-Bellmunt, O., Villafáfila-Robles, R.: A review of energy storage technologies for wind power applications. Renew. Sustain. Energy Rev. **16**(4), 2154–2171 (2012)
8. Pousinho, H.M.I., Mendes, V.M.F., Catalão, J.P.S.: A hybrid pso-anfis approach for short-term wind power prediction in portugal. Energy Convers. Manage. **52**(1), 397–402 (2011)
9. Khatib, T., Mohamed, A., Sopian, K.: A review of solar energy modeling techniques. Renew. Sustain. Energy Rev. **16**(5), 2864–2869 (2012)
10. Akkaya, E.: Anfis based prediction model for biomass heating value using proximate analysis components. Fuel **180**, 687–693 (2016)
11. Yang, X.-S., Cui, Z., Xiao, R., Gandomi, A.H., Karamanoglu, M.: Swarm Intelligence and Bio-inspired Computation: Theory and Applications. Newnes, Oxford (2013)
12. Hassanien, A.E., Alamry, E.: Swarm Intelligence: Principles, Advances, and Applications. CRC Press, Boca Raton (2015)
13. Gandomi, A.H., Alvai, A.H.: Krill herd: a new bio-inspired optimization algorithm. Commun. Nonlinear Sci. Numer. Simul. **17**(12), 4831–4845 (2012)
14. He, Z., Wen, X., Liu, H., Du, J.: A comparative study of artificial neural network, adaptive neuro fuzzy inference system and support vector machine for forecasting river flow in the semiarid mountain region. J. Hydrol. **509**, 379–386 (2014)
15. Ahmed, K., Hafez, A.I., Hassanien, A.E.: A discrete krill herd optimization algorithm for community detection. In: 2015 11th International Computer Engineering Conference (ICENCO), pp. 297–302. IEEE (2015)
16. Boyacioglu, M.A., Avci, D.: An adaptive network-based fuzzy inference system (anfis) for the prediction of stock market return: The case of the istanbul stock exchange. Expert Syst. Appl. **37**(12), 7908–7912 (2010)
17. Ghose, D.K., Panda, S.S., Swain, P.C.: Prediction and optimization of runoff via ANFIS and GA. Alexandria Eng. J. **52**(2), 209–220 (2013)

Optimize BpNN Using New Breeder Genetic Algorithm

Maytham Alabbas[1], Sardar Jaf[2(✉)], and Abdul-Hussein M. Abdullah[3]

[1] Department of Computer Science, College of CS and IT,
University of Basrah, Basrah, Iraq
ma@alumni.manchester.ac.uk
[2] School of Engineering and Computing Sciences,
Durham University, Durham, UK
sardar.jaf@durham.ac.uk
[3] Department of Computer Science, College of Science,
University of Basrah, Basrah, Iraq
abdo60_2004@yahoo.com

Abstract. In this paper, the ability of genetic algorithms in designing artificial neural network (ANN) is investigated. The multi-layer network (MLN) is taken into account as the ANN structure to be optimized. The idea presented here is to use the genetic algorithms to yield contemporaneously the optimization of: (1) the design of NN architecture in terms of number of hidden layers and of number of neurons in each layer; and (2) the choice of the best parameters (learning rate, momentum term, activation functions, and order of training patterns) for the effective solution of the actual problem to be faced. The back-propagation (BP) algorithm, which is one of the best-known training methods for ANNs, is used. To verify the efficiency of the current scheme, a new version of the breeder genetic algorithm (NBGA) is proposed and used for the automatic synthesis of NN. Finally, several problems of the experiment were taken and the results show that the back-propagation neural network (BpNN) classifier improved the current scheme has higher accuracy of classification and greater gradient of convergence than other classifiers, which have been proposed in the literature.

Keywords: Breeder genetic algorithm · Back-propagation network · Artificial neural network · Multi-layer neural network

1 Introduction

The back-propagation (BP) algorithm, despite having proved useful in a number of problems, still presents a certain range of difficulties as a network structure, convergence, calculation time, and teaching method.

There is neither theoretical result nor even a satisfactory empirical rule suggesting how a network should be dimensioned to solve a particular problem. Should the network use one hidden layer or more? How many neurons should there be on the hidden layer? What is the relationship between the number of training examples and the number of classes to separate these examples into? What should be the overall size of the network?

© Springer International Publishing AG 2017
A.E. Hassanien et al. (eds.), *Proceedings of the International Conference on Advanced Intelligent Systems and Informatics 2016*, Advances in Intelligent Systems and Computing 533, DOI 10.1007/978-3-319-48308-5_36

Several researchers used guesswork or trial and error to determine the number of layers, number of neurons in each layer, and the connection neurons for a certain problem. Also, many of the network parameters (learning factor, momentum factor… etc.) were determined empirically. This approach needs a long time to obtain good results, so genetic algorithms are used for solving these difficulties.

The rest of this paper is organized as follows: Sect. 2 will describe the breeder genetic algorithm and a brief explanation of the neural network will be given in Sect. 3. Section 4 explains the current scheme and in Sect. 5 results of our experiments are compared with those obtained by other researchers' results for the same problems through computer simulations. Finally, Sect. 6 presents the conclusion of this paper.

2   Breeder Genetic Algorithm

The breeder genetic algorithm (BGA) was designed by Heinz Mühlenbein in Germany at the beginning of the 1990s. It lies somehow in between the genetic algorithms (GA) and evolution strategies (ESs), in the sense that they borrow from each of them some basic ideas. The basic scheme for BGA is illustrated in Fig. 1 [1].

Procedure BGA;
  Begin
    t=0;
    Initialize randomly P(t) with N individuals;
    While (termination criterion not fulfilled) do
        Evaluate goodness of each individual;
        Save the best individual in the new population;
        Select the best T% individuals;
        For I= 1 to N-1 do
          Select randomly two elements within the best T% in P(t);
          Recombine them so as to obtain one offspring;
          Perform mutation on the offspring;
          Insert it in P'(t);
        End For;
        P(t+1)=P'(t);
        t=t+1;
        Update variable for termination;
      End while;
  End;

Fig. 1. Breeder genetic algorithm (BGA).

3   Back-Propagation Neural Network

Neural Networks (NNs) are algorithms for optimization and learning based loosely on concepts inspired by researches conducted on the nature of the brain [2]. An NN is simply a set of interconnected individual computation elements called neurons. In the case of the multi-layer neural networks (MLNs), the neurons are arranged in a series of layers. A layer is usually a group of neurons, each of which is connected to all neurons in the adjacent layer [3].

The back-propagation neural network (BpNN) is MLN. During the learning phase; input patterns are presented to the network in some sequence. Each training pattern is propagated forward layer by layer until an output pattern is computed. The computed output is then compared to the desired one and an error value is determined. The errors are used as input to feedback connections from which the adjustment is made to the synaptic weights layer by layer in a backward direction. The backward linkages are used only for the learning phase, whereas the forward connections are used for both the learning and the operational phases [4].

4 The Current Approach

The basic idea is to provide an automatic technique to define the most appropriate NN structure for a given problem. The optimization of MLN, which must be trained to solve a problem, is characterized by the need to determine: the architecture (the number of hidden layers (NHL) and number of neurons (HN_i) for each layer i), the activation function (Lf_i) to be used for each layer i, the momentum term (α), the initial temperature (t_0), the initial learning term (η_0), the bias cell (b), and the order of training patterns that are being presented during training ($PatOrd$). Choosing a suitable initial value for these parameters is a fundamental decision faced by all NN's users, but it is a problem. Often, the choice of these parameters can have a significant impact on the effectiveness of the NN. The choice needs to be tuned for efficiency.

New breeder genetic algorithm (NBGA) has been used to determine the appropriate set of parameters listed above. NBGA is an updated version of BGA (see Fig. 2) which is characterized by the following properties compared with BGA:

1. Its ability to determine the population size depending on the value of T, which represents the best individuals which are selected.
2. It has an efficient selection method, which prevents the repetition of the selected parents, and this implies to increase the diversity of the population.
3. It is capable of extending the population size by the Eq. 1. to move the best k individuals to the next generation instead of the best individual only.

$$N = \frac{T(T-1)}{2} + k, 1 \leq k \leq T, \tag{1}$$

4. Its ability to manipulate chromosomes with different length, which ensures solving different problems.

- Encoding

The NN is defined by "genetic encoding" in which the genotype is the encoding of the different characteristics of MLN and the phenotype is the MLN itself.

```
Procedure NBGA;
  Begin
    t=0;
    Input T & k; // where 1≤k≤T
```
$$N = \frac{T(T-1)}{2} + k$$
```
    Initialize randomly P(t) with N individuals;
    While (termination P(t) criterion not fulfilled) do
        Evaluate goodness of each individual;
        Save the best  k  individual(s) in the new population;
        Select the best  T  individuals;
        For i=1 to T-1 do
          For j=i+1 to T do
            Recombine chromosome i  & j in P(t) to obtain one offspring;
            Perform mutation on the offspring;
            Insert it in P' (t);
          End For;
        End For;
        P(t+1)=P' (t)
        t=t+1;
        Update variable for termination;
    End While;
  End
```

Fig. 2. New breeder genetic algorithm (NBGA).

The NBGA considered the chromosome structure $C = \{G_1,...,G_8\}$ as reported in Fig. 3.

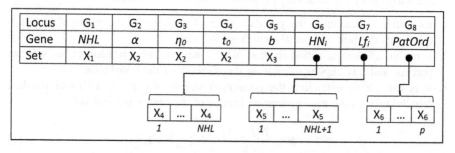

Fig. 3. The hierarchical chromosome representing an NN.

The loci are defined within the following subsets:

$X_1 \in \{1, ..., 4\}$

$X_2 \in (0, 1)$

$X_3 \in \{0 = \text{without bias}, 1 = \text{with bias}\}$

$X_4 \in \{1, .., M\}$, where M: maximum number of neurons.

$X_5 \in \{1 \equiv f_1, 2 \equiv f_2, 3 \equiv f_3, 4 \equiv f_4\}$, where:

Sigmoid function (f_1):

$$f_1(x) = \frac{1}{1 + e^{-kx}}, k > 0 \tag{2}$$

Tanh function (f_2):

$$f_2(x) = \frac{e^{\beta x} - e^{-\beta x}}{e^{\beta x} + e^{-\beta x}} \tag{3}$$

Hyperbolic Tanh function (f_3):

$$f_3(x) = \frac{1 - e^{-\beta x}}{1 + e^{-\beta x}} \tag{4}$$

Semi-linear function (f_4):

$$f_4(x) = \begin{cases} 1 & if\ x > 1 \\ x & if -1 \leq x \leq 1 \\ -1 & if x < -1 \end{cases} \tag{5}$$

X_6 = the set $\{1, .., p\}$ in permutation form, where p: number of training patterns.

- The fitness function

To evaluate the goodness of an individual, the network is trained with a fixed number of patterns and then evaluated according to Eq. 6 to determine parameters. The following function is used:

$$f = k_1 E + k_2 \frac{ep}{epmx} + k_3 \sum_{i=1}^{NHL} \frac{HN_i}{M} \tag{6}$$

Where,
E: (Mean Square Error) the error result from BpNN,
ep: current epoch,
$epmx$: maximum epochs,
NHL: number of hidden layers,
HN_i: number of neurons in hidden layer i,
M: maximum neuron,
$k_1, k_2, k_3 \in (0, 1)$ and $k_1 + k_2 + k_3 = 1$.

- Crossover

The uniform crossover (UX) [5] is used on the genes $(NHL, \alpha, \eta_0, t_0)$ because of its ability to yield an offspring more different from his parents and this implies to population variety. Depending on the value of NHL, which is inherited from the offspring, the UX used for NH and Lf genes in the case where NHL of parents is different. There are two cases, as shown in Fig. 4.

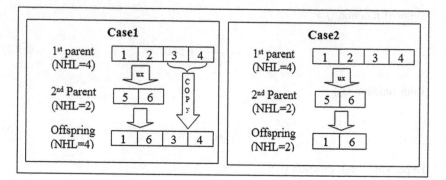

Fig. 4. UX for different chromosome length.

For the last part (*PatOrd*), multi-cycle crossover (MCX) is used because of its capability to prevent the repetition of a certain value for more than one gene. MCX is an updated version of cycle crossover (CX) [6] that builds offspring from ordered individuals by identifying cycles between two parents. To form the offspring, the cycles are copied from the respective parents. The basic scheme for MCX is explained in Fig. 5.

```
Procedure MCX(P₁,P₂);
  Begin
    i=1;
    j=2;
    C₁ & C₂ are empty chromosomes
    While (child chromosome has empty position) do
      x=random position from empty positions in the child
      While (C₁(x) is empty) do
        C₁(x) = Pi(x)
        C₂(x) = Pj(x)
        x = the position of the gene Pj(x) in Pi(x)
      End while;
      Swap(i,j)
    End while;
  Return C₁&C₂;
```

Fig. 5. Multi-cycle crossover (MCX).

A chromosome structure (genotype) for instance can look like the chromosome in Fig. 6.

• Mutation

A certain value is added (or deleted) from the genes (α, η_0, t_0) in certain probability in order to keep their values in the range (0, 1). If one of these values exceeds the range, the operations repeated until it becomes at the proper value. In addition, the mutation

NHL	α	η_0	t_0	b	HN_1	HN_2	Lf_1	Lf_2	Lf_3	PatOrd			
2	0.5	0.9	0.7	0	4	4	1	3	1	2	3	1	4

Fig. 6. A chromosome structure (genotype) for the current technique.

operator 1 m [5] is used for the genes (Lf, HN) because each of (Lf, HN) has minimum and maximum value. For the part (PatOrd), the mutation operator 2 m [5] is used to prevent presenting training pattern more than once to the network. The gene (NHL) is left unchanged, because any change on this gene will change the value of Lf and HN.

5 Results

In order to check the effectiveness of the current technique, the following problems were chosen because they allow us to compare our results with the previously published results. The tables below show the strength of the current scheme compared with the previous works. The following parameters were selected for each test:

- Maximum number of hidden neurons (M) = 5,
- Number of maximum learning trails (epmx) = 500,
- Number of genetic trails = 2000,
- Accepted error = 0.025,
- A number of best chromosomes move to the new population (k) = 1.

5.1 Artificial Problem

- Bit parity Problem

    Input cell = 6, output cell = 1, T = 7, number of patterns = 64. The result is shown in Table 1.

Table 1. The result of the current scheme compared with other works for 6 bit-Parity problem

Method	Purpose	Pop. size	No. of Gen. Trails	Structure (# connections)	No. of Learning Trail	Error
Yu [7]	Train	Not Reported	Not Reported	6-8-5-1 (49)	4000	0.075
Najim [8]	Design	100	300	6-7-1 (39)	4000	0.526
Al-Fadhly [9]	Design	25	215	6-7-1 (49)	319	0.053
Current Scheme	Train Design	22	250	6-6-1 (49)	1823	0.023

The parameters are: $\alpha = 0.6$, $\eta_0 = 0.429$, $t_0 = 0.29$, $b = 1$. Activation function: Tanh, Sigmoid. Pattern order: 1, 2, 3, 55, 51, 26, 7, 8, 9, 10, 11, 12, 13, 14, 15, 16, 17, 6, 32, 20, 21, 22, 23, 24, 25, 18, 27, 38, 29, 30, 31, 19, 48, 34, 35, 36, 37, 28, 39, 40, 45, 42, 43, 44, 41, 46, 47, 33, 49, 50, 5, 52, 53, 58, 4, 56, 57, 54, 59, 61, 62, 63, 64.

5.2 Realistic Problem

The artificial problem presented in Sect. 5.1 is not large enough to challenge the effectiveness of the current scheme. A number of realistic problems that consist of a real world data are used for this purpose in this section. The results of two of these problems are stated below.

- Breast Cancer

Diagnosis of breast cancer: try to classify a tumor as either benign or malignant based on cell descriptions gathered by microscope examination. Input attributes are for instance the clump thickness, the uniformity of cell size and cell shape, the amount of marginal adhesion, and the frequency of bare nuclei.[1]

Input cell = 9, output cell = 2, T = 10, number of patterns = 350. The result is illustrated in Table 2.

Table 2. The result of the current scheme compared with other works for Breast Cancer problem.

Method	Training set	Testing set	Structure	Learning percentage	Generalization percentage
Prechelt [10]	350	349	9-4-2-2	*Not Reported*	94.25 %
Engelbrecht [11]	450	170	9-8-2	97.6 %	95.7 %
Khaudeyer [12]	250	433	9-18-2	100 %	96.3 %
Current Scheme	350	349	9-6-2	100 %	96 %

- Iris Plants

This problem is a common benchmark problem in pattern recognition and classification studies. This dataset contains 150 instances of four attributes (sepal length, sepal width, pental length, pental width) from each of three classes (setosa, versicolor, virinica). The first class is separated from others clearly, while the other two classes overlap slightly. This dataset is publically available from the UCI Repository of Machine Learning Databases and Domain Theories.[2]

---

[1] This dataset is publically available at ftp://ftp.ira.uka.de/pub/neuron/proben1.tar.gz.

[2] This dataset is publically available at http://ftp.ics.uci.edu/pub/machine-learning-databases/iris/.

Table 3. The result of the current scheme compared with other works for Iris Plants problem.

Method	Training set	Testing set	Structure	Learning percentage	Generalization percentage
Swain [13]	75	75	4-3-1	*Not Reported*	96.66 %
Weihong [14]	90	60	4-25-3	100 %	96.67 %
Current Scheme	75	75	4-7-3	100 %	97.1 %

Input cell = 4, output cell = 3, T = 10, number of patterns = 150 (50 instances in each of three classes). The result is shown in Table 3.

As it can be seen from the tables above, the current scheme results have been compared with the results of other methods in the literature that provided to solve the same problems. These results show that the current scheme is very efficient and takes less time to solve the problems.

6 Conclusions and Future Work

In this paper, a new improved version of breeder genetic algorithm (BGA), which is called new breeder genetic algorithm (NBGA), was introduced. NBGA was used in designing MLN network to yield contemporaneously the optimization of the design of a neural network architecture in terms of the number of hidden layers and number of neurons in each layer, and the choice of the best parameters (learning rate, momentum term, activation functions order of training patterns, and the order of training patterns) for the effective solution of the actual problem to be addressed. The BpNN algorithm was used as a classifier. NBGA added new useful characteristics to BGA to increase the diversity of the population and manipulate the chromosomes with different length, which ensures solving different problems.

The current scheme was tested and the results were compared with previously published results on solving the same problems. The experimental results presented in this paper have demonstrated the effectiveness of NBGA for BpNN optimization. They have also proved the strength of NBGA in terms of solution quality and speed of conversion.

Based on the findings in this work, two further research tasks have been identified: (i) GA-based ANN model, construction and optimization, is computation intensive and could take quite a long time to process. In order to improve the performance of the presented scheme in terms of execution efficiency, we will work on a low-cost general-purpose graphics-processing unit (GPGPU), specifically, the NVIDIA graphics card, to adopt the ANN model training and validation; and (ii) integrate the artificial bee colony (ABC) algorithm [15] with NBGA algorithm. First, the ABC will be applied to derive an optimal set of initial weights from enhancing the accuracy of ANNs. Then, these weights will be used as the starting points for the NBGA evolution procedure.

References

1. Stoica, F., Boitor, C.: Using the breeder genetic algorithm to optimize a multiple regression analysis model used in prediction of the mesiodistal width of unerupted teeth. Int. J. Comput. Commun. Control **9**, 62–70 (2014)
2. Heaton, J.: Deep Learning and Neural Networks. CreateSpace Independent Publishing Platform (2015)
3. Souza, A., Soares, F.: Neural Network Programming with Java. Packt Publishing Ltd. (2016)
4. Rashid, T.: Make Your Own Neural Network. CreateSpace Independent Publishing Platform (2016)
5. Jacobson, L., Kanber, B.: Genetic Algorithms in Java Basics. Springer, New York (2015)
6. Simon, D.: Evolutionary Optimization Algorithms. Wiley, Berlin (2013)
7. Yu, X.-H., Chen, G.-A.: Efficient backpropagation learning using optimal learning rate and momentum. Neural Netw. **10**, 517–527 (1997)
8. Najim, S., Al-Sharibini, M.: Enhancement neural networks design by general genetic algorithm. Basrah J. Sci. **1**, 46–54 (2003)
9. Al-Fadhly, A.: A study of a neuro-genetic system performance. Department of Computer Science, Ph.D. thesis, University of Basrah (2004)
10. Prechelt, L.: Proben1: a set of neural network benchmark problems and benchmarking rules (1994)
11. Engelbrecht, A., Cloete, I.: Selective learning using sensitivity analysis. In: The 1998 IEEE International Joint Conference on Neural Networks, Proceedings of the 1998 IEEE World Congress on Computational Intelligence, Anchorage, Alaska, USA, vol. 2, pp. 1150–1155. IEEE (1998)
12. Khaudeyer, R.: Hybrid approaches: neuro-fuzzy and geno-neuro-fuzzy hybrid system for solving some classification and functions approximation problems. Department of Computer Science, Ph.D. thesis, University of Basrah (2003)
13. Swain, M., Kumar Dash, S., Dash, S., Mohapatra, A.: An approach for IRIS plant classification using neural network. Int. J. Soft Comput. (IJSC) **3**, 79–89 (2012)
14. Weihong, Z., Shunqing, X.: Optimization of BP neural network classifier using genetic algorithm. Intell. Comput. Evol. Comput. AISC **180**, 599–605 (2013)
15. Karaboga, D., Gorkemli, B., Ozturk, C., Karaboga, N.: A comprehensive survey: artificial bee colony (ABC) algorithm and applications. Artif. Intell. Rev. **42**, 21–57 (2014)

An Innovative Method for Key Frames Extraction in News Videos

Ibrahim A. Zedan$^{(\boxtimes)}$, Khaled M. Elsayed, and Eid Emary

Faculty of Computers and Information, Cairo University, Cairo, Egypt
{i.zedan,k.mostafa,e.emaryi.zedan}@fci-cu.edu.eg

Abstract. The process of key frames extraction is facing human subjectivity. The objectives of key frames extraction are mainly two objectives. The first objective is minimizing the number of extracted key frames. The second objective is maximizing the visual representation of a video shot that the key frames should have. There is a tradeoff between the two objectives. Also the needs of individuals are varied. Some individuals favor the compression ratio and others favor the representation level. Many works depend on a subjective ground truth of the key frames for assessing the performance of key frames extraction methods. In this paper we propose a key frame extraction system that takes a confidence level as input from the user to satisfy the different needs. We evaluate our system using fair measures which are the compression ratio and the fidelity that are directly related to the confidence level.

Keywords: Key frames extraction · News videos · Compression ratio · Fidelity

1 Introduction

Video key frame extraction is one of the basic issues in video indexing and retrieval [1], which aims to outline an entire video or a video shot to a small number of stationary key frames for visual content summarization [2]. Key frame is a representative frame mirrors the video shot main content and gives a succinct access to the video content [1–4]. To decrease the amount of video data indexing, key frames are utilized as video streams indexes instead of the original video streams [3]. Generally in the key Frame Extraction methods video is partitioned into shots and key frames are selected from these shots [1]. From each video shot, one or many key frames can be selected based on the content complexity [5, 19].

Matching, and clustering of video shots is an essential task for searching and retrieval in digital video libraries. In the matching process, it is computationally troublesome to incorporate each frame in the shot so it is desirable to process a reduced form of the video shots. Shot key frames are used as a reduced representation of shot [6] so video key frames extraction facilitates the capability to match, and cluster video shots that have similar visual contents.

In this paper, we introduce a key frames extraction system in news videos based on detecting cuts over ranges of frames. Dominant colors image representation is utilized. The difference in order of the dominant colors representation between two frames is

© Springer International Publishing AG 2017
A.E. Hassanien et al. (eds.), *Proceedings of the International Conference on Advanced Intelligent Systems and Informatics 2016*, Advances in Intelligent Systems and Computing 533, DOI 10.1007/978-3-319-48308-5_37

utilized as dissimilarity measure. A neural network is trained to differentiate between two classed: similar frames, and different frames. After extracting the key frames of the entire video, our neural network is utilized again to detect similarity and eliminate the duplicate key frames.

The rest of the paper is organized as follows; an overview of the key frames extraction methods is given in Sect. 2. A detailed description of the proposed system is given Sect. 3. Experimental results are shown in Sect. 4. Section 5 presents our conclusion.

2 Key Frames Extraction Methods

Key frames extraction methods can be assembled into two approaches: cluster-based approach and sequential-based approach [7].

2.1 Cluster-Based Approach

In the cluster-based approach, similar frames are grouped in the same cluster. Before adding a new frame into a specific cluster, the similarity between this frame and all the clusters centroids are calculated. The frame is added to the cluster that has the maximum similarity. If the maximum similarity value is less than the clustering threshold, it implies that this frame is not sufficiently close to be added into the cluster and new cluster is created [5]. After obtaining all the clusters, the frame that is closest to each cluster centroid is selected as a key frame.

Many works consider the clustering-based approach as in [5, 7]. In [5] the similarity of two frames is computed as the intersection of their 2D HS color histograms. In [7] image color histogram is refined using spatial feature based on the fact that pixels in the same color are contiguous or non-contiguous to each other. Similarity between frames is measured using the Euclidean distance between its spatial color histograms. In [8] video frames are divided into blocks. The inter-frame similarity is based on the information entropy on the block level. Hierarchical clustering is utilized to get initial clustering that is refined by k-means. Clustering algorithms have a trouble related to acquiring general clustering parameters that controls the clustering density [3–5]. As the clustering similarity threshold increases, the number of clusters also increases [5]. This approach don't give careful consideration to the shots temporal information [1, 7].

2.2 Sequential-Based Approach

In the sequential-based approach, models based on temporal information and visual features are utilized to determine key frames with thresholds [7]. Sequential-based approach can be further arranged into four methodologies: sampling-based, video shot based, content analysis based, and motion based.

Sampling-Based Methods. Earlier works in video key frames extraction automatically extracts key frames by sampling the video frames uniformly or randomly with a specific time intervals [2]. This technique faces a problem of information loss in small time shots [9]. Also might bring duplicate key frames in video shots with little changes [10].

Video Shot Based Methods. In the shot based method, video is partitioned into shots. Key frames are chosen by several traditional methods. Key frames are selected such as the first/last frame in the shot [1], the frame whose pixel value similar to the shot average value, or the frame whose histogram similar to the shot average histogram [3, 4]. In [11] the video is segmented into shots. Video Shot is classified as static or dynamic. In case of a dynamic shot, the frame with the maximum histogram difference is declared as key frame. In case of a static shot, the middle frame is declared as key frame. These methods have the benefit of simple computation complexity and the number of key frames is limited to one [5]. These methods not consider the content complexity of a video shot so video shot with many changes can't be totally described [3]. It is suitable for video shots with low camera motion [4].

Content Analysis based Methods. These methods extract key frames taking into account the change degree of the frames visual information [9]. When visual information changes significantly; a key frame is selected [3]. The color histogram is a well-known solution since it is computed very easily and insensitive to little changes in camera perspective or image orientations or background noises. In [3] the first k frames that their inter-frame histogram difference exceeds the average value are returned as key frames. The color histogram does not give any spatial information [2, 7] and this problem is handled in [1, 2, 4]. In [4] video frames are segmented into 9 grids. Each grid is associated with different weight. In [2] a temporally maximum occurrence frame (TMOF) is constructed. A weighted distance is calculated between the TMOF and all shot frames. From the weighted distance curve, Key frames are extracted at the peaks. In [1] video frames are divided into blocks. Each block color mean is computed. The color mean difference between the current frame and the last extracted key frame is computed. A block is changed if its color mean difference greater than the block threshold. In each frame, the number of changed blocks is counted. A frame is selected as a key frame if its changed blocks count is greater than the global threshold.

Many works consider the content analysis based approach as in [6, 9, 10, 12]. In [9] the entropy is used as a local feature. The entropies of all grey levels are computed and sorted descending. The grey levels that their entropies sum exceeds a threshold of the frame global entropy are selected. The consecutive frames are compared with the last extracted key frame using only the selected grey levels. If the difference is greater than a threshold; a new key frame is selected. In [6] the dissimilarity between each frame and the last extracted key frame is computed as the normalized sum of absolute difference of the luminance projection values. If the dissimilarity value is greater than a certain threshold a key frame is chosen. In [10, 12] each video frame is divided into blocks and vector quantization is applied to its blocks. Frame difference is calculated as the distance between the codebooks of each pair of consecutive frames. The frames that their difference greater than a threshold are considered as key frames. Many works utilize the frequency representation of video frames to extract key frames as in [13, 14]. In [13] the DCT is utilized. In [14] the discrete wavelet transform is adopted to combine time with frequency. These methods can choose number of key frames proportional to the content change degree in a video shot. But its inconvenience is that it is not handle camera motion well [3, 4].

Motion-Based Analysis Methods. Motion-based extracts key frames based on the scene temporal dynamics. The common procedure is applying motion estimation method followed by applying a motion metric [9]. Motion estimation sets up the inter-relationship of adjacent frames. The popular motion estimation methods are: pixel-recursive algorithms, block-matching algorithms, bayesian method, and the optical flow method. Panning, tilting, and zooming are examples of camera motions. In the panning-like shots, frames with minimum overlap are extracted as key frames. In the zooming-like shots, at least the first and last frames in the shot will be extracted as key frames to outline the focused and global views of the scene [5]. Motion-based algorithms have a trouble related to computation complexity [1, 3, 4].

3 Proposed System

In our proposed system we introduce a method for key frames extraction in news videos. After extracting the key frames of the entire video, key frames are compared with each other. Key frame is removed from the set of key frames if it is similar to other key frame. Our idea for detecting key frames and deciding the similarity between key frames is based on training of a neural network that able to differentiate between two classes: similar frames and different frames (abrupt cuts). The proposed system is shown in Fig. 1.

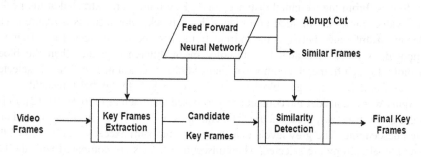

Fig. 1. The block diagram of the proposed system

In [15] we introduced a new feature for describing the video frames we called it as "dominate colors". We represent the video frame by the order of its colors sorted from the highest frequency to the lowest frequency. For each video frame, we truncate the bottom 25 % of the three color components as it is the candidate caption area and the patterns of caption insertion and removal can disrupt the key frames extraction process as shown in [16]. Then we generate the dominate colors of each color component using quantization step = 4. Colors quantization is involved in the process of dominate colors extraction to absorb the lighting conditions that result in variation of the pixels color intensities of the same object. The RGB representation of the video frame is the concatenation of the dominate colors of the three components.

The dissimilarity between two images is defined as a vector contains the difference in order for each color. If two images are similar so its order of dominate colors representations will be similar. Also if two video frames are different so a large difference of the dominate colors order will occur.

A feed forward neural network is trained to classify the video frames to abrupt cut frames (different frames) or similar frames. The neural network is trained by the feature vectors of the consecutive frames dissimilarity. Samples of similar frames correspond to consecutive frames belong to the same video shot. Samples of different frames correspond to samples of abrupt cuts or samples of two consecutive frames belong to different shots. We follow the suggested network structure in [17]. In [17] it is proved that two hidden layer feed forward network with the below structure significantly decreases the required number of hidden neurons while preserving the learning capability with any small error. Let M be the No of output classes = 2 and Let N be the length of dissimilarity feature vector. The number of neurons in the two hidden layers is as follows:

$$\text{No of neurons in the first hidden layer} = \text{sqrt}((M+2)*N) + 2*\text{sqrt}(N/(M+2)) \quad (1)$$

$$\text{No of neurons in the second hidden layer} = M*\text{sqrt}(N/(M+2)) \quad (2)$$

Given a video shot we generate its key frames as shown in Fig. 2. The first frame in the video shot is always considered as key frame and treated as a reference frame. We calculate the dissimilarity vector between the reference frame and its consecutive frames and simulate the neural network with that dissimilarity vector. If the neural network detected a cut so a new frame is added to the key frames list and treated as new reference. As long as no cut detected the reference frame is compared with its successive frames till a cut detected. When simulating the neural network it actually gives two outputs denoted as out1 corresponds to cut output and out2 corresponds to similarity output. The neural network is trained using tansig activation function so the ideal output for different frames is [1, −1] and for similar frames is [−1, 1]. We introduce a measure for the neural network confidence level for detecting key frames as the difference between the outputs values divided by the ideal range as follows:

$$\text{Key Frame Confidence Level} = (\text{out1} - \text{out2})/2 \quad (3)$$

We require a threshold to ensure the neural network confidence level in detecting key frames and this threshold effect is traced. Also to ensure that the proposed system will not generate false key frames due to camera flashes. So if a cut detected between reff and ind we verify this candidate key frame by simulating the network with the dissimilarity between the reff and ind+2 as flashes effect are introduced in maximum in two successive frames.

After extracting the candidate list of key frames of the entire video, a duplicate key frames elimination process is performed as described in Algorithm 1. We check the similarity between each key frame and the remaining key frames using our neural network. We take into account the probability of matching a certain key frame with multiple ones. A key frame is matched with the one that the neural network gives the max similarity. The main purpose of detecting similarity between key frames is to detect the studio shots, and any repeated parts in news video like introduction, ending,

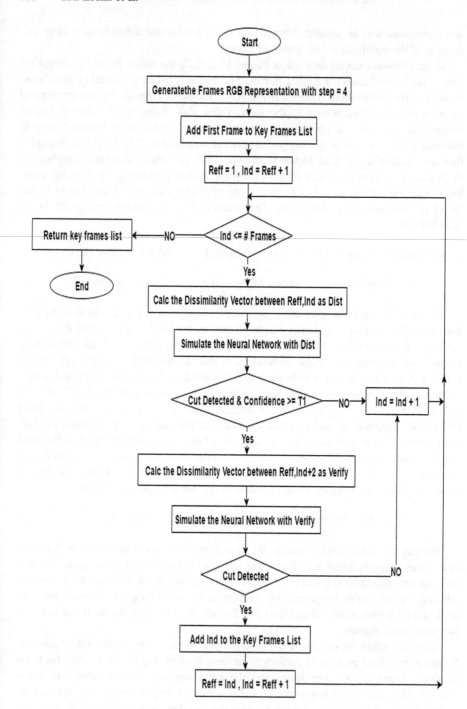

Fig. 2. Key frames extraction process

and separator sequences. Similar to the idea of confidence level in detecting key frames, we define the similarity score between two video frames as follows

$$\text{Similarity score} = (\text{out2} - \text{out1})/2 \tag{4}$$

After establishing the best matches of key frames, a thresholding process can be applied to require a minimum similarity score to decide a duplicate key frame.

Algorithm 1. Duplicate Key Frames Elimination Process

1: Generate the RGB representation of the candidate key frames using step=4
2: **For** Each candidate key frame **do**
3:    Calculate the dissimilarity feature vectors between the selected key frame and the remaining key frames
4:    Simulate the neural network with the dissimilarity feature vectors
5:    **If** a match detected **then**
6:       Calculate the similarity score and record the match
7:    **End If**
8:    **If** multiple matches exist **then**
9:       Retain only the best match that has the maximum similarity score
10:    **End If**
11: **End For**
12: Exclude the matched key frames with similarity score >= T2
13: Output the remaining key frames

---

4 Experimental Results

We tested the key frames extraction system with RGB colored video shots collected form YouTube. The video shots cover many shots types with different camera motion types as shown in Table 1.

To evaluate the performance of the key frames extraction system, we use two measures: the compression ratio, and the fidelity [2]. Fidelity measure compares every key frame with all the video shot frames. High fidelity means that the extracted key frames give great description of the video shot visual content.

$$\text{Compression ratio} = 1 - (\# \text{ key frames}/\# \text{ shot frames}) \tag{5}$$

We let a video shot (S) containing N_1 frames be

$$S = \{F(n)|n = 1, \ldots, N_1\} \tag{6}$$

And the key frames set (KF) of shot S be

Table 1. Test shots characteristics

Shot name	Shot type	Camera motion	Objects motion	# Frames
Shot 1	Studio	No	No	417
Shot 2	Sport	No	Yes	76
Shot 3	Report	Zoom out	No	267
Shot 4	Report	Zoom in	No	123
Shot 7	Sport	Panning	Yes	88
Shot 8	Report	Tilting – Zoom in	Yes	72
Shot 19	Report	No	Yes	57
Shot 21	Report	Panning – Tilting	Yes	150
Shot 24	Report	Panning	Yes	46
Overall				1296

$$KF = \{F_{KF}(n_1), F_{KF}(n_2), \ldots, F_{KF}(n_{KF}) | 1 \leq n_i \leq N_1\} \tag{7}$$

Each frame $F(n)$ in the shot S is compared with the entire set of key frames KF (N_2 Frames) and the minimum distance is retained. The distance between a video frame $F(n)$ and the key frames set KF is computed as

$$d(F(n), KF) = min_j\{diff\left(F(n), F_{KF}(n_j)\right)\} j = 1, 2, \ldots, N_2 \tag{8}$$

The distance $d(S, KF)$ between the video shot S and the key frames set KF is defined as the maximum value across the shot frames $F(n)$

$$d(S, KF) = max_n\{d(F(n), KF)\} | n = 1, \ldots, N_1 \tag{9}$$

The *Fidelity* measure is computed as

Table 2. Key frames extraction evaluation

Shot name	Confidence level = 0.5			Confidence level = 0.95		
	# Key frames	Fidelity	Com. ratio	# Key frames	Fidelity	Com. ratio
Shot 1	1	0.82	0.99	1	0.82	0.99
Shot 2	3	0.91	0.96	1	0.84	0.99
Shot 3	8	0.58	0.97	5	0.51	0.98
Shot 4	1	0.93	0.99	1	0.93	0.99
Shot 7	7	0.80	0.92	5	0.71	0.94
Shot 8	7	0.83	0.90	4	0.83	0.94
Shot 19	2	0.79	0.96	1	0.79	0.98
Shot 21	6	0.78	0.96	3	0.77	0.98
Shot 24	2	0.84	0.96	1	0.77	0.98
Overall	37	0.81	0.95	22	0.77	0.97

$$Fidelity(S, KF) = MaxDiff - d(S, KF) \tag{10}$$

Where the distance *diff* between two frames F_1, F_2 is defined as the complement of the histogram intersection similarity measure. As histogram intersection (HI) take values in the range [0,1] so *MaxDiff* will be equal to one.

$$diff(F_1, F_2) = 1 - \frac{HI_R(F_1, F_2) + HI_G(F_1, F_2) + HI_B(F_1, F_2)}{3} \tag{11}$$

$$HI(F_1, F_2) = \frac{\sum_{i=1}^{256} \min(H_1(i), H_2(i))}{width * height} \tag{12}$$

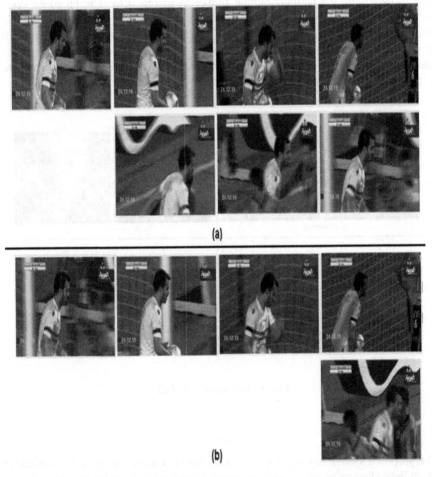

(a)

(b)

Fig. 3. Key frames extracted from Shot 7 [panning] (a) with confidence level = 0.5 (b) with confidence level = 0.95

Shown in Table 2 the evaluation of the key frames extraction process with confidence level equal 0.5 and 0.95. it is clear that when the confidence level increase the fidelity decrease and the compression ration increase as when the confidence level increase the number of extracted key frames will be reduced. Also for static shots, whatever the used confidence level the same number of key frames extracted as in shot 1 since it represent a studio shot. Figure 3 shows the extracted key frames of sport shot. The proposed system extracts 7 key frames with confidence level = 0.5 and only 5 key frames with confidence level = 0.95.

The main purpose of detecting the similarity between the extracted candidate key frames is to ignore the duplicate ones and to locate the repeated frames sequences like studio shots, beginning/ending sequences and the stories separators. Shown in Fig. 4(a) two identical key frames 1, 3248 corresponds to studio shot. The proposed system matches the two key frames with similarity score = 0.966. Also the proposed system is able to determine the partially similar frames as in Fig. 4(b). The proposed system locate two similar key frames with similarity score = 0.44. Actually a certain key frames can be matched with several frames so we always match with the frame that give the maximum similarity score as shown in Fig. 4(c) and (d). In Fig. 4(c) frame 6 matched with frame 977 with similarity score = 0.723 and in Fig. 4(d) it also matched with frame 1910 with similarity score = 0.948.

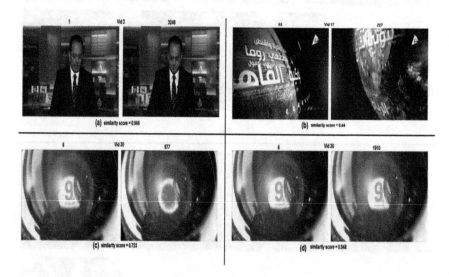

Fig. 4. Key frames similarities

5   Conclusion

The proposed system extracts key frames by detecting cuts between ranges of frames. Flash light elimination is necessary to avoid detecting false key frames. The proposed system takes from the user a confidence level as input. When the confidence level increase the fidelity decrease and the compression ratio increase so the proposed system add the ability to extract key frames with different levels of abstraction. Also the proposed system detects the similarity between video frames in order to detect the repeated parts in the news video. Repeated parts in news videos correspond to studio shots, the begging/ending sequences, or the separators. Locating the studio shots and the stories separators can be utilized to semantically segment the news video into stories units.

References

1. Cao, C., Chen, Z., Xie, G., Lei, S.: Key frame extraction based on frame blocks differential accumulation. In: 24th Chinese Control and Decision Conference, pp. 3621–3625, Taiyuan, 23–25 May 2012
2. Sun, Z., Jia, K., Chen, H.: Video key frame extraction based on spatial-temporal color distribution. In: International Conference on Intelligent Information Hiding and Multimedia Signal Processing, pp. 196–199, Harbin, 15–17 August 2008
3. Liu, H., Meng, W., Liu, Z.: Key frame extraction of online video based on optimized frame difference. In: 9th International Conference on Fuzzy Systems and Knowledge Discovery, pp. 1238–1242, Sichuan, 29–31 May 2012
4. Liu, H., Pan, L., Meng, W.: Key frame extraction from online video based on improved frame difference optimization. In: 14th International Conference on Communication Technology, pp. 940–944, Chengdu, 9–11 November 2012
5. Zhuang, Y., Rui, Y., Huang, T.S., Mehrotra, S.: Adaptive key frame extraction using unsupervised clustering. In: Proceedings of International Conference on Image Processing, ICIP 1998, vol. 1, pp. 866–870, Chicago, IL, 4–7 October 1998
6. Yeung, M.M., Liu, B.: Efficient matching and clustering of video shots. In: International Conference on Image Processing, pp. 338–341, Washington, 23–26 October 1995
7. Chang, J., Hu, R., Wang, Z., Hang, B.: Extracting key frames for surveillance video based on color spatial distribution histograms. In: 10th Pacific Rim Conference on Multimedia, pp. 1005–1010, Bangkok, 15–18 December 2009
8. Liu, H., Hao, H.: Key frame extraction based on improved hierarchical clustering algorithm. In: 11th International Conference on Fuzzy Systems and Knowledge Discovery, pp. 793–797, Xiamen, 19–21 August 2014
9. Mentzelopoulos, M., Psarrou, A.: Key-frame extraction algorithm using entropy difference. In: Proceedings of the 6th ACM SIGMM International Workshop on Multimedia Information Retrieval, pp. 39–45, New York (2004)
10. Thepade, S.D., Patil, P.H.: Novel keyframe extraction for video content summarization using LBG codebook generation technique of vector quantization. Int. J. Comput. Appl. **111**(9), 49–53 (2015)
11. Dhagdi, T.S., Deshmukh, P.R.: Keyframe based video summarization using automatic threshold & edge matching rate. Int. J. Sci. Res. Publ. **2**(7), 1–12 (2012)
12. Thepade, S.D., Patil, P.H.: Novel video keyframe extraction using KPE vector quantization with assorted similarity measures in RGB and LUV color spaces. In: International Conference on Industrial Instrumentation and Control, pp. 1603–1607, Pune, 28–30 May 2015
13. Thepade, S.D., Tonge, A.A.: Extraction of key frames from video using discrete cosine transform. In: International Conference on Control, Instrumentation, Communication and Computational Technologies, pp. 1294–1297, Kanyakumari, 10–11 July (2014)

14. Sharma, C.: Key frame extraction using wavelet transforms – a video summarization technique. Int. J. Adv. Res. Comput. Sci. Manage. Stud. 2(8), 207–213 (2014)
15. Zedan, I.A., Elsayed, K.M., Emary, E.: Abrupt cut detection in news videos using dominant colors representation. In: The 2nd International Conference on Advanced Intelligent Systems and Informatics (AISI), Cairo, 24–26 October 2016
16. Zedan, I.A., Elsayed, K.M., Emary, E.: Caption detection, localization and type recognition in Arabic news video. In: The 10th International Conference on Informatics and Systems Proceedings (INFOS 2016), Cairo, Egypt, 9–11 May 2016
17. Huang, G.: Learning capability and storage capacity of two-hidden-layer feed forward networks. IEEE Trans. Neural Netw. 14(2), 274–281 (2003)
18. Zawbaa, H.M., El-Bendary, N., Hassanien, A.E., Kim, T.: Event detection based approach for soccer video summarization using machine learning. Int. J. Multimed. Ubiquit. Eng. (IJMUE) 7(2), 1–18 (2012)

A Neural Network Approach for Binary Hashing in Image Retrieval

Mohamed Moheeb Emara[1(✉)], Mohamed Waleed Fahkr[2],
and M.B. Abdelhalim[1]

[1] Arab Academy for Science Technology and Maritime Transport,
College of Computing and Information Technology, Cairo, Egypt
mohamed.moheeb90@gmail.com, mbakr@ieee.org
[2] Arab Academy for Science Technology and Maritime Transport,
College of Engineering and Technology, Cairo, Egypt
waleedf@aast.edu

Abstract. Online and cloud storage has become an increasingly popular location to store personal data that led to raising the concerns about storage and retrieval. Similarity-preserving hashing techniques were used for fast storing and retrieval of data. In this paper, a new technique is proposed that uses both randomizing and hashing techniques in a joint structure. The proposed structure uses a Siamese-Twin architecture neural network that applies random projection on data before being used. Furthermore, Particle Swarm Optimization and Genetic Algorithms are used to fine-tune the Siamese-Twin neural network. The proposed technique produces a compact binary code with better retrieval performance than other hashing randomizing technique that varies from 2 % to 5 %.

Keywords: Neural network · Genetic algorithms · Similarity preserving hashing · Random projection

1 Introduction

In the past few years, the use of online and cloud storage has increased exponentially since it facilitates the storing of personal information such as face images and fingerprint. Consequently, two main concerns are raised; namely the ability to store and retrieve data efficiently.

One of the approaches that are used to efficiently store and retrieve data is the similarity preserving hashing function, in which each image feature vector is converted to a binary code where similar data has similar binary codes [1–3]. The binary code is very efficient in terms of storing the data since the space usage is very low and therefore the retrieval becomes fast. It is divided into different families like random hashing methods [4–6] and machine learning techniques [7–9]. The problem with randomizing hashing, it needs long codes to produce higher accuracy. A new direction was taken to produce shorter codes with higher accuracy using learning-based hashing methods. One of the most famous techniques of this direction is Binary Reconstructive Embedding (BRE) [2], Minimal Loss Hashing (MLH) [3] and Spectral Hashing (SP) [10].

© Springer International Publishing AG 2017
A.E. Hassanien et al. (eds.), *Proceedings of the International Conference on Advanced Intelligent Systems and Informatics 2016*, Advances in Intelligent Systems and Computing 533, DOI 10.1007/978-3-319-48308-5_38

In [1] locality sensitive hashing (LSH) is one of the randomized hashing techniques family. It uses a hash function that map the similarity from original space to the same hashing buckets with high probability. It depends on a random hyperplane with a Gaussian independent distribution. The hyperplane is used to produce similar code for similar inputs. There are different similarity criteria used like cosine similarity and Jaccard similarity [11].

In [2] binary reconstructive embedding is introduced where the Euclidean distance between inputs in the input space is calculated, and then the hamming distance between binary codes in the hamming space is calculated, the loss function used as a hash function learning is based on minimizing the error between the difference of the two spaces.

In [3] minimal loss hashing is introduced where a structural SVMs was proposed that apply an online algorithm with latent variables. It uses a loss function that takes into consideration hamming distance and binary quantization.

In [10] spectral hashing is introduced. It was noticed that the problem of finding good binary codes clearly resembles the graph portioning problem. It aims to keep neighbors in input space as neighbors in the hamming space.

This paper proposes a binary hashing technique that creates discriminative binary codes that are compact. It is applied on a Context Based Image Retrieval system (CBIR) where the search is done according to the image content similarity.

Siamese-Twin Random Projection Neural Network (STRPNN) architecture is proposed which is composed of two identical random projections with nonlinear hard-threshold neurons with adjustable bias. The Random Projection Neural Network (RPNN) executes extensive random projection on the input feature space, and the hard threshold produces the binary code. Both Genetic algorithm and particle swarm optimization are used to both select an optimal sparse number of neurons to ensure a short binary code, as well as to adjust the thresholds and the weights of the selected neurons.

The paper organization is as follows: Sect. 2 is a detailed explanation of the STRPNN. Section 3 is a detailed explanation of the GA-PSO tuning of the STRPNN. Section 4 explains the supervised and unsupervised versions of the STRPNN followed by a comparison between normal STRPNN and GA-PSO tuned STRPNN as well as a comparison between GA-PSOSTRPNN and hashing techniques. Section 5 is a conclusion about the proposed technique.

2 Siamese Twin Random Projection Neural Network (STRPNN)

The original idea of a Siamese structure was introduced in [12]. The basic idea is to find optimal weight that produce small distance between the output if the inputs are from the same category and large distance if the inputs are not.

The proposed structure for the Siamese twin Random Projection Neural Network (STRPNN) is shown in Fig. 1. It is composed of two identical RPNN structures with hard-threshold neurons. The STRPNN can work in both supervised and unsupervised image retrieval systems, where the training consists of selecting the best neurons, and tuning the thresholds and weights of the neurons. It converts feature data into secure

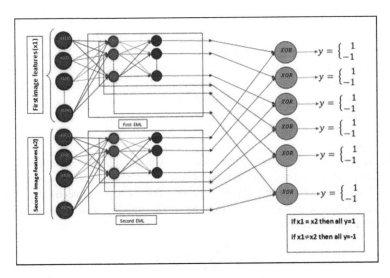

Fig. 1. Siamese Twin Random Projection Neural Network

binary codes where the objective is to produce similar codes for similar images and vice versa.

STRPNN goes through two main stages: the training stage and the binary code selection stage. In the first stage, the Euclidean distance is calculated between each feature vector (query) and the remaining training items, to select the closest (positive) and farthest (negative) match. In case of the supervised RPNN the positive class item has to be from the same class of the query while the negative class item has to be from different class than the query.

The query could have one or more positive and negative matches. The Target vector T is the vector that holds the values of the query compared to negative and positive matches. Target vector is composed of the $T_i i = 1 : N$ values where $T_i = 1$ if the 2 images are similar enough, and $T_i = -1$ if the 2 images are significantly different. RPNN consists of J neurons; the input for RPNN is the feature vector from the training data multiplied by random projection. The neurons outputs can be thought of as a vector Y which is given by for the twin neural networks:

$$Y_1 = f(\Phi * X_1) \tag{1}$$

$$Y_2 = f(\Phi * X_2) \tag{2}$$

Where X_1 and X_2 are the input feature vectors to the twin neural networks and f is the hard-thresholding neuron. Each neuron also has an adjustable threshold which is considered as an extra input feature with a value of 1. For the j_{th} neuron, the output is:

$$y_{ij} = f(\phi_j * X_k) \tag{3}$$

Where ϕ_j is the part of the random matrix Φ that is connected to the j_{th} neuron. Twin Random Projection Neural Network consists of two RPNN during the training phase; the first one is for the query image feature and the second RPNN is for checking the similarity with the query. To check for similarity the output of the first and second RPNN's are XORed: if they are similar the output (Z) is 1 and if they are not the output (Z) is −1:

$$if \; y_{1j} = y_{2j} \; Z_j = 1, \; and \; if \; y_{1j} \neq y_{2j} \; Z_j = -1 \qquad (4)$$

The next stage is the binary code selection. The outputs of the neurons over the whole training data are used to determine which of the neurons are more correlated to the target. The top M neurons (M the size of binary code produced) that have the highest correlation with the targets over the whole training data are kept: For each neuron calculate the following:

$$S_j = \sum_{i-1}^{N} y_{ij} * T_i \qquad (5)$$

Thus, if y_{ij} (the output of the jth neuron for the ith training example) has the same sign as the target (T) then this adds to the score of this neuron, while if the neuron output is different than the target then this will discount from its score. Finally, the scores of all the neurons are sorted and the top M kept.

Even though this FAST technique is very quick and does not need solving of optimization problems, it suffers from an obvious drawback: Some of the remaining neurons may be identical or highly correlated which will result in redundancy and sub-optimal sparsity. This can be resolved by applying some enhancement to the algorithm by checking the remaining neurons correlation and removing the redundant ones. The GA solves this problem as explained in the next section.

3 GA and PSO Tuning of the STRPN

The Genetic Algorithm (GA) was introduced by John Holland [13]. It applies Darwinian evaluation theory in searching for optimal solution for a given problem. Particle Swarm Optimization (PSO) was introduced by Kennedy and Eberhart [14]. It was inspired by the behavior and movement of animals in nature like birds flock. It was simplified and applied also in searching for optimal solution for a given problem.

Particle Swarm Optimization and Genetic Algorithms are used with the training of the STRPNN to improve the distinction of the produced binary codes and therefore improve the performance. The PSO works on improving the hard-limiter neuron thresholds. As for the GA, it works on selecting the best neurons to create the best sparse binary code. The number of neurons selected is according to the binary code length required.

PSO depend on velocity and position of each particle, velocity and position are update at each iteration to find optimal solution, Eq. (6) is for updating velocity and

Eq. (7) is used to update position found in [14] where c_1, c_2 and ω are parameters set by the user and r_1 and r_2 are random number generated in the range $[0 - 1]$.

$$V_{id}^{t+1} = w.V_{id}^{t} + c_1.r_1\left(P_{id}^{t} - X_{id}^{t}\right) + c_2.r_2\left(P_{gb}^{t} - X_{id}^{t}\right) \tag{6}$$

$$X_{id}^{t+1} = X_{id}^{t} + V_{id}^{t+1} \tag{7}$$

3.1 PSO Tuning of Thresholds

The PSO is used in tuning the hard-limiter neurons' thresholds. The function of hard-limiter neurons in RPNN is to converts input data to binary code. It depends on threshold: if the weighted sum is less than the threshold the output is -1 otherwise it is equal to 1. The PSO is used to find the best threshold for each neuron. The PSO consists of N particles; each particle consists of a random vector. Each random value in the vector represents a threshold for each neuron. The optimization is done on the random vector to find the best possible threshold suited for each neuron.

$$if \ y_j \geq P_j \ then \ f(Y) = 1 \ else \ f(Y) = -1 \tag{8}$$

The fitness function starts by using the initialized particles as weights for hard limited binary neurons (Eq. (8)). The vector P is the particle of N values. The vector has the same size as number of neurons in the RPNN, if the value of y_j is less than P_j value the output is -1 otherwise it is 1. The P vectors value are improved after each iteration using PSO. The next step is XORing the output of twin RPNN (explained in Sect. 2) using Eq. (4). To select the top M neurons to produce the binary code Eq. (5) is used (explained in Sect. 2). The next step is to convert the validation data using trained RPNN with each particle as hard-limiter to binary code. The output of the top M neurons is used in producing the validation binary code. The next step is to compare the distinctive of validation data in the original space with the binary codes produced in binary space. For this task a class matrix that contains the class of validation data compared with training data is produced in the original space and another is produced for binary codes in the Hamming space and then they are matched together as shown in Eq. (9):

$$MA = ||V - TD|| \tag{9}$$

Where *MA* have rows equal to number of training data and columns equal to the number of validation data. The intersection between the columns represents the Euclidian distance between training item and validation item. V is the validation data and TD is the training data

$$L = binarized \ (MA) \ according \ to \ TH \tag{10}$$

Where *TH* is calculated by sorting the Euclidian distance between the training data followed by selecting top *C* values and averaging it. *L* is the binary matrix where the element is 1 when the training data and validation data are from the same class, and is 0 it is not from the same class.

Equations (9) and (10) are used one more time to produce a class matrix for the Hamming space using the same value of C. The next step is to match the two matrices together using XNOR, as shown in Eq. (11).

$$f(P_i) = \text{sum}(\overline{MA \oplus MB}) \tag{11}$$

Where $f(P_i)$ represent the fitness function for each Particle P_i and is the sum of the matching pairs. MA is the class matrix for original data and MB is the class matrix for binary codes.

$$G = \sum_{j=1}^{m} \max(f(P_1), f(P_2), \ldots, f(P_n)) \tag{12}$$

In Eq. (12) m is the maximum number of iteration done by PSO, $f(P_n)$ is the sum of matching pairs produced by each particle calculated in Eq. (11). The basic idea for matching the two matrices used with PSO or GA, this to ensure that the binary code produced has its equivalent pair in the feature data.

3.2 Genetic Algorithm Sparse Neurons Selection

The GA is used to select the best neurons to create binary code. The problem with the FAST technique is some of the unselected neurons maybe more significant than the selected neurons. This is due to the fact that selecting the most correlated neurons doesn't ensure that best performance, least correlated neurons when combined with other neurons can produce a better distinct code therefore increasing the performance. The FAST technique selects the top M neurons that have the highest correlation with the targets over the whole training data. The GA overcomes this problem by exploring the different combination of neurons to find the most significant neurons. Each chromosome in the GA has same size as number of neurons, the 1s in the chromosome represent the index of the selected neurons to produce binary code otherwise the 0s represent the unselected neurons.

The GA RPNN population is initialized by four chromosomes. Each chromosome is a binary vector where 1's represent the selected neurons and 0's represent the unselected neurons. The first chromosome is initialized by the output of the FAST technique in case it acquires the best performance, the other three are initialized randomly according to the required binary code (if the required code is M the number of 1's should be exactly equal to M).

The Genetic algorithm is used with RPNN to replace the FAST technique and it uses a similar fitness function like the one used with PSO, instead of using the top M ranked neurons a randomly selected M neurons are selected to produce the binary validation data then calculate class matrix for both features and binary training and validation data.

The sum of matching pairs is also calculated using Eq. (11). The difference here is that the matching pairs reflects the quality of each chromosome in the Population. The reason for using GA is to find the best set of neurons of size M that produce the best distinct binary code. In Eq. (12) m is the max number of iteration done by GA, $f(P_n)$ is the sum of matching pairs for each chromosome.

For the selection phase the *roulette wheel* is used. In the crossover, the order 1 crossover is used to prevent any increase or decrease in the required code size as shown in Fig. 2., it is done by creating random permuted vectors as shown in Fig. 2(a), the value in the permuted vector are converted into 1's and 0's in the equivalent binary chromosomes. So if the binary code required like in Fig. 2 are 4 bits, the numbers from one to four is converted to 1's otherwise the numbers are converted to 0's. This is to make sure that value in the two parents (permutated vector) are the same in the equivalent binary chromosome for example the value 1, 2, 3 and 4 in Fig. 2(a) in both parents have the same value in the equivalent binary chromosomes in Fig. 2(b) equal to 1, therefore the number of 1's for sure will not increase or decrease. The order one crossover [15] is then applied on those two vectors to produce the new two children. If a value is moved from the random permuted vector (the parent) to the new child. The index of this value is taken and the same move is applied for the equivalent binary chromosome shown in Fig. 2(b). In the mutation phase, a swap takes place between two randomly selected bits from the newly created children. The last step is the elimination of the least two fit chromosomes and replacing them with the new produced children.

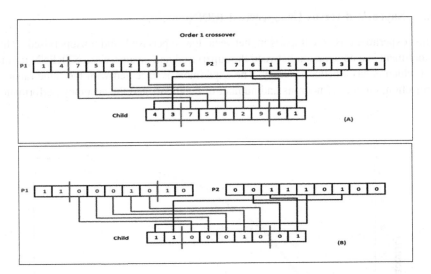

Fig. 2. The use of order 1 to maintain the size of binary code (4 bits binary code)

The technique used in pervious section is a smaller version of the GA-PSO RPNN, it doesn't take a lot of time in training and is suitable to use in application that has tight time constrain compared to GA-PSO RPNN. The two techniques only takes time in the training processes but takes average time in executing on testing data. The GA-PSO RPNN produce more secure and accurate binary codes since it has to extra layers of fine tuning and randomization using PSO and GA.

4 Experiments

4.1 Dataset Description

In following experiments, the COREL 1 k dataset is used. It has 10 categories and each category has 100 images. The COREL dataset is used in the first part of the experiments using STRPNN without GA-PSO. In the second part, the small datasets reported in [2] are used, namely the Labelme and MNIST. Each dataset consists of 5000 image feature vectors.

4.2 Feature Extraction

The features used on COREL database is a concatenation of indexed color histogram, Discrete Cosine Transform, Color Histogram, GIST and SURF-VLAD with vector lengths of 64, 64, 192, 512 and 1000 respectively. The formed concatenation feature is than reduced using PCA to a 25 dimension vector. Each image in the Labelme dataset is represented by a 512 GIST vector and each image in the MNIST dataset is represented by a 784 GIST vector [2].

4.3 Supervised versus Unsupervised RPNN

This experiment is a comparison between the supervised and unsupervised. This experiment is a comparison between the supervised and unsupervised STRPNN to find out which is has a better accuracy. The experiment is executed with the same random projection, number of neurons and number of neighbors and the accuracy performance

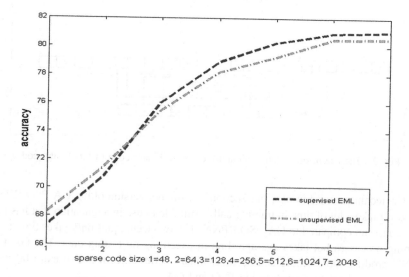

Fig. 3. Unsupervised vs. supervised for 8000 neurons

is shown in Fig. 3. The accuracy of unsupervised RPNN is better than supervised up to code length of 128 but the supervised overcomes unsupervised for longer codes. This experiment proves that the STRPNN can work on both supervised and unsupervised data with almost the same accuracy.

4.4 Comparing Hashing Techniques and GA-PSOSTRPNN

This subsection discusses a comparison between PSO and GA tuned STRPNN and other hashing techniques for compact binary codes using 2 dataset, namely MNIST and Labelme [5]. Each dataset is divided into 3000 images as training, 1000 for validation (used in the PSO and GA fitness function) and 1000 images as testing data randomly. The Euclidean distance is calculated for training data and itself, the 50 nearest data items are selected and averaged. The neighbors are the items that are less than average (ground truth neighbors), the PCA is applied on mean centered data and the top 40 PCA features are kept.

In this experiment, the RPNN used has 200 neurons with one hidden layer. PSO uses velocity update Eq. (5) that has ω (Inertia weight) equal to 0.9, c_1 equal to 1 and c_2 equal to 2 [14]. The GA algorithm has a mutation rate of 0.05 and crossover rate of 0.6 [13].

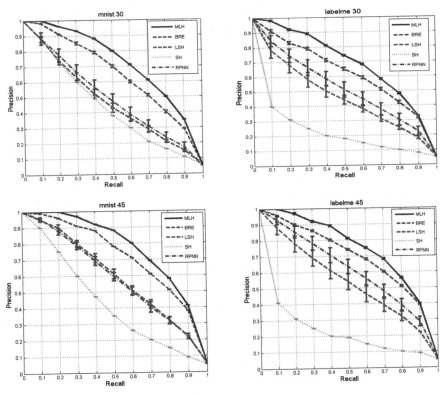

Fig. 4. Hashing techniques compared to RPNN [3]

The precision and recall curve found in [6] is used to compare between STRPNN and other state of the art hashing techniques. The precision recall curve is produced after varying the radius R from zero to "q", which is the radius value where the recall is equal to one. The experiments are executed 10 times and results shown represent the average with standard deviation as error bars, as shown in Fig. 4.

The results show that the RPNN is clearly better in precision than SH and LSH in almost the four precision and recall curves, but still need improvement to overcome the BRE and MLH.

4.5   Comparison Between Normal RPNN and GA-PSO RPNN

To conduct this experiment a precision and recall curve for 50 bits using COREL dataset concatenated features. The COREL dataset is divided into 500 images for training and validation and 500 images for testing. The same settings for normal RPNN and GA-PSO RPNN are used with the same random projection. The RPNN uses 200 neurons with one hidden layer and the settings for PSO and GA are the same as the previous subsection. In Fig. 5, the result shows that at recall 0.5 the precision of normal RPNN is improved by 15 % using PSOGA RPNN. The PSO and GA improved the average precision of normal RPNN by about 17.5 %.

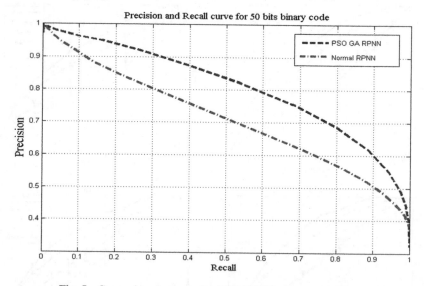

Fig. 5. Comparison between GA-PSO RPNN and normal RPNN

5   Conclusion

In this paper, a Siamese-Twin Random Projection Neural Network (STRPNN) is proposed for the fast retrieval and efficient storing of data, in which a combination of neurons are selected to produce the required binary code. A fine-tune PSO and GA are applied in the selecting process. STRPNN can also be used in both supervised and unsupervised data modes based on the availability of labeled data. When it is compared to other hashing techniques it overcomes SH with a precision that varies from 2 % to 20 % in average and LSH with a precision that varies from 2 % to 10 % in average, but it needs more improvement to overcome MLH and BRE [16] which will require the training of the random matrix weights for the selected neurons, and this is a future direction for research.

References

1. Gionis, A., Indyk, P., Motwani, R.: Similarity search in high dimensions via hashing. VLDB **99**(6), 518–529 (1999)
2. Kulis, B., Darrell, T.: Learning to hash with binary reconstructive embeddings. In: Advances in Neural Information Processing Systems, pp. 1042–1050 (2009)
3. Norouzi, M., Fleet, D.J.: Minimal loss hashing for compact binary codes. In: mij, 1, p. 2 (2011)
4. Lv, Q., Josephson, W., Wang, Z., Charikar, M., Li, K.: Multi-probe LSH: efficient indexing for high-dimensional similarity search. In: The 33rd International Conference on Very Large Data Bases, pp. 950–961. Endowment (2007)
5. Bawa, M., Condie, T., Ganesan, P.: LSH forest: self-tuning indexes for similarity search. In: The 14th International Conference on World Wide Web, pp. 651–660. ACM (2005)
6. Dong, W., Wang, Z., Josephson, W., Charikar, M., Li, K.: Modeling LSH for performance tuning. In: 17th ACM Conference on Information and Knowledge Management, pp. 669–678. ACM (2008)
7. Liu, W., Wang, J., Kumar, S., Chang, S.F.: Hashing with graphs. In: Proceedings of the International Conference on Machine Learning, Bellevue, WA, USA, pp. 1–8 (2011)
8. Gong, Y., Kumar, S., Verma, V., Lazebnik, S.: Angular quantization based binary codes for fast similarity search. In: Advances in Neural Information Processing Systems, Cambridge, MA, USA, vol. 25, pp. 1205–1213. MIT Press (2012)
9. Norouzi, M., Fleet, D.J., Salakhutdinov, R.R.: Hamming distance metric learning. In: Advances in Neural Information Processing Systems, Cambridge, MA, USA, pp. 1070–107. MIT Press (2012)
10. Weiss, Y., Torralba, A., Fergus, R.: Spectral hashing. In: Advances in Neural Information Processing Systems, pp. 1753–1760 (2009)
11. Charikar, M.S.: Similarity estimation techniques from rounding algorithms. In: The Thiry-Fourth Annual ACM Symposium on Theory of Computing, pp. 380–388. ACM (2002)
12. Chopra, S., Hadsell, R., LeCun, Y.: Learning a similarity metric discriminatively, with application to face verification. In: IEEE Computer Society Conference on Computer Vision and Pattern Recognition, CVPR 2005, vol. 1, pp. 539–546. IEEE (2005)

13. Holland, J.H.: Adaptation in Natural and Artificial Systems: An Introductory Analysis with Applications to Biology, Control, and Artificial Intelligence. University Michigan Press, Cambridge (1975)
14. Eberhart, R.C., Kennedy, J.: A new optimizer using particle swarm theory. In: The Sixth International Symposium on Micro Machine and Human Science, vol. 1, pp. 39–43 (1995)
15. Grefenstette, J., Gopal, R., Rosmaita, B., Van Gucht, D.: Genetic algorithms for the traveling salesman problem. In: The First International Conference on Genetic Algorithms and Their Applications, pp. 160–168. Lawrence Erlbaum, New Jersey (1985)
16. Wang, J., Liu, W., Kumar, S., Chang, S.F.: Learning to hash for indexing big data—a survey. IEEE **104**(1), 34–57 (2016)

An Integrated Smoothing Method for Fingerprint Recognition Enhancement

Muhammad Khfagy[1,2](✉), Yasser AbdelSatar[1], Omar Reyad[2,3], and Nahla Omran[1]

[1] Faculty of Science, South Valley University, Qena, Egypt
{Mohamd.khafagy,Yasser.abdelsatar,Nahlaa.fathy}@sci.svu.edu.eg
[2] Faculty of Science, Sohag University, Sohag, Egypt
ormak4@yahoo.com
[3] Faculty of Electronics and Information Technology,
Warsaw University of Technology, Warsaw, Poland

Abstract. Fingerprint identification systems are one of the most well-known and publicized biometrics because of the inherent ease in acquisition, the numerous sources (ten fingers) available for collection, and their established use by law enforcement and immigration. These systems rely on the unique biological characteristics of individuals to accurately verify their identities. To get reliable and accurate verification results, these systems need high quality images. The quality of the fingerprint image is obtained by using noise-free images during the pre-processing and filtering stages. In this paper, we proposed an integrated smoothing method (ISM) for fingerprint image recognition enhancement based on a linear combination of three different filtering techniques named median filter (MF), Wiener filter (WF) and anisotropic diffusion filter (ADF). This combination is made by using two coefficient parameters (α, β) with different values to enhance the quality of images and remove the unwanted distortion or noise that affect a fingerprint recognition system. The ISM is applied in the pre-processing stage to get a noise-free fingerprint image with high accuracy factor. We used the benchmarking FVC2004 and FVC2006 databases to test our method and the Wilcoxon signed-rank test (**W**) and the peak signal-to-noise ratio (PSNR) for results evaluation. The experimental results indicate that the proposed ISM improves the performance of the fingerprint identification significantly.

Keywords: Biometric authentication · Fingerprint recognition · Median filter · Wiener filter · Anisotropic diffusion filter

1 Introduction

Biometric authentication systems are based on the ways in which individuals can be uniquely identified through one or more distinguishing biological traits, such as fingerprints, hand geometry, earlobe geometry, retina and iris patterns, voice

© Springer International Publishing AG 2017
A.E. Hassanien et al. (eds.), *Proceedings of the International Conference on Advanced Intelligent Systems and Informatics 2016*, Advances in Intelligent Systems and Computing 533, DOI 10.1007/978-3-319-48308-5_39

waves, DNA (Deoxyribonucleic acid) and signatures [1]. Fingerprint identification system is one of the most popular biometric technologies and the performance of a fingerprint image matching algorithm depends heavily on the quality of the input fingerprint images [2,12].

A fingerprint refers to the flow of ridge patterns in the tip of a finger [3]. The ridge patterns flow exhibits anomalies in local regions of a fingertip and it's position and orientation that affect the matching process in fingerprints. The advantages of the fingerprint images are the using of relatively easy, low-cost scanners and the matching process can be more efficient in the case of small dimensional images. A good quality fingerprint image has around 40 to 100 minutiae. Minutiae are a point of interest in a fingerprint tip such as ridge bifurcation and ridge ending. Ridge bifurcation is a single ridge that divides into two ridges while ridge ending is a ridge that ends abruptly as shown in Fig. 1.

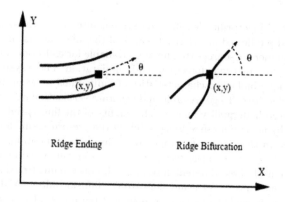

Fig. 1. Ridge bifurcation and ridge ending

Images are those data types that widely used in various areas such as [4–6]. Noisy and poor quality captured images which lead to false generation and loss genuine of minutiae are a challenge in any fingerprint image recognition system. Noise in fingerprint images may occur due to variations in skin and impression conditions such as scars, humidity, dirt, and non-uniform contact with the fingerprint capture device which corrupts the clarity of the ridge structures. The main goal includes building a trustworthy data model to represent randomly oriented lines of the fingertip and finding ways to compare the models with the highest accuracy in minutiae matching stage [7].

Any identification system design should consists of two major steps which is: enrollment stage and verification stage [8]. The enrollment stage includes capturing a fingerprint image from a sensor device then an extractor identifies the features of the fingerprint image and store them into an enrolling template in system database. In the verification stage, matching algorithms are used to compare previously stored templates of fingerprints against the candidate's fingerprints for authentication purposes.

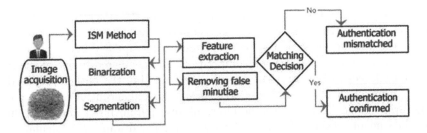

Fig. 2. Block diagram of the proposed fingerprint recognition system

In this paper, we proposed an integrated smoothing method (ISM) for fingerprint image recognition enhancement in the pre-processing stage to enhance noisy and poor quality images to acquire good quality images which result in increasing the performance of the minutiae extraction algorithm as shown in Fig. 2. The proposed ISM uses a combination of three main filters; median filter (MF), Wiener filter (WF) and anisotropic diffusion filter (ADF) with the addition of two coefficient parameters named α and β to remove noise models that affect fingerprint image recognition which result in an improvement of the matching results in the fingerprint verification process.

The rest of the paper is organized as follows: In Sect. 2, we present some preliminaries about the related work, MF, WF and ADF. The proposed method is presented in Sect. 3. The experimental results are given in Sect. 4, while conclusions and future work are made in Sect. 5.

2 Preliminaries

2.1 Related Works

Recently, several attempts for image enhancement in fingerprint image pre-processing stage has been proposed in literature. In [9], a fast fingerprint enhancement algorithm is presented, which can adaptively improve the clarity of ridge and valley structures of input fingerprint images based on the estimated local ridge orientation and frequency. The performance of the image enhancement algorithm is evaluated using the goodness index of the extracted minutiae and the accuracy of an on-line fingerprint verification system. An adaptive pre-processing method is proposed in [10], which extracts five features from the fingerprint images, analyzes image quality with clustering method, and enhances the images according to their characteristics. This method improves the performance of the fingerprint identification significantly as indicated by the experimental results. In [11], an efficient pre-processing algorithm is proposed to achieve good vertical orientation and high ridge curvature area around the core point for fingerprint authentication and analysis. The algorithm is implemented in two stages. The process of obtaining the vertical oriented fingerprint image is carried out in the first step and then followed by core point detection of a fingerprint. A Bat algorithm for gray scale fingerprint image contrast enhancement is proposed in [13]. The purpose is for the Bat algorithm to map the gray level distributions for contrast enhancement ends. To assess the approach, the enhancement process is evaluated on low quality images from the FVC2000 (fingerprint verification competition) and compared to one of the traditional related-work contrast-based enhancers. In [14], the author proposed a filter construction based on Log-Gabor theory for enhancement of high noisy fingerprint image. To construct the filter, all filter parameters is initialized first and the sine and cosine differences along with angular distance are calculated to overcome the wrap around effect. Then, the convolution with FFT (fast Fourier transform) of input noisy image is done.

2.2   The Median Filter

The median filter (MF) is a nonlinear operation often used to remove noise while preserving the sharp high-frequency edges [15]. Edges are of critical importance to the visual appearance of fingerprint images. The MF also eliminating noise, especially isolated noise spikes and is useful in reducing salt and pepper noise and high/low impulse-type noise with minimal degradation or loss of detail in the image. Each output pixel contains the median value in a 3×3 neighborhood around the corresponding pixel in an input image of size $[m\ n]$. MF put the image with 0's on the edges, so the median values for points within one-half the width of the neighborhood ($[m\ n]/2$) of the edges might appear distorted.

2.3   The Wiener Filter

The Wiener filter (WF) is a 2-D adaptive noise removal filtering operation that has two separate parts, an inverse filtering part and a noise smoothing part [16]. WF does not only perform the deconvolution by inverse filtering but also removes the additive noise. WF minimizes overall the mean square error (MSE) in the process of inverse filtering and noise smoothing. WF is a linear estimation of the original image and the approach based on a stochastic framework. The orthogonal principle implies that the WF in Fourier domain can be expressed as follows:

$$W(f_1, f_2) = \frac{H^*(f_1, f_2) S_{xx}(f_1, f_2)}{|H(f_1, f_2)|^2 S_{xx}(f_1, f_2) + S_{\eta\eta}(f_1, f_2)} \tag{1}$$

where $S_{xx}(f_1, f_2), S_{\eta\eta}(f_1, f_2)$ are power spectra of the original image and the additive noise respectively and $H(f_1, f_2)$ are the blurring filter.

2.4   The Anisotropic Diffusion Filter

Anisotropic diffusion filter (ADF) also known as Perona-Malik diffusion, is a technique aiming at reducing image noise without removing significant parts of the image content, typically edges, lines or other details that are important for the interpretation of the image [17]. Conductance function, the gradient threshold parameter, and the stopping parameter form a set of parameters which define the behavior and the extent of the diffusion. These process smoothed regions while preserving, and enhancing, the contrast at sharp intensity gradients. The basic equation of anisotropic diffusion equation is:

$$\frac{\partial I(x,y,t)}{\partial t} = div[g(\|\nabla I(x,y,t)\|)\nabla I(x,y,t)] \qquad (2)$$

where t is the time parameter, $I(x,y,0)$ are the original image and the gradient image with time t while $g(.)$ is the conductance function which satisfies $\lim_{x\to 0} g(x) = 1$ so that the diffusion is maximal within uniform regions, and, $\lim_{x\to\infty} g(x) = 0$ so that the diffusion stopped across edges.

3   The Proposed Integrated Smoothing Method

In this section, we propose an integrated smoothing method (ISM) to filtering out blurring troubles and smoothing as well as removing noise types while respecting the region in all boundaries and small structures within the fingerprint image without affecting minutiae specifics. The ISM is a combination of three filters (MF, WF, ADF) using two coefficient parameters α and β. The proposed ISM contains two sub-methods which represent ISM_1 and ISM_2 respectively.

$$ISM_1 = \alpha \times WF + MF \qquad (3)$$

$$ISM_2 = ISM_1 + \beta \times ADF \qquad (4)$$

The ISM_1 is the addition of MF with the result from multiplication of parameter α by WF as defined in Eq. (3). The ISM_2 is the addition of ISM_1 itself with the result from multiplication of parameter β by ADF as defined in Eq. (4). The two coefficient parameters are running various values, $\alpha = \{0.1, 0.2, \cdots, 0.9\}$ and $\beta = \{0.1, 0.2, \cdots, 0.9\}$, to enable appropriate estimation for the whole method. The two sub-methods are verified using the two benchmarking FVC2004 dataset [18] and FVC2006(DB2_A) dataset [19] which contains various samples of fingerprints.

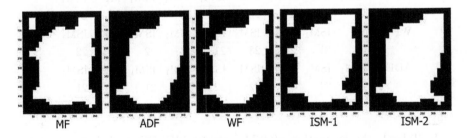

Fig. 3. Region mask image with different filtering techniques effect

In Fig. 3 is shown filtering techniques effect to extract the region of interest (ROI) of a sample fingerprint image from FVC2006(DB2_A) dataset. The ROI is a portion of an image that you want to filter or perform some other operation on. ROI is defined by creating a binary mask, which is a binary image that is the same size as the image you want to process. In the region mask image, the pixels that define the ROI are set to 1 and all other pixels set to 0. From Fig. 3, it is proven that the proposed ISM could locate the position of reference point of all fingerprints type more effectively and precisely.

4   Experimental Results and Statistical Analysis

To test our method, we use FVC2004 dataset which reduced to 160 fingerprint images for 40 fingers, 4 samples per each finger and FVC2006(DB2_A) dataset contains 1600 fingerprint images for 140 fingertips, 12 samples per each finger. Then, we obtained the matching value for each fingerprint image by computing the average results of the all test samples (4 or 12) per each finger. For statistical analysis, two methods are used here, the Wilcoxon signed-rank test ($\mathbf{W}$) and the peak signal-to-noise ratio (PSNR) to evaluate the matching results significance.

4.1   Wilcoxon Signed-Rank Test ($\mathbf{W}$)

The $\mathbf{W}$-test is a non-parametric statistical hypothesis test used when comparing two related samples, matched samples, or repeated measurements on a single sample to assess whether their population mean ranks differ [20]. We evaluate the matching results for each filter separately as a pre-processing tool to enhance the fingerprint image clarity. Then each filtered fingerprint is enrolled in the recognition stage. The final scored accuracy in the verification stage is evaluated via the $\mathbf{W}$-test to investigate the scored accuracy significance.

Table 1. An evaluation of ISM_1 against all other filters by $\mathbf{W}$-test

α value	0.1	0.2	0.3	0.4	0.5	0.6	0.7	0.8	0.9
MF	ISM_1	ISM_1	ISM_1	ISM_1	ISM_1	ISM_1	ISM_1	MF	ISM_1
	21	24	21	22	22	21	21	21	21
WF	WF	ISM_1	ISM_1	ISM_1	ISM_1	ISM_1	ISM_1	WF	ISM_1
	22	21	20	23	21	21	22	23	22
ADF	ISM_1	ISM_1	ISM_1	ISM_1	ISM_1	ISM_1	ISM_1	ADF	ISM_1
	21	24	21	22	22	21	21	21	21

In Table 1, the evaluation of ISM_1 against MF, WF and ADF using the $\mathbf{W}$-test for FVC2004 dataset is presented. The first row represent α values as $\alpha = \{0.1, 0.2, \cdots, 0.9\}$. The comparison of MF with ISM_1 is given in the second row which showing that ISM_1 is better than MF at all α values except when $\alpha = 0.8$ according to the numbers ($\mathbf{W}$=21) given in this row. These numbers represent the **W**-test values for the winner filter in each comparison at each α value. In the third row, the comparison of WF with ISM_1 is given and the **W**-test values shown that WF is winner only when $\alpha = (0.1, 0.8)$ according to the numbers ($\mathbf{W}=22, 23$) given in this row. The same is done for ADF in the fourth row.

In Table 2, the evaluation of ISM_1 against MF, WF and ADF using the **W**-test for FVC2006(DB2_A) dataset is presented. The first row represent α values as $\alpha = \{0.1, 0.2, \cdots, 0.9\}$. The comparison of MF with ISM_1 is given in the second row which showing that ISM_1 is better than MF at most of α values except when $\alpha = (0.1, 0.8)$ according to the numbers ($\mathbf{W}=80, 80$) given in this row. These numbers represent the **W**-test values for the winner filter in each comparison at each α value. In the third row, the comparison of WF with ISM_1 is given and the **W**-test values shown that WF is winner only when $\alpha = (0.1, 0.8)$ according to the numbers ($\mathbf{W}=73, 82$) given in this row. The same is done for ADF in the fourth row.

Table 2. An evaluation of ISM_1 against all other filters by **W**-test

α value	0.1	0.2	0.3	0.4	0.5	0.6	0.7	0.8	0.9
MF	MF	ISM_1	ISM_1	ISM_1	ISM_1	ISM_1	ISM_1	MF	ISM_1
	80	75	88	90	94	98	79	80	75
WF	WF	ISM_1	ISM_1	ISM_1	ISM_1	ISM_1	ISM_1	WF	ISM_1
	73	78	85	92	97	89	79	82	80
ADF	ADF	ISM_1	ISM_1	ISM_1	ISM_1	ISM_1	ISM_1	ISM_1	ISM_1
	70	76	79	75	85	87	77	81	77

From Table 2, it is found that the best matching results according to **W**-test values is when $\alpha = (0.3, 0.5, 0.6)$. So, we use these values to evaluate the ISM_2 accuracy for all β values against all other filters including the proposed ISM_1 for the fingerprint image enhancement whole method.

In Table 3, the evaluation of ISM_2 against MF, WF, ADF and ISM_1 using the **W**-test for FVC2006(DB2_A) dataset is presented. The first row represent β values as $\beta = \{0.1, 0.2, \cdots, 0.9\}$. The comparison of MF with ISM_2 at $\alpha = 0.3$ is given in the second row which showing that ISM_2 is better than MF at most of β values except when $\beta = (0.1, 0.3, 0.9)$ according to the numbers ($\mathbf{W}=84, 78, 76$) given in this row. These numbers represent the **W**-test values for the winner filter in each comparison at each β value. In the third row, the comparison of WF with ISM_2 is given and the **W**-test values shown that WF is winner only when $\beta = (0.1, 0.3)$ according to the numbers ($\mathbf{W}=74, 79$) given in this row. The same is done for all other filters at the selected α values.

Table 3. An evaluation of ISM_2 against all other filters by **W**-test

	β value	0.1	0.2	0.3	0.4	0.5	0.6	0.7	0.8	0.9
$\alpha = 0.3$	MF	MF	ISM_2	MF	ISM_2	ISM_2	ISM_2	ISM_2	ISM_2	MF
		84	73	78	71	82	72	73	71	76
	WF	WF	ISM_2	WF	ISM_2	ISM_2	ISM_2	ISM_2	ISM_2	ISM_2
		74	80	79	78	85	74	86	80	74
	ADF	ADF	ISM_2	ADF	ISM_2	ISM_2	ISM_2	ISM_2	ISM_2	ADF
		74	76	75	75	80	71	70	71	75
	ISM_1	ISM_1	ISM_1	ISM_1	ISM_1	ISM_1	ISM_1	ISM_1	ISM_1	ISM_1
		105	91	91	81	72	84	71	77	89
$\alpha = 0.5$	MF	MF	ISM_2	MF	MF	MF	ISM_2	ISM_2	ISM_2	MF
		73	81	102	96	71	75	75	70	81
	WF	ISM_2	ISM_2	WF	WF	ISM_2	ISM_2	ISM_2	ISM_2	WF
		72	78	99	93	75	84	84	79	76
	ADF	ISM_2	ISM_2	ADF	ADF	ADF	ISM_2	ISM_2	ISM_2	ADF
		76	79	90	95	72	75	75	73	78
	ISM_1	ISM_1	ISM_1	ISM_1	ISM_1	ISM_1	ISM_1	ISM_1	ISM_1	ISM_1
		96	84	115	108	87	78	78	81	97
$\alpha = 0.6$	MF	ISM_2	MF	MF	MF	MF	MF	MF	ISM_2	MF
		84	84	83	92	78	73	77	73	77
	WF	ISM_2	WF	WF	WF	WF	WF	WF	ISM_2	WF
		84	79	79	79	73	71	78	83	78
	ADF	ISM_2	ADF	ADF	ADF	ADF	ADF	ADF	ISM_2	ADF
		87	76	75	82	73	71	74	82	74
	ISM_1	ISM_1	ISM_1	ISM_1	ISM_1	ISM_1	ISM_1	ISM_1	ISM_1	ISM_1
		85	110	108	104	100	91	95	83	95

4.2   Peak Signal-to-noise Ratio (PSNR)

PSNR is the ratio between the maximum possible value (power) of the image and the power of distorting noise that affects the quality of its representation [6]. PSNR is usually used in image processing as a consistent image quality metric and the greater PSNR, the better the output image quality. The performance of proposed ISM enhancement method is evaluated on the basis of PSNR and measure values obtained are plotted in the graph shown in Fig. 4 for FVC2006(DB2_A) dataset. The PSNR values determine the efficiency of the denoising procedures. It is evident from the graph that mean PSNR values for ISM_2 at $\beta = 0.8$ is about 44 and is higher than that of all other filters. Added to that, PSNR values for ISM_1 (mean PSNR = 36) and ISM_2 (mean PSNR = 44) is better than that of the other filters. Thus the results clearly explain that the ISM enhancement method is well suited for many kinds of noisy images.

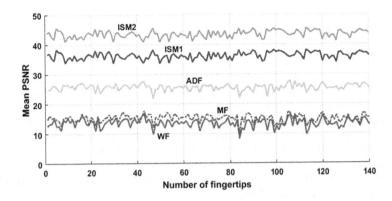

Fig. 4. PSNR for MF, WF, ADF, ISM₁ and ISM₂

5   Conclusions and Future Work

In this work, we proposed an integrated smoothing method (ISM) for image enhancement to increase the image quality used in fingerprint image recognition systems. The ISM is a linear combination of MF, WF, and ADF filtering techniques using coefficient parameters (α, β) to increase the image quality and remove many types of noise in digital images. This combination result in two sub-methods (ISM₁, ISM₂) which are considered in the pre-processing stage to increase the clarity of the fingerprint image, convert poor quality image to good quality image, and prepare the image for features extraction stage. Various performance metrics such as **W**-test and PSNR are used to evaluate and estimate each and every fingerprint enhancement technique. After demonstrating the aforementioned methods, we concluded that the enhancement of an image depends upon the coefficient parameter values used in this methods, while previous figures, tables, and graphs indicate that ISM₁ at most of α and β values is performing better in terms of **W**-test and PSNR since its **W**-test matching values and PSNR values has high values against all other filters. Also, it has been observed that, ISM₂ at $\beta = 0.8$ has the highest PSNR values and is performing better against MF, WF and ADF only at the small values of α for fingerprint images which has different kinds of noise. The future scope of this work is to evaluate ISM method by other denoising filters for other biometrics, image processing and computer vision that need the clarity of images to remove noise types that affects the image quality.

References

1. Clancy, T.C., Kiyavash, N., Lin, D.G.: Secure smart card based fingerprint authentication. In: Proceedings of the 2003 ACM SIGMM Workshop on Biometrics Methods and Application, pp. 45–52, New York (2003)

2. Ratha, N.K., Karu, K., Chen, S., Jain, A.K.: A real-time matching system for large fingerprint databases. IEEE Trans. Pattern Anal. Mach. Intell. **18**(8), 799–813 (1996)

3. Zhao, Q., Zhang, D., Zhang, L., Luo, N.: Adaptive fingerprint pore modeling and extraction. Pattern Recogn. **43**, 2833–2844 (2010)

4. Maltoni, D., Maio, D., Jain, A.K., Prabhakar, S.: Handbook of Fingerprint Recognition. Springer, Heidelberg (2009)

5. Reyad, O., Kotulski, Z.: Image encryption using koblitz's encoding and new mapping method based on elliptic curve random number generator. In: Dziech, A., Leszczuk, M., Baran, R. (eds.) MCSS 2015. CCIS, vol. 566, pp. 34–45. Springer, Heidelberg (2015). doi:10.1007/978-3-319-26404-2_3

6. Gonzalez, R.C., Woods, R.E.: Digital Image Processing, 3rd edn. Prentice-Hall Inc., Upper Saddle River (2006)

7. Kumar, B.A., Joshi, B.K.: A review paper: noise models in digital image processing. Signal Image Process. Inter. J. (SIPIJ) **6**(2), 63–75 (2015)

8. Jain, A.K., Hong, L., Pankanti, S., Bolle, R.: An identity-authentication system using fingerprints. Proc. IEEE **85**(9), 1365–1388 (1997)

9. Hong, L., Wan, Y., Jain, A.K.: Fingerprint image enhancement: algorithm and performance evaluation. IEEE Trans. Pattern Anal. Mach. Intell. **20**(8), 777–789 (1998)

10. Yun, E.K., Cho, S.B.: Adaptive fingerprint image enhancement with fingerprint image quality analysis. Image Vis. Comput. **24**, 101–110 (2006)

11. Gnanasivam, P., Muttan, S.: An efficient algorithm for fingerprint preprocessing and feature extraction. Procedia Comput. Sci. **2**, 133–142 (2010)

12. Hassanien, A.E.: Hiding iris data for authentication of digital images using wavelet theory. Pattern Recogn. Image Anal. **16**(4), 637–643 (2006)

13. Bouaziz, A., Draa, A., Chikhi, S.: Bat algorithm for fingerprint image enhancement. In: 12th International Symposium on Programming and Systems (ISPS), pp. 1–8. IEEE (2015)

14. Neeti, K., Khicha, A.: Image enhancement based on log-gabor filter for noisy fingerprint image. In: Satapathy, S.C., et al. (eds.) ICT4SD 2015 Volume 1. AISC, vol. 408, pp. 553–559. Springer, Singapore (2016)

15. Wu, C., Shi, Z., Govindaraju, V.: Fingerprint image enhancement method using directional median filter. In: Biometric Technology for Human Identification, SPIE 5404, pp. 66–75 (2004)

16. Jin, F., Fieguth, P., Winger, L., Jernigan, E.: Adaptive wiener filtering of noisy images and image sequences. In: Proceedings of IEEE International Conference on Image Process, vol. 3, pp. 349–352 (2003)

17. Tsiotsios, C., Petrou, M.: On the choice of the parameters for anisotropic diffusion in image processing. Pattern Recogn. **46**(5), 1369–1381 (2013)

18. Maio, D., Maltoni, D., Cappelli, R., Wayman, J.L., Jain, A.K.: FVC2004: third fingerprint verification competition. In: Zhang, D., Jain, A.K. (eds.) ICBA 2004. LNCS, vol. 3072, pp. 1–7. Springer, Heidelberg (2004). doi:10.1007/978-3-540-25948-0_1

19. Cappelli, R., Ferrara, M., Franco, A., Maltoni, D.: Fingerprint verification competition 2006. Biom. Technol. Today **15**, 7–9 (2007)

20. Wilcoxon, F.: Individual comparisons by ranking methods. Biom. Bull. **1**(6), 80–83 (1945)

RBG-CD: Residue Based Genetic Cancer Diagnosis

Mohamed A. Mahfouz$^{(\boxtimes)}$

Department of Computer and Systems Engineering, Faculty of Engineering,
Alexandria University, Alexandria, Egypt
m.a.mahfouz@gmail.com

Abstract. This research study is one of many computational methods that try to deal with huge gene expressions output from microarrays experiments in order to study different biological processes at the gene expression level. The proposed technique termed RBG-CD is a microarray based approach for cancer diagnosis in which the analysis of gene expression deviation from its normal composition is used to classify incoming sample. RBG-CD is based on new dissimilarity score called star residue (SR) which is similar to the mean square residue (MSR) that is usually used in biclustering problems. While MSR measures the coherence between rows and columns in a bicluster (sub-matrix of the input dataset), the introduced SR measures the coherence between a column representing a gene and other columns in a bicluster. RBG-CD starts by computing overestimated number of genes as representatives. The set of representatives and their corresponding biclusters should cover the whole set of genes. The representative is chosen in a way such that it has maximum relevancy with the target label vector and minimum SR with its corresponding bicluster. Then only the reduced set of genes (representatives) is used to compute a finally reduced set of overlapping biclusters in similar way which are able to distinguish between samples from different classes. Distinguishing Biclusters for class t has much lower residue on samples of class t than on samples from other classes. The proposed technique has been tested on several datasets and an improved performance has been achieved compared to several previously reported gene expression based cancer detection techniques.

Keywords: Cancer diagnosis · Classification · Gene expressions · Biclustering · Dimension reduction

1 Introduction

Cancer Diagnosis by manual examinations using traditional microscopic examination suffers from the dependency on the visual assessment of pathologists and has inter/intra observer variation in grading [3, 14, 18, 20]. In this approach, the microscopic analysis of samples by observing the level of organizational changes in tissues is used to assign cancer grade for incoming sample.

Accurate cancer diagnosis is critical step for cancer treatment. There are two major approaches for automating the process of cancer diagnosis which are texture analysis and genetic profile analysis. In texture analysis-based techniques, usually the contrast

© Springer International Publishing AG 2017
A.E. Hassanien et al. (eds.), *Proceedings of the International Conference
on Advanced Intelligent Systems and Informatics 2016*, Advances in Intelligent
Systems and Computing 533, DOI 10.1007/978-3-319-48308-5_40

between texture of normal and malignant parts of biopsy images are used for cancer detection on different body parts such as breast [21] and colon [4]. Several statistical and morphological features have proven to be good discriminators. Object-oriented (OO) texture analysis-based techniques [12], incorporates background information of the organization of normal and malignant tissue into the diagnostic process. Focus of few studies is to perform hyper-spectral analysis on biopsy images, and classify samples into different classes [1].

The second promising category of automated diagnostic methods of cancer detection is based on the analysis of thousands of human gene expressions to detect genetic alterations responsible for cancer. The input dataset for such methods is a matrix in which rows represent samples and columns represents genes (usually very large). Several studies worked as a feature selection technique to discover discriminating genes from the initial gene pool without actually classifying the sample into normal or malignant such as [2]. Genetic algorithms (GA) with the k-nearest neighbors (KNN) classifier [22] achieved much higher accuracy than [2]. The K-Nearest Neighbors (KNN) algorithm is one of the simplest instance-based learning algorithms suitable for multi-class problems but suffers high classification time. In [8] the performance of different distances that can be used in the KNN algorithm is evaluated. In [7] an evolutionary algorithms based method, is proposed for detection of colon cancer in which mutual information [10] is used to measures redundancy among genes and t-statistic were employed to minimize redundancy amongst selected genes in the future selection phase, after that, the genetic algorithm along with decision trees were used for classification. The results show that the combination of mutual information and genetic techniques is promising compared to others.

Several research studies are based on support vector machine (SVM). SVM is non-probabilistic binary classifier trying to find an optimal separating hyper plane having maximal distance to any data vector. In [19] a Genetic Algorithm (GA) was employed to select an optimal combination of linear SVM base classifiers. Top scoring pair method is employed to select 50 gene pairs and SVM classifiers were trained on them, a classification accuracy of 90 % is reported. Also Neural Network (NN) techniques are applied successfully to cancer diagnosis as in [16] by using wavelet transformation for reduction of feature space along with Probabilistic NN (PNN) to classify colon cancer data. Ensemble learning [11] is another approach to classification in which multiple classifiers are used, resulting multiple decisions, and a final decision is extracted from them. Therefore the decision of ensemble classifier is wrong only when most of them vote for the wrong class.

This research study introduces a measure for correlation between a gene and other genes called star residue. Star residue is similar to the mean square residue which is used in biclustering [13] but here it is not computed for the whole matrix representing a bicluster but computed for a certain gene compared to other genes in another bicluster. Star residue is used in tackling both the dimension reduction step and the classification step. The proposed technique start by computing overestimated number of genes as a reduced set using gene expressions belonging to all classes then uses this overestimated number of genes to reach a finally reduced set of genes that are able to distinguish between samples from different classes using gene expression from different classes separately in computing the star residue. The proposed technique was able to reduce the number of genes and still retaining higher class prediction accuracy. Several experiments are carried in order to validate the proposed algorithm and it was able to accurately divide the gene based samples into respective classes by considering the final reduced set of selected genes.

The remainder of this paper is organized as follows. Section 2 presents related concepts and commonly used abbreviations and symbols. Section 3 discusses the proposed techniques in detail. Section 4 discusses experimental results. Finally Sect. 5 concludes the paper and highlights future research directions.

2   Preliminaries

This section provides a supporting material to help the reader to better understand next sections. Commonly used abbreviations and symbols are listed in Table 1.

Table 1. Abbreviation and symbols used in the text

RBG-CD	The proposed classification scheme
X	Dataset comprising S samples size $n \times m$
S	Set of rows (samples) its size is n
G	Set of columns (genes) its size is m
S_t	Set of samples in X corresponds to class t, i.e. S_{normal} refers to normal samples.
T	Target label vector t = $\{t_1, t_2, ..., t_S\}$ corresponds to S samples in **X**
X^t	Dataset comprising samples of class t in **X**
n_t	Number of samples in X^t
SR(g_i, (I, J))	Star residue of genes g_i with set of columns J on the same set of rows I
x_{ij}	j^{th} feature value of the ith sample in the dataset X
α, β	Two thresholds used in controlling the number of reduced biclusters and distinguishing biclusters respectively

2.1   Gene Expression Data

Gene expressions are measured using DNA microarray technology by comparing genes expressions of one cell inhibited under certain condition (sample c) to gene expressions of another cell maintained in normal condition (sample n). In cancer diagnosis, each sample in the input dataset is labeled as malignant or normal in case of binary class or otherwise is given a cancer grade. Therefore, one gene expression means one feature in the feature vector. Likewise, one feature vector has multiple gene expressions of single gene on all samples. Entries of a gene expression matrix are real numbers representing gene expressions of each sample. The major difficulty of the analysis of this input data is the large number of genes (high dimensionality) with many irrelevant genes (noise) compared to very small number of samples.

2.2   Mean Square Residue(MSR)

Given a data matrix X with set of rows (samples) S of size n and set of columns (genes) G of size m a bicluster B = (I, J) where I subset of S and J subset of G such that each bicluster B satisfies some specific characteristics of homogeneity. Discovery of such biclusters is essential in revealing the significant connections in gene regulatory networks.

Mean squared residue [5] measures the coherence of the rows and columns in a bicluster and to quantify the difference between the value of an entry and the expected value of this entry predicted from the corresponding gene base, sample base, and the whole bicluster base as in (1). The residue of an entry x_{ij} in bicluster $B = (I,J)$ is

$$r_{ij} = x_{ij} - \frac{\sum\limits_{j \in J} x_{ij}}{|I|} - \frac{\sum\limits_{i \in I} x_{ij}}{|J|} + \frac{\sum\limits_{i \in I}\sum\limits_{j \in J} x_{ij}}{|I|\,|J|} \tag{1}$$

The mean square residue of the whole bicluster B is the average of the squared residue of its entries.

2.3   Star Residue

This paper introduces a measure for the coherence between a column j and other set of columns J ($j \notin$ J) on the same set of rows I. The star residue of column j can be defined as the average squared residue of its entries considering other columns in the bicluster (I, J) one by one. For example if we have a three columns bicluster (I,{1,6,9}) the star residue of another column 4 with this bicluster is the sum of squared residue of its entries in 3 biclusters (I,{1,4}), (I,{6,4}), (I,{9,4}) where I is the set of rows. The star residue of a column j with a bicluster (I, J) can be defined formally as follows:

$$SR(j,(I,J)) = \left(\sum_{k \in J} \sum_{i \in I} \left(x_{ij} - \frac{x_{ij} + x_{ik}}{2} - \frac{sum_j}{|I|} + \frac{sum_k + sum_j}{2|I|} \right)^2 \right) \Big/ (|J|\,\|\,|I|) \tag{2}$$

Where sum_j and sum_k are the sum of all entries belonging to I in columns j and k respectively. In computing SR for columns in the overestimated reduced set of genes, $|I|$ is fixed and equals n, number of samples. While in computing distinguishing biclusters for class t, $|I|$ equals n_t, the number of samples belong to class t. The complexity of computing the star residue SR can be reduced by keeping sum_j for each column j for each class.

2.4   Relevancy Measure

Similarity measures relative to correlation coefficient is one of the most common similarity measures in the field of bioinformatics. In this research study, the absolute Pearson correlation coefficient is used as a measure for the relevancy of a gene with the vector representing the target labels T. The mutual information [10] is well-founded measure for redundancy and can be used instead of absolute Pearson correlation. The absolute Pearson correlation coefficient of two random variables x and y is formally defined as follows:

$$ACorr(x, y) = |\frac{1}{n} \sum_{i=1}^{n} \left(\frac{x_i - \bar{x}}{\sigma_x}\right)\left(\frac{y_i - \bar{y}}{\sigma_y}\right)| \qquad (3)$$

In which $\bar{x}, \bar{y}$ are the sample mean of x and y respectively, while σ_x, σ_y are the sample standard deviation of x and y. It is a measure for how well a straight line can be fitted to a scatter plot of x and y. While the Pearson correlation coefficient fall between [−1, 1], the absolute Pearson correlation lies between [0, 1].

2.5 Performance Evaluation Measures

The quality of RBG-CD has been evaluated using well known performance measures such as accuracy, sensitivity, and specificity [9]. In order to calculate these measures number of true positive (TP), false positive (FP), true negative (TN), and false negative (FN) need to be computed. True negative and true positive are the count of correctly classified negative and positive samples respectively. False negative and false positive are the count of positive and negative samples, which are misclassified respectively.

$$Accuracy = \frac{TP + TN}{TP + FP + TN + FN} \qquad (4)$$

$$Sensitivity = \frac{TP}{TP + FN} \qquad (5)$$

$$Specificity = \frac{TN}{TN + FP} \qquad (6)$$

3 Proposed Classification Scheme

3.1 Classifier Construction Algorithm

This step has two main objectives, first to formulate a reduced feature vector for every sample without eliminating discriminative features (genes) which are necessary for classification. Second objective is to build distinguishing biclusters for each class. These biclusters are used in computing score for each class in the testing phase in the next section. As shown in Table 2, the algorithm starts by computing a degree for each gene depending on its relevancy to the target label vector computed using the absolute correlation coefficient presented above. After that, the genes are sorted according to their relevancy descending. In step 3, C_0 initially contains the complete list of sorted genes. In each iteration i, the top of C_0 is moved to the set of representative R as

Table 2. Proposed classifier construction algorithm

Input	X	: *Dataset of size n samles×m genes*
	T	: *Target label vector { $t_1,t_2,...,t_S$} corresponding to S samples in X*
	α	: *Dataset dependent threshold controls no. representatives*
	β	: *threshold in [0,1], controls no. distinguishing biclusters*
	maxn	: the maximum number of failed actions on distinguishing biclusters
Output:	R	: list of distinguishing biclusters for each class as a classifier

Begin (CGA)

1. $D(g_i)= ACorr(g_i,T)$ for $i=1,2,...m$ //*compute relevancy using abs. corr.*
2. Sort genes in descending order according to their values of $D(g_i)$.
3. Let C_0 be the ordered list of all genes which is computed in step 2
   $i = 0$; //*compute representatives and fill groups with* redundant genes
   while $C_0 \neq \{\}$ //i.e. $\neq \phi$
       $i = i + 1$;
       Move the top element of C_0 to the set of representative R as r_i
       $C_i=\{\}$

       for each gene $g_j \in$ C_0 such that $SR(r_i,(S, \{g_j\} \cup C_i)) <= α$

           Move g_j from C_0 to C_i **endfor endwhile**
4. **while** $R \neq \{\}$ // *randomized search to compute distinguishing biclusters*
       Select random r_i in R
       Let $f_i=\{\}$, *iter*=0
       While *iter<maxn*
           Select random $r_j \neq r_i$
           If r_j belongs to f_i then the action is deleting r_j otherwise it is adding r_j to f_i
           Let fb_i is f_i before the action and fa_i is f_i after the action
           Compute the gain of the action as follows:
           If $|SR(r_i, (S_m,fa_i))- SR(r_i, (S_n,fa_i))|> |SR(r_i, (S_m,fb_i))- SR(r_i, (S_n,fb_i))|$
               apply the action and update f_i
               Set *iter* = 0   Else   *iter=iter*+1   **endIf**
       endwhile
       Remove r_i from R
       If $|SR(r_i, (S_m,f_i))- SR(r_i, (S_n,f_i))| / |SR(r_i, (S_m,f_i))+ SR(r_i, (S_n,f_i))| > β$
           If $SR(r_i, (S_m,f_i)) < SR(r_i, (S_n,f_i))$
               Add r_i, f_i to distinguishing biclusters of the malignant class $B_{malignant}$
           Else
               Add r_i, f_i to distinguishing biclusters of the normal class B_{normal} **endIf endIf**
   endwhile
5. **Output** Classifier($B_{malignant}$,B_{normal})

End

representative for the new empty set C_i and termed as r_i. Any following gene g_j in the sorted list is added to C_i if the star residue of r_i with $\{g_j\} \cup C_i$ less than a threshold α. In step 4, randomized search [23] is used in identifying a biclusters for each representative in R. The difference between the star residue of the representative with f_i on the rows of normal samples and its star residue on malignant samples is maximized. The input parameter *maxn* controls the number of accepted failed trials before termination of randomized search step [23]. The representative is always moved out from R and the difference between its star residue with f_i on both normal and malignant is compared to input threshold β. If the bicluster is accepted then it is assigned to class t (normal or malignant) which has the lower star residue. The set of distinguishing biclusters of each class t denoted B_t is the output of the algorithm.

3.2   Predicting the Class Label for a Test Sample

After constructing a classifier as shown in Table 2, next comes the issue of testing data formulation. In this work, 10-folds Jackknife cross-validation scheme [9] has been applied in training and testing the proposed technique. In Jackknife test, data are divided into F partitions (folds). F–1 partitions participate in training, and the classes of the instances belonging to the remaining partition are predicted by the decision model based on the training performed on F–1 training partitions. This process is repeated F times and the class of each sample are identified. Data have been divided into 10 partitions. Each partition hosts S/10 instances except the last one that may contain less than S/10 instances. In the training phase only training data are used in computing distinguishing biclusters. As shown in Table 3, in order to predict the class of a test sample, the new star residue of each representative $r_i \in B_{normal}$ with the bicluster $(S_n \cup u_s, f_i)$ need to be computed. The absolute difference between newly computed star residue and previously stored star residue for (S_n, f_i) is added to *Ndiff* which is previously initialized to zero. Similarly *Mdiff* is computed for malignant class. The test sample is assigned to normal if *Ndiff is less than Mdiff* otherwise it is assigned to malignant.

Table 3. Proposed algorithm for predicting the class label for a test sample

Input: Classifier($B_{malignant}$, B_{normal}) :produced by the above algorithm in Table 2
               u_s : a test(unknown) sample
Output: The label of the class (Normal or Malignant)
Begin
     1. Set *Ndiff* and *Mdiff* to *zero*
     2. For each distinguishing bicluster B_i belongs to B_{normal} =(S_n, fi)
          $Ndiff = Ndiff + \|SR(r_i, (S_n, f_i)) - SR(r_i, (S_n \cup u_s, f_i))\|$
     3. For each distinguishing bicluster B_i belongs to $B_{malignant}$ =(S_n, fi)
          $Mdiff = Mdiff + \|SR(r_i, (S_n, f_i)) - SR(r_i, (S_m \cup u_s, f_i))\|$
     4. **if** *Ndiff* < *Mdiff* **then**
              return Normal
          else
              return Malignant
End

4 Experimental Results and Discussion

To validate the performance of the proposed RBG-CD for classification, it has been tested on two standard cancer datasets, namely KentRidge [2] and Stuart [17]. As shown in Table 4, KentRidge is colon cancer microarray data samples while Stuart is prostate cancer microarray data. Table 4 shows the part of the body, no. malignant samples, no. normal samples, total no. samples and no. genes respectively for each dataset.

Table 4. Gene expression datasets

Dataset	Type	Malignant	Normal	Total samples	No. genes
KentRidge	Colon	40	22	62	2000
Stuart	Prostate	38	50	88	8828

4.1 Performance Comparison with Related Work on KentRidge Dataset

To measure the efficacy of the proposed RBG-CD, accuracy, sensitivity, and specificity have been calculated and compared to the results of several related algorithms reported in [11] on KentRidge dataset. Table 5 shows the values of these measures for RBG-CD as well as the reported values for other algorithms. The results demonstrate better performance of RBG-CD in terms of higher accuracy, sensitivity and specificity compared to KNN, PNN, GA/KNN and DT. The input parameter α is found to be dataset dependent and can be estimated using histogram. α is set to 400 in all experiments on KentRidge in order to produce suitable set of representatives. The reported performance for RBG-Cd is average values on different values for $\beta = 0.95, 0.96, 0.97, 0.98, 0.99, 1$. The accuracy of RBG-CD ranges from 0.893–0.967. The complexity of the proposed algorithm is less than the ensemble SVM algorithm (GECC) which its performance highly dependent on the dimension reduction step. The reported value for GECC when using PCA but its performance ranges from 0.919–0.987 using different dimension reduction techniques other than PCA. Average results of RBG-CD are slightly less than NN but the black box nature of NN techniques gives an advantage for RBG-CD.

Table 5. Performance results on KentRidge dataset [11]

Algorithm	Performance		
	Accuracy	Sensitivity	Specificity
KNN	0.854	0.77	0.90
PNN	0.871	0.77	0.93
Decesion Tree (DT)	0.855	0.73	0.95
GA/KNN	0.890	0.89	0.88
NN	0.944	0.92	0.93
GECC (PCA)	0.919	0.90	0.95
RBG-CD	0.939	0.93	0.95

4.2   Performance Comparison with Related Work on Stuart Dataset

In this experimental study the performance of RBG-CD is compared to another set of algorithms KNN, Naïve Bayes, Random Forest, SVM and TC-VGC as shown on Table 6. The values reported for the first four algorithms are the average values reported in [15] for different ratio of selected features using two feature selection algorithms. The reported results for TC-VGC in [15] ranges from 0.838 to 0.983 depends on the procedure and the values used for the input parameters. Also the reported value for RBG-CD is the average values for different values of β ranges from 0.9–1 while α is set to 0.5 in all experiments used in calculating the reported value of RBG-CD on Stuart dataset. The accuracy of RBG-CD ranges from 0.855–0.991. TC-VGC [15] is based on computing correlation between pairs of genes and computing distinguishing pairs in similar manner to RBG-CD. However computing SR is much less expensive by keeping column sums for each class. Also considering the star relation between genes and the SR measure is novel approach in cancer diagnosis.

Table 6. Performance on Stuart dataset [15]

Algorithm	Accuracy
KNN	0.866
Naïve Bayes	0.898
Random Forest	0.816
SVM	0.890
TC-VGC	0.838-0.983
RBG-CD	0.923

5   Conclusions

The proposed RBG-CD technique employs a new measure (star residue) termed SR for producing a small set of biclusters that are able to distinguish between samples from different classes. The performance of RBG-CD has been validated on two gene expression datasets, and quite promising enhancement over several existing techniques in terms of Accuracy, Sensitivity and Specificity have been achieved without increase in classification time. The biological relevance [6] of the distinguishing biclusters produced by the proposed algorithm and a method for systematically estimating the input parameters is left as a future research.

References

1. Akbari, H., et al.: Detection of cancer metastasis using a novel macroscopic hyper-spectral method. In: Proceedings of SPIE (2012)
2. Alon, U., Barkai, N., Notterman, D.A., et al.: Broad patterns of gene expression revealed by clustering analysis of tumor and normal colon tissues probed by oligonucleotide arrays. Proc. Natl. Acad. Sci. **96**, 6745–6750 (1999)
3. Andrion, A., et al.: Malignant mesothelioma of the pleura: inter-observer variability. J. Clin. Pathol. **48**, 856–860 (1995)

4. Bauer, S., et al.: Integrated segmentation of brain tumor images for radiotherapy and neurosurgery. Int. J. Imaging Syst. Technol. **23**(1), 59–63 (2013)
5. Cheng, Y., Church, G.: Biclustering of expression data. In: Proceedings of Eighth International Conference Intelligent Systems for Molecular Biology (ISMB 2000), pp. 93–103 (2000)
6. Dupuy, A., Simon, R.M.: Critical review of published microarray studies for cancer outcome and guidelines on statistical analysis and reporting. J. Natl. Cancer Inst. **99**, 147–157 (2007)
7. Kulkarni, A., Kumar, N., Ravi, V., et al.: Colon cancer prediction with genetics profiles using evolutionary techniques. Expert Syst. Appl. **38**, 2752–2757 (2011)
8. Medjahed, S.A., Saadi, T.A., Benyettou, A.: Breast cancer diagnosis by using k-nearest neighbor with different distances and classification rules. Int. J. Comput. Appl. **62**(1), 1–5 (2013)
9. Mehdi, H., Chaudhry, A., Khan, A., et al.: Carotid artery image segmentation using modified spatial fuzzy C-means and ensemble clustering. Comput. Methods Programs Biomed. **108**, 1261–1276 (2012)
10. Peng, H., Long, F., Ding, C.: Feature selection based on mutual information: criteria of max-dependency, max-relevance, and min-redundancy. IEEE Trans. Pattern Anal. Mach. Intell. **27**(8), 1226–1238 (2005)
11. Rathore, S., Hussain, M., Khan, A.: GECC: gene expression based ensemble classification of colon samples. IEEE/ACM Trans. Comput. Biol. Bioinf. **11**, 1131–1145 (2014)
12. Rathore, S., Hussain, M., Khan, A.: A novel approach for colon biopsy image segmentation. In: Proceedings of Complex Medical Engineering Conference, pp. 134–139 (2013)
13. Madeira, S.C., Oliveira, A.L.: Biclustering algorithms for biological data analysis: a survey. IEEE Trans. Comput. Biol. Bioinf. **1**(1), 24–45 (2004)
14. Scholefield, J., et al. (eds.): Challenges in Colorectal Cancer, 2nd edn. Wiley-Blackwell, Oxford (2006)
15. Shin, E., Yoonb, Y., Ahna, J., Parka, S.: TC-VGC: a tumor classification system using variations in genes' correlation. Computer Methods and Programs in Biomed **104**(3), e87–e101 (2011)
16. Shon, H.S., Sohn, G., Jung, K.S., et al.: Gene expression data classification using discrete wavelet transform. In: Proceedings of International Conference on Bioinformatics & Computational Biology, pp. 204–208 (2009)
17. Stuart, R., Wachsman, W., Berry, C.C., Wang-Rodriguez, J., Wasserman, L., et al.: In silico dissection of cell-type-associated patterns of gene expression in prostate cancer. Proc. Natl. Acad. Sci. USA **101**, 615–620 (2004)
18. Thomas, G.D., et al.: Observer variation in the histological grading of rectal carcinoma. J. Clin. Pathol. **36**, 385–391 (1983)
19. Tong, M., Liu, K.H., Xu, C., et al.: An ensemble of SVM classifiers based on gene Pairs. Comput. Biol. Med. **43**, 729–737 (2013)
20. Young, A., Hobbs, R., Kerr, D. (eds.): ABC of Colorectal Cancer, 2nd edn. Wiley-Blackwell, Chichester (2011)
21. Yuan, Y., et al.: A sparse regulatory network of copy-number driven gene expression reveals putative breast cancer oncogenes. IEEE/ACM Trans. Comput. Biol. Bioinf. **9**(4), 947–954 (2012)
22. Li, L., Weinberg, C.R., Darden, T.A., et al.: Gene selection for sample classification based on gene expression data: study of sensitivity to choice of parameters of the GA/KNN method. Bioinformatics **17**(12), 1131–1142 (2001)
23. Rastrigin, L.A.: The convergence of the random search method in the extremal control of a many parameter system. Autom. Remote Control **24**(10), 1337–1342 (1963)

Distributed Topological Extraction Protocol for Low-Density Wireless Sensor Network

Walaa Abd-Ellatief[1]([⊠]), Hatem Abdelkader[2], and Mohee Hadhoud[1]

[1] Information Technology Department, Faculty of Computers and Information,
Menoufia University, Shibin Al Kawm, Egypt
walaaali285@gmail.com, mmhadhoud@yahoo.com
[2] Information Systems Department, Faculty of Computers and Information,
Menoufia University, Shibin Al Kawm, Egypt
hatem6803@yahoo.com

Abstract. Wireless sensor networks consist of a varying number of randomly deployed tiny sensors. One of its main applications is monitoring environmental phenomenons in remote and rugged places. In such applications, sensors are distributed with a helicopter passing over the area of study. This method of distribution causes different density subregions inside the network and unknown topology structure. Information is needed to be shared between sensors to enable them to detect and extract topological features about their random distribution. In this paper, we propose a technique which helps nodes to construct boundary cycles around each subregion of specified density-level. This cycles helps in extracting and describing the layout or the topology of the deployed network. Evaluation of our proposed technique shows that it uses less average node degree which equals 3 and achieves about 50 % decrease in energy consumption than Heuristic Boundary Cycles Finding Technique.

Keywords: WSN · Boundary nodes · Transmission range · Topology extraction

1 Introduction

Wireless Sensor Networks (WSNs) are networks with spatially distributed nodes called sensors. Sensors have good special characteristics which make WSNs differ from traditional wireless ad-hoc networks. These features allow it to be used in a wide range of discriminative applications. Sensor nodes are tiny ships called motes with a sensing, processing, and communication devices embedded on it. What makes it different is its sensing capabilities. It enables them to monitor physical phenomenons in real environments such as, but not limited to, pressure, temperature, and moisture. Examples of WSN applications are: (i) area monitoring where specific phenomenon has to be recognized such as air pollution [1, 2], forest fire detection [3, 4], disasters detection [5, 6], (ii) health-care [7, 8], (iii) civil services [9, 10]. One of the challenges in WSN is the used deployment method. According to the size of the application, one from two deployment methods can be used. The first one is planned deployment. It is used in small sized applications where a small number of sensors has to be deployed in the area

© Springer International Publishing AG 2017
A.E. Hassanien et al. (eds.), *Proceedings of the International Conference on Advanced Intelligent Systems and Informatics 2016*, Advances in Intelligent Systems and Computing 533, DOI 10.1007/978-3-319-48308-5_41

of interest. The location of these sensors has to be studied and analysed before deployment. This study helps in deciding the minimum number of sensors that is needed to satisfy the application requirements of coverage. The second deployment method is random deployment. It is used in large scale applications which specially serve in remote and rugged areas. In these situations, it is difficult for human to get through the area of the network. Therefore, helicopters are usually used to pass over these areas and scatter sensors. Random deployment method always suffers from holes in coverage. Because sensors are lightweight, it can be easily affected by air while dropping. Moreover, antennas may be poorly placed. So, it is always recommended to use large number of nodes, from hundreds to thousands, to compensate the unused nodes and achieve good coverage rate. However, this is not a good solution because of its high cost. As mentioned in [11], a 300×300 network, 95 % coverage rate can be achieved using 300 node while 100 % coverage rate can be achieved using 600 node. Extra 300 node for only more 5 % increase in coverage has to be considered. Therefore, random deployment with minimum number of nodes achieving desired coverage followed by relocation techniques which tries to solve coverage errors was proposed. However, all of them assume sensors are mobile to be able to move and optimize its location in the network [12]. Also, assuming all deployed nodes to be mobile is not a wise cost. Therefore, in this paper, we propose a topological extraction technique for low density randomly deployed WSN. The aim of topological extraction is to provide sufficient information to describe the layout of the network, i.e. the relation between points and subregions of high density or voids. After that, this information can be used by a mobile robot to place more nodes in holes or low coverage subregions to optimize the coverage in the networks. The remaining of the paper has been organized as follows: Sect. 2 shows the problem formulation of our work. Section 3 covers the related work section. In Sect. 4 we discusses our proposed technique. Simulation tests and evaluation of the proposed technique is presented in Sect. 5. Finally, we conclude our work in Sect. 6.

2   Problem Formulation

Topology extraction problem is a main issue in wireless sensor networks. From our review, it is always solved in a high density networks with a very large number of nodes that can achieve a relatively constant density all over the area of the network. In such networks, the shape of the network can be identified by finding the nodes with the highest degree which construct the skeleton of the topology. Added to that, when holes appear, it can be easily identified where the decrease in the degree can be easily observed. However, this assumption is not practically hold in real applications as the distribution of nodes is always random. By analysing nodes densities in randomly deployed networks, we can notice that there is no constant level of density if we divide the network into small units of area. This division yields areas with different densities starting from very high density to holes without nodes inside it. The network appears as if it is divided to a set of subregions. Each one with different density level of nodes as shown Fig. 1. The variation of density all over the network causes problems in the coverage and the connections between nodes. Therefore, in this paper, we focus on

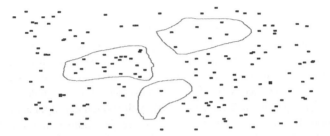

Fig. 1. Random distribution of wireless sensor network.

randomly deployed networks to extract spatial topological features which may help in additional redeployment process to overcome these shortages. Topological description for such networks can be a set of polygons, one for each subregion with specific density level. These subregions can be recognized by cycle of boundary nodes resides at the border of them.

Our proposed technique aims to define each subregion in randomly distributed WSN by defining nodes which border each subregion. Simple spatial analysis can be done on the neighboring information between nodes. This analysis aims to define the type of the node in its subregion. Node can be interior in its subregion or resides at the border of it. The identification of boundary nodes is a famous problem in WSN applications. Applications such as surveillance, and monitoring. To recognize these nodes, the proposed techniques depends on the adaptation of the transmission range value of each node to recognize boundary nodes sharing the same subregion with it. Each sensor node has Omni-directional antenna which enables communication within a maximum determined area. The transmission area is a disk centred at the node. The radius of the disk is the transmission range of this node. Two nodes within each other's transmission ranges are neighbors and can communicate directly. According to the node's transmission range, the number of neighbors can differ. In this paper, we propose the method upon which each node set its transmission range to a value that enable it to detect only neighbors of its subregion.

3 Related Work

In this paper we try to identify the shape of the network by describing it with a set of connected polygons that define each subregion with constant density. The aim is to collect sufficient information to recognize boundary nodes which define vertices of polygonal subregions and interior nodes inside it to extract the topology layout. Previous research made in this area tries to solve this problem such as in [13]. Kröller et al. proposed a topology extraction technique for large sensor networks. It is based on a sufficient nodes density. The technique is tested on a network with 60,000 sensor arranged along intersected streets. Nodes are deployed as strips representing streets and connected in the intersection of two streets. The technique constructs a flower-based graph of the connections between nodes along the street until it detects the boundary nodes of each street. The technique also identifies nodes which lies in the intersection cores between intersecting streets. Clusters are finally defined for intersections and streets. The average node's degree in this technique is about 20. In [14], Wang et al. proposed a technique to define the boundary nodes of the whole network area and recognize holes inside the network. This method depends on constructing the shortest path tree at one selected node. Paths of the tree are continuous with enough number of nodes achieving full connectivity. However, it diverges prior to a hole and meet again after it. The technique was tested on a network with enough density, i.e. average node's degree is about 10. In [15], Lederer et al. used a set of landmark areas in the network boundary with sufficient density to detect the shape of the network. Voronoi diagram for these landmarks and its dual combinatorial Delaunay complex is constructed. This technique figures out the correct network layout without flips where parts of network can fold over each other. Selected landmarks used to capture important topological information then it is used to reconstruct the network layout. The performance of this technique increases with higher nodes density and higher density of landmarks. The average node degree ranging from 6 to 10. In [16], Liu et al. proposed a skeleton extraction technique. The aim in this technique is to recognize nodes which figures the shape of the network. This is based on the idea that skeleton nodes has larger neighborhood size than others. The average node degree ranging from 7 to 10. From our review, all the previous techniques depends on high average node degree with a defined network structure and holes shapes which is not the case in randomly deployed networks. Figure 2 shows examples of used topologies in the previously proposed techniques.

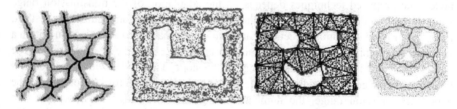

Fig. 2. Example of topologies used in previous work. [13–16]

4 Proposed Technique

In this section, the steps of our proposed technique are presented with the following assumptions: (i) The number of nodes is defines as N. (ii) The area of the network is defined as $W \times H$. (iii) Nodes are deployed using uniform random distribution. The aim is to define a set of nodes which are used as vertices of polygons to define subregions of the network. These polygons define areas according to the density of nodes inside it. The result is a set of empty subregions and others with high density nodes. To do that, we need to recognize the boundary node of each subregion. For this purpose, we use a border detection technique proposed by Shukla et al. in [17]. Many other techniques was proposed, but we choose this one because each node needs only its 1st hop neighbor's information to decide if it is interior or boundary one. It is used in a distributed algorithm and by one step it can differentiate between nodes as interior or boundary node. Node will be interior one if it is enclosed in a triangle of three neighbor nodes. Otherwise, it will be a boundary node. Shukla proposed his technique to define nodes which border the whole network. In our work, we need to define borders of each subregion, so we propose to adapt this border detection technique. We need to add a metrics that will be used as a stopping thresholds to border each subregion, not the whole area. This value can be the transmission range of each node. The transmission range of each node controls the number of neighbors that each node can detect. Therefore, it has an effect in the determination of the type of nodes. The question here is what is the value of the transmission range that has to be used to detect boundary nodes of subregions of the network? To test if the node is interior, it needs to recognize at least three neighbors in different directions which can construct a triangle around the test node.

To calculate the needed value of the transmission range, we use Eq. (1). It is proposed in by Bettstetter in [18] and relates between transmission range and the number of neighbors of each node. The degree of any node is described as $d(n)$ where n is the number of neighbors. Bettstetter assumed N randomly deployed nodes in an area $A = W \times H$. It is also assumed that a uniform random distribution with large N, a constant node density can be measured as $\rho = \frac{N}{A}$. Node's transmission radio range r_0 covers its transmission area $A_0 = \pi r_0^2$. The probability that the network has a minimum node degree $d_{min} \geq n$, is given by:

$$P(d_{min} \geq n) = \left(1 - \sum_{N=0}^{n-1} \frac{(\rho \pi r^2)^N}{N!} \cdot e^{-\rho \pi r^2}\right)^n \qquad (1)$$

In our protocol, the probability that the number of neighbor nodes is greater than or equal to 3 has to be maximized for a given value for number of N nodes and area size $(W \times H)$, where r changes between 1 to the maximum value of (W and H). To do that, $P(d_{min} \geq 3)$ must be greater than a specified threshold (Th) as follows:

$$\left(1 - \sum_{N=0}^{2} \frac{(\rho \pi r^2)^N}{N!} \cdot e^{-\rho \pi r^2}\right)^3 > Th \qquad (2)$$

The optimal value of Th is 1. A close value to 1 has been chosen for Th which is 0.999999. Therefore, using the last equation, the minimum value of r form the range $[1, max(W, H)]$ can be calculated. For example, if N = 100, and W = H = 500; the transmission range is in the range [1, 500]. Equation (1) suppose a uniform distribution of constant density achieved by using a large number of nodes that grantee enough constant density in each small unit area of the network. This contradicts with our assumption of initial low density distribution. Therefore, we propose to make a mediation for the resulted value of transmission range. This step is verified with simulation tests. In Table 1, Sample of different transmission range values to be applied for different network areas and different number of nodes. The value calculated in this step is the initial transmission range that will be used by all nodes to test if it is interior node or not.

Table 1. Values of transmission range for different network sizes

N	H = W	Min r
50	200	28
150	400	33
200	500	35
220	600	40

Defined boundary nodes of the subregions of different density in the network used to describe the layout of the topology. Each node broadcast advertisement message to its neighbors to announce its type. Upon receiving this information, each node constructs a neighboring table which contains all its 1st hop neighbors' type. Boundary nodes neighboring table contains additional information about its boundary neighbors. It has to know the direction to reach these neighbors to be able to construct the cycle which will surround this area. Figure 3 shows an example of simple subregion with nodes N1 and N2 as interiors and N3, N4, N5, N6 as boundary nodes. Constructed tables according to this subregion are shown in Tables 2, 3 and 4. After this, topological cycles construction procedures start by each boundary node. It summarizes this information to construct edges table which define all edges constructed by boundary nodes in the network.

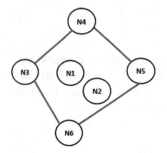

Fig. 3. Example of interior and boundary nodes constructing a subregion

Table 2. Neighboring table of interior node N1

Neighbours ID	Neighbours type
N2	Interior
N3	Boundary
N4	Boundary
N5	Boundary
N6	Boundary

From this table a polygons setup process is started to define all polygonal border lines for each subregion. The polygonal border lines are recognized by a loop of nodes which starts and ends at the same node. For example the polygon in Fig. 3 is N3-N4-N5-N6-N3. Given the estimated distance between each two adjacent boundary node and the direction from one to the other, each boundary node can detect its closed boundary cycle. Figure 4 shows an example scenario after detecting boundaries of all polygons that define regions of different density levels of the distributed nodes.

Table 3. Neighboring table of N3 boundary node

Neighbours ID	Neighbours type	Boundary neighbour directions
N1	Interior	
N2	Interior	
N4	Boundary	Clockwise
N6	Boundary	Counterclockwise

Table 4. Edges table of N3 boundary node

Edge ID	Starting node	Ending node
E1	N3	N4
E2	N3	N6

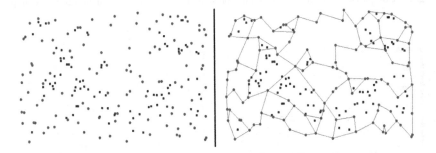

Fig. 4. The construction of boundary links describing each subregion.

5   Evaluation of the Proposed Technique

To evaluate our protocol, we compare it with the protocol presented in [19]. It proposed a Heuristic Boundary Cycles Finding technique. We choose this one because it assumed the least average degree used than all other techniques mentioned in the related work. The author assumed (5–9) average degree in the centralized scenario and (6–9) in the distributed case in low-density networks. This technique is based on choosing a seed node with the local highest degree among its neighbors. The technique starts at these node and begin to gradually augment a boundary cycle by adding more nodes to this cycles. These nodes has to be neither isolated with degree equals 0 nor an end-node with degree equals 1. This process is repeated until it hits the boundary of the network. If more than one cycle is constructed, merging process for these cycles begins. In this protocol all nodes are involved in the construction of the boundary cycles. On the contrary, our technique uses a (3) average degree all the time and the step of transmission range adaption selects a set of boundary nodes to define subregions. Only these nodes are involved in the construction of topological boundary cycles which causes a less number of messages and less energy consumption during the process of boundary cycles construction. Therefore, the benefit of our technique is the prediction of the nodes which will be help in the process of topology extraction in advance. To present the efficiency of our technique, we preview a sample scenario example in Fig. 5. We also measure the average number of exchanged messages and the average consumed energy by nodes involved in the construction process of the

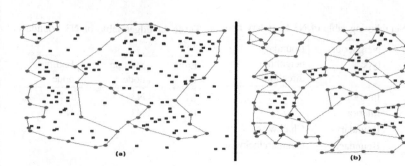

Fig. 5. Comparison between (a) technique presented in [19] and (b) our proposed technique.

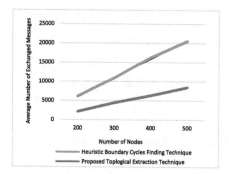

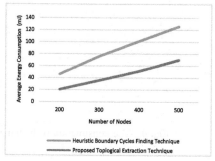

Fig. 6. Average number of exchanged messages for the construction of boundary cycles.

Fig. 7. Average energy consumption for the construction of boundary cycles.

boundary cycles. This values are measured for a 400 × 400 network with number of nodes ranging from 200 to 500 node. These values are viewed in Figs. 6 and 7.

6   Conclusion and Future Work

In this paper, we propose a simple technique to define the layout of low density randomly deployed network. This is done by defining boundary links that border polygonal subregions of similar density level resulted from the random distribution of sensors in the network. Our technique depends on node's transmission range adaptation by choosing a value which depends on the total number of distributed nodes and the total area of study. It is adapted to be able to test the type of each node in its own subregion. Each node can be interior or boundary one. Other techniques discussed in this paper depend on a high density of the distributed nodes. Some of them considers clear defined figure of holes, for example concave regions or separated holes not near to each other. These conditions of deployment are not feasible to many real applications. Future work includes the refinement of the constructed polygons to merge adjacent empty subregions and the use of defined boundary links to guide a redeployment process. This aims to redeploy nodes of high density subregions to low density subregions and holes.

References

1. Al-Ali, A., Zualkernan, I., Aloul, F.: A mobile GPRS-sensors array for air pollution monitoring. IEEE Sens. J. **10**(10), 1666–1671 (2010)
2. Khedo, K.K., Perseedoss, R., Mungur, A.: A wireless sensor network air pollution monitoring system. Int. J. Wireless Mob. Netw. **2**, 31–45 (2010)
3. Lloret, J., et al.: A wireless sensor network deployment for rural and forest fire detection and verification. Sensors **9**(11), 8722–8747 (2009)
4. Putra, E.H., Hariyawan, M.Y., Gunawan, A.: Wireless sensor network for forest fire detection. TELKOMNIKA (Telecommun. Comput. Electron. Control) **11**(3), 563–574 (2013)
5. Aziz, N.A.A., Aziz, K.A.: Managing disaster with wireless sensor networks. In: 13th International Conference on Advanced Communication Technology (ICACT). IEEE (2011)
6. Ohbayashi, R., et al.: Monitoring system for landslide disaster by wireless sensing node network. In: SICE Annual Conference. IEEE (2008)
7. Darwish, A., Hassanien, A.E.: Wearable and implantable wireless sensor network solutions for healthcare monitoring. Sensors **11**(6), 5561–5595 (2011)
8. Lee, Y.-D., Chung, W.-Y.: Wireless sensor network based wearable smart shirt for ubiquitous health and activity monitoring. Chem. Sens. Actuators B **140**(2), 390–395 (2009)
9. Ceriotti, M., et al.: Monitoring heritage buildings with wireless sensor networks: the Torre Aquila deployment. In: Proceedings of the 2009 International Conference on Information Processing in Sensor Networks. IEEE Computer Society (2009)
10. Laisheng, X., et al.: Research on traffic monitoring network and its traffic flow forecast and congestion control model based on wireless sensor networks. In: International Conference on Measuring Technology and Mechatronics Automation, ICMTMA 2009. IEEE (2009)
11. Senouci, M.R., Mellouk, A., Aissani, A.: Random deployment of wireless sensor networks: a survey and approach. Int. J. Ad Hoc Ubiquitous Comput. **15**(1–3), 133–146 (2014)
12. Sharma, V., et al.: Deployment schemes in wireless sensor network to achieve blanket coverage in large-scale open area: a review. Egypt. Inf. J. **17**, 45–56 (2015)
13. Kröller, A., et al.: Deterministic boundary recognition and topology extraction for large sensor networks. In: Proceedings of the Seventeenth Annual ACM-SIAM Symposium on Discrete Algorithm. Society for Industrial and Applied Mathematics (2006)
14. Wang, Y., Gao, J., Mitchell, J.S.: Boundary recognition in sensor networks by topological methods. In: Proceedings of the 12th Annual International Conference on Mobile Computing and Networking. ACM (2006)
15. Lederer, S., Wang, Y., Gao, J.: Connectivity-based localization of large-scale sensor networks with complex shape. ACM Trans. Sens. Netw. (TOSN) **5**(4), 31 (2009)
16. Liu, W., et al.: A general framework of skeleton extraction in sensor networks. IEEE Sens. J. **16**(4), 1103–1116 (2015)
17. Shukla, S., Misra, R.: Angle based double boundary detection in wireless sensor networks. J. Netw. **9**(3), 612–619 (2014)

18. Bettstetter, C.: On the minimum node degree and connectivity of a wireless multihop network. In: Proceedings of the 3rd ACM International Symposium on Mobile Ad Hoc Networking & Computing. ACM (2002)
19. Sitanayah, L., Datta, A., Cardell-Oliver, R.: Heuristic algorithm for finding boundary cycles in location-free low density wireless sensor networks. Comput. Netw. **54**(10), 1630–1645 (2010)

Particle Swarm Optimization with Random Forests for Handwritten Arabic Recognition System

Ahmed Sahlol[1,5]($\boxtimes$), Mohamed Abd Elfattah[2,5], Ching Y. Suen[3], and Aboul Ella Hassanien[4,5]

[1] Faculty of Specific Education, Damietta University, Damietta, Egypt
atsegypt@du.edu.eg
[2] Faculty of Computers and Information, Mansoura University, Damietta, Egypt
[3] Director of CENPRMI, Faculty of Engineering and Computer Sciences, Concordia University, Damietta, Canada
[4] Faculty of Computers and Information, Cairo University, Damietta, Egypt
[5] Scientific Research Group in Egypt (SRGE), Damietta, Egypt
http://www.egyptscience.net

Abstract. There are many problems facing the processing of a handwritten Arabic recognition system such as unlimited variation in human handwriting, similarities of character shapes and their position in the word. This paper presents a handwritten Arabic character recognition system based on Particle Swarm Optimization with random Forests. The main objective of the proposed system is to improve the recognition rate and reduce the feature set size. The proposed system is trained and tested by a well-known classifier; Random forests (RF) on CENPRMI dataset. The proposed optimization algorithm obtained promising results in terms of classification accuracy as the proposed system is able to recognize 91.66 % of our test set correctly, as well as, it reduced the computational time. When comparing our results with other related works we find that our results is the highest among other published results.

Keywords: Particle swarm optimization (PSO) · Optical character recognition (OCR) · Feature reduction · Optimization

1 Introduction

Arabic Alphabet is not only used by Arabic spoken but also by other languages like Persian, Kurdish, Malay and Urdu. Millions of people especially in Africa and Asia use it. The arabic alphabet has 28 basic characters. Sixteen Arabic letters have from one to three dots. Number and position of these dots differentiate between the otherwise similar characters (like ﺡ, ﺥ). Additionally, some

© Springer International Publishing AG 2017
A.E. Hassanien et al. (eds.), *Proceedings of the International Conference on Advanced Intelligent Systems and Informatics 2016*, Advances in Intelligent Systems and Computing 533, DOI 10.1007/978-3-319-48308-5_42

characters (like ڬ, أ) can have a zigzag-like stroke (Hamza ء). These dots and Hamza are called secondaries, and they are located above the primary character part as in (ALEF أ), or below like BAA ﻴ), or in the middle like JEEM ﺝ). Within a word, some characters connect to the preceding or following characters, and some do not connect. The shape of an Arabic character depends on its position in the word; a character might have up to four different shapes depending on whether it is isolated, connected from the right (ending form), connected from the left (beginning form), or connected from. Arabic characters do not have fixed size (height or width). Each character size varies according to its position in the word. There are many applications handwritten characters such as writer identification and verification, form processing, interpreting handwritten postal addresses on envelopes and reading currency amounts on bank checks.

Arabic characters do not have fixed size (height or width). Each character size varies according to its position in the word. There are many applications handwritten characters such as writer identification and verification, form processing, interpreting handwritten postal addresses on envelopes and reading currency amounts on bank checks. Particle swarm optimization (PSO) is a heuristic global optimization method and one of the well-known swarm intelligence and is based on the research of bird and fish flock movement behavior. PSO is widely used to solve the optimization and feature selection problems. In PSO, each solution is considered as a particle with specific characteristics (position, fitness, and speed vector) that defines the moving direction of the particle [1]. In PSO, each solution is considered as a particle which is defined by position, fitness, and a speed vector which defines the moving direction of the particle. For decades lots of research has been done to solve the problem of handwritten Arabic character recognition. Various approaches have been proposed to deal with this problem. Many techniques have been adopted in various ways to improve accuracy and efficiency. Abandah et al. [2] explored best sets of feature extraction techniques and studies the accuracy of well-known classifiers for Arabic characters. The principal component analysis (PCA) technique was used to select best subset of features out of a large number of extracted features. It was figured out that using more features did not substantially improve the accuracy. However, for features fewer than 25 features, a quadratic discriminant classifier was more accurate than the linear classifier. The highest recognition accuracy achieved was 87 % using LDA classifier. While in [3] five feature selection methods were presented to select feature subsets for recognizing handwritten Arabic characters. The used methods selected out of 96 extracted features. Subset sizes of 20 features or less are generally sufficient to achieve low classification error. The genetic algorithm method selected the best subsets of features compared with the other four methods. Also in [4] a feature extraction approach for handwritten Arabic characters was proposed. Efficient feature subsets were selected using multi-objective genetic algorithm. Various feature subsets were evaluated according to their classification error using the SVM. Selecting feature subset from the moments of the whole letter and the constituents of the letter yields around 9 % reduction in the classification error compared to selecting a feature subset from the moments of the whole body alone. Classification error of about 10 % was achieved when feature subsets were selected from the moment features and other efficient features. In [5] proposed approach based on hybrid method between PSO and Rough set for feature selection has been proposed, the used dataset was BCI Competition 2008 Dataset, the experimental result shows that the proposed approach has achieved the best ROC value. The overall accuracy of the proposed approach achieved the best results among the other the other methods when applying eleven test measurement (sensitivity, F-measure, accuracy, ...etc.). It was recommended the using this approach in another domain such as Big data, biological signals and Networks field. More details about the most recent Swarm, their Principles, Advances, and Applications can be found in [6].

This paper presents a hybrid character recognition system based on PSO Algorithm for feature selection to achieve better classification accuracy with minimum number of features instead of using the whole features. The organization of this paper is as follows: Sect. 2 presents the proposed approach. Experimental results with discussions are described in Sect. 4. Finally, conclusions and future work are provides in Sect. 5.

2   The Proposed Approach

Character recognition system has three main steps:

2.1   Preprocessing

Input data has to be prepared for further processing steps, this make extracted features more efficient in recognition. The first main step is binarization. Otsu method [7] is adopted because of its simplicity. The second step is noise removal. Noise comes during scanning the document or from the conversion the document to digital form or even during binarization. The following techniques were used for removing different kinds of noises:

1. **Median filtering** [8]: It is used to remove random noise like Pepper noise that can appear in a document image during the conversion process or caused by dirt on the document.
2. **Dilation** [9]: In dilation, a structuring element is used for expanding the shapes contained in the input image, therefore, if a pixel is set to foreground; it remains. However, if a pixel is set to background, but at least one of its eight neighbors is set to foreground, the pixel is converted to foreground. Nevertheless, if a pixel is set to background and none of its eight neighbors is set to foreground, the pixel remains set to background.
3. **Morphological noise removal:** Some morphological techniques is used like removing data pixels if they less than threshold, also removing a pixel if all adjacent or diagonal neighbors are zeros.

2.2 Feature Extraction

Since Arabic characters have unlimited variations, similarities of distinct character shapes, character overlapping and interconnections of neighboring characters so that need special kind of features. Moreover, number of Arabic characters share the same primary shape, but differ only in the presence/absence and location/number of dots, so to differentiate such character, dot information must be captured explicitly and structural features can naturally capture this information. The used Features can be categorized into:

Gradient. Gradient images are created from the original image (generally by convolving with a filter, one of the simplest being the Sobel filter [10] or Prewitt operator [11]) for this purpose. Each pixel of a gradient image measures the change in intensity of that same point in the original image, in a given direction. In this work the trained and the tested image were normalized by 128×128, this normalization scale was chosen because it achieves better results among the 64 64 or the 32×32 normalization scales.

Vertical and Horizontal Projections. Vertical projection profile is the sum of white pixels that is perpendicular to the y axis, horizontal projection is the sum of white pixels but it is perpendicular to the x axis. The character is traced horizontally along the x-axis using the sum of number of white pixels present in each row.

Right and Left Diagonal Projections of Each Part of the Four Triangular Character Parts. We started by splitting a character diagonally into four parts, and then the right and the left diagonal for each cropped part of the character have been got. Each character has been divided into four triangular and cropped each part by determining the boundaries for the last non-zero pixel, then the right and the left diagonal have been got for each triangle of the character.

Height to Width Ratio. The aspect ratio (height/width ratio) of the character is a useful feature because some Arabic characters are wider than others.

Number of Holes. It was supposed that this feature can give accurate results if all secondaries have been eliminated with the character correctly. So, only the main body of the character has been kept for this feature by using the algorithm in [12], then a tracing for the boundaries of holes inside the character has also been achieved by using Moore-Neighbor tracing algorithm modified by Jacob's stopping criteria [5].

Number of Secondaries. This feature recognizes the secondary component of a character and return with number of them (like Hamzas "ؤئ and dots). The connected component labeling techniques [13] has been used for that. The main body has been identified easily as it is usually the largest component so the second bigger connected components are considered as a secondary.

2.3 Position of Secondaries

As known, the position of a secondary is the only feature to distinguish between a character and another. Those groups of characters (ظ , ط , غ , ع , ـن , ـب , ـي , ـت) and (ح ' خ) can be only distinguished by machine or by human eye by the position of the position of the secondary component. By using connected component labeling techniques [12], a secondary component has been obtained, then, finally, the position of it has been obtained. Normalization of feature data: the Minmax normalization method [14] has been used because it is simple and more efficient in terms of time consuming.

2.4 Feature Selection

The main idea of feature selection is to choose a subset of input variables by eliminating features with little or no predictive information.

Feature Reduction Using PSO Algorithm. In PSO algorithm, particle swarm consists of N particles, and the position of each particle stands for the potential solution in D-dimensional space. The particle updates its positions and hence the obtained solutions are according to the following:

– Keeping its inertia; keeping its own position.
– Changing the condition according to its own most optimist position; named *pbest*.
– Changing the condition according to the swarm's most optimist position; social expertise named *gbest*.

The position of each particle in the swarm is affected both by the most optimum position during its movement and the position of the most optimist particle in its surrounding [1]. The repositioning of each particle is calculated using the Eq. (1).

$$X_{i,d}^{t+1} = X_{i,d}^t + V_{i,d}^{t+1}, \tag{1}$$

where X is a vector representing the particle position in the space, t is the iteration number, i is the particle number, $x_{i,d}^{t+1}$ is the updated position of dimension d for particle i, and $x_{i,d}^t$ is the current position of particle i at iteration t, $V_{i,d}^{t+1}$ is the modified velocity of a particle i. The velocity vector is calculated as given in Eq. (2):

$$V_{i,d}^{t+1} = V_{i,d}^t + c_1 rand_1^t \left(pbest_{i,d}^t - X_{i,d}^t \right) + c_2 rand_2^t \left(gbest_d^t - X_{i,d}^t \right), \tag{2}$$

where $V_{i,d}^t$ is the current velocity of dimension d for particle i at iteration t, $pbest_{i,d}^t$ is the most optimist position for particle i along dimension d at time t and $gbest_d^t$ is the most optimist position for the swarm at time t along dimension d, c_1, and c_2 represent the amount of loyalty and selfishness of particles. Usually, c_1 is equal to c_2 and they are equal to 2, $rand_1^t$ and $rand_2^t$ represent random fiction always uniformly randomized in the range $[0, 1]$.

According to wrapper approach, feature selection algorithm uses a machine learning technique as evaluation. In this work, we used the artificial neural network (ANN) with PSO algorithms to evaluate all the possible solutions. We use a simple ANN architecture with 7 hidden layers. ANN is used as fitness function to evaluate all the possible solutions during the optimization process of each algorithm. In this stage, ANN helps the optimization algorithms to reach to optimal solution (best selected feature set). In the training process, every particle represents one input feature. The training set is used to build ANN model to guide the optimization algorithms during the feature selection process.

3   Classification Phase

The feature vectors generated after the three different feature extraction algorithms were used as input to this stage. The main aim is to minimize the input features that will reduce the classification time as well as maximizing the classification performance. Finally, the output of the feature selection is given to classification phase to test the classification performance and computational time on the reduced feature set. The extracted features fed into the Random Forest (RF). An evaluation for features reduction ratio before and after applying the PSO is done. Figure 1 describes the proposed system overview.

4   Implementation Results and Discussions

CENPARMI dataset [15] is used to measure the efficiency of the proposed algorithm. The dataset contains several thousands of images of Arabic handwritten characters. The characters have been written by 328 writers. The samples were carefully selected to represent initial, medial, and final of the 28 Arabic character forms. In this work we work only with the 28 basic Arabic characters. Table 1 shows a collection of samples of isolated Arabic characters in CENPARMI.

The results of the used classifiers were evaluated using a well-known statistical Eq. (3); the overall accuracy:

$$Accuarcy = \frac{TP + TN}{TP + TN + FP + FN} \tag{3}$$

Where TP is the true positive samples, TN is the true negative samples, FP is the false positive samples and FN is the false negative samples. The proposed optimization approach is implemented by MATLAB, except Random forest which is implemented by Weka. To compare between the optimized feature sets and

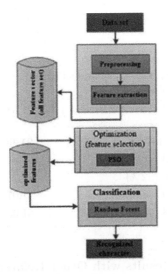

Fig. 1. The proposed OCR optimization approach

non-optimized in terms of time consumption, recognition rate and number of used features, see Table 2.

From Table 2, although the selected feature sets by PSO achieve slightly better results comparing with the whole feature set (91.66 % comparing with 91.58 %) as well as in terms of processing time (58.83 seconds comparing with 711.27 seconds). All the other PSO rounds didn't go less than 91 % of recognition rate.

Table 1. Variation in characters from dataset

Character	Variations			
ﻙ	ل	ﻙ	ﻙ	ﻙ
ﻩ	ﻪ	ﻪ	ﻪ	ﻪ
ﻥ	ﻥ	ﻥ	ﻥ	ﻥ
ﺯ	ﺯ	ﺯ	ﺯ	ﺯ
ﻕ	ﻕ	ﻕ	ﻕ	ﻕ

Table 2. Accuracy rate and running time for the whole feature set and five best PSO optimized feature sets by RF

Feature set	Features number (reduction)	RF (Parameters)	Accuarcy rate (%)	Time (sec)
The whole feature set	624	Trees = 100, seed = 1	91.58	71.72
PSO	330(47%)		**91.66**	58.83
	323 (48%)		91.44	74.29
	304 (52%)		91.13	64.29
	319 (48%)		91.07	61.45
	323 (48%)		91.03	64.10

4.1 Comparing the Results with Other Related Work

Table 3 summarizes the most recent work in Arabic handwritten characters. As seen, they are sorted according to the used classification approach, feature selection, time reduction and accuracy rate.

As seen in Table 3, the proposed algorithm based on PSO algorithm achieves the best recognition rate among the other state of the art on handwritten Arabic isolated character systems. A sufficient OCR system tends to have the best recognition rate to avoid misclassified characters. The main contribution of this research includes building of a new offline Arabic handwritten character recognition system based on the PSO optimization algorithm. The evaluation of our system has been done by applying the proposed algorithm on a standard dataset (CENPARMI) and classifying the selected features by a popular classifier (RF). The proposed method obtained competitive accuracy rate (91.66%).

Table 3. Comparisons between previous works and the proposed system

Previous work	Classifier	Feature selection approach	Time consumption achieved	Recognition rate (%)
G. A. Abandah et al. [4]	LDA	Not used	(Not used)	87
Ahmed.T. Sahlol [16]	FFD- Neural network	Not used	(Not used)	88
Ahmed.T. Sahlol [17]	SVM (rbf)	Not used	(Not used)	89.2
The proposed approach	RF	PSO	58.83 secs	91.66

5   Conclusion and Future Work

This paper presents an approach of improving the recognition rate and reduction of processing time for handwritten Arabic characters recognition. Each character has to pass through OCR steps, pre-processing, feature extraction and classification. Some important pre-processing operations were achieved. Some features were also extracted to overcome variations of characters shapes in a standard dataset (CENPARMI). To reduce the number of features extracted and to improve recognition rate, PSO optimization algorithm was used. The experimental results showed that PSO algorithm improved the classification rate (which measured by RF classifier) while at the same time, classification processing time was minimized comparing with that achieved by the whole feature set. When comparing the proposed system against another published works, the results proved advances in both computational time and classification accuracy over them.

In the future, we intend to work on up-to-date bio-inspired optimization algorithms to resolve feature selection to further minimize the number of features and maximize the classification accuracy. We may also examine the use of PSO for feature selection on data sets with a large number of features.

References

1. Eberhart, R.C., Kennedy, J.: A new optimizer using particle swarm theory. In: Proceeding of the Sixth International Symposium on Micro Machine and Human Science, Japan, pp. 39–43 (1995)
2. Abandah, G.A., Younis, K.S., Khedher, M.Z.: Handwritten arabic character recognition using multiple classifiers based on letter form. In: Proceedings of the 5th International Conference on Signal Processing, Pattern Recognition, and Applications (SPPRA), pp. 128–133 (2008)
3. Abandah, G.A., Malas, T.M.: Feature selection for recognizing handwritten arabic letters. Dirasat Eng. Sci. **37**(2), 1–21 (2011)
4. Abandah, G., Anssari, N.: Novel moment features extraction for recognizing handwritten Arabic letters. J. Comput. Sci. **5**(3), 226–232 (2009)
5. Udhaya Kumar, S., Hannah Inbarani, H.: PSO based feature selection and neighborhood rough setbased classication for BCI multiclass motor imagery task. Neural Comput. Appl. 1–20 (2011)
6. Hassanien, A.E., Alamry, E.: Swarm Intelligence: Principles, Advances, and Applications. CRC Taylor, Francis Group (2015). ISBN 9781498741064
7. Otsu, N.: A threshold selection method from gray-level histograms. IEEE Trans. Syst. Man Cybern. **9**(1), 62–66 (1979)
8. Jae, S.L.: Two Dimensional Signal and Image Processing. Prentice Hall PTR, Upper Saddle River (1990). 07458
9. Rosenfeld, A., Kak, A.: Digital Picture Processing, 1st edn. Academic Press, New York (1976). ISBN 10: 0125973608
10. Sobel, I.: An isotropic 33 gradient operator. In: Freeman, H. (ed.) Machine Vision for Three Dimensional Scenes, pp. 376–379. Academic Press, London (1990)

11. Prewitt, J.M.S.: Object enhancement and extraction. In: Lipkin, B., Rosenfeld, A. (eds.) Picture Processing and Psychopictorics, pp. 75–149. Academic, New York (1970)
12. Cheriet, M., Kharma, N., Liu, C., Suen, C.Y.: Character Recognition Systems: A Guide for Students and Practitioners. Wiley, New York (2007)
13. Zeki, A.M., Zakaria, M.S., Liong, C.Y.: Isolation of dots for Arabic OCR using voronoi diagrams. In: Proceedings of the International Conference on Electrical Engineering and Informatics, pp. 199–202 (2007)
14. Hann, J., Kamber, M.: Data Mining: Concepts and Techniques, 3rd edn. Morgan Kaufman, San Francisco (2011)
15. Alamri, H., Sadri, J., Suen, C.Y., Nobile, N.: A novel comprehensive database for Arabic off-line handwriting recognition. In: Proceedings of 11th International Conference on Frontiers in Handwriting Recognition, ICFHR, vol. 8, pp. 664–669 (2008)
16. Sahlol, A.T., Suen, C.Y., Elbasyoni, M.R., Sallam, A.A.: Investigating of pre-processing techniques and novel features in recognition of handwritten Arabic characters. In: Gayar, N., Schwenker, F., Suen, C. (eds.) ANNPR 2014. LNCS (LNAI), vol. 8774, pp. 264–276. Springer, Heidelberg (2014). doi:10.1007/978-3-319-11656-3_24
17. Sahlol, A.T., Suen, C.Y., Elbasyoni, M.R., Sallam, A.A.: A proposed OCR algorithm for cursive handwritten Arabic character recognition. J. Pattern Recogn. Intell. Syst. (PRIS) 2(1), 90104 (2014)

Group Impact: Local Influence Maximization in Social Networks

Ragia A. Ibrahim[1,3($\boxtimes$)], Hesham A. Hefny[1], and Aboul Ella Hassanien[2,3]

[1] Institute of Statistical Studies and Research (ISSR), Cairo University, Giza, Egypt
ragia.ibrahim@gmail.com, h.hefny@ieee.org
[2] Faculty of Computers and Information, Cairo University, Giza, Egypt
[3] Scientific Research Group in Egypt, Giza, Egypt
aboit@gmail.com
http://www.egyptscience.net

Abstract. Influence maximization defined as the problem of selecting influential small set of nodes that maximize influence spread over the social network. Influence maximization considered in number of domains, emergence situations, viral marketing, education, collaborative activities and political elections. In this paper, we propose Local Information Maximization LIM, considering group impact in terms of local propagation where the influencer(s) of each community has a direct effect on the nodes in the same community. We conduct experiments on synthetic data set and compare the performance of the LIM to various heuristics.

1 Introduction

The need for large information propagation can be shown in several situations: Emergency situations, viral marketing, education and research, collaborative activities and political events, such as in election. Emergency situations, such as in the great eastern Japan Earthquake. People behavior before and after the event can be inferred. Social media were used effectively during and after this earthquake, all revealed on a blog. What people did on and after the earthquake. Such a blog is one of the first blogs of people's actions a disaster [1]. In recent Syrian crises, social media used to organize medical supplies and donations from the Syrian diaspora, they uploaded videos in Arabic to YouTube that give advice to physicians inside Syria, and used a barcode system to track medical supplies to Syria [2].

In Cooperative activities such as what is going on the Arab spring, recent revolutions use social media to organize actions. Social events for attending an event such as musical party, Yoga class or books sale are done over social networks. In education, social media recently used to disseminate good exercise. Authors found a correlation between the inter-state total energy reduction during this campaign and between the amounts of inter-state online Twitter discussion [3].

Enermous amount of information flows through social networks, individuals communicate overs social networks various methods e.g. sharing posts, comments

© Springer International Publishing AG 2017
A.E. Hassanien et al. (eds.), *Proceedings of the International Conference on Advanced Intelligent Systems and Informatics 2016*, Advances in Intelligent Systems and Computing 533, DOI 10.1007/978-3-319-48308-5_43

on others post or private chat. This kind of interact may change individual state. State could be an opinion or an action, such as adopting new product or following specific political party. Affecting people state called influencing. RecentlyInfluence Maximization IM defined by Domingos and Richardson [4] thought of viral marketing or word of the-mouth as an epidemic spread. Marketing is not an independent action any more. It can be viewed as social network, customers decision of buying or not.

Individuals, relations and transmited data; are the main components that social networks formed around. Billions of people are interacting over virtual world forming communities, marketing product(s), spreading information and influencing followers decision(s). Lots of data are distributed over web on average in 2013, per minute, 216000 Instagram photos uploaded. Google performs 2 million searches and 72 h worth of video uploaded to YouTube within the space of 60 s. 278 000 tweets are tweeted and 20 million videos viewed on Flicker [5].

One of the most well known examples of influence propagation is Ted Williams called "GoldenVoice", Ted is a voice-over artist after being a homeless person. In January 3, 2011, he turned from a homeless man in Columbus, Ohio into a saved person, a reporter recorded a video for the homeless Ted. The video was re-posted to YouTube where it receives 11 million hits in months, and Williams received several job offers [6].

Also, On 16 November 2011, "WhyThisKolaveriDi" (English: Why This Murderous Rage, Girl?) was officially released. The song became the most searched YouTube video. Within a few weeks, YouTube honoured the video with a "Recently Most Popular" Gold Medal award and "Trending" silver medal award, by crossing 1.3 million hits in a short time [7].

Also in viral marketing Oreoa advertisement is a well known case. In February 2013 the Oreoa sandwich cookie company, tweeted during power failure in the Super Bowl XLVII, "Powerout? NoProblem, You can still dunk it in the dark". The blackout formed an adv. Opportunity, it has been retweeted more than 14, 000 times and the image on Facebook has gotten more than 20, 000 likes in around two days period. This was an example of maximum benefit gained with minimum cost since twitter is free but advertising industry's is much more expensive [8].

Other example, When Apple's launched an event for its new iPhones 5s earlier on Sep 2014, Nokia with a timely tweet pointed out the similarity between the newly unveiled iPhone 5s colour phones and its own range of bright designs. The two-word tweet from Nokia's UK account - "Thanks, #Apple" - associated with a picture stating "Imitation is the best form of flattery" quickly went viral and has been retweeted more than 38,000 times and more than 10, 000 likes [9].

Not much attention has been accord to the study of local information maximization problem. People adopt new behavior or information regarding mainly to the most influence neighbor can reach them, being in the same cluster mans more influence of affecting neighbors. This paper studies the relative importance between the vertex and its community from information propagation view. It studies the problem of maximization for the number of people that adopt information and provide efficient solutions with bounds.

Paper organization. Section 2 presents related work on information propagation maximization. Section 3 presents Problem definition and also points out Independent Cascade model. Section 4 our LIM model and the proposed algorithm for this model. We discuss results and future directions in Sect. 5.

2   Related work

A landmark study by Coleman, Katz, and Menzel [10], documented that for social contagion (diffusion of innovation) doctors decision to adopt a new drug influenced by people they were connected to, considering social and professional connections. They found that doctors that have not been named as a friend or as an advisor or disconnected doctors adoption rate in the beginning are much lower than linked doctors who has been named by others. Authors found that overtime the rate of adoption increased and the gap between rates decreased.

Identifying influential actors has received a great attention in recent research. For the influence maximization problem, given a probabilistic model of information propagation such as the Independent Cascade Model IC, a network, and a budget k, the objective is to select a set S of size k for initial activation so that the expected value of $f(S)$ is maximized for the hall graph.

Kempe et al. [11] proof that $f(S)$ is monotone, non-negative and submodular function. Therefore, the theorem of Nemhauser, Wolsey and Fisher 78 [12] where applied: if a function is non-negative, monotone and submodular, the greedy algorithm provides $(1-1/e) \cong 63\%$ approximation of the optimal value. Authors deduced that maximization problem could be solved via greedy algorithm within 63 % in linear time.

Evaluating $f(S)$ is an open question, but a very good estimation by simulating, repeating the diffusion process is often enough. Kempe et al., found that the system converges around 10 000. Kempe's solution most obvious disadvantages is taking enormous time.

Kimura et al. [13], used bond percolation and graph theory to determine the most influential nodes, instead of greedy algorithm, cause of greedy algorithm disadvantages. This method was used to solve influence maximization problem for both IC and LT models. Kimura et al. [13] shows that their method needs much less computation than the greedy hill- climbing algorithm. Using bond percolation method each edge (link) in graph **G** earns a state, that is determined by probability distribution. The state whether occupied or unoccupied, in other words active or not active.

Leskovec et al., enhanced the greedy algorithm by using lazy evaluation [14]. Leskovec et al., answered the question: Given information spread is happening, select set of nodes that detect this process effectively. Lescovec et al., addressed the problem of detecting outbreaks in networks, selecting few places or pipe joints to install sensors, in order to detect contamination as quickly as possible. The same concept could be applied to social network by selecting actors to observe for detecting virtuous spread. In addition, selecting few effective blogs to know what is the news spread over the network rather than reading full big blogs, select small effective set.

Cost-Effective Lazy Forward CELF, submodularity guarantees that the marginal gain decreases with the increase of solution size. The CELF optimization uses the submodularity feature of maximizing influence objective to reduce the number of evaluations on the influence spread of vertices estimation calls sufficiently, excluding the first iteration. However, the improved algorithm still takes long time.

Chen et al., illustrated [15] that Kempe et al., 2003 had a serious draw back. Monte Carlo simulations of the influence cascade were run for very large number of times with the purpose of getting accurate results. In addition, solution S is completely invisible for large graph over $500\,k$ edges. Chen et al., proposed scalable method thats faster than existing greedy algorithm for Maximum Influence Arborescence MIA model. Chen et al., invented a heuristic scalable MIA algorithm for IC on directed graph. Firstly the algorithm evaluates influence of each node when the seed set is empty, then the algorithm iterated two steps till finding k seeds.

Based on the community structure, Wang et al. [16] proposed a community-based greedy algorithm (CGA). Firstly: apply community detection to help influence maximization. Then influence maximization done mainly at local community level to speed up the computation. Authors show that is applicable to the large networks. Partitioning done by extending the information influence mechanism based on Independent Cascade model IC. The generated communities via step partition were small and dispersed; Combination is developed to combine these communities to get a better result. After dividing the network into communities Monte Carlo simulations are conducted within each community instead of the whole network to get the seed set of influential nodes [17].

3   Problem Definition

Influence maximizing problem aims to maximize the spread of influence to the largest number of users (nodes) overs social network, by targeting k initial set of nodes to adopt an innovation, information or influence spread. Selecting this set of nodes should let the innovation or product reach a large cascade. The optimization problem of selecting the most influential nodes is an NP-hard [11].

Influence function $f()$ defined as follows: for a set S of nodes, $f(S)$ is the expected number of active nodes at the end of the propagation process, assuming that S is the set of nodes that are initially active and the graphs are finite, thus the processes terminate in a number of time steps bounded by the total number of nodes n. From the spreader point of view, if the S is the set of initial adopters; given a budget k, where k is the number of set elements $k = |S|$ how large can $f(S)$ be, if a set S of size K initial adopters have been chosen. The objective is to activate the most larger amount of nodes is referred to as influence maximization [18].

Local Influence Maximization LIM aims to maximize influence locally. Thus, given a social graph $G(V, E)$ and a stochastic diffusion model on G. Partition G into sub graphs G_i, number of partitions of G denoted by I. Find a seed set $S_0 \subset V(G_i)$ with $| \sum_{i=1}^{I} S_0 | \leq k$, that maximizes $\sigma(S_0)$. Where $\sigma(S)$ is the expected number of nodes active at the end of the cascade. The aim is to compute $S^* \subseteq V$. The complexity of the proposed algorithm depends mainly on community detection used algorithm complexity in addition to number of nodes and edges in the induced sub graph I, number of selected k's from each community depends on the community relative importance, noticing that there might be very small sub groups where their marginal gain added to the total gain wont affect the spread. Thus ignoring very small group wouldn't harm the influence spread.

$$S = argmax_{S_0 \subset V(G_i), \sum_{i=1}^{I} S_0 | \leq k} \tag{1}$$

Determining the optimal number of k for each community depending on community relative importance is not considered before. Proposed approach considers the problem as posed in terms of local propagation, where the influencer(s) of each community has a direct effect only on the nodes in the same community. Considering social network represented as a directed graph $G(V, E)$ where V stands for Individuals and links (edges) $E \subset V \times V$ represents social ties. A function ρ associates a weight or probability with each tie $p(u, v)$, $\rho : E \rightarrow [0, 1]$. Influence from u to v perceives that u influenced v with probability $p(u, v)$ to do the same action. In local viral marketing, the maximization means selecting $V(seedset)$ that maximizes number of final adopters. A simple contagion could be presented via Independent Cascade model IC since each node gets active via one of its neighbors such as in viral of simple information.

3.1   Independent Cascade Model IC

The Independent Cascade IC model is a stochastic (prediction) model that depends on graph structure whether it is highly connected or sparse graph. Cascading means the behavior of being active or inactive that outcomes from inference (function) are based on neighbors or earlier individuals, thus cascading is based on the influence of neighbors that affect each node or individual to be earned one of two states whether active or inactive [19].

Firstly, seed nodes selected to be an active node (infectious). These seeds trigger information spread or the virus over the network. Then, each node or individual has got property to infect its neighbors according to their tie (edge). Thus IC uses diffusion probability associated with each edge and in each iteration each active node uses that probability once to activate its neighbors. Accordingly, for nodes (u,v) if u is active and v is not, the activation succeeds depending on nodes v and u and other nodes that already attempted to change v but failed. Once v is activated, it can activate its neighbors too [18,19].

4   Local Information Maximization LIM

Since most of real life networks follow power law distribution, it is not simple to find influencer actors; the skewed tail means that many actors have small degree all over the network despite of network huge size. Community based solution is naturally convening since socially people tend to be into groups that is translated in social into communities; those communities could be due to geographical location, financial level, level of education and so on. Thus, finding the most influencing actors in such community guarantees affecting actor on the same community.

The proposed algorithm is more realistic instead of calculating influence degree of the node (actor) for the whole graph, it would be computed for each community. In addition, selecting this community that presents relatively high marginal gain more than others, would reduce the cost. *birdswithfeatherflocktogether* friends adopt similar ideas and behavior, in each community there will be influencers that starts the adoption process than it propagates across the community. For example, we can see on social web pages, community of specific actor fans that propagate the actor news and events. Community of specific type of music lovers e.g. Jazz fans who shared news about Jazz parties and Jazz players even people who attend specific event or invite others to those events, they invite their friends who share the same properties.

Algorithm 1. LIM Algorithm

1: Input: Graph (G)
2: Output:selected seed set S
3: initialization $S = \phi$;
4: *Partition* G *into* i *subgraphs*
5: $RImportance(G_i)$
6: **for** $i = 1\ to\ k$ **do**
7:     **for** $v \in G_i$ **do**
8:         $v_{mg} = F(v_{ci})$
9:         $U_{Ci} = argmax(C_i)$
10:    **end for**
11:    $S \leftarrow \cup F(U_C i)$
12: **end for**
13: **return** S

$RImportance(G_i)$ function calculate the relative importance for each sub graph depending on graph connectivity. The more connected the sub-graph the higher percent of nodes in k will be adopted, where k represents the number of initial actors that will maximize $\sigma(S)$. $F(v_{ci})$ calculates each vertex v influence on its community C_i. $F(Ci)$ ranks selected k_i in each subgraph regarding marginal gain. The v_{mg} is node marginal gain, $RImportance(C)$ gets the heights nodes that have relatively higher importance in spreading information in the graph (G).

Community detection algorithms [20] applied to the proposed method is Walktrap, Walktrap is an approach based on random walks. The general idea is that if you perform random walks on the graph, then the walks are more likely to stay within the same community because there are only a few edges that lead outside a given community. Thus random walks will gravitate to stay within communities instead of bonding to other communities.

5   Experiments and Results

The proposed approach experimented over synthetic networks. Forest fire [21], Forest fire is a generative model that produces networks mimicking the structure of growing networks. New nodes attach to the network by burning through existing edges in epidemic fashion. As for the diffusion probabilities on the edges, each is chosen uniformly at random. The model produces realistic networks in terms of heavy-tailed degree distributions, community structure and other network properties. First graph NW_1 generated using the parameters: forward burning probability 0.3, and backward burning probability 0.25. The graph consists of 500 nodes and 1127 edges. Second graph NW_2 generated using the same parameters as in NW_1. The graph consists of 2000 nodes and 4965 edges, the graph generated using igraph library [22] (Fig. 1).

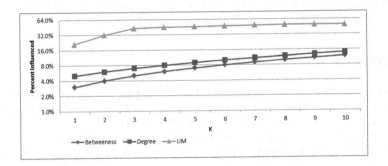

Fig. 1. Evaluating $LIM - NW_1$

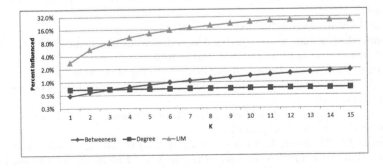

Fig. 2. Evaluating $LIM - NW_2$

Experiments show that community based approach for selecting seed set of nodes that maximize information propagation have high priority on heuristic approaches, especially on big graphs (Fig. 2).

6   Conclusion and Future Works

Results show that proposed LIM performs much better than heuristic methods, noticing that the success of the proposed method subject of many factors such as the graph structure and the graph connectivity, but the main benefit of using LIM is that: the more sparse subgraph results from the algorithm the more nodes ignored since very small subgroups wouldn't be considered relatively to other bigger subgraphs. On the other hand complexity selecting k nodes complexity reduced by omitting smaller subgraph from the greedy algorithm since their effect on the spread even if the influence probability guaranteed will not raise marginal gain in spreading influence over the network. The proposed algorithm may be extended to experiment more community detection approaches and applied on real social networks. In addition, this work could lead to examine different clustering approaches that are not based on community to examine information propagation maximization over social networks.

References

1. Sakaki, T., Toriumi, F., Matsuo, Y.: Tweet trend analysis in an emergency situation. In: Proceedings of the Special Workshop on Internet and Disasters, SWID 2011, New York, NY, USA, pp. 3: 1–3: 8. ACM (2011)
2. Echnology and community-centred humanitarian action (2014)
3. Cheong, M., Lee, V.: Twittering for earth: a study on the impact of microblogging activism on earth hour 2009 in Australia. In: Nguyen, N.T., Le, M.T., Świątek, J. (eds.) ACIIDS 2010. LNCS (LNAI), vol. 5991, pp. 114–123. Springer, Heidelberg (2010). doi:10.1007/978-3-642-12101-2_13
4. Domingos, P., Richardson, M.: Mining the network value of customers. In: Proceedings of the Seventh ACM SIGKDD International Conference on Knowledge Discovery and Data Mining, KDD 2001, New York, NY, USA, pp. 57–66. ACM (2001)
5. Doe, R.: Mail online @ONLINE, August 2014
6. Hollie, M.: Catching up with ted 'golden voice' williams @foxnews (2013)
7. Akbar, I.: Power out? No problem. commercial
8. Jadhav, P.: Why this kolaveri di is CNN's top song of the year (2013)
9. Clarke, C.: Nokia's 'thanks apple' taunt goes down as one of the most retweeted brand tweets ever (2013)
10. Coleman, J.S., Katz, E., Menzel, H.: Medical Innovation: A Diffusion Study. Bobbs-Merrill, New York (1966)
11. Kempe, D., Kleinberg, J., Tardos, É.: Maximizing the spread of influence through a social network. In: Proceedings of the Ninth ACM SIGKDD International Conference on Knowledge Discovery and Data Mining, KDD 2003, New York, NY, USA, pp. 137–146. ACM (2003)

12. Nemhauser, G.L., Wolsey, L.A., Fisher, M.L.: An analysis of approximations for maximizing submodular set functions-I. Math. Program. **14**(1), 265–294 (1978)
13. Kimura, M., Saito, K., Motoda, H.: Blocking links to minimize contamination spread in a social network. ACM Trans. Knowl. Discov. Data **3**(2), 9:1–9:23 (2009)
14. Leskovec, J., Krause, A., Guestrin, C., Faloutsos, C., VanBriesen, J., Glance, N.: Cost-effective outbreak detection in networks. In: Proceedings of the 13th ACM SIGKDD International Conference on Knowledge Discovery and Data Mining, KDD 2007, pp. 420–429 (2007)
15. Chen, W., Wang, C., Wang, Y.: Scalable influence maximization for prevalent viral marketing in large-scale social networks. In: Proceedings of the 16th ACM SIGKDD International Conference on Knowledge Discovery and Data Mining, KDD 2010, pp. 1029–1038 (2010)
16. Wang, Y., Cong, G., Song, G., Xie, K.: Community-based greedy algorithm for mining top-k influential nodes in mobile social networks. In: Proceedings of the 16th ACM SIGKDD International Conference on Knowledge Discovery and Data Mining, KDD 2010, New York, NY, USA, pp. 1039–1048. ACM (2010)
17. Lei, S., Maniu, S., Mo, L., Cheng, R., Senellart, P.: Online influence maximization (extended version). CoRR, abs/1506.01188 (2015)
18. Kleinberg, J.: Cascading behavior in networks: algorithmic and economic issues (2007)
19. Guille, A., Hacid, H., Favre, C., Zighedl, D.A.: Information diffusion in online social networks: a survey. SIGMOD Rec. **42**(2), 17–28 (2013)
20. Pons, P., Latapy, M.: Computing communities in large networks using random walks. J. Graph Algorithms Appl. **10**(2), 191–218 (2006)
21. Leskovec, J., Kleinberg, J., Faloutsos, C.: Graphs over time: densification laws, shrinking diameters and possible explanations. In: Proceedings of the Eleventh ACM SIGKDD International Conference on Knowledge Discovery in Data Mining, KDD 2005, New York, NY, USA, pp. 177–187. ACM (2005)
22. Csardi, G., Nepusz, T.: The igraph software package for complex network research. InterJournal Complex System 1695 (2006)

Controlling Rumor Cascade over Social Networks

Ragia A. Ibrahim[1,3]([✉]), Hesham A. Hefny[1], and Aboul Ella Hassanien[2,3]

[1] Institute of Statistical Studies and Research (ISSR), Cairo University, Giza, Egypt
ragia.ibrahim@gmail.com, h.hefny@ieee.org
[2] Faculty of Computers and Information, Cairo University, Giza, Egypt
aboit@gmail.com
[3] Scientific Research Group in Egypt, Giza, Egypt
http://www.egyptscience.net

Abstract. This work studies controlling rumor cascade over social networks problem. Previous research was mainly focused on removing nodes or edges to achieve the desired outcome. For this study, we firstly propose effective algorithms to solve rumor cascade controlling problem. Second, we conduct a theoretical study for our methods, including the hardness of the problem, the accuracy and complexity. Lastly, we conduct experiments on synthetic data set and compare results with local structure node measures.

Keywords: Graph mining · Social networks · Influence propagation · Simulation

1 Introduction

Social influence is the change in a person's behavior in respect of his or her relationships with other people. Social influence has been a widely practiced and used in social networks. Many applications, actions and events have been built based around the implicit notation of social influence between people. According to [1] Social networks could be classified as an explicit network where individuals and their relations are declared or implicit network. In explicit network diffusion could depend over aglobal decision as in herding behavior or local decision (depends on the neighbors choices) as in information cascade. Implicit network means that the graph is not explicitly constructed, but remain implicit. Incrementally construct nodes in the implicit graph as we search for a solution in the hope we can find a solution without ever generating all nodes. Diffusion of innovation and epidemics could be classified as propagation on implicit network.

Collective action integrates the network effect on both network global level and node local level, for examples organizing a protest, if huge numbers of people participate everyone in society will benefit, but on the other hand if few hundreds only show up then the demonstration will arrest them. Thus, In collective behavior the benefits occur only when enough number of users adapt specific opinion

© Springer International Publishing AG 2017
A.E. Hassanien et al. (eds.), *Proceedings of the International Conference on Advanced Intelligent Systems and Informatics 2016*, Advances in Intelligent Systems and Computing 533, DOI 10.1007/978-3-319-48308-5_44

or action. Individuals in such model will adopt the idea if at least k users of their neighbors adopt it [1]. Influence viral marketing early studies focused on person to person influence. Such as in hybrid seed corn in Iowa that were studied by Ryan and Gross (1943) [2], Drug adoption Coleman, Katz, Menzel (1966) and Hagerstrands (1970) [3]. Recently in February 2013 the Oreoaa sandwich cookie company, tweeted during a power failure in the Super Bowl XLVII is an example of maximum benefit gained with minimum cost [4].

The most known models to study influence are linear threshold model (LT) and independent cascade model (IC). Both are stochastic models, these model firstly pointed by [5]. In this work, The problem of minimizing a rumor studied under Independent Cascade (IC) model. Influence minimization computes the spread of the influence firstly, detect the rumor influence and selecting seed set of edges or nodes or both to minimize the cascade. The main contributions of the paper can be summarized as follows:

1. Immunization Algorithm: We propose effective and scalable algorithm to limit influence cascade overs social networks.
2. Proof and Analysis: We show the accuracy and the complex-ity of proposed Algorithm; the hardness of the problem.
3. Experimental Evaluations: Our evaluations synthetic network and compare results with heuristic local node metrics.

This paper starts with a brief introduction into information propagation on social networks; the rest of the paper is organized as follows. In Sect. 2, represents related work summary. Rumor influence propagation including an overview of Independent Cascade model IC and problem definition. In Sect. 3. Hardness and modularity of node immunization. In Sect. 4, Hardness and modularity of the proposed algorithm. In Sect. 5, The proposed algorithm introduced. Experimental results are presented and discussed in Sect. 6 before concluding and giving some future work suggestions in Sect. 7.

2 Related Work

Several studies have targeted influence propagation minimization via several methods, e.g. edge deletion, node deletion, computed cascade and blocking nodes. Eliminating cascade among network could be a priority especially during troubled status, such as epidemic spreads over water drinking network, false rumors and bad information cascade. The main advantage for edge removal versus node removal is: in removing nodes the node and all its linked edges are removed from the network on the other hand removing set of edges such as the inflow specific person on twitter resulting removing only the tie between two actors. Eliminating cascade via edges blocking has been studied by Kimura [6], proposed blocking links approach for IC model to minimize the spread of unwanted information (virus) over the network. In [6] Edge Bond percolation method was used to estimate the number of blocked links. By computing number of strongly connected components for a specific node u. In addition number of reachable nodes for the same node u.

Kimura, et al. (2009) conclude that blocking links between nodes with high the out-degrees is not certainly effective.

In addition, Kuhlman, et al. [7] focused on the Linear threshold model. They block edges on weighted and non weighted social networks. Kuhlman, et al. (2013) work differentiated between simple (interacts with one node) and complex (interact with at least two nodes) contagions, the authors conclude that edge blocking method is effective in simple and complex contagions, for both unweighted and weighted graphs.

Kuhlman [8] Used voter model in combination with Independent Cascade model IC. They assume that two opinions propagate over a network. In addition, they formalized a new parameter called minimum opinion control factor MOCF to control opening over a network. Given a graph G and k number of frozen nodes on specific opinion and the objective is to reach a desired minimum average opinion. The average opinion is calculated for the entire graph depending on all nodes opinion. The method applied into several synthetic and real graphs. The authors found that MOCF depends on network structure.

Callaway et al. [9] applied percolation model as the epidemic behavior of the network. In percolation model as in the spread of computer virus, the node gets infected if it was substible for the infection. Callaway et al. Allowed removing vertices from the network in an order that depends on their degree. They studied network resilience via two techniques firstly: removing random selected nodes or links. Secondly: targeting deletion of network nodes or links. The network follows random Poison degree destruction. These networks are unlike real world networks that frequently follow power law degree disruption. They examined site percolation for the general case where the infection probability is a random function. They conclude that the network is highly robust against random removal of nodes. On the other hand it is relatively easily broken, at least in terms of fraction of nodes removed, to the specific elimination of the most highly connected nodes [10].

Budak, et al. [11] shows that the influence function is non negative, submodular, and monotone, proposing two models for the simultaneous diffusion of two competing cascades. Budak aimed to minimize the spread of bad campaign, by minimizing the number of people adopting it or maximize good campaign.

Existing work in imminization have mostly focused on immunizing nodes before the rumor starts spreading. Tong et al. [12] proved that the epidemic threshold for a network is the inverse of the largest eigenvalue of the network adjacency matrix. The authors proposed an effective immunization strategy by maximizing that threshold using SIS (susceptible-infective-susceptible) epidemic diffusion model.

In addition, Tong, et al. [13] proposed edge removal methods named Net-Melt. NetMelt minimized influence cascade via minimizing the leading eigenvalue. Leading eigenvalue computed by computing the left and right eigenvectors corresponding to the left and right maximum eigenvalues of the network adjacency matrix. Note that the adjacency. This produces a left eigenvector u and a right eigenvector w. For an edge I, j, the product $u(I) \times w(j)$ is assigned as a weight to the edge. Edges with the greatest *eigenproduct* are selected as blocking edges.

Cohen et al. [14] propose immunization strategy which focuses on highly connected nodes for computer networks and populations networks with scale-free degree distributions. They observed that the probability of reaching a particular node by following a randomly chosen edge in a graph is proportional to the nodes degree, and it's more likely to find high degree nodes by following edges than by selecting nodes at random. They propose an immunization population technique of selecting a random person from that population and immune a friend (neighbor) of that person, then repeating the process. Cohen et al. validate their assumption by both analytic calculations, and by computer simulation that their strategy is strongly more effective than random immunization.

3   Rumor Influence Propagation

Principally, the social network is the sum of our living, changing relationships. A graph is a representation of these relationships. Thus, social networks can be represented as a graph $G = (V, E)$, where V is the set of nodes and E is the set of edges. Node u considered a neighbor to node $v \iff e_{v,u} \in E$, directed edge from v to u in E.

3.1   Independent Cascade IC Model

A simple contagion could be presented via Independent cascade IC model since each node gets active via one of its neighbors such as in a vial of simple information. IC model has been studied in different contexts [5,15,16]. The independent cascade IC model is a stochastic model that depends on graph structure weather it is highly connected or sparse graph. Cascading means the behavior of being active or inactive that outcomes from an influence function are based on neighbors or earlier individuals, thus cascading is based on the influence of neighbors that affect each node or individual to be earned in one of two states whether active or inactive [17].

Firstly, seed nodes selected to be an active node (infectious). These seeds trigger a rumor over the network. Then, each individual represented as a node has got a property to infect its neighbors, according to their tie represented as an edge. Thus, IC uses diffusion probability associated with each edge $p_{(v,u)} \in [0, 1]$; and in each iteration each active node uses that probability once to activate its neighbors. Accordingly, for nodes (v, u) if v is active and u is not, the activation succeeds depending on node v and other nodes that already attempted to change v but failed. Once a v activated node infected if connected to the initial active set via live (active) path, it can activate its neighbors too, thus the success probability depends on the order of activated neighbor. The process ends when no more nodes activate [17,18].

3.2   Problem Definition

Controlling Influence propagation requests, determining information type; rumor or not, then regarding its type spread or limit the contamination. The main goal of this work is to limit the spread of rumor influence. Influence minimization focuses on limiting the spread of misleading information over social networks. The problem can be formulated as follows: Given an infected directed social graph $G = (V, E)$, identified by links connections or Edges E and $|V| = N$ number of nodes consists of a set of vertex V. At time (t_0) number of vertices R infected, infection spread until the infection detected. The main aim is to minimize the contamination; the objective function could be one or more of the following: minimize further nodes to get influenced, maximize the number of uninfected nodes. This work focus on minimizing number of further infected nodes by immunizing selecting number of k nodes that's eliminated influence.

4   Hardness and Modularity of Node Immunization

Proposed approach depends on directed graph structure; immunizing expected rumor highly influence source does not change the graph structure since nodes or edges will not be deleted or added, immunizing provide an efficient soft approach to control influence.

Given a graph G, a set $\{r_1, r_2, r_3, ...r_n\}$ of initial activated nodes by rumor cascade R, $R \subseteq V$, and a budget k. S_0 is the initial set of nodes selected to limit cascade R by maximizing number of saved nodes by immunizing, $S \subseteq V \setminus R$. $f(S)$ represents number of expected saved nodes. node u is covered by $\sigma(S)$ if it is on the active path of initial set S.

$$S^* = argmax_{S \subseteq V \setminus R, |S_0|=k} \sigma(S_0) \tag{1}$$

Given budget k, a social graph $G(V, E)$ and a stochastic diffusion model on G, find a seed set $S_0 \subset V$ with $| S_0 |\leq k$, that maximize $\sigma(S_0)$. Where $\sigma(S)$ is the maximize number of lost incoming infected nodes. The aim is to compute $S^* \subseteq V$, where $\sigma \subseteq V \setminus R$ denotes the number of saved nodes from infected by rumor R nodes when the node set S is immunized.

Limiting rumor problem considered as an instance of the set cover NP- complete problem where If there exist a set k. More formally, given a universe set U and a family S of subsets of U, a cover is a subfamily $C \subseteq S$ of sets whose union is U. On the set covering decision problem, the question if there exist set of size k to cover. This is a decision problem and thus finding k nodes that saves all noninflected nodes is NP-hard. Limiting rumor problem is monotonicaly decreasing problem, means that, adding more nodes v to the set do not cause $\sigma(.)$ to decrease for all elements v and sets S. $f : 2^v \to R$ monotone if any subset $S \subseteq T \subseteq V$, $F(S) \leq F(T)$. Kempe et al. proved that influence function of IC said to be monotone and submodular [5].

A $\sigma(.)$ function is submodular function, it has diminishing returns, that means for all $v \in V$, adding a more node v to set S marginal gain can not exceed the marginal gain from adding node v to set T, where S is a subset of T. Formally, a set function $f : 2^v \to R$ is submodular if for all elements v and pairs of sets S, T; where $S \subseteq T$. $F(S \cup v - F(S) \geq F(T \cup v - F(T)$ [19].

Theorem 1. *Given A digraph G Immunizing set k of nodes to limit cascade spreading considered NP-hard problem.*

Proof. Regarding instance of NP-complete problems Set Cover problem, determining family $\mathscr{S}$ of subsets of universe set $\mathscr{U}$, where subfamily of sets whose union is $\mathscr{U}$. The aim is to find if there exist k subsets where there union cover the universe set.

Immunization is NP-hard, even if $P_{u,v} = 1$. given an instance of set cover problem, we define alike directed bipartite graph of $n+m+1$ nodes where $n = |V|$ and $m = |E|$. There exist a directed edge (i, j) whenever $u_i \in toS_j$. The set cover problem is the same as deciding set S^* of size k where $\sigma(S^*) \geqslant n+k$. If we are at time stamp t_0 and non of the node is infected yet, determining k nodes corresponding to set cover problem would prevent all the nodes to be infected.

The resulting graph structure after immunizing the nodes will not be changed, but the connectivity property doesn't hold. connection between nodes will be week since no rumor information really reaches node neighbors.

5 Greedy Immunization Algorithm

Proposed greedy based approach to limit rumors over social networks, Independent Cascade IC Model is adopted as the information diffusion model, reflecting greedy algorithm as in set cover problem, that is, the algorithm keeps selecting the set that covers most new elements and unite it to the previous solution, until their union gives the entire set. Since IC is stochastic model Monte Carlo simulation performed. We used Lazy Greedy algorithm CELF [20]. We scaled CELF into parallel processing. The benefit of multicore processor used while calculating node effectiveness for $\#NO_{sim}$ number of simulation times.

Algorithm 1. Greedy Immunization

1: Infected digraph (G) and k: number of initial nodes
2: Output:selected seed set S
3: initialization $S = \phi$, R: set of rumor nodes
4: **for** $i := 1$ to k **do**
5:    **for** $v \in V \setminus R \cup S_i$ **do**
6:       $\triangle_v = \sigma(V \setminus R \cup S_i) - \sigma(V \setminus R \cup S_i) \cup \sigma(v)$
7:       $S_i = S_i \cup argmax_{v \in V \setminus R \cup S_i} \triangle_v*$
8:    **end for**
9: **end for**
10: return S

Firstly the Digraph infected by a rumor, but after delay d rumor cascade caught. Immunization algorithm select set S of chosen nodes to be immunized. The $\triangle_v$ function returned the average marginal gain for each node v, the marginal gain calculated on each iteration for number of simulation times $\#NO_{sim}$. For each i in k, the node with highest marginal gain $\triangle_v*$ added to set S of chosen seed nodes.

6   Experimental Evaluation

Proposed immunization algorithm evaluated w.r.t several heuristics. The proposed approach performs better in the long run compared with heuristic approaches.

Regarding the algorithm complexity and huge real datasets, the proposed algorithm applied to syntahic data set that mimic real social networks. The synthetic network data has been created using Forest Fire Model. Forest Fire is a generative model that produces networks, mimicking the structure of growing networks. The model produces realistic networks in terms of heavy-tailed degree distributions, community structure and other network properties. Graph generated parameters are: forward burning probability 0.3, and backward burning probability 0.25. As for the diffusion probabilities on the edges, each is chosen uniformly at random. The graph consists of 2000 nodes and 4965 edges [21]. To evaluate the quality of the solution provided by the node Immunization algorithm, we compare the proposed algorithm against local structure graph metrics as defined in [22]. These heuristic can be described as selective k nodes that have the maximum local structure metric for each used heuristic. In these definition node j is the neighbor of node i. We used several groups of measures: Centrality matrices are as follows:

– **Average Outdegree AO**

$$AO = |[Outdegree(i) + Outdegree(j)]|/2 \tag{2}$$

– **Common Followers CF**

$$CF = Indegree(i) \cap Indegree(j) \tag{3}$$

– **Common Neighbors CNS**

$$CNS = [Outdegree(i) \cap Outdegree(j)] \cup [Indegree(i) \cap Indegree(j)] \tag{4}$$

– **Common Neighbors Metric CN**
  If $\nabla = Outdegree(i) \cap Indegree(j)$ and $\nabla' = Indegree(i) \cap Outdegree(j)$ then the Common Neighbors Metric is

$$CN = |\nabla| + |\nabla'| \tag{5}$$

- **Jaccard Coefficient JC**

$$JC = 0.5\frac{Outdegree(i) \cap Indegree(j)}{Outdegree(i) \cup Indegree(j)} + 0.5\frac{Indegree(i) \cap Outdegree(j)}{Indegree(i) \cup Outdegree(j)} \quad (6)$$

- **Adamic Adar Score AA**

  Mathematically the AA for two sets A and B can be represented as, $similarity(A, B) = 1/log(frequency of shared items)]$, from graph perspective.

$$AA = 0.5 \sum_{z \in \nabla} \frac{1}{log(d(z))} + 0.5 \sum_{z' \in \nabla'} \frac{1}{log(d(\bar{z}))} \quad (7)$$

- **Conservative Metric CM**

$$CM = 0.5 \sum_{z \in \nabla} \frac{1}{Outdegree(i) \cdot Outdegree(z)} +$$
$$0.5 \sum_{z \in \nabla'} \frac{1}{Outdegree(j) \cdot Outdegree(z)}$$

- **Conservative Attention Limited Metric CM_LA**

$$CM\_LA = 0.5[\sum_{z \in \nabla} \frac{1}{Outdegree(i) \cdot Outdegree(z) \cdot Indegree(j)} +$$
$$\sum_{z \in \nabla'} \frac{1}{Outdegree(i) \cdot Indegree(z) \cdot Outdegree(z) \cdot Indegree(j)}]$$

- **Non-Conservative Metric NCM**

$$NCM = 0.5[\sum_{z \in \nabla} 1 + \sum_{z \in \nabla} 1] = 0.5[|\nabla| + |\nabla'|] \quad (8)$$

- **Non-Conservative Attention Limited Metric NCM_AL**

$$NCM\_AL = 0.5[\sum_{z \in \nabla} \frac{1}{Indegree(z) \cdot Indegree(j)} + \sum_{z \in \nabla} \frac{1}{Indegree(z) \cdot Indegree(i)}] \quad (9)$$

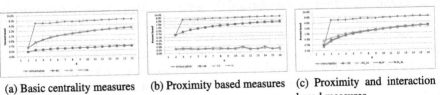

(a) Basic centrality measures   (b) Proximity based measures   (c) Proximity and interaction based measures

Fig. 1. Evaluation of immunization limiting delay $= 20\%$

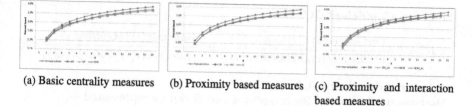

(a) Basic centrality measures   (b) Proximity based measures   (c) Proximity and interaction based measures

Fig. 2. Evaluation of immunization limiting delay $= 60\%$

Figure 1 represents the evaluation of immunization algorithm against each group of metrics. Figure 1a shows that Centrality metrics AO, CF and CNF perform similar to each other except for Average outdegree performs worse than others, obviously adding i and j neighbors to the metric enhance the performance rather than considering only outdegree for each of the. The greedy immunization performs much better, since it did not look to the very local neighbors but consider the whole graph.

Figure 1b represents a comparison between network proximity based measures: AA, CN and IC metric, as shown all of them have similar performance except AA performs better than IC metric, and CN. It seems that giving higher weight to common neighbors improve results its even close to the proposed immunization limiting algorithm. Figure 1c represents not only proximity based matrices, but proximity and interaction based matrices: CM, CM_LA, NCM and NCM_LA, the results show that proximity and interaction based metrics almost give similar results.

Figure 2, represents results when delay is 60%, results are very low and not effective compared with results when delay is 20%. Its clear that as long as detecting a rumor takes a long time as long as it is controlling it will be difficult. Immunization algorithm is slightly better in the all metrics used.

7   Conclusion and future works

In this paper, we presented Immunization limiting algorithm in social networks. We also showed that the problem is submodular and therefore a greedy algorithm guaranteed to provide a $(1 - 1/e)$ approximation. In addition the greedy algorithm is a polynomial time algorithm, but too costly. For that we used the Lazy greedy algorithm, it is performed by using priority queue data structure. In addition, we used parallel processing in the part of simulating the node marginal gain for each of them are independent. Also experimentally studied the performance of the Lazy greedy algorithm, comparing it with 10 different heuristics mainly represent different local node centrality measures. Also, In many cases the performance of the minimization Limiting algorithm performs better than heuristics.

Considering huge social network data, the proposed algorithm could be applied by firstly investigating the graph offline and determine influencers. Thus, when a rumor detected the proposed node immunizing applied for the top k influencers. In the second case considering graph dynamics and its changes over time a new method of dealing with online streamed data is a demand. In the future we intend to experiment scale proposed algorithm to work faster on several large social networks and apply it for real data sets.

References

1. Zafarani, R., Abbasi, M.A., Liu, H.: Social Media Mining: An Introduction. Cambridge University Press, Cambridge (2014)
2. Ryan, B., Gross, N.: The diffusion of hybrid seed corn in two iowa communities. Rural Sociol. **8**(1), 15–24 (1943)
3. Coleman, J.S., Katz, E., Menzel, H.: Medical Innovation: A Diffusion Study. Bobbs-Merrill, New York (1966)
4. Oreo Cookie. Power out? No problem. Commercial (2013)
5. Kempe, D., Kleinberg, J., Tardos, É.: Maximizing the spread of influence through a social network. In: Proceedings of the Ninth ACM SIGKDD International Conference on Knowledge Discovery and Data Mining, KDD 2003, pp. 137–146. ACM, New York (2003)
6. Kimura, M., Saito, K., Motoda, H.: Blocking links to minimize contamination spread in a social network. ACM Trans. Knowl. Discov. Data **3**(2), 9:1–9:23 (2009)
7. Kuhlman, C.J., Tuli, G., Swarup, S., Marathe, M.V., Ravi, S.S.: Blocking simple and complex contagion by edge removal. In: 2013 IEEE 13th International Conference on Data Mining, Dallas, TX, USA, 7–10 December 2013, pp. 399–408 (2013)
8. Kuhlman, C.J., Kumar, V.S.A., Ravi, S.S.: Controlling opinion propagation in online networks. Comput. Netw. **57**(10), 2121–2132 (2013). Towards a Science of Cyber Security Security and Identity Architecture for the Future Internet
9. Callaway, D.S., Newman, M.E.J., Strogatz, S.H., Watts, D.J.: Network robustness and fragility: percolation on random graphs. Phys. Rev. Lett. **85**(25), 5468–5471 (2000)
10. Albert, R., Jeong, H., Barabási, A.-L.: Error and attack tolerance of complex networks. Nature **406**, 378–382 (2000)
11. Budak, C., Agrawal, D., El Abbadi, A.: Limiting the spread of misinformation in social networks. In: Proceedings of the 20th International Conference on World Wide Web, WWW 2011, pp. 665–674. ACM, New York (2011)
12. Tong, H., Prakash, B.A., Tsourakakis, C., Eliassi-Rad, T., Faloutsos, C., Chau, D.H.: On the vulnerability of large graphs. In: 2010 IEEE 10th International Conference on Data Mining (ICDM), pp. 1091–1096. IEEE (2010)
13. Tong, H., Prakash, B.A., Eliassi-Rad, T., Faloutsos, M., Faloutsos, C.: Gelling, and melting, large graphs by edge manipulation. In: Chen, X.W., Lebanon, G., Wang, H., Zaki, M.J. (eds.) CIKM, pp. 245–254. ACM (2012)
14. Cohen, R., Havlin, S., Ben-Avraham, D.: Efficient immunization strategies for computer networks and populations. Phys. Rev. Lett. **91**, 247901 (2003)
15. Chen, W., Wang, C., Wang, Y.: Scalable influence maximization for prevalent viral marketing in large-scale social networks. In: Proceedings of the 16th ACM SIGKDD International Conference on Knowledge Discovery and Data Mining, KDD 2010, pp. 1029–1038 (2010)
16. David, E., Jon, K.: Networks, Crowds, and Markets: Reasoning About a Highly Connected World. Cambridge University Press, New York (2010)

17. Guille, A., Hacid, H., Favre, C., Zighed, D.A.: Information diffusion in online social networks: a survey. SIGMOD Rec. **42**(2), 17–28 (2013)
18. Kleinberg, J.: Cascading behavior in networks: algorithmic and economic issues. Algorithmic Game Theory **24**, 613–632 (2007)
19. Nemhauser, G.L., Wolsey, L.A., Fisher, M.L.: An analysis of approximations for maximizing submodular set functions-I. Math. Program. **14**(1), 265–294 (1978)
20. Leskovec, J., Krause, A., Guestrin, C., Faloutsos, C., Van Briesen, J., Glance, N.: Cost-effective outbreak detection in networks. In: Proceedings of the 13th ACM SIGKDD International conference on Knowledge Discovery and Data Mining, KDD 2007, pp. 420–429 (2007)
21. Csardi, G., Nepusz, T.: The igraph software package for complex network research. Inter J. Complex Syst. **1695**, 1–9 (2006)
22. Narayanam, R., Garg, D., Lamba, H.: Discovering signature of social networks with application to community detection. In: 2014 Sixth International Conference on Communication Systems and Networks (COMSNETS), pp. 1–7. IEEE (2014)

Medical Diagnostic System Basing Fuzzy Rough Neural-Computing for Breast Cancer

Mona Gamal Gafar[(✉)]

Faculty of Computers and Information, Information System Department,
KFS University, Kafr el-Sheikh, Egypt
Mona_gafar@fci.kfs.edu.eg,
Mona2004egypt@yahoo.com

Abstract. Medical diagnostic system is a branch in bioinformatics that is concerned with classifying medical records. Breast cancer is the most common deployed cancer in females worldwide. The main obstacle is the vagueness and ambiguity involving the breast cancer data. Human nature handles the vagueness and ambiguity easily. Therefore, doctors diagnose the patient condition using their expertise. Fuzziness and rough boundary theories simulate the human thinking. The fuzzy rough hybrids address the uncertainty in terms of membership degree of truth and lower and upper boundaries of fuzzy rough set theory. This research solves the diagnostic breast cancer problems via a proposed hybrid model of fuzzy rough feature selection and rough neural networks. The medical data is preprocessed by the fuzzy rough feature selection algorithm to remove unnecessary attributes. The reduced data set is applied to the rough neural network to learn the connection weights iteratively. The test data set are used to measure the proposed model accuracy and time complexities. Lower and upper approximations of the input features are weighted by input synapses learnt through training phase. The fuzzy rough proposed model design and implementation are declared. The experiments used WDBC and WPBC data sets from the UCI machine learning repository. The experimental results proved the fuzzy rough model ability to classify new instances compared with the conventional neural network.

Keywords: Data mining · Uncertain knowledge · Bioinformatics · RNN · FRFS

1 Introduction

Cancer is a fatal disease. Statistics of the National Cancer Institute [23] showed that breast cancer kills about 39,620 every year. Furthermore, 232,340 female breast cancers and 2,240 male breast cancers are detected. Analyzing and rearranging medical data is a field in computer science called biomedical information system [1]. Nowadays, doctors suffer from the amount of medical data they have to handle through curing and investigating diseases. Decision Support System (DSS) [2] are developed to facilitate accurate and fast decisions in poorly structured environment like medical field. The huge improvements in the Artificial Intelligence (AI) [3] algorithms create more feasible and enhanced DSS. Data are stored efficiently and knowledge extracted more accurately. The main obstacle is the problem oriented issue of the AI algorithms.

© Springer International Publishing AG 2017
A.E. Hassanien et al. (eds.), *Proceedings of the International Conference on Advanced Intelligent Systems and Informatics 2016*, Advances in Intelligent Systems and Computing 533, DOI 10.1007/978-3-319-48308-5_45

Hence, computer science researchers should investigate the AI algorithms used in medical diagnostics system to make these systems more trusted and help in lowering pressure on doctors.

In the medical environment, researchers are confused by the amount of relevant and irrelevant variables they have. The irrelevant attributes affect the medical data processing badly in terms of time consumption and accuracy rates. The medical data processing benefits the AI algorithms in removing redundant attributes. Fuzzy Rough Feature Selection (FRFS) [19] is an efficient and well examined approach in removing irrelevant attributes. Basing on the membership dependency degree between the conditional attributes and the classes attribute, the attributes with true contribution to the diagnostic system are selected. Fuzzy rough boundaries concepts push the feature reduction process towards handling uncertain medical data.

Diagnostic problems are estimated roughly by doctor's judgments. Lower and upper expectation of the attributes and the disease are common fields in any medical report. Rough Neural Networks (RNNs) [5–9] logically support doctors' decisions by recognizing rough patterns in the medical data. RNN handles uncertaininty via lower and upper approximation of rough set theory. The structure of RNNs is built by rough neurons. Two conventional neurons are imaginary connected to build the rough neuron. The lower and upper bounds of the input data are fed into the RNN for processing and producing the corresponding output. RNNs are feasible solutions for probabilistic and uncertain data than conventional ANN [4].

Researchers are very interested in analyzing the breast cancer disease medical data. Dheeba J., Singh N.A., Selvi S.T. investigated a new classification approach for detection of breast abnormalities in digital mammograms using Particle Swarm Optimized Wavelet Neural Network (PSOWNN) [10]. Moreover, Victor Balanica and his colleges differentiated between benign and malignant tumors based on the extracted speculation sets using intelligent neural networks [11]. Also, Azar A.T. and El-Said S. A. proposed a research with three classification algorithms, multi-layer perceptron (MLP), radial basis function (RBF) and probabilistic neural networks (PNN), which were applied for the purpose of detection and classification of breast cancer [12].

Other researchers like R.R. Janghel developed a system for prediction of breast cancer using Artificial Neural Network (ANN) models. The article implemented four models of neural networks namely Back Propagation Algorithm, Radial Basis Function Networks, Learning vector Quantization and Competitive Learning Network [13]. Karabatak M. and Ince M.C. implemented Association Rules (AR) for reducing the dimension of breast cancer database and Neural Networks (NN) for intelligent classification. The proposed AR + NN system performance is compared with NN model [14]. Furthermore, rough-neural computing is used to classify medical data in bioinformatics research in [25].

This research proposes a hybrid of FRFS and RNN for detecting benign and malignant tumors in WDBC and WPBC data sets. The FRFS removes redundant attributes using the membership dependency between the conditional attributes and the classes attribute. During the preprocessing phase, the weka data miner tool eliminates irrelevant features from the medical data sets using the FRFS algorithm. The RNNs benefit the rough theory to handle uncertaininty boundaries in medical data. The reduced data sets are fed into the RNNs which implement the Back-Propagation algorithm to learn the connection synapses. The RNN is supported with extra connecting weights, which are updated through learning phase, to prepare lower and upper approximations for the rough neuron. The hidden layer is consisted of rough neurons fully connected to the input layer. The output layer is a conventional neuron corresponding to the output class. After training, the RNN adjusts the connecting weights and is ready for classifying unseen cases. Through testing phase, network accuracy is measured by the rate of correctly classified instances and time complexity is defined. The rough sets, fuzzy rough feature selection and the rough neural networks are declared in Sect. 2 as theories. Section 3 proposes the hybrid model. The medical data sets and their results compared with other artificial intelligent algorithms working on the same data sets are presented in Sect. 4. Finally, Sect. 5 shows the conclusion of the research.

2 Theoretical Bases

2.1 Rough Set Theory

Rough sets [15, 16], presented by Professor Pawlak, define the concept of an upper and a lower approximation of a set. Features, in uncertain environment, are a disjoint of the two approximations. The lower and upper approximation $(\underline{P}X, \overline{P}X)$ of an attribute P over the target set $X \subseteq U$ are calculated by the following equations.

$$\underline{P}X = \{x | [x]p \subseteq X\} \tag{1}$$

$$\overline{P}X = \{x | [x]p \cap X \neq \Phi\} \tag{2}$$

Where $U = \{x1, x2 \ldots xn\}$ is a nonempty finite set of instances named the universe.

A rough set [17] defines an uncertain concept by lower and upper approximations which are precise concepts. These approximations are disjoint categories representing the regular attribute. Objects belonging to the same attribute category (lower or upper) are indistinguishable. Rough set analyzes the data without the need for any other external interfering. Rough sets generate classification rules from uncertain and ambiguous data efficiently.

2.2 Fuzzy Rough Feature Selection (FRFS)

The artificial intelligent algorithms improve the knowledge discovery process in multiple phases. The data reduction phase is critical in extracting knowledge from concrete data sets. The irrelevant and multiple attributes affect the time and space complexities badly. The reduction process discriminates the data sets size by eliminating the redundant attributes. Hence, the equivalent classes depend entirely on the core attributes. The main obstacle is finding the core attributes efficiently which increase the classification accuracy. For the uncertainty and ambiguity issues, the fuzzy rough attribute reduction (FRAR) is a competent and well examined algorithm. The dependency measure between the conditional attributes and the overall data set is reliable. Therefore, the FRAR Algorithm uses the dependency measure to allocate the core attributes out of the features set [19]. FRAR benefits the vagueness and indiscernibility of fuzzy and rough set theories respectively to calculate the dependency between the conditional fuzzy attributes and the classes attribute. The membership dependency degree is calculated using the membership function of each attribute and the membership degree of the objects in each class to find the best reduct. The dependency degree is defined over the positive region of the fuzzy rough variable. The attribute that increases the membership dependency degree is added to the reduct [19]. The algorithm is illustrated in Sect. 3.1.

2.3   Rough Neural Network

Rough neural networks [6–10] resemble traditional neural networks in training and connection mode. On the other hand, they vary in the structure of the neuron utilized in the network. The rough neural network replaces the conventional neuron by a pair of imaginary connected neurons to compose the rough neuron (upper and lower boundaries). A back propagation rough neural network is consisted of three layers. The input layer represents the attributes contained in the training data set. The input data feature is represented by a rough neuron (two conventional neurons imaginary connected for fast information exchange). The input features are multiplied by connection weights then fed into the lower and upper neurons of the input layer. These connection weights along with other weight synapses in the network are updated during the learning phase. The number of rough neurons in hidden layer is determined by the Baum-Haussler rule [18]:

$$Nhn = \frac{Nts * Te}{NI + No} \tag{3}$$

Where N_{hn} is the hidden neurons number, N_{ts} is the training samples number, Te is the tolerance error, N_I is the number of inputs, and N_o is the output number.

The hidden layer is consisted of rough neurons. The sub layers of lower neurons in the input layer and the hidden layer are fully connected. Similarly the sub layers of upper neurons are connected too. The activation of a rough neuron is a pair of upper and lower bounds, while the input of a conventional neuron is a single value. The mathematical equations of the input/output of the lower/upper rough neuron are:

$$ILn = \sum_{j=1}^{n} wLnjOnj \tag{4}$$

$$IUn = \sum_{j=1}^{n} wUnjOnj \tag{5}$$

$$OLn = Min(f(ILn), f(IUn)) \tag{6}$$

$$OUn = Max(f(ILn), f(IUn))$$ (7)

Where (ILn, OLn) is the input/output of the lower rough neuron,
(IUn, OUn) is the input/output of the upper rough neuron.
f is the sigmoid function.

$$f(x) = \frac{1}{1 + e^{-x\lambda}}$$ (8)

Where x is the neuron input, λ is a constant chosen according to experiments.

The output layer is consisted of one conventional neuron (the output of the classified pattern). The hidden and the output layers are fully connected. The rough neurons activation (O) is calculated according to Eq. 9. This activation is fed to the output layer as one value.

$$O = OUn + OLn$$ (9)

The previous equations are used to calculate the activation of neurons from the hidden layer to the output neuron. Equations 4, 5, 6 and 7 are used to calculate the lower and upper neurons activation from input layer to hidden layer.

3   The Proposed Hybrid Model of FRFS and RNN

The proposed model is a fuzzy rough hybrid system that is consisted of three main sub modules. The first sub module preprocesses that data set by eliminating the irrelative attributes in a fuzzy rough basis to handle the uncertainty nature of the medical data. The second sub module uses the rough neural network intelligence to learn from the uncertain reduced data set. After training, the rough neural network becomes the intelligent classifier of the unseen cases to predict their medical condition of the illness. The third sub module tests the accuracy of the intelligent classifier by the test data set. Figure 1 shows the main sub modules and their input/output structure.

The main modules of the proposed model are:

(1) **Preprocessing phase**: this process eliminates redundant attributes basing on the fuzzy rough feature reduction algorithm. The core attributes are defined by the membership dependency degree between the conditional attributes and the output attribute.
(2) **The classifier Training phase**: this module train the RNN algorithm using the reduced data set resulted from the data reduction phase.
(3) **The Testing phase**: this module computes the time and accuracy complexities during the classification process.

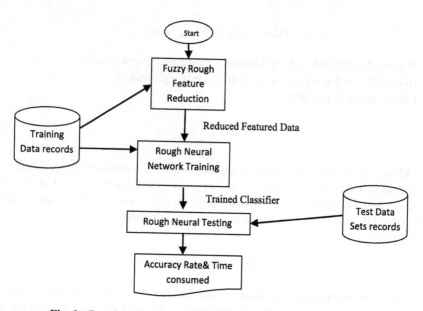

Fig. 1. Rough Neural Network model in classifying uncertain Data

3.1 Data Reduction Using Fuzzy Rough Feature Selection (FRFS)

The data reduction is essential in terms of reducing time and space complexity. But, the unconfirmed nature of the medical data is a major problem in intelligent systems. FRFS [19] is a reduction algorithm built for uncertainty handling via fuzzy logic. FRFS builds the core attributes using the membership dependency degree between the conditional attributes and the decision classes. The FRFS starts the core attributes as an empty set. Then it calculates the fuzzy membership dependency degree between the fuzzy variables (attributes) and the fuzzy equivalent classes, one attribute at a time. Finally it adds the attribute that maximizes the dependency degree to the core attribute set. This process is repeated until the dependency come to its highest level (usually 1). This process eliminates the problem of information loss caused by discretizing fuzzy attributes for dependency measuring.

This module is evaluated by the WEKA [24] data miner tool. WEKA is reliable and facilitates various knowledge extraction processes and search methods to be tested. One of the search methods used for the FRFS implementation in WEKA is the Best First algorithm [20]. It explores the search space by evaluating the most promising node with respect to a specified rule. It looks for the core attributes by greedy hill climbing [21] in addition to backtracking facility. Best first may search forward, backward, or in both directions. The output of that phase is the reduct (reduced data set features). The algorithm of the FRFS is declared in Fig. 2.

Input : Cond_ATT, training data set.
Output: CORE_ATT.

1: CORE_ATT← {} , $\gamma'_{optimal} = 0, \gamma'_{old} = 0$
2: do
       3: T= CORE_ATT
       4: $\gamma'_{old} = \gamma'_{optimal}$
       5: $\forall X \in (\text{Cond\_ATT} - \text{CORE\_ATT})$
       6: Calculate the fuzzy lower approximation of the fuzzy equivalence classes Fi ∈ U/T

$$\mu_{\underline{x}}(Fi) = \inf_{x \in U} \max\{1 - \mu_{Fi}(x), \mu(x)\}$$

       7: Calculate the fuzzy positive region of a fuzzy equivalence class Fi ∈ U/T

$$\mu_{posT}(F_i) = \sup_{x \in U|D}(\mu_{\underline{x}}(Fi))$$

       8: Calculate the membership degree of an object x to the fuzzy positive region of the fuzzy rough attribute T

$$\mu_{pos_T(D)}(x) = \sup_{x \in U|D} \min(\mu_{Fi}(x), \mu_{posT}(Fi))$$

       9: Calculate the fuzzy membership dependency degree

$$\gamma'_T(D) = \frac{\left|\mu_{pos_T(D)}(x)\right|}{|U|} = \frac{\sum_{x \in U} \mu_{pos_T(D)}(x)}{|U|}$$

       10: if $\gamma'_{R\cup\{x\}}(D) = \gamma'_T(D)$
           11: $T = \text{CORE\_ATT} \cup \{x\}$
           12: $\gamma'_{optimal} = \gamma'_T(D)$
      13: $\text{CORE}_{ATT} \leftarrow T$
    14: Untill $\gamma'_{optimal} = \gamma'_{old}$
15: Return CORE_ATT

Fig. 2. Fuzzy rough feature reduction algorithm

3.2 Building the Rough Neural Classifier Using Medical Data

This module implements the rough neural network structure. It applies the reduced training instances on RNN for the learning phase. The input data set is the crisp reduced data. The crisp features available have no rough data boundaries. The model is designed with two connections for computing the lower and upper approximation of the features values. The incoming data features will be weighted once by the input connection synapses of the lower boundary input layer neuron and the other by the connection synapses of the upper boundary neuron. These input connections will be updated during the weight update phase to learn the lower and upper boundaries of each input feature. The crisp data is directly used as input data without the need for further preparing of the lower and upper data approximation. The RNN structure is illustrated in Fig. 3.

The algorithm used to train the rough neural network is backpropagation [4]. The transfer function used in this model is the sigmoid function illustrated by Eq. 8. The algorithm teaches the network in two phases. The propagation phase applies the data records to the input layers and generates the output activations for all rough and

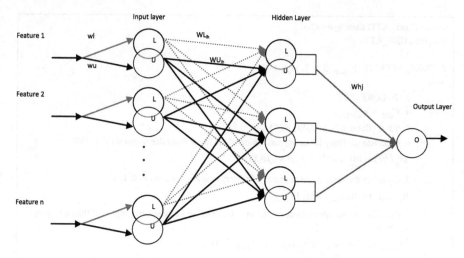

Fig. 3. The rough neural network structure

conventional neurons in the network. Furthermore, the algorithm computes the deltas (error rate) for all neurons basing on the training records actual output. The weight update phase scans all the weight connections and computes new weights according to Eq. 10.

$$new\ w_{ij}\ =\ old\ w_{ij} + (error * Derivative * learnRate * activation(input\ neuron))$$
$$(10)$$

The two phases are repeated until the accuracy of the rough neural network exceeds a predefined threshold (s satisfaction rate). The output of that module is a feasible trained rough neural network structure that is capable to classify new instances according to the inputs and connection weights (updated during the training by equations mentioned in Sect. 2.2). Actual algorithm for a 3-layer rough neural network (only one hidden layer) is illustrated in Fig. 4.

3.3 The Proposed Hybrid Accuracy Measure

The improvements achieved by the proposed hybrid model (FRFS + RNN) should be measured accurately. Most algorithms present improvements in terms of accuracy rate and time consumption. The testing module measures the proposed hybrid model accuracy rate by the number of correctly classified patterns in the test data set. The accuracy rate is the complement of the absolute error rate equation.

$$Accuracy = 1 - \frac{N_E}{N} \tag{11}$$

Input: reduced training set
Output: Network_weight_structure, Error_rate
    1 : Initialize Network_weight_structure (often randomly)
    2 : Do
       3: Foreach t in the training set
       //forward pass
       4: O $\leftarrow$ rough neural-net-output(network, t) ; //(actual output)
       5: T $\leftarrow$ target output for t
       6: error$\leftarrow$ (T - O) at the output units
       //backward pass
       7: Foreach wh in hidden_to_output
              8:delta_wh$\leftarrow$ error $*$ Derivative$*$ learnRate$*$ activation$($neuron$_h)$

       9: Foreach wli in inputLA_to_hiddenLA
         10: delta_wli$\leftarrow$ error $*$ Derivative$*$ learnRate$*$ activation$($neuron$_{li})$

       11: Foreach wui in inputUA_to_hiddenUA
         12:delta_wui$\leftarrow$ error $*$ Derivative$*$ learnRate$*$ activation$($neuron$_{ui})$

       14: Foreach wl & wu lower and upper approximation neuron in the input layer
         15: delta_wl$\leftarrow$ error $*$ Derivative$*$ learnRate$*$ activation$($neuron$_l)$

         16: delta_wu$\leftarrow$ error $*$ Derivative$*$ learnRate$*$ activation$($neuron$_u)$

       17: For all the weights in the network
           new w_{ij} = old w_{ij} + (error $*$ Derivative$*$ learnRate$*$ activation$($input neuron))
       18:
      19: Until all examples classified correctly or stopping criterion satisfied Return the network.
20: Return network_weight_structure, Error_rate

Fig. 4. The rough neural algorithm

Where N_E is the number of incorrect classified instances and N is the total number of the testing instances.

The time consumption is measured using the overall time used to train the network until convergence.

4   Experimental Results

The fuzzy rough proposed system is consisted of three sequential phases. The first phase removes the irrelevant attributes by the FRFS algorithm. The redundant attributes which do not contribute to the classification process are removed. The reduction process relies on the dependency membership degree between the conditional attributes and the classes attribute. The second sub module is RNN learning phase. It initializes the connection weights and feeds the reduced data features into the network until the optimal combination of synapses in the network are found. The output of that module is an intelligent classifier based on RNN. Through the test (final) phase, the accuracy rate that is measured by the absolute error rate of the network and the time complexity are determined.

The data preprocessing phase is implemented with the Weka data miner tool using the fuzzy rough feature selection evaluator and the Best First searching algorithm. The rough neural network structure, the backprobagation training algorithm and the testing are implemented in C# code. Through experiments, the learning rate and λ values are

Table 1. Description of the Breast Cancer data sets

Name of the data set	No of attributes	No of continuous attributes	No of categorical attributes	No of data records	No of classes
WDBC	32	31	1	569	2 classes M = malignant, B = benign
WPBC	34	33	1	198	2 classes R = recur, N = nonrecur

determined to be 0.6 and 7 respectively. The RNN is consisted of three layers. The input rough neurons layer takes the data from input attribute. The connection synapses before the input layer weight the crisp input and convert them to the lower and upper input approximation of each attribute. Rough neurons in the input and hidden layers are fully connected (lower to lower and upper to upper). The weights connections between the input and hidden layer are updated according to Eqs. 4, 5, 6 and 7. The hidden layer rough neurons are linked with the output neuron (conventional neuron) with one weight for each rough neuron (lower and upper) according to Eqs. 4, 5, 6, 7 and 9. The breast cancer data sets used in this research to train and test the RNN are from the UCI machine learning repository [22] and its properties are illustrated in Table 1. Features are computed from a digitized image of a fine needle aspirate (FNA) of a breast mass.

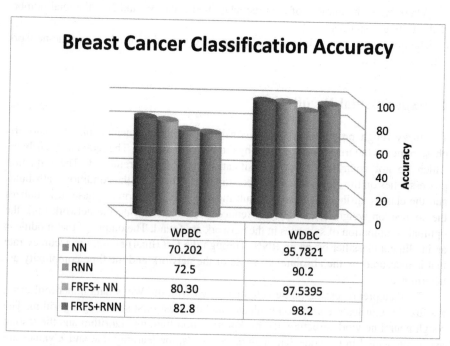

	WPBC	WDBC
NN	70.202	95.7821
RNN	72.5	90.2
FRFS+ NN	80.30	97.5395
FRFS+RNN	82.8	98.2

Fig. 5. FRFS + RNN and FRFS + NN accuracy for different data set sizes

They describe characteristics of the cell nuclei present in the image. The data sets records are separated into two equal sub sets (one for the training data and one for the test data).

The data sets are used to train and measure the accuracy for both the NN and the RNN. The experiments clarify that the RNN is more precise than the NN for the WPBC data set. The fuzzy rough feature selection enhanced the RNN with accuracy over 82 % and 98 % for the WPBC and the WDBC respectively. The fuzzy rough feature selection and NN classifier is evaluated using the Weka data miner tool. Figures 5 and 6, show the accuracy and time complexity comparisons for the WDBC and WPBC data sets respectively.

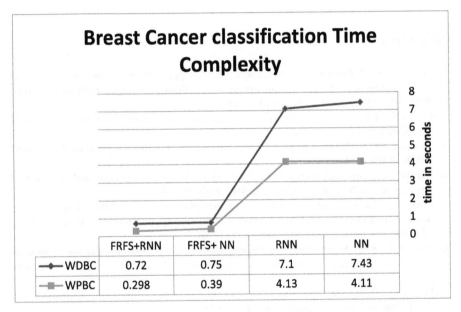

Fig. 6. FRFS + RNN and FRFS + NN time complexity for different breast cancer data sets

5 Conclusion

Medical data sets are full of uncertainty and ambiguity. Fuzzy Rough logic provides a suitable theoretical base to handle uncertainty in medical data classifications. The data reduction is a problem that faces the medical data classification. The fuzzy rough feature selection algorithm build the core attributes depending on the fuzzy membership dependency degree between the conditional attributes and the classes attribute. The core attributes are applied to the classification model to decrease time and space complexity. The rough neural networks mimic human thinking regarding lower and upper approximations of the rough set. They can be used perfectly to produce a medical data classification model.

This paper shows the ability of fuzzy Rough proposed model to reduce, learn and classify the medical data. It rough can be used as a medical diagnostic system. The proposed model simulates the human fuzzy thinking in classifying the input data records of the breast cancer data sets accurately. The experimental results showed that the proposed fuzzy rough modelmeets the effective levels of performance in terms of classification accuracy and time consumption. The comparisons are accomplished between the proposed fuzzy rough model (Fuzzy Rough Feature Selection and the rough neural network) and the neural network with and without feature reduction. The experiments used the Weka and rapid miner data mining tools to prepare and classify the data sets using multilayer perceptron classification as an example for neural networks.

References

1. Smolinski, G., Milanova, M.G., Hassanien, A.E.: Computational Intelligence in Biomedicine and Bioinformatics. Studies in Computational Intelligence. Springer, Heidelberg (2011). ISBN 10: 354070776X|13: 978-3540707769
2. Devlin, G.: Decision Support Systems: Advances. In-Tech Publishing, Croatia (2010). ISBN 9789533070698
3. Jackson, P.C.: Introduction to Artificial Intelligence. Courier Corporation, Chelmsford (1985). ISBN 048624864X
4. Suzuki, K.: Artificial Neural Networks: Methodological Advances and Biomedical Applications. InTech, Rijeka (2011). ISBN 13: 9789533072432
5. Lingras, P.J.: Rough neural network. In: Proceedings of the 6th International Conference on Information Processing and Management of Uncertainty in Knowledge-Based Systems (IPMU96), Granada, Spain, pp. 1445–1450 (1996)
6. Hassanien, A.E.: Rough neural intelligent approach for image classification: a case of patients with suspected breast cancer. Int. J. Hybrid Intell. Syst. **3**, 205–218 (2006). IOS press
7. Pal, S.K., Polkowski, S.K., Skowron, A.: Rough-Neuro Computing: Techniques for Computing with Words. Springer, Berlin (2002)
8. Peters, J.F., Liting, H., Ramanna, S.: Rough neural computing in signal analysis. Comput. Intell. **17**(3), 493–513 (2001)
9. Peters, J.F., Skowron, A., Liting, H., Ramanna, S.: Towards rough neural computing based on rough membership functions: theory and application. In: Rough Sets and Current Trends in Computing, pp. 611–618 (2000)
10. Dheeba, J., Singh, N.A., Selvi, S.T.: Computer-aided detection of breast cancer on mammograms: a swarm optimized wavelet neural network approach. J. Biomed. Inform. **49**, 45–52 (2014)
11. Balanica, V., Ioan, D., Luigi, P.: Breast cancer diagnosis based on speculation feature and neural network techniques. Int. J. Comput. Commun. Control **8**(3), 354–365 (2013)
12. Azar, A.T., El-Said, S.A.: Probabilistic neural network for breast cancer classification. Neural Comput. Appl. **23**(6), 1737–1751 (2013)
13. Janghel, R.R., Shukla, A., Tiwari, R., Kala, R.: Breast cancer diagnosis using Artificial Neural Network models. In: 3rd International Conference on Information Sciences and Interaction Sciences (ICIS), Chengdu, China, pp. 89–94 (2010)

14. Karabatak, M., Ince, M.C.: An expert system for detection of breast cancer based on association rules and neural network. Expert Syst. Appl. **36**(2), 3465–3469 (2009)
15. Pawlak, Z.: Rough sets. Int. J. Comput. Inform. Sci. **11**, 341–356 (1982)
16. Pawlak, Z., Grzymala-Busse, J., Slowinski, R., Ziarko, W.: Rough sets. Commun. ACM **38** (11), 89–95 (1995)
17. Pawlak, Z.: Rough Sets: Theoretical Aspects of Reasoning about Data. Kluwer Academic Publishing, Dordrecht (1991)
18. Baum, E., Huassler, D.: What size net gives valid generalization. Neural Comput. **1**, 151–160 (1989)
19. Jensen, R., Shen, Q.: New approaches to fuzzy-rough feature selection. IEEE Trans. Fuzzy Syst. **17**(4), 824–838 (2009)
20. Dechter, R., Pearl, J.: Generalized best-first search strategies and the optimality of A*. J. Assoc. Comput. Mach. **32**, 505–536 (1985)
21. Russell, S.J., Norvig, P.: Artificial Intelligence: A Modern Approach, 2nd edn. Prentice Hall, Upper Saddle River (2003). pp. 111–114. ISBN 0-13-790395-2
22. http://archive.ics.uci.edu/ml/
23. http://www.cancer.gov/
24. Hall, M., Frank, E., Holmes, G., Pfahringer, B., Reutemann, P., Witten, I.H.: The WEKA data mining software: an update. SIGKDD Explor. **11**(1), 10–18 (2009)
25. Saleh, A.A., Gamal, M.: An intelligent model in bioinformatics based on rough-neural computing. Int. J. Comput. Appl. **64**(2), 43–48 (2013)

Meta-Heuristic Algorithm Inspired by Grey Wolves for Solving Function Optimization Problems

Alaa Tharwat[1,3]([✉]), Basem E. Elnaghi[1], and Aboul Ella Hassanien[2,3]

[1] Faculty of Engineering, Suez Canal University, Ismailia, Egypt
engalaatharwat@hotmail.com
[2] Faculty of Computers and Information, Cairo University, Giza, Egypt
[3] Scientific Research Group in Egypt (SRGE), Cairo, Egypt
http://www.egyptscience.net

Abstract. In this paper, we suggest the use of Grey Wolf Optimization (GWO) algorithm to solve numerical optimization problems. GWO is compared with two well-known optimization algorithms namely, Bat Algorithm (BA) and Particle Swarm Optimization (PSO), to test the improvement in the accuracy of finding the near optimal solution and the reduction in the computational cost. Ten standard benchmark functions were applied to test the performance of the three optimization algorithms in terms of accuracy and computational cost. The experimental results proved that our proposed method achieved accuracy better than the other two algorithms and it reduced the computational cost and converged rapidly to the optimal solution.

Keywords: Numerical problems · Gray wolf optimization (GWO) · Bat Algorithm (BA) · Particle Swarm Optimization (PSO) · Bio-inspired optimization

1 Introduction

Searching for the optimal solution from a set of feasible solutions is an important topic to solve many problems in different applications such as segmentation [1], machine learning [2–5], and numerical problems [6]. Generally, an optimization problem includes minimizing or maximizing the value of a function by selecting the values of the input parameters. In other words, optimization algorithms search in the input space to reach the optimal solution. The dimension of the input space depends mainly on the number of input parameters of the optimization problem and the limits of these parameters. The optimal solutions consist of local and global solutions.

The optimization algorithms are divided into two main categories namely *deterministic* and *stochastic* algorithms. In deterministic algorithms, the same

© Springer International Publishing AG 2017
A.E. Hassanien et al. (eds.), *Proceedings of the International Conference on Advanced Intelligent Systems and Informatics 2016*, Advances in Intelligent Systems and Computing 533, DOI 10.1007/978-3-319-48308-5_46

set of solutions is generated if the iterations started from the same initial solution. On the other hand, in stochastic algorithms, different solutions are generated even if the iterations started from the same solution and hence the optimization algorithm converges to different optimal solutions in each run [7–9].

The recent studies in numerical analysis and applied mathematics appear to suggest that the algorithms that can guarantee convergence in a small time to the near optimal solution of a non-convex problem are called global numerical optimization problems. A variety of algorithms has been proposed to solve non-convex problems. Gilbert *et al.* used conjugate gradient method to solve nonlinear, non-convex, and global optimization problems. They used 26 functions to evaluate their proposed model. They found that the computational time for their algorithm was very high [10]. In contrast to gradient-based optimization algorithms, meta-heuristic algorithms optimize problems stochastically. In other words, the optimization process starts with the random solution(s) without no need to calculate the derivative of search spaces to find the optimum. For this reason, meta-heuristic algorithms are highly suitable for real problems with expensive or unknown derivative information. The bio-inspired optimization algorithms are used to solve numerical problems. Jorge Nocedal and Wright Stephen used Ant Bee Colony (ABC) optimization algorithm to solve numerical problems [11]. They evaluated their proposed algorithm on only three numerical functions with different dimensions, and they achieved excellent results. Pei-Wei Tsai *et al.* proposed a novel optimization algorithm called Evolved Bat Algorithm (EBA) to solve numerical optimization problems [6].

Grey Wolf Optimization (GWO) is one of the recent bio-inspired optimization algorithms. GWO has been used in different applications. For example, Jayakumar *et al.* used GWO in power system operation for allocating generation and heat outputs to the committed units [12]. Emary *et al.* used GWO to select the most discriminative features. They compared GWO algorithm with GA and PSO over a set of UCI machine learning data repository, and they proved that GWO is robust in comparison with GA and PSO. Seyedali Mirjalili used GWO to search for the weights to train Multi-Layer Perceptrons (MLPs). They compared GWO with PSO, Ant Colony Optimization (ACO), GA, Evolution Strategy (ES), and Population-based Incremental Learning (PBIL) and they found that GWO algorithm achieved classification accuracy lower compared with the other optimization algorithms [13].

In this paper, GWO algorithm is used to search for the global solution of numerical problems. The rest of this paper is organized as follows: Sect. 2 presents the fundamental of the optimization problem and GWO algorithm. Experimental results with discussions are presented in Sect. 3. Conclusions and future work are provided in Sect. 4.

2   Preliminaries

To begin with, in this section, we will provide a brief background on the optimization problem, and GWO algorithm.

2.1 Optimization Problem

The goal of any optimization algorithm is to search for the optimal solution from a set of feasible solutions. The optimal solution minimizes or maximizes the value of the objective function, f, by selecting input values from the search space, $S \in \mathcal{R}^d$, which includes all solutions, i.e. global or local, where d is the dimension of the search space. A feasible solution/region $s \in S$ is a region in which all solutions satisfy all constraints. Moreover, the dimension depends mainly on the number of parameters we need to optimize.

If the feasible region and the objective function are convex in a minimization problem, there is only one global solution $(x*)$; otherwise, there may be more than one local minima. A global minimum $x*$ is defined as a point that satisfies the following condition $f(x*) \leq f(x)$, $\forall x \in S$. In this paper, GWO algorithm is used to solve non-convex problems. The next section describes the details of GWO algorithm.

2.2 Grey Wolf Optimization (GWO)

Gray wolf optimization (GWO) is a new bio-inspired heuristic optimization algorithm that imitates the way wolves search for food and survive by avoiding their enemies [14]. Wolves are social animals that live in packs, and the pack size is from 5 to 12 wolves on average. The Grey wolves are species with very strict social dominant hierarchy of leadership. Grey wolves pack consist of four different categories as the following [2]:

1. *Alpha* (α) wolves (the leaders) consist of one male and one female. *Alpha* (α) wolves are mostly responsible for making decisions about hunting, sleeping place, time to wake, and so on. In addition, their decisions are dictated to the pack.
2. *Beta* (β) wolves are subordinate wolves that help the (α) in decision making or the other pack activities. The (β) wolves can be either male or female, and he/she is probably the best candidate to be the next (α) in case one of the (α) wolves passes away or becomes very old.
3. *Omega* (ω) wolves plays the role of scapegoat and (ω) wolves have to obey to all the dominant wolves. They are the last wolves that are allowed to eat.
4. *Delta* (δ) wolves have to submit to (α) and (β). *Scouts, sentinels, elders, hunters,* and *caretakers* belong to *delta* wolves. *Scouts* are responsible for watching the boundaries of the territory and warning the pack in case of any danger. *Sentinels* protect and guarantee the safety of the pack. *Elders* are the experienced wolves that used to be alpha or beta. *Hunters* have to help the alpha and beta wolves during the hunting process and providing the food needs for the pack. Finally, the *caretakers* are responsible for caring for the weak, ill, and wounded wolves in the pack.

Each grey wolf in the pack chooses its own position, continuously moving to a better spot and watching for potential threats. GWO is prepared with a threat probability, which mimics the incidents of wolves bumping into their enemies. When this happens, the wolf dashes a great distance away from its current position that helps break the deadlock of getting stuck in local optima. They travel when moving away from a threat are random for both the direction and distance, which is similar to mutation and crossover in GA. In addition, each wolf has a sensing distance that creates a sensing radius or coverage area generally referred to as visual distance. This visual distance is applied to the search for food (the global optimum), an awareness of their peers (in the hope of moving into a better position) and signs that enemies might be nearby (for jumping out of visual range). Once they sense that the prey is near, they approach quickly, quietly and very cautiously because they do not wish to reveal their presence [15].

In the mathematical model, $alpha(\alpha)$ is the fittest solution, $beta(\beta)$ and $delta(\delta)$ are the second and third best solutions, respectively. The hunting process is guided by α, β, and δ, while the ω follows the three candidates. The gray wolves are encircling the prey during the hunting process. The mathematical model of encircling behavior applied as in Eqs. (1) and (2) [16].

$$\vec{X}(t+1) = \vec{X}_p(t) + \vec{A}.\vec{D} \tag{1}$$

$$\vec{D} = |\vec{C}.\vec{X}_p(t) - \vec{X}(t)| \tag{2}$$

where t is the iteration number, $\vec{A}$, $\vec{C}$ are coefficient vectors, $\vec{X}_p$ is the prey position, and $\vec{X}$ is the gray wolf position.

The $\vec{A}$, $\vec{C}$ vectors are calculated as in Eqs. (3) and (4).

$$\vec{A} = 2\vec{A}.\vec{r_1} - \vec{a} \tag{3}$$

$$\vec{C} = 2\vec{r_2} \tag{4}$$

where components of $\vec{a}$ are linearly decreased from 2 to 0 over the course of iterations and r_1, r_2 are random vectors in $[0, 1]$.

α, β, and δ solutions obtained so far and oblige the other search agents (including ω) to update their positions according to the position of the best search agents. Therefore, the wolves' positions are updating as shown in Eqs. (5, 6, and 7).

$$\begin{aligned}
\vec{D_\alpha} &= |\vec{C_1}.\vec{X_\alpha} - \vec{X}|, \\
\vec{D_\beta} &= |\vec{C_2}.\vec{X_\beta} - \vec{X}|, \\
\vec{D_\delta} &= |\vec{C_3}.\vec{X_\delta} - \vec{X}|
\end{aligned} \tag{5}$$

$$\begin{aligned}
\vec{X_1} &= |\vec{X_\alpha} - \vec{A_1}.\vec{D_\alpha}|, \\
\vec{X_2} &= |\vec{X_\beta} - \vec{A_2}.\vec{D_\beta}|, \\
\vec{X_3} &= |\vec{X_\delta} - \vec{A_3}.\vec{D_\delta}|
\end{aligned} \tag{6}$$

$$\vec{X}(t+1) = \frac{\vec{X_1} + \vec{X_2} + \vec{X_3}}{3} \tag{7}$$

Table 1. The search space, name, and optimal solutions of the benchmark functions that were used in our experiments.

Function name	Formula	Search space	$f*$
Michalewicz's (f_1)	$f(x) = -\sum_{i=1}^{D} sin(x_i).(sin(\frac{i.x_i^2}{\pi}))^2.m$	$0 \leq x_i \leq \pi$	-1.801
De Jong's (f_2)	$f(x) = \sum_{i=1}^{D} x_i^2$	$-5.12 \leq x_i \leq 5.12$	0
Rastrigin's (f_3)	$f(x) = 10.n + \sum_{i=1}^{D}(x_i^2 - 10.cos(2.\pi.x_i))$	$-5.12 \leq x_i \leq 5.12$	0
Schwefel's (f_4)	$f(x) = \sum_{i=1}^{D} -x_i.sin(\sqrt{|x_i|})$	$-500 \leq x_i \leq 500$	-837.9658
Easom's (f_5)	$f(x) = -cos(x_1).cos(x_2).e^{((x_1-\pi)^2+(x_2-\pi)^2)}$	$-100 \leq x_i \leq 100$	-1
Goldstein-Price's (f_6)	$f(x) = (1 + (x_1 + x_2 + 1)^2 . (19-14.x_1+3.x_1^2-14.x_2+6.x_1.x_2+3.x_2^2)). (30 + (2.x_1 - 3.x_2)^2 . (18 - 32.x_1 + 12.x_1^2 + 48.x_2 - 36.x_1.x_2 + 27.x_2^2)]$	$-2 \leq x_i \leq 2$	3
Six-hump Camel Back (f_7)	$f(x_1, x_2) = (4 - 2.1.x_1^2 + x_1^4/3).x_1^2 + x_1.x_2 + (-4 + 4.x_2^2).x_2^2$	$-2 \leq x_1 \leq 2$ $-3 \leq x_1 \leq 3$	-1.0316
Alpine (f_8)	$f(x) = \sum_{i=1}^{D} |x_i sin(x_i) + 0.1x_i|$	$-10 \leq x_i \leq 10$	0
Bartels Conn (f_9)	$f(x) = |x_1^2 + x_2^2 + x_1x_2| + |sin(x_1)| + |cos(x_2)|$	$-500 \leq x_i \leq 500$	1
Beale (f_{10})	$f(x) = (1.5 - x_1 + x_1x_2)^2 + (2.25 - x_1 + x_1x_2^2)^2 + (2.625 - x_1 + x_1x_2^3)^2$	$-4.5 \leq x_i \leq 4.5$	0

A final note about the $\overrightarrow{a}$ is the updating parameter of the GWO that controls the trade-off between exploration and exploitation. $\overrightarrow{a}$ is linearly updated in each iteration to range from 2 to 0 as follows, $\overrightarrow{a} = 2 - t.\frac{2}{Max_{iter}}$, where t is the iteration number and Max_{iter} is the total number of iterations allowed for the optimization.

3   Simulation and Experimental Results

In this section, the GWO algorithm was evaluated using ten benchmark functions. GWO was compared with two well-known optimization algorithms, namely PSO and BA. All functions are classical benchmark functions that were used by many researchers [17]. All benchmark functions and their ranges are listed in Table 1. In all functions, the dimension was only two and all functions are minimization functions. The detailed descriptions of the benchmark functions are reported in [17]. Due to the random initialization and random walk of all algorithms, all algorithms run ten times. Table 2 shows the worst, best, and mean optimal solutions.

To allow a fair comparison of running times, all the experiments were conducted on a PC with an Intel Core i5-2400 CPU @ 3.10 GHz and 4.00 GB. Our implementation was compiled using MATLAB R2012a (7.14) running under Windows XP3.

3.1 Parameter Setting for GWO

In this section, the influence of the number of wolves and the number of iterations on the CPU time and deviation from the optimal solution of the proposed model were investigated.

Number of Wolves: The number of wolves needs to be sufficient to explore the space to find the global solution. In this section, the effect of the number of wolves on the deviation from the optimal solution and CPU time of the proposed model was investigated when the number of wolves ranged from 4 to 40 wolves. In this experiment, De Jong's function was used. The MSE and CPU time of this experiment are shown in Figs. 1a and b, respectively. From the figures, it is observed that increasing the number of wolves reducing the MSE, but more CPU time was needed. Moreover, the minimum MSE achieved when the number of wolves was more than 15.

Number of Iterations: The number of iterations also affects the performance of the proposed model. In this section, the effect of the number of iterations on the MSE and CPU time of the proposed model was tested when the number of iterations was ranged from 10 to 100. The MSE and CPU time are shown in Figs. 1c and d, respectively. In this experiment, De Jong's function was used. From the figure, it can be noticed that, when the number iterations were increased, the MSE was decreased until it reached an extent at which increasing the number of iterations did not affect the MSE. Moreover, the CPU time increased when the number of iterations was increased.

On the basis of the above parameter analysis and research results, the values of population size and a maximum number of iterations parameters of all algorithms were 15 and 50, respectively. In PSO, the value of the inertia factor (ω) was 0.5 and the acceleration factor was 0.1. In BA, The values of loudness, pulse rate, minimum frequency, and maximum frequency parameters were 0.5, 0.5, 0, and 2, respectively. For GWO algorithm, the values of the updating parameters (r_1 and r_2) ranged in $[0, 1]$.

A brief description of the ten functions is summarized as follows:

1. *Michalewicz's function*: is a multimodal[1], hence higher dimensions lead to the more difficult search. Figures 2(a) and 3(a) show the surface and contour plot of the Michalewicz's function, respectively.
2. *De Jong's function*: is the simplest test function, and it is also known as sphere model. It is continuous, convex and unimodal[2]. As shown in Figs. 2(b) and 3(b), the global minimum is located at the origin, i.e. (0,0).

---

[1] Multimodal function has at least two optimal solutions in the search space, indicating that these solutions have more than one local optima except the optimum global.

[2] Unimodal function has one solution to the function, which is the optimum global.

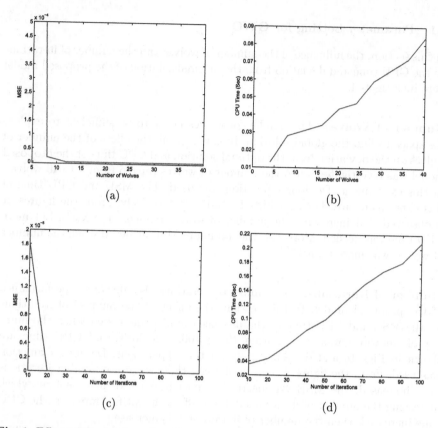

Fig. 1. Effect of the number of wolves and number of iterations on the MSE and CPU time for De Jong's function. (a) MSE of GWO algorithm using different numbers of wolves; (b) CPU time of the GWO algorithm using different numbers of wolves; (c) MSE of GWO algorithm using different numbers of iterations; (b) CPU time of the GWO algorithm using different numbers of iterations.

3. *Rastrigin's function*: is based on De Jong's function with the addition of cosine modulation to produce many local minima. Hence, Rastrigin's function is highly multimodal, and the location of the minima is regularly distributed. Figures 2(c) and 3(c) shows the surface and contour plot of Rastrigin's function.

4. *Schwefel's function*: the global minimum of this function is geometrically distant from the next best local minima. However, this function has many local minima and hence it is difficult to reach to global solution (see Figs. 2(d) and 3(d)). In other words, the search algorithms are potentially prone to convergence in the wrong direction.

5. *Easom's function*: is a unimodal test function, where the global minimum has a small area relative to the search space as shown in Figs. 2(e) and 3(e). The function was inverted for minimization.

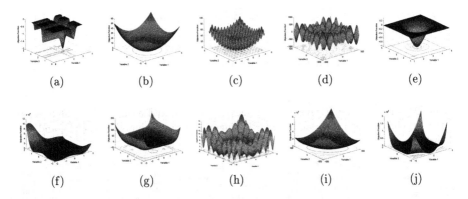

Fig. 2. Visualization of the surface of the benchmark functions that were used in our experiments (a) Michalewicz's function, (b) De Jong's function, (c) Rastrigin's function, (d) Schwefel's function, (e) Easom's function, (f) Goldstein-Price's function, (g) Six-hump Camel Back function, (h) Alpine function, (i) Bartels Conn function, (j) Beale function.

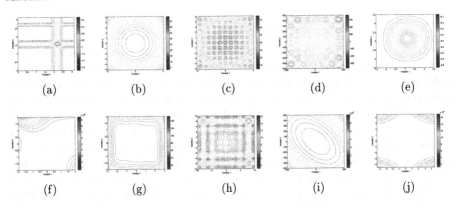

Fig. 3. Visualization of the contour plot of the benchmark function that were used in our experiments (a) Michalewicz's function, (b) De Jong's function, (c) Rastrigin's function, (d) Schwefel's function, (e) Easom's function, (f) Goldstein-Price's function, (g) Six-hump Camel Back function, (h) Alpine function, (i) Bartels Conn function, (j) Beale function.

6. *Goldstein-Price's function*: is a global optimization test function. Figures 2(f) and 3(f) show the Goldstein-Price's function. As shown, it has many local minimum solutions.
7. *Six-hump Camel Back function*: the two-dimensional Six-hump camel back function is a global optimization test function. This function has six local minima, and two of them are global minima as shown in Figs. 2(g) and 3(g).
8. *Alpine function*: has many local minima and only one global minimum as shown in Figs. 2(h) and 3(h).

9. *Bartels Conn function*: as shown in Figs. 2(i) and 3(i), the Bartels Conn function is simple function.
10. *Beale function*: Figs. 2(j) and 3(j) show the Beale function. As shown, the Beale function is multimodal with sharp peaks.

Table 2. The best, worst, mean ± standard deviation, and average CPU time (secs) of the benchmark functions and the CPU time of 10 runs using GWO, PSO, and BA algorithm.

Function	PSO				BAT				GWO			
	Best	Worst	Mean±SD	CPU Time	Best	Worst	Mean±SD	CPU Time	Best	Worst	Mean±SD	CPU Time
f_1	-1.801	-1.00	-1.240± 0.379	0.231	-1.747	-1.168	-1.550± 0.203	0.354	-1.801	-0.999	-1.676± 0.261	0.312
f_2	$4.495e-10$	$1.168e^{-07}$	$3.446e^{-08}± 3.829e^{-08}$	0.160	0.001	0.014	0.005± 0.005	0.31	$7.470e^{-09}$	$6.828e^{-06}$	$1.159e^{-06} ± 2.229e^{-06}$	0.218
f_3	0.995	9.958	3.453± 2.706	0.230	1.299	13.924	5.826± 4.275	0.325	0.004	4.977	1.635± 1.412	0.314
f_4	-693.424	-316.797	-521.743± 130.047	0.233	-798.521	-609.989	-723.706± 56.977	0.314	-718.506	-500.677	-620.121± 92.454	0.298
f_5	-1	-0.961	-0.994± 0.013	0.228	-0.999	-0.943	-0.982± 0.019	0.310	-0.999	-0.996	-0.999± 0.001	0.300
f_6	3	33.363	8.760± 12.111	0.531	3.001	69.120	20.781± 21.883	0.345	3.004	30.057	8.513± 11.341	0.366
f_7	-1.032	-0.712	-0.971±0.113	0.228	-1.028	-0.474	-0.913± 0.166	0.325	-1.032	-0.959	-1.024± 0.023	0.311
f_8	$1.564e^{-05}$	0.312	0.047± 0.104	0.128	0.430	0.203± 0.147	0.324	0.235	$4.675e^{-05}$	0.224	0.025± 0.070	0.300
f_9	1.001	96.746	19.663±32.848	0.228	1.949	263.928	70.121± 82.083	0.318	1.001	4.077	1.505± 1.021	0.291
f_{10}	0.0306	8.970	1.716± 2.596	0.229	0.011	0.919	0.286± 0.292	0.348	0.002	1.009	0.102± 0.319	0.312

From Table 1, we see that GWO achieved the best solution in five of the ten functions, i.e. 50 % of the total functions while BA achieved the best solution for only one function. According to the worst solutions, BA achieved the worst solutions in six of the ten functions, while both GWO and PSO algorithms achieved the worst solutions in only two functions. Moreover, in terms of the means of the optimal solutions, GWO achieved the best results. Further analysis of this point is that the mean of the optimal solutions using GWO was closer to the optimal solution than the other two algorithms, i.e. PSO and BA. In addition, GWO achieved the minimum standard deviation which reflects that all solutions that achieved using GWO were closer to the optimal solution than PSO and BA algorithms. In terms of computational time, from Table 2, PSO algorithm needs CPU time lower than GWO and BA while BA needs more CPU time. However, the best results obtained when GWO algorithm was used and the advance in the parallel computing and the super-computing could address this issue in the real-time implementation.

From these findings we can conclude that the performance of GWO is better than PSO and BA algorithms and GWO algorithm minimizes the deviation from the optimal solution.

4   Conclusions and Future Work

The main aim of this paper was to propose Grey Wolf Optimization (GWO) algorithm for solving numerical problems. The proposed algorithm is compared to Particle Swarm Optimization (PSO) and Bat Algorithm (BA) using different evaluation criteria on ten standard benchmark functions to test the three optimization algorithms' performance. The experimental results proved the capability of GWO to reach to the near optimal solution and decrease the computational time. In addition, the results proved that GWO algorithm is converged rapidly to the optimal solution and its performance significantly better than BA and PSO. On the basis of future performance, the proposed method will be assessed using more complex numerical functions and comparing with new bio-inspired algorithms.

References

1. Paulinas, M., Ušinskas, A.: A survey of genetic algorithms applications for image enhancement and segmentation. Inf. Technol. Control **36**(3), 278–284 (2015)
2. Emary, E., Zawbaa, H.M., Grosan, C., Hassenian, A.E.: Feature subset selection approach by gray-wolf optimization. In: Abraham, A., Krömer, P., Snasel, V. (eds.) AECIA 2014. AISC, vol. 334, pp. 1–13. Springer, Heidelberg (2015). doi:10.1007/978-3-319-13572-4_1
3. Gaber, T., Tharwat, A., Hassanien, A.E., Snasel, V.: Biometric cattle identification approach based on weber's local descriptor and adaboost classifier. Comput. Electron. Agric. **122**, 55–66 (2016)
4. Tharwat, A., Gaber, T., Hassanien, A.E.: Cattle identification based on muzzle images using gabor features and SVM classifier. In: Hassanien, A.E., Tolba, M.F., Taher Azar, A. (eds.) AMLTA 2014. CCIS, vol. 488, pp. 236–247. Springer, Heidelberg (2014). doi:10.1007/978-3-319-13461-1_23
5. Semary, N.A., Tharwat, A., Elhariri, E., Hassanien, A.E.: Fruit-based tomato grading system using features fusion and support vector machine. In: Filev, D., Jabłkowski, J., Kacprzyk, J., Krawczak, M., Popchev, I., Rutkowski, L., Sgurev, V., Sotirova, E., Szynkarczyk, P., Zadrozny, S. (eds.) Intelligent Systems'2014. AISC, vol. 323, pp. 401–410. Springer, Heidelberg (2015). doi:10.1007/978-3-319-11310-4_35
6. Tsai, P.W., Pan, J.S., Liao, B.Y., Tsai, M.J., Istanda, V.: Bat algorithm inspired algorithm for solving numerical optimization problems. Appl. Mech. Mater. **148**, 134–137 (2012). Trans Tech Publ
7. Yamany, W., Fawzy, M., Tharwat, A., Hassanien, A.: Moth-flame optimization for training multi-layer perceptrons. In: Proceedings of the 11th International Computer Engineering Conference (ICENCO), pp. 267–272. IEEE (2015)
8. Tharwat, A., Zawbaa, H., Gaber, T., Hassanien, A., Snasel, V.: Automated zebrafish-based toxicity test using bat optimization and adaboost classifier. In: Proceedings of the 11th International Computer Engineering Conference (ICENCO), pp. 169–174. IEEE (2015)
9. Yamany, W., Tharwat, A., Hassanin, M.F., Gaber, T., Hassanien, A.E., Kim, T.H.: A new multi-layer perceptrons trainer based on ant lion optimization algorithm. In: 2015 Fourth International Conference on Information Science and Industrial Applications (ISI), pp. 40–45. IEEE (2015)
10. Gilbert, J.C., Nocedal, J.: Global convergence properties of conjugate gradient methods for optimization. SIAM J. Optim. **2**(1), 21–42 (1992)
11. Karaboga, D.: An idea based on honey bee swarm for numerical optimization. Technical report, Technical report-tr06, Erciyes University, Engineering Faculty, Computer Engineering Department (2005)
12. Jayakumar, N., Subramanian, S., Ganesan, S., Elanchezhian, E.: Grey wolf optimization for combined heat and power dispatch with cogeneration systems. Int. J. Electr. Power Energy Syst. **74**, 252–264 (2016)

13. Mirjalili, S.: How effective is the grey wolf optimizer in training multi-layer perceptrons. Appl. Intell. **43**(1), 150–161 (2015)
14. Mirjalili, S., Mirjalili, S.M., Lewis, A.: Grey wolf optimizer. Adv. Eng. Softw. **69**, 46–61 (2014)
15. Hassanien, A.E., Alamry, E.: Swarm Intelligence: Principles, Advances, and Applications. CRC Press (2015). ISBN 9781498741064
16. Emary, E., Zawbaa, H.M., Hassanien, A.E.: Binary grey wolf optimization approaches for feature selection. Neurocomputing **172**, 371–381 (2016)
17. Liang, J., Suganthan, P., Deb, K.: Novel composition test functions for numerical global optimization. In: Proceedings of Swarm Intelligence Symposium (SIS), pp. 68–75. IEEE (2005)

Automatically Human Age Estimation Approach via Two-Dimensional Facial Image Analysis

Alaa Tharwat[1,3]($\boxtimes$), Basem E. Elnaghi[1,3], Ahmed M. Ghanem[1,3], and Aboul Ella Hassanien[2,3]

[1] Faculty of Engineering, Suez Canal University, Ismailia, Egypt
engalaatharwat@hotmail.com
[2] Faculty of Computers and Information, Cairo University, Giza, Egypt
[3] Scientific Research Group in Egypt (SRGE), Cairo, Egypt
http://www.egyptscience.net

Abstract. In this paper, the human age automatically estimated via two-dimensional facial image analysis. The exact age estimation is often treated as a classification problem while it can be formulated as a regression problem. In our research, a classification and regression models are proposed. The two proposed models are evaluated using the same database images and the same features. Due to a big difference between the number of samples in each class or age group, the two proposed models used the complete and missing data in different experiments. Moreover, we compared age estimation errors when (1) Age estimation is performed without discrimination between males and females (gender unknown); (2) Age estimation is performed in males and females separately (gender known). Conclusions and results of these proposed models are shown by extensive experiments on the public available FG-NET database.

Keywords: Facial age estimation · Classification · Regression · k-Nearest neighbor · Linear regression · Non-linear regression · Local binary patterns (LBP)

1 Introduction

Facial age estimation is a relatively new research topic in the area of facial image analysis. Compared with much other facial information, estimation of age has many challenges such as health, lifestyle, weather conditions, and a human gene. Another reason is that large aging databases are hard to collect. Moreover, the aging process can be accelerated or slowed down by a physical condition or lifestyle [1–5].

Age estimation can be considered either as a *classification* or a *regression* problem [6]. In the classification-based problem, the age group is estimated while in regression-based problem, the exact age can be estimated. Some earlier work

© Springer International Publishing AG 2017
A.E. Hassanien et al. (eds.), *Proceedings of the International Conference on Advanced Intelligent Systems and Informatics 2016*, Advances in Intelligent Systems and Computing 533, DOI 10.1007/978-3-319-48308-5_47

has been reported on different aspects of age progression and estimation. Some studies deal with the age estimation process as a classification problem. Kwon and Lobo, proposed an age classification method that used both the shape and wrinkles of a human face to classify input images into only one of the three age groups: babies, young adults, and senior adults [7]. Lanitis *et al.*, proposed a quadratic aging function that maps the *Active Appearance Model* (AAM) features of a face image to an age [8]. Moreover, they compared different classifiers for age estimation based on AAM features in their later work [9]. With AAM based face encoding, Geng et al., handled the age estimation problem by introducing an aging pattern subspace (AGES), which is a subspace representation of a sequence of individual aging face images [10]. Feng Gao and Haizhou Ai, used Gabor features as a face representation and the *Linear Discriminant Analysis* (LDA) to construct the age classifier that classifies human faces as a baby, child, adult, or elder people. In their proposed model, the images in the training set were labeled without the age information [11]. On the other hand, many studies deal with age estimation process as a regression problem. Guodong Guo *et al.*, introduced the age manifold learning scheme for extracting face aging features and have designed a locally adjusted robust regressor for learning and prediction of human ages [12]. Ni *et al.*, presented a multi-instances regression method in order to adopt the face images with noisy labels that were collected from Web image resources [13].

In this paper, a classification and regression models are proposed to estimate exact age from two-dimensional face images. Theoretically, classification affected by missing data than regression. Hence, in this paper, we compared between the classification and regression models when one class is neglected, i.e. missing data. In addition, the proposed models are evaluated when the genders are known or unknown. The rest of the paper is organized as follows: Sect. 2 describes some of the related work. Section 3 presents the proposed age estimation system. Experimental results and discussion are discussed in Sect. 4. Finally, concluding remarks are presented in Sect. 5.

2   Preliminaries

2.1   Local Binary Pattern (LBP) Features

LBP is one of the feature extraction methods that are used to extract local features from greyscale images [14]. In LBP, the LBP code is calculated for each pixel by comparing each pixel with its neighbors [14,15]. The LBP code is represented by a binary number and it is calculated as follows:

$$LBP_{P,R} = \sum_{i=0}^{P-1} s(g_i - g_c)2^i, \text{ where } s(x) = \begin{cases} 1, & x \geq 0 \\ 0, & x < 0 \end{cases}, i = 0, 1, \ldots, P-1 \quad (1)$$

where g_c is the gray level of the center pixel, g_i represents the gray levels of the neighbours of g_c, $LBP_{P,R}$ is the LBP code when the number of neighbors pixels is P, $R(R > 0)$ is the distance from the center to the neighboring pixels, i.e. radius, and s is the threshold function of x [14]. The LBP code is rotated until reach its minimum to make LBP robust against rotation as follows, $LBP_{P,R} = \min\{ROR(LBP_{P,R}, i)\}$, $i = 0, 1, \ldots, P - 1$, where $ROR(f, i)$ represents the circular bit-wise right shift on the f value i times. More details about LBP algorithm are reported in [14].

2.2 Facial Landmarks

The Regions Of Interest (ROI) in face images are called facial landmarks such as eyes, nose, lips and mouth. The landmarks are first located and then a template matching is used to accurately locate the location of the facial features. Cristinacce et al., have shown that precise landmarks are essential for a good face-recognition performance [16].

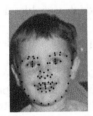  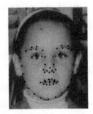 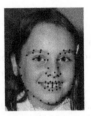

Fig. 1. Some images of the FG-NET database with landmarks.

In our model, 68 landmarks of each face image that attached into the database are used to extract the features from specific regions that are the most regions affected by age variation. Sample images with landmark annotations are shown in Fig. 1.

2.3 k-NN Classifier

k-Nearest Neighbor (k-NN) classifier is one of the most widely used classifiers. In k-NN classifier, an unknown pattern, x_{test}, is classified based on the similarity to the labeled/training samples by computing the distances from the unknown sample to all labeled samples and select the k-nearest samples as the basis for classification. The unknown sample is assigned to the class that has the most samples among the k-nearest samples. Hence, k-NN classifier algorithm depends on; (1) Integer k (number of neighbors) and changing the values of k parameter may change the classification result, (2) A set of labeled data, thus adding or removing any samples to the labeled samples will affect the final decision of k-NN classifier, and (3) A distance metric [17]. In k-NN, Euclidean distance is often used as the distance metric to measure the similarity between two samples as follows, $d(x_i, x_j) = \sum_{k=1}^{d}(x_{ik} - x_{jk})^2$, where $x_i, x_j \in \mathcal{R}^d$ and $x_i = \{x_{i1}, x_{i2}, \ldots, x_{id}\}$.

Table 1. Age group distribution of the facial images in FG-NET Database.

Age Range	Distribution (%)	# of Samples
0-9	37.03	371
10-19	33.83	339
20-29	14.37	144
30-39	7.88	79
40-49	4.59	46
50-59	1.5	15
60-19	0.8	8

2.4   Regression

Regression is used to build a relationship between a dependent variable, Y, and one or more independent variables, $X = \{x_1, x_2, \ldots, x_p\}$ as follows, $h_\beta = \beta_0 + \beta_1 x_1 + \beta_2 x_2 + \cdots + \beta_p x_p$, where p is the number of independent variables and β_i's are the parameters/weights or regression coefficients.

Given n patterns, hence $Y \in \mathcal{R}^{n \times 1}$ and $X \in \mathcal{R}^{n \times p}$. The intercept is denoted by β_0, i.e. where the line cuts the y axis while $\beta = \beta_1 \ldots, \beta_p$ represent the slope of the regressor. In gradient descent algorithm, the values of β are calculated to minimize the cost function ($J(\beta)$) as follows, $J(\beta) = \frac{1}{2n} \sum_{i=1}^{n} (h_\beta(x^{(i)}) - y^{(i)})^2$, where h_β is the hypothesis or regressor that is used to estimate the output value of $x^{(i)}$. The values of β are initialized randomly and iteratively converged to minimize $J(\beta)$ as follows, $\beta_i = \beta_i - \alpha \frac{\partial J(\beta)}{\partial \beta_i} \ \forall i = 1, 2, \ldots, n$, where $\frac{\partial J(\beta)}{\partial \beta_i} = \frac{1}{n} \sum_{i=1}^{n} (h_\beta(x^{(i)}) - y^{(i)}) x^{(i)}$ and $0 \leq \alpha \leq 1$ is the learning rate [18]. Hence, the general form is $\beta_j = \beta_j - \frac{\alpha}{n} \sum_{i=1}^{n} (h_\beta(x^{(i)}) - y^{(i)}) x^{(i)} \ \forall \ i = 1, \ldots, n$.

To avoid the problem of overfitting in regression, a regularization parameter, λ, is used as follows, $\frac{\partial J(\beta)}{\partial \beta_i} = \frac{1}{n} (\sum_{i=1}^{n} (h_\beta(x^{(i)}) - y^{(i)})^2 x^{(i)} + \lambda \beta_i)$, where λ is the regularization parameter and it is used to reduce the overfitting, which reduces the variance of the estimated regression parameters and increases the bias [18].

In non-linear regression, the objective or cost function is the same in linear regression, but the independent variables are in higher orders and hence the regressor will be non-linear.

3   Proposed Models

This section describes the proposed models, i.e. classification and regression, in detail. In the proposed models, the LBP algorithm and facial landmarks were used to extract features. The extracted features were then fused or combined through concatenating the feature vectors of the local features, i.e. LBP, and global features, i.e. facial landmarks, as shown in Fig. 2. There are two types of models, namely, classification and regression models.

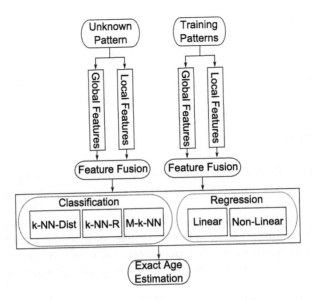

Fig. 2. Framework of the proposed Algorithm for exact facial age estimation.

3.1 Classification Models

In the classification models, three different methods were used to estimate the exact age. In the three methods, k-NN classifier was used to determine the class or age group of the unknown sample. The exact age still unknown and it will be calculated using one of the following methods.

k-**NN-Dist Method:** In this method, minimum distance classifier is used to match the unknown sample and all the trained samples belong to the class that has the highest number of the nearest neighbors.

Mean k-NN (M-k-NN) Method: Calculating the exact age using the class with a maximum number of the nearest neighbors only may avoid other correct or semi-correct classes. Hence, in this method, i.e. k-NN-Dist method, the exact age of the unknown sample was estimated by calculating the mean of the classes that were resulted from k-NN multiplying by the weight of each class as follows, $\sum_{i=1}^{n} w_i \times \mu_i$, where μ_i is the mean of each class and it is calculated as follows, $\mu_i = \frac{1}{m} \sum_{x_i \in C_i} x_i$ and w_i is the weight of each class and it represents the ratio between the number of the nearest neighbors from that class, C_i, to the total number of the nearest neighbors, k, as follows, $w_i = \frac{C_i}{k}$.

k-**NN-R Method:** In this method, the classification and regression techniques were used to calculate the exact age of the unknown sample. Simply, the samples of the class that has the maximum number of nearest neighbors that are used to build a regression model and then the exact age will be estimated.

3.2   Regression Models

In the regression model, there were two different methods were used to estimate the exact age, namely, linear and non-linear regression methods as shown in Fig. 2. The features of the training samples were used to build both regression methods and then the exact age will be estimated.

4   Experimental Results

In this section, different experiments were conducted to estimate the exact human age from two-dimensional face images. All experiments were applied on FG-NET Database [19]. The FG-NET Aging Database (Face and Gesture Recognition Research Network) has 1002 color and grey scale face images from 82 subjects. Each subject has a different number of images ranged from 6 to 18 face images at different ages. Each face image was manually annotated with 68 landmark points. The ages are distributed in a wide range from 0 to 69 as shown in Table 1.

The original face images in the database have many different backgrounds, clothes, hair, color, illumination, and orientation. Thus, it is necessary for each face image to crop and convert it to grey scale. Moreover, all images were resized to (64×64). In addition, due to a small number of images of subjects older than 40 in the database, only the first four age groups were used in our experiments.

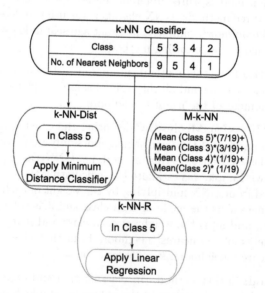

Fig. 3. Examples of M-k-NN, k-NN-Dist, and k-NN-R proposed age estimation system models.

4.1 Simulated Example

Figure 3 shows three examples to calculate the exact age in three of the proposed models, i.e. k-NN-Dist, M-k-NN, and k-NN-R. As shown, the unknown sample is classified using k-NN classifier when $k = 19$. As a result, the unknown sample has nine, five, four, and one nearest neighbors belong to the fifth, third, fourth, and second classes, respectively, and hence the unknown sample belongs to the fifth class which has the highest number of the nearest neighbors. However, the exact age still unknown and it will be calculated using one of the following methods.

k-**NN-Dist Method:** Figure 3 shows an example to calculate how the exact age is estimated using this method (k-NN-Dist). As shown, the unknown sample is matched with all training samples that belong to the fifth class that has the highest number of nearest neighbors and the age of the nearest sample is assigned to the unknown sample.

M-k-NN Method: In this method, the weight of each class is first calculated. The weight of each class is the ratio between the number of the nearest neighbors in that class to the total number of nearest neighbors. Hence, the weight of the fifth class is calculated as follows, $\frac{7}{19}$. Similarly, the weight of the third, fourth, and second classes are as follows $\frac{3}{19}$, $\frac{1}{19}$, and $\frac{1}{19}$, respectively. The exact age can be estimated as shown in Fig. 3.

k-**NN-R Method:** In this method, the samples of the fifth class that has the maximum number of nearest neighbors represent the training samples that are used to train the regression model to estimate the exact age of the unknown sample as shown in Fig. 3.

4.2 Real Data Experiments

In our experiments, Leave-One-Person-Out (LOPO) evaluation scheme is used. In each fold, all samples of a single person were used as the testing set and the remaining samples were used as the training set. To evaluate our experiments, Mean Absolute Error (MAE) was used. MAE is one of the most commonly used metric for age estimation and it is calculated as follows, $MAE = \frac{\sum |l_k - l_k^*|}{N}$, where l_k^* is the estimated age for the sample, l_k is the ground truth age of the sample and N is the total number of testing images [9]:

In the first experiment, the classification and regression models were used to estimate the exact age for unknown patterns. This experiment is divided into two sub-experiments. In the first one, the gender of the subject is ignored, i.e. gender unknown, while in the second sub-experiment, the subjects are divided into male and female groups and the unknown sample is matched with its group. The results of this experiment are summarized in Table 2.

In the second experiment, we proposed to evaluate the influence of losing one age group or class on the two proposed models. The third class, i.e. age group (20–29), which consists of 144 images, was neglected from the training stage. In this experiment, both classification and regression models were used. The results of this experiment are summarized in Table 3.

Table 2. A comparison between classification and regression models in terms of MAE using FG-NET database.

FG-Net Database			Classifier	Age Group				MAE
				0-9	10-19	20-29	30-39	
Classification	Unknown Gender		k-NN-Dist	3.9	5.7	11.8	22.4	6.6
			M-k-NN	7.2	3.2	8.2	18.6	6.2
			k-NN-R	3.2	4.8	11.2	21.2	5.8
	Known Gender	Male	k-NN-Dist	3.6	5.2	10.6	20.3	6.0
			M-k-NN	6.6	3.2	7.6	17.3	5.9
			k-NN-R	2.9	4.4	9.0	19.5	5.2
		Female	k-NN-Dist	5.2	5.2	13.0	23.8	7.2
			M-k-NN	9.3	2.6	8.8	19.5	7.0
			k-NN-R	4.8	4.6	12.8	23.4	6.8
Regression	Unknown Gender		Non-Linear $\lambda = 0$	6.3	3.9	6.9	16.1	5.8
			Non-Linear $\lambda = 100$	6.1	3.8	7.2	16.5	5.8
			Linear $\lambda = 0$	8.8	2.4	10.0	20.4	7.0
			Linear $\lambda = 100$	8.6	2.3	10.0	20.3	6.8
	Known Gender	Male	Non-Linear $\lambda = 0$	4.7	3.8	5.4	14.2	4.8
			Non-Linear $\lambda = 100$	4.5	3.7	5.6	14.4	4.8
			Linear $\lambda = 0$	7.6	2.5	10.1	20.4	6.5
			Linear $\lambda = 100$	7.6	2.5	10.1	20.4	6.5
		Female	Non-Linear $\lambda = 0$	6.5	4.7	7.4	14.6	6.2
			Non-Linear $\lambda = 100$	6.6	4.7	7.5	14.8	6.3
			Linear $\lambda = 0$	10.0	2.4	9.1	19.4	7.3
			Linear $\lambda = 100$	9.8	2.4	9.0	19.3	7.1

Table 3. A comparison between classification and regression models in terms of MAE, MAE change rate (%), and total MAE change using FG-NET database in case of missing one class.

FG-Net Database			Classifier	Age Group				MAE
				0-9	10-19	20-29	30-39	
Classification	Unknown Gender		k-NN-Dist	3.6	6.7	13	23.3	7.1
			M-k-NN	4.8	3.9	10.8	21.6	6.1
			k-NN-R	2.6	5.4	12.2	22.2	6.0
	Known Gender	Male	k-NN-Dist	3.4	6.8	12.0	21.7	6.8
			M-k-NN	4.7	3.9	10.1	21	5.9
			k-NN-R	3.0	5.1	10.0	20.9	5.7
		Female	k-NN-Dist	4.1	5.8	14.3	22.4	7.1
			M-k-NN	6.4	3.8	11.3	20.2	6.7
			k-NN-R	3.8	5.2	13.1	20.1	6.5
Regression	Unknown Gender		Linear $\lambda = 0$	5.7	3.7	8.5	18.4	5.9
			Linear $\lambda = 100$	4.2	3.9	11.2	21.7	6.0
			Non-Linear $\lambda = 0$	7.1	3.1	12	22.4	7.0
			Non-Linear $\lambda = 100$	7.3	3.2	12.1	22.4	7.2
	Known Gender	Male	Linear $\lambda = 0$	4.4	3.7	6.5	16.1	4.9
			Linear $\lambda = 100$	3.9	3.7	7.1	16.8	4.8
			Non-Linear $\lambda = 0$	6.0	3.5	12.2	22.6	6.7
			Non-Linear $\lambda = 100$	6.0	3.4	12.0	22.5	6.4
		Female	Linear $\lambda = 0$	5.9	4.7	7.8	15.5	6.1
			Linear $\lambda = 100$	5.9	4.6	8.0	15.9	6.1
			Non-Linear $\lambda = 0$	8.3	2.7	10.8	21.1	7.1
			Non-Linear $\lambda = 100$	8.2	2.6	10.7	21.5	7.3

4.3   Discussion

From Table 2 many notices can be seen. First, k-NN-R method in classification model achieved the lowest MAE among all methods in all classification models when the gender was known or unknown. Moreover, k-NN-Dist method achieved the worst results. The reason for that is k-NN-Dist depends mainly on one sample while the other two methods, i.e. M-k-NN and k-NN-R, depend on the nearest samples to the unknown sample. Second, non-linear regression method achieved MAE lower than linear regression method. The reason for this result is that the relation between the features of the images and the age of those images are non-linear and hence non-linear regression is suitable for this problem. However, there is no clear conclusion about the regularization parameter and this point needs more experiments. Generally, regression model achieved results better than classification model and the best result was 4.8 using the non-linear method and when the gender was known. Another positive finding is that the results of the gender known were better than the results of the gender unknown.

From Table 3 we can note that regression model achieved results better than classification model. Moreover, k-NN-R method and non-linear methods achieved the lower MAE in the classification and regression models, respectively. In addition, gender known results are better than gender unknown results and the MAE of male subjects was lower than female subjects. These findings are consistent with the findings from the first experiment. It is not surprising that the MAE was decreased when one class was removed. However, the difference between the MAE of the first and second experiment reflects the robustness of our two proposed models. In other words, the two proposed models achieved good results, despite removing one class. Moreover, the average changes of MAE of regression model were lower than the classification model and hence regression model deals with missing data better than classification model.

5   Conclusions and Future Work

In this paper, we have implemented the framework for facial age estimation. In this research, we have proposed two different models to estimate the facial age estimation. In the first model, classification model, k-NN classifier was used to determine the class of the unknown sample. In this model, three methods (M-k-NN, k-NN-Dist, and k-NN-R) were used to estimate the exact age. In the regression model, linear and non-linear regression methods were used to estimate the exact age of the unknown pattern. The implementation results illustrated that the regression model outperforms classification model. Moreover, the results demonstrated that the regression model deals with missing data better than classification model. Moreover, non-linear regression achieved results better than linear regression. In addition, k-NN-R method achieved the best results among all other methods in the classification model. Finally, the gender known results were better than gender unknown.

In the future, an optimization technique will be used to search for the optimal values of the regularization parameter of the linear and non-linear methods.

References

1. Tharwat, A., Mahdi, H., Hennawy, A.E., Hassanien, A.E.: Face sketch synthesis and recognition based on linear regression transformation and multi-classifier technique. In: Gaber, T., Hassanien, A.E., El-Bendary, N., Dey, N. (eds.) The 1st International Conference on Advanced Intelligent System and Informatics (AISI2015), November 28-30, 2015, Beni Suef, Egypt. AISC, vol. 407, pp. 183–193. Springer, Heidelberg (2016). doi:10.1007/978-3-319-26690-9_17

2. Gaber, T., Tharwat, A., Ibrahim, A., Snael, V., Hassanien, A.E.: Human thermal face recognition based on random linear oracle (RLO) ensembles. In: Proceedings of the International Conference on Intelligent Networking and Collaborative Systems (INCOS), pp. 91–98. IEEE (2015)

3. Tharwat, A., Mahdi, H., Hassanien, A.E., El Hennawy, A.: Face sketch recognition using local invariant features. In: Proceedings of 7th IEEE International Conference of Soft Computing and Pattern Recognition, pp. 464–473 (2015)

4. Gaber, T., Tharwat, A., Hassanien, A.E., Snasel, V.: Biometric cattle identification approach based on weber's local descriptor and adaboost classifier. Comput. Electron. Agric. **122**, 55–66 (2016)

5. Yamany, W., Fawzy, M., Tharwat, A., Hassanien, A.E.: Moth-flame optimization for training multi-layer perceptrons. In: 2015 11th International Computer Engineering Conference (ICENCO), pp. 267–272. IEEE (2015)

6. Fu, Y., Guo, G., Huang, T.S.: Age synthesis and estimation via faces: A survey. IEEE Trans. Pattern Anal. Mach. Intell. **32**(11), 1955–1976 (2010)

7. Kwon, Y.H., da Vitoria Lobo, N.: Age classification from facial images. Comput. Vis. Image Underst. **74**(1), 1–21 (1999)

8. Lanitis, A., Taylor, C.J., Cootes, T.F.: Toward automatic simulation of aging effects on face images. IEEE Trans. Pattern Anal. Mach. Intell. **24**(4), 442–455 (2002)

9. Lanitis, A., Draganova, C., Christodoulou, C.: Comparing different classifiers for automatic age estimation. IEEE Trans. Syst. Man Cybern. Part B: Cybern. **34**(1), 621–628 (2004)

10. Geng, X., Zhou, Z.H., Smith-Miles, K.: Automatic age estimation based on facial aging patterns. IEEE Trans. Pattern Anal. Mach. Intell. **29**(12), 2234–2240 (2007)

11. Gao, F., Ai, H.: Face age classification on consumer images with gabor feature and fuzzy LDA method. In: Tistarelli, M., Nixon, M.S. (eds.) ICB 2009. LNCS, vol. 5558, pp. 132–141. Springer, Heidelberg (2009). doi:10.1007/978-3-642-01793-3_14

12. Guo, G., Mu, G., Fu, Y., Huang, T.S.: Human age estimation using bio-inspired features. In: Proceedings IEEE Conference on Computer Vision and Pattern Recognition, (CVPR 2009), pp. 112–119. IEEE (2009)

13. Ni, B., Song, Z., Yan, S.: Web image mining towards universal age estimator. In: Proceedings of the 17th ACM International Conference on Multimedia, pp. 85–94. ACM (2009)

14. Ojala, T., Pietikäinen, M., Mäenpää, T.: Multiresolution gray-scale and rotation invariant texture classification with local binary patterns. IEEE Trans. Pattern Anal. Mach. Intell. **24**(7), 971–987 (2002)

15. Tharwat, A., Gaber, T., Hassanien, A.E.: Two biometric approaches for cattle identification based on features and classifiers fusion. Int. J. Image Min. **1**(4), 342–365 (2015)

16. Cristinacce, D., Cootes, T.F.: A comparison of shape constrained facial feature detectors. In: Proceedings of Sixth IEEE International Conference on Automatic Face and Gesture Recognition, pp. 375–380. IEEE (2004)

17. Tharwat, A., Ghanem, A.M., Hassanien, A.E.: Three different classifiers for facial age estimation based on k-nearest neighbor. In: Proceedings of the 9th International Computer Engineering Conference (ICENCO), pp. 55–60. IEEE (2013)
18. Duda, R.O., Hart, P.E., Stork, D.G.: Pattern Classification, 2nd edn. Wiley, New York (2012)
19. The fg-net aging database. http://www.fgnet.rsunit.com/

A Behavioral Action Sequences Process Design

Moataz Kilany[1,5(✉)], Aboul Ella Hassanien[2,3,5], Amr Badr[2], Pei-Wei Tsai[4], and Jeng-Shyang Pan[4]

[1] Faculty of Computers and Information, Minia University,
El-Minia, Egypt
moetaz.kilany@mu.edu.eg

[2] Faculty of Computers and Information, Cairo University,
Giza, Egypt

[3] Faculty of Computers and Information,
Beni Suef University, Beni Suef, Egypt

[4] Fujian Provincial Key Laboratory of Big Data Mining and Applications,
Fujian University of Technology, Fuzhou, Taiwan

[5] Scientific Research Group in Egypt (SRGE), Cairo, Egypt
http://www.egyptscience.net

Abstract. Modeling human actions in a format that is suitable for computer systems to understand is a target for behavior analysis systems. This work introduces a high level design for a behavioral analysis system based on action sequences. The design is introduced in terms of process modeling. System processes are presented in terms of a set of data flow diagrams (also known as DFDs) of multi levels. They represent the decomposition of all processing components required in such system and the data flows among them. The system is designed to receive structured information for human behaviors and actions and produce insights, predictions and classifications for personal and behavioral characteristics. Proposed process decomposition is introduced as a core step towards design and implementation of a behavior analyzer system.

This work also introduces an approach to model the contextual factors affecting personal activities. This should lead to a more precise behavioral models that can capture activities as well as considering contextual factors such as the surrounding physical environment, person committing the activity, surrounding culture, social, and religious norms. The targeted accuracy of modeling is meant by the precise evaluation of personal activities after taking all mentioned factors into account.

Keywords: Human behavior analysis · Prediction · Process modeling · Data flow diagrams

1 Introduction

Modern computer systems rely on advanced human interfaces designed to understand behavioral patterns and produce natural and intelligent response accordingly, human-machine interaction is a vital trend in this field [1]. Human behavior

© Springer International Publishing AG 2017
A.E. Hassanien et al. (eds.), *Proceedings of the International Conference on Advanced Intelligent Systems and Informatics 2016*, Advances in Intelligent Systems and Computing 533, DOI 10.1007/978-3-319-48308-5_48

analysis system aims to make human interactions effective, employing a behavioral model that preserves data structures for actions information and functions that perform behavioral analysis. The targeted output of such analysis here is to produce generic insights about action sequence patterns after application of psychological theories and logic relations among behavioral factors and modifiers. By generic we mean the level of extensibility and abstraction, where the core data elements are designed to abstract human behaviors of any type in any context, with defined rules that can be extended at any point of system lifetime. This in addition to a prediction component to foresee future behavioral sequences of actions and of human affects such as personal mood, attitude and emotions. A number of modeling ideas appeared in research efforts upon which we built our modeling approach. [2,3] discussed ideas for the representation of human actions.

Process modeling is an important step in system analysis and design. This work introduces a decomposition of all required logical processes inside the targeted behavior analysis system in terms of a set of data flow diagrams [4]. We start from context diagram through level one of decomposition. The proposed process model is a step towards design and implementation of a human behavior analysis system.

There exist many modeling efforts in former research such as, the identification of personal intentions [5], tracking human actions during driving [6]. All former efforts focused on a given context of actions and isolated many environmental variables. Such as limiting the location, the time and other important factors. A number of researchers focused on defining context-aware systems as for Cosimo Palmisano et al. [7], where context is defined as the customer intent of purchase in an e-commerce application. In [8], the location of a person, surrounding people identities and objects were considered for contextual analysis. Others relied on timing, season, and temperature [9].

This work is organized as follows; Sect. 2 Discusses the overview of system design, input, and output, Sect. 3 Shows the higher level of composition for the system, Sects. 4 and 5 show level 1 and level 2 of decomposition respectively, Sect. 6 discusses a rule-based approach for representing the context of human actions.

2 System Context

The following paragraphs introduce an overview of system in terms of input, output and processes.

Input Set:

- **Actor Information**, A history about targeted persons for analysis purposes such as a list of action sequences, contexts, locations, and other data elements.
- **Action Information**, Action is a main data element in proposed system, composed of three factors; a descriptive verb, a target of action, and an actor along with a set of temporal modifiers. This abstract definition enables the system to deal with any action whether physical, oral or mental.

- **Personality Classification**, Represents a set of ten factors employed by the big five aspect scales of the big five personality traits theory [10], each represented by a quantifiable variable. Those factors are neuroticism, agreeableness, conscientiousness, orderliness and extraversion. The five traits are decomposed into ten aspect scales which are targeted here.
- **Personal Attitude, Personal Rules, Location Rules**, Action context is also a major element in the system data model. It is the surrounding controlling factors that affect evaluation of a given behavior, a behavior can be treated as aggressive or abnormal in some environments, while not in others [11]. The context of a given action is represented in terms of a set of rules that, for each human action, define information for action and environment information, and a directed value (positive/negative) defining how much this action is tolerated or accepted in the given environment. A context can be a person, physical location, logical location, culture or a religion [12,13]. A set of rules each incorporates a positive value defining personal attitude towards the action target.
- **Location Information**, A set of items defining location of action either contextual location (work, home, street or others) or a physical location addressing the country, state, town, continent and coordinates. Such locations are assigned contextual rules to capture location-based context as discussed in previous section.
- **Behavior Classification**, A classification of behavior in terms of personal affects such as aggression, anger, enthusiasm and other.
- **Action Motivation**, A classification for action based on motivation. This work applied Alderfer's theory of motivation [14] which decomposes motivations into three factors; existence, relatedness and growth motivations.

Output Set: Most output data elements are previously defined in Sect. 2 and will be briefed here.

- **Personal Attitude (Rule Set)**, An updated attitude value.
- **Personality Classification**, An updated personality classification.
- **Behavior Classification**, An updated behavior classification.
- **Action Motivation**, An updated value for motivation.
- **Personal Mood**, An inference evaluating personal mood in terms of a directed weighted value. This inference depends on a set of logic rules that map a range of sequences of actions to an evaluation for personal mood.
- **Predicted Action Sequences**, A predicted action sequences set, based on pattern matching algorithms applied on a history of action sequences.
- **Predicted Mood**, A predicted mood based on the predicted sequences of actions and a set of inference rules.
- **Predicted Attitude**, A prediction for personal attitude towards an object based on pattern matching applied on patterns of actions (action sequences).

3   Level Zero Data Flow Diagram

This section shows the decomposition of internal processes along with data flows employing a level zero data flow diagram presented in Fig. 1 which identifies a set of six processes and 5 data stores responsible for data storage.

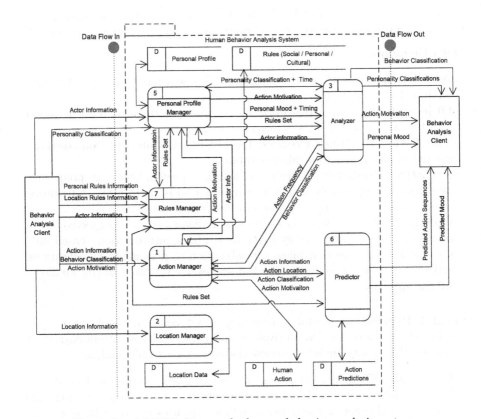

Fig. 1. Level-0 DFD diagram for human behavior analysis system

Internal Processes:

– **Action manager,** Responsible for actions information storage, retrieval and action sequences similarity check and pattern matching.
– **Location Manager,** Responsible for location information management discussed in Sect. 2.
– **Analyzer,** Incorporates logic and mathematical rules based on psychological theories and logic insights to analyze patterns of actions and make inferences about actions.
– **Profile Manager,** Manages action actors information as shown in Sect. 2.
– **Predictor,** Responsible for prediction logic, receives sequences of actions and applies a set of pattern matching algorithms to predict future sequences.

– **Rules Manager,** Responsible for storage and retrieval of rules that are the basis of actions context and personal attitude as discussed in Sect. 2.

4   Level One Data Flow Diagram

Presents a decomposition of level zero DFD. Most important decompositions discussed here are the analyzer and action predictor processes appearing in Fig. 2.

Data Elements:

– **Action sequences,** A set of predicted action sequences based on statistical analysis for action history.
– **Object of concern,** A real world object that is a target of a given action. Can be a real world item, a person or another action.
– **The Action Predictor Process,** The prediction component, composed of a statistical predictor and a rule based filter.
– **Statistical Predictor,** Makes use of action sequences history to predict the most probable sequences of actions, employs a set of pattern matching procedures applied on a large data base of action sequences [15].
– **Rule Based Filter,** Helps the statistical predictor narrowing expected results based on context. Also makes suggestions about future actions based on a set of inference rules that maps a combination of personality classifications, attitudes, and locations into expected sequences. Research in [16] applied a similar approach.

Level 1 Processes: Generates inferences about behavior classification, personality classification, action motivation and personal mood. It employs three processes; Personality classifier, action analyzer and attitude analyzer.

– **Personality Classifier.** Depends on a set of inference rules to identify the five personality aspect scales by combining the time factor along with personal mood, identified action motivations, action sequences and their frequencies.
– **Action Analyzer.** Identifies for a given action, a classification, motivation and action frequencies.
– **Attitude Analyzer.** Makes inferences on personal attitude based on classifications for the given action, personal attitude history and based on the history of action sequences and their context.

5   Level Two Data Flow Diagram

The next level of decomposition focuses on process 3.1 and 3.2, the personality classifier and action analyzer respectively. both processes are decomposed into seven functions presented in the following sections.

The Action Analyzer Process: Action analyzer employs three processes; The frequency calculator, motivation identifier and action classifier.

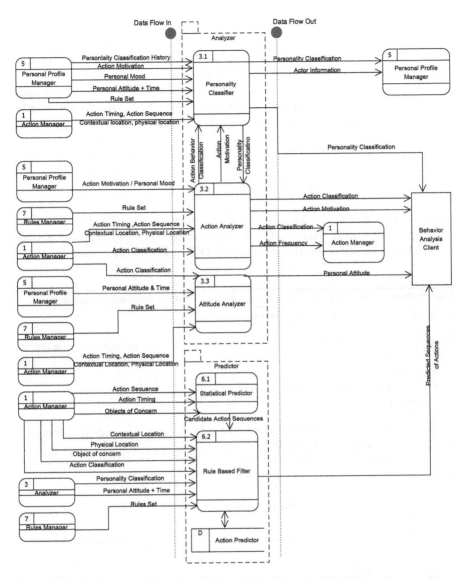

Fig. 2. Level 1 DFD diagram for human behavior analysis system

- **Frequency Calculator.** Responsible for matching new sequences of actions with action history based on a given affinity value to calculate how frequent a given single action or a sequence of actions is recurring on a given time unit. [17] and [15], introduced valuable ideas for defining behaviors and how to formally define and measure behavior frequency. Affinity measurement between action sequences is applied in all core processes such as frequency calculator, statistical predictor, and other action analysis elements. Multiple sequences of actions can lead to the same behavior. Here, we apply the Needlman-Wunsch algorithm that is being applied in many distance measurement applications [18]. Algorithm 1 employs the Needlman-Wunsch algorithm to measure distance or affinity between two given sequences of actions, where (**d**) is an initial value for first row/column of matrix. The algorithm builds a two dimensional scoring matrix (D) with dimensions (**N** ∗ **M**) where N, M are here the lengths of both action sequences applying the following rule:

$$D(i,j) = \max_{\substack{0 \leq i \leq N \\ 0 \leq j \leq M}} \begin{cases} D((i-1),(j-1)) + s(x_i, y_i) \\ D(i-1,j) + g \\ D(i,j-1) + g \end{cases} \tag{1}$$

where $(s(x_i, y_i))$ finds the equality of two single actions (x_i, y_i). Once the scoring matrix is calculated, the affinity among both sequences of actions (A_1, A_2) is identified by the value of cell $D(N, M)$ in scoring matrix.

Algorithm 1. Needleman-Wunsch algorithm - action sequences similarity

---

1: Input: Action sequence 1 (A_1) sequence 2 (A_2)
2: Output: A Scoring Matrix D
3: **for** $i \leftarrow 0$ to $Length(A_1)$ **do**   ▷ Initialize scoring matrix
4:     $D(i,0) \leftarrow (d * i)$
5: **end for**
6: **for** $j \leftarrow 0$ to $Length(A_2)$ **do**
7:     $D(0,j) \leftarrow (d * j)$
8: **end for**
9: **for** $i \leftarrow 0$ to $Length(A_1)$ **do**   ▷ Insertions - Deletions calculation Eq. (1)
10:     **for** $j \leftarrow 0$ to $Length(A_2)$ **do**
11:         Match $\leftarrow D(i-1, j-1) + S(A_1(i), A_2(i))$
12:         Delete $\leftarrow D(i-1, j) + d$
13:         Insert $\leftarrow D(i, j-1) + d$
14:         $D(i,j) \leftarrow$ max(Match,Insert,Delete)
15:     **end for**
16: **end for**

- **Motivation Identifier:** Combining patterns of action sequences with their contexts (in terms of rules as discussed before) to infer motivational factors about such sequences. [19] discusses how to computationally represent human motivations and actions. This research employs Alderfer's theory of motivation to similarly represent motivations in a computational manner.
- **Action Classifier:** Makes classification of actions on the basis of their frequencies and a comparison of action rule set with the context rule set to identify specific classifications such as aggressive, angry, or normal actions.

The Personality Classifier Process: Produces inferences about personality traits based on a number of different approaches, each one is applied by an internal process. The results of classifications are finally merged to produce an enhanced classification list. This employs four processes; (1) The rule set analyzer, (2) action history classification analyzer, (3) psychological affect analyzer, and (4) a personality classifier which merges all results and produces the final inferences.

Rule Set Analyzer: Applies a set of inference rules on the contextual rules of a given actor. Such contextual rules are to give indications about his personality traits.

Action History Classification Analyzer: Makes inferences of personality traits based on the history of action sequences and their classifications.

Psychological Affect Analyzer: Makes analysis based on the history of recorded motivations, personal mood, and attitude for a given actor.

Personality Classifier: Makes a merging process to the supplied lists of personality classifications and produce a refined list of the big five aspect scales.

6   Rule-Based Context Representation

As discussed before, a set of rules can define action tolerance with respect to some context to reflect a very high level of granularity in representing contextual effects. There are many examples of contexts; a person, location, time, contextual location, culture, religion, social group, human crowds and others.

6.1   Core Entities

In order to represent human actions in terms of a set of rules, we need to begin by defining the required data entities to be used by such rules later.

- **Action Sequence.** Represents a sequence of human activities and expressed by the following tuple.

$$a = \{(\{\hat{a_1}, \hat{a_2}, ..., \hat{a_n}\}, pr, l, c, pi, pm, o, v)\} \tag{2}$$

where, ($\hat{a_1}$) through $\hat{a_n}$ is a list of action characteristics. (**pr**), internal action sequence. Each action consists of an inner sequence of more trivial actions. (**l**), a list of action characteristics. (**c**), the context of action. (**pi**), personal profile information. (**v**), action verb. (**o**), the targeted object of the given action.
- **Action Context.** Any entity that can change action evaluation such as the time of action, actor, physical location, contextual location and the set of social, cultural and religious factors.
- **Action Evaluation.** Visualized in Figs. 3 and 4. Action is evaluated in terms of how much it is accepted in a given context and with a given criteria and represented by a directed value (positive / negative) identifying the acceptance of action along with the strength of such acceptance.

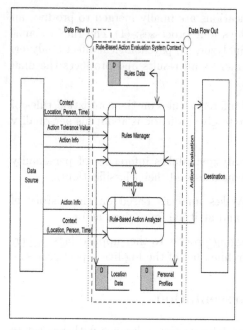

Fig. 3. Rule-based action evaluation system context

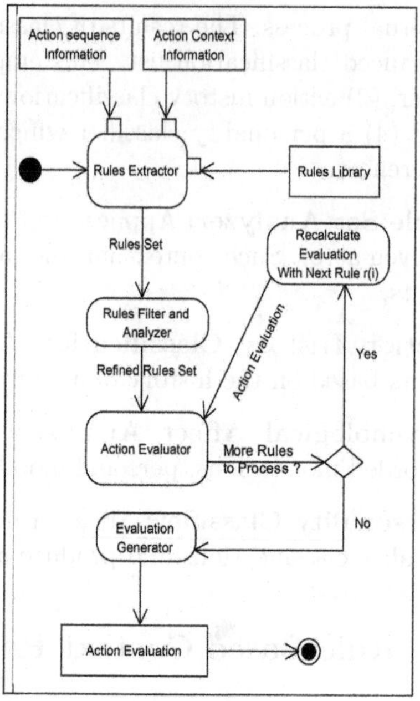

Fig. 4. Rule-based action evaluator component context

- **Activity Environmental Rules.** A personal action is controlled and evaluated on the basis of surrounding environment. Many research efforts discussed the effects of environment on human behavior as in [20] Here, the environment can be the person committing the action, location, time, or the set of social and cultural rules controlling action. A contextual rule is represented by a tuple of four elements; The action being evaluated (denoted by (**a**)), The context of action (denoted by (**c**)), and action tolerance value (denoted by (**t**)) that identifies the tolerance of action in the given environment using a directed value. Rule definition appears in 3.

$$r = (a, c, t) \tag{3}$$

7   Conclusion and Future Work

This work presented a design approach using a process model of a human behavior analysis system. The model presented a decomposition of effective internal processes along with assigned responsibilities. This system will be the basis of design and implementation of a human behavior analysis system to be integrated in later stages with a data model that illustrates the data model. The targeted model is expected to be a core component in real life applications that require prediction of human behavioral patterns such as recommendation systems that is capable of making product recommendations based on the analysis and prediction of customer actions, mood, attitude and other factors. In addition to smart systems that can predict human actions (driving actions for instance).

References

1. Cannan, J., Hu, H.: Human-machine interaction (HMI): a survey. University of Essex (2011)
2. Murakami, Y., Sugimoto, Y., Ishida, T.: Modeling human behavior for virtual training systems. In: Proceedings of the National Conference on Artificial Intelligence, vol. 20(1), p. 127. AAAI Press; MIT Press, Menlo Park, Cambridge, London (1999, 2005)
3. Luo, L., Zhou, S., Cai, W., Low, M.Y.H., Tian, F., Wang, Y., Xiao, X., Chen, D.: Agent-based human behavior modeling for crowd simulation. Comput. Anim. Virtual Worlds 19(3–4), 271–281 (2008)
4. Dennis, A., Wixom, B.H., Roth, R.M.: Systems Analysis and Design, 5th edn. Wiley, New York (2015)
5. Kim, S., Kavuri, S., Lee, M.: Intention recognition and object recommendation system using deep auto-encoder based affordance model. In: The 1st International Conference on Human-Agent Interaction (2013)
6. Pentland, A., Liu, A.: Modeling and prediction of human behavior. Neural Comput. 11(1), 229–242 (1999)
7. Palmisano, C., Tuzhilin, A., Gorgoglione, M.: Using context to improve predictive modeling of customers in personalization applications. IEEE Trans. Knowl. Data Eng. 20(11), 1535–1549 (2008)
8. Schilit, B.N., Theimer, M.M.: Disseminating active map information to mobile hosts. IEEE Netw. 8(5), 22–32 (1994)
9. Brown, P.J., Bovey, J.D., Chen, X.: Context-aware applications: from the laboratory to the marketplace. IEEE Pers. Commun. 4(5), 58–64 (1997)
10. DeYoung, C.G., Quilty, L.C., Peterson, J.B.: Between facets and domains: 10 aspects of the big five. J. Person. Soc. Psychol. 93(5), 880 (2007)
11. Gifford, R.: Environmental Psychology: Principles and Practice. Optimal Books, Colville (2007)
12. Brislin, R.: Understanding Culture's Influence on Behavior. Harcourt Brace Jovanovich, Fort Worth (1993)
13. Schultz, P.W., Nolan, J.M., Cialdini, R.B., Goldstein, N.J., Griskevicius, V.: The constructive, destructive, and reconstructive power of social norms. Psychol. Sci. 18(5), 429–434 (2007)
14. Alderfer, C.P.: An empirical test of a new theory of human needs. Organ. Behav. Human Perform. 4(2), 142–175 (1969)
15. Campuzano, F., Garcia-Valverde, T., Botia, J.A., Serrano, E.: Generation of human computational models with machine learning. Inf. Sci. 293, 97–114 (2015)
16. Hashimoto, K., Doki, K., Doki, S., Okuma, S.: Study on modeling and recognition of human behaviors by if-then-rules with hmm. In: 35th Annual Conference of IEEE Industrial Electronics: IECON 2009, pp. 3410–3415. IEEE (2009)
17. Skinner, B.F.: Science and Human Behavior. Simon and Schuster, New York (1951)
18. Needleman, S.B., Wunsch, C.D.: A general method applicable to the search for similarities in the amino acid sequence of two proteins. J. Mol. Biol. 48(3), 443–453 (1970)

19. Tooby, J., Cosmides, L., Sell, A., Lieberman, D., Sznycer, D.: 15 internal regulatory variables, the design of human motivation: a computational and evolutionary approach. In: Handbook of Approach and Avoidance Motivation, vol. 251. Lawrence Erlbaum, Mahwah (2008)
20. Remington, R., Boehm-Davis, D., Folk, C.: Determinants of human behavior. In: Introduction to Humans in Engineered Systems, pp. 105–112 (1961)

Hybrid Rough Set and Heterogeneous Ensemble Classifiers Model for Cancer Classification

Mohamed E. Helal[1(✉)], Mohammed Elmogy[2(✉)],
and Rasheed M. El-Awady[3]

[1] Faculty of Computers and Information, Information Systems Department,
Mansoura University, Mansoura, Egypt
`mehelal84@gmail.com`
[2] Faculty of Computers and Information, Information Technology Department,
Mansoura University, Mansoura, Egypt
`melmogy@mans.edu.eg`
[3] Faculty of Engineering, Electronics and Communications,
Mansoura University, Mansoura, Egypt
`actt_egypt@yahoo.com`

Abstract. The enhancing accuracy of machine learning algorithms is essential in making a high-performance model for the diagnosis of cancer diseases. Many studies have indicated that an individual classifier performance could be improved by ensemble classifiers. In this paper, we build ensemble classifiers of Support Vector Machine, C5.0, and Naive Bayes machine learning algorithms to assess their classification performances using Hepatitis C Virus (HCV) and Ovarian Cancer datasets. Experiments are done in three stages: First, the original features of the two datasets are reduced using the Rough set theory. Second, ensemble classifiers performances are calculated for two datasets. Third, performances of respective classifiers are measured. All the experiments are carried out and evaluated using three parameters: the percentage of accuracy, sensitivity, and specificity to measure performances of respective classifiers. Experimental results show that classification techniques performance have depended on the used dataset.

Keywords: Hepatitis C Virus (HCV) · Ovarian cancer · Feature selection · Classification · Ensemble classifier · Rough Set Theory (RST) · Support Vector Machine (SVM) · Naïve Bayes (NB) · C5.0

1 Introduction

Medical data contains many features where missing values and redundant data need to be discarded. In the medical area, one of the essential requirements for feature selection and classification is the ability to deal with inconsistent and imprecise information [1, 2]. Medical data analysis is a complex task because it requires information from the medical data set and also requires advanced techniques for processing, storing, and accessing

© Springer International Publishing AG 2017
A.E. Hassanien et al. (eds.), *Proceedings of the International Conference
on Advanced Intelligent Systems and Informatics 2016*, Advances in Intelligent
Systems and Computing 533, DOI 10.1007/978-3-319-48308-5_49

information from the data. Traditional techniques are not capable enough of creating optimal results from incomplete or redundant data through the analysis process [3].

To make the untreated data more suitable for further analysis, pre-processing steps should be applied. Some different techniques exist, with these techniques (1) feature selection, (2) dimensionality reduction, and (3) feature extraction. Machine learning algorithms work better when the dimensionality is reduced because it can reduce noise, eliminate irrelevant features, and can create more robust learning models. The main objective of machine learning techniques is to produce a model, which can be used to do classification, prediction, or any other similar task.

In the classification task, the data set is divided into training and testing sets. Classification process includes two stages that are known as training and testing stages. A good classification model should be appropriate the training set well and correctly classifies all the examples. If the test error rates of a classification model begin to increase although the training error rates decrease, then this is known as data overfitting. Therefore, feature selection can be used as a preprocessing step to solving this problem [4].

Classification of tumors is one of the significant steps to detect cancer. Tumors can be categorized into malignant and benign ones. The malignant tumors are named cancer and they more serious than benign ones. Early diagnosis needs an accurate and reliable diagnosis algorithm that helps experts to distinguish between different tumors without the need for surgical biopsy. On the other hand, feature selection technique helps to create a subset of the original features that are most significant in forecasting a particular outcome. The reduced feature set increases the performance of the classification task in comparison of applying the classification task on the original dataset [5]. The best feature selection algorithm tests all different combinations of features against measures, such as forecast accuracy and execution time [6]. It minimizes the dimension of data and time execution. Therefore, it can make good classification performance. The most important objective of feature selection is to improve model performance, prevent overfitting, and to present more efficient models.

Ensemble classifier is an effective technique, which is used to build a predictive classifier by integrating multiple classifiers to get an enhanced classification performance. This method is also called committee of learning machines. The idea of combining classifiers is to enhance the accuracy rate.

The rest of this paper is organized as follows. Section 2 contains related work that shows some of the previous and current work in medical diagnosis based on machine learning techniques. Section 3 talks about the proposed model. Section 4 shows results of applying different techniques and classification accuracy. Finally, the conclusion and future work are discussed in Sect. 5.

2 Related Work

There are different algorithms proposed by various researchers for the classification of various cancer diseases. Some of the existing techniques are presented in this section. For example, Badria et al. [9] designed a framework that uses Fuzzy-Rough Sets with SVM for ovarian cancer diagnosis. They used Fuzzy with RST as a pre-processing stage for feature selection. In addition, they used SVM as post-processing stage for classifying data. They applied this framework to two case studies. The first case is to find the connection between the Amino Acids and ovarian cancer. They discovered that the phenylalanine amino acid is the most important factor for discovering ovarian cancer in its early stage. The second case is to determine the connection between the amino acids and the tumor marker (CA-125). They found that the proposed model achieves moderate accuracy rate. In the future, they will use this hybrid model with larger datasets with additional attributes to investigate the main factors that affect the CA-125.

Helmy et al. [10] proposed ensembles and hybrid computational intelligence models for classification of bioinformatics data sets. They used individual, ensembles, and a hybrid of computational intelligence techniques to classify real bioinformatics data sets. Results showed that the classification accuracy of the hybrid and ensemble models is much better than the performance of the single models.

Pujari and Gupta [11] proposed a model that used feature selection technique with an ensemble model to enhance classification accuracy. They used Pearson Chi-square measure as a pre-processing step to select the most important features. Moreover, they combined the prediction of CART, CHAID, and QUEST classifiers as ensemble model. Results showed that the classification accuracy of ensemble models after applying feature selection is much better than the performance of the single models.

Srimani and Koti [12] presented a study to show that the performance of single classifier might be improved by ensemble classification approaches. They used correlation-based feature subset selection to evaluate the power of features to return the most related variable. They have used various base classifiers along with ensemble methods to evaluate the performance. The experiment results showed that ensemble classifiers cannot generalize doing enhancing the classification accuracy in all the cases.

Roy and Mohapatra [8] conducted a study that presents the comparison of different classification techniques. They investigated the performance of NB, SVM, and an ensemble SVM-KNN models for a set of MicroArray data. The experimental results showed that different classification techniques behave differently on several data sets according to nature and the size of their attributes. They showed that SVM-KNN ensemble model could improve the classification process in the medical field.

Dehzangi et al. [13] proposed an ensemble classifier, which is based on a combination of NB, SVM, Multi-Layer Perceptron, LogitBoost, and AdaBoost classifiers. They used five combinational policies to avoid complications. The highest result was achieved by using the majority voting policy for the employed classifiers. The proposed ensemble classifier achieved high accuracy rate greater by 3.4 % than the highest achieved result in the previous works using the same set of features.

Hota [7, 14] proposed predictive models for breast cancer. He used many classification techniques. The individually developed models are tested and combined to form ensemble model. He utilized (SVM & C5.0) and (Bayesian & C5.0) ensemble classifiers. Feature subsets are achieved by applying feature selection algorithm and models that are verified on these data sets. Results showed that ensemble model with selected subset features could be the best alternative to health care predictive model for the diagnosis of breast cancer.

Elsayed [15] built an ensemble model that combines three different data mining methods. C5.0 decision tree, Multilayer perceptron neural network, and linear discriminate analysis are used to construct an ensemble model for the problem of differential diagnosis of these Erythemato-Squamous diseases. A confidence-weighted voting scheme is used as a combination scheme for the proposed ensemble model. The classification performance of the proposed model was displayed using statistical accuracy, specificity, and sensitivity. The proposed Ensemble model achieved an accuracy of 98.23 %, which is very close to the work of Ubeyli [16].

In this section, some of the related works are listed above. Several authors proposed different algorithms for the prediction of cancer regions. Some of the existing techniques are using the feature selection as a pre-processing step to enhance classifier performance. In this paper, we propose a model that uses the RST as a feature selection technique to improve the performance of separate classifiers and ensemble classifiers.

3 The Proposed Model

This study focuses on integrating feature selection technique with individual and ensemble classifiers (NB, C5.0, and SVM). In this proposed model, the RST is employed as a feature selection technique and C5.0, SVM, and NB are employed as an ensemble model. The proposed model includes three stages as displayed in Fig. 1.

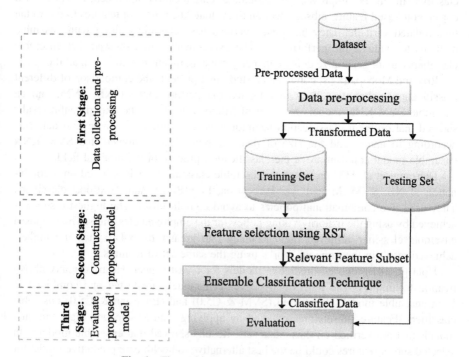

Fig. 1. The architecture of the proposed model.

First Stage: Data Collection and Preprocessing. In this paper, two different cancer data sets are involved. They can be considered as the 'input' of our proposed model. The first is the HCV data set, which is gathered from medical studies of a recently developed medication for HCV [17, 18]. The other data set is called 'Ovarian Cancer', which is collected and confirmed by the ethical institutional review board at Mansoura University. Data pre-processing is one of the most important steps which transforms data into a pattern that will be simpler and more efficient process. The dataset should be prepared to eliminate the redundancy, and check for the missing values. The data preparation process is often the mainly time-consuming and computational stage. In this step, data sets have been split into the training set, which is used to build the model and testing set that is used to evaluate the proposed model.

Second Stage: Building Proposed Model. In this model, RST is used as a feature selection technique to generate a subset of features from the original features that make machine learning easier and less time-consuming. After generating a subset, NB, C5.0, and SVM are separately used as classification techniques. Then, an ensemble classifier (NB + C5.0 + SVM) is proposed to classify the data. This study classifies data by using the combination of RST with different classification techniques and the combination of ensemble model with RST. Results are compared with other used techniques.

Third Stage: Evaluating the Proposed Model. In this model, to assess the performance of each classification technique and the proposed model, three statistical measures are used (Accuracy, Sensitivity, and Specificity).

The pseudo code of the proposed model:

Step 1: Collect data and transform it into a decision table (A). $A = (U, A \cup \{d\})$ Where U = {x1, x2, x3,....,xn} that is called Universe and A is a nonempty finite set of attributes, $\forall a \in A, a : U \rightarrow V_a$. The set V_a is called the value set of a.

Step 2: Split data into training and testing datasets.

Step 3: Find an optimal subset of features using a rough set on the training dataset.

- Generate the discretize table using training data set. Transform the original decision table $A = (U, A \cup \{d\})$ into new discrete decision table $A|_D = (U, A_D \cup \{d\})$ where AD = {a Da: an∈A}and Da is set of cuts. The table $A|_D$ is called the D-discretized table of A.
- Calculate reducts by using the exhaustive algorithm. The decision reduct is a set B $\subset$ A of attributes such that it cannot be reduced and IND(B) = IND(B). Where IND (B) is defined as: $_xIND(B)_y \Leftrightarrow \forall_{a \in B} a(x) = a(y)$, x,y∈U. If an (x,y)∈U × U belong to IND(B) then x and y are indiscernible by attributes from B.
- Minimize reducts by using shorten method using a 90 % shortening ratio without reducing accuracy and coverage measurement.

- Generate rules and remove rules with small support. $A|_D = (U, A_D \cup \{d\})$ discrete decision table every x∈U determines a sequence a1(x), a2(x), ... an(x); d1(x), d2 (x), ... dn(x) where {a1(x), a2(x), ... an(x)} = A and {d1(x), d2(x), ... dn (x)} = d. Decision rule induced by $A \rightarrow xd$
- Test the efficiency of using the rough set as feature selection.

Step 4:   Classify data by using the combination of (NB + SVM + C5.0) classifica-
           tion techniques with RS. $A_{D_{min}} = (U, A_D \cup \{d\})$ is minimized discrete
           decision table that contains optimal feature selected by Rough set. The
           optimal feature with Ensemble technique (NB + SVM + C5.0) is used to
           classify data.
Step 5:   Evaluate the performance of classification techniques by using accuracy,
           sensitivity, and specificity.

4   Experiment Results and Analysis

This section evaluates some hybrid techniques for classification, such as RS-NB, RS-SVM, RS-C5.0, and the proposed ensemble model. In this study, several experiments have been managed by two available data sets. A brief description of each data set was provided below.

4.1   Ovarian Cancer Data Set

It consists of 135 female patients with 18 attributes. They contain 113 positive cases and 22 negative cases. The decision attribute of this data set takes the values "positive" or "negative" that means positive or negative test for ovarian cancer, respectively. These attributes are listed in Table 1.

Table 1. The tested attributes in ovarian cancer data set.

Attribute Name	Attribute Description
Alanine, Arginine, Aspartic, Cystine, Lysine, Glutamine, Glycine, Histidine, Isoleucine, Leucine, Methionine, Phenylalanine, Proline, Serine, Threonine, Tyrosine, Valine	All Attribute are Numeric data type
Decision Class	−1 absent, 1 present of cancer

4.2   Hepatitis C Virus Data Set

It is composed of 119 cases. Each case is labeled by 28 attributes (5 categorical attributes and 23 numerical). The objective of the data set is to predict the attendance or non-attendance of HCV to the proposed medication. These attributes are listed in Table 2.

Table 2. The tested attributes in HCV data set.

No.	Attribute name	Attribute description	No	Attribute name	Attribute description
1	Sex	{Male, female}	15	Lesions	{0,1,2}
2	Source	blood transfusion, non-sterile tools by dentist or surgery	16	Serum Bilirubin	Regular range [0:1.1] mg/dL
3	SGPT (ALT)	Regular range [0:40] U/L	17	Blood Pressure	{Yes,No}
4	SGOT (AST)	Regular range [0:45] U/L	18	Serum Albumen	Regular range [3.5:5.1] g/dL
5	Ascites	{No, Mild, Ascites}	19	Serum Ferritin	Regular range [22:300]
6	Spleen	{Normal, Absent, Enlarged}	20	Fatigue	{Yes,No}
7	Portal vein	Natural diameter is 12 mm	21	Gasp	{Yes,No}
8	Constipation	{Yes,No}	22	Headache	{Yes,No}
9	Skin Colour	{Yes,No}	23	Nausea	{Yes,No}
10	PLT	Platelets regular range [150:450] cm/m	24	PCR	Quantitative analysis of the virus U/mL
11	WBC	Regular range [4:11]cm/m	25	Vertigo	{Yes,No}
12	Diarrhea	{Yes,No}	26	Vomiting	{Yes,No}
13	Appetite	{Yes,No}	27	Eye Colour	{Yes,No}
14	HGB	Range for a male [12.5:17.5] g/dL,Range for a female [11.5:16.5] g/dL	28	Decision Class	−1 absent, 1 present of HCV

In this study, machine-learning tools (RSES v2.2 [19] and Clementine 12.0.0.319 [20]) are used to achieve the proposed objectives. In Clementine, a confidence-weighted voting method is used as a combining method to create ensemble classifier. To analyze the performance of classification techniques the percentages of accuracy, sensitivity, and specificity are calculated. The accuracy parameter proposes a high value of accuracy rate for a classification model applied to a data set shows that the obtained classifier highly properly classifies the data set. On the other hand, the low value of accuracy rate for a classification model applied to a data set shows that the data set is less properly classified by the obtained classifier. The sensitivity parameter is used to measure the rate of positive class that is properly classified. Specificity parameter is used to measure the rate of the negative class that is properly classified.

During the experiments, the datasets were split into training and testing parts. The training set is used to construct the classifier and then use testing data to validate it. Then, the involved classification models are applied to generate the classifiers via

Table 3. The classification accuracy for different classification models.

Data Set	Classification performance parameters	RS + NB	RS + SVM	RS + C5.0	Proposed Model RS + (NB + SVM + C5.0)
3 Month dataset	Accuracy	96.77 %	93.55 %	96.77 %	96.77 %
	Sensitivity	96.78 %	96.67 %	96.78 %	96.78 %
	Specificity	0 %	0 %	0 %	0 %
6 Month dataset	Accuracy	83.87 %	77.42 %	61.29 %	80.65 %
	Sensitivity	83.87 %	85.18 %	81.82 %	83.33 %
	Specificity	0 %	25 %	11.1 %	0 %
9 Month dataset	Accuracy	64 %	64 %	56 %	64 %
	Sensitivity	0 %	0 %	40 %	0 %
	Specificity	64 %	64 %	65 %	64 %
Ovarian Cancer dataset	Accuracy	100 %	100 %	100 %	100 %
	Sensitivity	100 %	100 %	100 %	100 %
	Specificity	100 %	100 %	100 %	100 %

Table 4. The selected sub feature on the different datasets.

Dataset	No. features	Selected features	Accuracy	Coverage
3 Months Dataset	7	Sex -WBC - SGPT (ALT) - HGB - Fatigue - Serum Bilirubin – Gasp	93.3 %	100 %
6 Months Dataset	7	SGOT (AST) - HGB - Skin color - Serum Bilirubin - Serum Ferritin - SGPT (ALT) – Appetite	80 %	100 %
9 Months	13	Sex - Source - SGOT (AST)- SGPT (ALT) - HGB - Serum Bilirubin - Serum Albumen- Spleen - Serum Ferritin– PLT – Gasp – Fatigue - Skin Color	83.3 %	100 %
Ovarian Cancer	10	Serine – Methionine – Isoleucine – Leucine – Tyrosine – Phenylalanine – Histidine – Lysine – Arginine – Proline	100 %	100 %

Clementine 12.0.0.319 tools. The results are recorded regarding the percentage of accuracy rates, sensitivity, and specificity as shown in Table 3.

During the experiment, HCV dataset has three snapshots during treatment after three, six, and nine months. Therefore, data pre-processing, feature selection, and classification have been executed three times to estimate the progress of HCV cases. Otherwise, in Ovarian Cancer Data Set the same tasks are done one time.

In this paper, RSES tool is used as a feature selection technique. After splitting data set, discretize method is applied to generating cuts. After that, those cuts are used to generate the discretize table. Moreover, this table is used to calculate the reducts by using the exhaustive algorithm. This method generated 56 reduct sets. The following step is to minimize reducts by using shorten algorithm without reducing accuracy and coverage measurement. After this, the generated reducts are minimized to five. Then, generate rules algorithm are used to generate rules. This method produced 109 rules. The next task is to remove rules with support less than three. After this, the test set is used to test the efficiency of using the rough set as feature selection. Applying the previous steps in 6 months dataset, 9 months dataset, and Ovarian Cancer dataset, the results are displayed in Table 4.

The RS is applied on three snapshots of the first dataset and the ovarian cancer dataset. hybrids RS-NB, RS-SVM, RS-C5.0, and RS-(NB + C5.0 + SVM) are tested for the selected subset features. Table 3 shows that RS-NB, RS-C5.0, and proposed model have achieved high accuracy and sensitivity rates with the three months snapshot of HCV dataset, unlike the RS-SVM model. In the six months snapshot, RS-NB has higher accuracy rates than other techniques. Moreover, RS-SVM has higher sensitivity and specificity rates than others techniques with selected subset features. In the nine months snapshot, RS-NB, RS-SVM, and proposed model have achieved high accuracy rate, unlike RS-C5.0. On the other hand, RS-C5.0 has attained high sensitivity and specificity rates. In Ovarian Cancer Data Set, all models used in this paper have achieved 100 % accuracy, sensitivity, and specificity parameters.

5   Conclusion

Classification technique is used for identifying classes of unknown data. The classification of the tumor is one of the main steps to diagnose cancer. Feature selection plays a vital role in building intelligent classification systems. It can enhance classification performance. Ensemble classifier often has more accuracy performance than any of the individual classifiers in the ensemble. In this paper, we have offered a model for classifying HCV and Ovarian Cancer data sets based on RST as feature selection technique and combined it with ensemble classification technique. The proposed model tries to classify patients into two classes. The classification performances of the tested models are evaluated and compared to each other using three statistical measures; accuracy, specificity, and sensitivity. Experimental results show that ensemble model has not achieved high performance than other techniques on HCV dataset. In addition, different classification models behave differently on different datasets relying on the nature of their attributes and size. In future work, we intend to test our proposed system on different data sets. Also, we will test other classification techniques on tested data sets.

Acknowledgment. The authors would like to thank Prof. Dr. Farid Badria, Prof. of Pharmacognosy, Department and head of Liver Research Lab, Mansoura University, Egypt; and Prof. Dr. Hosam Zaghloul, Prof. at Clinical Pathology Department, Faculty of Medicine, Mansoura University, Egypt, for their efforts in this work.

References

1. http://www.cancer.org/cancer/cancerbasics/what-is-cancer. Accessed on 16 May 2015
2. Suhana, S.N., Mariyam, S.S.: Feature granularity for cardiac datasets using Rough Set. In: CSAE, 2011 IEEE International Conference on, Shanghai, pp. 346–352 (2011)

3. Durairaj, M., Sathyavathi, T.: Applying rough set theory for medical informatics data analysis. ISROSET - Int. J. Sci. Res. Comput. Sci. Eng. **1**(5), 1–8 (2013). ISSN: 2320-7639

4. Kourou, K., Exarchos, T.P., Exarchos, K.P., Karamouzis, M.V., Fotiadis, D.I.: Machine learning applications in cancer prognosis and prediction. Comput. Struct. Biotechnol. J. **13**, 8–17 (2015)

5. Lavanya, D., Usha, R.K.: Ensemble decision-making system for breast cancer data. Int. J. Comput. App. **51**(17), 19–23 (2012)

6. Vesterlund, J.: Feature selection and classification of cDNA microarray samples in ROSETTA, Uppsala University, Ph.D. thesis, April 2008

7. Hota, H.S.: Diagnosis of breast cancer using intelligent techniques. Int. J. Emerg. Sci. Eng. (IJESE) **1**(3), 45–58 (2013)

8. Roy, S., Mohapatra, A.: Performance analysis of machine learning techniques in microarray data classification. Int. J. Softw. Web. Sci. **4**(1), 20–25 (2013)

9. Badria, F.A., Shoaip, N., Elmogy, M., Riad, A.M., Zaghloul, H.: A framework for ovarian cancer diagnosis based on amino acids using fuzzy-rough sets with SVM. In: Hassanien, A.E., Tolba, M.F., Taher Azar, A. (eds.) AMLTA 2014. CCIS, vol. 488, pp. 389–400. Springer, Heidelberg (2014)

10. Helmy, T., Al-Harthi M. M., Faheem M. T.: Adaptive ensemble and hyper models for classification of bioinformatics datasets, AH-I10/GJTO, vol. 3 (2012)

11. Pujari, P., Gupta, J.B.: Improving classification accuracy by using feature selection and ensemble model. IJSCE **2**(2), 380–386 (2012)

12. Srimani, P.K., Koti, M.S.: Medical diagnosis using ensemble classifiers - a novel machine-learning approach. J. Adv. Comput. **1**, 9–27 (2013)

13. Dehzangi, A., Phon-Amnuaisuk, S., Dehzangi, O.: Enhancing protein fold prediction accuracy using an ensemble of different classifiers. Aust. J. Intell. Inf. Process. Syst. **26**(4), 32–40 (2010)

14. Hota, H.S.: Identification of breast cancer using ensemble of support vector machine and decision tree with reduced feature subset. Int. J. Innov. Tech. Explor. Eng. (IJITEE) **3**(9), 99–102 (2014)

15. Elsayad, A.M.: Diagnosis of erythemato-squamous diseases using ensemble of data mining methods. ICGST-BIME J. **10**(1), 13–23 (2010)

16. Ubeyli, E.: Multiclass support vector machines for diagnosis of erythemato-squamous diseases. Expert Syst. Appl. **35**(4), 1733–1740 (2008)

17. Badria, F.A., Eissa, M.M., Elmogy, M., Hashem, M.: Rough based granular computing approach for making treatment decisions of hepatitis C. In: 23rd International Conference on Computer Theory and Applications Iccta, Alexandria, Egypt, 29–31 October 2013 (2013)

18. Badria, F.A., Eissa, M.M., Elmogy, M., Hashem, M.: Hybrid rough genetic algorithm model for making treatment decisions of hepatitis C. In: 2nd International Conference For Engineering and Technology Icet, Guc, Cairo, Egypt, April 2014

19. http://logic.mimuw.edu.pl/~rses/. Accessed 25 June 2015

20. http://www.the-datamine.com/bin/view/Software/ClementineSoftwar. Accessed 02 April 2015

WEMA to Speed up NIDS Packet Header Detection Engine

Adnan A. Hnaif[✉]

Computer Networks Department,
Al-Zaytoonah University of Jordan, Amman 11733, Jordan
adnan_hnaif@zuj.edu.jo

Abstract. The traditional firewall provides the first level of defense for computer networks and prevents unauthorized people to access the internal networks from the external attacks. Thus, the Network Intrusion detection System (NIDS) is complementary to the firewall. One of the major functions of NIDS is to act as misuse detection. In This paper we used A Neural network with multi-connect architecture and Weighted Exact Matching Algorithm (WEMA) to enhance the speed of matching process between the incoming packets header and SNORT-NIDS rule sets.

Keywords: NIDS · Exact string matching algorithms · WEMA

1 Introduction

The speed of the internet has increased over the time, and the traditional firewall which swaps information between internet and intranet is insufficient to ensure security in computer networks. The firewall only provides the first level of defense for computer networks by preventing unauthorized people from accessing the information [1]. Therefore, we need a powerful and secure mechanism such as the Network Intrusion Detection system (NIDS) to enhance security.

There are several types of packet transport from internet to intranet and vice versa, and after the capturing process using any tool like snort [2], we need to filter each packet for the analyzing process to exclude the uninteresting packets to increase the system performance. As an example, a snort of 10/10/2000 has around 854 rules, 68 of these rules is only header without contents [3].

Experiments demonstrate that on high speed networks, software alone is insufficient to process all traffic on fast Internet connection [4]. Many attempts have been made to enhance hardware as an alternative to software. In spite of the fact that software has a slow speed it only performs lightweight processing on low size network links, compared to hardware which is faster and performs intensive processing on network traffic and supports much higher network output [5]. As we mentioned before the internet speed has increased over the time and because of the existence of intrusions people, who try to sneak away the system exploit the defects of software in a high speed link. Therefore, there is a need to build up another method to keep these intruders from getting to unapproved information.

© Springer International Publishing AG 2017
A.E. Hassanien et al. (eds.), *Proceedings of the International Conference on Advanced Intelligent Systems and Informatics 2016*, Advances in Intelligent Systems and Computing 533, DOI 10.1007/978-3-319-48308-5_50

2 Existing Works of String Matching Algorithms

Snort depends on pattern matching to decide the intruders, and the number of rules is developing every once in a while, thus, snort divides its rule sets into two dimensional connected records first: Rule Tree Nodes (RTNs) which hold the fundamental data of every principle, for example, source/destination address, source/destination port and protocols type (TCP, ICMP, UDP), second: Option Tree Nodes (OTNs) which hold the data for different choices that can be added to every standard, for example, TCP flags, ICMP codes and types, packet payload size and a major bottleneck for efficiency and packet contents.

These two structures are composed into chains where RTNs are hung from left to all right headers and the OTN's hang down from the RTNs [3].

2.1 Boyer-Moore Algorithm

The Boyer-Moore algorithm is viewed as a standout among the most famous exact pattern matching algorithms of many strings against a single keyword, and it is very fast in practice.

The first heuristic used is "bad character heuristic" which means to start working by comparing from right to left. If a character does not exist in the text to search for, then, it can be shifted forward "M":character where "M"is the length of the pattern.

The second heuristic used is "repeated substrings", which means to start working by comparing from right to left and if it matches, then it is continued, but if it does not match, then it looks for the next occurrence of a substring which has been matched before [3].

2.2 Quick Search Algorithm

The Quick Search algorithm is more improved adaptation of Boyer-Moore algorithm, but the Quick Search algorithm used only the "bad-character shift". The Quick Search algorithm is easy to implement and is quick in the practice for short and huge files [6].

3 The Proposed Matching Process

Our proposed technique is divided into three stages: classification of the header rule sets according to its protocols, learning stage and matching stage.

3.1 Classification of Header Rule Sets According to Its Protocols

Figure 1 and 2 shows the main classification of header rule sets according to its protocols, and the representation of some packet header respectively.

Protocol				S.A	S.P	D.A	D.P
0	0	0	0	...	...	...	...
0	0	0	1	...	...	...	...
0	0	1	0	...	...	...	...
0	0	1	1	...	...	...	...
..				..	..	..	..
0	1	0	0	...	...	...	...

Fig. 1. Classification of header rule sets according to its protocols

Protocol	Rep
any	00000000
TCP	00000001
ICMP	00000010
UDP	00000011
IP	00000100

S/D port	Rep.
any	00000000

Range of S.P/D.P	Rep. of the first 2 bits from left
No Range	00
Begin the range	01
End the range	10

Fig. 2. Representations for some fields of packet header

3.2 Learning Stage

This system will learn the header rule sets by using a neural network with a multi-connect architecture, which is described as an associative memory and a single-layer neural network.

A neural network with multi-connect architecture will learn a set of patterns pairs (associations), and can store a large set of patterns as memory. The memory will be excited during the recall process with a key pattern which contains a portion of the information about a particular member of a stored pattern set [7].

The process of learning creates eight learning weights for all possibilities of patterns matrices. Each matrix will be only of the size 3*3 (see Fig. 3).

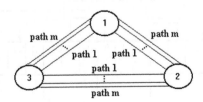

Fig. 3. NN with multi-connect architecture [4]

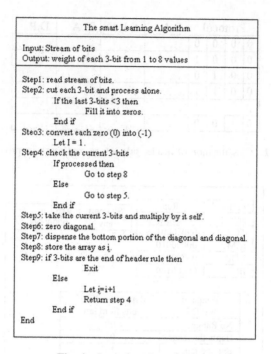

Fig. 4. Smart learning algorithm

As shown by Fig. 3, each path represents one learning weight matrix $(1 < m < 8)$, thus, we will change all vectors to the number of the net path [7], and the process of learning will be a one-time only (See Fig. 4).

In some cases, there is a collision coming about the two dimensional array which will be the same for two different values for example: [0 1 0] and [1 0 1]. These two different values have the same result when they are multiplied by itself. To solve this collision we take the sum of the each vector (index) and store only one of the two dimensional arrays with its index, see Fig. 5.

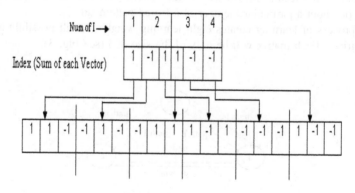

Fig. 5. Storing process

Ultimately, each header of rule sets (which is containing 106 bits of length), will be reduce to 35 bits of length by using a previous method. We divided each header into 3-bits 3-bits then converted each 3-bits into symmetric matrix weight (3-bits have 8 possibilities), thus we need only $3 \times 8 = 24$ bits of memory, therefore the memory exhausted is very low and the searching process is increasing.

3.3 Matching Stage

As a rule, existing packets header detection engine uses any exact string matching algorithm to recognize the intruders. These string matching algorithms calculations manage decimal, hexadecimal, and character values. The proposed algorithm deals with the weight values, and the header rule sets are converted in "N" weights (matrix "M").

The proposed algorithm applies a different technique to match the header rule set with the incoming packets header. this technique works as follows:

The matching process can be utilized to locate the accurate exact matching between the incoming packet header "P" and the header rule set "R.S" (matrix "M") as takes after:

1. Create array list "L" for every incoming packet header.
2. Determine the minimum value weight of pattern "P", which refers to the minimum number of occurrences of each value in matrix "M". If the minimum value weight is equivalent to zero, then stop the matching process, in light of the fact that the pattern "P" does not exist in the text "T". Else, go to step c.
3. Create the attempt matching process of the array list "L" by adding the minimum weight value, (which chose in step b) under the relating value position in the array list "L" (index [i]).
4. Compare the following and the past value in the pattern "P", which its indices are: index [i+1] and index [i−1], with the relating value of matrix "M", if both exist, then keep matching with index [i+2], and index [i−2] until achieving the end of the pattern "P" or until getting an exact matching.
5. If mismatch has happened, then read the following occurrence of the minimum value weight of the pattern "P", and repeat from step c [8].

4 Working Example

The implementation of the matching process has the following steps:

1. Convert the header rule sets into weight (matrix "M").
   Table 1 depicts the header rule set after the conversion into weight, where each row represents one header rule from the original header rule set. For efficient processing, and especially dealing with intrusions, the calculated weights will be rearranging in a descending order having an index of each group in the first column.
2. Create the array list "L" for the incoming packet header after finding its weight as shown by Fig. 5 (create the array list "L").

Table 1. Wight of the header rule sets

G	W₁	W₂	W₃	W₄	W₅	...	Wₙ
7	7	7	6	4	2	...	7
	7	6	5	1	5	...	2
6	6	4	2	3	6	...	1
	6	3	6	2	3	...	6
5	5	7	3	5	2	...	1
4	4	7	1	6	3	...	2
	4	5	3	2	6	...	3
3	3	7	4	3	2	...	4
2	2	6	3	4	1	...	3
1	1	6	4	7	2	...	3
	1	1	2	3	4	...	4
0	0	5	3	2	5	...	7

Where: G: group number, and W: weight

3. Determine the character in the incoming packet header, which has the minimum number of occurrences in group 6.
   Based on matrix "M", the minimum character occurrences in the incoming packet header is "2" (number 6 found 3 times, number 3 found 2 times, and number 2 found 1 time).
4. Create the first attempt of the searching process by selecting the first occurrence of the character "2", which is 3, under the corresponding character in the array list "L". Show Fig. 6 (first attempt).
5. Create the second attempt of the searching process by checking the index [i−1], and index [i+1] in the array list "L".
6. Reference to the matrix "M", "6" does not exist in the index [i−1]. See Fig. 6 (second attempt). Therefore, no match can get in this attempt, and then we will examine the next occurrence of the character "3", which are 4. See Fig. 6 (third attempt).
   As shown by Fig. 6 (third attempt), we can get an exact matching between the incoming packet header and one of the header rule set (full sequence: 6, 3, 6, 2, 3... and 6).

	Incoming packet header = 3623...6						
Index	1	2	3	4	5	...	n
Create the array list "L"	6	3	6	2	3		6
First attempt			2				
Second attempt		4≠6					
Third attempt	6	3	6	2	3		6

Fig. 6. Attempts of the searching process

5 Conclusion

In this work we have introduced a proposed technique to speed up the packets header detection engine. We separated the work into three stages, to be specific: classification of header rule sets, the learning stage and matching stage.

Acknowledgment. I would like to thank Al-zaytoonah University of Jordan for supporting this research paper.

References

1. Abedin, M., Nessa, S., Khan, L., Thuraisingham, B.: Detection and resolution of anomalies in firewall policy rules. In: Damiani, E., Liu, P. (eds.) Data and Applications Security 2006. LNCS, vol. 4127, pp. 15–29. Springer, Heidelberg (2006)
2. Snort – The open source network intrusion. Detection system. http://www.snort.org
3. Coit, C.J., Staniford, S., Mchlerney, J.: Towards faster string matching for intrusion detection or exceeding the speed of snort, pp. 367–373. IEEE (2001)
4. Xiang, Y.: Protecting information infrastructure from DDoS attacks by MADF. Int. J. High Perform. Comput. Netw. 4(5/6), 357–367 (2006)
5. Deri, L.: High-speed dynamic packet filtering. J. Netw. Syst. Manag. (ACM) **15**, 401–415 (2007)
6. Lecroq, C.C.: Handbook of Exact String matching Algorithm. King's College Publications, London (2004). ISBN 0954300645
7. Karem, E.: Alternative hopfield neural network with multi-connect architecture. Journal of College of Education, Computer Department, Al-mustansiryah university, Baghdad, Iraq (2004)
8. Hlayel, A.A., Hnaif, A.A.: An algorithm to improve the performance of string matching. J. Inf. Sci. (2014). doi:10.1177/0165551513519039. Accessed 14 Jan 2014

Content-Based Arabic Speech Similarity Search and Emotion Detection

Mohamed Meddeb[✉], Hichem Karray, and Adel M. Alimi

REGIM: REsearch Groups on Intelligent Machines, University of Sfax,
National Engineering School of Sfax (ENIS), BP 1173, 3038 Sfax, Tunisia
{dr.mohamed.meddeb, dr.hichem.karray,
Adel.alimi}@ieee.org

Abstract. System performance of speech emotion recognition is prejudiced by many factors, including the quality of speech samples, features and classifiers. The present study proposed a specific architecture of an automatic recognition system of Arabic speech emotions, recognized as: Anger, Happiness, Neutral, Sadness and Fear. To perform this system we select two approaches classifications. KNN (k-nearest neighbor's algorithm) and SVMs support vector machines multiclass. Good performances are carried out: The rate of cross-validation is over 92 % and the rate of recognition accuracy is about 93 %. The intelligent system outputs contribute to the addition of phonemes in Arabic speech, but it should be noted that it is related not only to the language but also to culture and environment. Corpus is performed without delimiter syntax and language, because our goal is to extract emotions not speech recognition. Even the silent sounds are processed as dynamic signal. To accomplish the emotions, we had created a REGIM_TES database as support and we developed a similarity module to compare features vectors to settle on emotions. The subjective evaluation of emotions is carried out by performing the subjective listening tests with listeners candid. Twelve listeners (6 women and 6 men) with ages ranging from 18 to 42 years participated in the experiment and each listener was presented with 550 expressions. We conclude that mistakes of the recognition system (objective) are significantly lower than those of human (subjective) evaluators. Finally, the emotion recognition system, allow research to detect a person's behavior and speech abnormalities.

Keywords: Speech · Arabic · Emotions · SVMs · Similarity · Intelligent · Phonemes

1 Introduction

We proposed in this paper the concept of emotional speech recognition system in Tunisian dialect. The proposed architecture is provided with an updated system of emotions database and a measuring similarities module.

This system allows providing assistance to Elderly person, with physical disabilities, intelligent television (ITV) and more other applications. We are interested in emotion recognition considering the utterances as the basic unit in their modeling. The use of this unit fits very well to the problem when the utterance does not convey one

© Springer International Publishing AG 2017
A.E. Hassanien et al. (eds.), *Proceedings of the International Conference on Advanced Intelligent Systems and Informatics 2016*, Advances in Intelligent Systems and Computing 533, DOI 10.1007/978-3-319-48308-5_51

emotion. However, sometimes this hypothesis is not checked, including statements from real conversations, where certain words or phonemes are more emotionally prominent than the rest of the statement. For instance, a ("نعم" = yes) or ("لا" = not) may be stronger emotionally than other words of the same utterance. Based on this observation, other units of analysis, such as words, syllables and phonemes were experienced by some researchers. However, constraints and significant challenges result from the implementation of the systems based on these units. Among these constraints, the need to provide the RAE system, not only with automatic speech recognition, but also with a system for temporal alignment between the system output of automatic speech recognition and phonemes or syllables of the utterance. In this paper, we propose the study of a new application for the recognition of emotion that overcomes the disadvantages of previous units. The new method is based on an independent segmentation of any system of speech recognition, which can be achieved in a very short time, and is therefore well suited for applications in real time. The latter method PROS (independent segmentation of speech recognition) has been used successfully in the field of language identification [1–3] and that of the speaker verification [4–7]. Independent segmentation method statements that we use in our system is based on the same technique used in the last two works cited [9, 11, 12].

2 Related Works

The subject of emotion digital processing via speech signal as an input is a new and active research topic. Consequently, there are several works that deal with Arabic speech emotion recognition and classification. One of these efforts is the research conducted by Khan et al. [2]. They attempted to classify Arabic sentences into either question or non-question sentences by segmenting them from continuous speech using intensity and duration features as well as by extracting the prosodic features from each sentence. Their approach achieved an accuracy of 75.7 %.

Al Dakkak et al. [5] developed an automated tool for emotional Arabic synthesis based on an automatic prosody rough generation model as well as the number of phonemes in a sentence. They claimed that their system's model worked successfully in tests. In a previous work [1] we extracted the acoustic features of pitch, intensity, formants, and speech rate and used them to classify five Arabic speech emotions: neutral, sad, happy, surprised, and angry, using twelve sentences read by six male and six female native Arabic speakers.

The content-based spectral and acoustic features are usually calculated for every short-time frame of sound based on the Short Time Fourier Transform (STFT) [7]. Typical timbral features include Spectral Centroid, Energy, Zero Crossings, Linear Prediction Coefficients, and Mel-Frequency Cepstral Coefficients (MFCCs). Among these timbral features MFCCs have been dominantly used in speech recognition. Logan [8] examines MFCCs for music modeling and music/speech discrimination. There has been much work on music style recognition, genre categorization, and similarity. Dannenberg, Thom, and Watson [10] demonstrate that machine learning can be used to build effective style classifiers for interactive performance systems. Classification using the K-Nearest Neighbor Model and the Gaussian Mixture Model. Lambrou et al. [6]

use statistical features in the temporal domain. While there has been much work on emotion recognition from speech [13, 14], there has been little work on automatic music emotion detection.

3 Description of the System

Figure 1 provides the structure of an intelligent emotional remote control [3] based on the architecture of the system of AER. The system is divided into four main modules. Each module is responsible for processing a data set in favor of the next section by presenting the most optimal and most useful acoustic vectors. The first (1) section is responsible for the acquisition of acoustic data, whether synthetic or real. The acquisition phase is a critical phase for the faithful and decisive results. The sound signal presented to the input of the system can be a speech or monologue. Modules (2,3,4) in Fig. 1 constitute the emotional intelligent remote control [3] can integrate the environment of ITV (Smart TV).

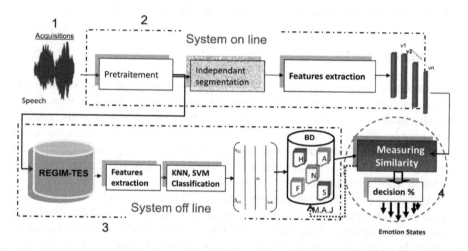

Fig. 1. Structure of remote intelligent emotional

The advent of digital technology has increased the number of channels of streaming media, so the individual cannot monitor the developments. Especially he becomes increasingly busy doing painful tasks along the day. For this reason and to provide him assistance in order to have comfortable life for older people living alone and also for motor handicaps, intelligent remote control can fill this gap and provide a moment of distraction in any case cited above. Admits in the acquisition phase the speech signal, performs the extraction of descriptors chosen and selected entry in our emotional database. The duration T ms is imposed on the size of the segments in order to have enough points for extraction Indeed, our system relates a database of emotional models built on a combination of descriptors to identify each emotion. Of course, the records have undergone a pretreatment according to the acoustic speech signal theory. Filtering and preprocessing phase solves the problem of terminals and the frame delimiters signal after segmentation. The extraction of descriptors is obviously on a chosen set well studied thus avoiding the slow performance of the automatic recognition system due to the presence of not decisive secondary parameters.

3.1 Pretreatment/Filtering

Performing pretreatments of acoustic speech signal aim to shaping and extracting parameters necessary to obtain an invalidating representation of this signal. Human being is sensitive to the bandwidth from 200 Hz to 5000 Hz. According to Shannon's sampling rate must be at least twice greater than the maximum useful frequency at least equal to 10 kHz but in our project we took 44,100 Hz sampling frequency. Quantization defines the number of bits wherein the scan will be performed and also to measure the signal amplitude at each sampling. If the number of bits is 16 bit, it will be 65536 amplitude value between -32768 and 32767 possible.

3.2 Segmentation

In Module (2) speech signals usually contain many areas of silence and noise. Therefore, in the speech analysis, it is first necessary to apply a method of silence suppression, to detect the "clean" speech. The best known method is that of Theodore Giannakopoulos for the detection of speech segments in the online part of the system. The general algorithm of the method is as follows:

- Extraction of two descriptors of the signal, the short-term energy and spectral centroid.
- For each descriptor, a threshold is dynamically estimated.
- The segments are formed by successive frames for which the values of the two descriptors are larger than the thresholds calculated.

3.3 Similarity Measure

This module compares the flow of descriptor vectors with reference models in the emotional database based on cosine technique to detect possible similarity toward the emotion class which can finally decide on the emotional state. This measurement can give every moment the emotional state of the conversation.

3.4 Integration of Similarity

Audio similarity aims to discover a concept that has the maximum degree of kinship to the sound application. To calculate the relationship, we adapt the concept of reasoning. Audio similarity is a linear space containing the semantic relation between the interrogation signals. The similarity between two semantic concepts and If Q is calculated with the cosine similarity measure: where and are the feature vectors of Q and Si, respectively. Unlike the usual distance measures, audio similarity is a computable space and provides an overview of the concepts in a semantic proximity between the concepts of influence. The vector chosen is a data recovery or response to the user request. Audio similarity is computed using the following equation:

$$Audiosimilarity = \frac{V_Q V_{Si}}{|V_Q||V_{Si}|}$$

3.5 Evaluation by Subjective Listening Tests

In this subsection, we describe the set of listening tests and present the results of evaluation.

3.5.1 Experimental Strategy Test

The subjective evaluation of emotions, happiness, fear, anger, neutral, sadness is carried out by performing the subjective listening tests with listeners candid. Twelve listeners (6 women and 6 men) with ages ranging from 18 to 42 years participated in the experiment and each listener was presented with 550 expressions. The test stimuli were presented in random order to eliminate all the consequential effects in decision making. Headphones were used and subjects were free to adjust the volume and listen to the word as many times as they wished, however once given the choice, they no longer have the right to go back. The evaluation is to listen to a set of six emotions randomly distributed in a structure that guarantees the transparency of the registration order not to influence the decision, the auditor is required to recognize the emotion whenever the word heard and check the box corresponding to the class to which it belongs word or phrase. The average test duration was 20 min per listener.

3.5.2 Results of the Tests

The average results for 12 listeners are presented in Tableau.19 and shown in percentage: N: Neutral, F: fear, A: Anger, S: Sadness, H: Happiness (Table 1).

The following confusion matrix presented in tabular form, allows us to deduce the average rates of recognition on the diagonal, the Anger emotion presents the best recognition percentage (76.48 %) followed by emotion Happiness (75.58 %), emotion Neutral (67.53 %), emotion Fear (65.56 %) and emotion Sadness (56.32).

3.6 Objective Evaluation by the System of Automatic Emotion Recognition

In this part, we perform the identification of six basic emotions presented in the test database consisting of 1,620 records with 168 words and 156 phrases with emotion.

Table 1. Confusion matrix of subjective listening tests

%	Neutral	Fear	Sadness	Happiness	Anger
Neutral	60.20	02.4	27.30	06.10	03.20
Fear	02,20	66,50	05,70	12,50	15,50
Sadness	25,50	09.40	61.70	00.20	00.40
Happiness	04,50	01.20	00.30	75,20	19,20
Anger	04,50	03,30	05.70	19,80	73,30

3.6.1 Classification Result

The classification of emotions speech using features (12 MFCC, LPC, Pitch, Energy, Power) is presented as a confusion matrix in Tableau.20, this matrix is used to evaluate the confusion between different emotions. Note that the emotion anger has the best recognition rate (95.42 %), followed by emotion Bonheur (90.33), emotions sadness, fear and neutral have accuracy rate around 64 % (Table 2).

Table 2. Confusion matrix of objective listening tests

%	Neutral	Fear	Sadness	Happiness	Anger
Neutral	58,42	13,08	20,42	07,12	07,80
Fear	06,50	63,33	12,17	08,14	8,10
Sadness	12,00	8,33	73,75	3,14	2,55
Happiness	05,33	03,25	0	90,33	2,56
Anger	01,17	02,83	0	01,41	95,42

3.6.2 Comparison of Objective and Subjective Results

The figure below shows the error rate obtained with the subjective and objective evaluations, respectively operated by listening and automatic emotion recognition of speech tests. Errors of the recognition system are significantly lower than those of human evaluators. The neutral emotions and fear is the only emotion that this mis-classification by the above error listening tests RAE automatic system.

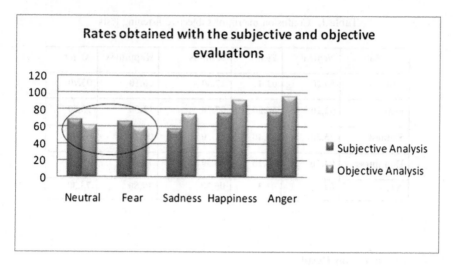

Fig. 2. Structure of remote intelligent emotional

Figure 2 shows the differences between objective and subjective assessments. Three emotions happiness, sadness and anger are remarkably better separable by our system of recognition by the human ear.

4 Discussion

System performance of emotion recognition of speech is influenced by many factors, including the quality of speech samples, the extracted features and classification algorithm. The initial work focused on the stage of acoustic analysis of the speech signal in order to ensure vectors sufficiently informative acoustic parameters. The most used feature sets are energy, Pitch, LPC, and MFCC. Other combination of emotional characteristics may get different rates of emotion recognition and the sensitivity of different emotional characteristics. As can be seen from the experiment, the rate of emotion recognition system that uses only the characteristics of the speech spectrum is slightly higher than that uses only prosodic features of speech. And the system which uses features of both spectral and prosodic has the best performance. The recognition rate of this energy use, pitch, LPC, MFCC is slightly lower than the energy use, features field MFCC and LPC. The set of descriptors is subject to separate emotional excitations, and then the group takes different forms and behaves differently organizing the following descriptors structures and singular groups. Therefore, yielded emotion model for each of the defined descriptor. Thus constructed models form the structure of the database emotions models, this will be the last base reference tool for real-time classification of emotions, where the real-time recognition of the subject's emotional behavior question. Thereafter, our choice is focused on two approaches classifications. The first classifier uses the kNN. While the second is based on SVM which belong lacking a priori models, called non- parametric approaches. Our work shows a good performance. The rate of cross-validation is high over 92 %, and the rate of recognition accuracy is about 93 %.

5   Conclusion and Perspective

In this paper, we presented the overall architecture of our system whose major purpose is to assist the user during the process of interaction. An experimental study was also set up to evaluate the performance of the proposed techniques. Our system corresponds to the intelligent emotional remote for iTV (Smart TV) kernel. It is in fact increase the functionality offered by our system techniques developed in our laboratory team REGIM to strengthen the system and lead to an intelligent access to television in general. In this study, we explored how people and computers identify emotions Arabic word. We applied the recognition from the Arabic words isolated using a protocol-independent text and speaker. The emotion model parameters are the result of the concatenation of the spectral and prosodic parameters. Several conclusions can be derived from the results, on the one hand, the decoding of emotional states is influenced by cultural, social and intellectual environment of the speaker, and on the other hand, the human being is unable to detect voice disorders caused by mental and emotional state. Through our emotion recognition system, we can detect a person's behavior or speech abnormalities. Will result allow nervous gents for example, to check whether they are equipped with a device for automatic recognition of emotion. The rate of recognition of emotions of the database emotional Arab REGIM_TES have reached almost 90 %, which represents a rate very interested viewed variability imposed on the system such as recognizing an independent mid- term and speaker.

Our research has shown the influence of Arabic on the prosodic and spectral descriptors. Among the contributions that we have outlined in this brief, the addition of phonemes in speech enabled suite has a sense of anger or happiness. This phenomenon has never been demonstrated before. This is specific to the Arabic language in Tunisian dialect, but it should be noted that it is related not only to the language but directly related to the culture and living environment. In our work we have adopted a method of independent segmentation delimiter syntax and language, because our goal is to extract emotions not to recognize the words. Even in the silent sound recording is processed in the same way as the rest of the signal. Recognition of emotions in speech has many useful applications. In human-machine interfaces, robots can learn to interact and recognize human emotions. For example, robotic animals, automated home lighting and interactive television (ITV) [4], must be able to understand not only the voice commands, but also other information and modify their actions accordingly. The advent of digital satellite television, redundancy peripheral airways by increasing the redundancy of programs and channels, make it very difficult choice of television programs that meet user preferences. A recommendation system for personalized television programs based on human behavior through the automatic recognition of emotion in speech and proposed as a solution to this problem in order to achieve a smart remote. In addition, through this remote digital TV we can estimate our interests and preferences by observing our behavior to watch and have a conversation on topics that may be of interest to us.

Emotion does not have a universal definition, especially the recognition and detection for each medium, each culture and its secret language, among other Arabic. In the future we propose to improve the basic REGIM_TES data, the actual recordings with sequences of spontaneous conversation. The combination of several classifiers may lead to improved precision, especially the method of the fuzzy logic to the real-time recognition. In particular, we must develop the system feedback (feedback) learning to update the database of emotion intelligent remote models. Work successor aim to introduce two new parameters, the jitter and shimmer representative micro variations of the fundamental frequency and energy, and to test their influence on the proposed recognition system respectively. Another issue concerns the variation of HMM states 8, 16, 32 and 64 and to note the effect of this variation on the recognition rate. Our system is smart emotional control is a multimodal system. Under the project of our laboratory, we believe in integrating the module neuro recognition and language pack to improve the state of the neutral emotion that marks an ambiguity for most gents and automatic systems, to improve the rate recognition of emotions in the Arab word.

References

1. Meddeb, M., Hichem, K., Alimi, A.: Automated extraction of features from Arabic emotional speech corpus. Int. J. Comput. Inf. Syst. Ind. Manag. Appl., IJCISIM **8**, 184–194 (2016). ISSN 2150-7988
2. Khan, O., Al-Khatib, W.G., Lahouari, C.: Detection of questions in Arabic audio monologues using prosodic features. In: Ninth IEEE International Symposium on Multimedia, ISM 2007, pp. 29–36 (2007)
3. Meddeb, M., Hichem, K., Alimi, A.: Intelligent remote control for TV program based on emotion in arabic speech. Int. J. Sci. Res. Eng. Technol. (IJSET) **1** (2014). ISSN: 2277-1581
4. Meddeb, M., BenAmmar, M., Alimi, A.: Towards a recommendation system for TV programs based on human behavior. In: The International Conference on Control, Engineering Information Technology, CEIT 2013, Proceedings of Engineering and Technology, vol. 4, pp. 180–182 (2013)
5. Al-Dakkak, O., Ghneim, N., Abou Zliekha, M., Al-Moubayed, S.: Prosodic feature introduction and emotion incorporation in an arabic TTS. In: 2nd International Conference on Information and Communication Technologies, ICTTA 2006, vol. 1, pp. 1317–1322 (2006)
6. Lambrou, T., Kudumakis, P., Speller, R., Sandler, M., Linney, A.: Classification of audio signals using statistical features on time and wavelet transform domains. In: Proceedings of the International Conference on Acoustic, Speech, and Signal Processing (ICASSP 1998), vol. 6, pp. 3621–3624 (1998)
7. Rabiner, L., Juang, B.H.: Fundamentals of Speech Recognition. Prentice-Hall, NJ (1993)
8. Logan, B.: Mel frequency cepstral coefficients for music modeling. In: Proceedings of the International Symposium on Music Information Retrieval, ISMIR (2000)
9. Meddeb, M., Hichem, K., Alimi, A.: Speech emotion recognition based on arabic features. In: 15th International Conference on Intelligent Systems Design and Applications (ISDA 2015), Marrakesh, Morocco, IEEE Conference, 14–16 December 2015
10. Dannenberg, R., Thom, B., Watson, D.: A machine learning approach to musical style recognition. In: International Computer Music Conference, pp. 344–347 (1997)

11. Meftah, A.H., Selouani, S.-A., Alotaibi, Y.A.: Preliminary arabic speech emotion classification. In: IEEE International Symposium on Signal Processing and Information Technology (2014)
12. Meftah, H., Selouani, S.-A., Alotaibi, Y.A.: Investigating speaker gender using rhythm metrics in Arabic dialects. In: 8th International Workshop on Systems, Signal Processing and their Applications (WoSSPA), pp. 347–350 (2013)
13. Petrushin, V.A.: Emotion in speech: recognition and application to call centers. In: Proceedings of the Artificial Neural Networks in Engineering 1999 (1999)
14. Polzin, T., Waibel, A.: Detecting emotions in speech. In: Proceedings of the CMC (1998)

Symptom-Diagnosis-Care: A Framework for a Collaborative Medical Chat Bot

Michael Fischer[✉]

Computer Science Department, Stanford University, Stanford, USA
mfischer@stanford.edu

Abstract. Medical chat bots could have the ability to deliver fast, high-quality information directly to people all over the world. A step preventing this from becoming a reality is training medical chat bots with reliable medical data. Medical books though, such as *The American Medical Association Family Medical Guide*, are well structured with a series of flowcharts to help users diagnose their symptoms by answering yes and no questions. In this paper we present and evaluate a tool for crowd workers to train medical chat bots using the information in books. To enable the collaboration of crowd workers, we present the symptom-diagnosis-care framework. The medical chat bot on the phone allows the information accessible to people that primarily use a mobile phone and are not medical professionals. By having the data on the phone, we are able to make the data actionable by integrating it with the computational capabilities and sensors of the phone.

Keywords: Chat bots · Medical bots

1 Introduction

Personal digital assistants, known as chat bots, are becoming popular as a way to interact with computers. With the rise of mobile computing there has been an increase in the number of chat bots. A problem with conversational chat bots are they don't have a HAL 9000 level of intelligence and there is ambiguity in how the chat bots interpret what the human is trying to convey. A user speaks to it, it speaks back, and it uses natural language, but it isn't general AI. How likely is it that the users can ask something that breaks the bot?

In this paper we create a chat bot that constrains what the user is able to respond with according to a set number of predefined responses. We design for the chat bot to take the lead in the conversation. The benefit of this is two-fold. The first benefit of this style of chat bot is that it limits the amount of opportunities for the user to input a response outside of the domain of experience of the chat bot. In essence, the chat bot is always able to set the context of the conversation so the user knows what to expect.

The second benefit is that the system asks questions and provides structure for patients that are unsure about how to diagnosis their problem. Given a free form text box, the user can go down a path that might not be helpful in diagnosing the problem.

The user might try to describe everything because they don't know the important things they should be focusing on in the domain. By asking specific questions, the

© Springer International Publishing AG 2017
A.E. Hassanien et al. (eds.), *Proceedings of the International Conference on Advanced Intelligent Systems and Informatics 2016*, Advances in Intelligent Systems and Computing 533, DOI 10.1007/978-3-319-48308-5_52

system is able to provide a landscape of the important questions in getting to the root cause of the problem. This helps to limit the number of red herrings that they will have to investigate.

We are limiting ourselves in this project to medical diagnosis. We customize our technical solution to this domain. Other approaches to creating chat bots have used a kitchen sink approach of machine learning to train on large volumes of chat logs to try and back out conversational models.

However, in the domain of medicine, we train the bot on medical literature. The information that goes into the system has a provenance and can be audited and corrected. With a statistical approach to building chat bots, it can be difficult to correct specific problems.

Moreover, in medicine, chat bots need to be updated as our knowledge of the domain evolves. Updating this system can be done in a narrow and intentional way, whereas a machine learning approach would require the system to be retrained.

Overall we try with this user interface to make a bot with a UI that works best for something very specific - where the user knows what they can ask. This limits the number of errors and allows for the fallback to a doctor if the bot goes into an area where it doesn't know the answer to a question.

2 Related Work

2.1 Expert Systems

An expert system is a computer program that is programmed to try and mimic a human expert's decision-making ability [1]. Expert systems try to solve complex problems by reasoning about knowledge represented in an if–then–else format.

MYCIN, an early example of an expert system, identified bacteria causing severe infections and recommended antibiotics [2]. The dosage adjusted for a patient's body weight. One of the difficulties of this type of system was the complexity of mapping out the data in the expert system. Another problem was the gathering of the knowledge needed to formulate the inference rules. However, gathering knowledge at scale is now possible. Wikipedia [3] is a successful example of how the complexity of developing an encyclopedia was overcome with crowd sourcing. The majority of Wikipedia is loosely structured which reduces the complexity of organizing the knowledge.

2.2 Script Theory

Script theory, based in cognitive psychology, is a theoretical framework used to explain how a physician's medical diagnostic knowledge is structured for diagnostic problem solving [4]. In script theory, the medical expert has an integrated network of prior knowledge that leads to an expected outcome. Medical experts apply pre-stored knowledge to place the patients' illnesses into a given class of diseases. After an illness is classified, the medical expert determines which treatments could work given the particulars of the patient.

2.3   Personal Digital Assistants

Achievements in language processing have created new assistants such as Apple's Siri, Google's Now, Microsoft's Cortana, and Amazon's Echo. Each can respond to a fixed number of automated queries. The tasks that they are doing are predefined [5]. When a user asks a question, a set number of parameters are filled into the application and it is then run. At present, these systems don't use structured data to complete a users' request.

The currently available commercial assistants are not domain specific. It thus makes it difficult to allow for experts in the field to structure their knowledge to be useful for such bots.

In this project we utilize the symptom-diagnosis-care framework to allow experts in the field to structure their knowledge. Our goal with providing this domain specific framework is to provide an interface for experts to work together to train a bot.

2.4   Crowd Sourced Conversational Personal Assistant

Crowd workers have been used to power real-time personal assistants. A user asks a questions and a human responds to it. This has shown to be useful for describing images [6], speech recognition [7], interface control [8], activity recognition [9], document editing [10], and vacation planning [11]. There are three problems with these human powered systems. Firstly, such systems rely on people to accomplish tasks. The second problem is the latency associated with humans doing a task. Users expect an instant response. The time it would take another human to complete many sets of tasks is too long. Thirdly, crowd sourced personal assistants have one person interacting with many people. There is no constant contact between the end-user and a single person on the other end. This means that maintaining context and state is difficult [12].

3   Data

A challenge with chat bots is how to acquire data. Generally there is a chicken and egg problem. Without data, no one will use the system. If no one is using the system, it is hard to get data.

Obtaining the time of domain experts for any software application is always difficult because domain experts, the doctors, are highly valued and in constant demand.

As a result of this problem, we looked toward tools and processes for systematic knowledge acquisition and for solutions that are scalable using parts of the medical infrastructure that already exist.

Since medical knowledge exists in the medical literature, one of the insights we had was not to ask medical experts to train the system using their knowledge.

Instead, we frame the problem so that someone without a medical degree can use a tool to train only a portion of the chat bot using the medical literature.

With this method, the task of teaching the chat bot becomes scalable. In order to make the project scalable, we had to add structure to the process of training the bot. Without structure it would be difficult to compound the efforts of each of the individual workers.

To bootstrap the problem of training the bot, we use the information contained in the existing medical literature. The idea to do this came from a discussion with a doctor that was describing the process that they learned in medical school and how they go about assisting patients.

The *American Medical Association Family Medical Guide* [13], is a medical reference that is comprehensive and targeted for non-medical professionals. It contains flowcharts to help readers find the possible cause of symptoms. We use the book as the basis to bootstrap the medical chat bot.

Through a series of questions and answers, the patient is led to a possible diagnosis or to other charts within the book (Fig. 1).

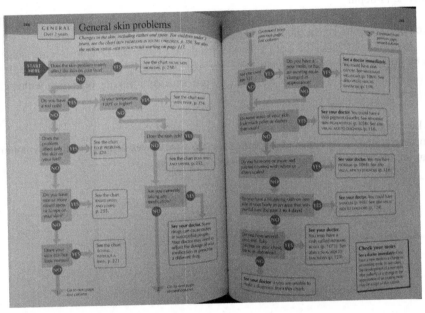

Fig. 1. An example of flowchart from the *American Medical Association Family Medical Guide*. The flowchart is used as the basis of the data to bootstrap the medical chat bot system. The flowchart works by asking a series of yes and no questions. There are many cases where the chart recommends the patient see a doctor either soon or immediately.

4 The Symptom-Diagnosis-Care Pattern

We developed a pattern to allow crowd workers to work together. This pattern provides a structure that allows interoperability between the work of crowd workers.

After doing a close reading of the medical literature, we structured our system to allow for three different types of information: symptoms, diagnoses, and care. Absent suitable categorizations, the knowledge is too amorphous for crowd workers. The structure provided by this framework is analogous to that of design patterns like

Model-View-Controller, which codified best practices for programmers developing large systems. In this section, we propose the Symptom-Diagnosis-Care pattern for programming medical chat bots. We describe the pattern and then explain its use in our system.

4.1   Symptom-Diagnosis-Care

The Symptoms-Diagnosis-Care pattern separates medical assistance into three stages. The first stage, Symptom, has a crowd worker enter a symptom as a question phrased to require either a yes or a no answer.

The Diagnosis stage gives a diagnosis and more information that the patient can browse in-depth.

The Care stage gives a number of recommended next steps for the patient. The recommendations are in the form of text and javascript programs.

Why should tasks be split into independent Symptoms-Diagnosis-Care stages? Why not let medical professionals have free rein? By having a framework for people to work within, our goal is to encourage re-use and intra-linking of nodes within the system.

4.1.1   Symptom

To determine if a user has a symptom they are asked a yes or no question. Symptoms can lead to other symptoms or to a diagnosis (Fig. 2).

Fig. 2. Symptoms are entered into the node builder tool. Symptoms are the nodes in the graph. An edge can be either a yes or a no. Symptoms lead to a diagnosis.

4.1.2   Diagnosis

The diagnosis section contains information for the user about the disease. Each diagnosis is given a title and supplementary text. The title appears alongside the Node UUID in a lookup table to allow crowd workers to make connections between their work. Each diagnosis can link to multiple care nodes.

A recommendation to see a doctor for further consultation is an acceptable diagnosis from the system (Fig. 3).

Fig. 3. Diagnosis nodes can include text or links to websites. Diagnoses are linked to one or many possible treatments.

4.1.3 Care

The care section contains both text and code. Each diagnosis can link to multiple care nodes. The crowd worker can fill in the care as it appears in the book. In addition, small javascript apps can be added to help the patient as part of the care. For example, a javascript app can remind a patient to take their medication every six hours. As another example, if a patient has too high blood pressure, an app could be used to notify the doctor if the blood pressure goes over a set limit (Fig. 4).

Fig. 4. Care nodes can include text and javascript code. The javascript code is used to run mini-apps that help the user improve their health. The chat bot is able to use run these programs for the user because the system runs on the mobile phone. An example care app could notify a patient to take aspirin twice a day.

5 System Flow

5.1 Crowd Worker Flow

A crowd-worker, who does not have a medical background, sees a photo of a flowchart. Their job is to translate the image into a semantic document.

This bootstrapping process could be replicated by a computer vision algorithm. We didn't choose this method because the tool used by the crowd worker can also be used by medical professionals to keep the content to the system up to date.

6 Chat Interface

To narrow the scope of the symptoms so that the correct dialogue can be shown, the system first shows a map of the body. The user then clicks on the area of the body where the symptoms are occurring. Each of the dialogues are categorized under one or more of the areas of the picture by an expert (Fig. 5).

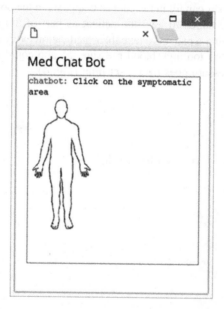

Fig. 5. The chat bot initially uses a triage system to quickly narrow the scope of the problem before going into a narrower set of yes and no questions. The triage portion of the chat uses a map of the body that the user clicks on.

After the user is within a specific dialogue, they are presented with a series of yes and no questions (Fig. 6).

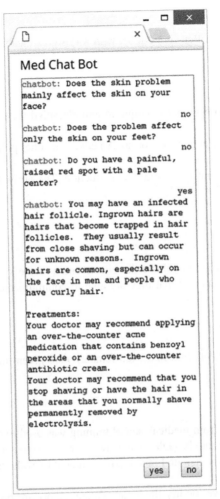

Fig. 6. A chat bot asks the user a series of yes and no questions. The yes and no question are designed to help the user find the possible cause of their symptoms. If the bot can't figure out the problem, it advises the patient to see a doctor.

7   Implementation

7.1   Frontend

The interface that the user uses is built to run in a web browser. The chat bot interface was built using jQuery, Bootstrap, and HTML. The flowchart builder was built using interact.js.

7.2   Backend

The system server is written in Node.JS with a Postgres database.

7.3   Data Structure

Each node is assigned a globally unique id and classified as: symptom, diagnosis, or care.

Symptom edges are either a yes or no. An edge can either point to another symptom or to a diagnosis.

Diagnosis nodes point to an array data structure. The array data structure contains care nodes.

8   Evaluation

We conduct a user study to evaluate the effectiveness of the system. The study evaluates how well end-users feel that they are able to communicate with the chat bot.

8.1   Participants

Twenty university students (ten male, ten female) were given the role of a patient.

8.2   Background

A technique borrowed from medical school training was used to evaluate the chat bot. Normally, patient actors are brought in and are given scripts to act out different medical and social situations. It is then up to the medical student to evaluate the patient actors to determine the diagnosis.

In our evaluation, we modify the setup. Patient actors are brought in and given a script to read. It is then up to the chat bot to give them a diagnosis. We measure how effective the system is by how accurately it is able to diagnosis the patients.

8.3   Method

We evaluated the interface with ten head and neck flowcharts. From these ten flowcharts there was a possibility of 78 different diagnoses.

Patient actors were given a 289 word script that described their symptoms which took about two minutes to read. After they were done reading the script, they were not allowed to look at it again.

9   Results

Of the twenty patient actors, the system found that the correct diagnosis for eighteen. For the patients that did not get the correct diagnosis, the chat bot recommended that the patient see a doctor because it was unable to make a diagnosis.

9.1   Post-Task Questionnaire

Post-task questions	1 (not very much) to 7 (very much so)
Were you able to describe your symptoms expressively to the chat bot?	5.6
Was the chat bot easy to communicate with?	6.2
Did you understand the questions the chat bot was asking?	6.4

9.2   Qualitative Assessment

At the end of the evaluation, the patient actors and the examiner had an informal talk to find out more about examine the usage of the system in more depth. Some key quotes:

> *"I would find this helpful when I'm traveling especially if it could be made to work on the phone without data"*

> *"Useful if for when I'm not sure if I should see a doctor or if it is something that I'm embarrassed about"*

> *"I'd enjoy using this as a way to learn when I'm bored. It would be especially interesting for learning if there were game mechanics built-in."*

> *"I liked how I only had to answer yes or no. It simplified the process and made it so I didn't have to type at all. It helped to also set my expectation as to what the system could do."*

10   Discussion

In this paper we develop the Symptom-Diagnosis-Care framework to allow crowd workers to collaborate in building a medical chat bot using the information in existing medical books.

The quantitative and qualitative feedback from patient actors showed that the system was effective in determining the correct diagnosis. When the chat bot was not able to make the correct diagnosis, it suggested the patient see the a doctor and did not give an incorrect diagnosis.

Patient actors found using the yes and no buttons effective for interacting with the system. They were regarded as simple and well suited for the form factor of the mobile phone.

Overall, these results show a promising research direction for further developing a collaborative medical chat bot for the mobile phone.

11 Future Work

11.1 Integration with Electronic Medical Records

The chat bot is in a position to have an understanding of the context of the patient and their symptoms. A patient will always have to answer similar questions over and over again to determine some diagnosis, as for example, family history or age. A feature that should be added to the system is a way to lookup patient information that the system already knows. This lookup feature we predict would come in the form of an electronic medical record. Ideally this medical record could be exported at any time to another system. Additionally, it could be made human readable so that a patient could review it.

11.2 Connected Devices

A computer should never ask a question it should know the answer to. Overall, some of the questions that the system may need to ask could be answered by connected devices. As such, a chat bot should have interfaces for such devices. If there is a blood pressure monitor cuff attached to the patient, a question about the person's blood pressure would not need to be asked. If the person is wearing a connected pedometer, that could be used as a signal to infer the patient's level of activity.

References

1. Jackson, P.: Introduction To Expert Systems, 3rd edn., p. 2. Addison Wesley (1998). ISBN 978-0-201-87686-4
2. Shortliffe, E.H., Buchanan, B.G.: A model of inexact reasoning in medicine. Math. Biosci. 23(3–4), 351–379 (1975)
3. Wikipedia contributors. "Wikipedia." Wikipedia, The Free Encyclopedia, January 2016
4. Charlin, B., Tardif, J., Charlin, H.P.A.: Scripts and medical diagnostic knowledge: theory and applications for clinical reasoning instruction and research. Acad. Med. 75(2), 182–190 (2000)
5. McGarry, C.: Here's everything Siri can do on Apple TV. http://www.macworld.com/article/3002099/home-players/heres-everything-siri-can-do-on-apple-tv.html
6. Bernstein, M.S., Teevan, J., Dumais, S., Liebling, D., Horvitz, E.: Direct answers for search queries in the long tail. In: CHI 2012, pp. 237–246 (2012)
7. Lasecki, W.S., Miller, C.D., Sadilek, A., AbuMoussa, A., Kushalnagar, R., Bigham, J.: Real-time captioning by groups of non-experts. In: UIST 2012, pp. 23–34 (2012)

8. Lasecki, W.S., Murray, K., White, S., Miller, R.C., Bigham, J.P.: Real-time crowd control of existing interfaces. In: UIST 2011, pp. 23–32 (2011)
9. Lasecki, W.S., Song, Y.C., Kautz, H., Bigham, J.P.: Real-time crowd labeling for deployable activity recognition. In: CSCW 2013, pp. 1203–1212 (2013)
10. Bernstein, M.S., Little, G., Miller, R.C., Hartmann, B., Ackerman, M.S., Karger, D.R., Crowell, D., Panovich, K.: Soylent: a word processor with a crowd inside. In: UIST 2010, pp. 313–322 (2010)
11. Zhang, H., Law, E., Miller, R.C., Gajos, K.Z., Parkes, D.C., Horvitz, E.: Human computation tasks with global constraints. In: CHI 2012, pp. 217–226 (2012)
12. Lasecki, W.S., Wesley, R., Nichols, J., Kulkarni, A., Allen, J.F., Bigham, J.P.: Chorus: a crowd-powered conversational assistant. Proceedings of the 26th Annual ACM Symposium on User Interface Software And Technology (UIST 2013), pp. 151–162. ACM, New York (2013)
13. Kunz, J.R.M.: The American Medical Association Family Medical Guide. Random House, New York (1982)

Informatics Track

Reducing Execution Time for Real-Time Motor Imagery Based BCI Systems

Sahar Selim[1(✉)], Manal Tantawi[2], Howida Shedeed[2], and Amr Badr[3]

[1] Computer Science Department,
Modern Academy for CS and MT, Maadi, Cairo, Egypt
s.selim@grad.fci-cu.edu.eg
[2] Faculty of Computer and Information Science,
SC Department, Ain Shams University, Cairo, Egypt
[3] Faculty of Computers and Information, CS Department,
Cairo University, Cairo, Egypt

Abstract. Brain Computer Interface (BCI) systems based on electroencephalography (EEG) has introduced a new communication method for people with severe motor disabilities. One of the main challenges of Motor Imagery (MI) is to develop a real-time BCI system. Using complex classification techniques to enhance the accuracy of the system may cause a remarkable delay of real-time systems. This paper aims to achieve high accuracy with low computational cost. Two public datasets (BCIC III IVa and BCIC IV IIa) were used in this study; to check the robustness of the proposed approach. Dimension reduction of input signal has been done by channel selection and extracting features using Root Mean Square (RMS). The extracted features have been examined with four different classifiers. Experimental results showed that using Least Squares classifier gives best results, compared to other classifiers, with minimum computational time.

Keywords: Brain Computer Interface · EEG signals · Channel Selection · Motor Imagery · Dimension Reduction

1 Introduction

For millions of people worldwide who suffer from severe physical disabilities as a result of accidents or diseases such as amyotrophic lateral sclerosis (ALS) and locked-in syndrome, and certain types of cerebral palsy, their ability to interact with environment is limited or even impossible. A BCI offers an alternative to natural communication and control for those people. Numerous BCI systems using EEG signals have been presented. BCI based on Motor Imagery is one of the most promising [1] particularly for paralyzed patients. In BCI systems, extracted features from raw EEG signal are often of high dimensionality which led to increase in time and space requirements of processed data [2]. This probably had bad impact on real-time systems. One way to solve this problem is to reduce dimensionality of input data by channel selection and extraction of most discriminative features. Researchers have examined different methods as common Spatial Pattern (CSP) for feature extraction. Zhang et al.

© Springer International Publishing AG 2017
A.E. Hassanien et al. (eds.), *Proceedings of the International Conference on Advanced Intelligent Systems and Informatics 2016*, Advances in Intelligent Systems and Computing 533, DOI 10.1007/978-3-319-48308-5_53

[4] selected the significant filter bands using a Sparse Filter Band Common Spatial Pattern (SFBCSP) algorithm. Shedeed et al. [3] examined the effect of normalizing data on extracted features using Wavelet Transform.

Root Mean Square (RMS) time-domain feature was used by Hamedi et al. [5] and Hindarto et al. [6] for dimension reduction. RMS feature was able to achieve discriminative information with low computational cost. For channel selection Kee et al. [7] used genetic algorithms for channel selection method for both P300 and motor imagery data. Siuly et al. [8] suggested a clustering technique for feature extraction. Six Statistical features (Minimum, Maximum, Mean, Median, Mode, and Variance) out of nine features were extracted from each sub-cluster data. The obtained features were fed into the Least Square Support Vector Machine (LS-SVM) for classification. In this paper, the issue of minimizing computational time has been addressed. The rest of the paper is organized as follows: Sect. 2 introduces the BCI system architecture, MI EEG datasets, feature extraction and classification techniques utilized. Section 3 presents the description of the proposed approach. The experimental results are provided in Sect. 4. Finally Sect. 5 draws the conclusions of the study.

2 Materials and Methods

2.1 BCI System Architecture

As shown in Fig. 1, a BCI system consists of signal acquisition, preprocessing, feature extraction and selection, classification (detection), and application interface. The brain signal is recorded by the signal acquisition component. The preprocessing component refines the recorded signal to increase the signal-to-noise ratio by using artifact reduction methods. Thus, the refined signal is input to feature extraction algorithm. The extracted significant features would be classified using suitable classifier [9].

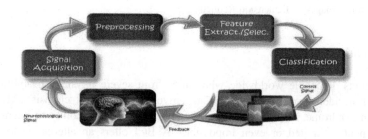

Fig. 1. Architecture of BCI systems

2.2 Data Acquisition

The proposed approach was evaluated using data of 14 subjects from two widely used datasets of BCI competitions. The two datasets are described below.

Data set IVa, BCI Competition III. BCI competition III data set IVa [10], contains EEG signals recorded from 5 subjects, performing imagination of right hand and foot. The EEG signals were recorded from 118 electrodes (as shown in Fig. 2) with sampling rate 100 Hz. Each subject has a training set and a testing set. The size of these sets is variant from one subject to another. The 280 trials are given as follows: the training set of subjects are 168, 224, 84, 56 and 28 'aa', 'al', 'av', 'aw' and 'ay' respectively. The rest of the 280 trials formed the testing set.

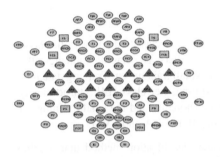

Fig. 2. Electrode montage corresponding to the international 10–20 system for BCIC III IVa dataset

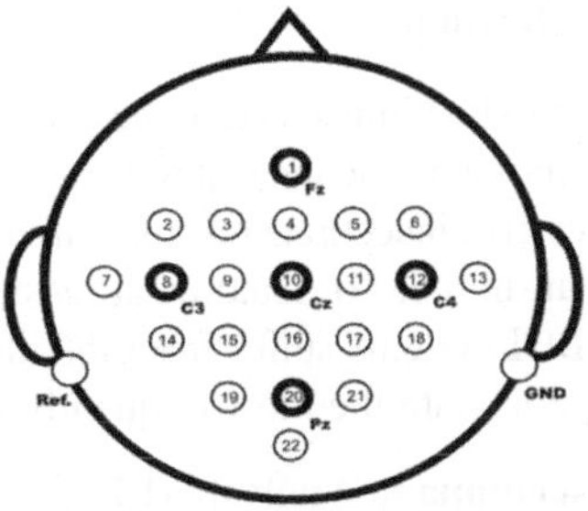

Fig. 3. Electrode montage corresponding to the international 10–20 system for BCIC IV IIa dataset

Data set IIa, BCI Competition IV. BCI competition IV data set IIa [11] recorded from 9 subjects performing imagination of left hand, right hand, foot and tongue. This dataset was recorded from 22 electrodes (as shown in Fig. 3) with sampling rate 250 Hz. In this study, we used MI EEG signals of left and right hand. Each subject has a training and a testing set containing 72 trials for each class.

2.3   Feature Extraction

From pattern recognition point of view [12], one of the tackling issues is how to represent the huge amount of the recorded EEG signals for classification. Effective features must be extracted from raw EEG signals. Discriminative and informative features may achieve better classification of different MI classes. These discriminative features should be computed without increasing in computational cost of BCI system [5]. Therefore, dimension reduction algorithms must be used to reduce amount of data input to classifier. Many studies investigated numerous statistical feature extraction techniques on different bio-signals as in [13].

Root Mean Square (RMS). "is modeled as amplitude modulated Gaussian random process" (Eq. (1)). The effectiveness of RMS feature have been investigated on MI movement classification by calculating the average power of the signal to get the Root Mean Square (RMS).

$$RMS_t = \sqrt{\frac{1}{N} \sum_{i=1}^{N} x_i^2} \tag{1}$$

Where t is the current Epoch, N is the length of the Epochs, x_i is the current point of signal and i is the index of the current point.

2.4   Classification

After feature extraction/selection the extracted features are classified into various classes of the corresponding mental states, using various classifiers. The classification algorithm would rather give inaccurate results with relatively small training set. Training data must be more than number of features at least 5 to 10 times per class [14]. Unfortunately, almost all BCI systems suffer from the high dimensionality and the small training set. Linear classifiers are the most frequently used in MI as LDA and SVM.

Linear Discriminant Analysis (LDA).   A linear classification method that requires fewer examples in order to obtain a reliable classifier output with very low computational requirements as compared to other classifiers. LD achieved good results in classifying different combinations of mental tasks. This classifier achieved remarkable success for motor imagery BCI systems. Meanwhile, LDA fails with data of non-Gaussian distributions with complex structure [15].

Support Vector Machine (SVM).   Many BCI applications uses the SVM linear classifier. SVM catches a hyper plane to separate the data sets. It broadens the gap between datasets to be able to classify them appropriately. The distance between the hyper plane and the nearest points from each class is maximized by the hyper plane, that are called as support vectors. This method increase the performance and reduce the learning model complexity. Thus, it provides notable generalization [15].

Least Square-Support Vector Machine (LS-SVM).   This process is a robust classification technique that proved success in BCI applications. It can converge to a global optimum, not to a local optimum. LS-SVM is faster than other machine learning techniques as it needs fewer random parameters. In addition to, the generalization provided by support vectors [16].

Least Square (LS).   The Least Squares Method determines what "best fit" to the data by using some calculus and linear algebra. It can accomplish great generalizations. The best fit could be found by *any* finite linear combinations of specified functions as an alternative of finding the best fit line [17].

3   Proposed Methodology

The proposed approach (shown in Fig. 4) is evaluated on two MI EEG datasets; dataset IVa of BCI Competition III and dataset IIa of BCI Competition IV. For both data sets, the EEG signals are further band-pass filtered between 10 Hz and 30 Hz using a 5th order Butterworth filter and zero centered to get mu (μ) and beta (β) frequencies needed for motor imagery. The filtered signal is segmented to extract epochs between 2.5 s to 5.5 s for dataset BCIC III IVa and between 3.5 to 5.5 for dataset BCIC IV IIa. Root Mean Square (RMS) algorithm is adopted to extract representative features from the original signals. Extracted features are used to compare between 4 linear classifiers LS-SVM, SVM, LS and LDA to find out best linear classifier that can accomplish best accuracy with minimum computational cost.

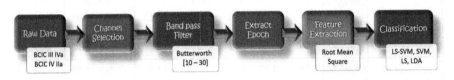

Fig. 4. Proposed BCI scheme

4   Experimental Results and Discussion

This section, discusses the results obtained from the experiments. Classification accuracy and computational time have been considered as the main metrics for the purpose of evaluation. This study aimed to reduce the execution time which is one of the challenges for real-time systems.

4.1   Channels Selection

Some researchers used the 8 centro-parietal channels (narrowly related to the mental tasks) "C3, Cz, C4, CP1, CP2, P3, Pz and P4". Wang et al. [18] selected 18 electrodes around the sensorimotor cortex including the channels "C5, C3, C1, C2, C4, C6, CP5, CP3, CP1, CP2, CP4, CP6, P5, P3, P1, P2, P4 and P6" based on the international 10-20 system. Their experimental results showed that the suggested electrodes are the best channels for getting the MI data. For BCIC III IVa, four combinations of channels have been tried: (1) Three channels (C3, Cz, C4), (2) Eight centro-parietal channels (C3, Cz, C4, CP1, CP2, P3, Pz and P4), (3) 18 channels taken from sensory motor area (C5, C3, C1, C2, C4, C6, CP5, CP3, CP1, CP2, CP4, CP6, P5, P3, P1, P2, P4 and P6), and (4) 25 channels [51–75].

Table 1 shows that using only 18 channels suggested by Wang et al. [18], gave almost same results with less execution time as in SVM. LS & LDA gave better accuracies with using the 18 channels (**78.77 % with less computational time** = 1.5 s instead of 1.9833 s of using the 25 channels).

For BCIC IV IIa most researchers used the 22 channels. This study examined channels closest to the 18 optimal channels of the Motor imagery area suggested by Wang et al. [18] for the dataset IVa of BCIC III. Out of the 22 channels of BCIC IV IIa, empirically 13 channels have been selected (C1, C2, C3, C4, Cz, CP1, CP2, CP3, CP4, CPz, P1, P2, Pz) located at (8, 9, 10, 11, 12, 14, 15, 16, 17, 18, 19, 20, 21) in Fig. 3.

Three combinations of channels have been tried for BCIC IV IIa:

Table 1. Results of examining 3, 8, 18, 25 channels for BCIC III IVa

Channels	Classifier	aa	al	av	aw	ay	Accuracy ± SD	Ex Time
C3 Cz C4	LS-SVM	62.50	80.36	53.57	82.59	68.65	69.53 ± 12.17	0.45
	SVM	61.61	78.57	56.63	82.14	82.54	72.30 ± 12.26	0.51
	LS	62.50	80.36	55.61	82.59	80.56	72.32 ±12.38	0.43
	LDA	62.50	80.36	55.61	82.59	80.56	72.32 ± 12.38	0.49
C3 Cz C4 CP1 CP2 P3 Pz P4	LS-SVM	63.39	87.50	58.16	87.05	55.95	70.41 ± 15.63	0.81
	SVM	61.61	87.50	63.78	85.27	72.22	74.07 ± 11.94	1.02
	LS	59.82	92.86	58.67	87.05	72.22	74.13 ± 15.53	0.79
	LDA	59.82	92.86	58.67	87.05	72.22	74.13 ± 15.53	0.86
C5 C3 C1 C2 C4 C6 CP5 CP3 CP1 CP2 CP4 CP6 P5 P3 P1 P2 P4 P6	LS-SVM	67.86	91.07	62.76	83.48	83.33	77.70 ± 11.88	1.52
	SVM	66.96	92.86	61.22	80.80	86.11	77.59 ± 13.20	1.73
	LS	69.64	89.29	59.18	88.84	86.90	**78.77** ± 13.65	1.50
	LDA	69.64	89.29	59.18	88.84	86.90	**78.77** ± 13.65	1.57
Channels [51 75]	LS-SVM	73.21	98.21	65.82	75.00	66.27	75.70 ± 13.23	2.00
	SVM	**74.11**	96.43	60.71	71.88	84.92	77.61 ± 13.59	2.30
	LS	73.21	98.21	56.12	82.14	72.62	76.46 ± 15.37	1.98
	LDA	73.21	98.21	56.12	82.14	72.62	76.46 ± 15.37	2.05

1. Three channels (C3, Cz, C4).
2. 13 channels taken from sensory motor area (C1, C2, C3, C4, Cz, CP1, CP2, CP3, CP4, CPz, P1, P2, Pz) closest to the 18 electrodes suggested by [18] for the BCIC III IVa.
3. 22 channels (Fz, FC3, FC1, FCz, FC2, FC4, C5, C3, C1, Cz, C2, C4, C6, CP3, CP1, CPz, CP2, CP4, P1, Pz, P2, POz).

As shown in Table 2 of BCIC IV dataset IIa, some classifiers as SVM has provided same accuracy when using either 13 or 22 channels. Results show that working with only 13 channels reduces execution time approximately ∼4 s which would be promising for real-time systems. LS & LDA gave almost same accuracies with using the 13 channels than using 22 channels (74.97% with less computation time = 9.33946 s instead of 13.38063 s obtained when using 22 channels). Results showed that dimension reduction and time reduction have been achieved with same results as when using much more channels.

4.2 Dimension Reduction

As mentioned in Sect. 1, Siuly et al. [8], used Statistical Features extraction techniques. They selected six features which gave best results out of nine. The six features (min, max, mean, median, mode, variance) were combined then they applied LS-SVM for classification. This study compared between using the six combined features and using RMS feature only.

Table 2. Results of examining 3, 13, 22 channels for BCIC IV IIa

Channels	Classifier	1	2	3	4	5	6	7	8	9	Accuracy ± SD	Ex Time
C3 Cz C4	LS-SVM	77.08	45.77	86.01	68.75	55.56	63.64	56.94	87.50	90.97	70.25 ± 16.05	3.12
	SVM	77.08	49.30	86.71	65.97	53.47	65.03	54.17	85.42	90.97	69.79 ± 15.81	3.53
	LS	76.39	52.11	86.01	66.67	53.47	65.03	55.56	84.72	90.97	70.10 ± 14.96	3.09
	LDA	76.39	52.11	86.01	66.67	53.47	65.03	55.56	84.72	90.97	70.10 ± 14.96	3.19
8 9 10 11 12 14 15 16 17 18 19 20 21	LS-SVM	93.06	57.75	86.71	68.06	47.22	66.43	74.31	97.92	70.83	73.59 ± 16.50	9.37
	SVM	92.36	58.45	86.71	70.14	56.94	60.14	68.06	98.61	82.64	74.89 ± 15.60	10.2
	LS	95.14	56.34	83.22	70.14	59.72	60.14	71.53	96.53	81.94	74.97 ± 15.10	9.34
	LDA	95.14	56.34	83.22	70.14	59.72	60.14	71.53	96.53	81.94	74.97 ± 15.10	9.45
Channels [1 22]	LS-SVM	89.51	59.86	94.41	75.00	49.30	66.43	63.89	94.44	84.72	75.28 ± 16.38	13.41
	SVM	91.61	56.34	93.71	70.83	53.52	63.64	61.81	97.22	85.42	74.90 ± 17.17	14.53
	LS	90.91	59.86	91.61	76.39	53.52	64.34	66.67	95.14	84.72	75.91 ± 15.40	13.38
	LDA	90.91	59.86	91.61	76.39	53.52	64.34	66.67	95.14	84.72	75.91 ± 15.40	13.5

BCIC III IVa. Only 18 channels have been used instead of 118 channels (C5, C3, C1, C2, C4, C6, CP5, CP3, CP1, CP2, CP4, CP6, P5, P3, P1, P2, P4 and P6). Each signal is input from 18 selected channels. Using RMS is able to achieve dimension reduction of input signal by representing the signal input from 18 channels by only one feature (RMS feature). Thus, having comparable results and less computation time as shown in Table 3. Results in Figs. 5 and 6 shows that using RMS feature outperforms the 6-features in both classification accuracies' of all classifiers and less execution time.

Table 3. Classification accuracy of using different feature extraction techniques for BCIC III IVa

Features	Classifier	aa	al	av	aw	ay	Accuracy	Ex Time
Mean, Median, Variance, Min, Max, Mode	LS-SVM	59.82	89.29	59.18	74.11	48.41	66.16	5.00
	SVM	59.82	89.29	59.18	74.11	48.41	66.16	5.68
	LS	57.57	85.36	56.72	67.32	62.59	65.91	4.99
	LDA	63.39	82.14	57.14	69.64	69.84	68.43	5.07
RMS	LS-SVM	67.86	91.07	62.76	83.48	83.33	77.70	1.52
	SVM	66.96	92.86	61.22	80.80	86.11	77.59	1.73
	LS	69.64	89.29	59.18	88.84	86.90	**78.77**	1.50
	LDA	69.64	89.29	59.18	88.84	86.90	**78.77**	1.57

BCIC III IVa. Thirteen channels have been selected instead of using 22 channels (C1, C2, C3, C4, Cz, CP1, CP2, CP3, CP4, CPz, P1, P2, Pz) denoted by (8, 9, 10, 11, 12, 14, 15, 16, 17, 18, 19, 20, 21). Each signal is input from 13 channels. Using RMS reduce dimension of 13 channels to one feature only. The results in Table 4 show that using only RMS feature outperforms the 6-features in both accuracies and execution

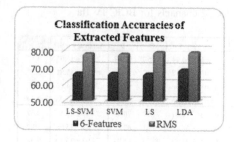

Fig. 5. BCIC III IVa classification accuracies of extracted features

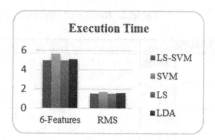

Fig. 6. BCIC III IVa execution time

Table 4. Classification accuracy of using different feature extraction techniques for BCIC IV IIa

Features	Classifier	1	2	3	4	5	6	7	8	9	Accuracy	Ex Time
Mean, Mode,	LS-SVM	78.47	54.93	79.02	69.44	53.47	62.24	57.64	92.36	78.47	69.56	19.33
Variance,	SVM	81.94	54.93	81.12	62.50	56.94	61.54	52.78	92.36	73.61	68.64	22.5
Min, Median,	LS	72.00	55.59	79.60	69.20	55.39	67.20	53.00	93.40	74.86	68.92	19.3
Max	LDA	77.08	55.63	75.52	66.67	56.94	67.13	58.33	93.75	75.69	69.64	19.43
	LS-SVM	93.06	57.75	86.71	68.06	47.22	66.43	74.31	97.92	70.83	73.59	9.367
RMS	SVM	92.36	58.45	86.71	70.14	56.94	60.14	68.06	98.61	82.64	74.89	10.20
	LS	95.14	56.34	83.22	70.14	59.72	60.14	71.53	96.53	81.94	74.97	9.34
	LDA	95.14	56.34	83.22	70.14	59.72	60.14	71.53	96.53	81.94	74.97	9.45

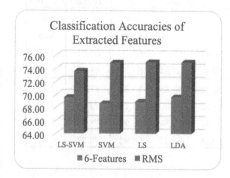

Fig. 7. BCIC IV IIa classification accuracies of extracted features

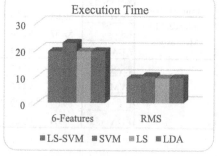

Fig. 8. BCIC IV IIa execution time

time. Figures 7 and 8 shows that execution time and accuracies of RMS is more promising than the 6-features.

All experiments have been performed using the MATLAB software package version 8.5 (R2015a) and run on a 2.60 GHz Intel(R) Core(TM)i7 CPU processor machine with 16.0 GB of RAM. The operating system on the machine was Microsoft Windows 10.

4.3 Discussion

Many complex non-linear algorithms have been proposed to deliver better performance. However, these methods may need much more computation time compared with conventional classifiers. The goal of this paper is to achieve high accuracy with low computational cost, which is a critical issue for real-time system.

One of the main challenges in motor Imagery BCI is the high dimensionality of input signal. Channel Selection and Dimension Reduction have been implemented to tackle this issue. The proposed approach was applied on two datasets to check its robustness.

For channel selection:

1. **BCIC III IVa.** Data of this dataset was recorded from 118 channels. Results proved that using the 18 channels suggested by Wang [18] can lead to almost same results as using more channels, but with less execution time.
2. **BCIC IV 2a.** Empirically, thirteen channels have been selected closest to the 18 optimal channels of the Motor imagery area suggested by Wang [18] for BCIC III IVa. Using the 13 selected channels achieved almost same results with less execution time which will be promising for real-time systems.

For dimension reduction, the six combined features (min, max, mean, median, mode and variance) proposed by Siuly et al. [8] have been compared with RMS feature as shown in Tables 3 and 4. The effectiveness of the extracted features has been evaluated by four linear classifiers LS-SVM, SVM, LS and LDA. Experimental results showed that RMS provided more distinctive information for MI task than the 6-combined features for both classification accuracies of all classifiers and minimal execution time. Least Square Classifier gave comparable results with other classifiers. Yet, it outperforms them as it gives less computational time.

5 Conclusion

The approach proposed in this study, implemented channel selection by minimizing number of input channels and dimension reduction by extracting effective features of input signal and thus reduces computational time which is important for real-time systems. Two datasets; BCIC III IVa and BCIC IV IIa have been used to evaluate the proposed approach. For BCIC IV IIa, most researchers used all the 22 channels of this dataset. To our knowledge, this study is the first one which has empirically selected the most effective 13 channels for this dataset.

For dimension reduction, the six combined features (min, max, mean, median, mode and variance) have been compared with RMS feature. Using RMS feature not only gave better classification accuracies for all classifiers than using the six combined features yet it's simpler with less computation time and reduced dimension. Among the four examined classifiers (LS-SVM, SVM, LS and LDA), LS classifier was the best classifier with minimum execution time, which would be useful for real-time systems. In BCI research, it's recommended to have a reasonable trade-off between accuracy and speed. The proposed approach showed that it's reliable with minimal computation time. Therefore, this study recommends using channel selection with RMS and LS Classifier for real-time systems. All experiments were implemented offline. In the future, this approach will be applied on a real-time BCI system and the multi-class classification problems will be addressed.

References

1. Pfurtscheller, G., Neuper, C.: Motor imagery and direct brain computer communication. Proc. IEEE Neural Eng. (2001)
2. Hassanien, A., Azar, A.: Brain-Computer Interfaces: Current Trends and Applications. ISRL, vol. 74. Springer, Heidelberg (2015). ISBN: 978-3-319-10977-0
3. Shedeed, H., Issa, M.: Brain-EEG signal classification based on data normalization for controlling a robotic arm. Int. J. Tomogr. Simul. 29(1), 72–85 (2016)
4. Zhanga, Y., Zhoub, G., Jina, J., Wanga, X., Cichock, A.: Optimizing spatial patterns with sparse filter bands for motor-imagery based brain-computer interface. J. Neurosci. (2015)
5. Hamedi, M., Salleh, S.-H., Noor, A.M. Mohammad-Rezazadeh, I.: Neural network-based three-class motor imagery classification using time-domain features for BCI applications. In: Proceedings of 2014 IEEE Region 10 Symposium, pp. 208–211 (2014)
6. Hindarto, M.H., Purnomo, M.H.: EEG signal identification based on root mean square and average power spectrum by using BackPropagation. J. Theor. Appl. Inf. Technol. 66(3), 782–787 (2014)
7. Kee, C.-Y., Ponnambalam, S.G., Loo, C.-K.: Multi-objective genetic algorithm as channel selection method for P300 and motor imagery data set. Neurocomputing (2015)
8. Siuly, S., Li, Y., Wen, P.P.: Clustering technique-based least square support vector machine for EEG signal classification. Comput. Methods Programs Biomed. 104(3), 358–372 (2011)
9. Pfurtscheller, G., Neuper, C., Birbaumer, N.: Motor cortex in voluntary movements: a distributed system for distributed functions. In: Vaadia, E., Riehle, A. (eds.) Human Brain-Computer Interface. Methods and New Frontiers in Neuroscience. CRC Press, Boca Raton (2005)
10. Müller, K.-R., Blankertz, B.: BCI Competition III Data set IVa. Berlin BCI group, University Medicine Berlin, Department of Neurology, Neurophysics Group (2005)
11. Brunner, C., Leeb, R., Müller-Putz, G., Schlögl, A., Pfurtscheller, G.: BCI Competition 2008 – Graz data set A. Graz University of Technology (2008)
12. Duda, R.O., Hart, P.E., Stork, D.G.: Pattern Classification, 2nd edn. John Wiley & Sons, New York (2001)
13. Khorshidtalab, A., Salami, M.J.E., Hamedi, M.: Evaluation of time domain features for motor imagery movements using FCM and SVM. In: Proceedings of Computer Science and Software Engineering, International Joint Conference on IEEE, pp. 17–22 (2012)

14. Al-ani, T., Trad, D.: Signal processing and classification approaches for brain-computer interface. In: Somerset, V.S. (ed.) Intelligent and Biosensors (2010)
15. Bashashati, A., Fatourechi, M., Ward, R.K., Birch, G.E.: A survey of signal processing algorithms in brain–computer interfaces based on electrical brain signals. J. Neural Eng. **4** (2), R32–R57 (2007)
16. Esen, H., Ozgen, F., Esen, M., Sengur, A.: Modelling of a new solar air heater through least-squares support vector machines. Expert Syst. Appl. **36**, 10673–10682 (2009)
17. Miller, S.J.: The Method of Least Squares. Mathematics Department, Brown University (2006)
18. Wang, S., James, C.J.: Extracting rhythmic brain activity for braincomputer interfacing through constrained independent component analysis. Comput. Intell. Neurosci. (2007). https://www.hindawi.com/journals/cin/2007/041468/abs/

CNN for Handwritten Arabic Digits Recognition Based on LeNet-5

Ahmed El-Sawy[1], Hazem EL-Bakry[2], and Mohamed Loey[1(✉)]

[1] Faculty of Computer and Informatics, Computer Science Department,
Benha University, Benha, Egypt
{ahmed.el_sawy,mohamed.loey}@fci.bu.edu.eg
[2] Faculty of Computer and Information Sciences, Information System Department,
Mansoura University, Mansoura, Egypt
helbakry5@yahoo.com

Abstract. In recent years, handwritten digits recognition has been an important area due to its applications in several fields. This work is focusing on the recognition part of handwritten Arabic digits recognition that face several challenges, including the unlimited variation in human handwriting and the large public databases. The paper provided a deep learning technique that can be effectively apply to recognizing Arabic handwritten digits. LeNet-5, a Convolutional Neural Network (CNN) trained and tested MADBase database (Arabic handwritten digits images) that contain 60000 training and 10000 testing images. A comparison is held amongst the results, and it is shown by the end that the use of CNN was leaded to significant improvements across different machine-learning classification algorithms.

1 Introduction

Recognition is an area that covers various fields such as, face recognition, image recognition, finger print recognition, character recognition, numerals recognition, etc. [1]. Handwritten Digit Recognition system (HDR) is an intelligent system able to recognize handwritten digits as human see. Handwritten digit recognition is an important component in many applications; check verification, office automation, business, postal address reading and printed postal codes and data entry applications are few examples [2]. The recognition of handwritten digits is a more difficult task due to the different handwriting styles of the writers.

Over the last few years, deep learning [3] are the most researched area in machine learning that model hierarchical abstractions in input data with the help of multiple layers. Deep learning techniques have achieved state-of-the-art performance in computer vision [4,5], big data [6,7], automatic speech recognition [8,9] and in natural language processing [10]. Although, increase in computing power has contributed significantly to the development of deep learning techniques, deep learning techniques attempts to make better representations and create models to learn these representations from large-scale data.

© Springer International Publishing AG 2017
A.E. Hassanien et al. (eds.), *Proceedings of the International Conference on Advanced Intelligent Systems and Informatics 2016*, Advances in Intelligent Systems and Computing 533, DOI 10.1007/978-3-319-48308-5_54

Deep learning have many architectures such as Convolution Neural Networks (CNN). CNN is a multi-layer feed-forward neural network that extract properties from the input data. CNN trained with neural network back-propagation algorithm. CNN have the ability to learn complex, high-dimensional, non-linear mappings from very large number of data (images). Moreover, CNN shows an excellent recognition rates for characters and digits recognition [11]. The advantage of CNN is that it automatically extracts the salient features which are invariant and a certain degree to shift and shape distortions of the input characters [12].

Feature extraction is an important key factor to make a successful recognition system. The recognition system requires that features distinguished characteristics among different labels while retaining invariant characteristics within the same labels. Traditional feature extraction that designed by hand is really a boring and time consuming task and cannot process raw images, but the automatic extraction techniques can restore and reconstruct features directly from raw images. Based on CNN [11] extract feature from trainable dataset can done automatically. So, the propose of the paper is using CNN to create deep learning recognition system for Arabic handwritten digits recognition.

The rest of the paper is organized as follows: Sect. 2 gives a review on some of the related work done in the area. Section 3 describes the motivation and proposed approach, Sect. 4 gives an overview of the dataset and results, and we list our conclusions and future work in Sect. 5.

2   Related Work

Various methods have been proposed and high recognition rates are reported for the recognition of English handwritten digits [13–15]. Niu and Suen [13] proposed recognize handwritten digits using Convolutional Neural Network (CNN) and Support Vector Machine (SVM). There Experiments have been conducted on MNIST digit database. They achieve recognition rate of 94.40 % with 5.60 % with rejection. Tissera and McDonnell [14] introduced a supervised auto-encoder architecture based on extreme machine learning to classify Latin handwritten digits based on MNIST dataset. The proposed technique can correctly classify up to 99.19 %. Ali and Ghani introduced Discrete Cosine Transform based on Hidden Markov models (HMM) to classify handwritten digits. They used MNIST as training and testing datasets. HMM have been applied as classifier to classify handwritten digits dataset. The algorithm provides promising recognition results on average 97.2 %.

In recent years many researchers addressed the recognition of text including Arabic. In 2011, Melhaoui et al. [16] proposed an improved method for recognizing Arabic digits based on Loci characteristic. Their work is based on handwritten and printed numeral recognition. The recognition is carried out with multi-layer perceptron technique and K-nearest neighbour. They trained there algorithm on dataset contain 600 Arabic digits with 200 testing images and 400 training images. They were able to achieve 99 % recognition rate on small database.

In 2008, Mahmoud [17] proposed a technique for the automatic recognition of Arabic handwritten digits using Gabor-based features and Support Vector Machines (SVMs). They used a medium database have 21120 samples written by 44 writers. The dataset contain 30 % for testing and the remaining 70 % of the data is used for training. They achieved average recognition rates are 99.85 % and 97.94 % using 3 scales & 5 orientations and using 4 scales & 6 orientations, respectively.

In 2014, Takruri et al. [18] presented three level classifier based on Support Vector Machine, Fuzzy C Means and Unique Pixels for the classification of handwritten Arabic digits. They tested the new algorithm on a public dataset. The dataset contain 3510 images with 40 % are used for testing and 60 % of images are used for training. The overall testing accuracy reported is 88 %.

In 2013, Pandi Selvi and Meyyappan [1] presented a method to recognize Arabic digits using back propagation neural network. The final result shows that the proposed method provides an recognition accuracy of more than 96 % for a small sample handwritten database.

In 2014, Majdi Salameh [19] proposed two methods about enhancing recognition rate for typewritten Arabic digits. First method that calculates number of ends of the given shape and conjunction nodes. The second method is fuzzy logic for pattern recognition that studies each shape from the shape, and then classifies it into the numbers categories. Their proposed techniques was implemented and tested on some fonts. The experimental results made high recognition rate over 95 %.

In 2014, AlKhateeb et al. [20] presented a system to classify Arabic handwritten digit recognition using Dynamic Bayesian Network. They used discrete cosine transform coefficients based features for classification. Their system trained and tested on Arabic digits database (ADBase) [21] which contains 70,000 Arabic digits. They reported average recognition accuracy of 85.26 % on 10,000 testing samples.

3 Proposed Approach

3.1 Motivation

Arabic digits recognition and different handwriting styles as well, making it important to find and work on a new and advanced solution for handwriting recognition. A deep learning systems needs a huge number of data to be able to make good decisions. In [1,16–18] they applied the algorithms on a small handwritten images, the problem is the small database of training and testing images. In [21] proposed a large Arabic handwriting digits database called (MADBase) with training and testing images. The proposed database of images gives us with large different handwriting styles. So, the use of MADBase database and deep learning lead to the suggestion of our approach.

3.2 Suggested Approach

Convolutional neural networks (CNN) are a class of deep models that were inspired by information processing in the human brain. In the visual of the brain, each neuron has a receptive field capturing data from certain local neighborhood in visual space. They are specifically designed to recognize multi-dimensional data with a high degree of in-variance to shift scaling and distortion.

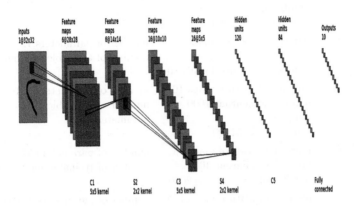

Fig. 1. Convolutional neural networks LeNet-5

CNN architecture is made of one input layer and multi-types of hidden layers and one output layer. The fist kind of hidden layers is responsible for convolution and the other one is responsible for local averaging, sub sampling and resolution reduction. The third hidden layers act as a traditional multi-layer perceptron classifier.

In this study LeNet-5 CCN architecture is used with an 8 layers including one input layer, one output layer, two convolutional layers and two sub-sampling for automatic feature extraction, two fully connected layers as multi-layer percep-tron hidden layers for nonlinear classification. The CNN architecture is shown in Fig. 1. The input image size is 32 × 32 with 1 channel (i.e. grayscale image).

The first convolutional layer C1 is a convolution layer with 6 feature maps and a 5 × 5 kernel for each feature map. There are inputs for each neuron of C1 planes, which is obtained from a 5 × 5 receptive field at the previous (input) layer. According to weights sharing strategy, all units in these feature maps use the same weights and bias to produce a linear location invariant filter to be applied to all regions of the input image. The sharing weights of this layer will be adapted during training procedure. This layer C1 has six different biases and six different 5 × 5 kernels including 156 trainable parameters, 4704 number of neurons and 122304 connections. The next layer is a sub sampling Layer S2 with six feature maps and a 2 × 2 kernel for each feature map.

In fact, after averaging the input samples in the receptive field of an out-put pixel, the result is multiplied and added by two trainable coefficients which

Table 1. CNN layers description for our approach

Lenet5 layers	Description	
Layer 1 [Input]	number of feature maps:1 number of neurons:0 number of connections:0	number of parameters:0 number of trainable parameters:0
Layer 2 [C1]	number of feature maps: 6 number of neurons:4704 number of connections:122304	number of parameters:156 number of trainable parameters:156
Layer 3 [S2]	number of feature maps:6 number of neurons:1176 number of connections:5880	number of parameters:12 number of trainable parameters:12
Layer 4 [C3]	number of feature maps:16 number of neurons:1600 number of connections:151600	number of parameters:1516 number of trainable parameters:1516
Layer 5 [S4]	number of feature maps:16 number of neurons:400 number of connections:2000	number of parameters:32 number of trainable parameters:32
Layer 6 [C5]	number of feature maps:120 number of neurons:120 number of connections:48120	number of parameters:48120 number of trainable parameters:48120
Layer 7 [F6]	number of feature maps:10 number of neurons:10 number of connections:1210	number of parameters:1210 number of trainable parameters:1210

are assumed to be similar for the output pixels of a feature map, but different in different feature maps. This layer has 12 trainable parameters and 5880 connections.

The third layer C3 is a convolution layer with 16 feature maps and a 5×5 kernel for each feature map acts as previous convolutional layer. This layer has 16 feature maps and each neuron of each output feature map connects to some 5×5 pixels areas at the previous layer S2. The next layer S4 is a sub-sampling layer with 16 feature maps and a 2×2 kernel for each feature map. Layer S4 has 32 trainable parameters and 2000 connections. Layer C5 is a convolution layer with 120 feature maps and a 6×6 kernel for each feature map which has 48120 connections and trainable parameters. Layer F6 is the last fully connected layer which selected 84 neurons. The final layer is the output layer that has 10 neurons for 10 digit classes. The output of each CNN layers for Arabic digit 4 illustrated in Fig. 2. Each CNN layers have number of feature maps, neurons, connections, parameters, trainable parameters. All this parameters descripe in Table 1.

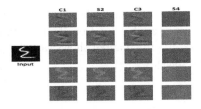

Fig. 2. Layers output for digit "4"

4 Experiment

4.1 Dataset

El-sherif and Abdleazeem released an Arabic handwritten digit database (ADBase) and modified version called (MADBase) [21]. The MADBase is a modified version of the ADBase benchmark that has the same format as MNIST benchmark [22]. ADBase and MADBase are composed of 70,000 digits written by 700 writers.

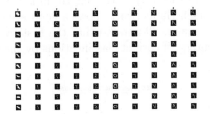

Fig. 3. Sample of MADBase benchmark training database

Each writer wrote each digit (from 0 −9) ten times. To ensure including different writing styles, the database was gathered from different institutions: Colleges of Engineering and Law, School of Medicine, the Open University (whose students span a wide range of ages), a high school, and a governmental institution. The databases is partitioned into two sets: a training set (60,000 digits to 6,000

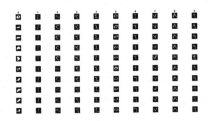

Fig. 4. Sample of MADBase benchmark testing database

images per class) and a test set (10,000 digits to 1,000 images per class). The ADBase and MADBase is available for free (http://datacenter.aucegypt.edu/shazeem/) for researchers. Figures 3 and 4 shows samples of training and testing images of MADBase database.

4.2   Results

In this section, the performance of CNN was investigated for training and recognizing Arabic characters. For the setting architecture, a convolutional layer is parameterized by the size and the number of the maps, kernel sizes, skipping factors. This section describes about our attempt to apply the CNN based Arabic digits classification. The experiments are conducted in MATLAB 2016a programming environment. LeNet-5 network is used for implementing CNN on Arabic digits on MADBase database. At first for evaluating the performance of CNN on Arabic digits, incremental training approach was used on the proposed approach. We started with 4 classes for training first and calculated the accuracy. Then number of classes is slowly increased. As shown in Fig. 5, the Root Mean Square Error (RMSE) of the proposed approach is .894 for training data and 1.105 fot testing data. The miss-classification rate is come down to 1 % for training and 12 % for testing shown in Fig. 6. Here the algorithm is trained for 30 iterations, but from iteration 12 itself the network shows a good accuracy.

In Fig. 7, confusion matrix illustrated that the first two diagonal cells show the number and percentage of correct classifications by the trained network. For

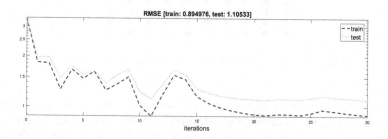

Fig. 5. Root Mean Square Error

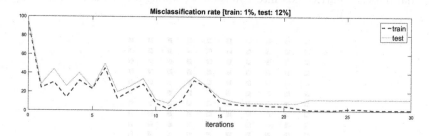

Fig. 6. Miss-classification rate

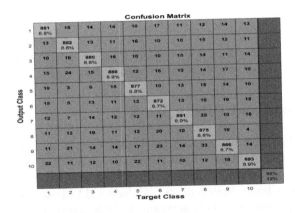

Fig. 7. Confusion matrix

Table 2. Comparison between proposed approach and other approach

Authors	Database	Training data / Testing Data	Misclassification error
Takruri et al. [18]	Public database	3510 images 60 % training digits 40 % testing digits	12 %
AlKhateeb et al. [20]	ADBase	60,000 training digits 10,000 testing digits	14.74 %
Majdi Salameh [21]	Fonts	1000 training digits 1000 testing digits	5 %
Melhaoui et al. [16]	Private database	600 images 400 training digits 200 testing digits	1 %
Pandi Selvi and Meyyappan [1]	Private database	Sample handwritten images are tested	4 %
Mahmoud [17]	Private database	21120 images 70 % training digits 30 % testing digits	0.15 % and 2.16 %
Our approach	MADBase	70000 images 60000 training digits 10000 testing digits	**1 % training 12 % testing**

example 881 of class (1) are correctly classified as class (1). This corresponds to 8.81 % of all 1000 test images of class (1). The column in the target class show the miss-classification of the class. Overall, 88 % of the predictions are correct and 12 % are wrong classifications.

Finally, in Table 2 shown the obtained results with CNN on MADBase database. It can be seen from Table 2 that the proposed approach have the large database and have the best miss-classification error. The results are better than the results reported in related work [1,16–18,20], although it is sometimes hard to compare, because previous work has not experimented with large database benchmark. The proposed method obtained 1 % miss-classification error on training data and 12 % miss-classification error on testing data.

5   Conclusion and Future Work

Handwritten Character Recognition for Arabic digits is an active research area which always needs an improvement in accuracy. This work is based on recognition of Arabic digits using convolution neural networks (CNN). The approach were tested on a large Arabic digits database (MADBase). As experimental results, the approach gives best accuracy in large database with 1 % training miss classification error rate and 12 % testing miss classification error rate. Our future work will be focusing on improving the performance of handwriting Arabic digits recognition using other improved deep learning techniques.

References

1. Selvi, P.P., Meyyappan, T.: Recognition of Arabic numerals with grouping and ungrouping using back propagation neural network. In: International Conference on Proceedings of Pattern Recognition, Informatics and Mobile Engineering (PRIME), pp. 322–327 (2013)
2. Mahmoud, S.: Recognition of writer-independent off-line handwritten Arabic (Indian) numerals using hidden Markov models. Sig. Process. 88(4), 844–857 (2008)
3. Lecun, Y., Bengio, Y., Hinton, G.: Deep learning. Nature 521(7553), 436–444 (2015)
4. Krizhevsky, A., Sutskever, I., Hinton, G.E.: ImageNet classification with deep convolutional neural networks. In: Pereira, F., Burges, C.J.C., Bottou, L., Weinberger, K.Q. (eds.) Advances in Neural Information Processing Systems, vol. 25, pp. 1097–1105. Curran Associates Inc., Red Hook (2012)
5. Afaq Ali Shah, S., Bennamoun, M., Boussaid, F.: Iterative deep learning for image set based face and object recognition. Neurocomputing 174, 866–874 (2016)
6. Zhang, Q., Yang, L.T., Chen, Z.: Deep computation model for unsupervised feature learning on big data. IEEE Trans. Serv. Comput. 9(1), 161–171 (2016)
7. Chen, X.W., Lin, X.: Big data deep learning: challenges and perspectives. IEEE Access 2, 514–525 (2014)
8. Cai, M., Liu, J.: Maxout neurons for deep convolutional and LSTM neural networks in speech recognition. Speech Commun. 77, 53–64 (2016)
9. Sainath, T.N., Kingsbury, B., Saon, G., Soltau, H., Mohamed, A.-R., Dahl, G., Ramabhadran, B.: Deep convolutional neural networks for large-scale speech tasks. Neural Netw. 64, 39–48 (2015)
10. Collobert, R., Weston, J., Bottou, O., Karlen, M., Kavukcuoglu, K., Kuksa, P.: Natural language processing (almost) from scratch. J. Mach. Learn. Res. 12, 2493–2537 (2011)

11. Maitra, D.S., Bhattacharya, U., Parui, S.K.: CNN based common approach to handwritten character recognition of multiple scripts. In: 2015 13th International Conference on Document Analysis and Recognition (ICDAR), pp. 1021–1025 (2015)

12. Yu, N., Jiao, P., Zheng, Y.: Handwritten digits recognition base on improved LeNet5. In: Proceedings of Control and Decision Conference (CCDC), 2015 27th Chinese, pp. 4871–4875 (2015)

13. Niu, X.-X., Suen, C.Y.: A novel hybrid CNNSVM classifier for recognizing handwritten digits. Pattern Recogn. **45**(4), 1318–1325 (2012)

14. Tissera, M.D., McDonnell, M.D.: Deep extreme learning machines: supervised autoencoding architecture for classification. Neurocomputing **174**, 42–49 (2016)

15. Ali, S.S., Ghani, M.U.: Handwritten digit recognition using DCT and HMMs. In: 2014 12th International Conference on Frontiers of Information Technology (FIT), pp. 303–306 (2014)

16. Melhaoui, O.E., Hitmy, M.E., Lekhal, F.: Arabic numerals recognition based on an improved version of the loci characteristic. Int. J. Comput. Appl. **24**(1), 36–41 (2011)

17. Mahmoud, S.A.: Arabic (Indian) handwritten digits recognition using Gabor-based features. In: International Conference on Proceedings of Innovations in Information Technology, IIT 2008, pp. 683–687 (2008)

18. Takruri, M., Al-Hmouz, R., Al-Hmouz, A.: A three-level classifier: fuzzy C means, support vector machine and unique pixels for Arabic handwritten digits. In: World Symposium on Proceedings of Computer Applications & Research (WSCAR), pp. 1–5 (2014)

19. Salameh, M.: Arabic digits recognition using statistical analysis for end/conjunction points and fuzzy logic for pattern recognition techniques. World Comput. Sci. Inf. Technol. J. **4**(4), 50–56 (2014)

20. Alkhateeb, J.H., Alseid, M.: DBN - based learning for Arabic handwritten digit recognition using DCT features. In: 2014 6th International Conference on Computer Science and Information Technology (CSIT), pp. 222–226 (2014)

21. Hafiz, A.M., Bhat, G.M.: Boosting OCR for some important mutations. In: Second International Conference on Advances in Computing and Communication Engineering (ICACCE), pp. 128–132 (2015)

22. Wu, H., Gu, X.: Towards dropout training for convolutional neural networks. Neural Netw. **71**, 1–10 (2015)

A Multilateral Agent-Based Service Level Agreement Negotiation Framework

Amira Abdelatey$^{(\boxtimes)}$, Mohamed Elkawkagy, Ashraf El-Sisi,
and Arabi Keshk

Computer Science Department, Faculty of Computers, Information,
Menofiya University, Shibin al Kawm, Egypt
{amira.mohamed,mohamed_shamseldeen,ashraf.elsisi,
arabi.keshk}@ci.menofia.edu.eg

Abstract. Negotiation gets more interesting With the increasing demand for discovering web services. Negotiation requires that the non-functional consumer requirements have to meet with providers. Conducting a negotiation between participants is the key issue for reaching an agreement between them. Service Level Agreements (SLA) plays an important role in service-based systems. Different researchers conduct different bilateral negotiation frameworks. Multilateral negotiation helps the consumer to get the best suitable provider not to get an agreement between agreed participants. This paper presents a multilateral SLA negotiation framework for non-functional requirements using three different functions with time-based strategy. That is for getting the best suitable provider for a consumer. Through the proposed framework, a model is defined to map attributes of participants to a main parameter used in decision-making model. A prototype of the proposed framework is implemented with conducting a multilateral negotiation scenario. The proposed framework reached the best-suitable provider, not an acceptable one.

Keywords: Negotiation · Web service · Service level agreement · Multi-agent · Quality of service

1 Introduction

Negotiation is an effective process especially for IT business [1]. Nowadays, most commercial transactions are applied through web services. A web service defines a unit of application that provides some business functionality or information to other application through the internet [2]. The business functionality of a web service is called functional requirement. In addition to a functional requirement, a web service has non-functional requirements. WS-Policy (WS-P) is capable of representing the non-functional properties of a web service [3]. Negotiation is a business interaction process between two (bilateral) or more (multilateral) participants to reach an agreement [4]. The participants communicate using a specific negotiation protocol, rules for exchanging messages until they reach a common agreement. Through the negotiation process, the Decision Making System (DMS) is used as the model controlling negotiation offers as described in details in the proposed framework. For automated

© Springer International Publishing AG 2017
A.E. Hassanien et al. (eds.), *Proceedings of the International Conference on Advanced Intelligent Systems and Informatics 2016*, Advances in Intelligent Systems and Computing 533, DOI 10.1007/978-3-319-48308-5_55

negotiation, multi-agents perform the negotiation process instead of the participants [5]. The process of negotiation comprises of the specifications of participants' attributes, exchanging proposals to conduct the actual negotiation process, The most usable strategy in a negotiation is the time-dependent strategy [6]. The different aspects of negotiation system are negotiation policy, negotiation protocol, and decision support system [7]. The set of rules that identify what actions will be taken when certain conditions happen is so-called policy. It is the way to describe all the rules in any business process. Different languages are used to describe the negotiation rules such as WS-Agreement [8]; WS- Negotiation [7]; WS-Policy [9]. In the proposed framework, we use WS-P for describing QoS requirements and constraints for all participants. For bargaining, different researchers use different negotiation protocols for negotiating about different terms and conditions between participants [10]. Due to the high performance of the Foundation for Intelligent Physical Agents (FIPA) protocol[1], we used it in this paper. In addition, it is used with an agent to dynamically controlling the negotiation process. Automated multilateral SLA negotiation framework for web services is proposed through this paper. In this framework, Multi-agents used to conduct the negotiation dynamically between each two participants. Different three time-based negotiation functions used as a negotiation tactic. In our framework, a novel relation is proposed to calculate the value of the conceding offers for negotiation.

The rest of the paper is organized as follows: Sect. 2 introduces the related work for web service negotiation. The proposed multilateral SLA negotiation agent-based framework is provided in Sect. 3. In Sect. 4, the practical scenario is conducted with the results. Conclusion and future work are included in Sect. 5.

2 Related Work

With the emergence of Service-Oriented Applications (SOA) technology, researchers focus on negotiation of web service participant requirements and exploit Negotiation Support System (NSS) systems [11–14]. Different negotiation specification languages have been proposed [7–9, 15]. Through our proposed framework, we extend WS-Policy [15] for specifying requirements and constraints of negotiable participants. Jennings et al. [4] defined the negotiation process from different prospects. They also discuss the main approaches used in negotiation such as heuristic search, game theory and argumentation based approaches with drawbacks of each approach. Conducting a negotiation system requires a decision-making model. Faratin et al. [16] introduced a formal model for negotiation with introducing decision functions such as exponential and polynomial for conducting automated negotiation process. Authors provided different negotiation strategies such as time-based strategy, resource-based strategy and imitative strategy. After that, Zulkernine et al. [17] proposed a middleware framework for web service SLA automated negotiation. This Middleware expresses business requirements, constraints, and attributes of different participants with WS-policy. They used time-based negotiation strategy for conducting the negotiation. This framework used an

[1] http://www.fipa.org/specs/fipa00029.

exponential function as a utility function because of its conceding pattern through their strategy. WS-policy used to specify constraints and requirements with the two participants for this automatic SLA negotiation framework [18]. A polynomial time based dependent function is used as a time-based dependent strategy for conducting the negotiation. Xiao et al. [19] propose a policy-based framework for supporting automated SLA bilateral negotiation for web services called "WebNeg". Authors apply polynomial decision time-based function as the utility function. The WebNeg used a genetic algorithm as a search-based approach for reaching acceptable solutions. Hashmi et al. [20] provide an automated negotiation among Web services. They use a genetic algorithm (GA) based approach for finding an acceptable solution. This approach depends on of GA operators over a time constraints. They negotiate over QoS attributes. GA is one of the methods used for automated negotiation.

As opposed to this, conducting the negotiation on multilateral SLA negotiation to select the best suitable participant from a list of suitable participants has not been considered so far. In addition, most researchers define the main parameter affecting the conceding curve of the negotiation offers as a static value. Through the proposed framework, we define this parameter dynamically with getting its value from desirability factor (DF) of reaching an agreement and weights of each attribute. Besides, we conduct a multilateral SLA automated negotiation.

3 Multilateral SLA Negotiation Framework

The proposed framework is extended from automated SLA negotiation framework [19]. The Negotiation Manager, which is responsible for getting the best suitable provider, is an added module within our proposed framework. The decision-making module is modified in our proposed framework. Automated SLA negotiation framework [19] conducts a bilateral SLA negotiation process between a consumer and a provider. Whilst, the proposed framework conducts a multilateral negotiation process between different participants to get the best-matched provider for a service. Multilateral SLA negotiation framework, shown in Fig. 1, takes as input the policy specification of the different participants. Policy specification of participants identifies the requirements and constraints of participants in the form of WS-P. Beside different attributes' ranges, a definition of the name, measuring unit of the attributes are provided in WS-P [18]. Through this paper, we will consider price, response time, and availability as a negotiable QoS attributes. Policy processor processes the policy specification and provides these specifications to factory agent module. Agent factory module based on FIPA Agent Communication Language (ACL) enables the creation of distributed-based runtime agent applications. Also, it connects agent pairs of participants to conduct bilateral negotiations as an independent process. Negotiation Manager is connected with factory agent module, which conducts bilateral negotiations of different participants, to receive the result of negotiation processes and select the best suitable provider. The provider that reaches an agreement faster than others is so-called "The best suitable provider". To conduct a negotiation process between agents, a DMS have to connect with the agent factory. The DMS contains the negotiation strategy, which is the backbone of the multilateral negotiation framework. Finally, the proposed

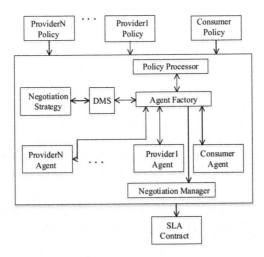

Fig. 1. Multilateral SLA negotiation framework

framework produces a SLA contract between the best suitable providers and the consumer. In any negotiation process, a negotiation tactic is the decision process that is based on a specific mathematical model to fit the values of attributes (price, availability, etc.) for participants [4]. Note that negotiation tactic varies depending on a negotiation problem. Three stages help for making the decision on behalf of the participants; (1) the cost-benefit model which is used to evaluate the cost-benefit of a given proposal. (2) the judging criteria that determine the decision accept, reject, or generate a new proposal. (3) The negotiation tactic which defines how to generate a new proposal. This paper considers a time-dependent model as a negotiation tactic. A utility function $v_j(x_j)$, defined in Faratin, etc. [16], used as cost-benefit model as well as computing the cost of each attribute. This utility function is represented as:

$$v_j(x_j) = \begin{cases} \frac{max_j - x_j}{max_j - min_j}, & vj(x_j) \, is \, decreasing \\ \frac{x_j - min_j}{max_j - min_j}, & vj(x_j) \, is \, increasing \end{cases}$$

Where, $x_j \in [min_j, max_j]$ for attribute j and $v_j(x_j)$ denotes the satisfaction level of the value of attribute j. To compute the universal function $v(x_j)$ of a proposal with a multi attribute, we need to consider the utility function value of each attribute $v_j(x_j)$ and the importance of each attribute w_j. So, it is computed as:

$$v(x_j) = \sum_{1 \le j \le n} w_j v_j(x_j), \, where \sum_{1 \le j \le n} w_j = 1$$

And the judging criteria is represented as:

$$I^a\left(t'^{,x^t_{b\to a}}\right) = \begin{cases} reject, & if\ t' > t^a_{max} \\ accept, & Ifv^a\left(x^t_{b\to a}\right) \ge \left(x^{t'}_{b\to a}\right) \\ x^{t'}_a, & otherwise \end{cases}$$

Where, $x^t_{b\to a}$ is the offer which agent a received from agent b at time t, $x^{t'}_{b\to a}$ is the offer which agent a should make to agent b at time t' and t^a_{max} is the maximum negotiation time for agent a. Time-dependent tactics vary the proposals as the time deadline varies. An agent a has to find a deal until t^a_{max}. A proposal $x^t_{a\to b}[j]$ for attribute j from agent a to agent b at time t, $0 \le t \le t^a_{max}$ is calculated as:

$$x^t_{a\to b}[j] = \begin{cases} min^a_j + \propto^a_j(t)\left(max^a_j - min^a_j\right), Ifv^a_j isdecreasing \\ min^a_j + (1-\propto^a_j(t))(max^a_j - min^a_j), Ifv^a_j isincreasing \end{cases}$$

Time-based tactic uses different functions (polynomial, exponential, sigmoid) depending on the time that can be parameterized. Faratin et al. [16] identified two functions polynomial and exponential. We define function sigmoid and apply it as a time-based dependent function. Polynomial and exponential are considered linear functions whilst a sigmoid function is a non-linear function. Time-dependent behavior can be realized in different ways by choosing the function $\propto(t)$ according to Table 1. The table contains three different functions, which are polynomial, exponential, and sigmoid. Constraints on function $\propto(t)$ are also provided in the table.

Table 1. Different alpha functions.

Time-based dependent functions	Function relation	The function $\propto(t)$ constraints
Polynomial	$\propto(t) = e^{\left(1-\frac{min(t,t_{max})}{t_{max}}\right)^\beta \ln k}$	It must satisfy the following constraints:
Exponential	$\propto(t) = k + (1-k)\left(\frac{min(t,t_{max})}{t_{max}}\right)^{(1/\beta)}$	• $0 \le \propto(t) \le 1$
Sigmoid	$\propto(t) = 1/(1+e^{-\beta*(t-t_{max})})$	• $\propto(t_0) = k$ • $\propto(t^a_{max}) = 1$

Throughout a negotiation, every offer has to satisfy the range of values for an issue. As the initial offer has to be in the range of acceptable values for an issue. A constant k is used to adjust the initial value. These functions depend on the parameter β. This parameter defines the conceding pattern of $\propto(t)$ curve. The conceding pattern of $\propto(t)$ curve depends on the parameter β. Figures 2, 3 and 4 represent the three functions polynomial, exponential, and Sigmoid. Through these three functions, we compute alpha $\propto(t)$ within time interval [0–1] with different beta β values 0.05, 0.5, 0, 1, 2, and 5. From Figs. 2 and 3, which represent a polynomial and an exponential function, the behavior of values of $\beta < 1$ is described as Boulware. While the behavior of function classes $\beta > 1$ is described as conceder. The difference of both function classes in conceder behaviour is that polynomial functions concede faster at the beginning. In the

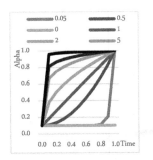

Fig. 2. Polynomial function curve

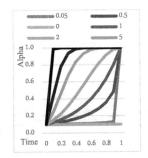

Fig. 3. Exponential function curve

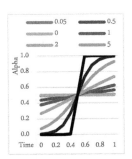

Fig. 4. Sigmoid function curve

Boulware behavior, polynomial functions also concede earlier. While as, the $\propto (t)$ curve of the two functions are not constant at any point, a sigmoid function is always constant except only around the middle of the curve. The sigmoid curve, shown in Fig. 4, has increasing steps around the middle (0.4 - 0.6) of time t. The curve is a constant at t_0 from the initial time till around the middle. Also, it is constant at t_{max} after the middle of t. Although the β parameter is the affecting value of the conceding pattern of time-based dependent functions, most frameworks define it as a constant value. We define a model that computes β dynamically from QoS attribute definition represented in WS-P. As we mentioned before in Sect. 3, WS-P contains attribute ranges, weight the of each attribute, and desirability factor (DF) to reach an agreement. In our model, the β parameter of each attribute is computed according to the DF and the weight of each attribute, which are the most important values affecting the offers conceding. Therefore, we can propose a novel relationship used to calculate the best value of β as following:

$$\beta = (1 - weight)/(1 - DF)$$

This relationship is represented by a chart in Fig. 5. It is validated with the substitution of different values of the DF and weights. When the DF increases and decreasing value of the weight, the offer of an attribute concedes faster. Conceding faster means a larger value of the β parameter. The β parameter is computed using values presented in negotiation preferences given in participant policy. With the same weight, the β parameter has a larger value with increasing DF. The participants of our conducted negotiation scenario have a price, a number of users and availability attributes. With computing the conceding pattern of these attributes, the β parameter of that attributes are "1, 1.6, and 1.4" respectively. This indicate that the conceding pattern of number of users concedes faster than other attributes as it has a greater β value than other attributes with the same desirability factor.

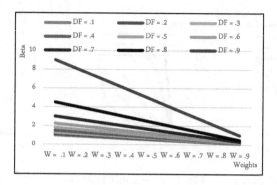

Fig. 5. The β parameter-computing model

4   Results and Practical Scenario

For demonstrating the proposed multilateral negotiation framework, we have implemented the prototype in Java. The scenario that we depend on is from a previous work [17]. In this prototype, Multi-agent JADE development environment is used to create and simulate communications between different participants. Through the implementation, a multilateral negotiation scenario between one consumer and three different providers is conducted. A web service is offered by three different providers; provider1 (Bronze), provider2(Silver), and provider3(Gold). Negotiation conducted with applying a win-win model, collaborative model, for different participants. Time-dependent three functions, which dependent on time, are applied to this scenario. Time, which is defined in the policy, represents iteration number from the total number of iterations. These three service providers offered three negotiable nonfunctional properties; price, number of users, and availability. Weights of the three providers for these QoS properties are shown in Table 2. Consumer offers with its attributes are shown in Table 3. The offers of providers and consumer show the Desirability Factor (DF) that describes the tendency of each participant to reach an agreement. The maximum time for negotiation is also defined in the offers of negotiated partners. The consumer negotiates these issues with each provider separately to get a satisfactory offer for both of them.

Table 2. Service Providers attributes

Providers	Negotiable issues	Range of values	Weights	DF	Time
Provider1 (Bronze)	Price	10–20$	0.5	0.5	200
	Number of Users	100–200	0.2		
	Availability	96–97.9 %	0.3		
Provider2 (Silver)	Price	20–30$	0.5	0.5	200
	Number of Users	100–350	0.2		
	Availability	97.9–98.9 %	0.3		
Provider3 (Gold)	Price	30–50$	0.5	0.5	200
	Number of Users	100–500	0.2		
	Availability	98.9–99.9 %	0.3		

Table 3. Consumer attributes

	Negotiable issues	Range of values	Weights	DF	Time
Consumer	Price	15–35 $	0.5	0.5	200
	Number of users	100–250	0.2		
	Availability	97–99 %	0.3		

Table 4. Consumer negotiation process with provider1

Polynomial time-based function

Provider1(Bronze)				Consumer		
Time	Price	Users	Avail	Price	Users	Avail
123	15.13	163.471	0.9679	24.8171	154.2154	0.9785
124	15.0825	163.868	0.9699	24.9114	153.6270	0.9784
125	15.035	164.263	0.970	Offer accepted		

Exponential time-based function

Provider1(Bronze)				Consumer		
Time	Price	Users	Avail	Price	Users	Avail
151	15.2	172.93	0.9699	24.919	138.48	0.9816
152	15.127	173.68	0.9711	25.062	137.40	0.9814
153	15.054	174.43	0.9720	Offer accepted		

Sigmoid time-based function

Provider1(Bronze)				Consumer		
Time	Price	Users	Avail	Price	Users	Avail
98	18.808	103.92	0.9611	17.384	244.12	0.9889
99	17.311	116.8	0.9638	20.378	224.80	0.9860
100	15	150	0.970	Offer accepted		

The proposals of the consumer and provider1 with polynomial, exponential and sigmoid time-based function are shown in Table 4. For polynomial function, the consumer agent accepts the proposal values"15.035, 164.2631547, and 0.97" at time 125 of negotiation. Accepting that proposal cause of the global utility value of provider1 proposal at time 125 is greater than the global utility value of the conceding offer "23.8743, 160.2376 and 0.9794" of the consumer at the same time. With an exponential function, the consumer accepts the proposal "15.054, 174.43, 0.9720" at time 153. With a sigmoid function, the consumer accepts the proposal "15, 150 and 0.970" at time 100. For the second provider2, Table 5 shows the offers of polynomial, exponential and sigmoid functions. The consumer accepts the first offer of the provider2 using the three different functions. The reason for the consumer to accept the first offer of that provider2 is that the range of QoS attributes of a provider2 is nearly sub of the range of attribute values of the consumer. For the third provider3, as shown in Table 6, the consumer accepted the provider3 offer at time 178 with a polynomial function. With an exponential function, the consumer accepts the offer at time 183. Moreover, with a sigmoid function the consumer accepts the offer at time 102. Note that, all the accepted proposals' values have to be in the range of values defined in

Table 5. Consumer negotiation process with provider2

Polynomial time-based function						
Provider2(Silver)				Consumer		
Time	Price	Users	Avail	Price	Users	Avail
0	29.667	117.86	0.979	Offer accepted		

Exponential time-based function						
Provider2(Silver)				Consumer		
Time	Price	Users	Avail	Price	Users	Avail
T = 0	29.667	117.86	0.979	Offer accepted		

Sigmoid time-based function						
Provider2(Silver)				Consumer		
Time	Price	Users	Avail	Price	Users	Avail
T = 0	30	100	0.979	Offer accepted		

policy by the opponent participant. That is beside the condition of accepting the proposal, which is the value of global utility value of a participant have to be greater than or equal to the global utility value of the opponent participant. Although the sigmoid function through different providers gets an agreement faster than the other two functions, it gets results that are not suitable as its offers are constants at the beginning and the end. Its offer is only variable around the middle with varying the offers faster than usual. After the negotiation process terminated, the negotiation manager chooses the best suitable provider from three providers. The negotiation manager chooses provider2 (silver) as the most suitable provider.

Table 6. Consumer negotiation process with provider2

Polynomial time-based function						
Provider3(gold)				Consumer		
Time	Price	Users	Avail	Price	Users	Avail
176	35.269	230.29	0.98998	30.238	109.45	0.9841
177	35.03	231.12	0.98999	30.458	108.69	0.9838
178	34.787	232.9	0.98980	Offer accepted		

Exponential time-based function						
Provider3(gold)				Consumer		
Time	Price	Users	Avail	Price	Users	Avail
181	35.184	210.29	0.9892	29.9086	125.1767	0.9840
182	35.088	211.58	0.98931	30.0029	124.6815	0.9839
183	34.992	212.87	0.98971	Offer accepted		

Sigmoid time-based function						
Provider3(gold)				Consumer		
Time	Price	Users	Avail	Price	Users	Avail
100	40	200	0.9894	25.0000	175.0000	0.9800
101	35.379	205.81	0.9897	27.0000	175.0000	0.9800
102	32.384	206.33	0.990	Offer accepted		

5 Conclusion and Future Work

This paper proposed a multilateral SLA agreement framework using multi-agent for supporting automated negotiation between a consumer and different providers. The QoS negotiable attributes for different participants specified using WS-policy. Besides, the negotiation strategy is provided with using three different utility functions. In the proposed framework, a model is provided for the parameter affecting the conceding offers of different participants. Moreover, finally, the negotiation manager chooses the best suitable provider for a consumer. As a future direction, we intend to adopt the decision-making process for participants using different negotiation strategies. In addition, we will apply negotiation technique with other nonfunctional requirements such as security and privacy.

References

1. Baber, W.W., Ojala, A.: Cognitive negotiation schemata in the IT industries of Japan and Finland. J. Intl. Technol. Inf. Manage. **24**(3), 87–104 (2015)
2. Liu, C., Liu, D.: QoS-oriented web service framework by mixed Programming Techniques. J. Comput. **8**(7), 1763–1770 (2013)
3. Weerawarana, S., et al.: Web services platform architecture: SOAP, WSDL, WS-policy, WS-addressing, WS-BPEL, WS-reliable messaging and more, Prentice Hall PTR (2005)
4. Jennings, N.R., et al.: Automated negotiation: prospects, methods and challenges. Group Decis. Negot. **10**(2), 199–215 (2001)
5. Okumura, M., Fujita, K., Ito, T.: An implementation of collective collaboration support system based on automated multi-agent negotiation. In: Ito, T., Zhang, M., Robu, V., Matsuo, T. (eds.) Complex Automated Negotiations: Theories, Models, and Software Competitions. SCI, vol. 435, pp. 127–144. Springer, Heidelberg (2012)
6. Dastjerdi, A.V., Buyya, R.: An autonomous time-dependent SLA negotiation strategy for cloud computing. Comput. J. (2015)
7. Hung, P.C., Li, H., Jeng J.-J.: WS-Negotiation: an overview of research issues. In: Proceedings of the 37th Annual Hawaii International Conference on System Sciences, 2004. IEEE (2004)
8. Andrieux, A., Czajkowski, K., Dan, A.: Web Services Agreement Specification (WS-Agreement), 17 March 2007 (2002)
9. Anderson, A.H.: An introduction to the web services policy language (wspl). In: Proceedings of the Fifth IEEE International Workshop on Policies for Distributed Systems and Networks (2004)
10. Chen, J.-H., et al.: Architecture of an agent-based negotiation mechanism. In: Proceedings of the 22nd International Conference on Distributed Computing Systems Workshops. IEEE (2002)
11. Lock, R.: Automated negotiation for service contracts. In: COMPSAC'06, The 30th Annual International Conference on Computer Software and Applications, 2006. IEEE (2006)
12. Koumoutsos, G., Thramboulidis, K.: Towards a knowledge-base for building complex, proactive and service-oriented e-negotiation systems. In: International MCETECH Conference on e-Technologies. IEEE (2008)

13. Mukun, C.: Multi-agent automated negotiation as a service. In: The 7th International Conference on Service Systems and Service Management (ICSSSM). IEEE (2010)
14. Ogawa, R., Orgill, D.P.: Mechanobiology of cutaneous wound healing and scarring. In: Gefen, A. (ed.) Bioengineering Research of Chronic Wounds. SMTEB, vol. 1, pp. 31–42. Springer, Heidelberg (2009)
15. Bajaj, S., et al., Web Services policy 1.2-framework (WS-policy). W3C Member Submission, (2006)
16. Faratin, P., Sierra, C., Jennings, N.R.: Negotiation decision functions for autonomous agents. Robot. Auton. Syst. **24**(3), 159–182 (1998)
17. Zulkernine, F., et al.: A policy-based middleware for web services SLA negotiation. In: ICWS IEEE International Conference on Web Services. IEEE (2009)
18. Ludwig, H.: Web services QoS: external SLAs and internal policies or: how do we deliver what we promise? In: Workshops, 2003, Proceedings of Fourth International Conference on Web Information Systems Engineering. IEEE (2003)
19. Xiao, Z., et al.: A policy-based framework for automated service level agreement negotiation. In: ICWS IEEE International Conference on Web Services (2011)
20. Hashmi, K., et al.: Automated negotiation among web services. In: Web Services Foundations, pp. 451–482. Springer (2014)

An Adaptive Load-Balanced Partitioning Module in Cassandra Using Rendezvous Hashing

Sally M. Elghamrawy[(✉)]

MISR Higher Institute for Engineering and Technology, Mansoura, Egypt
sally@mans.edu.eg

Abstract. With the rapid growth and use of social networks, the appearance of Internet technology and the advent of the cloud computing a need for new tools and algorithms is appeared to handle the challenges of the big-data. One of the key advances in resolving the big-data challenges is to introduce scalable storage systems. NoSQL databases are considered as efficient big data storage management systems that provide horizontal scalability. To ensure scalability of the system, data partitioning strategies must be implemented in these databases. In this paper, an Adaptive Rendezvous Hashing Partitioning Module (ARHPM) is proposed for Cassandra NoSQL databases. The main goal of this module is to partition the data in Cassandra using rendezvous hashing with proposing a Load Balancing based Rendezvous Hashing (LBRH) algorithm for guaranteeing the load balancing in the partitioning process. To evaluate the proposed module, Cassandra is modified by embedding the APRHM partitioning module in it and a number of experiments are conducted to validate the load balancing of the proposed module by using the Yahoo Cloud Serving Benchmark.

Keywords: Cassandra · NoSQL database · Rendezvous hashing · Partitioning · Consistent hashing · Load balancing

1 Introduction

Over the past decade, information and internet technology had grown rabidly, resulting in data explosion, and the traditional data processing applications became not suitable for handling these data. The term Big Data [1] appeared with its associated techniques and technologies to manage the massive amounts of datasets. Big data had become related to all aspects of human life. The researchers found that there is a vital need to address the challenges of storing Big Data. The relational database RDBMS can't handle big data. NoSQL [2] (meaning 'not only SQL') databases are used for storing big data due to its ability to expand easily according to the data scale. The data in NoSQL database must be distributed over heterogeneous servers. NoSQL databases provide shared-nothing horizontal scalability. To ensure this scalability, data partitioning strategies must be implemented. Sharding is the horizontal partitioning of data, it means the ability to shard the database and then distribute data stored in each shard. Range, random and hashing partitioning techniques are presented for NoSQL databases to partition data on different nodes. Some NoSQL databases like Apache HBase [3]

© Springer International Publishing AG 2017
A.E. Hassanien et al. (eds.), *Proceedings of the International Conference on Advanced Intelligent Systems and Informatics 2016*, Advances in Intelligent Systems and Computing 533, DOI 10.1007/978-3-319-48308-5_56

uses range partitioning, while the most common NoSQL databases like Google Big-Table [4] and Cassandra [5] used by Facebook adopt in consistent hashing [6] as their partitioning strategies.

The basic consistent hashing used in Cassandra presents a load balancing problem because it ignores the nature of nodes being assigned the data to them, it only depends on blind hashing. There have been great efforts done to adopt inconsistent hashing to solve load balancing problem [7, 8]. Unlike consistent hashing, Rendezvous hashing (HRW) uniformly distributes records of database over nodes using specific hash function. In this paper an Adaptive Partitioning module is proposed based on Rendezvous Hashing (ARHPM) for Cassandra databases, with proposing an associated Load Balancing LBRH algorithm to guarantee the load balancing in the partitioning process. To enhance the speed of the partitioning process, a spooky hash function is used in ARHPM that proved to be twice the speed of the Murmur hash function used by Cassandra consistent hashing. In addition, using a virtual hierarchical structure with the Highest Random Weight (HRW) [9] algorithm reduce time taken by consistent hashing in precomputing and tokens storage processes.

The rest of the paper is organized as follows. Section 2 covers related partitioning studies. Section 3 demonstrates the proposed Adaptive Rendezvous Hashing Partitioning module, the Load Balancing algorithm based on Rendezvous Hashing LBRH is shown too. Cassandra performance is evaluated in Sect. 4 showing the effect of the load balancer sub module in ARHPM, and the results are compared against recent work's results.

2 Related Work

There are enormous interests of researchers in investigating the partitioning strategies of NoSQL databases due to its impact on the performance of systems. Hash, range and hybrids between hash and range partitioning are the most common strategies used by NoSQL systems. The types of data partitioning are summarized by Ata Turk et al. [10], showing the advantages and disadvantages of each type. A number of researchers have developed different methods to enhance in Cassandra performance and in partitioning strategies of different NoSQL databases: Ata Turk et al. [10] proposed data partitioning method based on hypergraph that is constructed to correctly model multi-user operations. Abramova et al. [11] analyzed the scalability of Cassandra by testing the replication and data partitioning strategies used. Lakshminarayanan [7] proposed an adaptive partitioning scheme for consistent hashing that effect in the heterogeneity of the systems. X. Huang et al. [12] proposed a dynamic programming algorithm in the consistent hashing ring based on imbalance coefficient for Cassandra cluster to calculate the position of the new coming node. Ramakrishnan, et al. [13] proposed a processing pipeline using the Random Partitioner of Cassandra to allow any non-Java executable make use of the NoSQL and allowing offline processing for Cassandra. Zhikun et al. [14] proposed a Hybrid Range Consistency Hash (HRCH) partition strategy for NoSQL database to improve the degree of processing and the data loading speed. Most of these approaches either consider a scalability test of the NoSQL database or enhancing in consistent hashing of Cassandra. Cassandra had three basic partitioning strategies: Random Partitioning, ByteOrdered and Murmur3 partitioning. To the best of our knowledge, it's the first attempt to replace the whole partitioning module of Cassandra with the rendezvous hashing technique and test their scalability, load balancing, and timing of partitioning process.

3 The Proposed Adaptive-Rendezvous Hashing Partitioning Module (ARHPM)

This section's main goal is to demonstrate the proposed Adaptive-Rendezvous Hashing partitioning module **ARHPM** using Highest Random Weight (HRW) algorithm based on the Hash Based virtual hierarchies [9], in partitioning Cassandra. In addition, a proposed load balancing algorithm is presented. The proposed **ARHPM** presents a load balancing scheme to ensure that any exchange of load or capabilities between nodes will ultimately reduce the large potential value and will lead the system to be in a more balanced state, as will be shown in details. The implementation of the proposed partitioning module is based on load balancer and rendezvous hash partition using spooky hash function. The goal of the proposed partitioning module is to improve Cassandra in the following: (1) Manage the load of forward request. (2) Balance the load among each node by taking in consideration the node's storage capacity using Load Balancing based Rendezvous (HRW) Hashing algorithm **LBRH**. (3) Enhance the timing of hashing by using the spooky hash function and using the rendezvous hashing that dispense of using the virtual servers in Cassandra. The proposed ARHPM module consists of three main sub-modules, as shown in Fig. 1.

The Load Request Manager Sub Module: Its main goal is to control the requests coming from different nodes, it consists of: **Workload analyzer**: It is used to analyze

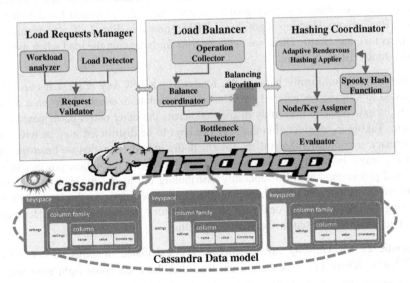

Fig. 1. The proposed Adaptive-Rendezvous Hashing partitioning module (ARHPM)

the workload in each node in the system and caches them, and also caches the most frequently appeared node locations in a specific Cassandra cluster. Based on *Heuristic 1*: Suppose that there is a key with name K and a set of nodes N = $\{n_1, n_2, n_3, \ldots n_i\}$. Each node n ∈ *N* is defined in terms of: =<CAPACITY(n), ACTIVE DEGREE (n), DEPENDENCY (n)>, where capacity(n) indicates the ability of the node to accept number of keys and queries. Active degree(n) indicates the number of finished queries. Dependency(n) is concerned of how many times the node asks the help of another node. **Load detector**: manages the load requests and checks the ability for managing this load. **Request validator**: It validates the requests sent to the system and parsing the query if it contains any unrecognizable format.

The Load Balancer Sub Module: Cassandra's current load balancing [20] depends on transferring most of keys requests of the most loaded node, this operation does not guarantee node balancing. As a result, many attempts were proposed to build load balancers in front of Cassandra to ensure the balancing [7, 8, 12], but this load balancer itself can be the bottleneck in the system. In the proposed module, a load balancer will process in a different way by using the benefits of current Cassandra load balancing policy without transferring any data. The load balancer manages the balancing in the partitioning process without affecting the performance. **The operation collector** can record the nodes participated and the operations which are done in the system and the load statuses for each node collected in the previous sub module. Unlike the consistent hashing techniques that categorize the nodes to be allocated based on their distance between each other which result in non-uniform distribution. In other hand, using the rendezvous hashing (HRW) will make each node has the opportunity to equally receive the key K, which result in uniform the load among nodes. The operation collector main goal is to make each node exchange information about one another. Each node will be associated with its corresponding load. **Balance Coordinator**: it decides each node's role in the balancing procedure according to the list created from operation collector. The balance coordinator's main goal is to implement the proposed Load Balancing based Rendezvous Hashing algorithm **LBRH**, shown in Fig. 2. It also resolves all conflicts that may occur. There are two scenarios for balancing the load when using the rendezvous (HRW) hashing:

Scenario 1: To uniformly balance the load when a new key request needed to be allocated to specific node based on balancing algorithm. *Scenario 2*: When a node is down, its load must be uniformly distributed across the other nodes participated in the system. Taking in consideration that: (a) *The* keys to be distributed may be have more importance or popularity than others. (b) *The* nodes in both scenarios are heterogeneous in many ways: its ability to proceed operations, network capabilities, power consuming, and processing time. So randomly distributing the load may lead to bottleneck problems or insufficient performance, so the process of balancing the load when partitioning the database, must deal with all the attributes for node and keys. Heuristic 2 illustrate some terms used in the LBRH algorithm.

Heuristic 2: The capacity (n) in heuristic 1 is calculated in terms of < $LD_i^{Cur}(n)$, $LD_i^{MaX}(n)$>. Where LD_i^{Cur} is the current load that the node holds right now and the LD_i^{MaX} is the maximum load that the node i can hold calculated based on each node

Load Balancing based Rendezvous Hashing (LBRH)
Input: Node Tuples's ID $(Node_i(CAPACITY(n_i), ACTIVE\ DEGREE(n_i), DEPENDENCY(n_i))$ and node (n_i)'s neighbors' in same zone according to HRW Output: A List of the distributed load assigned to which node (LD_{DS_j}, n)

1.	Supervise $(\sum_{i=1}^{x}(n_i)\ in\ n_i) \in C_j$
2.	For Each n in $\sum_{i=1}^{x}(n_i)$
3.	Calculate $LD_i^{Cur}((n_i)$
4.	Check load (n_i)
5.	IF $LD_i^{Cur} > T_{\mathcal{HLD}}$ IS OVERLOAD THEN INTENSE-LOAD()
6.	INTENSE-LOAD ( )
7.	{ aa : Register→ Pending_Relaod_List()
8.	LD_i^{Cur}=fragment ($\mathcal{LD}$)
9.	If $LD_i^{MaX} < T_{\mathcal{HLD}}$
10.	Choose specific (n(i))
11.	Split (n(i))
12.	Partition (n(i))
13.	Bid_Bonus ($LD_i^{MaX}, LD_{i+1}^{MaX}, LD_{i-1}^{MaX}$)
14.	Else Go to aa
15.	Combine (n)
16.	Create (LD_{DS_j}, n) LIST }
17.	If $LD_i^{Cur} < T_{\mathcal{LLD}}$ is overload then LOW_KEY-LOAD()
18.	Low_KEY-LOAD( )
19.	{
20.	$LD_i^{Cur}=LD_i^{Cur} - T_{\mathcal{LLD}+i}$
21.	For each CC_{LD}^{ji} in C_j
22.	Bid_Bonus($LD_i^{Cur}, LD_{i+1}^{Cur}, LD_{i-1}^{Cur}$)
23.	Minimize (φ)
24.	End For
25.	Create LD_{DS_j}, n) LIST }

Fig. 2. The Load Balancing based Rendezvous Hashing (LBRH)

feature. There are two main phases for the process: **First phase**: Let $(n_i \in N)$ be the set of the participating nodes in the cell C_j in HRW structure [11]. And CC_{ji} be the set of cell coordinator for each cell responsible of a number of node i. Each CC_{ji} has a load denoted by CC_{LD}^{ji}, the cell coordinator's main goal is to manage the load between the nodes in this cell based on each node's LD_i^{Cur} and LD_i^{MaX}, to categorize the nodes to Intense load, Moderate Load, and Low-Key load, as will be shown in the LBRH algorithm. By using two threshold values as a high value threshold $T_{\mathcal{HLD}}$ and low value threshold $T_{\mathcal{LLD}}$. To allocate the key, it must assign a popularity factor $((\mathcal{PK}))$. While the load is assigned to a specific node, the load balancing algorithm will be activated to ensure that the load is balanced in the allocating process by minimizing φ, shown in Eq. (1). **Second phase**: Each node in the cell has neighbor's node, the main goal of the LBRH algorithm is to predict the neighbor node performance in allocating the load by using a bid-bonus algorithm proposed in Fig. 3.

The Bid bonus algorithm used by LBRH
Bid_Bonus( )
1.   For Each n_i(Cap) in $C_i$do
2.   CreatBID(n_i)
3.   If bid_i (Cap)=trustybids[i] then
4.   If Expct{bid_i (Cap)} > Higest(Cap) then
5. bid_i (Cap) → $higestbidlist[i]$
6.   *Else if*
7.   Recieved {bid_i (Cap)} > Higest(Cap) then
8. bid_i (CCap) → $higestbidlist[i]$)
9.   For each bid_i (Cap)in $higestbidlist[i]$
10.   if bid_i (Cap)=trustybids[i]then
11.   Sort (bid_i (Cap)) → $sortedbidlist[i]$
12.   sendnew (bid_i (Cap))
13.   For each bid_i (Cap) in $sortedbidlist[i]$
14.   Get HigestPairBids (bid_i (Cap)) → $HighestPair[i]$
15.   If $HighestPair[i]$ > < $\mathcal{T}_{\mathcal{LLD}}$ then
16.   For first bid_i (Cap, AD, Dep) in $HighestPair[i]$
17.   Get (bid_i (AD))
18.   If n_i-ActDgree (bid_i(AD)> n_i-ActDgreebid_{i+1} (AD)
19.   then bid_i (AD)) → $FinalHighestPair[i]$
20.   If n_i-ActDgree(bid_i (AD)< n_i-ActDgree(bid_{i+1}(AD)
21.   then bid_{i+1} (AD) → $FinalHighestPair[i]$
22.   Else
23.   Get(n_i-Dependency (bid_i (Dep))
24.   If n_i-Dependency (bid_i (Dep)< n_i-Dependency (bid_{i+1} (Dep) then
25. bid_i (Dep)) → $FinalHighestPair[i]$
26.   If n_i-Dependency (bid_i (Dep)> n_i-Dependency (bid_{i+1} (Dep) then
27. bid_{i+1} (Dep)) → $FinalHighestPair[i]$
28.   Else
29.   RandomPair(bid_i (Cap, AD, Dep)) → $FinalHighestPair[i]$

Fig. 3. The Bid bonus algorithm used by LBRH

$$\varphi = \sum_{i=1,n}^{j=1,m} CC_{LD}^{ji} - \frac{\sum_{j=1}^{m} CC_{LD}^{ji}}{\sum_{i=1}^{n} LD_i^{MaX}} \tag{1}$$

Bottleneck Detector: The Balance Coordinator module contacts this sub-module when the load is distributed across different node. The involved node's information in this distribution is sent to bottleneck detector. Then the detector has two main roles: *First*, it detects any bottlenecks that might occur in the load distribution process. Second, it updates the list by detecting the node statues after applying the load balancing algorithms. It detects if a node is involved in the load balancing process, if yes, then its capacity and active degree are changed, so the list must be refreshed.

Hashing Coordinator Sub Module: It is the core module in APRHM. Its main goal is to implement the adaptive HRW hashing algorithm for the partitioning process. The hashing module consists of four main submodules: **Adaptive rendezvous**

(HRW) Hashing applier: In this sub-module, the virtual hierarchies' skeleton on HRW [9] is used. It partitions keys of the Cassandra database using rendezvous hashing on the nodes distributed in a sufficient way to guarantee balanced partitioning. The nodes on a Cassandra cluster are divided into a number of Rendezvous Geographic Zones ($\mathbb{R}\mathcal{G}_\mathbb{Z}$) [16], and the data are distributed according to the virtual hierarchical design in each zone. **Spooky Hash Function**: The default hash function used by the original partitioning module in Cassandra is the Murmur Hash [17]. APRHM implements the non-cryptographic Spooky Hash [18] function for the required nodes and keys in Cassandra. One of the main reasons of using spooky hash instead of Murmur hash is that later proved to be half the speed of spooky hash on x86-64. **Node/key assigner**: Implements the process of assigning the hashed key to the node yielding the highest weight. **Evaluator**: The evaluator is used in every node of the rendezvous hash range to ensure the accurate allocation of keys to nodes.

4 Performance Evaluation

A number of experiments are conducted on Cassandra to evaluate the improvement of the proposed partitioning module APRHM. The experiments have been divided into two main parts for the evaluations, as follows: (1) Evaluates the performance of the LBRH algorithm implemented in the load balancer sub module. (2) The response time of Cassandra with APRHM is tested under different environments. The Apache Cassandra version 3.4 is modified by embedding the APRHM partitioning module in it. The standard Yahoo! Cloud Serving Benchmark [15] is installed YCSB 0.1.4. In the experiments, two clusters of 8 nodes/cluster are used. The experiment configurations are shown in Table 1. The nodes are created by using the VMware vSphere system [19]. Run at Ubunt10 system.

Table 1. Cassandra cluster nodes specifications

Number of Cassandra clusters:	2	**Hadoop 2.6.4 configurations**	**Cassandra 3.4 configurations**
Number of nodes per cluster:	8	dfs.replication.max: 512	Replication Factor: 1, Heap
		dfs.blocksize: 128MB	size: 1GB
Node specification:		Dual-core Intel Core i5-4200U at 2.6GHZ, 8G of RAM, 200GB disks and 1GB Ethernet, run. 200 GB disk space. 8 million records database	

First Experiment: The performance of the load balancing sub module embedded in our partitioning (APRHM) module is evaluated, and compares it against two partitioners used in original Cassandra [11, 12]. Workload C of YCSB is used in load mode to upload the data to a cluster of 8 nodes. Figure 4 shows the performance of Cassandra with ByteOrder partitioner, the load is condensed in a one or two nodes which result in uneven distribution of data. Figure 5 demonstrates that using Murmur in Cassandra the load is almost distributed on the 8 nodes but not uniformly. However, Fig. 6 demonstrates a uniformly balanced performance of Cassandra between the 8 nodes

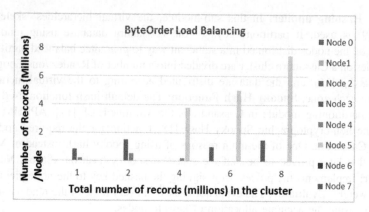

Fig. 4. The performance of Cassandra with Byteorder partitioner

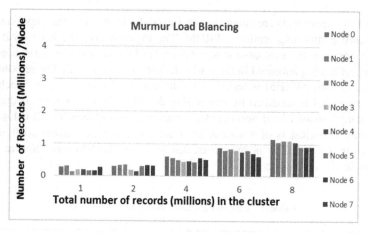

Fig. 5. The performance of Cassandra with Murmur partitioner

when using the proposed APRHM. This comparison validated the Load Balancing based Rendezvous Hashing algorithm LBRH.

Second Experiment: The execution time of assigning the records to the nodes in APRHM is compared with standard Cassandra ByteOrdered, Random, and Murmur3 [11–13] partitioning when using READ workload c in YCSB. Figure 7 shows that APRHM is the fastest partitioner when varying number of loaded records.

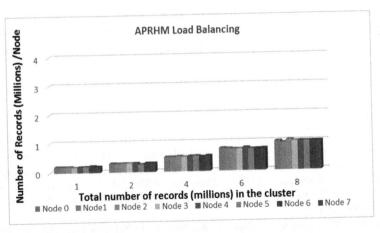

Fig. 6. The performance of Cassandra with APRHM partitioner

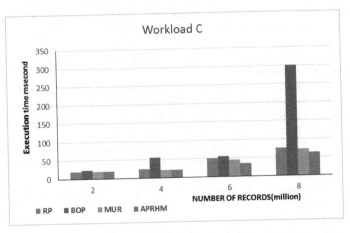

Fig. 7. The execution time of different partitioners with workload C

5 Conclusions and Future Work

An Adaptive Rendezvous Hashing Partitioning ARHPM module is proposed in this paper to enhance performance of Cassandra databases. ARHPM partitions the data in Cassandra database using rendezvous hashing that uses the spooky hash function in a random way to make each node have the same chance to receive the keys, which result in uniformly distribute the load. In addition, a Load Balancing based Rendezvous (HRW) Hashing LBRH algorithm for guarantee load balancing with heterogeneous nodes. ARHPM module proved the ability to balance the load upon nodes with different storage, capabilities and load. ARHPM adapted to the load variations of YCSB with maintaining Cassandra performance. The comparative experiments showed that the load in APHRM is balanced more uniformly than the two default partitioners of standard Cassandra. And the ARHPM timing for distributing the load is faster than different partitioners of standard Cassandra. As a future work, an intention to use bloom filter method on every node of the rendezvous hash range to enhance the timing of hashing.

References

1. Demchenko, Y., Membrey, P., Grosso, P., de Laat, C.: Addressing big data issues in scientific data infrastructure. In: First International Symposium on Big Data and Data Analytics in Collaboration (BDDAC 2013). Part of The 2013 International Conference on Collaboration Technologies and Systems (CTS 2013), 20–24 May 2013, San Diego, California, USA
2. Benzaken, V., Castagna, G., Nguyen, K., Siméon, J.: Static and dynamic semantics of NoSQL languages. SIGPLAN Not. 48(1), 101–114 (2013)
3. HBase Development Team. HBase: BigTable-like structured storage for Hadoop HDFS [EB/OL], 20 March 2013. http://wiki.apache.org/hadoop/Hbase/
4. Chang, F., Dean, J., Ghemawat, S., Hsieh, W.C., et al.: BigTable: a distributed storage system for structured data. In: Proceedings of the 7th OSDI, pp. 205–218. ACM, Seattle (2006)
5. Lakshman, A., Malik, P.: Cassandra: a decentralized structured storage system. Oper. Syst. Rev. 44(2), 35–40 (2010)
6. Karger, D., Lehman, E., Leighton, T., Panigrahy, R., Levine, M., Lewin, D.: Consistent hashing and random trees: distributed caching protocols for relieving hot spots on the world wide web. In: Proceedings of the Twenty-Ninth Annual ACM Symposium on Theory of Computing, 1997, pp. 654–663. ACM, New York (1997)
7. Srinivasan, L., Varma, V.: Adaptive load-balancing for consistent hashing in heterogeneous clusters 2015. In: 15th IEEE/ACM International Symposium on Cluster, Cloud and Grid Computing (2015)
8. Byers, J., Considine, J., Mitzenmacher, M.: Simple load balancing for distributed hash tables. In: Kaashoek, MFrans, Stoica, Ion (eds.) IPTPS 2003. LNCS, vol. 2735, pp. 80–87. Springer, Heidelberg (2003)
9. Yao, Z., Ravishankar, C., Tripathi, S.: Hash-based virtual hierarchies for caching in hybrid content-delivery networks (PDF), CSE Department, University of California, Riverside, Riverside, CA, 13 May 2001. Accessed 15 November 2015
10. Turk, A., Oguz Selvitopi, R., Ferhatosmanoglu, H., Aykanat, C.: Temporal workload-aware replicated partitioning for social networks. IEEE Trans. Knowl. Data Eng. 26(11), 2832–2845 (2014)
11. Abramova, V., et al.: Testing cloud benchmark scalability with cassandra. In: 2014 IEEE 10th World Congress on Services (2014)
12. Huang, X., Wang, J., Zhong, Y., Song, S., Yu, P.S.: Optimizing data partition for scaling out NoSQL cluster, 20 September 2015 in Wiley Online Library (wileyonlinelibrary.com). doi:10.1002/cpe.3643
13. Ramakrishnan, L., et al.: Processing cassandra datasets with hadoop-streaming based approaches. IEEE 2015 Trans. Serv. Comput.
14. Chen, Z.: Hybrid range consistent hash partitioning strategy–a new data partition strategy for NoSQL database. In: 2013 12th IEEE International Conference on Trust, Security and Privacy in Computing and Communications (2013)

15. Cooper, B.F., Silberstein, A., Tam, E., et al.: Benchmarking cloud serving systems with YCSB. In: Proceedings of SoCC. ACM, Indianapolis (2010)
16. Seada, K., Helmy, A.: Rendezvous regions: a scalable architecture for service location and datacentric storage in large-scale wireless networks. In: Proceedings of the 18th International Parallel and Distributed Processing Symposium, IPDPS 2004
17. Yuki, K.: "Digest::MurmurHash", GitHub.com. Accessed 18 Mar 2015
18. Jenkins, B.: SpookyHash: a 128-bit noncryptographic hash. Accessed 29 Jan 2012
19. Server Virtualization with VMware vSphere|VMware India. www.vmware.com. Accessed 08 Mar 2016
20. https://datastax.github.io/python-driver/api/cassandra/policies.html. Accessed 4 Jan 2016

3D Mesh Segmentation Based on Unsupervised Clustering

Dina Khattab[1]([⊠]), Hala M. Ebeid[1], Ashraf S. Hussein[2],
and Mohamed F. Tolba[1]

[1] Faculty of Computer and Information Sciences,
Ain Shams University, Cairo 11566, Egypt
dina.reda.khattab@gmail.com, fahmytolba@gmail.com,
hala@cis.asu.edu.eg
[2] Faculty of Computer Studies, Arab Open University, Headquarters,
13033 Kuwait City, Kuwait
ashrafh@acm.org

Abstract. 3D mesh segmentation is considered an important process in the field of computer graphics. It is a fundamental process in different applications such as shape reconstruction in reverse engineering, 3D models retrieval, and CAD/CAM applications, etc. It consists of subdividing a polygonal surface into patches of uniform properties either from a geometrical point of view or from a perceptual/semantic point of view. In this paper, unsupervised clustering techniques for the 3D mesh segmentation problem are introduced. The K-means and the Fuzzy C-means (FCM) clustering techniques are selected for the development of the proposed clustering-based 3D mesh segmentation techniques. Since the mesh faces are considered the main element, the clustering technique is applied to the dual mesh. The 3D Euclidean distance is used as the distance measure to compute matching between mesh elements. Based on empirical results on a benchmark dataset of 3D mesh models, the FCM-based mesh segmentation technique outperforms the K-means-based one in terms of accuracy and consistency with human segmentations.

Keywords: 3D mesh segmentation · Unsupervised clustering · K-means · FCM

1 Introduction

In the field of Computer Graphics, 3D models are used to describe the shape of a 3D space object that they represent. Different tools are used to represent these shapes, for example, the surface of curve modeling, constructive solid geometry, voxels, etc. They are used to create primitive shapes that are grouped together to obtain the final model. The most popular way to represent the geometric model is the 3D polygon mesh.

A polygon mesh is defined as a collection of connected polygons (i.e. bounded planar surfaces of different sizes) which approximate the surface of any 3D object. Polygons can be triangles, quadrangles or any kind of polygons. A polygon mesh M consists of a triple of three different kinds of mesh elements $\{v, e, f\}$; where v is the set of vertices, e is the set of edges and f is the set of faces.

© Springer International Publishing AG 2017
A.E. Hassanien et al. (eds.), *Proceedings of the International Conference on Advanced Intelligent Systems and Informatics 2016*, Advances in Intelligent Systems and Computing 533, DOI 10.1007/978-3-319-48308-5_57

The information describing the mesh elements consists of the mesh connectivity, mesh geometry and mesh attributes. The mesh connectivity (also referred as topology) describes the incidence relationship between the mesh elements. The incidence relations specify -for each face- the vertices and edges on the bounding loop, for each edge the end vertices and the faces to which the edge is incident, and for each vertex the incident edges and faces. The mesh geometry defines the coordinate values of vertices in 3D space which make up the model. The mesh attributes (also referred as photometry) are such as color, normal and texture coordinates at the vertices that describe the visual appearance of the mesh at rendering time. The problem of mesh segmentation consists of decomposing the polygonal surface of mesh M into a set of different regions S as a connected set of faces or vertices of uniform properties, where $S = \{M_0,...,M_{k-1}\}$.

Mesh segmentation is considered an important process in the field of computer graphics. It is a fundamental process in different applications such as shape reconstruction in reverse engineering, 3D models retrieval, and CAD/CAM applications, etc. Several recent works have studied the problem of 3D mesh segmentation [1–4].

The authors in [5, 6] have introduced the use of the unsupervised clustering techniques of K-means [7] as hard clustering and Fuzzy C-means (FCM) [8, 9] as a Soft/Fuzzy clustering for the segmentation of 2D images. Their work was based on modifying the classic semi-automatic GrabCut [10] segmentation technique for an automatic one. The proposed work of this paper extends the use of the two clustering techniques for the problem of 3D mesh segmentation. Evaluations on a proper benchmark dataset of 19 models are carried out. The proposed segmentation techniques achieved acceptable segmentation results compared to human segmentations.

In the following, Sect. 2 presents the related work to the different approaches of 3D mesh segmentation and introduces the most popular segmentation techniques. Section 3 presents the proposed work achieved for segmenting 3D meshes. Section 4 illustrates the algorithms of K-means and FCM clustering techniques used throughout the paper. Section 5 illustrates the achieved experimental results while presenting the benchmark of the 3D models and the metrics used for evaluating the segmentation techniques. Conclusions and future work are presented in Section 6.

2   Related Work

The 3D mesh segmentation techniques can be categorized as Region Growing techniques, Hierarchical Clustering techniques, Iterative Clustering techniques, Spectral Segmentation techniques, Skeleton Extraction Based Segmentation techniques, Interactive techniques, and Learning Segmentation techniques. Concerning the region growing techniques, they start by selecting a seed element and grow by adding successively compatible elements e.g. vertices which satisfy a given criterion. Then the growing process is repeated with a new seed element each time the previous growing is interrupted. The algorithm stops when all the seed elements are visited.

Zeckerberger et al. [11] have proposed a segmentation method which is based on a region growing algorithm. The algorithm starts by computing the dual graph of the input mesh. Each vertex of the graph represents a face of the mesh, and the edges which connect the vertices represent the adjacency relation between the faces. Then, the algorithm randomly selects a vertex to continue by collecting faces which form a convex patch (or convex region). The process of computing a new patch is launched each time the previous one is interrupted because of the violation of the convexity.

In Lavoué et al. [12] work, the curvature is first calculated for all vertices of the mesh and classified into several clusters. A region growing mechanism then extracts connected regions (associated with similar curvature), starting from several seed-facets.

In the hierarchical clustering, each segment is initially represented by a unique mesh element. Then each pair of adjacent segments is assigned a merging score based on a given criterion e.g. the boundary shape or the region curvature. After that, the merging process is implemented according to the increasing order of scores. Some parameters such as the region size are used to control the size of the clusters and decide if they should continue the merging operation or not.

Attene et al. [13] proposed a segmentation algorithm based on fitting primitives. The set of primitives includes plan, sphere, and cylinder. The algorithm generates a binary tree of segments where each segment is fitted to one of the employed primitives. To this end, all possible pairs of adjacent segments are considered (initially each pair of the segment is represented by two adjacent facets), and the pairs that are fitted well with one of the defined primitives, form a new segment.

In the iterative clustering approach, an iterative process begins by K clusters, each one with a representative centroid, where K is a predefined number representing the needed number of segments. Then, each mesh element is added to the closest cluster according to a certain distance function. This distance between the vertex and the each class centroid is measured based on a certain criterion. Once all vertices are added to the different clusters, the centroids are updated, and the iterative process is repeated until the centroids stop changing.

Lai et al. [14] proposed a clustering-based iterative segmentation algorithm dedicated for large models with high connectivity. They first generate a mesh hierarchy suitable for segmentation using a feature sensitive re-meshing algorithm. Then, they apply a clustering algorithm to segment the mesh. To increase the robustness of their method against geometric variations on the regions of the mesh, they introduced a metric which allows efficiently computing the distance between faces for clustering. The metric definition includes geodesic distance, integral invariants related to averaged normal curvature [15], and statistical measures of these invariants characterizing local properties such as geometric texture.

In the spectral segmentation approach, the combinatorial graph partitioning problem is converted into a geometric partitioning problem. In this problem, any mesh is represented as a graph with two A and D matrices, where A is the adjacency matrix and D is the degree matrix. Then, the Laplacian L of the graph is computed as $L = D - A$. The Laplacian eigenvectors are computed, and the graph is embedded in the R^K space using the first K eigenvectors.

Liu et al. [16] have proposed a segmentation algorithm that makes use of spectral space partitioning. They defined a symmetric affinity matrix $W \in R^{n \times n}$ such that for each i, j, $0 \leq W_{ij} \leq 1$ encodes the probability that two faces i and j can be grouped in the same segment. The matrix allows avoiding grouping facets which are separated by concave regions. The W matrix can be seen as a graph adjacency matrix A. The spectral analysis of the W matrix (using its first k largest eigenvectors) creates a partitioning which makes a segmentation of the mesh. The authors in [17] have proposed another algorithm which improves the results of their previous one [16] by making use of contour analysis besides the spectral analysis.

In the Skeleton Extraction Based Segmentation approach, an approximate skeleton of the input mesh is computed, and then each critical joint of the skeleton will correspond to a segment. In the segmentation algorithm proposed by Tierny et al. [18], the skeleton is used to restrict the object core and to identify the junction surfaces. Thus, the resulting segmentation is a coarse one which is refined following a hierarchical schema based on the topology of the model. The authors developed their own algorithm that allows computing an enhanced topological skeleton. The algorithm seeks to follow both of topological and geometrical variations of mesh contours computed using a geodesic mapping function.

Another work by Au et al. [19] was developed that differs in the skeleton computation. Their work was base on mesh contraction where a Laplacian smoothing is applied on the mesh to iteratively contract its geometry. The contraction allows removing details from the input mesh and leads to a zero-volume mesh. The conversion is carried out by removing all the collapsed faces from the zero-volume mesh. In order to perform this removal, a sequence of edge-collapse operations is applied using a coast function.

The approach of interactive techniques requires user interaction, where the user draws on the 3D mesh a set of sketches which corresponds to future segments. Each set of vertices traversed by a sketch is assigned a unique label. Then, a queue is filled with the unlabeled vertices where distance is measured between each unlabeled vertex and its direct labeled neighbors. An iterative process is applied where the queue vertex with the minimum distance is assigned to the closest segment. The process stops when the queue is empty.

Generally, the main difference between the proposed methods relies on the definition of the criterion to decide how to merge the mesh elements. For Example, Wu et al. [20] use the angular distance, while Fan et al. [21] use the Gaussian mixture and shape diameter function.

In the Learning Segmentation approach, the problem is defined as an optimization problem where an objective function is learned using a set of geometric features from a collection of labeled training meshes of ground truth segmentations. These ground truth segmentation databases have given the opportunity to quantitatively analyze and learn the mesh segmentations. The work proposed by Kalogerakis et al. [22] allows to simultaneously segment and label the input mesh. The problem consists in optimizing a Conditional Random Field (CRF). The algorithm has demonstrated its efficiency through the improvement of the results over the state-of-the-art of mesh segmentation.

3 Clustering Techniques

This section describes the algorithms of K-means clustering technique as hard segmentation approach and FCM clustering technique as a soft segmentation approach that are used throughout the paper.

3.1   K-means

K-means is one of the most popular and well used hard clustering algorithms. The standard algorithm was first proposed by Stuart Lloyd in 1982 [7]. The algorithm works as follows:

1. Select the number of desired clusters k and assign the k cluster centers randomly to different initial mesh vertices.
2. For the remaining mesh vertices, assign each mesh vertex to the cluster whose center is the closest i.e. the cluster which it is most similar to, using a measure of distance or similarity.
3. Recompute the cluster centers by getting the average coordinates of the mesh vertices that make up the cluster.
4. Go to step 2 until no more changes occur, i.e. the cluster centers remain the same, or a maximum number of iterations is reached.

3.2   FCM

Fuzzy C-means (FCM) [11, 12] is considered one of the famous clustering approaches for segmentation. The algorithm works as follows:

1. Initialize the fuzzy centroids C_k randomly for $K = 1, ...,nc$ where nc is the number of clusters.
2. Repeat for N iterations:
3. Compute the fuzzy membership $U^{(k)} = [u_{ij}]$, which is the membership of data point i to cluster j using

$$u_{ij} = \frac{\left(\frac{1}{\|x_i - c_j\|}\right)^{\frac{1}{m-1}}}{\sum_{j=1}^{nc}\left(\frac{1}{\|x_i - c_j\|}\right)^{\frac{1}{m-1}}} \tag{1}$$

4. Update the fuzzy centroids using the computed fuzzy memberships, where m is the fuzzy parameter and n is the number of data points.

$$c_j = \frac{\sum_{i=1}^{n}\left(u_{ij}\right)^m x_i}{\sum_{i=1}^{n}\left(u_{ij}\right)^m} \tag{2}$$

5. If $\|U^{(k)} - U^{(k-1)}\| < \varepsilon$, then STOP, else continue iterations.
6. Determine membership cutoff. For each data point x_i, assign x_i to cluster j where u_{ij} is the maximum among $U^{(k)}$ for $K = 1, ...,nc$, or $u_{ij} > \alpha$

4 Proposed Work

A benchmark for 3D mesh segmentation is used for quantitative evaluation of the proposed clustering-based 3D mesh segmentation techniques. The benchmark includes 3D meshes from the Watertight Track of the 2007 SHREC Shape-based Retrieval Contest provided by Daniela Giorgi [23]. The dataset contains 380 models spread evenly among a broad set of 19 object categories including human bodies, furniture, mechanical CAD parts, animals, tools, etc. These models exhibit different shape variability within each category and are represented in a form that most segmentation algorithms are able to use as input.

All models of the benchmark dataset are considered as closed manifold triangular meshes with only one connected component. A mesh M is called manifold if every vertex v has a neighborhood homeomorphic to a disk or half a disk. A closed mesh is the one that does not contain any boundary edges which has only one adjacent face f.

In addition to the 3D mesh models, the benchmark dataset includes ground truth segmentations that are collected using manual segmentation from human subjects. Figure 1 shows one example of the human ground truth segmentations provided for each category.

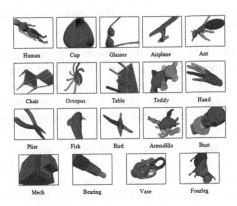

Fig. 1. Examples of the human ground truth segmentations provided for 19 categories of the benchmark dataset for 3D mesh models. Colors distinguish the different segments.

Since the process of 3D mesh segmentation is interested in clustering the mesh faces into different segments, the mesh face is considered the pivot mesh element for the clustering technique. For this reason, the algorithm starts by computing the dual graph of the input mesh. In the dual representation, each vertex of the original mesh is considered as a face in the dual mesh. Two vertices in the dual mesh are connected by an edge if their original faces were adjacent.

In order to achieve meaningful comparisons, the number of initial cluster centroids of both implementations of the K-means and FCM is set to the exact number of K segments equal to the corresponding human ground truth segmentation. These K cluster centroids are selected randomly among the set of dual mesh vertices. The Euclidean distance is used as a measure of distance between all remaining mesh vertices and the cluster centroids. Equation 3 shows the formula used for computing the Euclidean distance d between two vertices p and q in the 3D space.

$$d(p,q) = \sqrt{(p_1 - q_1)^2 + (p_2 - q_2)^2 + (p_3 - q_3)^2} \tag{3}$$

5 Results and Discussions

In the carried out experiment of applying the clustering-based 3D mesh segmentation techniques, one object of each dataset category is used for evaluation. This results in a total of 19 different meshes. As the results of both the K-means and the FCM clustering techniques are affected by the initial selection of centroids, five different implementations of each technique are applied to each mesh model and the average result is selected.

For quantitative evaluations of the proposed techniques, two well-known performance accuracy measures [24–26] are used from literature. These metrics are appropriate for measuring the degree of refinement, similarity and consistency and provide meaningful comparisons between segmentation results and ground truth images [24, 27]. These metrics include:

1 *The Rand Index (RI)* [25], counts the fraction of pairs of pixels whose labels are consistent between the computed segmentation and the ground truth [28]. The RI range is [0, 1], where the higher value is better. In other words, the score of 'zero' indicates the labeling of the test image is totally opposite to the ground truth segmentation and 'one' indicates that they are the same on every pixel pair.
2 *The Global Consistency Error (GCE)* [24], measures the degree of overlap of regions. In other words, it measures the extent to which one segmentation can be viewed as a refinement of the other [28]. These related segmentations are considered to be consistent since they represent the same natural image segmented at different scales. The GCE range is [0, 1], where the lower value is better.

Table 1 illustrates the achieved results of K-means-based and FCM-based 3D mesh segmentation techniques for the selected dataset of mesh models including the RI and GCE accuracy measures. The last row shows the average RI and GCE calculated for the entire dataset over all categories for each technique. The second column shows the number of segments used for segmenting each mesh model. Figure 2 shows the visual segmentation results for both techniques for the 19 mesh models.

It can be observed from Table 1 that the FCM-based 3D segmentation technique managed to increase the average rate of RI and decrease the average rate of GCE on the dataset used. Since the higher RI value is the best and the lower GCE value is best, it is concluded that the FCM-based segmentation technique outperforms the K-means-based one. This illustrates the higher efficiency of the FCM-based segmentation technique in terms of achieving more accurate segmentations compared to human segmentations. According to the achieved results, the FCM-based managed to

Fig. 2. Visual 3D segmentation results achieved for the 19 category models. The first row of each model shows the results of the K-means-based clustering while the second row shows the results of the FCM-based clustering technique

Table 1. Experimental results of the clustering-based 3D mesh segmentation techniques

Name	No. of segments	K-means		FCM	
		RI	GCE	RI	GCE
Human	6	0.193	0.201	0.163	0.163
Cup	2	0.023	0.194	0.499	0.177
Glasses	3	0.291	0.089	0.536	0.156
Airplane	5	0.352	0.367	0.336	0.286
Ant	11	0.145	0.170	0.061	0.029
Chair	6	0.108	0.105	0.214	0.123
Octopus	9	0.115	0.231	0.153	0.145
Table	5	0.098	0.182	0.101	0.013
Teddy	8	0.195	0.398	0.038	0.050
Hand	6	0.282	0.315	0.406	0.185
Plier	3	0.106	0.066	0.247	0.077
Fish	5	0.523	0.221	0.441	0.268
Bird	4	0.106	0.292	0.197	0.156
Armadillo	7	0.141	0.391	0.106	0.195
Bust	5	0.171	0.489	0.233	0.237
Mech	2	0.038	0.209	0.038	0.034
Bearing	4	0.057	0.147	0.303	0.115
Vase	10	0.266	0.503	0.420	0.071
Fourleg	3	0.337	0.355	0.338	0.027
Avg.		**0.187**	**0.259**	**0.254**	**0.132**

increase the average rate of the RI accuracy measure by 138 % and decrease the average rate of the GCE accuracy measure by 50.9 %.

6 Conclusions and Future Work

The paper introduced the development of 3D mesh segmentation techniques that are based on the use of the unsupervised clustering of K-means and FCM for segmentation. The proposed techniques were evaluated against a dataset of 19 different models representing different categories from the benchmark of SHREC with human ground truths. The Rand Index (RI) measure and the Global Consistency Error (GCE) are used for evaluating the segmentation results.

The FCM-based technique managed to achieve an average RI rate of 0.254 and an average GCE rate of 0.132 compared to 0.187 and 0.259 achieved by the K-means-based clustering technique respectively. On other words, it managed to increase the RI rate by 138 % while decreasing the GCE rate by 50.9 %. These results showed that the FCM-based mesh segmentation technique outperformed the K-means-based technique in terms of accuracy and consistency. This work of clustering-based 3D segmentations would be considered a step toward applications of mesh deformations and reconstruction for future work.

References

1. Wang, H., Lu, T., Au, O., Tai, C.: Spectral 3D mesh segmentation with a novel single segmentation field. Graph. Models **76**(5), 440–456 (2014)
2. Guha, S.: 3D mesh segmentation using local geometry. Int. J. Comput. Graph. Anim. **5**(2), 37 (2015)
3. Gu, M., Duan, L., Wang, M., Bai, Y., Shao, H., Wang, H., Liu, F.: An improved approach of mesh segmentation to extract feature regions. PLoS ONE **10**, 10 (2015)
4. Jia, H., Zhang, J.: Extract segmentation lines of 3D model based on regional discrete curvature. Int. J. Sig. Process. Image Process. Pattern Recogn. **9**(1), 265–274 (2016)
5. Khattab, D., Ebied, H.M., Hussein, A.S., Tolba, M.F.: A comparative study of different color space models using FCM-based automatic GrabCut for image segmentation. In: Gervasi, O., Murgante, B., Misra, S., Gavrilova, M.L., Rocha, A.M.A.C., Torre, C., Taniar, D., Apduhan, B.O. (eds.) ICCSA 2015. LNCS, vol. 9155, pp. 489–501. Springer, Heidelberg (2015). doi:10.1007/978-3-319-21404-7_36
6. Khattab, D., Ebeid, H.M., Hussein, A.S., Tolba, M.F.: Clustering-based Image Segmentation using automatic GrabCut. In: Proceedings of the 10th International Conference on Informatics and Systems (INFOS 2016), Cairo (2016)
7. Lloyd, S.: Least squares quantization in PCM. IEEE Trans. Inf. Theor. **28**(2), 129–137 (1982)
8. Dunn, J.C.: A fuzzy relative of the ISODATA process and its use in detecting compact well-separated clusters. J. of Cybern. **3**(3), 32–57 (1973)
9. Bezdek, J.C.: Pattern Recognition with Fuzzy Objective Function Algorithms. Kluwer Aca-demic Publishers, Norwell (1981)

10. Rother, C., Kolmogorov, V., Blake, A.: GrabCut: interactive foreground extraction using iterated graph cuts. ACM Trans. Graph. **23**(3), 309–314 (2004)
11. Zuckerberger, E., Tal, A., Shlafman, S.: Polyhedral surface decomposition with applications. Comput. Graph. **26**(5), 733–743 (2002)
12. Lavoué, G., Dupont, F., Baskurt, A.: A new CAD mesh segmentation method, based on curvature tensor analysis. Comput. Aided Des. **37**(10), 975–987 (2005)
13. Attene, M., Falcidieno, B., Spagnuolo, M.: Hierarchical mesh segmentation based on fitting primitives. Vis. Comput. **22**(3), 181–193 (2006)
14. Lai, Y.-K., Zhou, Q.-Y., Hu, S.-M., Martin, R.R.: Feature sensitive mesh segmentation. In: Proceedings of the ACM Symposium on Solid and Physical Modeling, pp. 17–25. ACM (2006)
15. Manay, S., Hong, B.-W., Yezzi, A.J., Soatto, S.: Integral invariant signatures. In: Pajdla, T., Matas, J(George) (eds.) ECCV 2004. LNCS, vol. 3024, pp. 87–99. Springer, Heidelberg (2004)
16. Liu, R., Zhang, H.: Segmentation of 3D meshes through spectral clustering. In: Proceedings of the 12th Pacific Conference on Computer Graphics and Applications, pp. 398–305. IEEE (2004)
17. Liu, R., Zhang, H.: Mesh segmentation via spectral embedding and contour analysis. Comput. Graph. Forum **26**(3), 385–394 (2007)
18. Tierny, J., Vandeborre, J.-P., Daoudi, M.: Topology driven 3D mesh hierarchical segmentation. In: Proceedings of IEEE International Conference on Shape Modeling and Applications (SMI 2007), pp. 215–220. IEEE (2007)
19. Au, O.K.-C., Tai, C.-L., Chu, H.-K., Daniel, C.-O., Lee, T.-Y.: Skeleton extraction by mesh contraction. ACM Trans. Graph. (TOG) **27**(3), 44 (2008). ACM
20. Wu, H.-Y., Pan, C., Pan, J., Yang, Q., Ma, S.: A sketch-based interactive framework for real-time mesh segmentation. In: Computer Graphics International (2007)
21. Fan, L., Liu, K.: Paint mesh cutting. Comput. Graph. Forum Wiley Online Libr. **30**(2), 603–612 (2011)
22. Kalogerakis, E., Hertzmann, A., Singh, K.: Learning 3D mesh segmentation and labeling. ACM Trans. Graph. (TOG) **29**(4), 102 (2010)
23. Giorgi, D., Biasotti, S., Paraboschi, L.: SHREC: Shape retrieval contest: Watertight models track (2007). http://watertight.ge.imati.cnr.it/
24. Martin, D., Fowlkes, C., Tal, D., Malik, J.: A database of human segmented natural images and its application to evaluating segmentation algorithms and measuring ecological statistics. In: 8th IEEE International Conference on Computer Vision (ICCV), vol. 2, pp. 416–423 (2001)
25. Unnikrishnan, R., Hebert, M.: Measures of similarity. In: 7th IEEE Workshops on Application of Computer Vision, p. 394. IEEE (2005)
26. Meilă, M.: Comparing clusterings: an axiomatic view. In: Proceedings of the 22nd International Conference on Machine learning, pp. 577–584. ACM (2005)
27. Peng, B., Zhang, L., Zhang, D.: A survey of graph theoretical approaches to image segmentation. Pattern Recogn. **46**(3), 1020–1038 (2013)
28. Yang, A.Y., Wright, J., Ma, Y., Sastry, S.S.: Unsupervised segmentation of natural images via lossy data compression. Comput. Vis. Image Underst. **110**(2), 212–225 (2008)

Affective Dialogue Ontology for Intelligent Tutoring Systems: Human Assessment Approach

Samantha Jiménez[1][(✉)], Reyes Juárez-Ramírez[1], Victor Castillo Topete[2], and Alan Ramírez-Noriega[1]

[1] Universidad Autónoma de Baja California, Mexicali, Mexico
{samantha.jimenez,reyesjua,alan.david.ramirez.noriega}@uabc.edu.mx
[2] Universidad de Colima, Colima, Mexico
victorc@ucol.mx

Abstract. Researchers has focused on integrate the dialogue to learning systems, such as Intelligent Tutoring Systems. To date, the researchers have focused on make more intelligent, accuracy and faster this systems. However, they become cold. So, it is important that these systems consider affective aspects when they bring feedback to the students because affective aspects and motivation have been shown to improve learning in face-to-face scenarios. This work introduces a novel ontology which models aspects of student, tutor and dialogue to bring an affective feedback. We conducted a human assessment approach to evaluate the proposed ontology. We design an instrument derived from ontological components (concepts and relations). We collected 24 responses of professors of undergraduate programs. Most of the domain experts, more than the 70 %, were agree with the relations proposed in the affective dialogue ontlogy proposed in this work. The results suggest that the proposal will serve as useful guidelines for the design of Intelligent Tutoring Systems, particularly to dialogue modules.

Keywords: Affective dialogue · Ontology · Intelligent tutoring systems

1 Introduction

Dialogue is essential to the learning process independently of the type of tutoring, educational level, face to face or with a computer tools in distance [10]. In classrooms the most used type of dialogue is the speech, where feedback is given immediately [16]. When students interact with a computational tool the dialogue is commonly written [1]. Regular communication between tutor and student has a positive effect on student's motivation [2]. There are some advantages of a frequent dialogue between these two actors (tutor and student) [5]:

1. The teacher will understand better how the presentations of the topic are received and the teacher can adjust the lectures to the audience.

© Springer International Publishing AG 2017
A.E. Hassanien et al. (eds.), *Proceedings of the International Conference on Advanced Intelligent Systems and Informatics 2016*, Advances in Intelligent Systems and Computing 533, DOI 10.1007/978-3-319-48308-5_58

2. The student may receive elucidation of a particular point and benefit from the rest of this particular lesson, where he or she might otherwise be lost. The question might be of benefit to a number of other students who had not thought of it.

Although the benefits of the dialogue in different learning contexts, sometimes students do not ask questions (doubts) to tutors, so that, tutor does not know when to assist student. The absence of dialogue during the learning process causes problems such as low motivation in the student and poor academic performance [1,16,19].

In recent years, education has been promoted the independent study and distance learning through computational tools [18,25]. This is where the Intelligent Tutoring Systems (ITS) are a viable option to support the student in the independent study. However, as we mentioned above, it is important that these systems provide feedback to student [9]. Feedback is crucial in the learning process [16]. On time feedback can help learners tackle problems, get to know the quality of their work, based on which learners can reflect and adjust learning ways so as achieve the purpose of effective learning [29]. For that reason, the integration of dialogue into ITS became a priority in order to improve student's motivation and learning performance [12,27].

When ITS emerged, they provided effective support to students taking into account factors such as cognition and learning style of the student. However, ITS do not know emotional aspects of student during the tutoring process, which are also important factors during the learning process [14]. As we mentioned earlier, in traditional tutoring, human tutors detect the emotional state of the student and provide feedback according to the detection in order to positively impact the student performance [20]. This led to the integration of affective aspects to ITS.

To date, research works have focused on addressing the first part of this process, detecting student's emotional states during the tutoring [6–8,15]. It should be noted that research about this area has progressed in providing to computers the ability of recognize the emotional state of the student such as human tutor does. These systems became faster, accuracy, intelligent; but they became colder. The question arises *how ITS should provide affectivity to the student?*

According to the latter, this paper purposed an affective dialogue ontology for ITS. The ontology suggests bringing affectivity using words, gestures and tone of voice. It is well known that the affectivity are strong related to motivation and these factors improve the learning process [2,28].

Taking into account that an ontology is a semi-formal model (that could be validated), we design a conceptual model using an ontology for representing the affective dialogue in learning. The purposed ontology will allow us to support the dialogue module design for ITS.

This paper is organized as follows. Section 2 describes the affective dialogue ontology's design. Section 3 shows the experimental design. Section 4 summarizes the results of this study. Finally, Sect. 5 presents our conclusions.

2   Affective Dialogue Ontology

Based on a systematic literature review in the affective learning domain, we identified some elements that ITS need to considered. First, some qualities of the student that ITS need to take into account to bring an affective tutoring. The, it is important to select which aspects of ITS should be considered and how should be the dialogue to provide affectivity to the student.

The design of affective dialogue ontology was guided by a methontology [11]. In a previous work we explain the complete method used to construct the ontology and mathematical analysis of the ontology [17]. However, it is important to note that we evolve the ontology presented in the previous work. Table 1 presents the terminology concepts extracted from the literature [2,3,13,19,24,26].

Table 1. Terminology glossary.

Student	ITS	Dialogue
Motivation	Empathy	Gestures
Objectives	Friendly attitude	Voice tone
Learning style	Immediacy	Words
Cognition	Respect	Affectivity
Emotional intelligence	Individual Support	Speech acts
		Grammar
		Phrases

Student is the person who acquires knowledge [19], and is the principal actor of the learning process. In order to achieve an affective learning is important to consider student's motivation, objectives, cognition, learning style, and the emotional intelligence. The motivation is important for the student because is the reason to act or to accomplish learning [26]. In the same way, the student needs to establish goals or objectives. Also, the ability to acquire, store, retrieve and use knowledge (cognition) [24] is necessary into the learning process. If this learning is conceived taking into account the learning style of the student, preferences for learning, it is possible that the learning will be more effective. Finally, the ability if perceive, recognize and use emotions is needed by both actors, student and ITS.

ITS must be sensitive to these aspects of the student. In a human tutoring, the tutor takes into account student's needs, and then he interacts, with the student, in an appropriate and friendly manner. However, the current ITS are no sensible systems. For that reason, we suggest some aspects of affective learning that need to be added to these systems. In a perfect scenario, ITS must be friendly and respectful to the student, the student needs individual support with immediacy and the tutor expressed empathy through the dialogue.

Our propose was to divide the dialogue into speech acts, grammar, phrases and words, in order to bring affectivity and to reduce the dialogue complexity. Speech acts are a theory that assumes the language is an action and when people say something, do something [22]. In addition, each speech act must has grammar to structure sentences. The speech acts, grammar, phrases, gestures and voice tone must has affectivity.

We integrated all the aspects mentioned above into an ontology, the graphical representation if this ontology is shown in Fig. 1.

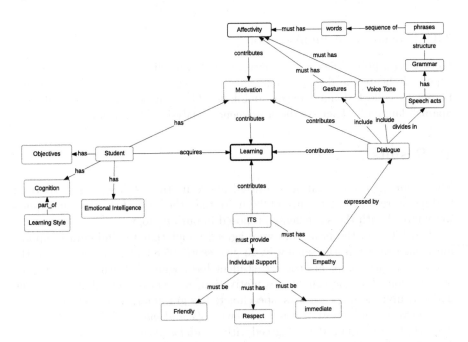

Fig. 1. Affective dialogue ontology.

3   Experimental Design

Ontology evaluation is a crucial activity during the ontology life cycle [30]. One of the most well known ontology assessment approaches is the one achieved by domain experts such as [30] suggested. We conducted an evaluation using this approach.

3.1   Sample

The questionnaire was completed by 24 domain experts. The domain experts were professors teaching in engineering undergraduate programs in Autonomous University of Baja California and Technological Institute of Tijuana. The 24 professors were from México. Study participation was voluntary.

3.2 Instrument Design

Zemmouchi et al. [30] purposed a methodological baseline for ontology assessment by humans (domain experts and end-users) via the usage of a questionnaire which were derived from ontology components.

First we design the instrument derived from affective dialogue ontology. A statement for every two concepts liked by a relationship was generated. For example, in a relation such as $C_1 \overrightarrow{uses} C_2$, the statement is: C1 uses C2. The instrument has 22 statements.

3.3 Procedural Administration

We used a 3-point Likert scale (agree, neutral and disagree) where the domain experts indicate the degree of agreement with each statement. We use Google Forms to collect domain expert opinions. The questionnaire was applied in Spanish language. Then we did a frequency analysis of the results.

4 Results

The results of the evaluation are listed above in Table 2. The most important findings were that in 20 items, of 22, more than the 70 % of the domain experts were agreed with the statements proposed in our ontology.

The 100 % of the professors were agreed about that the individual support must has respect. In the same way, 23 professors (95.83 %) were agreed that the following relationships are correct: student has cognition, learning style is part of cognition, dialogue contributes motivation, motivation contributes learning, dialogue divides in speech acts, speech acts has grammar, were correct.

The 87.5 % of professors were agreed with *the dialogue contributes in learning*. Also, the 95.83 % of professors agreed with *the dialogue contributes in motivation*. The 70.83 % of the professors were agreed that *the affectivity contributes in motivation*. And finally, the 95.83 % of the professors were agreed with that *motivation contributes learning*.

The ontology hypothetically establish what relations must exist between a pair of elements, but in real scenarios, some conditions are different. Answers with a low percentage suggest that the experts are not agreed with the statement or maybe they considered that this statement do not represent the reality, based on in their experience. For example, only the 58.33 % of the professors think that the student has learning objectives. This represents a chance for doing improvements in ITS, it is important to promote the inclusion of learning objectives in the design of ITS in order eradicate this deficiency that affects in real scenarios.

In the same way, the 66.66 % of the teachers believe that the individual support must be immediate. It is possible that the professors think it is not necessary that the feedback is in real time. This makes sense in traditional scenarios, where teachers can advise students in the next class session, however, in the context of computer-based tutoring systems, it is better to provide feedback in real time.

Table 2. The domain expert's responses.

Questions	Agree	Neutral	Disagree	Agree (%)
Student acquires learning	21	2	1	87.5
Student has motivation	17	2	5	70.83
Student has cognition	23	0	1	95.83
Student has objectives	16	6	2	66.66
Student has emotional intelligence	17	4	3	70.83
Learning style is part of cognition	23	1	0	95.83
ITS brings individual support	17	5	2	70.83
Individual support must be friendly	21	2	1	87.5
Individual support must has respect	24	0	0	100
Individual support must be immediate	14	8	2	58.33
ITS contributes learning	21	3	0	87.5
Dialogue contributes learning	21	3	0	87.5
Dialogue contributes motivation	23	0	1	95.83
Affectivity contributes motivation	17	4	3	70.83
Motivation contributes learning	23	1	0	95.83
ITS must has empathy expressed by dialogue	18	6	0	75
Dialogue divides in speech acts	23	1	0	95.83
Dialogue includes gestures	21	2	1	87.5
Dialogue includes voice tone	22	1	1	91.66
Speech acts has grammar	23	0	1	95.83
Grammar structure phrases	21	2	1	87.5
Words, gestures and voice tone must has affectivity	18	4	2	75

According, to these results we can establish that the affective dialogue ontology proposed in this work could be a based guidelines to implement dialogue modules in ITS. Also, it could provide some affective aspects of the tutor that must be considered to develop a new generation of ITS.

5   Implementation

Our affective ontology is focused on the tutoring system response. The objective is to integrate the affective feedback in tutoring systems. For our implementation, the students' actions are important because student starts the interaction in a turn n, however the proposal is focused on the tutoring system response (turn $n + 1$).

The n turn represents when a student is studying a topic and he/she wants to ask something he/she did not understand. In this implementation, the *Knowledge category (Remembering)* of Bloom's taxonomy [4] were considered. Knowledge category is used when a student defines, describes, identifies or knows concepts of a topic. For that reason, in the turn n the student ask a definition of a concept.

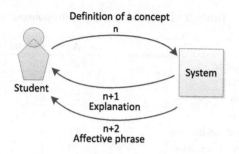

Fig. 2. Dialogue process.

The affective dialogue model is focused on the systems responses to the student when he asks a definition. In the turn $n + 1$ the system gives to the student an explanation of the concept ($d6$). Then, the turn $n + 2$ is a system act again; the tutoring system generate an affective phrase, in this case an encouragement phrase to the student in order to motivate him.(see Fig. 2).

As we mentioned earlier the main propose is the response to student, but the interpretation of the students' dialogue can be avoided. We propose to use a simple technique of auto-completion, which is a widely deployed facility in systems that require user input [21,23]. This allows student to avoid unnecessary typing. The auto-completion mechanism recommends the user a set of suggestions; these suggestions are meant to assist the user in finding what he wants quickly and also which information is available in ITS.

We suggest using the architecture show in Fig. 3 and a MVC pattern (Model-View-Controller) for the implementation.

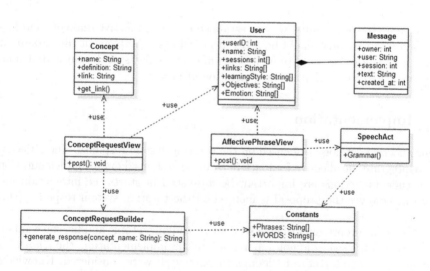

Fig. 3. Class diagram of our affective dialogue model propose.

The Model is represented by the databases of the system. The class *Concept* has the concept and their definitions and links of the learning material to structure the explanation. User class saves the users' information and Message has the conversation history. In the View, we propose to design a dialog box like a chat, and here is where the student see the messages from ITS and send his messages. We have a controller to each Speech Act (for example, Greeting, Congratulation, Goodbye, Encouragement). In each class for each speech act we have the grammar to structure the phrases. The Constants class is a static Model, it has predefined *Words as Phrases*.

6 Conclusions and Future Work

In this paper, we introduced our affective dialogue ontology based on a literature review. This ontology attempts to answer the research question mentioned earlier *how ITS should provide affectivity to the student?* We purposed a model that describes how can we bring affectivity through the dialogue to the student.

Based on the domain expert opinions, we deduce that it is important to consider some aspects of students such as: motivation, cognition, emotional intelligence and learning style. Also, the responses of the experts suggests that a dialogue module for ITS based on speech acts, gestures, voice tone and affective phrases will be an option to bring affectivity to students. This work could be a start point to define guide lines on the design of dialogue modules for ITS.

Moreover, the experts were agreed with include empathy, individual support, immediacy, friendliness and respect to ITS. It is important to mentions that most of the domain experts were agreed with the dialogue and the affectivity contributes to motivation which eventually will impact learning.

We have two statements were the percentage of agreement is less than 70 %. The domain experts thought that students has not objectives and it is not necessary to bring an immediacy support to them. However, maybe we have to go deeper in these aspects to improve our proposal.

It is well known that ITS cannot replace human tutors. However, we can improve the user experience with these systems making them more effective and affective taking into account the elements of the purposed ontology.

This work is the basis for some lines for future research, such as:

- Designing an ontology-based guidelines for affective ITS design.
- Implementing the affective dialogue module using this approach. The integration of this module to ITS.
- An evaluation from the students point of view.

References

1. Alcalá, M.D.S.P.: Afectos, aprendizaje y virtualidad. México, primera edn. (2012)
2. Angelaki, C., Mavroidis, I.: Communication and social presence: the impact on adult learners' emotions in distance learning. Euro. J. Open, Distance E-learn. **16**(1), 78–93 (2013)

3. Armour, W.: Emotional intelligence, student engagement, teaching practice, employability, ethics curriculum. Invest. Univ. Teach. Learn. **8**(2004), 4–10 (2012)
4. Bloom, B.: Taxonomy of Educational Objectives, Handbook I: The Cognitive Domain, 2nd edn. Addison Wesley Publishing Company, New York (1984)
5. Case, A.: Students who are reluctant to ask questions - Articles - UsingEnglish.com (2012). http://www.usingenglish.com/articles/students-who-are-reluctant-to-ask-questions.html
6. D'Mello, S., Olney, A., Williams, C., Hays, P.: Gaze tutor: a gaze-reactive intelligent tutoring system. Int. J. Hum. Comput. Stud. **70**(5), 377–398 (2012)
7. D'Mello, S.K., Graesser, A.: Multimodal semi-automated affect detection from conversational cues, gross body language, and facial features. User Model. User-Adap. Interact. **20**(2), 147–187 (2010)
8. Duffy, M.C., Azevedo, R.: Motivation matters: interactions between achievement goals and agent scaffolding for self-regulated learning within an intelligent tutoring system. Comput. Hum. Behav. **52**, 338–348 (2015)
9. Elizabeth, K., Michael, D., Mladen, A., James, C., Carolina, N.: Learner characteristics and feedback in tutorial dialogue. In: Proceedings of the Third ACL Workshop on Innovative Use of NLP for Building Educational Applications, pp. 53–61, June 2008
10. Feller, S.: The nature of dialog. In: Proceedings of the Workshop at SIGGRAPH Asia, WASA 2012, p. 73. ACM Press, New York, USA, November 2012. http://dl.acm.org/citation.cfm?id=2425296.2425309
11. Fernández-López, M., Gómez-Pérez, A., Juristo, N.: Methontology: from ontological art towards ontological engineering. Assessment SS-97-06, pp. 33–40 (1997)
12. Forbes-Riley, K., Rotaru, M., Litman, D.J.: The relative impact of student affect on performance models in a spoken dialogue tutoring system. User Model. User-Adap. Interact. **18**(1–2), 11–43 (2007). http://link.springer.com/10.1007/s11257-007-9038-5
13. García, B.: Las dimensiones afectivas de la docencia. Rev. Digit. Univ. **10**, 1–14 (2009)
14. Garrett, P., Young, R.F.: Theorizing affect in foreign language learning: an analysis of one learner's responses to a communicative portuguese course. Mod. Lang. J. **93**(2), 209–226 (2009)
15. Grafsgaard, J.F., Wiggins, J.B., Boyer, K.E., Wiebe, E.N., Lester, J.C.: Embodied affect in tutorial dialogue: student gesture and posture. In: Lane, H.C., Yacef, K., Mostow, J., Pavlik, P. (eds.) AIED 2013. LNCS (LNAI), vol. 7926, pp. 1–10. Springer, Heidelberg (2013). doi:10.1007/978-3-642-39112-5_1
16. Yangin, I.: Importance of feedback in teaching, communication and information systems for learning. Komunikacija i kultura online **1**(1), 309–317 (2010)
17. Jimenez, S., Juarez-Ramirez, R., Castillo, V.H., Ramírez-Noriega, A.: An affective learning ontology for educational systems. In: Rocha, A., Correia, A.M., Adeli, H., Reis, L.P., Teixeira, M.M. (eds.) New Advances in Information Systems and Technologies. AISC, vol. 444, pp. 1117–1126. Springer, Heidelberg (2016)
18. Juárez-Ramírez, R., Navarro-Almanza, R., Gomez-Tagle, Y., Licea, G., Huertas, C., Quinto, G.: Orchestrating an adaptive intelligent tutoring system: towards integrating the user profile for learning improvement. Procedia - Soc. Behav. Sci. **106**, 1986–1999 (2013)
19. Kopp, K., Britt, M., Millis, K., Graesser, A.: Improving the efficiency of dialogue in tutoring. Learn. Instr. **22**(5), 320–330 (2012)

20. Kort, B., Reilly, R., Picard, R.W., Media, M.I.T.: Affective model of interplay between emotions and learning- reengineering educational pedagogy-building a learning companion. In: Proceedings of IEEE International Conference on Advanced Learning Technologies, pp. 43–46, Madison, WI (2001)
21. Matani, D.: An O (k log n) algorithm for prefix based ranked autocomplete. English, pp. 1–14 (2011)
22. Moldovan, C., Rus, V., Graesser, A.C.: Automated speech act classification for online chat. In: CEUR Workshop Proceedings, vol. 710, pp. 23–29 (2011)
23. Nandi, A., Jagadish, H.V.: Effective phrase prediction. In: Proceedings of the 33rd International Conference on Very Large Data Bases, pp. 219–230 (2007). http://dl.acm.org/citation.cfm?id=1325851.1325879
24. Navarro, M.R.: Procesos cognitivos y aprendizaje significativo. Comunidad de Madrid Consejeria de Educación, Madrid, España (2008)
25. Pagano, C.M.: Los tutores en la educación a distancia. Un aporte teórico. Revista de Universidad y Sociedad del Conocimiento **4**(2), 1–11 (2008)
26. Rica, U.D.C., Pedro, S., Oca, M.D., Rica, C.: The Emotional Intelligence, its importance in the learning process. Educación **36**(1), 1–24 (2012)
27. Tetreault, J., Litman, D.: Using reinforcement learning to build a better model of dialogue state. In: EACL, pp. 289–296 (2006). http://acl-arc.comp.nus.edu.sg/archives/acl-arc-090501d3/data/pdf/anthology-PDF/E/E06/E06-1037.pdf
28. Tooman, T.: Affective learning : activities to promote values comprehension. In: Soultice Training, pp. 10–14 (2013)
29. Wang, D., Han, H., Zhan, Z., Xu, J., Liu, Q., Ren, G.: A problem solving oriented intelligent tutoring system to improve students' acquisition of basic computer skills. Comput. Educ. **81**, 102–112 (2015). http://www.sciencedirect.com/science/article/pii/S0360131514002231
30. Leila, Z.G., Réda, G.A., Bouguerra, M.: Position paper: a new approach for human assessment of ontologies, Université, U.M.B.B., October (2015)

Hybrid Randomized and Biological Preserved DNA-Based Crypt-Steganography Using Generic N-Bits Binary Coding Rule

Ghada Hamed[(⊠)], Mohammed Marey, Safaa El-Sayed Amin,
and Mohamed Fahmy Tolba

Faculty of Computer and Information Sciences, Ain Shams University, Cairo, Egypt
ghadahamed@cis.asu.edu.eg, {mohammedmarey,safaa_amin007}@hotmail.com,
fahmytolba@gmail.com

Abstract. In this paper, a blind crypto-stego technique is introduced using the cryptography and steganography concepts to achieve a double layered secured system. The proposed method consists of two phases: First is the data conversion to DNA using the proposed generic N-bits binary coding rule that leads to lower cracking probability proved by some comparative studies. Then, DNA and amino acids Playfair is applied to encrypt the DNA of the message resulting in ambiguity. In the second phase, the cipher text is placed with the ambiguity using 3:1 placement strategy. Then they are shuffled to be hidden in DNA at random positions generated by using a true real random number seed that is obtained from the atmospheric noise, thereby achieving very low cracking probability. The proposed technique is a blind preserved one as it achieves zero modification for the generated protein without extra data.

Keywords: Information security · Hybrid technique · Cryptography · Steganography · Playfair cipher · DNA · Amino acids · Binary coding rule · Cracking probability · True random number

1 Introduction

The information capacity is growing significantly as well as its level of importance, sensitivity and transformation rate. Consequently, high secured systems are required for thriving networks [1]. The most common and widely used methods in the computer security fields are cryptography and steganography [2–5]. Cryptography is the science of converting some data to an incomprehensible format that becomes unintelligible to a third party [6–8]. Steganography is the art of hiding sensitive data like text, images etc. in a different media such as image, video and audio in order to prevent attracting attention that the data is there [9,10]. The data hiding technique should preserve a minimum change in the characteristics of the original media after covering the data to conceal its existence [11–13]. Deoxyribonucleic acid (DNA) has been proposed for many

© Springer International Publishing AG 2017
A.E. Hassanien et al. (eds.), *Proceedings of the International Conference
on Advanced Intelligent Systems and Informatics 2016*, Advances in Intelligent
Systems and Computing 533, DOI 10.1007/978-3-319-48308-5_59

computational algorithms. A remarkable property of DNA is its vast storage capacity as one gram of DNA is known to store about 10^8 terra-bytes. However like every data storage device, DNA requires protection through a secured algorithm.

The rest of the paper is organized as follows: Sect. 2 is some related work then Sect. 3 discusses the proposed randomized hybrid crypto-stego technique using DNA. The algorithm's security is measured and analyzed in Sect. 4. In Sect. 5, the experimental results are given and discussed followed by the conclusions in Sect. 6.

2   Related Work

Wang, et al. brought back Clellands idea in [13] of using microdot to send half of the encrypted message, while the other half is sent via public channel [14]. Shiu et al. in [17], proposes three data hiding methods based on DNA: First, the insertion method where bits from the message are inserted in separated positions within a selected DNA sequence which results in expanding the length of the original sequence. In the complementary pair method, the longest complementary pairs in a DNA sequence is detected to hide the message parts before their positions which results in expanding the original sequence's length too. Finally, the substitution method is implemented by substituting some of the DNA nucleotides with others based on the message bits. GUO et al. in [9], 2012 introduce a data hiding scheme based on substituting the repeated bases only of a DNA sequence to minimize the modification rate.

Torkaman et al. in [3], introduce an innovative approach to improve hybrid cryptography that uses DNA steganography which is a combination of public and private keys to change the secret information. This is achieved by concealing the secret session key which is transferred among sender and receiver throughout an unsecured channel and this makes attackers not aware of exchanging the session key. Mitras and Aboo in [2], propose a new data hiding approach that uses DNA properties. This is achieved through two stages: The first stage is encrypting the secret message by using the RSA encryption algorithm, then in the second phase, the cipher secret message is hidden in it using the complementary rule where the complementary DNA sequence is the sequence on the other strand of DNA directly opposite a specified sequence.

3   The Proposed Approach

In the proposed work, a blind hybrid crypto-stego technique is achieved through 2 main stages at the sender and their reverse at the receiver which are:

3.1   Phase A1: Data Encryption - Sender's Side

In this paper, the secret message is encrypted after converting it into DNA format then to amino acids using the table in [15] derived from the standard universal table of amino acids and DNA codons. When, the message is converted from DNA to amino acids, there should be something that refers to the index of each DNA codon corresponding to each amino acid for successful decryption phase. These indices are defined as AMBIG which refer to ambiguity. In [2, 3, 6, 15, 16], the naive binary coding rule (BCR) maps each 2 bits to one DNA nucleotide. Then, the first choice for nucleotide A can be within 4 options; 00, 01, 10 or 11. Then, the overall unrepeated permutations for the rule = 4! = 24.

The binary coding rule can be generalized by representing each n base with N bit such that N = 2n. This is declared by the following example: Given A, AA, AAA, ..., (A...A) which are the DNA bases where A is repeated n times in the last term. Their corresponding binary bits are: 00, 0000, 000000, ..., (00...00), where 00 is repeated 2n times in the last term. As the example shown in Table 1, a 4-bits binary coding rule is proposed that maps each 4 bits of the binary message to $N/2$ DNA bases. Since AA can be assigned by 0000 or 0001 or ... or 1111 so it has 16 options. The next choice is for AC that will have 15 binary codes remaining after removing those assigned to AA. So, the overall unrepeated permutations = 16*15*..*2*1 = 16! that lead to low cracking probability as discussed in the security analysis section.

Table 1. Proposed 4-bits binary coding rule

DNA nucleotides	Binary representation	DNA nucleotides	Binary representation
AA	0000	GG	1000
AC	0001	GA	1001
AG	0010	GC	1010
AT	0011	GT	1011
CC	0100	TT	1100
CA	0101	TA	1101
CG	0110	TC	1110
CT	0111	TG	1111

Algorithm A1 shows the formal applied encryption algorithm.

Algorithm A1. Data encryption algorithm

Input: Secret message M and secret key K.
Step 1: Map M to binary using 8-bits coding to form M_{bin}.
Step 2: Convert M_{bin} to DNA using N-bits BCR in M_{DNA} then to amino acids in M_{AA} getting the ambiguity in AMBIG.
Step 3: Encrypt M_{AA} by applying DNA and amino acids Playfair using K and by applying the conventional rules of the main Playfair to form the cipher amino acids in CipherM$_{AA}$.
Step 4: Map CipherM$_{AA}$ to DNA using N-bits BCR in CipherM$_{DNA}$.
Output: Encrypted DNA text (CipherM$_{DNA}$) and ambiguity (AMBIG).

3.2   Phase A2: Steganography Phase - Sender's Side

The data hiding method is based on the LSBase method in [16] to obtain 0 biological modification after the steganography process. LSBase method is based on hiding a bit per LSBase of each codon in a DNA. First, the formed $CipherM_{DNA}$ is converted into binary ($CipherM_{bin}$) using the N-bits binary coding rule. The AMBIG is converted to binary ($AMBIG_{bin}$) to be ready for the hiding process. Since the maximum number of codons corresponding to an amino acid is 4, indexed from 0 to 3 so it can be represented as a maximum 2 bits. The $CipherM_{DNA}$ and $AMBIG_{bin}$ are hidden in arbitrary way in a DNA sequence using a true random number seed instead of using a pseudo-random. The true random seed comes from the atmospheric noise to be within 1 and the sequence's length results in lower cracking probability. Figure 1 shows an example for the 3:1 ratio of $CipherM_{DNA}$ to $AMBIG_{bin}$ that are hidden in a randomized way to avoid sending extra data as a separator between the ambiguity and the data. Algorithm A2 shows the formally applied steganography algorithm.

	m1	m2	m3	a1	m4	m5	m6	a2
$CipherM_{bin}$	0	0	1		1	1	0	
$AMBIG_{bin}$				1				1
S	CAC_3	TAT_6	GTA_9	TAA_{12}	AAA_{15}	CTA_{18}	ATT_{21}	CAA_{24}
RP	15	12	3	24	9	6	18	21
$CipherM_{bin}$	m3	m5	m4	m2	m1	m6	a2	a1
S`	CAC	TAC	GTG	TAA	AAA	CTA	ATC	CAG

Fig. 1. The hiding process of M_{bin} and $AMBIG_{bin}$ according to random positions (RP) generated using a true random seed number, m1, .., m6, .. are the message bits, a1 and a2 are the ambiguity bits, S is the original sequence before the steganography while S^* is the fake one after it

---

Algorithm A2. Data hiding algorithm

Input: $cipherM_{DNA}$, AMBIG, a selected DNA reference sequence S, a true random seed TRS within $[1, |S|]$.
Step 1: Convert ($cipherM_{DNA}$) into binary format using N-bits binary coding rule to form $cipherM_{bin}$ and AMBIG to binary in $AMBIG_{bin}$.
Step 2: Generate $|cipherM_{DNA}| + |cipherM_{bin}|$ random numbers within $[0, |S|]$ using TRS.
Step 3: Shuffle $cipherM_{bin}$ and $AMBIG_{bin}$ in $RandData_{bin}$ according to the generated number randoms leads to hiding the data in arbitrary way.
Step 4: Hide $RandData_{bin}$ in S in the LSBase of each codon by applying the proposed 3:1 data hiding strategy to get S^*.
Output: The S^* that contains $cipherM_{bin}$ and $AMBIG_{bin}$.

3.3 Phase B1: Data Extraction - Receiver's Side

The extraction process is simply the inverse of the embedding algorithm where S^* generated in algorithm 2 from step 4, is divided into codons. Using the true random seed, the random numbers of the data positions are generated to extract the shuffled hidden bits from the LSBases at those numbers using the 3:1 hiding strategy. Each three extracted consecutive bits by the LSBase method are added to the message and the next bit is added to the ambiguity of the secret message till $cipherM_{bin}$ and $AMBIG_{bin}$ are completely extracted from S^*. Finally $cipherM_{bin}$ is converted back to DNA using the N-bits binary coding rule in $cipherM_{DNA}$ and $AMBIG_{bin}$ is mapped to its corresponding decimal in AMBIG.

3.4 Phase B2: Data Decryption - Receiver's Side

Decryption is the inverse of the encryption phase. First, the $cipherM_{DNA}$ is converted to amino acids in $CipherM_{AA}$ to apply the Playfair cipher on it. M_{AA} is obtained by using the secret key that's used by the sender in the encryption process. Then, use each ambiguity number in AMBIG generated from the extraction phase with each amino acid to retrieve its corresponding codon in M_{DNA}. After that M_{DNA} is converted into binary using N-bits binary coding rule. To get the plain text of the secret message, the binary sequence extracted is converted to ASCII by using the 8-bits coding and finally the original form of the secret data is extracted.

4 Security Analysis of the Proposed Approach

Cracking probability, the probability of hacking an algorithm, is the most important parameter that indicates the security degree of an algorithm. In terms of security, to crack a steganography algorithm, some fundamental information including the used reference sequence, the binary coding rule, LSBase substitution rule and the positions sequence of the secret data must be known by an intruder to be able to extract the plain text of the secret data. Each following subsection gives a statistical study to show how each one contributes in decreasing the attacker's successful guess probability in an obvious way. This information is:

4.1 DNA Reference Sequence

There are 163 million DNA reference sequences available publicly. So, the probability of an attacker to make a successful guess to the correct chosen reference sequence (DNA_{Ref}) is:

$$P(DNA_{Ref}) = \frac{1}{163 * 10^6}.$$

(1)

4.2 Binary Coding Rule

The proposed N-bits representative binary coding rule represents each n nucleotides by N bits such that n = 2N, i.e. each DNA nucleotide base is mapped to two binary bits such that 2 N bits are used to represent N bases. Since there are 4 bases (A, C, G and T), the number of the possible combinations from n DNA bases are 4^n couples. For example, if it is the 4-bits binary coding rule that assigns 4(N) bits to each 2(n) nucleotide bases, it means that; AA can be represented by '0000', '0001', '0010', ..., '1111', so, it has 16 choices. If '0000' is selected to represent AA then AC can be represented by the remaining ones: '0001', '0010', ..., '1111', so, it has 15 options to be assigned to and so on till 16 pairs of nucleotides have been assigned a binary representation. So, the number of all possible rules for 4-bits binary coding rule is 16*15*14* .. *1 = 16!. Hence generically, the total number of required guesses to get the actual used N-bits binary coding rule to map each n DNA bases is (number of BCRs):

$$Number\_of\_BCRs = 4^n * (4^n - 1) * (4^n - 2) * .. * 3 * 2 * 1 = (4^n)!. \quad (2)$$

Given N which is the number of bits used to represent n DNA bases, then it can be described as the following:

$$N = 2n. \quad (3)$$

Since, the likelihood of making correct guess to the applied rule (BCR) is:

$$P(BCR) = \frac{1}{(4^n)!}. \quad (4)$$

From (3) and (4), it is derived that:

$$P(BCR) = \frac{1}{(2^N)!}. \quad (5)$$

4.3 The Least Significant Base Substitution Rule

LSBase hiding approach is applied by substituting the pyrimidine base by 'U' to encode the secret bit '0' or 'C' to encode '1'. But it can also encode '0' by C and '1' by U and the same for the purine base. So, briefly the '0' secret bit can be encoded by substituting the pyrimidine either by 'U'or 'C'. If it is selected to be substituted by 'U' then 'C' , it will be used to substitute the pyrimidine base to encode '1'. So the number of possibilities is 2*1 guesses and the same goes for the purine base. So, the probability of making successful guess for the substituted nucleotides is:

$$P(N) = \frac{1}{4}. \quad (6)$$

4.4   The Random Positions Sequence of the Hidden Secret Bits

Since the data is hidden in an arbitrary way in a DNA sequence(S) according to random numbers generated using a true random seed within $[1, |S|]$. So the probability of making successful guess for the seed and the random numbers sequence respectively is:

$$P(N) = \frac{1}{|S|} * \frac{1}{(\frac{|S|}{3})!}. \tag{7}$$

Therefore, using the proposed randomized hybrid technique, the total probability of an attacker to make a successful guess (SG) is very small as the following shows:

$$P(SG) = \frac{1}{163 * 10^6} * \frac{1}{(4^n)!} * \frac{1}{4} * \frac{1}{|S|} * \frac{1}{(\frac{|S|}{3})!}. \tag{8}$$

From the derived algorithm's cracking probability, it is concluded that as n increases, the search level for the used DNA reference sequence, binary coding rule and the substitution rule increases and consequently the cracking probability decreases.

5   Experiments and Comparison

A set of experiments are carried out to evaluate the performance of the proposed scheme in terms of multiple parameters. There are three parameters to be taken into consideration other than the cracking probability parameter in the DNA based steganography field. The first parameter is the capacity which is the total length of the modified DNA reference sequence after the secret message and ambiguity are hidden within it. The second parameter is the payload which is the remaining length of the DNA sequence after extracting the nucleotides of the original one. The third parameter is the bpn parameter which is the number of hidden bits per one nucleotide.

In this section, the main substitution method in [17] is compared with the proposed substitution algorithm. They were tested on Intel(R) Core (TM) i5-3230M CPU @ 2.60 GHz personal computer with 6 GB RAM. NCBI database [18] was used to get the data set of the real sequences. The message M used is of size 21 KB and the secret key for encryption is 'SECURITY'. The true selected seed is 1000 and 4-bits binary code rule is used in the implementation. Table 2 displays the experimental results in terms of capacity, payload and bpn parameters to evaluate the system's performance when 20 kB of data are used to be hidden in the eight reference sequences. It is obvious from the results in Table 2 that as the length of a DNA sequence increases, i.e. as the number of the sequence's bases increases, the capacity increases. Since only the LSBase of each codon of the reference sequence is used to hide one bit to preserve the biological functions, so it is concluded that the sequence's (S) capacity is:

$$Capacity = \frac{1}{3} * |S|. \tag{9}$$

Table 2. The results obtained using the proposed scheme to hide 20 KBs within the tested DNA sequences

Locus	Capacity (Bits)	Payload	$bpn = \frac{|M|+|A|}{C}$	Sender side (Time in seconds)	Receiver side (Time in seconds)
AC166252	46,685	0	3.7	0.49	0.32
AC168901	58,488	0	3.0	0.56	0.42
AC168907	60,032	0	2.9	0.62	0.42
AC153526	61,862	0	2.8	0.70	0.42
AC168897	60,077	0	2.9	0.68	0.43
AC167221	63,042	0	2.7	0.58	0.45
AC168874	63,271	0	2.7	0.60	0.44
AC168908	66,622	0	2.6	0.8	0.47

As shown in Table 2, since data hiding is based on the substitution, the payload is zero, which means that the fake sequence length is not expanded after the steganography, which avoids drawing the attackers' attention. The bpn is within the interval [2.6, 3.7] and the proposed scheme has an acceptable embedding capacity distributed on both the message and ambiguity bits. Finally, the time required to hide all the bits of the message and ambiguity is 0.5 s. It is shown from Table 2 that the capacity and the execution times are affected by the length of the sequence used for hiding i.e. the execution time is directly proportional to the length.

Table 3 compares the main features of the introduced scheme, the main substitution in [17] and the LSBase method in [16]. From the comparison criteria listed in Table 2, it was shown that the drawbacks in the 2 methods are overcome by the proposed one.

Table 3. A comparison between the proposed, main substitution and the Main LSBase methods

Comparison critera	Proposed method	Main substitution method [17]	LSBase method [16]
Capacity	$C = \frac{1}{3} * |S|$	$C = |S|$	$C = \frac{1}{3} * |S|$
Preserve DNA biological functions	Yes	No	Yes
Encryption Phase	Exists	Not exist	Not exist
Blind (Yes/No)	Yes	No	Yes
Payload	$P = 0$	$P = 0$	$P = 0$
Cracking probability	$CP = \frac{1}{(163*10^6)*((4^N)!)} * \frac{1}{(4)*(S)*(\frac{|S|}{3})!}$	$CP = \frac{1}{(163*10^6)*(6)}$	$CP = \frac{1}{(163*10^6)*(4)}$

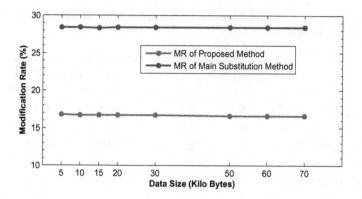

Fig. 2. Modification rate (MR) of the proposed technique against the main substitution one

Figure 2 shows the modification rate caused by the introduced approach and the main substitution method [17]. The graph proves that the proposed one achieves less modification rates than [17]. While this modification is considered biologically zero, since the substitution leads to the same protein sequence and consequently the same functions and characteristics of the original one are kept.

6   Conclusion

In this paper, a data hiding method is proposed by combining the means of cryptography and steganography as well as achieving a double layer security on the system. The first contribution is the generic binary coding rule that represents each n nucleotides with 2n (N) bits instead of representing each 1 nucleotide with only 2 bits which increases the number of rules from 4! to 4^n!, strengthen the algorithm's security. The second contribution is the random placement such as the message bits are not hidden in a continuous way within the DNA, as they are separated by ambiguity bits and in arbitrary way based on a true random seed. The random seed is selected from the atmospheric noise to be true random which leads to a very low cracking probability compared with other algorithms. The third contribution is the blindness property, as the embedded data can be extracted without the need of the original DNA sequence. Also, using the LSBase in hiding the secret message, the DNA biological functions are preserved as only the LSB of each codon is substituted and leads to the same protein sequence that leads to a zero payload which avoids attracting attention to the fake sequence. So, recovering the message and the ambiguity as well in a correct form is actually a complicated process for any third part.

References

1. Hamed, G., Marey, M., El-Sayed, S., Tolba, F.: DNA based steganography: survey and analysis for parameters optimization. In: Hassanien, A.-E., Grosan, C., Tolba, M.F. (eds.) Applications of Intelligent Optimization in Biology and Medicine. Intelligent Systems Reference Library, vol. 96, pp. 47–89. Springer, Switzerland (2015). ISSN: 1868–4394

2. Mitras, B.A., Abo, A.K.: Proposed steganography approach using DNA properties. Int. J. Inf. Technol. Bus. Manag. **14**(1), 96–102 (2013)

3. Torkaman, M.R.N., Kazazi, N.S., Rouddini, A.: Innovative approach to improve hybrid cryptography by using DNA steganography. Int. J. New Comput. Architectures Appl. (IJNCAA) **2**(1), 224–235 (2012)

4. Kayarkar, H., Sanyal, S.: A survey on various data hiding techniques and their comparative analysis. ArXiv preprint arXiv **5**(3), 35–40 (2012)

5. Hamed, G., Marey, M., El-Sayed, S., Tolba, F.: Hybrid technique for steganography based on DNA with N-Bits binary coding rule. In: 7th International Conference on Soft Computing and Pattern Recognition (SoCPaR), Unpublished manuscript, Japan. IEEE (2015)

6. Skariya, M., Varghese, M.: Enhanced double layer security using RSA over DNA based data encryption system. Int. J. Comput. Sci. Eng. Technol. (IJCSET). **4**(06), 746–750 (2013)

7. Siper, A., Farley, R., Lombardo, C.: The rise of steganography. In: Proceedings of Student/Faculty Research Day, CSIS, Pace University (2005)

8. Sharma, M.K., Upadhyaya, A., Agarwal, S.: Adaptive steganographic algorithm using cryptographic encryption RSA algorithms. J. Eng. Comput. Appl. Sci. **2**(1), 1–3 (2013)

9. Guo, C., Change, C., Wang, Z.: A new data hiding scheme based on DNA sequence. Int. J. Innovative Comput. Inf. Control **8**(1), 139–149 (2014)

10. Thakar, S., Aggarwal, M.: A review-steganography. Int. J. Adv. Res. Comput. Sci. Softw. Eng. **3**(12) (2013)

11. Nosrati, M., Karimi, R., Hariri, M.: An introduction to steganography methods. World Appl. Program. **1**(3), 191–195 (2011)

12. Das, S., Das, S., Bandyopadhyay, B., Sanyal, S.: Steganography and steganalysis: different approaches. Int. J. Comput. Inf. Technol. Eng. (IJCITAE), **2**(1) (2008)

13. Clelland, C.T., Risca, V., Bancroft, C.: Hiding messages in DNA microdots. Nature **399**, 533–534 (1999)

14. Wang, Z., Zhao, X., Wang, H., Cui, G.: Information hiding based on DNA steganography. In: 4th IEEE International Conference on Software Engineering and Service Science (2013)

15. Sabry, M., Hashem, M., Nazmy, T., Khalifa, M.E.: A DNA and amino acids-based implementation of playfair cipher. Int. J. Comput. Sci. Inf. Secur. **8**(3), 129–136 (2010)

16. Khalifa, A.: LSBase: a key encapsulation scheme to improve hybrid crypto-systems using DNA steganography. In: 8th International Conference on Computer Engineering and Systems (ICCES), Cairo, Egypt, pp. 105–110 (2013)

17. Shiu, H.J., Ng, K.L., Fang, J.F., Lee, R.C., Huang, C.H.: Data hiding methods based upon DNA sequences. Inf. Sci. **180**(11), 2196–2208 (2010)

18. National Center for Biotechnology Information. http://www.ncbi.nlm.nih.gov/

Parallel Implementation of Super-Resolution Based Neighbor Embedding Using GPU

Marwa Moustafa[2(✉)], Hala M. Ebeid[1], Ashraf Helmy[2],
Taymoor M. Nazamy[1], and Mohamed F. Tolba[1]

[1] Faculty of Computer and Information Sciences,
Ain Shams University, Cairo, Egypt
{halam, fahmytolba}@cis.asu.edu.eg
[2] Data Reception, Analysis and Receiving Station Affairs,
National Authority for Remote Sensing and Space Science, Cairo, Egypt
{marwa, akhelmy}@narss.sci.eg

Abstract. Neighbor Embedding (NE) is one of the powerful manifold techniques used in super-resolution, but due to the complexity of manifold learning algorithms, their vast computation times are very challenging. The Locally Linear Embedding (LLE) algorithm requires massive computing power, especially when applied to satellite images. In this paper, we propose a parallel implementation of a super-resolution method based on the locally linear embedding algorithm using the CUDA multi-thread model. Locally linear embedding algorithm is implemented on NVIDIA NVS 5200 M with 1 GB RAM, which is based on the optimized Fermi architecture of the GF108 chip and offers 96 cores. Experiment results show that the proposed method based on manifold learning outperforms significantly state-of-art methods. The proposed super-resolution method achieves an acceleration from $11 \times$ to $163 \times$ depending on the image size.

Keywords: Manifold learning · Super resolution · CUDA

1 Introduction

Super-resolution (SR) is a signal processing technique that substitutes the low frequency in the low-resolution (LR) input image, with their high frequency (HR) depending on prior knowledge. Over the past two decades, various super-resolution methods had been proposed which are grouped into classical interpolation methods and example-based methods. In this paper, we focus on the example learning based method where the high-frequency details can be predicted from prior information learned from the training set. The goal of the single image super-resolution method is to infer high-resolution details that are missing in the original low-resolution image and can't be retrieved by simple interpolation.

In [1, 2], example-based learning was introduced to SR work firstly by Freeman in 2002. The high-resolution image is estimated from a low-resolution input image with the prior knowledge estimated from a Markov network model for the training set examples, which means the pairs between the training low-resolution image and their

© Springer International Publishing AG 2017
A.E. Hassanien et al. (eds.), *Proceedings of the International Conference
on Advanced Intelligent Systems and Informatics 2016*, Advances in Intelligent
Systems and Computing 533, DOI 10.1007/978-3-319-48308-5_60

corresponding high resolution one. Single image super-resolution methods estimate the HR image based on two relationships: Low–low relation and Low–high relation. Low–low relation (LL-relation) discloses the compatibility between the input image and the training set samples. Low–high relation (LH-relation) is estimated by learning algorithms used on the training set samples [3].

Different probability models have been investigated to increase the flexibility and effectiveness of the learning phase by using machine learning statistical models. Inspired by Freeman's work, Chang et al. [4] introduced a novel super-resolution method based on locally linear embedding (LLE). An improvement was proposed by Chan et al. [5], where histogram matching was used to choose more relative training images to the input test image and the new combination of feature selection and different neighborhood sizes learning. In [7], the manifold learning method used to reconstruct high-resolution images from low input images by a linear weight summation of its several most similar training LR patches. Then, recover the HR patch by using the same linear summation of the corresponding training HR patches to reduce the computation calculations. Only the patches with larger variance, which means with high-frequency components, are selected for super-resolution procedures.

Sparse representation had been used to enhance the prior knowledge used in both the training and reconstruction steps. In [8], the histogram of oriented Gradient (HoG) is employed in the feature selection to characterize the local geometric structure then a sparse neighbor selection was proposed for the SR reconstruction. In [9], the authors proposed an improvement for local linear embedding (LLE). The SR method was improved by using a new feature selection vector which was obtained using the discrete cosine transform coefficients of the norm luminance. This new feature selection preserves the neighborhood well. To reduce the reconstruction time, each image patch is classified into flat or not, according to its variance; then it uses different methods to solve and the reconstruction time is greatly reduced. In [10], a hierarchical support vector machine (HSVM) to learn representative features of both training and test LR image patches is presented. Then, a sparse manifold assumption is cast on training patch features to find the appropriate number of local neighbors for each LR input patch. The obtained local neighbors are more real and feasible than the strict assumption in NE.

In this paper, we propose an acceleration of the locally linear embedding algorithm to be employed in an example-based super resolution method. The algorithm is implemented using CUDA API, and the performance is evaluated and compared with the implementation of the same algorithm on a single CPU and the state-of-the-art algorithm such as; the bi-cubic interpolation algorithm and nearest neighbor embedding algorithm. Also, the Multi-spectral Landsat-8 scene is used during the experiments. Various experiments are conducted to evaluate the effectiveness and the performance of the proposed method towards different parameters such as; patch size, the number of neighbors, robustness to the noise and the size of the training examples.

The rest of the paper is organized as follows. In Sect. 2, the super-resolution problem is formulated, and a utilization of manifold learning methods are introduced. In Sect. 3, acceleration to the proposed method was introduced using general computing CUDA. In Sect. 4, experimental results are reported, including feature representation and training set. Finally, Sect. 5 gives the conclusion of this paper.

2 Proposed Example-Based Super-Resolution Framework

In this section, the single image super-resolution problem is formulated. Consider the forward image model [10], let the high-resolution image $Y \in \mathbb{R}^N$ represents a vector $Y = [y_1, y_2, \ldots, y_N]^T$, where $N = L_1 N_1 \times L_2 N_2$, L_1 and L_2 is the down sampling factors in the image observation model and N_1 and N_2 are the height and width. The low-resolution image can be denoted as $= [x_1, x_2, \ldots, x_M]^T \in \mathbb{R}^M$ where $M = N_1 \times N_2$. The forward image model represents the observed low-resolution images from warping, blurring and sub-sampling the high-resolution image.

Generally, a single image example-based super resolution method includes two phases: Training and reconstruction phases. In the training phase, several HR images and their corresponding statistically similar LR images are collected to input test image.

From the single image super-resolution algorithm based on neighbor embedding proposed in [1, 4–7, 11, 17], we propose an improved SR algorithm based on the locally linear embedding algorithm as illustrated in Fig. 1. The proposed method algorithm is summarized follows:

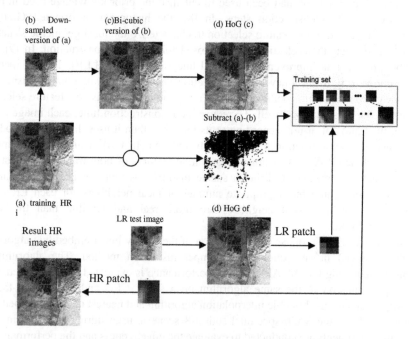

Fig. 1. Proposed example based Super-resolution framework.

[1] Training set setup: The original high-resolution images are blurred and down-scaled by a factor of 2 to obtain a low resolution. The resulting low-resolution images are scaled up using bi-cubic interpolation to achieve the mid-resolution, which lacks the high-resolution features.

[2] Feature extraction: Low features are extracted by utilizing the Histogram of Gradient (HoG) on the mid-resolution images. High features (lost components) are generated by subtracting the original high-resolution image and the mid resolution image (interpolated HR image).

[3] Vectorization: For each LR-HR image pair in the training set, a partition into a sequence of image patches with two pixels overlapped in a raster scan order is done and generates a training LR feature patch set with its corresponding HR feature patch set.

[4] Reconstruction Iteration: The initial HR image is produced using bi-cubic interpolation for the LR input image. First, the LR image is transformed to the mid frequency image using the same procedure as in the training phase. Then, the LLE is adopted for each patch in the input image; we iteratively carry out the steps of Locally Linear Embedding as illustrated in Algorithm 1 to obtain HR patches and merge them to obtain the HR image. For the overlapped regions between the adjacent patches, averaging fusion is used to obtain the final pixel value.

[5] De-blurring: Finally, we performed the TV based de-blurring on the estimated HR image. Then, the final HR image is produced by using the Iterative Back Projection procedure.

Algorithm 1. Local Linear Embedding (X: input LR image, K: Number of nearest neighbors, Y: Output LR image) **[12]**

1- For each patch $x_i \in X$, finds its K-nearest neighbors.

2- Compute the weights matrix w_{ij} that is best the linearly reconstruct x_i from its neighbors and minimizes the following cost function:

$$\varepsilon(w) = \sum_i \left| x_i - \sum_j w_{ij} x_j \right|^2$$

The sum of the coefficient of w_{ij} must be equal to 1: $\sum w_{ij} = 1$

3- Find the d-dimensional embedding vectors y_i by using the weights w_{ij}, which minimizes the following cost function

$$\varphi(Y) = \sum_i \left| y_i - \sum_j w_{ij} y_j \right|^2$$

---

3   GPU Implementation

Classical implementation of LLE became quite expensive for high dimensional datasets especially satellite images. Therefore, there is a real need to reduce the execution time of the algorithm by suggesting a faster algorithm implementation using affordable parallel architectures such as graphics processing units (GPUs). The GPU architecture [13] abides the Single Instruction Multiple Thread (SIMT) execution model.

We propose a parallel implementation of the LLE algorithm by using Compute Unified Device Architecture (CUDA) API [12] and CUBLAS library [14]. CUBLAS is a CUDA implementation of classical BLAS functions (vector/matrix operations). Five kernels are proposed to reduce the execution times. Line 1 in Algorithm 1 implements the k-nearest neighbors' algorithm. First, the kernel computes the Euclidean distance that is re-written to involve the matrix additions and multiplication. The addition and subtraction kernels are implemented to compute the norm using CUDA kernels. The fourth kernel implements the sorting algorithm had been proposed in [15]. Fifth kernel implements line-2 in Algorithm 1 that computes the weights matrix to best linearly reconstruct x from its neighbors is re-written to implement on the GPU. Line 3, in Algorithm 1, is implemented as serial C-MEX file function of Matlab to accelerate the computation of eigenvalues and eigenvectors. Figure 2 shows the proposed parallel implementation for the local linear embedding method.

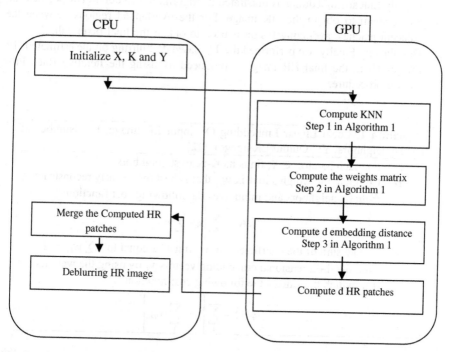

Fig. 2. Flow chart of the proposed local linear embedding method on CPU/GPU processing schemes.

4   Experimental Results

In this section, the consistency and stability of the proposed method were experimentally tested using multispectral satellite image. In Sect. 4.1, the hardware and software configurations are presented. Section 4.2 introduces a detailed description of the dataset. Finally, experimental results are reported in Sect. 4.3.

4.1 Hardware Configuration

The platform used to perform these experiments is an optimized Fermi based architecture called NVIDIA NVS 5200 M. This architecture, GF117, has a GF108 chip and offers 96 cores, 16 TMUs, and 4 ROPs. Each core is clocked twice as fast as the rest of the graphics chip. Table 1 defines the detailed hardware specification used in our experiments.

Table 1. Hardware specification used in the experiment.

Host specification	CPU	Intel ® Core (TM) i7-3740QM
	Memory	4 GB, 800 MHz DDR2
	Operating system	Windows 7 x64
Device specification (GPU)	GPU	NVIDIA NVS 5200 M @ 625 MHz, 96 cores; 1024 MB memory size
	Shared memory per block	16 KB
	Compiler	Visual Studio 2012
	CUDA runtime version	CUDA v 6
	Max # of threads per block	512
	Warp size	32

4.2 Dataset

The dataset used to perform these experiments is the Landsat-8 image [16]. NASA launched on 11 February 2013 and recently it became available free of charge [16]. It provides moderate resolution imagery, from 15 m to 100 m. Landsat 8 operates in the visible, near-infrared, short wave infrared and thermal infrared spectrums. The Landsat-8 scene is located in Egypt at a latitude of $21°40'17.69''$N and longitude of $35°58'21.47''$E. In our experiments, a true-color image acquired on the 2014/02/15 is used. Fourteen 512×512 pixels were cropped from the original Landsat-8 image to be used in the training phase as shown in Fig. 3. Five windows of sizes 64×64, 128×128, 256×256, 512×512, and 1024×1024 were cropped from the original Landsat-8 image to use in the reconstruction phase as shown in Fig. 4. The cropped sub-images were cropped to represent the diversity of features such as water, deserts, urban, roads and agriculture areas.

4.3 Results

To validate the performance of the proposed locally linear embedding using the single image super-resolution, we compared our parallel algorithm implementation with other algorithms including; bi-cubic interpolation algorithm and nearest neighbor embedding algorithm [4]. The low-resolution images corresponding to the fourteen training

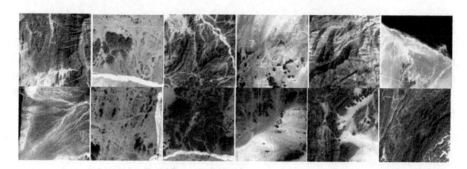

Fig. 3. Samples of images used in the training set.

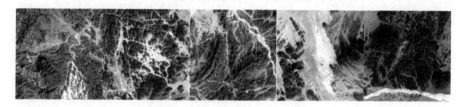

Fig. 4. Five windows images of different size utilized in the reconstruction phase.

high-resolution images are generated by down sampling (factor 2) using bi-cubic interpolation and for all experiments; the magnification factor is set to be 2. The size of the low-resolution patch is set to be 3 × 3, the number of overlapping pixels is 2, and the number of nearest neighbors is 5.

During our experiments, some preprocessing has been done; we considered the natural color combination (4, 3, 2) to the RGB bands. We converted images from the RGB color model into the YIQ color model. The super resolution method is applied only to the luminance values (Y channel). I and Q channels were enlarged to the required size using the bi-cubic interpolation. The feature vector is generated using a Histogram of oriented gradients (HoG) of the luminance.

The performance of the proposed method is evaluated using the Peak Signal to Noise Ratio (PSNR). Table 2 shows a comparison between the proposed sequential CPU and the proposed parallel GPU solutions against the bicubic interpolation method, Freeman et al. [1] and Chang et al. [4] methods using the PSNR. Figure 5 shows the reconstructed image for two images of the five test images using different super-resolution algorithms. Visually, the HR reconstructed images show that more details are well estimated by the proposed method than those estimated by the nearest neighbor based algorithms. Table 2 shows no considerable change in accuracy between the sequential and the parallel implementations of the proposed method. Besides, the proposed method outperforms state-of-the-art methods.

Different experiments are carried out to investigate the noise impact to the proposed method using *image 03* of the size of 256 × 256. The image is corrupted by adding different Gaussian noise filters (with zero mean and variance $\sigma = 5$, $\sigma = 10$, $\sigma = 20$, and $\sigma = 50$). Figure 6 shows the PSNR values corresponding to different

Table 2. PSNR (dB) values for the HR recovered images using different SR methods

Images	Bicubic interpolation	Freeman et al. [1]	Chang et al. [4]	CPU proposed	GPU proposed
Image 01	28.06	28.56	28.87	29.14	29.14
Image 02	27.29	27.34	27.64	27.83	27.83
Image 03	25.31	25.64	25.85	26.13	26.13
Image 04	30.15	30.28	30.33	30.69	30.69
Image 05	29.08	29.12	29.74	29.92	29.92
Average	**27.978**	**28.188**	**28.486**	**28.742**	**28.742**

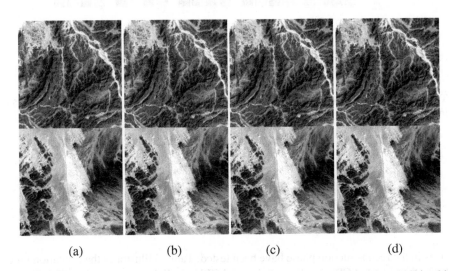

(a)   (b)   (c)   (d)

Fig. 5. Reconstructed HR images for two images of the 5 test images using (a) bi-cubic interpolation, (b) Freeman et al. [1] method, (c) Chang et al. [4] method, (d) GPU proposed method

super-resolution algorithms using different Gaussian noise variance. One can observe from Fig. 6 that the PNSR values for all SR algorithms declined over the increase of the noise. In the bi-cubic and Chang methods, the quality of the HR image changes marginally at the noise variance ($\sigma = 5$, $\sigma = 10$, $\sigma = 20$), while the quality decreased slightly from 2.6 % at $\sigma = 20$ to 8.6 % at $\sigma = 50$ for the bi-cubic method and from 6.8 % at $\sigma = 20$ to 10 % $\sigma = 50$ for the Chang method. In the Freeman method, the quality of the recovered image declined dramatically from 8.7 to 10.1 % for noise variance $\sigma = 20$ and $\sigma = 50$ respectively. In the proposed method, the PNSR values regularly but only slightly change from one noise variance to another. The PNSR values dropped by 7.9 % in $\sigma = 50$. Overall, the proposed method outperforms other algorithms in terms of PNSR.

To study the speedup of the proposed GPU implementation compared to sequential CPU implementation, we conducted different experiments to compare the time of the training phase and reconstruction by using the optimal parameters. Different image

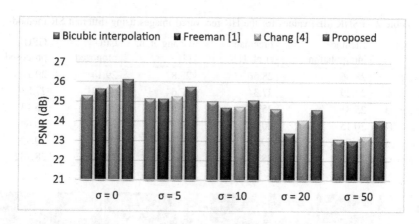

Fig. 6. Performance of the proposed SR method using different noise rates.

Table 3. Execution times (ms) for the proposed method on CPU and GPU platform for different image size.

Image size	CPU	GPU	Speedup
64 × 64	5105.93	500.41	10.20
128 × 128	10858.9	904.76	12.00
256 × 256	12680.25	845.02	15.01
512 × 512	15941.61	520.98	30.60
1024 × 1024	68135.47	418.27	162.90

sizes in the reconstruction phase have been tested. Table 3 illustrates the execution time for both sequential CPU implementation and GPU parallel implementation. The CUDA implementation execution time is speeded up by 11 × to 163 × approximately for small and large image sizes respectively during the reconstruction phase.

5   Conclusion

In this paper, we present acceleration for a learning based method for single image super resolutions and the locally linear embedding algorithm is used in the training set. We use the histogram of the gradient (HoG) to generate a training set that is most similar to the input images. Our current implementation was implemented on three main kernels to improve the speed up of the performance. The bottleneck of the LLE is the matrix manipulation. We adopted the CUBLAS library for matrix inverse and the calculation of matrix-vector multiplication modules; we utilize the shared memory in the matrix-addition and subtraction kernels.

Different experiments are conducted using a multi-spectral Landsat-8 satellite image to show the effectiveness of the proposed algorithm, and also demonstrates the value of the manifold method which utilized in the single image super-resolution.

Fourteen sub-images were clipped by a window of size 512×512 pixels. The experimental results show that our GPU implementation with CUBLAS speeds up from $11 \times$ to $163 \times$ faster. However, the accuracy of the HR images using sequential CPU and parallel GPU implementation are almost identical, and both outperform the state-of-art method (bi-cubic interpolation).

References

1. Freeman, W.T., Jones, T.R., Pasztor, E.C.: Example-based super-resolution. IEEE Comput. Graph. Appl. **22**(2), 56–65 (2002)
2. Huang, J.B., Singh, A., Ahuja, N.: Single image super-resolution from transformed self-exemplars. In: IEEE Conference on Computer Vision and Pattern Recognition (CVPR), pp. 5197–5206 (2015)
3. Tang, Y., Yan, P., Yuan, Y., Li, X.: Single-image super-resolution via local learning. Int. J. Mach. Learn. Cybern. **2**(1), 15–23 (2011)
4. Chang, H., Yeung, D.-Y., Xiong, Y.: Super-resolution through neighbor embedding. In: Proceedings of the IEEE Computer Society Conference on Computer Vision and Pattern Recognition (2004)
5. Chan, T.-M., Zhang, J., Pu, J., Huang, H.: Neighbor embedding based super-resolution algorithm through edge detection and feature selection. Pattern Recogn. Lett. **30**(5), 494–502 (2009)
6. Dong, C., Loy, C.C., He, K., Tang, X.: Image super-resolution using deep convolutional networks. IEEE Trans. Pattern Anal. Mach. Intell. **38**(2), 295–307 (2016)
7. Taniguchi, K., Ohashi, M., Han, X.-H., Iwamoto, Y., Sasatani, S., Chen, Y.-W.: Example-based super-resolution using locally linear embedding. In: 6th IEEE International Conference on Computer Sciences and Convergence Information Technology (ICCIT) (2011)
8. Gao, X., Zhang, K., Tao, D., Li, X.: Image super-resolution with sparse neighbor embedding. IEEE Trans. Image Process. **21**(7), 3194–3205 (2012)
9. Yu, T., Gan, Z., Zhu, X.: Novel neighbor embedding supe resolution method for compressed images. In: International Conference on Image Analysis and Signal Processing (IASP) (2012)
10. Liao, Z., Shuyuan, Y., Jiren, Z., Licheng, J.: Single image super-resolution via learned representative features and sparse manifold embedding. In: International Joint Conference on Neural Networks (IJCNN) (2014)
11. Dang, C., Radha, H.: Fast image super-resolution via selective manifold learning of high-resolution patches. In: IEEE International Conference on Image Processing (ICIP), pp. 1319–1323 (2015)
12. Roweis, S.T., Saul, L.K.: Nonlinear dimensionality reduction by locally linear embedding. Science **290**(5500), 2323–2326 (2000)
13. Owens, J.D., Luebke, D., Govindaraju, N., Harris, M., Krüger, J., Lefohn, A.E., Purcell, T.J.: A Survey of General-Purpose Computation on Graphics Hardware. Computer Graphics Forum. Wiley Online Library, Hoboken (2007)
14. Nvidia, C.: Cublas library. NVIDIA Corporation 15, p. 27, Santa Clara, California (2008)
15. Garcia, V., Debreuve, E., Barlaud, M.: Fast k nearest neighbor search using GPU. In: IEEE Computer Society Conference on Computer Vision and Pattern Recognition (2008)

16. Knight, E.J., Kvaran, G.: Landsat-8 operational land imager design, characterization and performance. Remote Sens. **6**(11), 10286–10305 (2014)
17. Moustafa, M., Ebied, H.M., Helmy, A., Nazamy, T.M., Tolba, M.F.: Parallel super-resolution reconstruction based on neighbor embedding technique. In: Gervasi, O., Murgante, B., Misra, S., Gavrilova, M.L., Alves Coutinho Rocha, A.M., Torre, C., Taniar, D., Apduhan, B.O. (eds.) ICCSA 2015. LNCS, vol. 9156, pp. 134–143. Springer, Heidelberg (2015). doi:10.1007/978-3-319-21407-8_10

Ultrasound Image Enhancement Using a Deep Learning Architecture

Mohamed Abdel-Nasser[1]([⊠]) and Osama Ahmed Omer[1,2]

[1] Department of Electrical Engineering, Faculty of Engineering,
Aswan University, Aswan 81542, Egypt
egnaser@gmail.com, omer.osama@aswu.edu.eg
[2] Department of Electronics and Communications Engineering,
College of Engineering and Technology, Arab Academy for Science,
Technology and Maritime Transport, Aswan 81516, Egypt

Abstract. Ultrasound images have been used for detecting several diseases such as kidney stones and breast tumors. However, ultrasound images suffer from speckle noise and several artifacts, thus degrading the quality of the images. In this paper, we propose a new method for enhancing the quality of ultrasound images. This method is a multi-frame super resolution approach, where it extracts a high resolution image from a set of low resolution images (down-sampled, blurred, and shifted versions of a high resolution source image). The critical step in multi-frame super resolution approaches is motion estimation, especially when there is noise in the images. To cope with this issue, we propose the use of a deep learning based method for motion estimation. Experimental results using synthetic and realistic sequences demonstrate that our proposed approach is feasible and effective for enhancing the quality of ultrasound images.

Keywords: Ultrasound imaging · Deep learning · Image enhancement

1 Introduction

Ultrasound (US) imaging is a safe, non-invasive and relatively cheap real-time screening method. US images have been used to detect several diseases such as abdominal aortic aneurysm, gallstones, kidney stones and breast cancer. Unfortunately, US images suffer from speckle noise, shadowing and other artifacts. US artifacts are the set of structures in ultrasound images that do not have corresponding anatomical structures [1]. Artifacts usually appear when an US image is displayed; they are produced by the physical properties of ultrasound themselves. Artifacts can be classified into four main categories: missing structures, degraded images, falsely perceived objects and structures with a mis-registered location [2]. Due to advances in hardware and image-forming algorithms, the

© Springer International Publishing AG 2017
A.E. Hassanien et al. (eds.), *Proceedings of the International Conference on Advanced Intelligent Systems and Informatics 2016*, Advances in Intelligent Systems and Computing 533, DOI 10.1007/978-3-319-48308-5_61

quality of ultrasound images has been improved since the 1970s. However, US imaging has a fundamental limit, which is the way that waves diffract as they travel. Indeed, we can distinguish between two objects in the resulting US image if the distance between them is more than half a wavelength [3]. Obviously, low quality US image may yield unwarranted clinical intervention. Image super-resolution (SR) methods may help to enhance the quality of US images at no cost.

In the literature, several methods were proposed for enhancing medical images. The study in [4] presented several advantages and challenges of applying the SR framework to applications in medical imaging. To enhance kidney US images, the study in [5] compared the performance of five common enhancement techniques: the spatial domain filtering, frequency domain filtering, histogram processing, morphological filtering and wavelet filtering.

The authors of [6] developed a SR ultrasound imaging method based on the phase-coherent multiple-signal-classification. The method accounts for the phase response of transducer elements for improving image resolution. Unlike the work in [6], we propose a multi-frame SR method, in which a high resolution (HR) image is extracted from a set of low resolution (LR) images. The most critical step in multi-frame SR approaches is the motion estimation process as the quality of the reconstructed image is sensitive to the accuracy of motion estimation. The artifacts and noise that exist in US images degrade the accuracy of most motion estimation methods. To cope with this problem, in this paper we propose the use of deep learning based approach for motion estimation. It includes convolution, max-pooling and rectification processes, which reduces the effect of noise and artifacts that exist in US images. When this motion estimation approach is used with the SR framework, it produces a high quality reconstructed US image.

The rest of this paper is organized as follows. Section 2 explains the proposed method. Section 3 presents the experimental results and the discussion. The conclusion is given in Sect. 4.

2 Methods

2.1 Extracting the HR Ultrasound Image

Given a set of LR images of the same region of interest, which are degraded versions of an HR image X, the following degradation process models the observation of LR images:

$$Y^k = DHF^kX + q^k \tag{1}$$

In Eq. 1, Y^k is the kth LR image (down-sampled, blurred, and shifted version of HR), X is the HR image, D is a down-sampling operator, H is a blurring operator, F is the motion operator and q^k is the noise in the kth frame, $k = 1, \ldots, N$, where N is the number of LR images.

As we can see in Eq. 1, the SR problem is an inverse problem, therefore the HR image $\hat{X}$ can be estimated from a set of LR images by minimizing the following cost function:

$$\hat{X} = arg \min_{X} \left[\sum_{k=1}^{N} \rho(Y^k, DHF^k X) + \lambda \Gamma(X) \right] \tag{2}$$

In Eq. 2, ρ is a similarity cost function, Γ is a regularization function and λ is a regularization parameter.

Table 1. Cost functions

Norm	Cost function $\rho(Y^k, DHF^k X)$
$\ell 1$	$\|Y^k - DHF^k X\|_1$
$\ell 2$	$\|Y^k - DHF^k X\|_2$
ℓ_{lor}	$\sum \log \left[1 + 0.5 \left(\dfrac{Y^k - DHF^k X}{T} \right)^2 \right]$

For the similarity cost function, $\ell 1$- and $\ell 2$- norms are widely used to fuse LR images [7,8]. In this study, we use the $\ell 1$-norm, $\ell 2$-norm and Lorentzian norm (ℓ_{lor}-norm) as a similarity cost function (see Table 1). The bilateral total variation (BTV) is used for regularization. It can be defined as

$$\Gamma(X) = \sum_{l=-p}^{p} \sum_{m=-p}^{p} \alpha^{|l|+|m|} \|X - S_x^l S_y^m X\|_1 \tag{3}$$

where α is a scalar weight ($0 < \alpha < 1$) used to apply a spatially decaying effect to the summation of the regularization term. S_x^l is a shifting operator in the horizontal direction by l pixels, whereas S_y^m is a shifting operator in the vertical direction by m pixels. In this paper, the motion operator (F) is estimated from the US LR images by using a deep learning-based motion estimation method (see Sect. 2.2), and the blurring operator is assumed to be Gaussian with a kernel size of 5×5. We used the steepest descent optimization method to optimize the cost function of Eq. 2. In the subsection below, we explain the motion estimation method that uses a deep learning architecture.

2.2 Deep Learning-Based Motion Estimation

To calculate the displacement between two ultrasound LR images, we propose the use of a deep learning akin architecture optical flow method—deepflow [9]. The main reason behind the use of this method is that it is robust against the noise and artifacts that exist in US images. Deepflow includes two main steps: correspondence extraction using a deep matching approach, and the calculation of optical flow using a large displacement optical flow cost function.

Deep matching. As we can see in Fig. 1, the deep matching method has an architecture similar to the one of deep convolutional neural network—CNN (convolution, pooling/aggregation, and rectification layers). The deep matching method samples 4×4 window at each pixel of the reference image and convolves it with the pixels of the template images, thus generating L response maps. Given a response map, the maximum response of each 8×8 window is returned (akin to max. pooling layer in the CNN) and then rectified using a power transform (akin to rectification layer in the CNN). The previous aggregation process is repeated N times (N controls the number of deep layers). Finally, the response maps of all levels are stacked into a pyramid.

Given the maps pyramid, the algorithm determines the local maxima in the response maps then it merges the correspondences extracted from all local maxima. The algorithm weights each correspondence according to three factors: similarity of the concerned atomic 4×4 patches, the level of the maximum in the pyramid and the value of the maximum. The algorithm retains the best correspondence (in terms of its weight) in every 4×4 non-overlapping block in both images. The final set of correspondences is the intersection of the retained correspondences in both image.

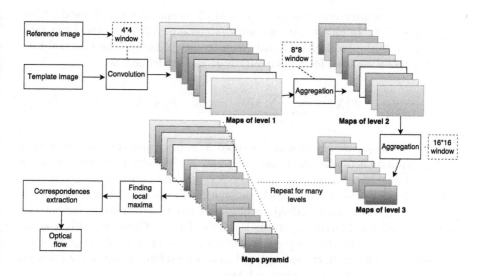

Fig. 1. Deep matching

Large Displacement Optical Flow (LDOF) Cost Function. Assume that I1 and I2 are two ultrasound images captured at times t and $t + 1$, respectively. LDOF attempts to find the global minimum of the following objective function in terms of the displacement vector w [10]:

$$E(w) = E(w)_{intensity} + \gamma E(w)_{gradient} + \alpha E(w)_{smooth}$$
$$+ \beta E(w, w1)_{match} + E(w1)_{descriptor} \quad (4)$$

In this expression, $w =: (u, v)^\mathsf{T}$ is the flow velocity vector, where u is the displacement in the horizontal direction and v is the displacement in the vertical direction, γ, α and β are tuning parameters. We experimentally set γ, α and β to 5, 30 and 300, respectively. $E(w1)_{descriptor}$ is an energy term that matches a set of descriptors extracted from the two US images (here, the results of deep matching were used). $E(w, w1)_{match}$ integrates the descriptor correspondences from a descriptor matching step into the variational approach, where $w1$ is an auxiliary variable.

$E(w)_{intensity}$ is an energy term that penalizes the deviation from the brightness constancy.

$$E(w)_{intensity} = \int_{\Omega} \Psi(|I_2(x + w(x)) - I_1(x)|^2)dx \tag{5}$$

where $\Psi(z^2) = \sqrt{z^2 + \epsilon^2}$ is used to deal with outliers ($\epsilon = 0.001$). $E(w)_{smooth}$ is a regularity constraint used to enforce a smooth optical flow field by penalizing the total variation (TV) in the optical flow as follows:

$$E(w)_{smooth} = \int_{\Omega} \Psi(|\nabla u(x)|^2 - |\nabla v(x)|^2)dx \tag{6}$$

$E(w)_{gradient}$ is used to add the gradient constancy constraint:

$$E(w)_{gradient} = \int_{\Omega} \Psi(|\nabla I_2(x + w(x)) - \nabla I_1(x)|^2)dx \tag{7}$$

In this equation, $\nabla(z)$ calculates the gradients of z in x- and y- directions of the image domain.

Indeed, the convolution, max-pooling and rectification layers eliminate the noise that exists in the US images, and thus it calculates accurate optical flow (the motion between each pair of US-LR images; F in Eq. 1). As a result, when this optical flow is used with the multi-frame SR framework, it gives a reconstructed US image of high quality.

3 Experimental Results and Discussion

3.1 Datasets

To validate the proposed method, we used B-mode image of cyst phantom generated by Field II simulator [11]. The cyst phantom was generated using the following settings: a linear scan of the phantom was performed with a 192 element transducer, and using 64 active elements with a Hanning apodization in transmit and receive. The element height was 5 mm, the width was a wavelength and the distance in x-direction between elements (*kerf*) was 0.05 mm. A single transmit focus was placed at 60 mm, and receive focusing was performed at 20 mm intervals from 30 mm from the transducer surface. More details about the cyst phantom can be found at http://field-ii.dk/?examples.html. The resulting image for 100,000 scatterers is shown in Fig. 2. We used the phantom image as

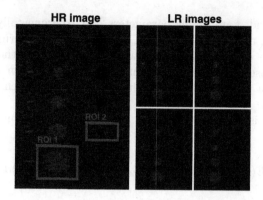

Fig. 2. The phantom image and samples of the generated LR images

a source HR images. We then generated N LR images using Eq. 1 (synthetic LR images).

We also used two realistic ultrasound sequences to validate the proposed method. Each sequence consists of a set of LR images. These sequences are collected from a clinical database of ultrasonic radio frequency strain imaging data that was created by the Engineering Department, Cambridge University. It is available at http://mi.eng.cam.ac.uk/research/projects/elasprj/.

3.2 Results

Parameter Settings. For the number of LR images N, we assessed several values (3, 5, 7, 9 and 11). The resolution increment factor was set to 2. The threshold T of the ℓ_{lor}-norm was 50. The spatial window size in the gradient of the bilateral-filter was set to 5×5. The size of the point spread function (PSF) was set to 5×5. This value was common to all frames. The exponential decay coefficient (α) and the steepest descent coefficient were set to 0.2 and 0.5, respectively. The maximum number of iterations was set to 10. However, when the cost function converged, the algorithm automatically stopped.

To asses the performance of the proposed method, we picked up a region of interest (ROI) around the biggest object (ROI1) and another around the smallest object (ROI2) in the phantom image (see Fig. 2, left). We then used the peak-signal-to-noise ratio (PSNR), and the structural similarity (SSIM) index to measure the similarity between a ROI extracted from the source (phantom) image and the corresponding ROI of the reconstructed HR image (an SSIM value of 1 refers to identical ROIs). Indeed, the best method for evaluating the proposed method is to measure how it enhances the regions of interest [12].

In Table 2, we show the PSNR and SSIM values of the proposed method for ROI1 with $\ell 1$-, $\ell 2$- and ℓ_{lor}- norms when varying the number of LR images. As shown, our method based on $\ell 1$-norm gave the best PSNR and SSIM values because this norm is more robust against noise. The best results is achieved with 9 LR images. Similarly with ROI2, $\ell 1$-norm and 9 LR images gave the best PSNR

Table 2. PSNR and SSIM values of the proposed method with cyst phantom (ROI 1)

Method/N	3		5		7		9		11	
	PSNR	SSIM	PSNR	SSIM	PSNR	SSIM	PSNR	SSIM	PSNR	SSIM
Proposed with $\ell 1$-norm	35.347	0.835	35.970	0.850	36.600	0.867	**36.694**	**0.870**	36.494	0.867
Proposed with $\ell 2$-norm	30.489	0.773	20.459	0.345	13.323	0.079	11.101	0.062	08.501	0.015
Proposed with ℓ_{lor}-norm	34.394	0.811	34.557	0.815	34.892	0.823	34.900	0.821	35.267	0.832

and SSIM values (see Table 3). Figure 3 shows the reconstructed HR images using the proposed method with $\ell 1$-, $\ell 2$- and ℓ_{lor}- norms, and with 9 LR images. As shown in Fig. 3, $\ell 1$-norm and ℓ_{lor}-norm give high quality images while $\ell 2$-norm produces low quality images because $\ell 2$-norm is sensitive to non-Gaussian noise. Indeed, the performance of the $\ell 1$- and $\ell 2$- norms was compared in [7], and the $\ell 1$-norm proved to be more robust, which coincides with our findings.

Table 3. PSNR and SSIM values of the proposed method with cyst phantom (ROI 2)

Method/N	3		5		7		9		11	
	PSNR	SSIM	PSNR	SSIM	PSNR	SSIM	PSNR	SSIM	PSNR	SSIM
Proposed with $\ell 1$-norm	36.146	0.845	35.986	0.836	35.881	0.830	**37.278**	**0.872**	36.252	0.836
Proposed with $\ell 2$-norm	32.601	0.753	24.202	0.357	16.670	0.043	11.110	0.0634	10.080	0.010
Proposed with ℓ_{lor}-norm	35.498	0.828	35.601	0.831	35.614	0.831	36.192	0.844	36.175	0.844

In Table 4 we also compared the proposed method with other alternatives where the motion parameters were estimated blindly from the LR images using Lucas-Kanade (LK) affine motion model [13]. Indeed, LK affine motion model has been widely used with SR methods. For example, the SR method proposed in [7] used $\ell 1$-norm and LK for extracting an HR image from a set of LR real-world images. In this study, we compared five methods ($\ell 1$-norm with LK, $\ell 2$-norm with LK, ℓ_{lor}-norm with LK, the method of [14], and *bicubic* interpolation) with the proposed method using the same procedure mentioned above. The authors of [14] used the SR method of Zomet et al. [15] to reconstruct HR images. They then used a nonlinear diffusion (NLD) algorithm for suppressing the noise in the restored images. As we can see in Table 4, the proposed method gave PSNR and SSIM values better than other methods. We also calculated the statistical

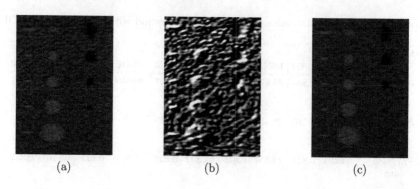

Fig. 3. The extracted HR image with 9 LR images, and using (a)ℓ1-norm, (b)ℓ2-norm and (c)ℓ_{lor}-norm

significance for the results of the SR methods using the *Wilcoxon signed rank test*. We found that the PSNR and SSIM values of the proposed method are not statistically significant than the values of other methods.

As ℓ1-norm and 9 LR led the proposed method to the best results, we used this configuration to extract an HR image from a realistic US sequence for a benign (Fig. 4) and malignant (Fig. 5) breast tumors. As shown, in the case of benign and malignant tumors, the extracted HR image has more details than the LR images (see the red arrow in the HR image of Figs. 4 and 5), and the extracted HR images also have good quality.

The average execution time of the deepflow method is less than 1.5 s. The whole SR algorithm takes less than 10 s to extract an HR image. The experiments were carried out using MATLAB on an Intel processor core 2 Quad at 2.5 GHz and 8 GB of RAM. Note that, in this paper we used a non-optimized version of the codes implemented in MATLAB language.

Table 4. Comparing the proposed method with five SR methods

Method	ROI1		ROI2	
	PSNR	SSIM	PSNR	SSIM
Proposed	**36.694**	**0.870**	**37.278**	**0.872**
ℓ1-norm with LK [7]	34.238	0.792	35.900	0.827
ℓ2-norm with LK [7]	33.651	0.779	33.623	0.758
ℓ_{lor}-norm with LK	34.185	0.806	35.292	0.823
SR with NLD [14]	35.263	0.846	36.580	0.863
Bicubic interpolation	34.100	0.810	35.230	0.820

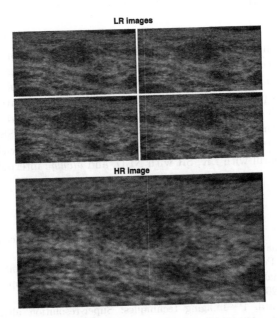

Fig. 4. Extraction of an HR image from a sequence of realistic LR-US images for a benign tumor (Color figure online)

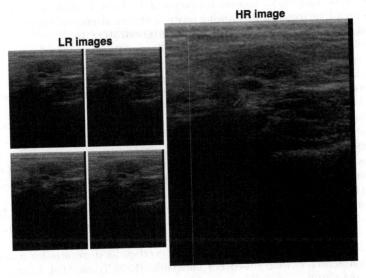

Fig. 5. Extraction of an HR image from a sequence of realistic LR-US images for a malignant tumor (Color figure online)

4 Conclusion

In this paper we proposed a new method for extracting an HR-US image from a sequence of LR ultrasound images. As the motion estimation process is sensitive to noise and artifacts, we proposed the use of a deep learning based approach for motion estimation. Indeed, the deep learning architecture of the motion estimation method (convolution, max-pooling and rectification layers) reduces the effect of noise that exists in the US images. We compared several cost functions, of which, ℓ1-norm with 9 LR images gave the best PSNR and SSIM values. Our method also outperformed other methods such as ℓ1-norm with LK, ℓ2-norm with LK, ℓ_{lor}-norm with LK, SR with NLD, and *bicubic* interpolation.

References

1. Ortiz, S.H.C., Chiu, T., Fox, M.D.: Ultrasound image enhancement: A review. Biomed. Signal Process. Control **7**(5), 419–428 (2012)
2. Sehmbi, H., Perlas, A.: Basics of ultrasound imaging. In: Sehmbi, H., Perlas, A. (eds.) Regional Nerve Blocks in Anesthesia and Pain Therapy, pp. 27–56. Springer, Heidelberg (2015)
3. Cox, B., Beard, P.: Imaging techniques: Super-resolution ultrasound. Nature **527**(7579), 451–452 (2015)
4. Robinson, M.D., Chiu, S.J., Farsiu, S.: New applications of super-resolution in medical imaging. In: Super-Resolution Imaging, pp. 384–412. CRC Press (2010)
5. Wan, M.H., Supriyanto, E.: Comparative evaluation of ultrasound kidney image enhancement techniques. Int. J. Comput. Appl. **21**(7), 15–19 (2011)
6. Huang, L., Labyed, Y., Hanson, K., Sandoval, D., Pohl, J., Williamson, M.: Detecting breast microcalcifications using super-resolution ultrasound imaging: A clinical study. In: SPIE Medical Imaging, pp. 86751O–86751O (2013)
7. Farsiu, S., Robinson, M.D., Elad, M., Milanfar, P.: Fast and robust multiframe super resolution. IEEE Trans. Image Process. **13**(10), 1327–1344 (2004)
8. Elad, M., Hel-Or, Y.: A fast super-resolution reconstruction algorithm for pure translational motion and common space-invariant blur. IEEE Trans. Image Process. **10**(8), 1187–1193 (2001)
9. Weinzaepfel, P., Revaud, J., Harchaoui, Z., Schmid, C.: Deepflow: Large displacement optical flow with deep matching. In: Proceedings of the IEEE International Conference on Computer Vision, pp. 1385–1392 (2013)
10. Brox, T., Malik, J.: Large displacement optical flow: descriptor matching in variational motion estimation. IEEE Trans. Pattern Anal. Mach. Intell. **33**(3), 500–513 (2011)
11. Jensen, J.A.: Field: A program for simulating ultrasound systems. In: 10th Nordicbaltic Conference in Biomedical Imaging, vol. 4, Supplement 1, Part 1, pp. 351–353 (1996)
12. Rohlfing, T.: Image similarity and tissue overlaps as surrogates for image registration accuracy: widely used but unreliable. IEEE Trans. Med. Imaging **31**(2), 153–163 (2012)
13. Bouguet, J.Y.: Pyramidal implementation of the affine Lucas Kanade feature tracker description of the algorithm. Intel Corporation **5**, 1–10 (2001)

14. Wang, B., Cao, T., Dai, Y., Liu, D.C.: Ultrasound speckle reduction via super resolution and nonlinear diffusion. In: Asian Conference on Computer Vision, pp. 130–139 (2009)
15. Zomet, A., Rav-Acha, A., Peleg, S.: Robust super-resolution. In: IEEE Conference on Computer Vision and Pattern Recognition, vol. 1, pp. I-645 (2001)

Prediction of Medical Equipment Failure Rate: A Case Study

Rasha S. Aboul-Yazeed$^{(\boxtimes)}$, Ahmed El-Bialy,
and Abdalla S.A. Mohamed

Systems and Biomedical Engineering Department,
Faculty of Engineering, Cairo University, Giza, Egypt
rashasaleh24@hotmail.com, abialy_86@yahoo.com,
amohamed@eng.cu.edu.eg

Abstract. Medical equipment is one of the important inputs required for the provision of efficient healthcare services. Following maintenance programs will make the equipment last longer, work more efficiently and reduces the likelihood of equipment failure during critical processing operations. Prediction of these failures affects the efficiency and enlarges the uptime of medical equipment, minimizes sudden failures and even can prevent it. Therefore, time series analysis using autoregressive model (AR) has been used to analyze failure rate data. AR model uses the past behavior of the system output to predict its behavior in the future. The mean squared error (MSE) between model output and real-life data was less than 0.1 %. Moreover, it succeeded to predict duration of next failures.

Keywords: Medical equipment maintenance · Time series analysis · AR model · Failure rate forecasting

1 Introduction

Medical equipment is designed to aid in the diagnosis, monitoring or treatment of medical conditions. Following the recommended maintenance program can reduce equipment failures [1]. Prediction of these failures affects the efficiency of health care delivery system, and enlarges the uptime of medical equipment to reach the optimal usage and minimizes sudden failures and even can prevent it. This can be achieved either by installing the necessary spare parts before the failure's occurrence, using a backup equipment or by purchasing new equipment in case the forecasting process predicts continuous failures that recommends equipment's retirement. As the core interest is patient safety, this will positively influenced by forecasting such failures that aid in decision making. The practical usage of equipment failure rate models is to find optimum equipment replacement policies, and to identify imperfect or hazardous repair, so that maintenance practice can be further investigated and enhanced.

Over the past few decades, failure rate prediction, also known as reliability prediction, of medical devices have been studied by many researchers at the design stage [2, 3]. But not many researchers dealt with the medical equipment's reliability prediction while they were used in hospitals. For example, the field data have been used to

A.E. Hassanien et al. (eds.), *Proceedings of the International Conference on Advanced Intelligent Systems and Informatics 2016*, Advances in Intelligent Systems and Computing 533, DOI 10.1007/978-3-319-48308-5_62

study the reliability prediction for Philips medical systems using power law and exponential law models. The aim was to model the failure patterns and to valuate models' prediction results. For some failure patterns; as each system has different failures number, the models couldn't fit individual systems that makes the predictive values lacking accuracy [4]. Similarly, medical imaging systems' field data have been utilized for failures prediction using non-homogeneous Poisson process and non-parametric Nelson-Aalen model [5]. Moreover, different failures type of a particular infusion pump could be classified to establish policies for analyzing field data at system and component levels [6]. The Cox proportional hazard model have been utilized to develop a model for failure events prediction based on a single event sequence collected from in-service equipment [7].

Failure rate models are also of intrinsic interest to mathematicians. Over the years, probability and operational researchers have published many papers on failure rates of repairable systems. Most work is concerned with formulating models of failure rate, as a function of system age t and the periods $t1...t_n$ at which failures occur and repairs are carried out. Policies for system replacement (such as replacement at a specified age) are then evaluated under the model [8]. Time series analysis for prediction can be a very useful tool in the field of medical equipment failure to forecast and to study the behavior of failure along time. This creates the possibility to give early warnings of possible equipment malfunctioning [9]. Proper care should be taken to fit an adequate model to the underlying time series. It is obvious that a successful time series forecasting depends on an appropriate model fitting. Various important time series forecasting models have been evolved in [10].

The aim of this paper is to produce a rigorous model for failure data analysis leading to the estimation of optimal approach for failure rate forecasting to aid in replacement decision-making.

The rest of this paper is organized as follows: Method description is in the next section where data analysis and AR model were introduced. Section 3 reports results and discussion. While conclusion is illustrated in Sect. 4.

2 Material and Method

Observing the failure history of medical equipment in order to anticipate the future failures was the main interest. This is referred to as forecasting or prediction.

The analysis is applied to the COULTER MAX.M hematology equipment, Beckman Coulter, Inc. used in the clinical laboratory at Dar Al-Fouad hospital that is a private hospital in Egypt. Failures history occurred with the MAX.M have been observed and scheduled along three years. The differences between two failures have been calculated.

2.1 Data Analysis

Consider the obtained data as a time series that is defined as a sequential set of data points, measured typically over successive times. The measurements taken during an event in a time series are arranged in a proper chronological order. The time series is represented here by the failure rate data. Failure rate is the time difference between two failures.

Time Series Stationary. Properties of stationary time series do not depend on the time at which the series is observed. A white noise series is stationary because it does not matter when it is being observed, it should look much the same at any period of time. Yet, some cases can be confusing, a time series with cyclic behavior (but not trend or seasonality) is stationary. That is because the cycles are not of fixed length, so we cannot assure where the peaks and troughs of the cycles will be before we observe the series [11]. Thereupon, the failure rate data are considered as a stationary time series.

In general, models for time series data can have many forms and represent different stochastic processes. One of the most widely used linear time series model in literature is Autoregressive (AR) [12] that is remarkably flexible at handling a wide range of different time series patterns [11] and appropriate when the time series is stationary. This Linear model has drawn much attention due to its relative simplicity in understanding and implementation [10]. Failure rate data analysis involves data preprocessing in which failure data were smoothed and interpolated before using the AR stochastic model for failure forecasting.

Data Preprocessing: Smoothing and Interpolation. Initially, data fitting has been utilized to find the curve which matches the failure rate data. Smoothing approach has been followed to overcome any abrupt changes in these data. One of the most widely used nonlinear correction techniques is the locally weighted scatterplot smoother (Lowess) technique that was first applied to microarray data [13, 14]. The Lowess is a data analysis technique for producing a smooth set of values from a time series that has been contaminated with noise that may causes missing or masking the true data value. This missing data can be found by linear interpolation [15]. Or, by way of explanation, the resulted smoothed data are non-uniformly space sampled that necessitates applying interpolation technique to achieve uniform sampled data. Compared to other non-parametric regression techniques, the Lowess is more robust in many types of scenarios [14]. Steps required to select the model structure are shown in Fig. 1.

This structure depends on two basic steps: (1) Model Order Selection; and (2) Parameters Estimation.

The model order and parameters were estimated so that to yield minimum mean squared error (MMSE) as in Eq. (1),

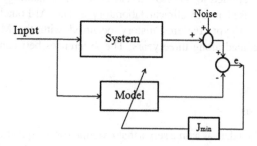

Fig. 1. Criterion for model structure selection

$$J_{min} = \left(\frac{1}{n}\right) \sum\nolimits_{k=1}^{n} e_k^2 \qquad (1)$$

Where:

J_{min}: MMSE;

n: No. of data points;

e: System to model error.

2.2 AR Model

AR model is a simple and effective method in time series modeling. It is a model which uses the past behavior of a variable to predict its behavior in the future. In other words, it is simply a linear regression of the current value of the series against one or more prior values of the series where the input is assumed to be white noise. It can easily handle messy data frequently seen in biological signals such as in heart rate variability studies [16]. Yet, in that instance, we have applied it to the smoothed, interpolated failure rate data. The most often seen form of the equation is a linear form as shown in Eq. (2):

$$Y_t = C + \sum\nolimits_{k=1}^{p} \varphi_k Y_{t-k} + \varepsilon_t \qquad (2)$$

Where:

Y_t: Dependent variable values at the moment t;

Y_{t-k} (k = 1, 2, ..., p): Dependant variable values at the moment $(t - k)$;

p: Order of AR model and written as AR(p);

$\varphi_1, \ldots \ldots \varphi_p$: Parameters or regression coefficients;

C is a constant; and ε_t is an error at moment t.

The constant term can be omitted for simplicity if the dependent variable Y_t has zero mean value. The Eq. (2) will become as in Eq. (3),

$$Y_t = \varphi_1 Y_{t-1} + \varphi_2 Y_{t-2} + \ldots + \varphi_p Y_{t-p} + \varepsilon_t \qquad (3)$$

Accordingly, the observed failure rate data have been normalized (i.e. having zero mean and unit variance) before being used to omit the constant term in the model representing equation.

Some constraints are necessary on the values of the parameters of this model in order that the model remains stationary. For example, processes in the first-order autoregression model AR(1) with $|\varphi 1| > 1$ is not stable. Generally, $|\varphi p|$ should be less than one to assure the stability of the model [17]. Therefore, it is important to use a stability test that is a fast and reliable numerical method based on calculating the reflection coefficients [18].

Reflection coefficients are the partial autocorrelation coefficients scaled by–1. They indicate the time dependence between Y_t and Y_{t-p} after subtracting the prediction based on the intervening p − 1 time steps [19].

Another constraint is that AR model's default input data are white noise (Gaussian distribution). Figure 2 shows the flowchart of the AR model estimation.

Selecting the Model Order. There is no straightforward way to determine the correct model order. As one increases the order, the mean square error generally decreases

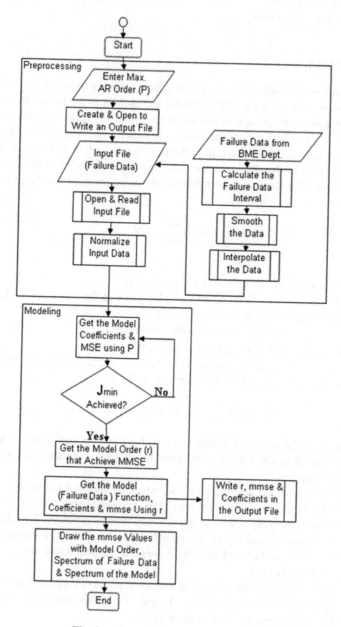

Fig. 2. AR model estimation flowchart

quickly up to some order and then become more slowly. An order just after the point at which the mean square error flattens out is usually an appropriate order [20].

Fitting Models. Models in general, after choosing the order (p), are fitted by least square regression to find the parameters' values which minimize the error term. The autoregressive model parameters can be estimated by minimizing the sum of squares residual with respect to each parameter. The aim is to find the smallest values of p which provides an acceptable fit to the data [21].

3 Results and Discussion

3.1 Best Fit Model

Model identification is one of the most challenging and distressing issues in AR modeling. As shown in Fig. 3, Starting with AR (20) to estimate the variation of MMSE values relative to the AR model order, there is no significant improvement in MMSE with increasing model order. The MMSE value is decreased then stabilized at order 2. Consequently, AR (2) has been used and the result is illustrated in Fig. 4.

Table 1 compares the accuracy of different model order selections. The absolute value of the reflection coefficient for AR (2) is almost equals 1. To avoid model instability, a higher model order AR (3) has been used. The output is depicted in Fig. 5.

Figures 3, 4 and 5 show the identification of the spectral density of the original failure data with the spectral density of the AR (20), AR (2), and AR (3) models, respectively. This is because MMSE values between different models' output and real-life data were less than 0.1 %, as illustrated in Table 1.

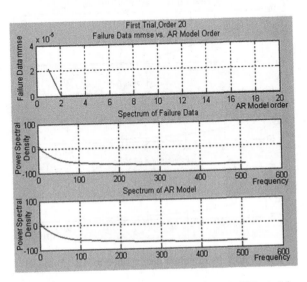

Fig. 3. Results of failure data applied to AR (20) model

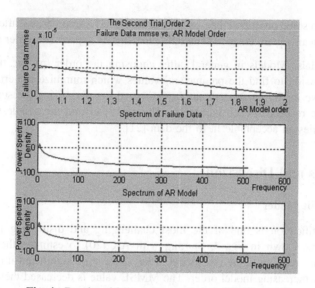

Fig. 4. Results of failure data applied to AR (2) model

Table 1. Influence of model order on MMSE

	1st trial	2nd trial	Last trial
Model order	20	2	3
MMSE	0.00060878	0.00067152	0.00067033
AR model reflection coefficient	−0.03315649	−0.9938938	−0.06470016

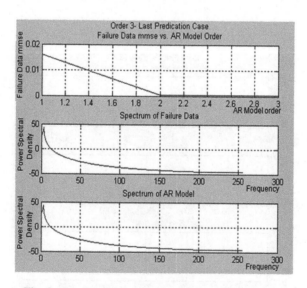

Fig. 5. Results of failure data applied to AR (3) model

It is clear that higher-order structures needs more computations especially for testing its stability which needs the determination of a polynomial roots of order 20; and consequently, the effect of its parameterization is much more difficult to generalize. The results for the AR (3) model showed that its parameter values could imply the same spectral pattern and smaller number of parameters represent an adequate fit for those data. Therefore, AR (3) model has been chosen to be used for failure rate forecasting.

3.2 Failure Rate Forecasting

Based on the main objective of this research is to forecast the period of next failure, whereas the prediction accuracy depends on the available set of failure rate data (n). These data have been evaluated over an interval from (n + 1) to (n + m) using AR (3). The results of prediction are illustrated in Fig. 6. It is noticed from this figure that failures will occur almost every two days.

This result can be considered in the replacement decision-making process as an alarm to put a plan for the equipment replacement as soon as possible.

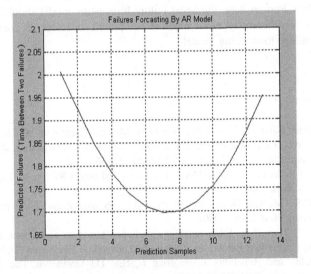

Fig. 6. The forecasted failure data

4 Conclusion

Observing the failure history of medical equipment in order to anticipate the future failures is referred to as forecasting. This can affect the medical equipment efficiency and minimize sudden failures. Time series analysis for prediction is a vital tool in the field of medical equipment failure to forecast and to study the behavior of failures along time. This creates the possibility to give early warnings of possible equipment malfunctioning and help in medical equipment replacement decision-making process.

Considering our model compared to others; both failures prediction models in [4] have given wrong predications for future failures when a failure pattern of 900 days has been used due to limited number of failures. Yet, in our model, we applied interpolation to the smoothed failure data to overcome this problem by reaching large number, uniform sampled data to achieve accurate prediction results.

Moreover, a failure prediction model based on Cox model could be implemented in [7]. However, recent publications suggest a growing interest in the quality of Cox applications [22]. On the other hand, the AR model is remarkably flexible at handling a wide range of different time series patterns [11], and it has drawn much attention due to its relative simplicity in understanding and implementation [10]. As well, it could define a highly rigorous failure forecasting model with MMSE less than 0.1 % and succeeded to represent failure rate data of Max.M equipment. The model could predict the failure rate occurrence every two days that persuades decision-making of equipment replacement.

In future, time series analysis using different stochastic models such as MA, ARMA, and GARCH should be applied to the failure rate data and the outcome of each model should be compared so that to achieve the best model that well represent the failure rate data. Furthermore, the finest failure forecasting model should be applied to the field data of different types of medical equipment to assess its contribution as a generalized failure forecasting model.

References

1. WHO Library Cataloguing-in-Publication Data: Medical Equipment Maintenance Program Overview. Who medical device technical series. World Health Organization (2011). http://apps.who.int/medicinedocs/documents/s21566en/s21566en.pdf
2. Fries, R.C.: Reliable Design of Medical Devices, 3rd edn. CRC Press, Taylor & Francis Group, New York (2005)
3. Dhillon, B.S.: Medical Device Reliability and Associated Areas. CRC Press, Taylor & Francis Group, New York (2000)
4. Roelfsema, S., Ion, R.A.: Early reliability prediction based on Field Data. In: Eindhoven: Technische Universiteit Eindhoven, Technische Universiteit Eindhoven (TUE). Capaciteitsgroep Quality and Reliability Engineering (QRE). TU Eindhoven. Fac. Bedrijfskunde: afstudeerverslagen (2004) (in Dutch)
5. Ion, R.A, Sonnemans, P.J.M., Wensing, T.: Reliability Prediction for Complex Medical Systems. In: Reliability and Maintainability Symposium, pp. 368–373 (2006)
6. Sharareh, T., Dragan, B., Andrew, K.S.: Reliability Analysis of Maintenance Data for Medical Devices. Quality and Reliability Engineering International, Wiley online library (2010)
7. Zhiguo, L., Shiyu, Z., Suresh, C., Crispian, S.: Failure event prediction using the cox proportional hazard model driven by frequent failure signatures. IIE Trans. **39**, 303–315 (2007)
8. Ascher, H., Feingold, H.: Repairable Systems Reliability, Modeling, Inference, Misconceptions and Their Causes. Marcel Dekker, New York (1984)

9. Thissen, U., Van, B.R., De, W.A.P., Melssen, W.J., Buydens, L.: Using support vector machines for time series prediction. Chemometr. Intell. Lab. Syst. **69**, 35–49 (2003)
10. Adhikari, R., Agrawal, R.K.: An Introductory Study on Time Series Modeling and Forecasting. LAP Lambert Academic Publishing, Germany (2013)
11. Rob, J.H., George, A.: Forecasting: principles and practice. OTexts (2012). https://www.otexts.org/fpp
12. Hipel, K.W., McLeod, A.I.: Time Series Modelling of Water Resources and Environmental Systems. Elsevier, Amsterdam (1994)
13. Box, G.E.P., Jenkins, G.M., Reinsel, G.C.: Time Series Analysis: Forecasting and Control. Wiley, San Francisco (2008)
14. John, A.B., Sampsa, H., Anna-Kaarina, J., Henrik, E., Sanjit, K.M., Jaakko, A.: Optimized LOWESS normalization parameter selection for DNA microarray data. BMC Bioinf. **5**(1), 1 (2004). BioMed Central Ltd.
15. Robert, B.N.: Introduction to Instrumentation and Measurements, 2nd edn. CRC Press, Taylor & Francis, New York (2005)
16. Burr, R.L., Cowan, M.J.: Autoregressive spectral models of heart rate variability. practical issues. J. Electrocardiol. **25**(Suppl), 224–233 (1992). US National Library of Medicine. National Institute of Health
17. Lauer, A., Wolff, I.: Improved autoregressive (AR) signal modeling for FDTD resonance estimation. In: Microwave Symposium Digest. IEEE MTT-S International, pp. 255–258. IEEE Press, Boston (2000)
18. Julius, O.S.I.: Introduction to Digital Filters with Audio Applications. Center for Computer Research in Music and Acoustics (CCRMA). Stanford University (2007)
19. MathWorks, Signal Processing Toolbox: Documentation (R2016a). Accessed August (2016)
20. Zhigang, Z., Guohua, C.: Adaptive predictive based on equal-dimension and new information for the hydraulic mechanism of wave motion compensating platform. In: Applied Mechanics and Materials, pp. 236–242. Trans Tech Publications, Switzerland (2010)
21. Brockwell, P.J., Davis, R.A.: Time Series: Theory and Methods, 2nd edn. Springer, New York (1991)
22. Carine, A.B., Gaëtan, M.G., Marc, D.: Christine, T.D.L., Véronique, B., Simone, M.P.: Variables with time-varying effects and the cox model: some statistical concepts illustrated with a prognostic factor study in breast cancer. BMC Med. Res. Methodol. **10**(1), 1 (2010). BioMed Central Ltd.

Predicting Learner Performance Using Data-Mining Techniques and Ontology

Alla Abd El-Rady$^{(\boxtimes)}$, Mohamed Shehab, and Essam El Fakharany

Arab Academy for Science, Technology & Maritime Transport, Cairo, Egypt
alla.abdelrady@gmail.com, melemam9@gmail.com,
essam.fakharany@gmail.com

Abstract. The high rates of learners' dropout and failures in different courses that are offered by many universities and educational institutions through the use of e-learning or online learning systems have been a serious concern. Analyzing and studying learners' learning data in order to predict their future performance can support both tutors and e-learning systems to determine learners' progress or status and spot those with low performance. Thus they can offer learners with personalized learning resources and activities designed to each one in order to maximize their learning outcomes and overcome their learning gaps. This paper presents a methodology that uses semantic web technologies as well as data mining techniques to predict learners' future performance based on data produced by learners through their interaction with LMS (Learning Management System) and social networks.

Keywords: E-learning · Semantic web · Ontologies · SWRL · Education data mining · Social networks · LMS

1 Introduction

Over the last years there has been a quick change in the web technologies and internet. This quick change has driven changes in various fields such as economy and education. In education, web technologies play an essential role in changing the way of learning as well as the learning procedures. They change the way of how learning contents are delivered to learners and prompt a new way of learning called e-learning, online learning or distance learning.

E-learning is generally referred to as "web based learning" [1]. What is unique about e-learning is that it facilitates the learning process through the use of computers, networks and web technologies to ease delivery, sharing and management of different e-learning resources and activities to learners. It also aids learners to learn in ways that are much faster, easier and cheaper than the traditional way of learning in order to enhance their performance as well as to improve their final outcomes [1].

Today people use e-learning systems as a normal way of learning. However with the rapid growth of e-learning systems or online learning systems there has been a rising concern over number of problems related to learners such as the high dropout rates [2, 3] and failure to get a degree or pass a course. A lot of researches and studies

© Springer International Publishing AG 2017
A.E. Hassanien et al. (eds.), *Proceedings of the International Conference on Advanced Intelligent Systems and Informatics 2016*, Advances in Intelligent Systems and Computing 533, DOI 10.1007/978-3-319-48308-5_63

in the last years are concentrating on how to solve these problems through the analysis of learning process in order to improve it.

In general the massive amount of data produced by learners through their learning process is an important asset to different educational stakeholders such as tutors, parents and learners themselves. Highlighting different features and characteristics of learners' data is an important process as it can help tutors in getting better image of learners' progress so they can adapt their learning process according to learners' needs [4].

Different e-learning systems have begun to use educational data mining techniques to analyze learners' data. Therefore tutors can offer those learners personalized and adapted learning activities designed for each one to enhance their final performance.

Moreover, a lot of educational studies and researches have begun to develop personalized e-learning systems using semantic web technologies such as ontology and semantic web rule languages (SWRL) [5]. Utilizing both semantic web technologies and educational data mining techniques in developing e-learning systems is an essential part in the process of analyzing learners' data as they enrich each other by interacting over the time [6].

In this paper we propose a methodology that uses semantic web technologies such as ontology and SWRL as well as data mining techniques to predict learners' final performance according to their learning data. The data is derived from learners' engagement with both learning management system (LMS) and social networks (Facebook). This methodology can help and support both tutors and e-learning systems to analyze the learners' performance and determine those who are underachieving and have high possibility to fail, so they can offer them extra educational materials and activities to avoid their failure.

2   Background Knowledge

E-learning systems, data mining techniques and ontology are three areas of interest in building our system. Brief description of each one is given in this section.

2.1   E-learning

E-learning is a web based learning which introduces a new environment and a new way of learning through the use of internet, interactive multimedia and different web technologies. It's defined as "interactive learning which permits learning through the deployment of computers as an education medium" [7]. Using e-learning, learners can easily start learning independently from others.

E-learning technologies have been changed and evolved over the last years by utilizing internet and web technologies. They can be classified into e-learning 1.0, e-learning 2.0 and e-learning 3.0 [8].

In e-learning 1.0 the learning contents are kept and viewed online so learners can easily access different learning entities and resources. Moving forward to learning 2.0 it became more advanced by allowing learners to access diverse learning materials passively. It also allows learners to share their own beliefs through writing comments or notes. Finally in e-learning 3.0 learning process and environment can be adjusted or personalized through the use of sematic web technologies [9].

2.1.1   E-learning and Social Networks

Lately, terms like "Social Media" and "Social networks" became so widespread. It gained a strong position in different educational studies and researches as they are being accessed and used by most tutors and learners frequently. Researchers have begun to explore the opportunities and challenges of using social media in educational systems [10].

Social networks are used by different learners for communicating each other as well as for discovering, sharing and exchanging knowledge. Facebook, Twitter, LinkedIn, Instagram, MySpace and Google+ are the most common social networks used around the world. There are over than 1.65 billion active Facebook users [11]. Because of Facebook popularity between learners and tutors we have nominated it for our study as a source of learners' social data. Learners' activities at course Facebook groups such as number of comments, shares, posts and likes can be used in determining those with active engagement and interaction with course which consequently can be used in predicting leaner's future performance.

2.2   Educational Data Mining Techniques

Data Mining (DM) is referred to as the method of analyzing massive amount of raw data from different perceptions and perspectives for the purpose of discovering novel, non-trivial, understandable patterns and valuable information [11].

Educational Data Mining (EDM) is defined as the process of analyzing and studying raw data derived from educational systems in order to discover and extract valuable information and useful knowledge. This knowledge can be presented to diverse participants or stakeholders such as tutors, learners, system developers, parents and educational researchers [12].

Data mining techniques have been categorized and classified from different views and perspectives. Ryan Baker [13] has classified it into: "Prediction, Clustering, Relationship mining, Distillation of data for human judgment and Discovery with models".

For the time being there is an increasing interest in employing data mining techniques in e-learning systems [14]. Digging the vast amount of data made by learners through their learning process and extracting knowledge from it can support tutors in identifying learners' progress [15].

2.3   Ontology

Basically, Semantic Web is defined as an extension of the existing web. Information in semantic web is given a well-defined meaning which can permit both people and computers to work together. It supports people in getting accurate answers to their inquiries by allowing different web agents to reason the multiple web resources and contents [16].

Ontology is an important and essential part of semantic web layer cake. It is defined as "Explicit specification of conceptualization" [17]. It is broadly used in artificial intelligence, knowledge engineering and computer science related applications. These applications are related to different fields such as knowledge discovery, e-commerce and education.

Ontology has been well utilized in e-learning in diverse fields for the purpose of representing learning domain and learner profile, personalizing and recommending learning materials, evaluating learning process and planning course syllabus and contents [17]. It offers a way of discovering and extracting new knowledge through the use of its inference mechanism such (reasoner). Semantic web rule language (SWRL) is used to extend ontology reasoning capabilities. As it is generally used to express different types of relationships and conditions that cannot be expressed or defined using ontological reasoning only.

3 Proposed Methodology

The main purpose of this study is to build a methodology that can be used to predict learners' performance by analyzing data generated from their interaction with both learning management system and Facebook groups. In this study both data mining techniques and semantic web technologies are being used.

3.1 Dataset Description

In this study an educational data of 140 learners with different variables that highlight learners' different aspects was analyzed. The data set was obtained from "UCI Machine Learning Repository" which is an online data sets repository. Modifications were made on it in order to meet our requirements.

The dataset contains variables about learner age, address, sex, family members, average time spent on learning, number of previous failures, learner activities (curriculum related or unrelated), attended sessions, exercises grades, midterm grade and final grade. It also contains variables about average number of comments, posts and likes submitted by leaner on course Facebook groups.

Our objective is to analyze the data and then classify learner into one of two classes "Pass" or "Fail" based on his/her final grade.

3.2 Methodology Architecture

Figure 1 shows the architecture of our proposed methodology. It consists of two main phases. In the following a description of each phase is given.

Phase 1: The main objective of this phase is to select and implement different data mining techniques in order to build our predictive model which predicts learner's future performance based on learner's data. Initially the data was collected and prepared for the analysis process through the using of different preprocessing and filtering

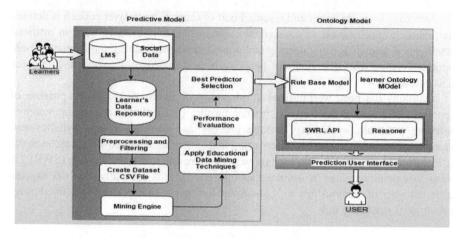

Fig. 1. Overall methodology architecture

techniques in order to remove any duplication in data and handle any missing values which could cause errors later.

The dataset was created as CSV file format (comma separated values file format which is supported by WEKA data mining tool) and then different classification techniques were chosen to be applied by utilizing WEKA data mining tool.

3.2.1   Applying Classification Techniques

Mainly the type of data and the problem domain affect the choice of the classification techniques. Twelve different data mining classification techniques form different groups were chosen to be implemented and evaluated through the use of WEKA data mining tool:

- Functions Family: SMO(Support Vector classifier) and Simple Logistics
- Bayesian Network Family: Naïve Bayes and Bayes Net
- Decision Tree Family: J48, Simple Cart, Random forest, and ADTree
- Rule Based Family: JRIP and OneR
- Lazy Classifiers Family: IB1 and KStar.

These classification techniques are frequently used by researchers in multiple educational researches and studies. They have high potential to yield to good results and high accuracy. Furthermore, these classification techniques used different methodologies in building their prediction models which can increase the chance of finding a prediction model with high accuracy and fewer errors. The results of classification techniques are given in Sect. 4. The output of this phase is the prediction rules which will be mapped and stored in rule base model.

Phase 2: The objective of phase2 is to build learner ontology model and develop our inference engine through the use of ontology reasoning and SWRL rules. The classification results which are the output of phase 1 will be mapped as the input of phase2.

Learner data and prediction rules extracted from the data mining output in phase1 is used to build learner ontology and build our rule base model which is the base of ontology inference engine. This phase consists of the following components:

Learner ontology model which is a key component of phase 2 is used to represent and model learner's learning data such as learner's personal information (gender, age, address) through the identification of different sets or classes, object properties or variables and relationships between these sets. OWL (web ontology language) is a language which is used for ontology encoding. It is used to describe and represent knowledge of each set. Figure 2 shows a part of learner ontology model implemented using Protégé software tool which is the software used for the implementation of phase2.

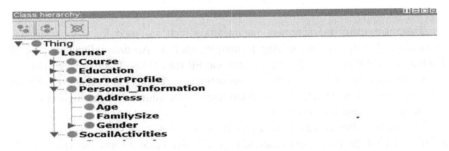

Fig. 2. Example of learner ontology implemented in Protégé

Rule base model stores the prediction rules which were produced from data mining results in phase 1. These rules can be driven for example from decision tree rules. As in the decision tree each branch could have a set of leaves. Each leaf represents a classification (prediction) rule. The following is an example of a developed prediction rule which illustrates some key performance indicators (KPIs) such as leaner midterm grade, session grade and average number of comments made by learner on course group at Facebook:

Example Rule 1: If (Midterm_Grade >= 10) && (SessionGrade_2 >= 2.25) && (AVG_No_Comments > 2) then Pass.

These rules then will be transformed and then represented as Semantic rules through the use of **SWRL API.** Using both SWRL prediction rules along with ontology reasoning will increase and extend the **ontology reasoning** capabilities by allowing creation of multiple conditional rules which cannot be expressed using ontology relationships only.

Both Semantic rules generated using SWRL API and ontology reasoning mechanism work as our inference engine as they enrich and complete each other. By implementing this methodology the predication of learner's final performance can be carried out directly from ontology online.

4   Experimental Results

The performance evaluation of different data mining techniques is given in this section. We used confusion matrix to evaluate the different data mining techniques [18]. Table 1 shows the confusion matrix used for this study.

Table 1. Confusion matrix

Data label	Correctly classified	Incorrectly classified
Correct	TP (True Positive)	FN (False Negative)
Incorrect	FP (False Positive)	TN (True Negative)

There are many measurement parameters which are used for the performance evaluation of different data mining techniques such as Accuracy, Precision, Recall, F-Measure, TP Rate (True Positives Rate) and FP Rate (False Positive Rate) [18, 19].

In our experiment we used those measurement parameters to evaluate 12 different classification techniques in order to determine the technique with the highest classification accuracy and fewest errors.

The training dataset was divided into 10-folds using cross validation to be used as testing data for the evaluation process. Figure 3 and Table 2 show the values of different measurement parameters. Random forest records the best values among other techniques except for FP Rate measurement. The results show that Simple Logistics, SMO, Bayes Net, Random Forest, JRip and ADtree record the best value for FP Rate respectively.

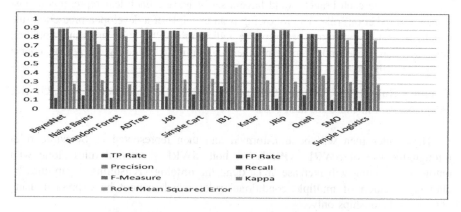

Fig. 3. Results of different classification techniques.

In general, results of different data mining techniques depend on data to be classified and on its distribution. In our Experiment, Random Forest gives best accuracy (91.36 %) followed by Simple Logistics and SMO with accuracy (89.9281 %), then

Table 2. Classification techniques results

	Accuracy	TP Rate	FP Rate	Precision	Recall	F-Measure	Kappa	Root Mean Square Error
Bayes Net	89.2086	0.892	0.12	0.893	0.892	0.892	0.7687	0.2776
Naïve Bayes	87.0504	0.871	0.149	0.871	0.871	0.871	0.7213	0.3217
Random Forest	91.3669	0.914	0.124	0.915	0.914	0.912	0.8095	0.2734
AD Tree	88.4892	0.885	0.141	0.884	0.885	0.884	0.7502	0.2882
J48	87.7698	0.878	0.145	0.877	0.878	0.877	0.7357	0.3342
Simple Cart	86.3309	0.863	0.17	0.862	0.863	0.862	0.7021	0.3425
IB1	74.8201	0.748	0.261	0.758	0.748	0.751	0.4732	0.5018
Kstar	85.6115	0.856	0.141	0.862	0.856	0.858	0.6978	0.3321
JRip	89.2086	0.892	0.128	0.892	0.892	0.892	0.7668	0.3179
OneR	84.8921	0.849	0.17	0.85	0.849	0.849	0.6761	0.3887
SMO	89.9281	0.899	0.116	0.899	0.899	0.899	0.7832	0.3174
Simple Logistics	89.9281	0.899	0.108	0.9	0.899	0.9	0.785	0.2914

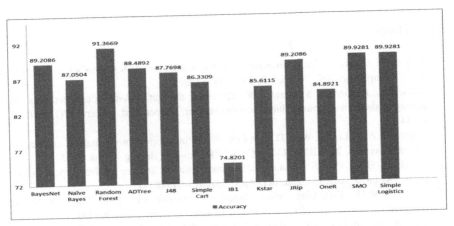

Fig. 4. Accuracy results of different classification techniques.

Bayes Net and JRip with same accuracy (89.2086 %), AD Tree with accuracy (88.4892 %), then J48 with accuracy (87.76 %) as shown in Fig. 4.

In our experiment, random forest technique shows the best result in predicting learner's class label. Random forest is built from various decision tree techniques. It uses the majority vote technique to predict the class label. Usually combining results of multiple techniques using an ensemble way give better results compared to using single technique for prediction.

5   Conclusion and Future Work

This paper proposed a methodology which can be used to predict learners' future performance and identify learners with high risk to drop out course or fail in the final exam. The prediction of learners' future performance can help learners to be aware of their progress as well as tutors to improve their teaching procedures in order to engage underachieving learners in an appropriate learning process.

The data produced through learner engagement and interaction with both e-learning systems and social networks was analyzed through the utilization of different classification techniques in order to extract prediction rules and valuable knowledge. The extracted prediction rules along with learner's data were used to build learner ontology model and rule base model with the use of semantic web technologies.

The results of this study show that some variables have direct impact on learner performance such as average number of comments, midterm grade, study time, age and gender. In the future more variables could be examined in order to obtain prediction with fewer errors and higher accuracy. Also we could categorize learner's comments into positive and negative comments. Moreover we could consider the type of errors made by each learner in different tests in order to enrich our analysis and identify learner's weak points.

References

1. Surjono, H.D.: The design of adaptive e-learning system based on student's learning styles. Int. J. Comput. Sci. Inf. Technol. **5**, 2350–2353 (2011)
2. Yukselturk, E., Ozekes, S., Türel, Y.K.: Predicting dropout student: an application of data mining methods in an online education program. Eur. J. Open Dist. e-Learn. **17**(1), 118–133 (2014)
3. Dewan, M.A.A., Lin, F., Wen, D., Kinshuk.: Predicting dropout-prone students in e-learning education system. In: UIC-ATC-ScalCom-CBDCom-IoP, Beijing, China (2015)
4. Baradwaj, B.K., Pal, S.: Mining educational data to analyze students' performance. (IJACSA) Int. J. Adv. Comput. Sci. Appl. **2**(6), 63–69 (2011)
5. Qwaider, W.Q.: E-learning system based on semantic web technology. In: Second International Conference of E-learning and Distance Learning, Riyadh (2011)
6. Kazi, A., Kurian, D.T.: An ontology based approach to data mining. Int. J. Eng. Dev. Res. **2** (4), 3394–3397 (2014)
7. Titthasiri, W.: A comparison of e-learning and traditional learning: experimental approach. In: International Conference on Mobile Learning E-society and E-learning Technology (ICMLEET), Singapore, November 2013
8. Chung, H., Kim, J.: An ontological approach for semantic modeling of curriculum and syllabus in higher education. Int. J. Inf. Educ. Technol. **6**(5), 365–369 (2016)
9. Weber, P., Rothe, H.: Social networking services in e-learning. In: Proceedings of World Conference on E-learning in Corporate, Government, Healthcare, and Higher Education (2012). https://www.researchgate.net/publication/235975162. Accessed 11 May 2016
10. ZEPHORIA Digital Marketing: The Top 20 Valuable Facebook Statistics, April 2016. https://zephoria.com/top-15-valuable-facebook-statistics/. Accessed 24 May 2016
11. Srivastava, J., Srivastava, A.K.: Understanding linkage between data mining and statistics. Int. J. Eng. Technol. Manage. Appl. Sci. **3**(10), 4–12 (2015)
12. Lakshmi Prabha, S., Mohamed Shanavas, A.R.: Educational data mining applications. Oper. Res. Appl. Int. J. (ORAJ) **1**(1) (2014)
13. Elaal, S.A.E.A.: E-learning using data mining. Chin. Egypt. Res. J. (2011)

14. Prakash, B.R., Hanumanthappa, M., Kavitha, V.: Big data in educational data mining and learning analytics. Int. J. Innov. Res. Comput. Commun. Eng. **2**(12), 7515–7520 (2014)

15. Romero, C., Ventura, S., Espejo, P.G., Hervás, C.: Data mining algorithms to classify students. In: Proceedings of the 1 st International Conference on Educational Data Mining, Montreal, Quebec, Canada, pp. 20–21 (2008)

16. Kolovski, V., Galletly, J.: Towards e-learning via the semantic web. In: International Conference on Computer Systems and Technologies – CompSysTech 2003 (2003)

17. Al-Yahya, M., George, R.: A. Alfaries:"Ontologies in e-learning: review of the literature. Int. J. Softw. Eng. Appl. **9**(2), 67–84 (2015)

18. López, V., del Río, S., Benítez, J.: Cost-sensitive linguistic fuzzy rule based classification systems under the MapReduce framework for imbalanced big data. Sciencedirect Fuzzy Sets Syst. **258**, 5–38 (2014)

19. Gupta, D.L., Malviya, A.K., Singh, S.: Performance analysis of classification tree learning algorithms. Int. J. Comput. Appl. **55**(6), 0975–8887 (2012)

A Random Forest Model for Mental Disorders Diagnostic Systems

Horeya Abou-Warda[1], Nahla A. Belal[1(✉)], Yasser El-Sonbaty[1], and Sherif Darwish[2]

[1] College of Computing and Information Technology, Arab Academy for Science, Technology and Maritime Transport, P.O. Box 1029, Alexandria, Egypt
h.abouwarda@gmail.com, {nahlabelal,Yasser}@aast.edu
[2] Department of Dual Diagnosis, El-Mamoura Hospital for Psychiatric and Addiction Treatment, Alexandria, Egypt
shd203@yahoo.com

Abstract. Data mining has established new applications in medicine over the last few years. Using mental disorders diagnostic systems, data possession, and data analysis has been of enormous succor for clinicians to recognize diseases more precisely, especially when dealing with overlapping mental symptoms. In this study, random forests used to predict a number of mental disorders and drug abuse. Two datasets were used, the first dataset consists of substances abuse patients derived from United States, and the second dataset consists of psychotic patients collected from Egypt. Random Forest classification technique was used to increase the accuracy rate of mental disorders prediction systems. The exploratory data mining analysis produced two models of the random forest algorithm with two groups of patients at different risks for substances abuse and psychotic diseases. This study presents a new data mining approach to enhance the diagnosis of mental disorders. This approach also serves therapeutic interventions.

Keywords: Data mining · Mental disorders · Psychotic diseases · Random forest · Substance abuse

1 Introduction

Data Mining refers to extracting knowledge from large amounts of data [1]. It enables us to explore and analyze the large patterns in the data by statistical means and artificial intelligence [2]. The data mining technique is used to discover hidden patterns and knowledge that are embedded in the behavior of the data, and it is widely applied by clinicians in medicine. The ultimately utilized techniques are association rules [3], classification [4], and clustering [5].

Psychiatry is the branch of medicine that deals with the science and practice of treating mental, emotional, or behavioral disorders. Analysis of data and decision making is a crucial step, especially with mental illnesses, due to the existence of several, sometimes overlapping, illness types, such as depression, bipolar disorder, schizophrenia, and illness subtypes, such as bipolar I and II. Many times the analysis and decision making depends on the mood of the Psychologist. In addition to that, the

© Springer International Publishing AG 2017
A.E. Hassanien et al. (eds.), *Proceedings of the International Conference on Advanced Intelligent Systems and Informatics 2016*, Advances in Intelligent Systems and Computing 533, DOI 10.1007/978-3-319-48308-5_64

human error cannot be avoided, given that diagnosis is the first step from a set of curative actions that are developed in order to save patients' life or to ameliorate their health; an error at this level can have dramatic outcomes. This necessity has led to birth data mining. These techniques are broadly utilized in medical activity because of their relevant advantages. It is lucid that technology cannot replace human expertise in this point of medical assistance; it only gives support through medical systems that are able to select or to generate relevant data for clinicians.

The DSM (Diagnostic and Statistical Manual of Mental Disorders) [6], which is the eventual resource of diagnostic standard for all mental disorders, includes substance abuse as a mental illness because addiction converts the brain in essential tendencies, staggering a person's normal hierarchy of needs and desires and replacing new priorities connected with the drug. Thus, producing compulsive manners that override the capability to control impulses despite the consequences are similar to lineament of other mental illnesses.

Data mining is very valuable in diagnosing numerous life menacing diseases at an early phase; it is effectively utilized in many medical missions, for examples, in intensive care medicine analysis [7], breast cancer screening [8], and diagnosis of ischaemic heart disease [9]. Horvitz-Lennon et al. [10], in their paper on the general health need of adults with severe mental illness, explain that we need to totally cuddle information technology and its power for improving service capacity and evolve a superior information infrastructure for the patient's care. Data mining also plays an important role in psychology. Dipnall and Poorna [11] utilized data mining to detect biomarkers associated with depression, while Sumathi et al. [12] applied it for prediction of children mental health problems.

The Random Forest classifier [13] is a data mining method, which has been shown to offer superior classification performance than other innovative algorithms [14]. These properties have made Random Forest increasingly popular in the last few years, especially in the field of mental diseases [15, 16].

In this research, the random forest classification technique combined with impute missing values learner [17], to predict missing values in the dataset, is utilized in the mental disorders procedure and decision making process. The study has been performed by classifying addiction patients based on substances abuse, and classifying psychotic patients according to different psychopathologies symptoms (attributes). Using data mining techniques enables the uncovering of hidden data which cannot be revealed using normal brain imaging [18, 19] techniques. Brain imaging [20, 21] allows clinicians to see inside the patient brain, without any surgical intervention, and understand the relationships between specific areas of the brain and what function they serve. It also enables clinicians to locate the areas of the brain that are affected by neurological disorders. However, due to the complexity of brain structure and function, brain imaging is not sufficient to output the diagnosis as presented in this study, as with brain imaging alone there could be an overlap between disorders, or even inability to identify the disorder without all the features discussed and used in this paper.

2 Datasets and the Proposed Random Forest Model

This section starts by discussing the datasets used in the study and then the random forest model presented.

2.1 Datasets

The study includes two groups of mental disorders patients. One group of substance abuse patients, a total of 405,393 patients [22] who were admitted to substance abuse treatment in 2012, Center for Behavioral Health Statistics and Quality (CBHSQ), Substance Abuse and Mental Health Services Administration (SAMHSA), United States Department of Health and Human Services. Patients were included if they had been diagnosed of Substance Use Disorder, SUDs, either alcohol dependence, opioid dependence, cannabis dependence, or cocaine dependence.

The other group is a group consisting of 1800 Psychotic patients[1] collected from the psychiatric division of El-Mamoura hospital for psychiatric and addiction treatment in Egypt, between September 1, 2015 and November 30, 2015. The study was approved by the local ethics committee, carried out in accordance with the ethical standards of the Egyptian General Secretariat of Mental Health and Addiction Treatment, Ministry of Health and Population (Ratification 1860||8/17), and was registered at isrctn.com. The extracted data provide useful association between different symptoms that are in some way linked together according to the different psychopathologies of psychotic patients. Data collection was performed by structured psychiatric interview and examination, and clinical psychiatric scales, evaluation of health status and functioning, neuropsychological tests, and personality tests were applied to all patients. Patients must be free of psychoactive substances. Patients were contained if they had been diagnosed with either major depression, drug-induced psychosis, schizophrenia, schizoaffective disorder, obsessive – compulsive disorder, bipolar disorder - manic episode, or bipolar disorder - mixed episode.

2.2 The Proposed Model

This research studies the structure of a mental disorders predictive model in three stages [23], namely, data pre-processing and attribute (variable) inspection, classification model construction by the random forest classifier, and classification efficacy rapprochement, optimum predictive model determination, and diagnosis classification rules extraction.

The random Forest is an ensemble classifier that consists of many decision trees and outputs the class that is the mode of the class's output by individual trees. A random forest classifier takes as an input the training dataset of N records, which is a set of cases (patients) and their associated class (diagnosis). The classifier then creates

---

[1] Trial Registration—Psychotic diseases dataset used in this study came from isrctn.com, identifier: ISRCTN75534069.

decision trees, where each tree is structured based on the number of training cases (patients) be N, and the number of variables in the classifier be M. The number m of input variables to be used to define the decision at a node of the tree; m should be much less than M, and a training set for each tree. For each node of the tree, randomly choose m variables on which to base the decision at that node. Calculate the best split based on these m variables in the training set. For prediction, a new case is pushed down the tree. It is assigned the label (diagnosis) of the training case in the terminal node it ends up in. This procedure is iterated over all trees in the ensemble, and the average vote of all trees is reported as the random forest prediction.

Data processing is performed using RapidMiner [24]. The following variables were used to predict substances abuse for the first dataset: age, gender, patient's specific Hispanic origin, education, employment status, living arrangement, number of previous treatment episodes the patient has received in any drug or alcohol program, primary, secondary, and tertiary substances used by the subject, their route of administration, and if injection was reported as primary, secondary, or tertiary route of administration, frequency of use, and age at first use, number of substances reported at admission from primary, secondary, and tertiary substances of abuse, and classifies patient's substance abuse type.

The variables used to predict psychotic diseases for the other dataset were age, gender, governorate, marital status, employment status, and if psychopathologies symptoms [6] were reported, as samples like (stupor, ambitendency, dystonia, psychogenic movements, avolition, social phobia, agoraphobia, manneristic speech, flight-of-ideas, illusions, delusions of persecution, eating disorders, body dysmorphic disorder), and classifies patient's psychotic disease type.

Prediction systems and models were generated using machine learning algorithms [25] provided by RapidMiner. The original number of patients (n) before the pre-processing phase for substances abuse dataset were $n = 1,749,767$ with only 405,393 diagnosed patients, while for psychotic diseases dataset $n = 1,800$ diagnosed cases. In the pre-processing phase, Principal Component Analysis [26] (PCA) was used with variance threshold parameter $= 0.95$ for attributes reduction procedure by selecting 24 attributes only from 63 for the substances abuse dataset, while minimizing the number of attributes from 89 to 79 for the psychotic diseases dataset. The attributes removed were ineffective attributes, like health insurance of patients in the substances abuse dataset and hypochondriasis symptom in the psychotic diseases dataset. Impute missing values also applied to predict the values missing for attributes, in order to use the two whole datasets, since examples are too precious to squander through model generation [27].

Random forest classifier for multi-classification [28] (more than two classes) was used, for substances abuse dataset before impute missing values was applied, total $n = 77,790$ (alcohol dependence $= 22,090$ cases, opioid dependence $= 35,441$ cases, cannabis dependence $= 11,385$ cases, cocaine dependence $= 8,874$ cases), after impute missing values was applied, the total became $n = 405,393$ cases (alcohol dependence $= 163,012$, opioid dependence $= 140,082$ cases, cannabis dependence $= 64,921$ cases, cocaine dependence $= 37,378$ cases).

For the psychotic diseases dataset, impute missing values was not applied because this dataset was collected without missing values, total cases $n = 1,800$ (major depression $= 300$ cases, drug-induced psychosis $= 200$ cases, schizophrenia $= 650$

cases, schizoaffective disorder = 100 cases, obsessive compulsive disorder = 50 cases, bipolar disorder - manic episode = 400 cases, bipolar disorder - mixed episode = 100 cases).

The parameters utilized for both datasets (for best accuracy according to the experiments carried out): Number of trees to build = 3, Allowed maximal depth of the trees = 20, Minimal leaf size = 2 examples, Minimal size for split = 4 examples, Ten-fold cross-validation [29] was chosen as a performance measure in evaluating models based on the number of class learners.

3   Results

The resulted random forest classification accuracy values applied on substances abuse dataset were equal to 87.72 % without impute missing values learner, and 92.15 % with impute missing values learner. After adding impute missing values learner, classification efficiency was increased by 4.43 %. Applying random forest classifier on psychotic diseases dataset, accuracy was equal to 100 % without impute missing values learner, possibly due to the lack of noisy data in the 1,800 cases, unlike 405,393 cases of the substances abuse dataset. Precision and recall [30] were used to evaluate the effectiveness of information retrieval models. Precision means how many patients were diagnosed rightly, while recall means, how many relevant patients for a specific disease were diagnosed (see Table 1, 2 and 3). The data mining analysis (random forest classifier with impute missing values learner) created two models of those with substances abuse disorders or psychotic disorders and identified the most important factors (attributes) in these two separate groups of patients.

Table 1. Precision and recall rates of random forest model **WITHOUT** impute missing values learner for substances abuse patients (**Accuracy 87.72 %**)

	True opioid dependence	True cannabis dependence	True alcohol dependence	True cocaine dependence	Class precision
Pred. opioid dependence	32611	648	886	360	94.51 %
Pred. cannabis dependence	639	9519	785	1353	77.42 %
Pred. alcohol dependence	1497	885	19631	681	86.50 %
Pred. cocaine dependence	694	333	788	6480	78.12 %
Class recall	92.01 %	83.61 %	88.87 %	73.02 %	

3.1   Substances Abuse Patients' Model

The most relevant attribute in this model was the number of substances reported at patient admission from primary, secondary, and tertiary substances of abuse.

Table 2. Precision and recall rates of random forest model **WITH** impute missing values learner for substances abuse patients (**Accuracy 92.15 %**)

	True alcohol dependence	True cocaine dependence	True opioid dependence	True cannabis dependence	Class precision
Pred. alcohol dependence	155309	3087	4822	3787	93.00 %
Pred. cocaine dependence	2766	29060	2385	1115	82.26 %
Pred. opioid dependence	2266	1032	130792	1598	96.39 %
Pred. cannabis dependence	2671	4199	2083	58421	86.71 %
Class recall	95.27 %	77.75 %	93.37 %	89.99 %	

Specifically, single substance abuse patients were significantly more likely to be diagnosed by the mental disorder of alcohol dependence. In addition, double substances abuse patients had the mental disorder risk of cocaine dependence versus those on three substances abuse who had the risk of cannabis dependence.

3.2   Psychotic Patients' Model

In those with a psychotic disorder, stereotypy symptom was the strongest attribute of classification. Particularly, patients with the stereotypy symptom were diagnosed as schizoaffective disorder. Patients with clang associations symptom were identified as bipolar disorder (mixed episode), while those with hoarding symptom were identified as obsessive-compulsive disorder (OCD) patients. Major depression disease risk reliably identified patients with neurasthenia and chronic fatigue syndrome symptoms. In schizophrenia, patients had avolition symptom, versus patients with decreased arousal symptom were more likely to be diagnosed by drug-induced psychosis disease.

4   Discussion

To our knowledge, the present study is the first using an advanced random forest algorithm which is based on a collection of decision trees, especially suited for non-linear and complex data structures, and combined with impute missing values learner, applied on collected dataset of psychotic patients from Egypt parallel to substances abuse patients dataset from the United States for constructing mental disorders diseases predictive models. The datasets included demographic characteristics, substances abuse and psychopathologies symptoms aspects, and diagnoses. This paper showed that the ability of the addiction model to predict diagnosis based on substances abuse was high, ten-fold cross-validation accuracy of 92.15 %, and the psychotic diseases model based on psychopathologies symptoms obtained a ten-fold cross-validation accuracy of 100 %. This was not surprising, considering that the

Table 3. Precision and recall rates of random forest model **WITHOUT** impute missing values learner for psychotic patients (**Accuracy 100 %**)

	True manic_episode	True schizophrenia	True depression	True drug-induced_psychosis	True OCD	True mixed_episode	True schizoaffective_disorder	Class precision
Pred. manic_episode	400	0	0	0	0	0	0	100.00 %
Pred. schizophrenia	0	650	0	0	0	0	0	100.00 %
Pred. depression	0	0	300	0	0	0	0	100.00 %
Pred. drug-induced_psychosis	0	0	0	200	0	0	0	100.00 %
Pred. OCD	0	0	0	0	50	0	0	100.00 %
Pred. mixed_episode	0	0	0	0	0	100	0	100.00 %
Pred. schizoaffective_disorder	0	0	0	0	0	0	100	100.00 %
Class recall	100.00 %	100.00 %	100.00 %	100.00 %	100.00 %	100.00 %	100.00 %	

amount of noisy data of patients in the addiction model was higher than in psychotic diseases model, with a higher amount of missing values, which decreases the accuracy during the classification process.

From the substances abuse model, alcohol dependence was considered as the highest substance abuse consumption, by usual route of administration as oral, 50.9 %, frequency of use as daily, 45.1 %, and age at first use between 15 and 17 year old, 27.2 %. In psychotic diseases model, the most prevalent three psychiatric diseases were Schizophrenia, 36.1 %, major depression, 16.7 %, and drug-induced psychosis, 11.1 %. Figure 1 shows the methodology of the study, which includes two diagrams of the study procedures from collected datasets followed by pre-processing phase, applying random forest classifier and impute missing values learner respectively for creating the more accurate mental disorders diagnostic models.

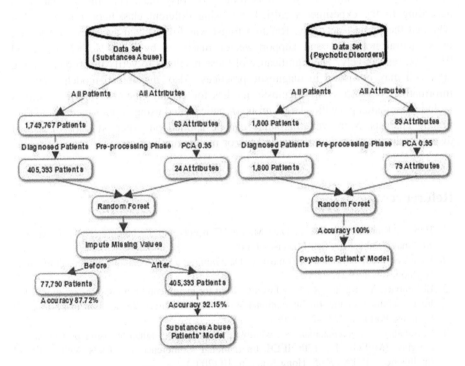

Fig. 1. The methodology of the study

The present study is subject to a number of limitations. Random forest classifier may have low performance to a single pruned decision tree when very large numbers of noisy data are present [31], but this is easily conquered by changing in parameters. Also, the number of psychotic patients collected in Egypt was much lower compared to the number of addiction patients of the other dataset from the United States, but all psychotic inpatients in psychiatric division of El-Mamoura hospital were taken into consideration in this study.

5 Conclusion

A combination of data mining techniques using a machine learning algorithm provided an effective systematic approach for diagnosing mental disorders. The use of this novel methodology will yield more accurate results in mental disorders diagnostic systems. This methodology highlights the effectiveness of implementing hybrid data mining methodologies, by comparing random forest classifier with and without impute missing values learner on two datasets of mental disorders problems. It is evident from the results that the random forest classifier combined with impute missing values learner produce more accurate results than without impute missing values learner.

The classification task in this study could have been applied by other classifiers, such as decision tree, neural networks, or support vector machines. The preferences of random forest classifier are that it is robust to noisy data, and missing data, and according to the experiments carried out using different classifiers, random forest obtained the highest accuracy. Random forest was followed in accuracy by decision trees, neural networks, and support vector machines by 92.10 %, 92.01 %, and 91.04 %, respectively. The significance of these results is to verify that a data mining approach may be useful in diagnostic practices. Also, the trained models could be informative in constructing diagnostic models for mental illnesses. Future directions include more attempts with a larger group of patients and using various methods of data mining analysis. Moreover, this study can be applied on expert systems of mental disorders diagnostic systems as a desktop or mobile application.

References

1. Han, J., Kamber, M., Pei, J.: Data Mining: Concepts and Techniques, 2nd edn. Morgan Kaufmann Publishers, San Francisco (2006)
2. Papageorgiou, E., Kotsioni, I., Linos, A.: Data mining: a new technique in medical research. Hormones **4**, 210–212 (2005)
3. El-Sonbaty, Y., Kashef, R.: New fast algorithm for incremental mining of association rules. In: Proceedings of the 4th International Workshop on Pattern Recognition in Information Systems, Porto, pp. 275–281 (2004)
4. El-Sonbaty, Y., Neematallah, A.: Multiway decision tree induction using projection and merging (MPEG). In: 17th IEEE International Conference on Tools with Artificial Intelligence - ICTAI 2005, Hong Kong, p. 10 (2005)
5. El-Sonbaty, Y., Ismail, MA., Farouk, M.: An efficient density based clustering algorithm for large databases. In: 16th IEEE International Conference on Tools with Artificial Intelligence, Florida, pp. 673–679 (2004)
6. American Psychiatric Association: Diagnostic and statistical manual of mental disorders: DSM-5. In: American Psychiatric Association, Washington, DC, 5th edn. (2013)
7. Ganzert, S., Guttmann, J., Kersting, K., et al.: Analysis of respiratory pressure-volume curves in intensive care medicine using inductive machine learning. Artif. Intell. Med. **26**, 69–86 (2002)

8. Ronco, A.L.: Use of artificial neural networks in modeling associations of discriminant factors: towards an intelligent selective breast cancer screening. Artif. Intell. Med. **16**, 299–309 (1999)
9. Kukar, M., Kononenko, I., Grošelj, C., et al.: Analysing and improving the diagnosis of ischaemic heart disease with machine learning. Artif. Intell. Med. **16**, 25–50 (1999)
10. Horvitz-lennon, M., Kilbourne, A.M., Pincus, H.A.: From silos to bridges: meeting the general health care needs of adults with severe mental illnesses. Health Aff. **25**, 659–669 (2006)
11. Dipnall, J.F., Pasco, J.A., Berk, M., et al.: Fusing data mining, machine learning and traditional statistics to detect biomarkers associated with depression. PLoS ONE **11**(2), e0148195 (2016)
12. Sumathi, M.R., Poorna, B.: Prediction of mental health problems among children using machine learning techniques. Int. J. Adv. Comput. Sci. Appl. **7**(1), 552–557 (2016). doi:10.14569/ijacsa.2016.070176
13. Breiman, L.: Random forests. Mach. Learn. **45**, 5–32 (2001)
14. Qi, Y., Bar-joseph, Z., Klein-seetharaman, J.: Evaluation of different biological data and computational classification methods for use in protein interaction prediction. Proteins **63**, 490–500 (2006)
15. Lebedev, A.V., Westman, E., Van Westen, G.J.P., et al.: Random Forest ensembles for detection and prediction of Alzheimer's disease with a good between-cohort robustness. NeuroImage Clin. **6**, 115–125 (2014)
16. Pflueger, M.O., Franke, I., Graf, M., Hachtel, H.: Predicting general criminal recidivism in mentally disordered offenders using a random forest approach. BMC Psychiatr. **15**(1), 1 (2015)
17. Rahman, M.M., Davis, D.N.: Machine learning-based missing value imputation method for clinical datasets. In: Yang, Gi-Chul, Ao, Sio-long, Gelman, Len (eds.) IAENG Transactions on Engineering Technologies, pp. 245–257. Springer, Dordrecht (2013)
18. Cetin, M. S.: New approaches for data-mining and classification of mental disorder in brain imaging data, Dissertation (2015)
19. Savitz, J.B., Rauch, S.L., Drevets, W.C.: Clinical application of brain imaging for the diagnosis of mood disorders: the current state of play. Mol. Psychiatr. **18**(5), 528–539 (2013)
20. Doehrmann, O., Ghosh, S.S., Polli, F.E., et al.: Predicting treatment response in social anxiety disorder from functional magnetic resonance imaging. JAMA Psychiatr. **70**(1), 87–97 (2013)
21. Kambeitz, J., Kambeitz-Ilankovic, L., Leucht, S., et al.: Detecting neuroimaging biomarkers for schizophrenia: a meta-analysis of multivariate pattern recognition studies. Neuropsychopharmacology **40**(7), 1742–1751 (2015)
22. United States Department of Health and Human Services: Substance Abuse and Mental Health Services Administration. Center for Behavioral Health Statistics and Quality. Treatment Episode Data Set – Admissions (TEDS-A). ICPSR35037-v1. Inter-university Consortium for Political and Social Research [distributor], Ann Arbor, 5 July 2014. http://doi.org/10.3886/ICPSR35037.v1
23. Cabena, P., Hadjinian, P., Stadler, R., et al.: Discovering Data Mining: From Concept to Implementation. Pearson Education, Upper Saddle River (1997)
24. Mierswa, I., Wurst, M., Klinkenberg, R., et al.: Yale: rapid prototyping for complex data mining tasks. In: The 12th ACM SIGKDD International Conference on Knowledge Discovery and Data Mining, Philadelphia, pp. 935–940 (2006)
25. Witten, I.H., Frank, E.: Data Mining: Practical Machine Learning Tools and Techniques, 2nd edn. Morgan Kaufmann Publishers, San Francisco (2005)

26. Johnstone, I.M., Lu, A.Y.: On consistency and sparsity for principal components analysis in high dimensions. J. Am. Stat. Assoc. **104**, 682–693 (2009)
27. Roecker, E.B.: Prediction error and its estimation for subset-selected models. Technometrics **33**, 459 (1991)
28. Ibrahim, N., Belal, N., Badawy, O.: Data mining model to predict Fosamax adverse events. Int. J. Comput. Inf. Technol. **3**, 936–941 (2014)
29. Kohavi, R.: A study of cross-validation and bootstrap for accuracy estimation and model selection. In: Proceedings of the 14 International Joint Conference on Artificial Intelligence, Montréal, pp. 1137–1145 (1995)
30. Davis, J., Goadrich, M.: The relationship between precision-recall and ROC curves. In: Proceedings of the 23rd international conference on Machine learning - ICML 2006, New York, pp. 233–240 (2006)
31. Segal, MR.: Machine learning benchmarks and random forest regression. Center for Bioinformatics and Molecular Biostatistics, University of California, San Francisco (2003)

Performance Analysis of Biased Localization of Heterogeneous Nodes Combined with Pure LEACH Routing Protocol

Ahmed Lotfy[1(✉)], Amr A. Awamry[2], and M.B. Abdelhalim[1]

[1] College of Computing and Information Technology,
Arab Academy for Science, Technology and Maritime Transport, Cairo, Egypt
ahmed_lotfy@aast.edu, mbakr@ieee.org
[2] Benha Faculty of Engineering, Benha University, Cairo, Egypt
Amr.awamry@bhit.bu.edu.eg

Abstract. Energy efficiency and network lifetime are the one of the most important issues in Wireless Sensor Network (WSN). As well as, they are mandatory for low-cost solutions to various real world challenges. Sensor nodes depend on the power of its battery which has a significant effect on their performance. Clustering and localization of sensor nodes are one of the efficient techniques to increase the network lifetime. This paper discusses a new clustering protocol scheme is proposed and evaluated; the new scheme depends on hybrid nodes localization formation to improve the existing Low Energy Adaptive Clustering Hierarchy (LEACH) protocol. The LEACH static clustering is based on the heterogeneous routing protocol. Hence, enhancing allocation improves a number of transmission data rate and node lifetime to achieve the network longevity in WSN. Results are simulated using Matlab simulation tools; the proposed scheme shows 15 % improvement in WSN total network lifetime and 15 %–49 % enhancement in the total number of packets transmitted.

1 Introduction

Wireless sensor network (WSN) is a set of specialized transducers with a communication infrastructure. It is composed of hundreds of sensor nodes which sense the physical field in terms of temperature, humidity, light, sound, vibration,... etc. [1]. The main function of the sensor node is to gather the data and information from the sensing elements and send it to the end user via Base station (BS). These sensor nodes can be deployed in many applications. The current wireless sensor network is working on the problems of minimal power communication, sensing, energy storage, and computation.

Communication protocols great affect the performance of WSNs by an evenly distribution of energy load, decreasing their energy consumption thus increase their lifetime [2]. Therefore, many kinds of efforts have been done for developing energy-efficient transmission protocols for wireless sensor networks.

A clustering technique is one of the most effective techniques which enhance the sensor network to work efficiently. It also decreases the depletion of energy and hence it increases the lifetime of the sensor. The main role of cluster head is sending data

© Springer International Publishing AG 2017
A.E. Hassanien et al. (eds.), *Proceedings of the International Conference on Advanced Intelligent Systems and Informatics 2016*, Advances in Intelligent Systems and Computing 533, DOI 10.1007/978-3-319-48308-5_65

communication between sensor nodes and the BS efficiently. So the cluster head is more preferable than the other nodes because it produce higher energy as well as it performs the data aggregation.

In this paper, a new scheme is presented. In this scheme using the heterogeneous clustering protocol is used which focuses on localizing the nodes for improving entire Network lifetime. The paper also crystallizes a comparison between the existing clustering protocols in terms of simulation and evaluation, the results will show that the proposed scheme outperformed LEACH [3] in terms of energy consumption and network lifetime.

The rest of paper is planned as follows. In Sect. 2, the background of the existing works is presented in relation with the routing and clustering protocols for wireless sensor networks. In Sect. 3, the design of LEACH is presented. Section 4 it presents our scheme. Section 5, shows evaluated performance of the proposed scheme using Matlab simulator. Finally, several concluding statements are given in Sect. 6.

2   Related Work

In latest years intelligent WSNs, as an important monitoring technology, have been utilized to collaboratively and comprehensively obtain the target data through a large scale of sensors deployed in a target area and self-organized sensor networks [4].

It applied to various fields, including military [5]. Transportation [6], environmental monitoring [7], the agriculture monitoring [8, 9], forest fire detection [10, 11], volcano [12] and earthquake [13], measurements, and hazards detection [14].

Many approaches had been adopted by the researchers to protect the reduction of the node energy as much as possible. LEACH [3] proposed by Heinzelman is one of the first hierarchical or clustering-based protocols in which Cluster heads are randomly selected. Others advancements are proposed for LEACH are Advanced LEACH (AD-LEACH) [15], Hybrid Energy-Efficient Distributed clustering (HEED) [16], Stable Election Protocol (SEP) [17], Distributed Energy-Efficient Clustering (DEEC) [18] and Enhanced Parallel Cat Swarm Optimization (EPCSO) [2]. Energy efficient routing techniques have been proposed to route data within the sensor network. They also decrease duplicated packets and reduce total number of hops to deliver the data. Eventually, they consume network energy in a way that needs to be improved [19–21].

A clustering-based protocol utilizes randomized rotation the cluster head (CH) so that it can distribute the energy load among the sensors in the network [22]. LEACH protocol uses localized management to enable scalability and robustness of the dynamic network and combines data fusion into the routing protocol in order to increase the aggregate of information that must be transmitted to the base station. It rearranges the network cluster head dynamically and periodically, consequently, it is difficult to rely on long lasting node-to-node trust relationships to make the secure connection protocol. LEACH assumes that every node can directly connect to the base station by transmitting high power sufficiently.

The hierarchical structure based protocols, have been recognized to be energy efficient protocols of wireless sensor network (WSN), can be divided into more than one category, cluster based such as LEACH, HEED, and tree based such as TEEN [23].

In a hierarchical cluster-based protocol, the sensor nodes are classified into clusters with one CH (selected among the sensor nodes) in each cluster.

It is the responsibility of the CH to gather the data from its associated nodes. It may aggregate the data, and then forward it to the BS. No node is selected as a permanent CH to distribute the energy consumption load equally among the nodes. The data transmission process is divided into rounds and each round is divided into two phases; first phase is the Setup phase followed by the Steady State. Each sensor knows when each round starts by using a synchronized clock. In each round, a new CH is selected randomly and rotationally. Each random CH selection establishes a protocol which is proposed in research literature so far. Each protocol is based mostly on their own means of CH selection criteria, clustering parameters, and data transfer and aggregation mechanisms. All these mechanisms are meant to improve one or more parameters like energy consumption, packet delay and the throughput of the network [16–19, 23, 24]. The energy efficiency parameters of these protocols have been evaluated and compared.

3 LEACH Protocol

3.1 Radio Dissipation Model

Heinzelman's first order model is used for transmission of data in low noise environment. That l-bit message is transmitted over a distance d as in as shown in Fig. 1 [3, 25].

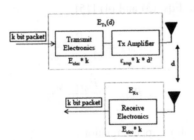

Fig. 1. Radio energy dissipation mode [3, 25]

3.2 Cluster Head (CH) Selection

LEACH is the one of the most popular cluster based routing technique. It is an energy-efficient communication protocol. The sensor nodes are divided periodically into several clusters. For each cluster, a sensor node is selected as a CH. Thus, LEACH executes a periodically randomized rotation of the CH nodes. The operations of LEACH are generally separated into two phases: first phase is the Setup phase followed by the Steady State. In the Setup phase, CHs are selected and clusters are organized. In the steady-state phase, the data transmissions to the BS take residence. The role of the CH is allocated by the node which gets a random number between 0 and 1. If the number is less than the threshold values (n), the node becomes a CH for the current round:

$$T(n) = \begin{cases} \dfrac{p}{1 - p * \left(r \, mod \, \frac{1}{p}\right)} & if \, s_in \in G \\ 0 & otherwise \end{cases} \qquad (1)$$

P is the predetermined percentage of CHs ($P = 10$ %), r is the current round, and G is the set of nodes that have not been selected as CHs in the last $1/P$ rounds. Using this threshold, each node will be a CH at some round within $1/P$. After the election of CH nodes, each ordinary node will determine the optimal CH to join in terms of minimum energy required for transmission [15].

3.3 Probability of Heterogeneous Nodes

From the Eq. (2) the optimal probability of a node becomes a cluster head P_{opt}, pi is the reference value in heterogeneous network and the reference value of node each other according to its initial energy value (2) [15].

$$p_i = P_{opt} \frac{E_i(r)}{\overline{E}(r)} \qquad (2)$$

Considering two level heterogeneous networks are used in this research. P_{nrm} is the reference to normal nodes and P_{adv} is the references to advanced nodes Modified values are produced to generate Eqs. (3) and (4) [15].

$$P_{nrm} = \frac{P_{opt}}{(1 + a * m)} \qquad (3)$$

$$P_{adv} = \frac{P_{opt(1+a)}}{(1 + a * m)} \qquad (4)$$

Equation (5) is a result of upper equations

$$P_i = \begin{cases} \dfrac{P_{opt} E_i(r)}{(1 + a * m)\overline{E}(r)} & if \, s_i = normal \, node \\ \dfrac{P_{opt(1+a)} E_i(r)}{(1 + a * m)\overline{E}(r)} & if \, s_i = advanced \, mode \end{cases} \qquad (5)$$

4 Proposed Scheme

This section describes the details of the proposed scheme algorithm. Figure 2 shows a flowchart of the proposed algorithm process. The initialization of the network environment to imitate a real application of the required parameters, including network environment and node specification. Consequently, network architecture is constructed

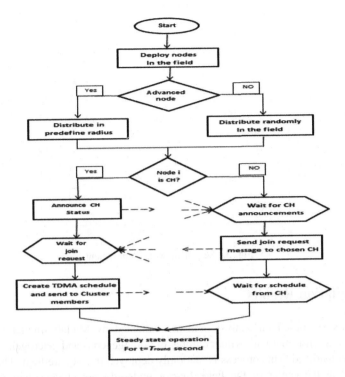

Fig. 2. Flowchart of proposed scheme

and the nods are deployed according to their types. During the cluster head selection phase, nodes decide whether they should take the role of a cluster head or not. Thus, every node is initialized to be a cluster head or not and does not have an associated cluster at the beginning of a cluster head setup phase. After electing Cluster Heads, a node decides to associate a cluster according to the strength of the signal. The proposed scheme uses a similar clustering algorithm with LEACH to form Cluster Heads.

Energy model as assumed is used for selection the cluster head mechanism for WSN of n nodes with an initial area size of $X \times Y$ m2 which has been prepared by using Matlab simulation. The BS is placed in the middle of the area to minimize the power consumed by the sensor nodes. The model can compute the energy consumed in the cluster head selection phase. More than one type of nodes in the network such as normal and advanced nodes. Eo is the initial energy for the normal nodes, m = 0.1 is the fraction of the advanced nodes and they are supporting energy harvesting. The additional energy factor between advanced and normal nodes a = 1 which means there are 10 % advanced nodes contain more energy than the normal nodes. The deployment strategy depends mainly on the type of sensors as well as the advanced nodes in the predefined radius around the BS. The rest normal nodes randomly are deployed in a given region as shown in Fig. 3.

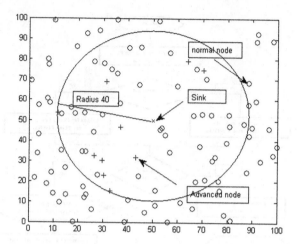

Fig. 3. Sensor network deployment

5 Simulation Results

The results are based on analysis and are validated by Matlab simulations. In the simulation, a sensor network with n number of homogeneous and heterogeneous sensor nodes is considered. This process is randomly deployed in a square field. The only BS is located at the center of the field. Sensor nodes do not change their place after deployment. Sensor normal nodes have a limited energy while the advanced node has energy more than normal node. The transmission ranges of sensor nodes are adjustable according to the distance from the BS. Eventually, all nodes send the data packets to the BS in a cycle time; Table 1 [25] shows the radio parameters used in this simulation.

Table 1. Simulation parameters

Parameters	Value
Sensing field	100 m² × 100 m²
Number of nodes	100
Number of advanced nodes	10
Number of normal nodes	90
TX energy	50/nj/bit
Rx energy	0.0013/pj/bit
Initial energy	0.5 j
Packet size	4000 Bits

Figure 4. represents the effect of changing radius of advanced node distribution on total life nodes, where are **Rxx** mains the radius = **xx** m², and it also shows an increase in the network lifetime and data packet, from the above figure it's clear that when R = 40 m² gives the highest lifetime.

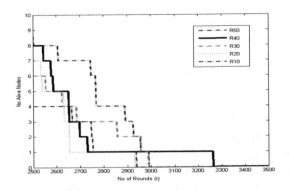

Fig. 4. Changing the number of alive nodes with different radiuses

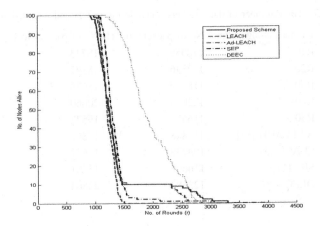

Fig. 5. Comparison between proposed scheme, Ad-LEACH, LEACH, SEP & DEEC

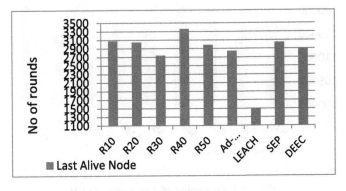

Fig. 6. Network lifetime

Table 2. First dead node and network lifetime after 3500 round

Parameters-algorithms	Round first dead node	Round last alive node
R10	929	3081
R20	956	3051
R30	919	2736
R40	935	3361
R50	1005	2989
Ad-LEACH [15]	1072	2842
LEACH [3]	1020	1497
SEP [17]	989	3050
DEEC [18]	1344	2894

Table 3. Total number of data packets transmission to CH and received at the BS

Parameters-algorithms	No of packets to CH	No of packets to BS
R10	118056	18743
R20	117346	18397
R30	118159	18676
R40	122056	20660
R50	116570	19577
Ad-LEACH [15]	117828	17866
LEACH [3]	109739	12671
SEP [17]	116695	13781
DEEC [18]	123583	22564

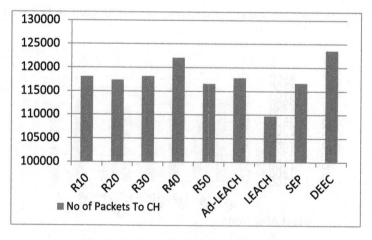

Fig. 7. Total number of data packets to CH

A comparison has been made between proposed scheme, Ad-LEACH, LEACH, SEP and DEEC will be discussed next.

Figures 5, 6 and Table 2, show the improvement in network lifetime. It was aforementioned before that the best results were found in case on radius 40 m². The improvement percentage in network lifetime at least 15 % with Ad-LEACH, SEP and DEEC and 120 % with LEACH.

Table 3 contains the total number of data packets transmitted to CH and received from the BS. These results are represented in Figs. 7, 8 and 9.

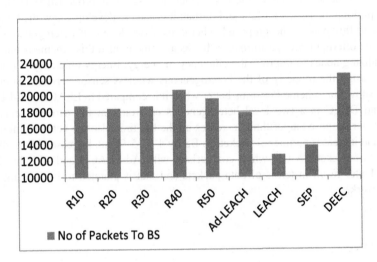

Fig. 8. Total number of data packets to BS

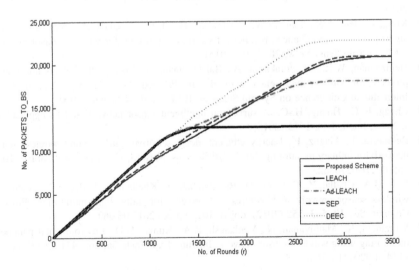

Fig. 9. Total number of packets to BS

In Figs. 7 and 8 the X-axis shows the number of packets to CH and BS respectively against radius (R) and other selected algorithms. While deploying the advanced nodes in radius around the BS as proposed scheme. Raised percentage is 15 % in a number of packets for BS and 4 % for CH. The best results found in case of the circle in radius = 40 m^2 as shown in Fig. 9.

6 Conclusion

Localization nodes are one of the keys which are used to improve LEACH algorithm and to prolong the network lifetime. For that, it is used to develop a new method which is used in this paper. The proposed scheme uses two levels of heterogeneous nodes which are offering better performance. It was aforementioned this scheme is uniformed by localizing nodes to improve the efficiency of energy. Energy efficiency is improved by increasing the property of the heterogeneous wireless sensor network. Simulation results show that scheme Performs better when it is compared to LEACH, Ad-LEACH, SEP and DEEC in a two level heterogeneous environment for wireless sensor networks. Our proposed algorithms increases all the life time also the network lifetime by 15 % and increases number of packets to BS by 4 %–5 % and increases 15 %–49 % in number of packets to CH while comparison with Ad-LEACH and SEP but DEEC shows better result in terms of data packets and that is planned to be investigated in future work.

References

1. Zhu, Y., Hua, K.: The development of mobile wireless sensor networks: a survey. In: Technological Breakthroughs in Modern Wireless Sensor Applications, p. 257, 31 March 2015
2. Kong, L., Pan, J.-S., Tsai, P.-W., Vaclav, S., Ho, J.-H.: A balanced power consumption algorithm based on enhanced parallel cat swarm optimization for wireless sensor network. Int. J. Distrib. Sens. Netw. **2015**, 10 (2015)
3. Heinzelman, W.R., Chandrakasan, A., Balakrishnan, H.: Energy-efficient communication protocol for wireless microsensor networks. In: Proceedings of the 33rd Annual Hawaii International Conference on System Sciences, IEEE, p. 10, 4 January 2000
4. Chang, F.-C., Huang, H.-C.: A survey on intelligent sensor network and its applications. J. Netw. Intell. **1**(1), 1–15 (2016)
5. Bekmezci, I., Alagoz, F.: Energy efficient, delay sensitive, fault tolerant wireless sensor network for military monitoring. Int. J. Distrib. Sens. Netw. **5**, 729–747 (2009). doi:10.1080/15501320902768625
6. Kafi, M.A., Challal, Y., Djenouri, D., Bouabdallah, A., Khelladi, L., Badache, N.: A study of wireless sensor network architectures and projects for traffic light monitoring. Procedia Comput. Sci. **10**, 543–552 (2012). doi:10.1016/j.procs.2012.06.069
7. Vairamani, K., Mathivanan, N., Venkatesh, K.A., Kumar, U.D.: Environmental parameter monitoring using wireless sensor network. Instrum. Exp. Tech. **56**, 468–471 (2013). doi:10.1134/S0020441213040118

8. Qu, Y., Zhu, Y., Han, W., Wang, J., Ma, M.: Crop leaf area index observations with a wireless sensor network and its potential for validating remote sensing products. IEEE J. Sel. Top. Appl. Earth Obs. Remote Sens. **7**(2), 431–444 (2014). Chang, F.C., Huang, H.C., 12

9. Marino, P., Fontan, F., Dominguez, M., Otero, S.: An experimental ad-hoc WSN for the instrumentation of biological models. IEEE Trans. Instrum. Meas. **59**(11), 2936–2948 (2011)

10. Farah, C., Schwaner, F., Abedi, A., Worboys, M.: Distributed homology algorithm to detect topological events via wireless sensor networks. IET Wireless Sens. Syst. **1**(3), 151–160 (2011)

11. Lei, Z., Jianhua, L.: Distributed coverage of forest fire border based on WSN. In: 2010 2nd International Conference on Industrial and Information Systems (IIS), vol. 1, pp. 341–344, July 2010

12. Lara, R., Benitez, D., Caamano, A., Zennaro, M., Rojo-Alvarez, J.: On real-time performance evaluation of volcano monitoring systems with wireless sensor networks. IEEE Sens. J. **15**(6), 1–9 (2015)

13. Tan, R., Xing, G., Chen, J., Song, W.-Z., Huang, R.: Quality-driven volcanic earthquake detection using wireless sensor networks. In: 2010 IEEE 31st Real-Time Systems Symposium (RTSS), pp. 271–280, November 2010

14. Jan, M., Habib, Q., Irfan, M., Murad, M., Yahya, K., Hassan, G.: Carbon monoxide detection and autonomous countermeasure system for a steel mill using wireless sensor and actuator network. In: 2010 6th International Conference on Emerging Technologies (ICET), pp. 405–409, October 2010

15. Iqbal, A., Akbar, M., Javaid, N., Bouk, S.H., Ilahi, M., Khan, R.D.: Advanced LEACH: a static clustering-based heteroneous routing protocol for WSNs: arXiv preprint, arXiv:1306.1146, 5 June 2013

16. Younis, O., Fahmy, S.: HEED: a hybrid, energy-efficient, distributed clustering approach for ad hoc sensor networks. IEEE Trans. Mob. Comput. **3**(4), 366–379 (2004)

17. Kumari, R.S., Chithra, A., Devi, M.B.: Efficient 2-level energy heterogeneity clustering protocols for wireless sensor network. Indian J. Sci. Technol. **9**(8) (2016)

18. Shaji, M.: Distributed energy efficient heterogeneous clustering in wireless sensor network. In: 2015 Fifth International Conference on Advances in Computing and Communications (ICACC), pp. 130–134. IEEE, 2 September 2015

19. Hussain, S., Matin, A.W.: Energy efficient hierarchical cluster-based routing for wireless sensor networks. In: Jodrey School of Computer Science Acadia University Wolfville, Technical report, pp. 1–33, Nova Scotia, Canada (2005)

20. Baranidharan, B., Shanthi, B.: A survey on energy efficient protocols for wireless sensor networks. Int. J. Comput. Appl. **11**(10), 35–40 (2010)

21. Al Ameen, M., Islam, S.M., Kyungsup, K.: Energy saving mechanisms for MAC protocols in wireless sensor networks. Int. J. Distrib. Sens. Netw. **2010** (2010)

22. Liu, M., Cao, J., Chen, G., Wang, X.: An energy-aware routing protocol in wireless sensor networks. Sensors **9**(1), 445–462 (2009)

23. Manjeshwar, A., Agrawal, D.P.: TEEN: a routing protocol for enhanced efficiency in wireless sensor networks. In: Null, p. 30189a. IEEE, 23 April 2001

24. Sankarasubramaniam, Y., Akyildiz, I.E., McLaughlin, S.W.: Energy efficiency based packet size optimization in wireless sensor networks. In: Proceedings of the First IEEE, 2003 IEEE International Workshop on Sensor Network Protocols and Applications, pp. 1–8. IEEE (2003)

25. Abdellah, E., Benalla, S.A.I.D., Hssane, A.B., Hasnaoui, M.L.: Advanced low energy adaptive clustering hierarchy. (IJCSE) Int. J. Comput. Sci. Eng. **2**(07), 2491–2497 (2010)

Effective Selection of Machine Learning Algorithms for Big Data Analytics Using Apache Spark

Manar Mohamed Hafez[1]([✉]), Mohamed Elemam Shehab[2],
Essam El Fakharany[3], and Abd El Ftah Abdel Ghfar Hegazy[4]

[1] Information System Department, Arab Academy for Science,
Technology and Maritime Transport, Cairo, Egypt
manar.aast@gmail.com
[2] Arab Academy for Science, Technology and Maritime Transport, Cairo, Egypt
melemam9@gmail.com
[3] College of Computing and Information Technology,
Arab Academy for Science, Technology and Maritime Transport, Cairo, Egypt
essam.fakharany@gmail.com
[4] Dean of College of Computing and Information Technology,
Arab Academy for Science, Technology and Maritime Transport, Cairo, Egypt
hegazyabdelfatah@gmail.com

Abstract. Big Data appears with not only the increasing size of data but also complex and different processing and analytical tools. This research aims to compare some selected machine learning algorithms on datasets of different types and sizes using Apache spark tool in order to make a fair judgment about which one is the best fitting in. The algorithms were compared based on few parameters including mainly accuracy and training time. The algorithms were applied on three datasets of different fields: marketing, packing and statistics, and security datasets. The findings of this experiment show that the decision tree algorithm is the most suitable algorithm for marketing and security datasets. Additionally, logistic regression algorithm had the highest accuracy for packing and statistics dataset.

Keywords: Big Data · Apache spark · Machine learning algorithms · Decision tree · Naïve Bayes · Logistic regression · Gradient boosted trees · Random forest

1 Introduction

Big Data is a new attractive technology trend in science, industry, and business. Big Data has been becoming correlated to almost all aspects of human activities from just recording events to research, design, production, and digital services or products delivery to the final user. It applies to data sets of extreme size (e.g., terabytes, petabytes, exabytes, and zettabytes). Besides, it means the size of the data to be of large amount, it also refers to the complexity of the data increasing, which is not possible to be processed or analyzed by a conformist methods or tools [1]. Big Data is very

© Springer International Publishing AG 2017
A.E. Hassanien et al. (eds.), *Proceedings of the International Conference on Advanced Intelligent Systems and Informatics 2016*, Advances in Intelligent Systems and Computing 533, DOI 10.1007/978-3-319-48308-5_66

complex and contains not only structured data, but also it contains unstructured data, such as text, video, pictures, etc. Those are beyond the ability of manual techniques and commonly used software tools to capture, manage, and process within a tolerable timeframe. In general, Big Data exhibits different characteristics like volume, variety, velocity, value, variability, and veracity, and it extends another characteristic as complexity [2].

In modern world, data analysis has become a necessary part of any research. Moreover, it can provide some significant insights and hidden patterns from the dataset. The process of estimating data using an analytical tool to observe the components of data is becoming more and more important. Nowadays, there are lots of machine learning algorithms, which can be applied to datasets to get necessary results of data analysis easily [3]. The researchers aim to find a suitable combination of the sizes and types of datasets to reach the efficiency and the accuracy of Big Data analytics. For solving the problem of existence of different sizes and types of datasets and algorithms performance comparison, this research seeks out obtaining some serial as well as parallel implementations of certain machine learning algorithms. Hence, parallel implementation of machine learning algorithms was developed to decide what algorithm is best suited to work in order to make a fair judgment and to know which one is the best to fit in with the data. One of the tools that can be used for this purpose is sparks MLlib. This library in spark has some parallel implementations of certain machine learning algorithms.

Three different datasets were used in the experiment: an online marketing, packing and statistics and security datasets. Online marketing is a small dataset while packing and statistics. Additionally, security dataset is a large dataset. With Apache spark tool, different algorithms are deployed like NaiveBayes, Decision Tree, Random forest, Logistic Regression and Gradient Boosted Trees. Comparisons performed to evaluate various classification algorithms on the same dataset. And comparisons were done to compare machine learning algorithms on various data. This study is important because it helps to decide which one of algorithms classifications is the best to fit in with this data set. Performance evaluation depends on many standards, but the main important ones are accuracy and training time. Accordingly, this helps in deciding which algorithm is the most suitable for the high accuracy and less consumed time during building the model.

This paper is organized into five sections: initially, the paper is an introduction to the first section. The second section discusses the related works and the literature review. In the third section discusses the methodology of the proposed model. Results are discussed in the experimental works are assumed in the fourth section, whereas the fifth section introduces the conclusion and recommended future work.

2   Related Works

There were some papers and articles on the general topic of Big Data in different areas, such as business, government, artificial intelligence, traffic, health, science, engineering, genomics, social media, psychology, and mobility [4–8].

The researchers in all paper had similar descriptions of Big Data and provided a specific detailed information about how Big Data is different in the various areas. The different trends or topics, security, clouds, storage, architectures, IT governance, and privacy, had been covered by researchers as standalone topics only and provided guidance on how to manage each of the individual topics with regard to Big Data [9–11].

After researching in this area, it is noticed that no one makes a fair judgment to know which machine learning algorithm is best to fit in for this Big Data. There are some comparisons, which have been done on different types and sizes of datasets in Big Data mining tool.

3 Methodology

The main objective of this study is to improve the quality of prediction processes in different sizes and types of the datasets. This comparative study seeks out a fair judgment and knows which algorithm is best to fit in with the introduced dataset. The study moved on certain phases described in the framework of the experimentation as shown in Fig. 1. The framework includes many phases: data collection, data preprocessing and selection, transformation phase, selection of Big Data tool, selection of programming language, and selection of data mining algorithms depending on the dataset binary classification or multi-classification.

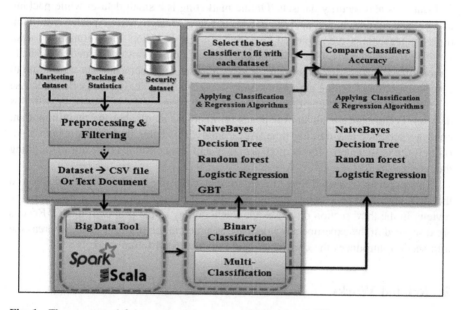

Fig. 1. The conceptual framework of the comparative study of different size and type dataset with different classifiers.

3.1 The Proposed Model

While processing the four phases, this study starts with data collection, data preprocessing, and filtering to clean, integrate, and transform the data. Then, the required
fields for data mining are selected by the programming language. The data was
transformed into a certain file format, which is acceptable by the Big Data mining tools.
The Big Data mining tools were different data mining algorithms based on the dataset
binary classification or multiclass which are tested by Scala programming language.

Data Collection and Selection Phase

In this experiment, there are three datasets; the first data set is a small data set
obtained from extremely reputational online shopping activity which sells only via
online. The dataset is collected by an online ordering log file for three months. The
dataset contains 304 instances and 26 attributes, four classes and 69 features. The
second data set is a medium dataset source from UCI/Adult [12]. The original adult
data set has fourteen features, among which six are continuous and eight are categorical. In this data set, the continuous features are discretized into quintiles, and each
quintile is represented by a binary feature. The dataset consists of 45222 instances and
122 attributes, and 123 features. Finally, the third dataset is a large dataset which is
obtained from The Mineta Transportation Institute (MTI) Database [13]. The dataset
consists of 456613 instances of five classes, 42 attributes, and 164 features. For all the
datasets, feature extraction, and feature selection should be done properly.

Data Preprocessing, Imputation and Transformation Phase

In this step, the dataset goes through many stages, such as data cleaning, data
integration, and data transformation. The gathered data was saved as excel spreadsheets
or text documents. The cleaning process is required in order to analyze the data based
on selected classifier algorithms, Data deletion technique was used for data imputation
where records with missing values were eliminated, inconsistent data was corrected,
outliers were identified, and duplicate data was removed. The data was exemplified by
numbers and stored in the form of a CSV or txt file so it can be introduced to the data
mining tool.

Selection of Big Data Mining Tool

Apache Spark is an open computing source cluster framework. It has been taken on
a new level of popularity in the last year. It is a fast and general-purpose cluster
computing system. It provides a high-level APIs in Java, Scala, Python and R, and an
optimized engine that supports general execution graphs. It also supports a rich set of
higher-level tools including Spark SQL for SQL and structured data processing, MLlib
for machine learning, GraphX for graph processing, and Spark streaming [14].

Spark loads the data into HDFS on disk and implements a parallel execution system
with in-memory optimization. The dataset was split into two parts: the first part is a
training set which was 70 % used to train the model of the real dataset, and the second part
was a 30 %, it was used as a testing set to train the model. Whereas, Apache spark loads
data onto HDFS on local system on disk and runs in parallel on all the cores in the CPU.

Selection of Data Mining Algorithms Phase

The selected machine learning algorithms depend on the dataset binary classification or multiclass. Accordingly, a comparison was made between the five different
classification algorithms, such as NaiveBayes, decision tree, random forest, logistic

regression, and gradient boosted trees, in order to evaluate the performance of the algorithms by several binary classification evaluation and regression metrics. Additionally, the performance of the algorithms by several multi-classification evaluation metrics and regression metrics highlights a difference between four different classification algorithms: NaiveBayes, decision tree, random forest, and logistic regression, and it explains algorithms in Sect. 3.2.

Time Consumed for Each Process

Cleaning and organizing data consumed about 55 % of the process time. Collecting data sets come second at 19 %, meaning data spent around 75 % of their time on preparing and managing data for analysis. Other data spent consumed almost 25 % of their the time on refining algorithms, mining data, building training sets as shown in Fig. 2.

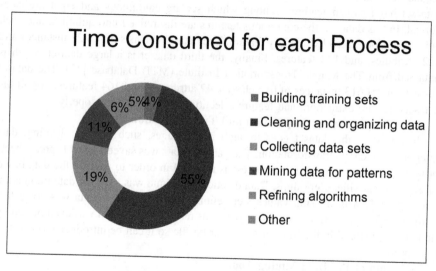

Fig. 2. Time consumed for each process

3.2   Machine Learning Algorithms

Big Data mining techniques can be segregated by their different model functions, representation, preference criterion, and algorithms. The main function of the model lies on its classification and a brief overview of several classification algorithms that have been used in this study [15, 16].

Naïve Bayes is a simple multiclass classification algorithm with the assumption of individuality between every pair of features. It can be used to train professionally, within a single pass of the training data and use it for prediction.

A decision tree is a successful and popular method for the machine learning task of classification and regression. It's usually used since they are easy to be interpreted, handled categorical features, extended to the multiclass classification setting.

Random forest is a collection of decision trees. It's one of the most successful machine learning models for classification and regression. It runs well on large datasets, but it is comparatively slower than other algorithms.

Logistic regression is a regression model in everywhere, the dependent variable is categorical. It's used to evaluation the probability of a binary response based on one or more predictor features.

Gradient-boosted tree (GBT) is a mixture of decision trees. It is training the decision trees in order to reduce the loss function. Alike decision trees, It's handled by some categorical features and a wide range of the multiclass classification setting.

3.3 Evaluation

The comparison of the different machine learning algorithms mentioned in Sect. 3.2 is based on the following measured parameters [15, 16]:

- **Training Time:** As an accomplished machine learning algorithms are measured. Time is taken to build the model is called training time. This varies on the implementations of the algorithms.
- **Accuracy:** It is the percentage of the correctly classified instances which are the total number of predictions that were correct.
- **Precision:** It is the element of the identified items that are correct and used to measure how well the proposed algorithm matches with the ground truth. It's likewise known as Positive Predictive Value (PPV).
- **Recall:** It is another measure used to compute how the proposed algorithm matches the ground truth. Recall, or sensitivity or consistently True Positive Rate (TPR), which is a measure of the number of true positives relative to the sum of the true positives and the false negatives.
- **F-measure:** It is sensitivity, an overall measure of how well we have been able to classify the ground truth foregrounds and backgrounds.
- **A receiver operating characteristic (ROC), or ROC curve:** It's a graphical plot that shows the performance of a binary classifier system as discernment threshold is diverse.
- **Area under Precision-Recall Curve (PRC):** Tuning the prediction threshold will change the precision and recall of the model and it is an imperative part of model optimization. It plots (precision, recall) points for diverse threshold values; while ROC, curve plots (recall, false positive rate) points.
- **Mean squared error (MSE):** It is the variance of the estimator and what is estimated. It is a risk function, consistent with the estimated value of the squared error loss or quadratic loss.
- **Root mean square error (RMSE):** It is often used to measure the variances between values predicted by a model or an estimator and the values actually detected.
- **Mean absolute error (MAE):** It is a quantity used to measure how close forecasts or predictions are to the ultimate results.

- **Coefficient of determination:** Represented R^2 or r^2 and pronounced R-squared are a number that indicates the quantity of the variance in the dependent variable that is predictable from the independent variable.
- **Explained variation (EV):** The proportion of the measured model accounts for the variation of a given dataset. Often, variation is quantified as variance.

From the previously mentioned measured parameters, we can compare the accuracies provided by all the algorithms on a dataset. Here, the focus is mainly on comparing major parameters like accuracy and training time in order to decide which machine learning algorithm is better suited for a selected type of data.

4 Experimental Work

Objectives of the experiment are to compare the performances of the machine learning algorithms when deployed different sizes and types datasets in terms of the training time, accuracy, true positive rate (TPR), precision, recall, F-measure, PRC area, ROC area, weighted precision, weighted recall, weighted F-measure, weighted TPR, weighted FPR, MSE, RMSE, R-squared, MAE and explained variance. This evaluation helps to decide which machine learning algorithm is best suited for a selected type of data. Results section has been presented for three experimental results. In all experimental results, the dataset is analyzed using different classifiers. They have been deployed Big Data mining tool Apache Spark version 1.5.0. These experiments were carried out on an HP Laptop using UBUNTU operating system version 15.10 which has been installed on VMware Workstation Pro version 12 having the following specifications and the number of the core processors 1 with Intel® Core ™ i7-5500U CPU, 2.40 GHz Processor, and 16.0 GB RAM.

4.1 First Dataset

Table 1 shows a comparison between four different classification algorithms, which are naivebayes, decision tree, random forest, and logistic regression. Evaluate the performance of the algorithms by several multi-classification evaluation metrics and regression metrics. The decision tree algorithm had the highest value of accuracy at 89.29 %, while naivebayes algorithm had the lowest value of accuracy at 31.03 %, although NaiveBayes algorithm had the lowest value of training time at 1.1 s, the decision tree algorithm had the highest value of training time at 2 s. Decision Tree had the highest value of precision at 0.8929, Decision Tree and Random forest had the highest value of recall at 0.893 and 0.735, respectively, and Decision Tree had the highest value of F-Measure at 0.893, Decision Tree and Logistic regression had the highest value of PRC Area at 0.9941 and 0.9678, respectively, and Decision Tree and Random forest had the highest value of ROC Area at 0.9523 and 0.8281.

In addition, Table 2 shows that Decision Tree had the highest value of weighted precision at 0.9018, Decision Tree and Random forest had the highest value of weighted Recall at 0.8929 and 0.7353, respectively, and Decision Tree had the highest value of weighted F-Measure at 0.877, Decision Tree and Random forest had the

Table 1. Comparison of several different classifiers using online shopping agency dataset.

Classifier	Accuracy (%)	TPR	Precision	Recall	F-Measure	PRC	ROC	Time (s)
NaiveBayes	31.03 %	0.311	0.3104	0.311	0.3105	0.8865	0.5220	1.1
Decision Tree	89.29 %	0.893	0.8929	0.893	0.893	0.9941	0.9523	2
Random forest	73.53 %	0.735	0.7353	0.735	0.73529	0.8917	0.8281	1.9
Logistic regression	68 %	0.677	0.6765	0.677	0.67647	0.9678	0.7243	2

highest value of weighted TPR at 0.893 and 0.7353, respectively, and Decision Tree had the lowest value of weighted FPR at 0.082. Random forest had the lowest value of MSE at 0.0693, Random forest and Logistic regression had the lowest value of RMSE at 0.2632 and 0.6807, respectively, and Random forest had the highest value of R-squared at 0.964, Random forest and Decision Tree had the lowest value of MAE at 0.089 and 0.289, respectively, and Logistic regression and Decision Tree had the lowest value of explained variance at 0.581 and 0.648.

Table 2. Comparison of another measured parameter.

Classifier	Weighted precision	Weighted recall	Weighted F-measure	Weighted TPR	Weighted FPR	MSE	RMSE	R^2	MAE	EV
NaiveBayes	0.5047	0.3104	0.2356	0.3104	0.2939	1.3608	1.166	−0.944	0.948	0.687
Decision Tree	0.9018	0.8929	0.877	0.893	0.082	0.5063	0.7116	0.2184	0.289	0.648
Random forest	0.7402	0.7353	0.728	0.7353	0.142	0.0693	0.2632	0.964	0.089	0.723
Logistic regression	0.6994	0.67647	0.6706	0.6765	0.1861	0.4634	0.6807	0.3764	0.414	0.581

Finally, these results of efficiency analysis show that the best classification algorithm was decision tree algorithm, which had the highest accuracy of 89.29 % in 2 s followed by random forest at 73.53 % by taking 1.9 s. Hence, these results can help several online retail shops by selecting the optimal classification algorithm which was decision table suitable to datasets related to the online retail shopping by building a complete recommended model to help the users to identify their needs from the represented product on e-commerce websites.

4.2 Second Dataset

Table 3 shows a comparison between five different classification algorithms, which are NaiveBayes, decision tree, random forest, logistic regression, and gradient boosted trees. Evaluate the performance of the algorithms by several binary classification evaluation and regression metrics.

Show that in Table 3, Logistic regression algorithm had the highest value of accuracy at 85.00 %, while NaiveBayes algorithm had the lowest value of accuracy at

81.67 %. Instead of NaiveBayes algorithm had the lowest value of training time at 3.9 s followed by logistic regression algorithm at 4.6 s. Logistic regression had the highest value of precision at 0.85, Logistic regression and Gradient Boosted Trees had the highest value of recall at 0.848 and 0.845, respectively, and Logistic regression had the highest value of F-Measure at 0.8478, Logistic regression and NaiveBayes had the highest value of PRC Area at 0.7243 and 0.7184, respectively, and NaiveBayes and Decision Tree had the highest value of ROC Area at 0.8060 and 0.7915.

Table 3. Comparison of several different classifiers using the census bureau dataset.

Classifier	Accuracy (%)	TPR	Precision	Recall	F-Measure	PRC	ROC	Time (s)
NaiveBayes	81.67 %	0.816	0.82	0.816	0.8158	0.7184	0.8060	3.9
Decision Tree	84.19 %	0.842	0.8419	0.842	0.8418	0.6502	0.7915	6.8
Random forest	82.21 %	0.822	0.8222	0.822	0.823	0.6226	0.6715	9.7
Logistic regression	85.00 %	0.848	0.85	0.848	0.8478	0.7243	0.7650	4.6
Gradient Boosted Trees	84.48 %	0.845	0.8448	0.845	0.84477	0.6905	0.7350	7.4

As shown in Table 4 show that Decision Tree had the highest value of weighted precision at 0.856, Logistic regression and Gradient Boosted Trees had the highest value of weighted recall at 0.847 and 0.844, respectively, and Decision Tree had the highest value of Weighted F-Measure at 0.847, Logistic regression and Gradient Boosted Trees had the highest value of Weighted TPR at 0.847 and 0.844, respectively, and NaiveBayes had the highest value of Weighted FPR at 0.222.

Table 4. Comparison of another measured parameter.

Classifier	Weighted precision	Weighted recall	Weighted F-Measure	Weighted TPR	Weighted FPR
NaiveBayes	0.840	0.815	0.823	0.815	0.222
Decision Tree	0.856	0.841	0.847	0.841	0.271
Random forest	0.809	0.822	0.805	0.822	0.447
Logistic regression	0.843	0.847	0.844	0.847	0.305
Gradient Boosted Trees	0.836	0.844	0.837	0.844	0.364

In addition, Table 5 show that Decision Tree had the lowest value of MSE at 0.115, Decision Tree and Random forest had the lowest value of RMSE at 0.340 and 0.340, respectively, and Logistic regression had the highest value of R-squared at 0.084, Logistic regression and Gradient Boosted Trees had the lowest value of MAE at 0.147 and 0.165, respectively, and Decision Tree and Random forest had the lowest value of Explained variance at 0.180 and 0.181.

Table 5. Comparison of some measured parameters.

Classifier	MSE	RMSE	R^2	MAE	EV
NaiveBayes	0.186	0.431	−0.001	0.186	0.221
Decision Tree	0.115	0.340	−0.727	0.231	0.180
Random forest	0.116	0.340	−0.970	0.239	0.181
Logistic regression	0.147	0.383	0.084	0.147	0.182
Gradient Boosted Trees	0.165	0.406	−0.078	0.165	0.183

This study used Big Data mining to help to achieve the high accuracy in different classification algorithms. The results show that the best classification algorithm was logistic regression algorithm, which had the highest accuracy of 85.00 % by taking 4.6 s followed by gradient boosted trees at 84.48 % by taking 7.4 s.

These results can help various packing and statistic (Census Bureau) by selecting the optimal classification algorithm which was the decision table suitable to datasets related to packing and statistic by building a complete recommended model to determine whether or not a person makes over 50K a year or less.

4.3   Third Dataset

Table 6 presents a comparison between four different classification algorithms, which are NaiveBayes, decision tree, random forest, and logistic regression. Evaluate the performance of the algorithms by several multiclass classification evaluation and regression metrics. As seen in Table 6 show that the decision tree algorithm had the highest accuracy of 96.73 % while NaiveBayes algorithm had the lowest value of accuracy of 82.33 %. The lowest value of training time had a random forest algorithm at 13.7 s; however, logistic regression algorithm had the maximum value of training time of 44.8 s. Decision Tree had the highest value of precision at 0.967, Decision Tree and Logistic regression had the highest value of recall at 0.967 and 0.961, respectively, and Decision Tree had the highest value of F-Measure at 0.967, Decision Tree and Logistic regression had the highest value of PRC Area at 0.999 and 0.997, respectively, and Decision Tree had the highest value of ROC Area at 0.966.

In addition, The Table 7 show that Decision Tree had the highest value of Weighted Precision at 0.966, Decision Tree and Logistic regression had the highest value of weighted recall at 0.9673 and 0.961, respectively, and Decision Tree had the highest value of weighted f-Measure at 0.963, Decision Tree and Logistic regression had the highest value of weighted TPR at 0.967 and 0.961, respectively, and Decision

Table 6. Comparison of several different classifiers using MTI dataset.

Classifier	Accuracy	TPR	Precision	Recall	F-Measure	PRC	ROC	Time (s)
NaiveBayes	82.33 %	0.8234	0.823	0.8234	0.823	0.997	0.799	14.3
Decision Tree	96.73 %	0.967	0.967	0.967	0.967	0.999	0.966	15.5
Random forest	94.70 %	0.947	0.947	0.947	0.947	0.993	0.691	13.7
Logistic regression	96.14 %	0.961	0.962	0.961	0.961	0.997	0.895	44.8

Tree had the lowest value of weighted FPR at 0.015, and NaiveBayes had the highest value of weighted FPR at 0.213. Random forest had the lowest value of MSE at 0.053, Random forest and Decision Tree had the lowest value of RMSE at 0.231 and 0.2456, respectively, and Random forest had the highest value of R-squared at 0.972, Logistic regression and Decision Tree had the lowest value of MAE at 0.057 and 0.071, respectively, and NaiveBayes had the lowest value of explained variance at 1.691.

Table 7. Comparison of another measured parameter.

Classifier	Weighted precision	Weighted recall	Weighted F-measure	Weighted TPR	Weighted FPR	MSE	RMSE	R^2	MAE	EV
NaiveBayes	0.836	0.823	0.816	0.823	0.213	1.070	1.034	0.469	0.413	1.691
Decision Tree	0.966	0.9673	0.963	0.967	0.015	0.060	0.246	0.969	0.071	2.016
Random forest	0.95	0.947	0.928	0.947	0.033	0.053	0.231	0.972	0.073	2.015
Logistic regression	0.956	0.961	0.95	0.961	0.024	0.114	0.338	0.945	0.057	2.009

Finally, the dataset used Big Data mining to help to achieve the highest accuracy in the different classification algorithms. The results show that the best classification algorithm fit in with our dataset was decision tree algorithm, which had the highest accuracy of 96.73 % by taking 15.5 s followed by logistic regression at 96.14 %. Although the total time taken to build the model is at 44.8 s, the lowest accuracy classification was NaiveBayes by 82.33 % by taking 14.3 s.

These results can support MTI database on terrorist and serious criminal attacks against public surface transportation, which is giving transit officials, security planners, and policy maker's important insights into public surface transportation threats, such as terrorist attacks against trains and buses. MTI database provides some information on more than 3,000 attacks against public transit over the last 40 years for helping taking decisions in the upcoming situations.

5 Conclusion and Future Work

This paper investigated using different machine learning algorithms on datasets with different types and sizes by Big Data mining tool "Apache spark". This comparison can lead to a fair judgment about which algorithm is the best fit with each field. The algorithms were applied on three datasets: marketing, packing and statistics, and security datasets. The algorithms were compared based on few parameters including mainly accuracy and training time. It was observed that there is a direct relationship between execution time in building the model and the volume of data records and attributes size of the datasets. The finding of this experiment shows that Decision Tree algorithm is the most suitable one for the marketing and security dataset. Additionally, logistic regression algorithm had the highest value of accuracy in packing and statistic dataset. The study can conclude that Decision Tree and logistic regression algorithms have higher classification accuracy over the compared algorithms.

The experiment was implemented locally on specific datasets. In future work, enhancements will be done on the algorithm with the highest accuracy putting into consideration saving time. This will be implemented on a cluster and deployed on Big Data from different sources.

References

1. Raj, P., Raman, A., Nagaraj, D., Duggirala, S.: The high-performance technologies for big and fast data analytics. In: Raj, P., Raman, A., Nagaraj, D., Duggirala, S. (eds.) High-Performance Big-Data Analytics, pp. 25–66. Springer, Heidelberg (2015)
2. McAfee, A., et al.: Big data. The management revolution. Harvard Bus. Rev. 90(10), 61–67 (2012)
3. Han, J., Pei, J., Kamber, M.: Data Mining: Concepts and Techniques. Elsevier, Amsterdam (2011)
4. Chen, M., Mao, S., Liu, Y.: Big data: a survey. Mobile Netw. Appl. 19(2), 171–209 (2014)
5. Chen, Z., Liu, B.: Topic modeling using topics from many domains, lifelong learning and big data. In: Proceedings of the 31st International Conference on Machine Learning (ICML 2014) (2014)
6. Agrawal, D., Budak, C., El Abbadi, A., Georgiou, T., Yan, X.: Big data in online social networks: user interaction analysis to model user behavior in social networks. In: Madaan, A., Kikuchi, S., Bhalla, S. (eds.) DNIS 2014. LNCS, vol. 8381, pp. 1–16. Springer, Heidelberg (2014). doi:10.1007/978-3-319-05693-7_1
7. Chen, H., Chiang, R.H., Storey, V.C.: Business intelligence and analytics: from big data to big impact. MIS Q. 36(4), 1165–1188 (2012)
8. Solanas, A., et al.: Smart health: a context-aware health paradigm within smart cities. IEEE Commun. Mag. 52(8), 74–81 (2014)
9. Schell, R.: Security—a big question for big data. In: 2013 IEEE International Conference on Big Data. IEEE (2013)
10. Assunção, M.D., et al.: Big Data computing and clouds: trends and future directions. J. Parallel Distrib. Comput. 79, 3–15 (2015)

11. Cuzzocrea, A.: Privacy and security of big data: current challenges and future research perspectives. In: Proceedings of the First International Workshop on Privacy and Security of Big Data. ACM (2014)
12. Machine Learning Repository. http://mlr.cs.umass.edu/ml/. Accessed 4 Jan 2016
13. Mineta Transportation Institute. https://www.dhs.gov/publication/mineta-transportation-institute-mti-database. Accessed 4 Jan 2016
14. Zaharia, M., et al.: Fast and interactive analytics over Hadoop data with Spark. USENIX: login 37(4), 45–51 (2012)
15. Ratner, B.: Statistical and Machine-Learning Data Mining: Techniques for Better Predictive Modeling and Analysis of Big Data. CRC Press, Boca Raton (2011)
16. Suthaharan, S.: Big data classification: problems and challenges in network intrusion prediction with machine learning. ACM SIGMETRICS Perform. Eval. Rev. 41(4), 70–73 (2014)

Mammogram Classification Using Curvelet GLCM Texture Features and GIST Features

Syed Jamal Safdar Gardezi[1,2], Ibrahima Faye[1,2(✉)], Faouzi Adjed[1,2,4], Nidal Kamel[1,3], and Mohamed Meselhy Eltoukhy[5]

[1] Centre for Intelligent Signals and Imaging Research,
Universiti Teknologi PETRONAS, Tronoh, Malaysia
jamalgardezi@gmail.com, ibrahima_faye@petronas.com.my
[2] Department of Fundamental and Applied Sciences,
Universiti Teknologi PETRONAS, Tronoh, Malaysia
[3] Department of Electrical and Electronics Engineering,
Universiti Teknologi PETRONAS, Tronoh, Malaysia
[4] Laboratoire IBISC EA 4526,
Université D'Evry Val d'Essonne, 91020 Evry, France
[5] Computer Science Department, Faculty of Computers and Informatics,
Suez Canal University, Ismailia 41522, Egypt

Abstract. This paper presents a feature fusion technique that can be used for classification of ROIs in breast cancer into normal and abnormal classes. The texture features are extracted using geometric invariant shift transform and statistical features from the curvelet grey level co-occurrence matrices. First classification accuracy of both methods were evaluated independently. Later, feature fusion is done to improve the classification performance. Support vector machine classifier with polynomial kernel was implemented using 2×5 folds cross validation. Fusion of features produces better results with accuracy of 92.39 % as compared to 77.97 % and 91 % for GIST and CGLCM respectively.

Keywords: Classification · Mass patches · Texture · GIST · Curvelet

1 Introduction

Breast cancer is amongst one of the most diagnosed cancer in women around the world. According to recent reports on breast cancer by World Health Organization (WHO) agency for cancer research IARC (International Agency for Research on Cancer) more than 522,000 death cases of women due to cancer were reported in 2012 only [1]. In Malaysia the statistics show that one out of nineteen Malay women is at risk of being diagnosed for breast cancer during their life time. The women with age group 40–60 years have the highest prevalence rate of 64 % [2]. Unfortunately 56 % of these cases were already in advanced cancer stages [3]. Early detection plays a significant role in reducing the causality rates. To diagnose the breast cancer, mammography screening programmes are one the most effective and widely used tool [4]. Computer aided diagnosis (CAD) systems assist the clinicians for diagnosis of breast

© Springer International Publishing AG 2017
A.E. Hassanien et al. (eds.), *Proceedings of the International Conference on Advanced Intelligent Systems and Informatics 2016*, Advances in Intelligent Systems and Computing 533, DOI 10.1007/978-3-319-48308-5_67

cancer and reduce the computational complexity issues [5]. However due to hetero-geneous nature of breast tissues, irregular and subtle signs of abnormalities the existing CAD systems still require improvements in detection and diagnosis [6].

In the literature, many researchers have used texture and statistical features for mammogram classification. Oliver et al. [7] used 322 ROIs from the MIAS dataset in their study. They employed local binary pattern technique to obtain the local texture information from each ROI. The local texture was computed using four different neighborhoods and radii values. From this local texture information the grey level co-occurrence matrices (GLCM) at $0°$, $45°$, $90°$ and $135°$ orientations were computed and 216 GLCM features were obtained. The classification was performed using four different classifier such as K nearest neighbors (KNN), Fisher discriminant analysis (FDA), decision tree and Support vector machine (SVM) with leave one out strategy. The performance in each case was compared with other classifiers. The results showed that SVM produced better results amongst all, with classification accuracy of 78 %. Eltoukhy et al. [8] presented a feature extraction technique to extract the biggest coefficients of curvelet and wavelet. The performance of each feature matrix was evaluated using the Euclidian distance classifier. The results show that curvelet pro-duced higher classification i.e. 94.07 % as compared to wavelet transform (90.05 %). Eltonsy et al. [9] presented a study using concentric layers morphology to determine the mass and normal tissues. The regions with higher concentration of concentric layers were identified as suspicious regions. Their method produced maximum sensitivity of 92 % for malignant cases and 61.6 % for benign cases. The study utilized 540 mammogram images obtained from the Digital Database for Screening Mammography (DDSM) dataset. Chan et al. [10] used spatial gray-level dependence (SGLD) matrices using some region-based algorithm and extracted eight texture features from these matrices. In their study the dataset was divided in two parts, one with the biopsy-proven masses i.e. 168 mass and 504 ROIs containing normal breast tissue were used. The method recorded area under the receiver operative characteristic (ROC) curve was 0.84 and 0.82 for training and test sets respectively. Rangayyan et al. [11] used spiculation index and fractional concavity two new shape factors for the classifying the manually segmented mass ROIs. The combination of spiculation index, fractional concavity with compactness yielded an accuracy of 81.5 % for classifying benign versus malignant tissues.

The current paper presents feature fusion technique that can provide significant texture information to improve the classification performance in breast cancer diag-nosis. To achieve this task, features from curvelet and GIST descriptor are fused together and performance metrics are computed. The rest of the paper is organized as follows: Sect. 2 presents a brief theoretical background of Curvelet GLCM and GIST descriptor. In Sect. 3 experimental work is discussed, which includes description of dataset, feature extraction and fusion methodology and classifier used. The results are discussed in Sect. 4, while the conclusion is presented in Sect. 5.

2 Preliminaries

2.1 Curvelet Grey Level Co-occurrence Matrices (CGLCM)

The curvelet transform introduced by Candes and Donoho [12] to efficiently represent the edge details of images. The curvelet transform follows the scaling law i.e. width $\sim$ length^2 and has superior directional representations as compared to other multiresolution transforms such as wavelet or ridgelets. Generally there are two curvelet implementations namely: the un-equispaced Fast Fourier transform and the warping technique [13]. While the grey level co-occurrence matrix (GLCM) [14] is constructed from the co-occurring intensity pixel values at certain offset in an image. Significant statistical features can be computed from the GLCM matrices, these features are capable to discriminate different tissue types in mammograms. The GLCM can be computed in horizontal, vertical or diagonally ($i.e.\ \theta = 0°, 45°, 90°\ and\ 135°$) directions.

In the current study, we implement the methodology of [15] which utilized 128×128 ROIs to compute the curvelet grey level co-occurrence matrices (CGLCM). However, in the current study the size of ROI patches under study are not fixed but depends upon the abnormality involved.

2.2 Geometric Invariant Shift Transform (GIST)

The geometric invariant shift transform (GIST) operator was developed by Oliva et al. [16] for recognition of real world scenes and characterization. The method is based on spatial envelop which means, the low dimensional spatial representation of an image are computed. The gist descriptor is widely used for natural scene recognitions and facial recognition applications. The operator is based on five spatial properties i.e. degree of naturalness, degree of openness of images, degree of roughness of the image textures, degree of expansion and ruggedness.

3 Experimental Work

3.1 Dataset

The current study utilizes images from the Image retrieval in Medical Applications (IRMA) framework [17] and the Mammographic Image analysis Society (MIAS) database [18]. The IRMA is a collaborative project initiated by Aachen University to collect mammogram images from different database combined in one bigger dataset. The ground truth for both dataset were prepared by experts radiologists, and provide the information on abnormality types, tissues densities and radii measures for masses as well. In the current study 513 mammogram ROIs patches were used with 289 abnormal and 224 normal class patches. The patches are extracted from the original images based on ground truth information provided in these datasets. The size of patches varies with the size of abnormality present in each mammogram e.g. Fig. 1. While in case of normal mammogram patches are extracted randomly with patch size no less than 40×40.

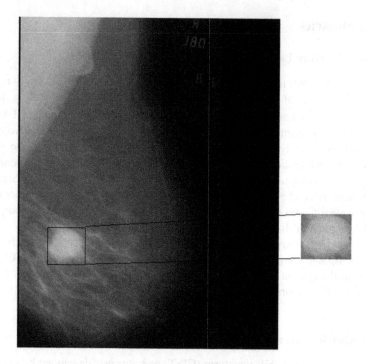

Fig. 1. Extraction of ROI patches

3.2 Methodology for Feature Extraction

Once the ROI patches are obtained GIST descriptor is applied. The motivation in this work is to use the spatial properties of GIST operator to extract the texture information from the ROI patches. It can also work for non-square images thus also suits for ROI patches with variable sizes. To compute the texture feature the descriptor uses 8 orientations with block size of 4×4. From each ROI patch 288 texture features were computed.

Similarly, two level curvelet decompositions are applied on the ROIs patches to compute the curvelet coefficients at sub-band level. The curvelet produces nine coefficient matrices for the two levels i.e. one matrix at level one and eight matrices at level 2. For each matrix a curvelet grey level co-occurrence (CGLCM) matrix is computed. Hence in total we obtain nine CGLCM matrices.

For feature extraction from the CGLCM, nine statistical features Eq. (1–9) were computed. Thus, from each ROI patch 81 statistical features were extracted i.e. 9 for first level and 72 from the eight matrices in second level decomposition of curvelet as can be seen in Fig. 2. Here it should be noted that CGLCM had different size for each patch as the patch size depended upon the abnormality itself.

$$Energy = \sum_{m=0}^{g-1} \sum_{n=0}^{g-1} p(m,n)^2 \tag{1}$$

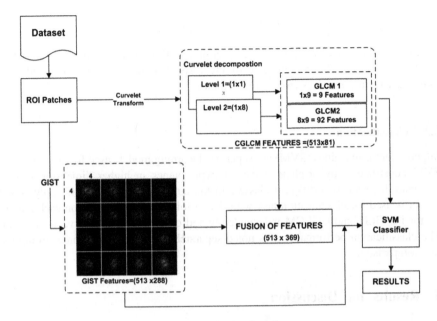

Fig. 2. Proposed methodology

$$Contrast = \frac{1}{(g-1)^2} \sum_{m=0}^{g-1} \sum_{n=0}^{g-1} (m-n)^2 * p(m,n) \qquad (2)$$

$$Homogeneity = \sum_{m=0}^{g-1} \sum_{n=0}^{g-1} \frac{p(m,n)}{1+(m,n)} \qquad (3)$$

$$Correlation = \frac{\sum_{m=0}^{g-1} \sum_{m=0}^{g-1} mnp(m,n) - \mu_x\mu_y}{\sigma_x\sigma_y} \qquad (4)$$

where $\mu_x = \sum_{m=0}^{g-1} m \sum_{n=0}^{g-1} p(m,n)$, $\mu_y = \sum_{m=0}^{g-1} n \sum_{n=0}^{g-1} p(m,n)$

$\sigma_x = \sum_{m=0}^{g-1} (m-\mu_x)^2 \sum_{n=0}^{g-1} p(m,n)$, $\sigma_x = \sum_{m=0}^{g-1} (m-\mu_x)^2 \sum_{n=0}^{g-1} p(m,n)$

$$Mean = \sum_{i=2}^{g-1} i.P_{m+n}(i) \qquad (5)$$

$$Moment = \sum_{m} \sum_{n} \{p(m,n)\}^2 \qquad (6)$$

$$Variance = \sum_{m} \sum_{n} (m-\mu)^2 p(m,n) \qquad (7)$$

$$Entropy = -\sum_{m=0}^{g-1} \sum_{n=0}^{g-1} p(m,n).log(p(m,n)) \qquad (8)$$

$$Max\ Prob(i) = max\{p(m, n)\} \tag{9}$$

In the last step features form GIST descriptor and CGLCM are combined to obtain the fusion matrix. The performance of the fused feature matrix is compared with individual feature matrices from both methods.

3.3   Classification

Support vector machines (SVM) are supervised learning models used for classification. SVM constructs a hyper-plane or set of hyper-planes in higher dimension to discriminate between two different classes. In the current study SVM with polynomial kernel is used for classification task with 2×5 folds cross validation on each feature vector i.e. GIST and CGLCM independently and on the fused feature vector as well. The classification performance was noted separately for each case in order to draw a fair comparison.

4   Results and Discussion

In the current study the texture features from the ROI patches are computed and performance measures such as accuracy, sensitivity and specificity are measured in Eqs. 10–12.

$$accuracy = \frac{\text{true positives (TPs)} + \text{true negatives (TNs)}}{\text{true positives (TPs)} + \text{true negatives (TNs)} + \text{false positives(FPs)} + \text{false negatives(FNs)}} \tag{10}$$

$$sensitivity = \frac{\text{true positives (TPs)}}{\text{true positives (TPs)} + \text{false negatives(FNs)}} \tag{11}$$

$$specificity = \frac{true\ negatives(TNs)}{true\ negatives(TNs) + false\ positives(FPs)} \tag{12}$$

First, a feature matrix of size 513×288 is obtained using the GIST descriptors on each ROI. Using the 10 fold cross validation it produces a maximum accuracy of 77.97 % as shown in Table 1.

Table 1. Classification results for GIST texture features

Method	Accuracy	Sensitivity	Specificity
GIST	77.97 %	79.93	75.45 %
CGLCM	91.03 %	91 %	91.07 %
GIST + CGLCM	92.40 %	92.39 %	92.41 %

In the second step the nine statistical texture features are computed from CGLCM. The CGLCM at two levels of curvelet decompositions having 9 matrices give a feature vector of 513 × 81in sizes. The results in Table 1 show the CGLCM features produce superior results as compared to GIST features. This better performance is evident that curvelet has superior representations of edges and small scale details.

In the last step, both feature matrices are fused to obtain a fused feature matrix of size 513 × 369. Since both the techniques are using the same images thus fusion of feature can be done. The resultant fused matrix is presented to SVM classifier with polynomial kernel function, with 10 fold of cross validations produces better results in terms of accuracy, sensitivity as well as specificity as can be seen in Table 1. The combination of both features improves the accuracy by 14.43 % in terms of accuracy as compared to classification results by GIST descriptor and 1.4 % higher than CGLM as well. Similarly, the sensitivity and specificities are also increased significantly.

The comparison of proposed method with other existing methods that have used same dataset is presented in Table 2. The performance of our method is comparable with existing techniques as can be seen below in Table 2.

Table 2. Comparison with existing methods

Method	Dataset	Classifier	Accuracy %
Agrawal et al. [19]	MIAS	SVM	89 %
Vaidehi and Subashini [20]	MIAS	Adaboost, Neural network and sparse representation classifier (SRC)	93.75 %
Oliveria et al. [21]	IRMA	SVM	89.6 %
Deserno et al. [22]	IRMA	SVM with Gaussian kernel	80 %
Esener et al. [23]	IRMA	SVM, kNN, Decision tree and FLDA	94.67 %
Proposed Method	MIAS and IRMA	SVM	92.40 %

5   Conclusion

This paper presents a fusion based feature extraction method for mammogram ROI classification. The method uses two state of the art feature extraction methods such as curvelet and GIST descriptors. GLCM matrices are computed at the curvelet sub band level and significant statistical features are calculated. Similarly the spatial texture features using the GIST operator are also extracted from each ROI. Features from each method are separately evaluated using SVM classifier and their performances were recorded. To further improve the classification performance both features are fused to obtain a feature matrix of 513 × 369. The results show that feature fusion enhances the performance produced better classification accuracy i.e. 92.39 % as compared to CGLM (91 %) and GIST (77.97 %).

Acknowledgements. The work is supported by the international grant 0153AB-E13.

References

1. Latest world cancer statistics Global cancer burden rises to 14.1 million new cases in 2012: Marked increase in breast cancers must be addressed. In: International Agency for Cancer Research (IARC),World Health Organisation (WHO), Lyon/Geneva (2013)
2. Pride Foundation. http://pride.org.my/breast-cancer/ (2016)
3. Global Cancer Facts & Figures, 3rd edn. American Cancer Society, Atlanta (2015)
4. Penhoet, E.E., Petitti, D.B., Joy, J.E.: Saving Women's Lives: Strategies for Improving Breast Cancer Detection and Diagnosis. National Academies Press, Washington (DC) (2005)
5. Sampat, M.P., Markey, M.K., Bovik, A.C.: Computer-aided detection and diagnosis in mammography. In: Handbook of Image and Video Processing, vol. 2, no. 1, pp. 1195–1217 (2005)
6. Christoyianni, I., Dermatas, E., Kokkinakis, G.: Fast detection of masses in computer-aided mammography. Sign. Process. Mag. IEEE **17**(1), 54–64 (2000)
7. Oliver, A., Llado, X., Marti, R., Freixenet, J., Zwiggelaar, R.: Classifying mammograms using texture information. In: Medical Image Understanding and Analysis, pp. 223–227 (2007)
8. Eltoukhy, M.M., Faye, I., Samir, B.B.: A comparison of wavelet and curvelet for breast cancer diagnosis in digital mammogram. Comput. Biol. Med. **40**(4), 384–391 (2010)
9. Eltonsy, N.H., Tourassi, G.D., Elmaghraby, A.S.: A concentric morphology model for the detection of masses in mammography. IEEE Trans. Med. Imaging **26**(6), 880–889 (2007)
10. Chan, H.-P., Wei, D., Helvie, M.A., Sahiner, B., Adler, D.D., Goodsitt, M.M., Petrick, N.: Computer-aided classification of mammographic masses and normal tissue: linear discriminant analysis in texture feature space. Phys. Med. Biol. **40**(5), 857 (1995)
11. Rangayyan, R.M., Mudigonda, N.R., Desautels, J.L.: Boundary modelling and shape analysis methods for classification of mammographic masses. Med. Biol. Eng. Comput. **38** (5), 487–496 (2000)
12. Candes, E.J., Donoho, D.L.: Curvelets, multiresolution representation, and scaling laws. In: International Symposium on Optical Science and Technology, pp. 1–12. International Society for Optics and Photonics (2000)
13. Candès, E., Demanet, L., Donoho, D., Ying, L.: Fast discrete curvelet transforms. Multiscale Model. Simul. **5**(3), 861–899 (2006)
14. Haralick, R.M.: Statistical and structural approaches to texture. Proc. IEEE **67**(5), 786–804 (1979)
15. Gardezi, S.J.S., Faye, I., Eltoukhy, M.M.: Analysis of mammogram images based on texture features of curvelet sub-bands. In: Fifth International Conference on Graphic and Image Processing, 906924-906924-906926. International Society for Optics and Photonics (2014)
16. Oliva, A., Torralba, A.: Modeling the shape of the scene: a holistic representation of the spatial envelope. Int. J. Comput. Vis. **42**(3), 145–175 (2001)
17. Lehmann, T.M., Gold, M., Thies, C., Fischer, B., Spitzer, K., Keysers, D., Ney, H., Kohnen, M., Schubert, H., Wein, B.B.: Content-based image retrieval in medical applications. Methods Inf. Med. **43**(4), 354–361 (2004)
18. Suckling, J., Astley, S., Betal, D., Cerneaz, N., Dance, D.R., Kok, S.-L., Parker, J., Ricketts, I., Savage, J., Stamatakis, E., Taylor, P.: The mammographic image analysis society digital mammogram database exerpta medica. Int. Congr. Ser. **1069**, 375–378 (1994)
19. Agrawal, P., Vatsa, M., Singh, R.: Saliency based mass detection from screening mammograms. Sig. Process. **99**, 29–47 (2014). doi:10.1016/j.sigpro.2013.12.010
20. Vaidehi, K., Subashini, T.: Automatic characterization of benign and malignant masses in mammography. Procedia Comput. Sci. **46**, 1762–1769 (2015)

21. De Oliveira, J.E., Deserno, T.M., Araújo, A.A.: Breast lesions classification applied to a reference database. In: 2nd International Conference: E-Medical Systems, pp. 29–31. Citeseer (2008)
22. Deserno, T.M., Soiron, M., de Oliveira, J.E.E., de A Araujo, A.: Towards computer-aided diagnostics of screening mammography using content-based image retrieval. In: 2011 24th SIBGRAPI Conference on Graphics, Patterns and Images (Sibgrapi), pp. 211–219. IEEE (2011)
23. Esener, I.I., Ergin, S., Yuksel, T.: A new ensemble of features for breast cancer diagnosis. In: 2015 38th International Convention on Information and Communication Technology, Electronics and Microelectronics (MIPRO), pp. 1168–1173. IEEE (2015)

Fingerprint Identification Using Hierarchical Matching and Topological Structures

M. Elmouhtadi[1(✉)], S. El fkihi[1,2], and D. Aboutajdine[1]

[1] LRIT – CNRST URAC29, Faculty of Science,
University Mohammed V, Rabat, Morocco
meryem.mouhtadi@gmail.com, aboutaj@fsr.ac.ma
[2] RIITM, ENSIAS, University Mohammed V, Rabat, Morocco
elfkihi.s@gmail.com

Abstract. Fingerprint identification is one of the most popular and efficient biometric technique used for improving automatic personal identification. In this paper, we will present a new indexing based method, in the first step, on estimation of singular point considered as an important feature in the fingerprint using a directional file. In the second step, a hierarchical Delaunay triangulation was applied on the minutiae around the singular point extracted. Comparison of two fingerprints was calculated by introducing the barycenter notion so as to ensure the exact location of the similar triangles. We have performed extensive experiments and comparisons to demonstrate the effectiveness of the proposed approach using a challenging public database (i.e., FVC2000) which contains small area and low quality fingerprints.

Keywords: Fingerprint indexing · Delaunay triangulation · Barycenter · Singular point

1 Introduction

Biometric technologies are already widely disseminated in numerous large-scale nation-wide projects [13, 15]. Fingerprint indexing and matching is among the most widely used biometric technologies, which is playing a major role in automated personal identification systems deployed to enhance security all over the world. Even without the advent of environmental noise, two same fingerprint impressions are not guaranteed to be identical due to variability in displacement, rotation, scanned regions, and non-linear distortion. Displacement, rotation, and disjoint detected regions are obviously due to the differences in the physical placement of a finger on a scanner.

In the literature, various methods of fingerprint matching exist, Such As correlation-based matching, texture descriptor, minutiae points based matching [1, 2]. Among these, minutiae based algorithms are the most widely used [3, 4]. Most methods of matching consist of relating two minutiae points. This decreases the performance of systems when one has transformations, then the minutiae in rotation will not be aligned and the high-level minutiae can be neglected. Furthermore the complexity of matching becomes more increased considering the number of extracted minutiae which increases the computing time and reduces the result quality.

© Springer International Publishing AG 2017
A.E. Hassanien et al. (eds.), *Proceedings of the International Conference on Advanced Intelligent Systems and Informatics 2016*, Advances in Intelligent Systems and Computing 533, DOI 10.1007/978-3-319-48308-5_68

In this article, we present a new method based on the extraction of singular point which represents the central point of the image, using the directional Field to locate it, then we'll focus on a block of 100*100 pixels around the extracted singular point. The selected block is an interval that contains the high quality minutiae which the probability of losing them still minimum for any type of transformation. In a second step, we will apply the Delaunay triangulation of minutiae points. We apply a process matching first by extracting similar triangles, and second extracting the barycenters of similar triangles extracted in the first, and then reapply the Delaunay triangulation and the extraction of similar triangles. The aim of this method is to ensure the right location of the matched triangles, which means a good matched minutiae points. The strengths of this system are: (1) The extraction of the singular point involves the reduction of the minutiae aligned at matching, which implies reducing time and complexity, (2) Location of the singular point at the center makes it robust to the loss in case of contour missing. This increases the recognition performance, (3) This method is robust to zoom because it is interested only in the corners of triangles when searching similarity of triangles, (4) The change in orientation of the incoming image does not affect the identification results. This approves the performance of using triangulation, and (5) Considering the barycenters of the similar triangles, ensures that the relevant triangles found are located in the same places in the both compared impression.

The rest of this paper is organized as follows: Sect. 2 presents analyzes the topological structure of fingerprint and present the extraction of the singular point. Section 3 details the proposed hierarchy triangulation Delaunay for matching. Section 4 presents experience and results. Conclusions are drawn in Sect. 5.

2 Singular Point Extraction

Generally, Fingerprint structure is classified into two categories: global characteristics and local ones. The local characteristics are minutiae, defined by end ridge and bifurcation. The global characteristics are the singular points (core and delta), obtained at the points when ridges change their orientation which influencing largely the directional field. They determine the topological structure and type of fingerprint, allowing a classification of fingerprints into six main classes; defined by Henry [2, 9] (Refer to Fig. 1).

The location of Core point at center makes him a high level singular point. Whatever the finger position on the sensor, we will get the center of the impression in

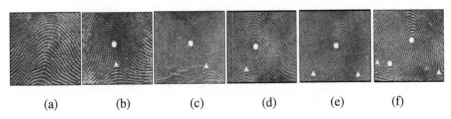

| (a) | (b) | (c) | (d) | (e) | (f) |

Fig. 1. (a) arch, (b) tented arch, (c) left loop, (d) right loop, (e) whorl, and (f) whorl (twin loop).

the required image. Contrary to the delta, it can be missed in a bad position or in the case of missing parts. Different approaches were based on the extraction of a core point [5, 6], as a useful point for obtaining the directional registration, classifying fingerprint and matching. Singular point detection and directivity ridge estimation achieve to use the Directional field. A number of methods have been proposed to estimate the fingerprint orientation field [3, 7, 8]. For this the gradient based method was adopted by many approaches [6].

In this work, we propose to estimate the orientation ridges using a directional filed. Using a gradient approach we compute the image derivative in the x and y directions to get the gradient vector defined as:

$$\begin{pmatrix} G_x I(x,y) \\ G_y I(x,y) \end{pmatrix} = \begin{bmatrix} \frac{\partial I(x,y)}{\partial x} \\ \frac{\partial I(x,y)}{\partial y} \end{bmatrix} \tag{1}$$

Orientation was estimated by using least square contour alignment method using a window $W(3,3)$ (Eqs. 2 and 3). The advantage is that this method gives continuous values [5]. With calculating gradient at pixel level, the orientation of ridge stays orthogonal to average phase angle of changes pixels value indicated by gradients. Then orientation of an image block is determined by averaging the square gradients $\begin{pmatrix} G_{s,x} I(x,y) \\ G_{s,y} I(x,y) \end{pmatrix}$ to eliminate the orientation ambiguity. Figure 2 presents an overview of extracting singular point steps.

$$\begin{pmatrix} G_{s,x} I(x,y) \\ G_{s,y} I(x,y) \end{pmatrix} = \begin{bmatrix} \sum_W G_x^2 - \sum_W G_y^2 \\ 2 \sum_W G_x G_y \end{bmatrix} = \begin{bmatrix} G_{xx} - G_{yy} \\ 2 G_{xy} \end{bmatrix} \tag{2}$$

Then the gradient orientation θ was estimated as:

$$\theta = \frac{1}{2} \nabla \left(G_{xx} - G_{yy}, 2 G_{xy} \right) \tag{3}$$

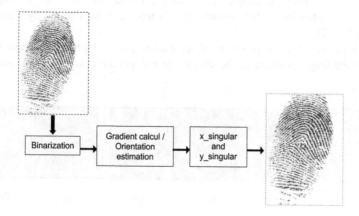

Fig. 2. Overview of singular point extracting

With $\nabla(x, y)$ defined as (Eq. 4):

$$\nabla(x, y) = \begin{cases} tan^{-1}\left(\frac{x}{y}\right) & x > 0 \\ tan^{-1}\left(\frac{x}{y}\right) + \pi & for \quad x < 0 \, and \, y \geq 0 \\ tan^{-1}\left(\frac{x}{y}\right) - \pi & x < 0 \, and \, y \geq 0 \end{cases} \tag{4}$$

As an important region, pixels around the singular point present a high level point on a fingerprint impression, for this we will focus to extract minutiae around the singular point extracted. Many approach approved that the high number of minutiae extracted can decrease the result matching especially when use minutiae features, because it can generate a bad point and then a false feature. We propose in this work to considering a bloc of (100*100) pixels around the singular point extracted, this reduce number of minutiae extracted and help to minimize time and complexity matching (Fig. 3).

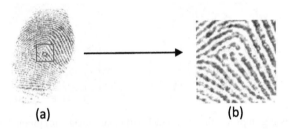

(a)   (b)

Fig. 3. (a) Original fingerprint with the extracted singular point, (b) The selection bloc of (100*100) pixels around singular point.

3 Hierarchical Delaunay Triangulation

Delaunay triangulation is a matching method based produces triangles starting from the center point of the fingerprint image, forming an arc with the closest points to cover all the minutiae points. This method provides good results in terms of complexity [9]. Delaunay triangulation [4, 10] consists in generating the largest possible number of triangles formed by the extracted minutiae points. The problem is that the number of the generated triangles may be a weakness for these kinds of algorithms. Indeed, as the set of extracted minutiae may contain false ones, so for each false minutia, multiple false triangles will be generated and the result quality will decrease. Thus, some approaches [10, 11] propose to reduce the number of the generated triangles, but this is not efficient as it can eliminate useful points.

Considering two sets of minutiae I (input image) and T (template image). The main problem of the triangulation methods is that they can find two similar triangles from I and T, but are not formed by the correspondent minutiae; (i.e. triangle nodes do not have the same position in the two considered impression). In the first step of our approach, we apply a preprocessing method to each image in order to extract minutiae that will be classified into a vector of ridge and bifurcation (see Fig. 4). Especially, we use the Otsu binarization [10] for obtaining the binary image. Zhang's skeletization approach was used for thinning [11]. Then for extracting minutiae composed by bifurcation and end of ridges, the Cross Number based method was used on a neighborhood of 8 pixels [12]. Then, based on the obtained minutiae vector (ridge and bifurcation), the different possible triangles will be formed applying a Delaunay triangulation (DT) function so as to obtain all possible triplets of the extracted points.

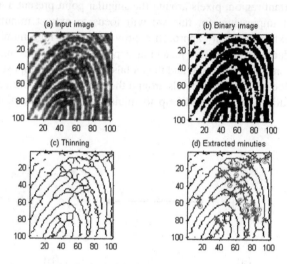

Fig. 4. Preprocessing steps: (a) the input block, (b) binary image, (c) thinned image, (d) Present the extracted minutiae, end ridges are presented by circles.

As a comparison and matching step, we define the similarity between the entry fingerprint image and the others in a database. We calculate the three angles of each triangle in DT. Among the existing methods we use the Akashi theorem (Eqs. 5, 6, and 7) to define the three angles α, β, and γ of each triangle ABC in DT.

$$\alpha = arccos(\frac{b^2 + c^2 - a^2}{2bc}) \tag{5}$$

$$\beta = arccos(\frac{a^2 + c^2 - b^2}{2ac}) \tag{6}$$

$$\gamma = arccos(\frac{a^2 + b^2 - c^2}{2ab}) \tag{7}$$

For each triangle of DT identified by its three angles (α, β, and γ) we look all the triangles in the second image DT2 that have the same angles. This similarity research method will always identify the fingerprint, even if the change of orientation and zoom image, since the triangles do not change angles in these cases. The similar triangles will be saved in Similar_DT.

The problem here is that we may find similar triangles that are not formed by the same minutiae. Indeed, even it exist a homothetic transformation between two triangles this does not guarantee that they are really similar because they can be in two different positions in the compared fingerprints. To solve this problem and to take into consideration the location of similar triangles, the next step of our method is to extract the barycenter of each triangle given in Similar_DT (Fig. 5).

Fig. 5. Delaunay triangulation of minutiae extracted on the chosen bloc [100*100]

Barycenter is calculated by using the mean of A_i, B_i, and C_i, nodes of each triangle $\Delta_i(A_i, B_i, C_i)$ given in Similar_DT. The result is saved in P_center(x_center,y_center). To keep the topological structure of minutiae, we apply the Delaunay Triangulation DT_Similar_DT of points saved in the vector P_center, then we measure the similarity between DT_Similar_DT of the entry fingerprint and DT2_Similar_DT2 of the compared fingerprint. This method will ensure improvement the similarity decision of triangles and so the identification of fingerprints. Finally, probability identification (P) of each fingerprint compared to the database images is defined as follows in Eq. 8:

$$P = \frac{|Similar\_barycenter|}{|DT\_similar\_DT|} \tag{8}$$

|DT_similar_DT|: number of similar triangles obtained using Delaunay tria gulation of barycenter points. And |Similar_barycenter|: number of similar triangles obtained using Delaunay triangulation of barycenter points.

4 Results and Discussion

To evaluate the performance of the proposed method, we use the benchmarking dataset Db1_a from FVC2000 database [14]. In the first we generate the file of blocs (100*100) pixels, obtained after extraction of the singular point. Given two different fingerprint **I1** and **I2**, Figs. 6 and 7 shows the Delaunay triangle (i.e.; DT and DT2)

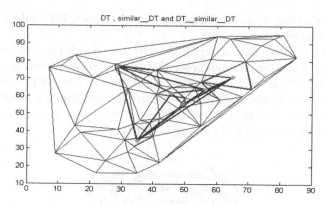

Fig. 6. Extracted similar triangles in DT (red) and DT_similar_DT (green) (Color figure online)

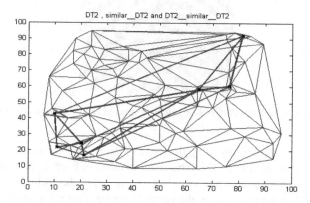

Fig. 7. Extracted similar triangles in DT (green) and DT_similar_DT (red) (Color figure online)

extracted in the both images, the similar triangles founded (i.e.; similar_DT and similar_DT2), generated barycenter and then the extracted similar triangles formed by barycenters (i.e.; DT_similar_DT and DT2_similar_DT2). We calculate the probability P of matching for an input image and others in the database.

Comparing a query image (I10) and others selected fingerprint from the database, Table 1 shows that the new approach is more accurate comparing with the other methods; matching using a simple Delaunay triangulation applied on the entry minutiae (SDT), matching with simple Delaunay triangulation around the singular point extracted (SPSDT). We obtain a similarity of 100 % only for the query image (I10) by using our proposed method and for all the others non similar fingerprints we obtain 0 % as a similarity rate. In the opposite, in the other methods, there are many false detection rates (rates non equal to 0 %) that are determined. They gives other relevant images with similarity rates 100 %, which decrease the efficiency. The average processing time for each fingerprint is around 0.098 s.

Table 1. Matching results

Compared image	Proposed method	SDT	SPSDT
I1	14 %	100 %	30 %
I2	0 %	15 %	0 %
I3	5 %	36 %	9 %
I4	0 %	56 %	0 %
I5	0 %	42 %	3 %
I6	12 %	62 %	0 %
I7	0 %	54 %	12 %
I8	0 %	27 %	0 %
I9	0 %	35 %	0 %
I10	100 %	100 %	100 %

5 Conclusion

In this paper we have present a new fingerprint matching approach based on the extraction of the singular point, and a hierarchical Delaunay triangulation based on triangles features. As an important point, singular point is considered as a high quality fingerprint pattern with a low probability of loss in case of the missing part on a fingerprint caption. The extraction of singular point helps to approve a high matching performance when considering minutiae on his neighbor. A Delaunay matching method was applied on the minutiae extracted around the singular point. As to ensure the best correspondence of similar triangles, we calculate hierarchically similarity of triangles formed by barycenters point of the extracted triangles in the first time.this as to ensure the topological distribution of aligned minutiae. The main experiences approved the efficiency and advantages of proposed method.

References

1. Kumar, D.A., Begum, T.U.S.: A comparative study on fingerprint matching algorithms for EVM. J. Comput. Sci. Appl. **1**(4), 55–60 (2013)
2. Jain, A., Pankanti, S.: Fingerprint classification and matching. In: Bovik, A. (ed.) Handbook for Image and Video Processing. Academic Press, London (2000)
3. de Boer, J., Bazen, A.M., Gerez, S.H.: Indexing fingerprint databases based on multiple features (2001)
4. Zaeri, N.: Minutiae-based fingerprint extraction and recognition, biometrics. In: Yang, J. (ed.) InTech (2011). DOI:10.5772/17527. http://www.intechopen.com/books/biometrics/minutiae-based-fingerprint-extraction-and-recognition
5. Bazen, A.M., Gerez, S.H.: Systematic methods for the computation of the directional fields and singular points of fingerprints. IEEE Trans. Pattern Anal. Mach. Intell. **24**(7), 905–919 (2002)
6. Zhou, J., Chen, F., Gu, J.: A novel algorithm for detecting singular points from fingerprint images. IEEE Trans. Pattern Anal. Mach. Intell. **31**(7), 1239–1250 (2009)

7. Liu, M., Jiang, X., Chichung Kot, A.: Efficient fingerprint search based on database clustering. Pattern Recogn. **40**(6), 1793–1803 (2007)

8. Ratha, N.K., Karu, K., Chen, S., Jain, A.K.: A real-time matching system for large fingerprint databases. IEEE Trans. Pattern Anal. Mach. Intell. **18**(8), 799–813 (1996)

9. U. S. F. B. of Investigation, The science of fingerprints: classification and uses. United States Dept. of Justice, Federal Bureau of Investigation (1979)

10. Sezgin, M., Sankur, B.: Survey over image thresholding techniques and quantitative performance evaluation. J. Electr. Imaging **13**(1), 146–165 (2004)

11. Parker, J.: Algorithms for Image Processing and Computer Vision, ser. IT Pro. Wiley, Indianapolis (2010)

12. Munoz-Briseno, A., Alonso, G., Palancar, J.H.: Fingerprint indexing with bad quality areas. Expert Syst. Appl. **40**(5), 1839–1846 (2013)

13. Hassanien, A.E.: Hiding iris data for authentication of digital images using wavelet theory. Int. J. Pattern Recogn. Image Anal. **16**(4), 637–643 (2006)

14. Maio, D., Maltoni, D., Cappelli, R., Wayman, J.L., Jain, A.K.: FVC2000: fingerprint verification competition. IEEE Trans. Pattern Anal. Mach. Intell. **24**(3), 402–412 (2002)

15. Hassanien, A.E.: A copyright protection using watermarking algorithm. Informatica **17**(2), 187–198 (2006)

Energy-Aware Topology Construction and Maintenance for Energy Harvesting Sensor Networks

Tien-Wen Sung[1], Chao-Yang Lee[2], You-Te Lu[3], Liang-Cheng Shiu[4], and Fu-Tian Lin[5(✉)]

[1] Fujian Provincial Key Laboratory of Big Data Mining and Applications
and College of Information Science and Engineering,
Fujian University of Technology, Fuzhou, China
twsung@fjut.edu.cn
[2] Department of Electrical Engineering,
National Cheng Kung University, Tainan, Taiwan
nelsorn@gmail.com
[3] Department of Information and Communication,
Southern Taiwan University of Science and Technology, Tainan, Taiwan
yowder@mail.stust.edu.tw
[4] Department of Computer Science and Information Engineering,
National Pingtung University, Pingtung, Taiwan
shiu@mail.nptu.edu.tw
[5] Department of Electrical Engineering,
Tung Fang Design University, Kaohsiung, Taiwan
lft@mail.tf.edu.tw

Abstract. This work designs a topology control approach which can stratify the perpetual energy supply to extend the system lifetime in energy harvesting sensor networks. Topology control is a well-known and energy-efficiency method that aims to reduce the energy consumption and prolong network lifetime in many research fields. The proposed perpetual and distributed topology control (PDTC) algorithm aims to make the harvesting ambient energy usefully and ensure network sustainability, and performs in each sensor which includes two phases, topology construction phase and topology maintenance phase. First, in topology construction phase, each sensor decides a most suitable parent node with maximal working time and adjusts the traffic generating rate to achieve the system sustainability. In the topology maintenance phase, this work adopts a topology maintenance algorithm to trigger the topology construction algorithm and then re-build a new network topology needed. The experimental results demonstrate the superiority of the PDTC algorithm in terms of energy efficient, network lifetime, and system sustainability.

Keywords: Topology construction · Topology maintenance · Network lifetime · Energy harvesting · Sensor network

© Springer International Publishing AG 2017
A.E. Hassanien et al. (eds.), *Proceedings of the International Conference on Advanced Intelligent Systems and Informatics 2016*, Advances in Intelligent Systems and Computing 533, DOI 10.1007/978-3-319-48308-5_69

1 Introduction

Energy harvesting has been applied to deal with the finite lifetime issue in WSNs [1]. Energy harvesting techniques can enable sensors to replenish energy from ambient sources, such as solar, wind, vibrations, foot strike, and finger strokes [2, 3]. If the harvesting ambient energy is large and continuously available, a sensor node can be powered perpetually. Since the harvesting ambient energy is stochastic, and energy collected from an ambient source is low. As a result, the battery of a sensor may exhausted and implies finite lifetime, because the energy availability constraint, which requires the energy consumption to be less than the energy stored in the battery. Thus, energy harvesting sensor network needs take into account an energy management strategy to management the energy consumption and the energy replenishment process of nodes. This work applies the topology control technology for energy management strategy in energy harvesting sensor network. Topology control is an effective method which offers the potential for improving the energy efficiency and prolonging the system lifetime [4]. This study proposed perpetual and distributed topology control algorithm, PDTC, aims to provide high throughput and system lifetime perpetually in energy harvesting sensor network.

2 Related Works

Eu *et al.* [5] designed an opportunistic routing protocol (EHOR) for energy harvesting wireless sensor networks. EHOR take into account link information and ambient energy sources at each node in order to select the best candidate region to forward a data packet. Sharma *et al.* [6] developed energy management policies that are throughput optimal and minimize the mean delay in the queue. Zhang *et al.* [7] developed a wireless sensor network architecture capable of supporting time-critical cyber-physical-systems using energy harvesting. It includes the harvesting aware speed selection algorithm which can maximize the minimum energy reserve.

Due to low recharging rates and the dynamics of ambient energy such as solar power, power supply at sensor nodes may exhausted. The energy harvesting sensor networks necessary to employ an energy management strategy for each sensor. Topology control is an effective method to extend system lifetime and control the topology in WSN [8, 9]. There exists number of distributed topology control algorithms addressing topology control issues where are mainly concerned about the reduction of energy consumption. Ding *et al.* [10] applied connectivity-based Partition Approach (CPA) for topology control with the aim of reducing energy consumption of sensors. The adaptive partitioning scheme guarantees connectivity for the backbone network and divides nodes based on their measured connectivity instead of guessing connectivity based on their positions. A distributed energy-efficient topology control (DETC) algorithm [11] has developed only for home machine-to-machine (M2M) networks. Recently, home networks are rapidly developing to including mobile phones, electronic appliances. The DETC algorithm can perform only in home M2M networks and apply the various sensor types to meet application requirements and achieve the objective of prolonged lifetime with reduced energy cost. Sethu *et al.* [12] developed Step Topology

Control (STC) reduces energy consumption while preserving the connectivity avoids the use of GPS devices or estimations of distance and direction. Liu *et al.* [13] proposed an opportunity-based topology control, called CONREAP, and focus on the reachability-preserving problem in a practical way. The CONREAP has the guaranteed network reachability which is based on reliability theory and the significantly reduced energy cost. Dahnil *et al.* [14] devised a topology-controlled adaptive clustering (TCAC) algorithm to increase the network's lifetime while maintaining the required network connectivity. The TCAC algorithm ensures elected cluster heads to adjust their power level by integrating a transmission power control algorithm to achieve optimal degree and maintain throughout of network operation. Rizvi *et al.* [15] presented a Connected Dominating Set (CDS) based topology control algorithm, A1, which provides better sensing coverage in an energy efficient manner. The A1 uses far less messages for the topology construction phase and achieves better connectivity under topology mainte-nance. These topology control algorithm didn't consider the harvesting ambient energy to power, and only function for a limited amount of time. This work designs a novel distributed topology control with harvesting ambient energy supply in energy harvesting energy sensor network. The proposed perpetual and distributed topology control (PDTC) algorithm aims to construction of energy-efficient topology, make the har-vesting ambient energy usefully, and ensure WSN sustainability.

3   Topology Construction and Maintenance

The power model in this study has adopted from [16], the amount of energy consumption to transmit a bit can be represented as $E_{tx} = E_{elec} + \epsilon_{amp} d^{\alpha}$ and the amount of energy consumption to receive a bit can be obtained by $E_{rx} = E_{elec}$. The E_{elec} represents the electronics energy, ϵ_{amp} is obtained by the transmitter amplifier's efficiency, d denotes the communication distance, and α represents the path loss exponent. Two network scenario parameters have been defined for network description. The network scenario parameters include the traffic generation rate r_i for each sensor and the distances d_{ij} between sensor i and sensor j. The goal of this study is to support a sustained operation for a sensor network. Thereto, sensors should not consume more energy than they can collect. This limited denotes that the traffic generation rate r_i is constrained by the amount of energy sensors can collect and the remaining energy of sensors. In the lifetime model, we adopt as a common lifetime definition, which is widely utilized in the sensor network research field, the time when the first sensor dies [17]. According to [16], we follow its' definition for topology control as an iterative process. The topology control method of our proposed scheme includes topology construction phase and topology maintenance phase. In the topology construction phase, a topology is established in the network while maintaining connectivity. As soon as the topology construction phase establishes the topology, the topology maintenance phase must start working. During topology main-tenance phase, nodes monitor the status of the topology and trigger a new topology construction phase when appropriate. Over the lifetime of the network, we expect this cycle to repeat many times until the energy of the network is depleted.

   To allow the sensors send their sensed data to the sink, we need to construct a sensor tree whose root is the sink and all leaf nodes are the sensors. Each sensor constructs a sub-graphic and chooses a parent node. The sensed data flow from sensors to the sink along the tree. Sensors perform both duties of sensing and forwarding data for other sensors at the same time. Thus, we can construct a sensor tree. Our distributed topology construct algorithm performs in each sensor, and it includes tree steps. First, we compute the upper bound on the working time and maximum traffic generation rate of the sensor. Second, we determine the energy-efficient parent node of each sensor without compromising the obtained maximal working time. Finally, we adjust the traffic generation rate of the sensor to achieve the sustained operation.

The proposed algorithms are based on the condition of an energy harvesting sensor network. In the first step for topology construction process, this work use linear programming (LP) technique to find the maximal working time for sensor v of a sub-graphic. This work adopt as a common lifetime definition the time when the first sensor dies. The working time definition is the same as the lifetime, which is when the first sensor dies in a sub-graphic. Let L_v denote the working time of sub-network in v's view, and l_v denotes the working time of node v. Table 1 shows the parameters used in the algorithm and the pseudo codes of the algorithm is showed in Fig. 1. Each sensor runs the topology construction algorithm, and attempts to find a parent node to build a sub-graphic with maximum working time.

Table 1. Parameters used in the algorithm.

Symbol	Description
N_v	The neighbor set of sensor v
λ_{tx}	Energy cost for per packet transmission
λ_{rx}	Energy cost for per packet reception
e_v^{init}	Initial energy of sensor v
e_v^{rm}	The remaining energy of sensor v
π_v	Amount of energy replenish by sensor v per time slot
t_{iv}	Amount of traffic that sensor i forwards to sensor v
r_v	Traffic generation rate of sensor v
t_{vp}	Amount of traffic that sensor v forwards to parent node p
e_v	Total energy consumption for sensor v per time slot
ω_v	Amount of energy available for sensor v per time slot

Once the maximum working time from a sensors' sub-graphic has been obtained, this sensor starts performing the topology maintenance scheme. Topology maintenance is defined as the process that restores, rotates, or recreates the network topology from time to time when the current one is no longer optimal. In sensor networks, many applications apply a many-to-one traffic pattern. This traffic pattern may cause the energy imbalance problem or hot spot problem. Nodes in this hot spot must forward a large amount of traffic, and often die at a very early stage. Thus, the energy imbalance problem is the limitation of network lifetime. In this study, a topology maintenance mechanism triggers the building of a new topology, consumes a node's energy balance, and increases network lifetime, when energy is not useful. Our topology maintenance

Function **Topology Construction Algorithm** at node v in G:

1. Calculate the upper bound on l_v and r_v by using LP formulation
2. *do*
3.   Broadcast *beacon* message to neighbor set N_v
4.   Receive *beacon* messages from each neighbor set N_v
5. $G_v \leftarrow (V_v, E_v)$
6.   Find Candidate Set C_v
7.   *for* $i \leftarrow 1$ to $|C_v|$ *do*
8.   Calculate $l_i, \forall i \in C_v$
9. $p_v \leftarrow \underset{i \in C_v}{\arg \max} \, l_i$
10. $r_v^* \leftarrow \dfrac{e_v^{rm}}{e_v^{init}} r_v$
11.   *end for*
12. $L_{v,t} = min\left(\underset{i \in N_v}{min} \, l_i, l_v\right)$
13. $\sigma \leftarrow L_{v,t} - L_{v,t-1}$
14.   *if* $\sigma \leq \delta$ *then*
15.   Trigger to **Topology Maintenance Algorithm**
16.   *end if*
17. *while*()

Fig. 1. Topology construction algorithm

process uses a dynamic topology maintenance technique, which triggering the topology construction mechanism and recreating a new sub-graph when necessary. Our topology maintenance scheme defines a migration rule to determine the time to trigger the topology construction scheme. Eq. (1) is adopted to perform topology maintenance process in this work and $\Phi(N)$ is the parent node selected. Each sensor determines when it should transit the topology maintenance to topology construction. If any sensor has no remaining energy, the system is end. The pseudo codes is of the algorithm is shown in Fig. 2.

$$\Phi(N) = \begin{cases} \Phi(N-1), & e_v^{rm} > 0 \cap \varepsilon_v < \theta \\ i, & e_v^{rm} > 0 \cap \varepsilon_v < \theta \\ \emptyset, & e_v^{rm} = 0 \end{cases} \tag{1}$$

4   Performance Evaluations

The performance of the proposed PDTC algorithm is evaluated and compared with that of the generalized probabilistic topology control algorithm (BRASP) and energy efficient topology control algorithm (A1) in this section. The BRASP algorithm determines the minimal transmission power for each sensor and the network reachability satisfies a threshold-based constraint. A1 is a connected dominating set based algorithm for topology control, and it provides a better coverage by an energy efficient approach.

Function **Topology Maintenance Algorithm** at node (v):

1. Broadcast and receive *beacon* message to neighbor set $N(v)$
2. If receive *reconstruct* message from neighbor $N(v)$
3.   Trigger to **Topology Construction Algorithm**
4. If $e_v^{rm} > 0$
5. {
6.   If $\varepsilon_v > \theta$
7.   {
8.   Send *reconstruct* message to children nodes of v
9.   wait until a pre-specified timeout
10.   Trigger to **Topology Construction Algorithm**
11.   }
12. }
13. Else
14.   *System End*

Fig. 2. Topology maintenance algorithm

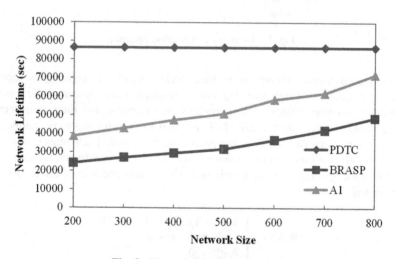

Fig. 3. Network lifetime comparison

We simulated a scenario in which nodes were randomly deployed in a 100 m × 100 m terrain. It is assumed that the sink node is located at the left-bottom corner in the sensing field. The communication radius R_c and initial energy level of each node are set to 15 m and 30 KW, respectively. The energy consumption model was based on the Berkeley mote. The energy costs of 0.075 W when transmitting, 0.03 W when receiving, and 0.025 W when listening. Figure 3 shows the simulation results of system lifetime. The lifetime was measured in term of the round when the first node dies. The proposed PDTC can achieve the maximum bound of system lifetime by using topology construction process and topology maintenance process. This is because the

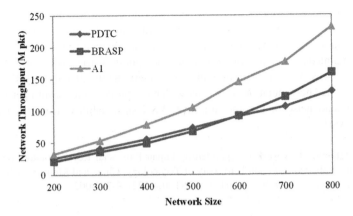

Fig. 4. Network throughput comparison

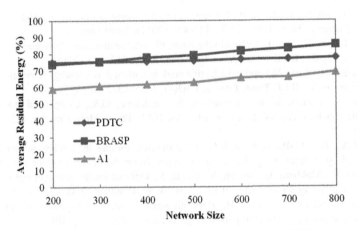

Fig. 5. Average residual energy comparison during 24 h

BRASP algorithm is only includes the topology construction process and both of BRASP and A1 algorithms didn't consider recharging. Figure 4 and Fig. 5 show the simulation results of network throughput and average residual energy respectively. It can be find that less network throughput per unit time can prolong the lifetime, because the energy consumption of packet transmit and receive is decreased.

5 Conclusion

This paper proposed a topology control algorithm, which is perpetual and distributed topology control in energy harvesting sensor network, for the construction of energy-efficient topology and ensure WSN sustainability. The proposed algorithm enables sensors to construct and maintain energy-efficient links. In the construction phase, sensors can autonomously find and determine a parent node with maximum working time of a subgraph by using linear programming technique. Then, sensors can adjust its traffic generating rate to achieve the sustained operation and trigger the topology maintenance phase. In the maintenance phase, it tries to balance the energy consumption of sensors through migration and determine the time to trigger a topology reconstruction. The simulation results show that the proposed algorithm performs well and can extend network lifetime, improve WSN sustainability in energy-harvesting sensor networks.

Acknowledgment. This work is supported by Fujian Provincial Key Laboratory of Big Data Mining and Applications, Fujian University of Technology, China and Ministry of Science and Technology, Taiwan under grant number MOST 104-2221-E-272-002.

References

1. Sudevalayam, S., Kulkarni, P.: Energy harvesting sensor nodes: survey and implications. IEEE Commun. Surv. Tutor. **13**(3), 443–461 (2011). Third Quarter
2. Yang, J., Wu, X., Wu, J.: Optimal scheduling of collaborative sensing in energy harvesting sensor networks. IEEE J. Sel. Areas Commun. **33**(3), 512–523 (2015)
3. Michelusi, N., Zorzi, M.: optimal adaptive random multiaccess in energy harvesting wireless sensor networks. IEEE Trans. Commun. **63**(4), 1355–1372 (2015)
4. Poduri, S., Pattem, S., Krishnamachari, B., Sukhatme, G.S.: Using local geometry for tunable topology control in sensor networks. IEEE Trans. Mob. Comput. **8**(2), 218–230 (2009)
5. Eu, Z.A., Tan, H.-P., Seah, W.K.G.: Opportunistic routing in wireless sensor networks powered by ambient energy harvesting. Comput. Netw. **54**(17), 2943–2966 (2010)
6. Sharma, V., Mukherji, U., Joseph, V., Gupta, S.: Optimal energy management policies for energy harvesting sensor nodes. IEEE Trans. Wirel. Commun. **9**(4), 1326–1336 (2010)
7. Zhang, B., Simon, R., Aydin, H.: Harvesting-aware energy management for time-critical wireless sensor networks with joint voltage and modulation scaling. IEEE Trans. Ind. Inf. **9** (1), 514–526 (2013)
8. Aziz, A.A., Sekercioglu, Y.A., Fitzpatrick, P., Ivanovich, M.: A survey on distributed topology control techniques for extending the lifetime of battery powered wireless sensor networks. IEEE Commun. Surv. Tutor. **15**(1), 121–144 (2013). First Quarter
9. Ren, H., Meng, M.Q.-H.: Game-theoretic modeling of joint topology control and power scheduling for wireless heterogeneous sensor networks. IEEE Trans. Autom. Sci. Eng. **6**(4), 610–625 (2009)
10. Ding, Y., Wang, C., Xiao, Li: An adaptive partitioning scheme for sleep scheduling and topology control in wireless sensor networks. IEEE Trans. Parallel Distrib. Syst. **20**(9), 1352–1365 (2009)
11. Lee, C.-Y., Chu-Sing, Y.: Distributed energy-efficient topology control algorithm in home M2M networks. Int. J. Distrib. Sens. Netw. **2012**, 1–8 (2012)
12. Sethu, H., Gerety, T.: A new distributed topology control algorithm for wireless environments with non-uniform path loss and multipath propagation. Ad Hoc Netw. **8**(3), 280–294 (2010)
13. Liu, Y., Zhang, Q., Ni, L.M.: Opportunity-based topology control in wireless sensor networks. IEEE Trans. Parallel Distrib. Syst. **21**(3), 405–416 (2010)

14. Dahnil, D.P., Singh, Y.P., Ho, C.K.: Topology-controlled adaptive clustering for uniformity and increased lifetime in wireless sensor networks. IET Wirel. Sens. Syst. **2**(4), 318–327 (2012)
15. Rizvi, S., Qureshi, H.K., Khayam, S.A., Rakocevic, V., Rajarajan, M.: A1: an energy efficient topology control algorithm for connected area coverage in wireless sensor networks. J. Netw. Comput. Appl. **35**(2), 597–605 (2012)
16. Labrador, M.A., Wightman, P.M.: Topology Control in Wireless Sensor Networks with a Companion Simulation Tool for Teaching and Research. Springer, The Netherlands (2009)
17. Chang, J., Tassiulas, L.: Energy conserving routing in wireless ad hoc networks. In: Proceedings of IEEE INFOCOM (2000)

Serious Educational Games' Ontologies: A Survey and Comparison

Ahmed M. Abou Elfotouh[1(✉)], Eman S. Nasr[2],
and Mervat H. Gheith[3]

[1] Information Systems Technology Department,
Institute of Statistical Studies and Research (ISSR), Cairo, Egypt
ahmed.fotouh@yahoo.com
[2] Independent Researcher, Cairo, Egypt
nasr.eman.s@gmail.com
[3] Computer Science Department,
Institute of Statistical Studies and Research (ISSR), Cairo, Egypt
mervat_gheith@yahoo.com

Abstract. Serious Educational Games (SEGs) are games not having a mere purpose of entertainment. They benefit from the main characteristics of games, such as engagement and immersiveness to achieve pedagogic objectives. In spite of the promising results of SEGs reported in the literature, their analysis and design still require complex tasks that incorporate game design activities within an educational context. Ontologies that include concepts, relations, and governing rules for both games and education domains could offer an approach to solve such problem. An ontology, as a domain modeling tool, could be used as a meta-model to guide a SEG designer, in addition to bridging the communication gap between the game design and pedagogic domains. This paper presents a survey of available ontologies for SEGs in the literature, in addition to comparing them. We managed to find only two SEGs' ontologies and a meta-model in the literature, and hence presented and compared them. After presenting the survey, and result analysis and general comparison, we followed an ontologies' comparison method called OntoMetric for further evaluation of the current SEGs' ontologies. Our research results revealed that SEGs' ontologies in the literature have two main diverse perspectives. One perspective intensively focuses on the game domain concepts, and the other perspective focuses on the pedagogic domain concepts. In addition, there is little proof that a comprehensive web-based SEGs ontology, which is characterized by completion, consistency, and reusability exists.

Keywords: Domain modeling · Model driven development · Ontology comparison · Serious educational games

1 Introduction

Serious Educational Games (SEGs) are games not having a mere purpose of entertainment [1]. They benefit from the main characteristics of games, such as engagement and immersiveness to achieve pedagogic (teaching) objectives. In addition, they

© Springer International Publishing AG 2017
A.E. Hassanien et al. (eds.), *Proceedings of the International Conference on Advanced Intelligent Systems and Informatics 2016*, Advances in Intelligent Systems and Computing 533, DOI 10.1007/978-3-319-48308-5_70

capture players' (learners') concentration for a long time, offer them practicing what they have learnt, and solve the cost, time, and space barriers [2]. SEGs, as a field of research, emerged from using games as an educational tool because of the similarity between learning processes and playing games [3].

In spite of the promising results of SEGs reported in the literature, e.g. [4, 5], their analysis and design still require complex tasks that incorporate game design activities within an educational context [2]. An integrated and comprehensive model is required that includes both domains' concepts to guide SEGs' analysts and designers [6]. Hence, ontologies which include concepts, relations, and governing rules for both games and education domains integrated together could offer an approach to solve such problem. An ontology, as a domain modeling tool, could be used as a meta-model to guide a SEG designer, in addition to bridging the communication gap between the game design and pedagogic domains. An ontology is defined to be the "formal, explicit specification of a shared conceptualization" [7]. This means presenting knowledge in a shared and structured form which facilitates concepts' communication and analysis [8]. One of the main benefits of building domain ontologies is eliminating ambiguity that arises during the system analysis process. The ambiguity is a consequence of the misinterpretation of concepts' semantics among different stakeholders [8], which might lead to building a wrong solution for a problem.

This paper presents a survey and a comparison of available ontologies for SEGs in the literature. We managed to find only two SEGs' ontologies and a meta-model in the literature, and hence presented and compared them. For the survey, we partially follow Kitchenham et al.'s [9] guidelines for conducting a Systematic Literature Review (SLR) to answer our three formulated Research Questions (RQs). After presenting the survey, and result analysis and general comparison, we followed an ontologies' comparison method called OntoMetric, developed by Lozano-Tello et al. [10], for further evaluation of the current SEGs' ontologies. Our research results revealed that SEGs' ontologies in the literature have two main diverse perspectives. One perspective intensively focuses on the game domain concepts, and the other perspective focuses on the pedagogic domain concepts. In addition, there is little proof that a comprehensive web-based SEGs' ontology, which is characterized by completion, consistency, and reusability exists.

The rest of this paper is organized as follows. Section 2 presents the research method. Section 3 discusses the results' analysis and a general comparison. Section 4 presents the results of using the OntoMetric ontology comparison method for further evaluation, and, finally, Sect. 5 gives the conclusion and future work.

2 Research Method

We partially follow Kitchenham et al.'s guidelines [9] for conducting an SLR to answer our three formulated RQs; conducting a complete SLR remains to be one of our future work. This section gives the research goal and questions, as well as the search strategy, which also includes our inclusion and exclusion criteria.

2.1 Research Goal and Questions

The goal of this study is to identify, analyze, and compare available SEGs' ontologies. To achieve this goal, the following RQs were formulated:

- RQ1: Are there currently available SEGs' ontologies?
- RQ2: What are the concepts used in each, and what are their corresponding meanings?
- RQ3: What ontology engineering approach was used to develop each?

2.2 Search Strategy

Our search strategy consisted of: (1) identifying the English online digital libraries to search, (2) preparing the search keywords based on our RQs, (3) determining the inclusion and exclusion criteria, and (4) applying the search keywords to each online digital library identified in step 1. The online digital libraries we used are: Science-Direct, ACM digital library, IEEE Xplore, and Springer Link. The search keywords are: serious game, serious games ontology, and serious games meta-model. While this research is concerned with SEGs' ontologies, the search keywords extend the inclusion criteria to serious games meta-models. One of the reasons for this extension is the close relation between an ontology and a meta-model from the perspective that both are used to describe and analyze the relations between concepts [7]. However, we acknowledge that there are differences, such as an ontology is a description model in the problem domain, while a meta-model is a prescriptive model of a modeling language in the solution domain [7]. Another reason is the lack of availability of comprehensive, complete, and integrated ontologies in the literature that discuss the SEGs domain.

Our inclusion criteria are: (1) papers published in the English language, (2) papers focusing on SEGs' ontologies, and (3) papers answering at least one RQ. Other than that, papers were excluded. The literature contains a lot of game ontologies, e.g. [11], and models, e.g. [12], but they are not included in our study as we focus on the SEGs' domain. The search keywords were used with each digital library, then we filtered the search results using our inclusion criteria. The filtering was done through reviewing the retrieved studies' titles, abstracts, and keywords. In order to formalize the information extraction process, we applied a predefined information form. The information form consisted of four components, namely: (1) study basic data (title and publication), (2) brief description (study objective, ontology methodology applied, and summarization), (3) glossary of concepts and (4) general comments. The next section gives a brief discussion of the extracted information.

3 Result Analysis and General Comparison

This section presents our result analysis and general comparison. The result analysis subsection lists the selected studies about SEGs' ontologies, in addition to a brief discussion of each. This will be followed by giving a general comparison that presents additional findings.

3.1 SEGs' Ontologies Result Analysis

This subsection gives answers to RQ1 and RQ2. The selected studies that match our inclusion criteria are two ontologies and a meta-model listed in Table 1; they span the period between 2011 and 2014. Each study is given an ID that will be used in the rest of the paper to refer to. Table 1 gives the study ID, title (followed by its reference citation number in this paper), and publishing year.

Table 1. Selected studies.

ID	Title	Year
S1	Game Content Model: An Ontology for Documenting Serious Game Design [13]	2011
S2	Towards Ontology of Gameplay: Application to Game Based Learning Systems [14]	2014
S3	Developing a Meta-Model for Serious Games in Higher Education [15]	2012

S1 introduces a complex SEGs ontology that aims to develop a high-level authoring environment to facilitate the development of SEGs for the nontechnical domain expert (teacher). It presents a SEG as a set of sections, and associates each section with a pedagogic event indicator. The pedagogic event indicator relates each section to one or more pedagogic event from the Gagne's nine events of instructions [16]. S1's ontology gives ten key concepts. These key concepts are related to each other, and are also related to thirty-one detailed classes by the use of a composition relationship. We only list here the key concepts with their exact definitions by the authors, see Table 2, due to space constraints.

Generally, S1 focuses on game concepts more than pedagogic concepts. It tries to address many aspects of SEGs' development. The authors dived into more technical details that have a lot of information and organized the ontology in such a way that might lead to the confusion of domain experts. The notation used are an oval with a white background to present key concepts, an oval with a grey background to present a class, and rectangle with grey background to present a property. However, there is an inconsistent use of notations; for instance, the avatar (an object in the game controlled by game players and represent them) presented by a grey rectangle as a property. In spite of, it is a class that has characteristics/properties. The precise distinction between key concept and class is not clarified, which leads to more ambiguity, such ambiguity has a negative relationship with the ontology reusability, such as reusability for ontology aligning, or merging.

S2 proposes an application of an ontology for gameplay on Game-Based Learning Systems (GBLS). The researchers in this paper aim to tackle one of the most game designers' problems, gameplay specifications. They introduce a formal model to solve understanding and communication issues of gameplay. Gameplay means the interaction done between the game player (learner in SEGs case) and the game. In simplified meaning how the game flows, what actions can the player do (called mechanics in the game design domain), what are the rules that govern these actions, what are consequences of these actions (dynamics). The process of gameplay specifications occurs traditionally in natural language which is difficult to be interpreted and traced either by

Table 2. S1's ontology concepts.

Concept	Definition
Game event	A game event is the happening associated to a game scenario.
Game objective	[T]he goal associated to a game scenario. It is presented using a goal condition.
Game objects	They are virtual things that populate the game world and can be designed to have a combination of abilities such as decision making, moving, acting and responding to surroundings and game player's input simulating their existence in the game world.
Game player	Game player is the user of the game application who provides inputs to the game system.
Game presentation	It is a virtual canvas form a game menu, a game notification or a cut-scene to present information.
Game rule	A game rule states the relationship between game objects and game world.
Game scenario	It is a description of a situation which require game player to overcome a number of challenges in order to achieve the defined objectives.
Game simulation	It is a mechanism that recreates scenarios virtually for game-play to take place.
Game structure	It provides the form and organises game into segments oflinked game presentations and game simulations.
Game theme	It describes most of the art requirements related the game through expressive written text.

human or machine. S2 introduces four parts of an ontology, namely: rules ontology for GBLS, actions of play ontology, game environment ontology, and gameplay bricks ontology. These ontology parts were presented in a Unified Modeling Language (UML) class diagram. S2 limits the introduced concepts on class's names only for simplicity as claimed, Table 3 illustrates part of their concepts with mostly exact authors' definitions.

Table 3. S2's ontology concepts.

Concept	Definition
Rule	Fundamental concept that clarify the difference between game and play.
Action	Responses player can do when facing a situation.
Condition.	Two type of condition Success and fail conditions.
Pedagogic goal	The goal of the game with respect to pedagogic (cognitive, affective, and psychomotor) to achieve.
Game environment.	No formal definition just mentioned (Character, non-player character).

S2 classifies rules according to Reyno et al. [12], and Frasca [17]. The rules classified into meta-rule, manipulation rule, and goal rule. The pedagogic goal, which introduced in S2, inherited from goal rule, then classified into a cognitive, affective, and psychomotor goal. The concepts used were not completely well defined, this leads to ambiguity either in understanding or communicating the domain concepts.

S3 gives a meta-model for SEGs in higher education. It presents a meta-model of three core parts, namely: external entities, educational game elements, and traditional game components. This model focuses mainly on the educational elements which are presented by the SoftWare Engineering Body of Knowledge (SWEBOK) as a curriculum reference. Then it incorporates the knowledge from SWEBOK with a learning taxonomy -that classifies the knowledge level- in a challenge. Table 4 lists S3's concepts and the exact definitions by the authors.

Table 4. S3's meta-model concepts.

Concept	Definition
Character	It includes the Protagonist (student player) as well as a variety of Non-Player Characters (NPC).
Challenge	It holds the formative educational opportunities in the game and has two modalities: expression and conditions.
Context	It situates the player/protagonist in an immersive world with a narrative providing the framework for game challenges.
Mechanics	It provides fundamental, low-level game resources such as character repository, music, and graphics.
Learning taxonomy	Each topic then has a Bloom's taxonomy ability indicating required proficiency.
SWEBOK	Software Engineering Body of Knowledge

S3 reduces the game concepts into character, context, and mechanics, meaning by mechanics the game resources and storage for game manipulations and player records. While the same concept defined in game design and development domain as "The core mechanics consist of the data and the algorithms that precisely define the game's rules and internal operations" [18]. As a result of such misusing of concepts, an ambiguity exists, that leads to imprecise analysis, specification, and communication of domain knowledge.

3.2   General Comparison

This sub-section presents a general comparison to compare the selected studies according to five criteria as follows. First, whether the authors of the studies consider their work as an ontology or meta-model. Second, the number of concepts. Third, the methodology used to develop ontology or meta-mode (which answers RQ3). Fourth, whether the ontology or meta-model provide an ontology validation or not. Fifth, whether a recognized pedagogic method is presented within the ontology or meta-model or not. Table 5 gives the result of our general comparison for the selected studies.

Table 5. General comparison of the selected studies.

Study Criterion	S1	S2	S3
Ontology/Meta-model	Ontology	Ontology	Meta-model
Number of concepts	63	44	52
Ontology developing methodology	Noy et al.'s [19] seven steps method	Common steps existing in different methodologies	–
Ontology validation	No	Yes	No
Pedagogic indicator	Gagne's [16] nine events of instructions	Bloom learning taxonomy [20]	Book Of Knowledge, A revision of Bloom learning taxonomy [21]

As illustrated in Table 5, S1 has the largest number of concepts, which might indicate that the others are more general and miss more elaboration of concepts such as the game presentation and the game theme concepts. This is probably a consequence of the purpose of S1, which is developing a high-level SEGs authoring environment, while S2 focuses on gameplay only. As for S3, it elaborates its concepts by listing valid values for each, which is the main meta-model objective [19].

S1 claims that it follows Noy et al.'s guides [19], while S2 mentions that it follows general steps for developing an ontology namely: (1) definition of ontology purpose, (2) conceptualization, (3) formalization, and (4) validation. S2 has two important steps that are not addressed by S1; formalization, and validation. Validation provides a good justification of the developed ontology.

Finally, the pedagogic indicator; SEGs embeds an educational content introduced in a game format, so it is important to consult a well-defined or proven pedagogic method. S2 uses the first version of Bloom's taxonomy of educational objectives [20], which was introduced in 1956. S1 mentions Gagne's [16] nine events of instructions, which is newer than the one used by S2. S3 uses a revision of Bloom's learning taxonomy [21], which is the newest one. In addition, S3 integrates it with SWEBOK.

4 Using OntoMetric Ontology Comparison Method

The OntoMetric method is an ontology comparing and evaluating method developed by Lozano-Tello et al. [10]. The method is detailed in a PhD thesis in Spanish [22], which we didn't manage to read because of the language barrier. Therefore, we used only the first publication we mentioned. The reason for choosing this ontology comparison or evaluation method among others in the literature is its close relation to both of our purpose of the research and RQs. The OntoMetric method compares and evaluates ontologies in five dimensions. Each dimension has a set of factors. Each factor has a set of characteristics. Each characteristic has a distinct value scored as very low, low, medium, high, or very high. The five dimensions are Content, Methodology, Language, Tools, and Cost. Unfortunately, one of the major challenges to applying the OntoMetric method is that it does not explain in details how its characteristic values are calculated. It is a descriptive measure. Another challenge is the manual work that had to be done to measure each ontology's characteristics, as none is digital nor web-based.

In this research, we consider only the Concept and Relation factors of the Content dimension. These factors support also in answering RQ2. The Concept factor analyses concepts in an ontology by six characteristics. These characteristics assure the importance of providing concise and consistent description of concepts in Natural Language (NL) as well as with formal specification. Table 6 gives the comparison results for the Concept factor of S1 and S2 ontologies. Both have reasonable essential concepts, but neither the pedagogical concepts nor the assessment methods are well presented. S1 introduces a higher-detailed description in NL that might make it more appropriate for reusability. S2 neither introduces well-described concepts in NL nor

Table 6. OntoMetric comparison: concept factor.

Study		
Concept factor characteristic	S1	S2
Essential concepts	Medium	Medium
Essential concepts in superior levels	High	High
Concepts properly described in NL	Very high	Medium
Formal specification of concepts coincides with NL	Medium	Low
Attributes describe concepts	Medium	Low
Number of concepts	Medium	Medium

provides detailed attributes. The more attributes provided for a concept the more specification and distinction exists.

Table 7 gives the Relations factor of the Content dimension, which checks to what degree an existing ontology's relations are satisfying the domain requirements.

Table 7. OntoMetric comparison: relations factor.

Study		
Relations factor characteristics	S1	S2
Essential relations	Medium	Medium
Relations relate appropriate concepts	High	High
Formal specification of relations coincides with NL	Medium	Low
Arity specified (relation degree)	Medium	High
Formal properties of relations	Low	Low
Number of relations	Medium	High

Although the two studies, S1 and S2, provide the essential relations, they do not propose an adequate formal specification for them.

5 Conclusion and Future Work

In spite of the promise that SEGs introduce, there is a lack in the literature for studying their analysis and design. In this paper, we presented a survey of the literature by partially following the SLR approach and evaluated two existing SEGs' ontologies and a meta-model in literature. Our research results revealed that SEGs' ontologies in the literature have two main diverse perspectives. One perspective focuses on game concepts neglecting educational details, that neither well-defined pedagogical concepts nor assessment methods are well integrated. The other one emphasizes on educational concepts rather than balancing required concepts. In addition, a comprehensive and consistent descriptive model (ontology) has to be developed first in a problem domain before attempts to propose a prescriptive or a solution model such as meta-model or a model. The required ontology has to be web-based, shared, and reusable for SEGs' community, in order to support the era of the semantic web. It should integrate well-defined shared concepts in both domains.

For our future work, we will work on developing a more comprehensive and digital SEGs' ontology. Our ultimate aim is to reach the automatic SEGs generation by having a model–driven requirements engineering approach for SEGs. We intend to first follow published approaches of game ontology. Then we will try to integrate current ontologies to form the basis for our new future ontology.

References

1. Abt, C.: Serious Games. Viking Press, New York (1970)
2. Bellotti, F., Berta, R., De Gloria, A.: Designing effective serious games: opportunities and challenges for research. Int. J. Emerg. Technol. Learn. (iJET) **5**, 22–35 (2010)
3. Gee, J.P.: Learning and games. In: Katie, S. (ed.) The Ecology of Games: Connecting Youth, Games, and Learning, pp. 21–39. The MIT Press, Cambridge (2008)
4. Boyle, E.A., Hainey, T., Connolly, T.M., Gray, G., Earp, J., Ott, M., Lim, T., Ninaus, M., Ribeiro, C., Pereira, J.: An update to the systematic literature review of empirical evidence of the impacts and outcomes of computer games and serious games. Comput. Educ. **94**, 178–192 (2016)
5. Mayer, I., Bekebrede, G., Harteveld, C., Warmelink, H., Zhou, Q., Ruijven, Tv, Lo, J., Kortmann, R., Wenzler, I.: The research and evaluation of serious games: toward a comprehensive methodology. Br. J. Educ. Technol. **45**(3), 502–527 (2014)
6. Dormans, J.: The effectiveness and efficiency of model driven game design. In: Herrlich, M., Malaka, R., Masuch, M. (eds.) ICEC 2012. LNCS, vol. 7522, pp. 542–548. Springer, Heidelberg (2012). doi:10.1007/978-3-642-33542-6_71
7. Corcho, O., Fernández-López, M., Gómez-Pérez, A.: Ontological engineering: principles, methods, tools and languages. In: Calero, C., Ruiz, F., Piattini, M. (eds.) Ontologies for Software Engineering and Software Technology, pp. 1–48. Springer-Verlag, Berlin (2006)

8. Castaneda, V., Ballejos, L., Caliusco, M., Galli, M.: The use of ontologies in requirements engineering. Glob. J. Res. Eng. **10**(6), 2–8 (2010)
9. Keele, S.: Guidelines for Performing Systematic Literature Reviews in Software Engineering. Software Engineering Group School of Computer Science and Mathematics Keele University, and Department of Computer Science University of Durham, Durham (2007)
10. Lozano-Tello, A., Gómez-Pérez, A.: Ontometric: a method to choose the appropriate ontology. J. Database Manage. **2**(15), 1–18 (2004)
11. Guo, H., Trætteberg, H., Wang, A.I., Gao, S.: PerGO: an ontology towards model driven pervasive game development. In: Meersman, R., et al. (eds.) OTM 2014 Workshops. LNCS, vol. 8842, pp. 651–654. Springer, Berlin (2014). doi:10.1007/978-3-662-45550-0_67
12. Reyno, E., Carsí, J.Á.: A platform-independent model for videogame gameplay specification. In: Proceedings of the 4th Digital Games Research Association Conference (2009)
13. Tang, S., Hanneghan, M.: Game content model: an ontology for documenting serious game design. In: Proceedings of the IEEE 4th International Conference on Developments in E-systems Engineering (DeSE), Dubai, UAE (2011)
14. Raies, K., Khemaja, M.: Towards ontology of gameplay: application to game based learning systems. In: Proceedings of the Second AIM Research Day on Serious Games and Innovation (2014)
15. Longstreet, C.S., Cooper, K.M.L.: Developing a meta-model for serious games in higher education. In: Proceedings of the IEEE 12th International Conference on Advanced Learning Technologies (ICALT), Rome, Italy, (2012)
16. Gagne, R.: The Conditions of Learning and Theory of Instruction, 2nd edn. Holt, Rinehart and Winston, New York (1970)
17. Frasca, G.: Simulation versus narrative. In: Perron, B., Wolf, M.J.P. (eds.) The Video Game Theory Reader, pp. 221–235. Routledge, London (2003)
18. Adams, E.: Fundamentals of Game Design, 2nd edn. Pearson Education Inc., Upper Saddle River (2010)
19. Noy, N., McGuinness, D.: Ontology Development 101: A Guide to Creating Your First Ontology. Technical report, SMI-2001-0880, Knowldege Systems Laboratory, Stanford University (2001)
20. Bloom, B.S., Engelhart, M.D., Furst, E.J., Hill, W.H., Krathwohl, D.R.: Taxonomy of Educational Objectives: The Classification of Educational Goals, Handbook I: Cognitive Domain, 1st edn. Longmans, New York (1956)
21. Anderson, L.W., Krathwohl, D.R., Airasian, P.W., Cruikshank, K.A., Mayer, R.E., Pintrich, P., Raths, J., Wittrock, M.C.: A Taxonomy for Learning, Teaching, and Assessing: A Revision of Bloom's Taxonomy of Educational Objectives, 1st edn. Longman, New York (2001)
22. Lozano-Tello, A.: Métrica de Idoneidad de Ontologías, Ph.D. thesis, University of Extremadura, Madrid (2002)

Enhancing Web Page Classification Models

Fayrouz Elsalmy[1]([⊠]), Rasha Ismail[2], and Walid AbdelMoez[1]

[1] College of Computing and Information Technology, Arab Academy for
Science, Technology and Maritime Transport, PO Box 1029, Alexandria, Egypt
Fayrouzelsalmy@gmail.com, walid.abdelmoez@gmail.com
[2] Ain Shams University, Cairo, Egypt
rashaismail@cis.asu.edu.eg

Abstract. Web page classification has a crucial role in web mining. The massive amount of data available on the web makes it so important to build web page prediction models. We aim to build classification models that classify new instances depending on existing labeled web documents. This paper investigates the effect of the two powerful ensemble methods called stacked generalization- also known as stacking- and random forest in web page classification context. In this paper, we suggest to enhance the predictive power of the web page classification models by stacking ensemble method. Random forest, stacking with multi-response model trees and four different base learners (Naïve Bayes, J4.8, IBK and FURIA) are used. Datasets are obtained from DMOZ (Open Directory Project). This paper provides an empirical study on the existing supervised classifiers and ensemble learning methods in web page classification context. It introduces that constructing ensembles of heterogeneous classifiers with stacking has higher predictive power than the individual classifiers, boosting and random forest for web page classification.

Keywords: Stacking with Multi-response model trees · Web document classification · Ensemble learning · Multiple classifiers

1 Introduction

The enormous volume of information accessible from the internet makes it of a great value to be used in web mining research area. The concept of applying machine learning methods on web contents to reveal hidden patterns is called web mining. It has three subclasses: web content mining, web structure mining and web usage mining [1, 2]. In this paper, we are discussing web content mining which aims to uncover helpful information from online contents, data or documents to improve finding, filtering and searching for information [2].

Classifying web pages is the process of categorizing web pages to a pre-known category based on pre-defined labelled data. The excessive amounts of information and the rapidly changing environment of the web are the reasons that make it difficult to manually classify web pages. Thus, automatically classifying web pages is essential to several retrieval tasks on the web, such as the construction, development and maintenance of web directories, improvement of search quality results, enhancement of

© Springer International Publishing AG 2017
A.E. Hassanien et al. (eds.), *Proceedings of the International Conference on Advanced Intelligent Systems and Informatics 2016*, Advances in Intelligent Systems and Computing 533, DOI 10.1007/978-3-319-48308-5_71

quality in question answering systems, development of focused web crawling, web content filtering, assisted web browsing and contextual advertising [3].

The ensemble of classifiers approach is an active research area in inductive Machine Learning. An ensemble classifier is combining individual decisions from several classifiers in some way to classify new examples. The objective behind combining of classifiers is to improve the accuracy of a single classifier [4].

The most frequently used ensembles are bagging [5], boosting [6] and the less widely used, stacking [7]. Thus, we choose to examine the stacking in the context of web page classification. In this paper, we are seeking to improve the predictive power of the web page classification models by stacking ensemble method. Therefore, two ensemble learning algorithms were applied Stacking with multi-response model trees [8] and Random forest [9] and four different base learners which are statistical classifier (naive Bayes Multi-nominal algorithm), an instance-based classifier (K-nearest neighbor algorithm), a decision tree classifier (C4.5 algorithm) and a rule-based classification algorithm (FURIA algorithm).

The paper is organized as follows. In Sect. 2, previous work is introduced. In Sect. 3, the backgrounds on areas needed in the analysis of the web page classification models are discussed. In Sect. 4, our proposed ensemble method, datasets and the experimental procedures are presented. Section 5, the experiment results are showed. The conclusion and future work are presented in Sect. 6.

2   Related Work

Previous work has proposed several web page classification models. They are introduced in this section.

Onan [10] used feature selection methods and four ensembles learning algorithms (Boosting, Bagging, Dagging and Random subspace). His results showed that consistency–based feature selection with Adaboost and naïve Bayes algorithms achieved 88.1 % accuracy. In addition, Bagging and Random Subspace showed better performances than base learning algorithms. In addition, Onan [11] has applied Immunos-1 and Immunos-99 classifiers in order to investigate the predictive power of the artificial immune system based methods for web page classification. The results showed that Immunos-1 outperformed the other examined algorithms with 92.4 % accuracy. The experiments used 50 data sets that were obtained from DMOZ (Open Directory Project).

Moreover, Cobos et al. [12] used cuckoo search and k-means clustering approaches. The experimental result was 93 % accuracy. In addition, they used suffix tree clustering that yields 88 % accuracy. Moreover, Bisecting k-means and lingo were applied and their results were 84.4 % and 83.4 % respectively.

Sun et al. in [13] proposed a model using support vector machines with text and context features. The experimental results indicated that the support vector machine performs well in web page classification, especially with context based features.

Haruechaiyasak et al. in [14] suggested to use a fuzzy association method-based approach to handle the ambiguity with the use of the same word/vocabulary.

The results indicated that fuzzy approach yields better results for web page classification with an average accuracy of 81.5 %.

The experimental results presented by Džeroski and Zenko [15] showed that choosing model trees with stacking as a meta level classifier is a very suitable choice, Regardless to the number of base learners used.

Marath et al. [16] have dealt successfully with large- scale datasets as the small-scale datasets for web page classification. They addressed several challenges with the large-scale datasets such as the imbalanced distribution of the class category.

3   Background

In this section, we give a brief background on areas needed in the analysis of the web page classification models. We will explain the classification algorithms, and the ensemble methods being used.

3.1   Classification Algorithms

This section describes the classifiers used in the experimental study. Several classifiers are used from a different set of learning algorithms. Naive Bayes classifier which is a probabilistic classifier, C4.5 which is a decision tree based classifier. K-nearest neighbor which is an instance-based algorithm and FURIA (Fuzzy Unordered Rule Induction Algorithm).

Naive Bayes classifier. Naïve Bayes supervised classifier depends on applying the probability Bayes' theorem. It gives an easy way to deal with any number of classes and attributes [17]. It works fast and well in a wide variety of problems in spite of its "naïve" simple model. The main assumption in Naïve Bayes classification is that the probabilities of attributes having specific values are independent of each other, given a particular class. In reality, this assumption may violate. Also, the results by Naïve Bayes classifier is not affected by small amounts of noise data. The feature independence assumption adopted may degrade the classifier's performance. High correlation between features may increase undesired bias in classification. Moreover, the removal of the redundant attributes can increase the Naïve Bayes classifier's results.

K-nearest neighbor algorithm. The K-nearest neighbor algorithm is an instance based learning algorithm. This algorithm classify instances based on the similarity to the k closest neighbor's classification [18]. K = 1 is used in our experiment.

Every instance has a single point in n-dimensional space. Based on the majority voting of k nearest neighbors of the instance, the algorithm will detect its class label accordingly. Euclidean distance is conducted to assess the similarity between neighbors. The algorithm works well with large training sets.

C4.5 algorithm. Decision trees use inductive inference. Greedy algorithm is used as a basic decision tree induction algorithm. It constructs top-down recursive divide-and-conquer manner decision trees. Decision trees algorithm uses the entire training data to generate a test node for the best feature, which has the maximum information for classification process. Accordingly, top down induction of decision trees splits up the tuples and their values. Also, the classifier stops generation, if the subset tuples are of the same class or if a certain threshold is reached to stop generation. The algorithm uses the information gain as a heuristic to split the data into classes. In each round, the feature with the maximum information gain is selected as the test feature for a given set of data [17].

FURIA algorithm. FURIA (Fuzzy Unordered Rule Induction Algorithm) is a classifier that extends over the Ripper algorithm by learning fuzzy rules and unordered rule sets [19]. It needs an unordered rule sets scheme. A set of rules are obtained for every class by following one-versus-another classes approach. Although, this approach may generate incomplete lists. As a result, the algorithm has an effective rule-stretching approach, which extends minimal generalization-based rule stretching approach by taking the order of the learned antecedents. As the pruning mechanism degrades the method's results, so it is not used.

3.2 Our Proposed Ensemble Methods

Combining multiple classifiers instead of a single classifier can improve the performance of classification significantly [20]. In this section, we describe the ensemble methods we applied: random forest, the stacking framework and stacking with model trees.

Random Forest. Random Forests classifier has been proposed by Breiman [9]. The classifier has a group of tree predictors that have the ability to produce response when given a set of predictor values.

The response of each tree is determined by an independently selected set of predictor values and with the same spread for the entire trees in the forest. Forest trees are built independently. They participate in a voting approach to have a final class prediction. Using tree ensembles can lead to significant improvement in prediction accuracy. This approach is fast, can handle noise, does not over fit and gives explanation and visualization of its output.

Stacking framework. Stacking is considered one of the first ensemble learning methods. It combines heterogeneous base classifiers, which can belong to a completely different family of machine learning algorithms. This is done by the help of the meta-classifier, that takes as its input the output value of the base classifiers [8, 21].

Moreover, the most important issue in stacking approach is the choice of the meta-level classifier. We choose to examine stacking with multi response model trees [15, 22] at the meta-level, as it outperformed other stacking methods as shown in [23–26].

Level-1 base classifiers train level-0 training dataset as shown in Fig. 1. Their outputs are used as inputs to train a layer-2 meta-classifier. The aim is to check if the training data have been learned properly.

Stacking with multi-response model trees. Stacking with model trees is an alternative to Multi-response Linear Regression (MLR) for learning at the meta-level. Model trees at the meta-level outperformed MLR for classification via regression as [28] has shown especially for domains with continuous attributes. Model tree induction was used instead of linear regression.

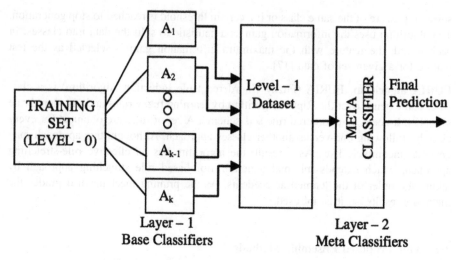

Fig. 1. Multiple classifiers based on stacking [27]

4 Experiments and Evaluation

This section describes the experimental procedure, environment and the evaluation measures used to conduct the experiments.

4.1 Experimental Environment

The experiments are conducted using the open-source platform called WEKA (Waikato Environment for Knowledge Analysis) machine learning toolkit version 3.7.11. It implements different machine learning algorithms in JAVA.

A 10-fold cross-validation method is utilized. The original dataset is divided into 10 folds. Then, the training and testing process is repeated 10 times. The testing uses one fold for validation every time, and the training uses the rest of the folds. At the end of the process, average results for 10-fold testing are determined.

4.2 Datasets

50 datasets of DMOZ (Open Directory Project) are used [29] to measure the performance of the classifiers and ensemble methods. The minimum sum of the documents is 104 and maximum sum of the documents is 164. The topics vary from 3 to 10. The minimum number of attributes is 495, and the maximum number of attributes is 822.

The datasets are available in a pre-processed format, which means that special characters and stop words were removed, lower case filtered and Porter stemming was applied. The datasets are available at http://artemisa.unicauca.edu.co/~ccobos/wdc/wdc.htm.

4.3   Evaluation Measures

Accuracy. To assess the performance of classification algorithms and ensemble learning methods, classification accuracy is selected as an evaluation measure. Classification accuracy (ACC) is the proportion of correctly classified instances obtained by the classifier over the total number of instances as given by the following equation:

$$ACC = \frac{TN + TP}{TP + FP + FN + TN} \tag{1}$$

Where *TN* presents the number of true negatives, *TP* presents the number of true positives, *FP* presents the number of false positives and *FN* presents the number of false negatives [10].

Wilcoxon signed ranks test. The Wilcoxon signed-ranks test is a non-parametric statistical test. It is used to compare the performances of two classifiers for the same data set. It is used to examine the null hypothesis that the distribution's median is equal to a certain value. In other words, it compares the population mean ranks differences between the two classifiers [30].

5   Experimental Results

Our experimental results are shown in this section. Figure 2 depicts the classification accuracies of the ensemble method stacking with model trees and the four base-level classifiers. Stacking with model trees yields more performance than each base classifier individually. It has 91.2 % classification accuracy.

Figure 3 shows a comparison between the two proposed ensemble-learning methods. We choose to classify with 500 random trees and set the default parameters of random forest. As shown in the figure stacking with model trees outperformed random forest. Wilcoxon test is used to measure the significance difference between the two classifiers. The result showed a P-value less than the significance level chosen, P-value = 0.035 < 0.05. Therefore, candidate classifiers differ significantly and not due to chance. The diversity of nature among the chosen base classifiers has led to better results of stacking when compared to random forest.

Table 1 shows the Comparison of classification accuracies obtained from DMOZ Open Directory Project dataset with classification algorithms from the literature. Several ensemble learning methods and classification algorithms were used.

As a result, the aim of the paper was to show that stacking with model trees was significantly better than the other ensemble learning approaches; random forest, boosting. The table also refers to Onan [11] that has suggested to use Immunos-1 classifier as it showed a higher average accuracy percentage of 92.4 %.

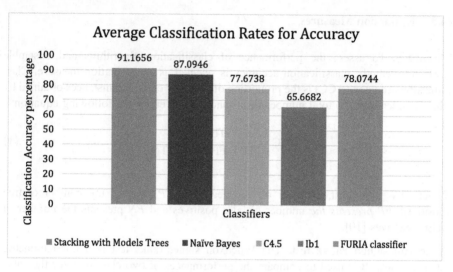

Fig. 2. Average classification rates for algorithms

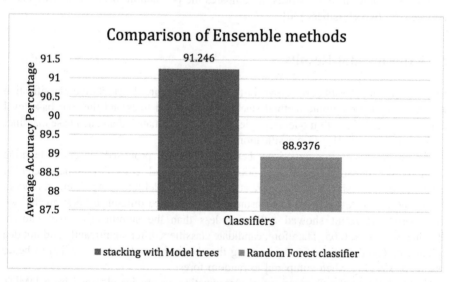

Fig. 3. Average accuracy between stacking with model trees and random forest

As a result, we applied Wilcoxon test to measure the significance difference between Immunos-1 and Stacking with model trees. With significance level 0.01, the result showed P-value equals 0.0479 > 0.01. Thus, the null hypothesis is not rejected, which means that the difference among the classifiers is not significant, and due to chance.

Table 1. Classification accuracy rates comparisons

Reference	Method	Accuracy %
Onan [11]	Immunos-1	92.4
our proposed work	**Stacking with model trees, Naïve Bayes, C4.5, IBk, FURIA**	**91.2**
Onan [10]	Adaboost, Naïve Bayes, consistency-based feature selection	88.1
our proposed work	**Random Forest**	**88.3**
Onan [11]	Immunos-99	77.8

6 Conclusion and Future Work

Web page classification has several benefits. It can help to make relevant and fast selection of information from large directories. Also, it is important for analytical and marketing companies. This paper revealed the literature review that introduces web page classification models. The paper also includes a comparison of three papers used classification algorithms on the same datasets from DMOZ. Our research problem is to find better ways for web page classification models. Thus, we have investigated the effect of two powerful ensemble-learning methods: stacking with multi-response models trees and random forest. We suggested an approach to improve the predictive power of the web page classification models by stacking ensemble method. Our results show that stacking with multi-response model trees outperforms random forest and the other existing ensemble methods examined in previous studies. We are planning to use hybrid clustering and classification methods simultaneously seeking to improve the predictive power of web page classification models.

References

1. Fürnkranz, J.: Web Mining. In: Maimon, O., Rokach, L. (eds.) Data Mining and Knowledge Discovery Handbook, pp. 891–920. Springer, Heidelberg (2005)
2. Kosala, R., Blockeel, H.: Web mining research: a survey. SIGKDD Expl. **2**, 1–15 (2000)
3. Qi, X., Davison, B.D.: Web page classification: features and algorithms. ACM Comput. Surv. **41**(2), 1–31 (2009). Article 12
4. Dietterich, T.: Machine learning research: four current directions. AI Mag. **18**(4), 97–136 (1997)
5. Breiman, L.: Bagging predictors. Mach. Learn. **24**(2), 123–140 (1996)
6. Freud, Y., Schapire, R.E.: A decision-theoretic generalization of on-line learning and an application to boosting. In: Ito, T. (ed.) TPPP 1994. LNCS, vol. 907, pp. 23–37. Springer, Heidelberg (1995)
7. Wolpert, D.: Stacked generalization. Neural Netw. **5**, 241–259 (1992)
8. Dzeroski, S., Zenko, B.: Is combining classifiers with stacking better than selecting the best one? Mach. Learn. **54**, 255–273 (2004)
9. Breiman, L.: Random forests. Mach. Learn. J. **45**, 5–32 (2001)

10. Onan, A.: Cassifier and feature set ensembles for web page classification. J. Inf. Sci. **42**(2), 150–165 (2015)
11. Onan, A.: Artificial immune system based web page classification. In: Silhavy, R., Senkerik, R., Oplatkova, Z.K., Prokopova, Z., Silhavy, P. (eds.) Software Engineering in Intelligent Systems, pp. 189–199. Springer, Berlin (2015)
12. Cobos, C., Munoz-Collazos, H., Urbano-Munoz, R., Mendoza, M., Leon, E., Herrera-Viedma, E.: Clustering of web search results based on the cuckoo search algorithm and balanced Bayesian information criterion. Inf. Sci. **281**, 248–264 (2014)
13. Sun, A., Lim, EP., Ng, WK.: Web classification using support vector machine. In: Proceedings of the 4th International Workshop on Web Information and Data Management, pp. 96–99. ACM, New York (2002)
14. Haruechaiyasak, C., Shyu, M.L., Chen, S.C., Li, X.: Web document classification based on fuzzy association. In: Proceedings of COMPSAC 2002, pp. 487–492. IEEE, New York (2002)
15. Džeroski, S., Ženko, B.: Stacking with multi-response model trees. In: Roli, F., Kittler, J. (eds.) MCS 2002. LNCS, vol. 2364, pp. 201–211. Springer, Heidelberg (2002)
16. Marath, S.T., Shepherd, M., Milios, E., Duffy, J: Large-scale web page classification. In 2014 47th Hawaii International Conference on System Sciences (HICSS), pp. 1813–1822 (2014)
17. Ratanamahatana, C., Gunopulos, D.: Feature selection for the naive Bayesian classifier using decision trees. Appl. Artif. Intell. **17**(5), 475–487 (2003)
18. Aha, D.W., Kibler, D., Albert, M.K.: Instance-based learning algorithm. Mach. Learn. **6**, 37–66 (1999)
19. Huhn, J., Hullermeier, E.: FURIA: an algorithm for unordered fuzzy rule induction. Data Mining Knowl. Disc. **19**(3), 293–319 (2009)
20. Dietterich, T.G.: Ensemble methods in machine learning. In: Kittler, J., Roli, F. (eds.) MCS 2000. LNCS, vol. 1857, pp. 1–15. Springer, Heidelberg (2000)
21. Wolpert, D.H.: Stacked generalization. Neural Networks. **5**, 241–259 (1992)
22. Seewald, A.K.: How to make stacking better and faster while also taking care of an unknown weakness. In: Nineteenth International Conference on Machine Learning, pp. 554–561 (2002)
23. Merz, C.J.: Using correspondence analysis to combine classifiers. Mach. Learn. **36**, 33–58 (1999)
24. Ting, K.M., Witten, I.H.: Issues in stacked generalization. J. Artif. Intell. Res. **10**, 271–289 (1999)
25. Todorovski, L., Džeroski, S.: Combining multiple models with meta decision trees. In: Zighed, D.A., Komorowski, J., Żytkow, J.M. (eds.) PKDD 2000. LNCS (LNAI), vol. 1910, pp. 54–64. Springer, Heidelberg (2000)
26. Seewald, A.K., Fürnkranz, J.: An Evaluation of Grading classifiers. In: Hoffmann, F., Adams, N., Fisher, D., Guimaraes, G., Hand, D.J. (eds.) IDA 2001. LNCS, vol. 2189, pp. 115–124. Springer, Heidelberg (2001)
27. Nagi, S., Bhattacharyya, D.K.: Classification of microarray cancer data using ensemble approach. Netw Model Anal Health **2**(3), 59–173 (2013)
28. Frank, E., Wang, Y., Inglis, S., Holmes, G., Witten, I.H.: Using model trees for classification. Mach. Learn. **32**(1), 63–76 (1998)
29. DMOZ Open Directory Project Dataset. http://www.unicauca.edu.co/~ccobos/wdc/wdc.htm
30. Demsar, J.: Statistical comparisons of classifiers over multiple data sets. J. Mach. Learn. Res. **7**, 1–30 (2006)

Artifact Elimination in Neurosciences

Thea Radüntz[1]([✉]), Mohamed A. Tahoun[2], Mohammed A-Megeed[3],
and Beate Meffert[4]

[1] Federal Institute for Occupational Safety and Health,
Unit 3.4 'Mental Health and Cognitive Capacity',
Nöldnerstr. 40-42, 10317 Berlin, Germany
`raduentz.thea@baua.bund.de`
[2] Faculty of Computers and Informatics, Suez Canal University, Ismailia, Egypt
[3] Faculty of Computer and Information Sciences,
Ain Shams University, Abbasiya, 1156, Cairo, Egypt
[4] Department of Computer Science, Humboldt-Universität zu Berlin,
Rudower Chaussee 25, 12489 Berlin, Germany

Abstract. Artifact elimination is a central issue in neurosciences. A method that has established itself as an important part of EEG analysis is the application of independent component analysis (ICA). It decomposes the multi-channel EEG into linearly independent components (ICs) that then can be classified as artifact or EEG signal component. However, classification of the ICs still requires visual, time-intensive inspection by experts.

In order to develop an automated artifact elimination method, we apply several classification algorithms on feature vectors extracted from ICA components via image processing algorithms. We compare their performance with the ratings of experts and identify range filtering as a feature extraction method with great potential. Range images classified with artificial neuronal networks yield accuracy rates of 95.5 %. The results are very promising regarding automated IC artifact recognition.

Compared to existing automated solutions the proposed method has the main advantage that it is not limited to a specific number or type of artifact. Furthermore, it is an automatic, real-time capable, and practical tool that reduces the time-intensive manual selection of ICs for artifact removal.

Keywords: Electroencephalogram (EEG) · Signal processing · Image processing · Pattern recognition · Classification

1 Introduction

The electroencephalogram (EEG) is a multi-channel signal of neuronal brain activity. Due to its temporal resolution, the electroencephalography is an excellent and widely used technique for investigating human brain functioning. Unfortunately, it is commonly contaminated by artifacts from various other sources

© Springer International Publishing AG 2017
A.E. Hassanien et al. (eds.), *Proceedings of the International Conference on Advanced Intelligent Systems and Informatics 2016*, Advances in Intelligent Systems and Computing 533, DOI 10.1007/978-3-319-48308-5_72

that produce undesired alterations. Artifacts arise from various physiological and non-biological interferences such as eye movements, blinks, muscle activity, heartbeat, high electrode impedance, line noise, and electric devices. Generally they are linearly summed to the cerebral signal and can be present in all electrodes with different proportions according to electrodes' position. Hence, an important processing step previous to the EEG interpretation is the elimination of every kind of artifacts.

A widely applied but insufficient approach is the discarding of all artifact contaminated EEG segments. This loss of experimental data becomes especially problematic if only a few epochs are available and artifacts such as blinks or movements occur too frequently. Furthermore, this approach is inappropriate when dealing with the continuous EEG, real-time brain-computer interface (BCI) applications, and online mental state monitoring [7]. Although there exist numerous other artifact reduction techniques [3] most of them are unsuitable for real-time applications [7].

A method which has established itself as an important part of EEG analysis is the application of independent component analysis (ICA) [7,8]. The idea central to this method is that the EEG can be decomposed into its linearly independent source components (ICs), which are then visually examined, and classified as artifact or EEG signal components. Artifact components are removed and the remaining EEG signal components are projected back to the original time domain. This procedure yields the reconstruction of an artifact-free EEG.

However, a limitation of the method is that it requires human raters for classifying the ICs. There are some works addressing this problem but they are limited to specific artifacts [7,10]. Hence, there is a further need to develop fully automated machine classification for automatic artifact elimination able to deal with all kinds of artifacts.

The idea underlying our method is inspired from the pattern similarity of IC images for each type of artifact that is clearly illustrated in Fig. 1. The similarities exist despite different EEG channel configurations and form the basis for the visual classification of the ICs by experts. The 2D scalp maps are called topoplots. They are generated by interpolating the ICA mixing matrix columns onto a fixed-size grid (in our case: 51 × 63). This interpolation step permits the classification of images derived from ICs composed of varying column lengths of the mixing matrix.

The decision if an IC is an artifact or EEG signal component is taken by the experts according to the topoplots' image pattern and their know-how. This ability of humans to perceive and understand the image for further decision making builds the core of our new method. For enhancing the recognition performance of our machine classification we extract image features from the topoplot images. We apply several image processing algorithms commonly used in the context of 2D object recognition. Finally, we use different classification methods and compare the results.

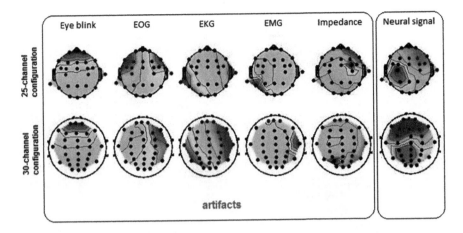

Fig. 1. Examples of similar IC artifact patterns for two different EEG channel configurations. First row: IC images from a 25-channel configuration; second row: IC images from a 30-channel configuration.

2   Material and Experiments

The data acquisition took place in the shielded lab of the Federal Institute for Occupational Safety and Health in Berlin. The sample consisted of 57 people in paid work and high variability with respect to cognitive capacity and age (age range 30–62, 31 female, 26 male). The experiment was fully carried out with each subject in a single day. During the investigation the subjects were asked to solve nine cognitive tasks. Furthermore, there was a short period of relaxation at the beginning and at the end of the experiment.

Thereby, the EEG was recorded using electrode caps and an amplifier from BrainProducts GmbH and their BrainRecorder software. We captured signals from 25 electrodes placed at positions according to the 10–20-system and recorded with reference to Cz and at a sample rate of 500 Hz.

3   Methods

We implemented a MATLAB [9] toolbox consisting of three main modules for pre-processing, feature generation, and classification. Figure 2 summarizes this processing pipeline that is described in-depth in the following.

3.1   Pre-processing

After filtering the EEGs with a bandpass filter (order 100) between 0.5 and 40 Hz, independent component analysis (Infomax algorithm) was applied to the 25-channel signals, with the aim of decomposing them into 25 ICs [2,8]. Each of the 25-element column vectors of the 25×25 ICA mixing matrix W^{-1} was interpolated onto a 51×63 grid using the inverse distance method of [16]. They form

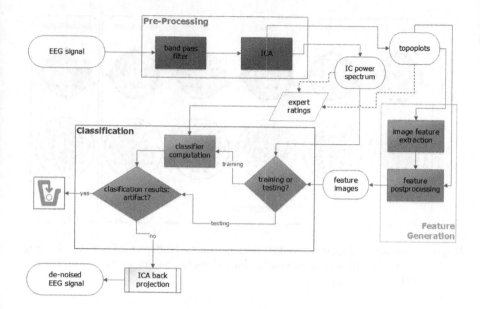

Fig. 2. Processing pipeline

the 25 two-dimensional scalp projection maps, here referred to as topoplots. The first plot on the left of Fig. 3 illustrates the interpolation on a fine, rectangular cartesian grid. The 25 computed ICs are characterized by their activations (the sources' time course), their activity power spectrum calculated with FFT, and their 2D scalp component maps (interpolated from the columns of the mixing matrix).

3.2   Feature Generation

The feature generation module consists of algorithms for image feature extraction. Furthermore, it comprises post-processing algorithms that can be applied to the generated features but also directly to the topoplot images. For the sake of convenience, we skip the detailed description of the post-processing module and refer the interested reader to a previous publication [14].

Image Feature Extraction. Image feature extraction generates features from the topoplot that can be generally categorized in three groups: texture features, gradient and range features, and geometric features.

Local Binary Pattern (LBP). LBP is a powerful feature for 2D texture classification first described in 1994 [11] and refined later in [12]. The LBP feature values are computed by building the differences between each pixel and its neighbors $P \in \{8, 16, 24\}$ within a prescribed radius R. The relation of the central pixel to its neighbors is binary coded. Hence, in the case of $P = 8$ neighbors, we obtain an 8-digit binary number that can be converted into a decimal number. This constitutes the new value of the central pixel.

A rotation-invariant alternative of the LBP method is described by [13]. Here, the digits of the binary number are shifted until a minimal value is created. Another variant is the uniform LBP, where thresholding is used to replace rare feature values with a single value. A combination of both results in a rotation invariant, uniform LBP feature.

Huang et al. introduced the 3D LBP method, which computes four distinct LBP layers [6]. Here coding is done not only for whether the central pixel is less or greater than its neighbor but also for the value of the differences themselves. In a four bit word the sign of the difference is tracked in the first bit and its value in the remaining three bits. In each LBP layer (sign layer, first bit layer, second bit layer, third bit layer) we set the corresponding decimal value at the position of the central pixel.

We implemented and tested all LBP variants described here.

Gradient Images, Range Filter, and Laplacian. We also evaluated the recognition performance of several gradient features. The horizontal Sobel operator (HSO) is obtained by spatial convolution with the filter mask

$$HSO = \begin{pmatrix} -1\ 0\ 1 \\ -2\ 0\ 2 \\ -1\ 0\ 1 \end{pmatrix} \tag{1}$$

and detects vertical edges.

The large horizontal gradient operator (LHG) detects horizontal range differences over a greater horizontal distance. Heseltine et al. tested a variety of image processing techniques for face recognition and identified the LHG operator as the most effective surface representation and distance metric for use in application areas such as security, surveillance, and data compression [4]. In consideration of this, we decided to use their recommended size of 5 and compute the LHG via the filter mask

$$LHG = \begin{pmatrix} -1\ 0\ 0\ 0\ 1 \end{pmatrix} \tag{2}$$

We also tested the range filter, which takes the difference between the maximal and minimal value for each topoplot image's pixel in a 3×3 neighborhood [1]. In addition, we applied the Laplacian operator as a second order feature.

Gaussian Curvature. The Gaussian curvature belongs to the geometric features and is defined as follows. The curvature K of a point on a surface is the product of the principal curvatures k_1 and k_2:

$$K = k_1 \cdot k_2 = \frac{1}{r_1} \cdot \frac{1}{r_2} \tag{3}$$

with main curvature radius r_1 and r_2.

We implemented the Gaussian curvature of the topoplot surface points as a further feature and tested its classification accuracy [5].

After the application of these operations for feature extraction we obtain images of the same size as the original topoplots. We refer to them as feature images. Examples of these are illustrated in Fig. 3.

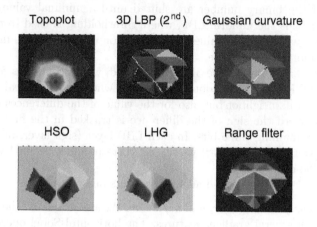

Topoplot   3D LBP (2nd)   Gaussian curvature

HSO   LHG   Range filter

Fig. 3. Examples of the feature images used.

3.3 Classification

We employed four frequently-used classification methods mentioned in numerous research works. These are: linear discriminant analysis (LDA), logistic regression (LgR), support vector machines (SVM), and artificial neuronal network (ANN).

LDA and LgR are linear classifiers. Compared to each other LgR is less affected by outliers due to a small loading while LDA includes the outliers in the computation of the covariance matrix. Hence, we assume LgR to be more robust in its performance although it might be less precise than LDA. As a large margin classifier we employed SVM with a nonlinear kernel (Gaussian radial basis function). Finally, we implemented an ANN with one hidden layer and trained it with back propagation [15] using 100 iterations.

We used visual ratings from two experts to train classifiers for different image features and the topoplots. Firstly, we identified a suitable subset of features and trained the different classifiers on 60 % of our data set. By this, we selected the best parameter for each classification method and feature (Table 1). Subsequently, we compared the trained classifiers to select the one with the best performance. This trained classifier automatically classifies then the ICs. Components classified as EEG components are projected back, artifact components are discarded from subsequent processing. In this way we reconstruct an artifact-free EEG signal.

4   Results

We evaluated our artifact removal algorithm by means of a close inspection of the classification results and the performance time. While the first is standard

for the assessment of recognition performance, the timing performance accounts
for the real-time feasibility of the algorithm.

4.1   Classification Results

Based on the results described in our previous paper [14] we focus on six selected
image features: the horizontal Sobel operator (HSO), the large horizontal gradient operator (LHG), the range filter, the second layer of the 3D Local Binary
Pattern (3D LBP (2nd)) and Gaussian curvature, as well as the raw topoplots.
In addition, we integrate the five conventional EEG frequency bands computed
from the IC activations in the classification process.

For these features we obtained different recognition rates in respect to the
classification method used (Table 1). The topoplots achieved their best performances in combination with ANN (93.2 %) and logistic regression (92.3 %). The
3D LPB (2nd) features performed similarly good for these two classification
methods (94.1 %, 92.9 %). The accuracy rates of the gradient images for logistic
regression, SVM, and ANN ranged between 92.2 % and 93.4 % and yield their
best results for the combination of HSO with SVM. The Gaussian curvature provides good recognition rates of 93.7 % for the SVM but also 93.1 % for the ANN.
Although the range images provide very good results with the SVM (94.6 %),
their best accuracy rates of 95.5 % were obtained in combination with ANN

Table 1. Recognition rates (%) for optimal parameter (in brackets) for each classification method and feature (N: number of eigenvectors, λ: regularization parameter, C:
SVM regularization, σ: kernel width, units: number of neurons for the hidden layer).

IC power spectra &	LDA % (N–1)	LgR % (λ)	SVM % (C/σ)	ANN % (λ/units)
Topoplots	86.9 (376)	92.3 ($2.5 \cdot 10^{-3}$)	88.4 (2.5/0.5)	93.2 ($1.8 \cdot 10^{-2}$/10)
3D LBP (2nd)	84.3 (376)	92.9 ($2.5 \cdot 10^{-3}$)	89.8 (2.5/4,5)	94.1 ($1.8 \cdot 10^{-2}$/10)
LHG	89.3 (376)	92.3 ($2.5 \cdot 10^{-3}$)	92.3 (1.5/2.8)	92.8 ($2.5 \cdot 10^{-3}$/3)
HSO	89.7 (376)	92.2 ($2.2 \cdot 10^{-1}$)	93.4 (2.6/0,8)	92.7 ($5.5 \cdot 10^{-4}$/4)
Range filter	90.9 (376)	92.8 ($1.8 \cdot 10^{-2}$)	94.6 (3.5/1.1)	95.5 ($5.0 \cdot 10^{-2}$/3)
Gaussian curvature	90.5 (376)	92.5 (2.7)	93.7 (1.5/0.4)	93.1 (2.7/3)

4.2   Performance Time

In respect to the real-time feasibility of the algorithm we tested our system on an
Intel Core i5-3320M Processor (2.6 GHz) with 8 GB DDR3-SDRAM (2×4 GB).
The input signal was a 25-channel EEG of 94.34 s (500 Hz sample rate). We
choose such a short signal because computation time of the ICA depends on
the signal length and is the most time-consuming operation of the processing
pipeline. Hence, regarding system's performance time, it can be seen as a bottleneck. At the same time we must consider that identification of N components
requires more than kN^2 data sample points (N denotes the number of EEG
channels, N^2 is the number of weights in the ICA unmixing matrix, and k is a
multiplier that increases as the number of channels increases).

Hence, in our case the pre-processing module needs approx. 5 s for filtering, performing ICA, and generating a topoplot. The computation of one feature image (here range filtering) and the corresponding frequency bands takes only 15.7 ms. Classification of one component with SVM as well as with ANN needs less than 1 ms. System training can be done offline and thus does not need to be considered for the evaluation of real-time performance.

5 Discussion

In this paper we described a novel approach for robust, automated EEG artifact rejection inspired by computer vision. Existing methods are either not fully independent from a user [7,10] or limited to a few artifacts [17]. However, real-world EEG contains an array of artifacts with unknown properties. Therefore, we developed an automated method capable of distinguishing between the pure EEG signal and all types of artifacts. There is no comparable method known to the authors.

Our method is based on machine classification of ICs as artifact or EEG signal via features gained with image processing algorithms. We implemented, tested, and validated several image features, e.g. range filter, local binary patterns, and geometric features. Linear discriminant analysis, logistic regression, SVM, and ANN were applied for classification. Accuracy rates were calculated for assessing the features' achievement potential. Our novel approach for real-time and fully automated artifact elimination reaches recognition rates between 89.13 % and 95.5 %. Its best recognition performance is gained for features derived from range images together with the IC frequency bands and in combination with ANN. Furthermore, the outlined performance times of the modules demonstrate that the system is real-time capable.

In summary, the presented method is an automatic, real-time capable, and practical tool that reduces the time-intensive manual selection of ICs for artifact removal. However, additional tests on further data have to be carried out. Their results will be presented in a separate paper.

Acknowledgements. We would like to thank Dr. Sergei Schapkin, Dr. Patrick Gajewski, and Prof. Michael Falkenstein for selection of the battery's tasks. We would like to thank Mr. Ludger Blanke for technical support during the timing tests for the tasks. In addition, we would like to thank Ms. Xenija Weißbecker-Klaus, Mr. Robert Sonnenberg, Dr. Sergei Schapkin, and Ms. Marion Freyer for general task testing and for conducting the laboratory experiments. Furthermore, we would like to thank Ms. Marion Freyer for daily operational support and our student assistant Jon Scouten for computational support and proofreading. A special thanks go to Dr. Gabriele Freude for her general project support.

More information about the project where our EEG data was acquired can be found under http://www.baua.de/de/Forschung/Forschungsprojekte/f2312.html?nn=2799254.

References

1. Bailey, D.G., Hodgson, R.M.: Range filters: localintensity subrange filters and their properties. Image Vis. Comput. **3**(3), 99–110 (1985). http://www.science direct.com/science/article/pii/0262885685900587
2. Delorme, A., Makeig, S.: EEGLAB: an open source toolbox for analysis of single-trial EEG dynamics. J. Neurosci. Methods **134**, 9–21 (2004)
3. Fatourechi, M., Bashashati, A., Ward, R.K., Birch, G.E.: EMG and EOG artifacts in brain computer interface systems: a survey. Clin. Neurophys. **118**(3), 480–494 (2007)
4. Heseltine, T., Pears, N.E., Austin, J.: Three-dimensional face recognition: a fishersurface approach. In: Campilho, A.C., Kamel, M.S. (eds.) ICIAR 2004. LNCS, vol. 3212, pp. 684–691. Springer, Heidelberg (2004). http://dx.doi.org/10.1007/978-3-540-30126-4_83. 01 Sept 2014
5. Horn, B.K.P.: Extended gaussian images. Proc. IEEE **72**(12), 1671–1686 (1984)
6. Huang, Y., Wang, Y., Tan, T.N.: Combining statistics of geometrical and correlative features for 3D face recognition. In: Proceedings of British Machine Vision Conference, pp. 879–888 (2006)
7. Jung, T.P., Makeig, S., Humphries, C., Lee, T.W., Mckeown, M.J., Iragui, V., Sejnowski, T.J.: Removing electroencephalographic artifacts by blind source separation (2000)
8. Makeig, S., Bell, A.J., Jung, T.P., Sejnowski, T.J.: Independent component analysis of electroencephalographic data. In: Advances in Neural Information Processing Systems, pp. 145–151. MIT Press (1996)
9. MathWorks: MATLAB (R2012b). www.mathworks.com
10. Mognon, A., Jovicich, J., Bruzzone, L., Buiatti, M.: ADJUST: an automatic EEG artifact detector based on the joint use of spatial and temporal features. Psychophysiology **48**(2), 229–240 (2011). doi:10.1111/j.1469-8986.2010.01061.x. ISSN: 1469–8986
11. Ojala, T., Pietikainen, M., Harwood, D.: Performance evaluation of texture measures with classification based on kullback discrimination of distributions. In: Proceedings of the 12th IAPR International Conference on Pattern Recognition, Computer Vision and Image Processing, vol. 1, pp. 582–585 (1994)
12. Ojala, T., Pietikainen, M., Harwood, D.: A comparative study of texture measures with classification based on featured distributions. Pattern Recogn. **29**(1), 51–59 (1996)
13. Ojala, T., Pietikainen, M., Maenpaa, T.: Multiresolution gray-scale and rotation invariant texture classification with local binary patterns. IEEE Trans. Pattern Anal. Mach. Intell. **24**(7), 971–987 (2002)
14. Radüntz, T., Scouten, J., Hochmuth, O., Meffert, B.: EEG artifact elimination by extraction of ICA-component features using image processing algorithms. J. Neurosci. Methods **243**, 84–93 (2015). http://www.sciencedirect.com/science/article/pii/S0165027015000370
15. Rumelhart, D.E., Hinton, G.E., Williams, R.J.: Learning representations by back-propagating errors. Nature **323**(6088), 533–536 (1986). http://www.iro.umontreal.ca/~vincentp/ift3395/lectures/backprop_old.pdf. Accessed 7 Dec 2014
16. Sandwell, D.T.: Biharmonic spline interpolation of GEOS-3 and SEASAT altimeter data. Geophys. Res. Lett. **14**(2), 139–142 (1987). http://dx.doi.org/10.1029/GL014i002p00139. Accessed 26 Nov 2014
17. Viola, F.C., Thorne, J., Edmonds, B., Schneider, T., Eichele, T., Debener, S.: Semi-automatic identification of independent components representing EEG artifact. Clin. Neurophys. **120**(5), 868–877 (2009)

Composing High Event Impact Resistible Model by Interactive Artificial Bee Colony for the Foreign Exchange Rate Forecasting

Pei-Wei Tsai[1,2], Li-Hui Yang[3], Jing Zhang[1,2], Yong-Hui Zhang[1,2], Jui-Fang Chang[3(✉)], and Vaci Istanda[4]

[1] College of Information Science and Engineering, Qingdao, China
peri.tsai@gmail.com, jing165455@126.com,
zyh@fjut.edu.cn
[2] Fujian Provincial Key Laboratory of Big Data Mining and Applications,
Fujian University of Technology, Fuzhou, China
[3] Department of International Business,
National Kaohsiung University of Applied Sciences, Kaohsiung, Taiwan
lohasgugu1988@yahoo.com.tw, rose@kuas.edu.tw
[4] Council of Indigenous Peoples, Executive Yuan, Taipei, Taiwan
biungsu@yahoo.com.tw

Abstract. Taiwan is an isolated island located in the south East Asia. Since Taiwan is lack of nature resources, thus, a huge part of the economy is export-oriented. To stimulate the economy to grow and activate the international trading, the Free Trading Agreement (FTA) is an activator to allow larger quantity of trading over the world. The foreign exchange rate plays the major role affecting the trade surplus in the export-oriented economic system. Hence, a stable and accurate foreign exchange rate forecasting model is important for the economic activity participants. In this paper, the event study method is used to examine 3 international trading related events including the Economic Cooperation Framework Agreement (ECFA), the Taiwan-Japan Bilateral Investment Arrangement (BIA), and the Agreement between Singapore and the Separate Customs Territory of Taiwan, Penghu, Kinmen and Matsu on Economic Partnership (ASTEP) signed between Taiwan and other participants. The foreign exchange rate forecasting models are built by the time-series methods and the computational intelligence method, namely, the Generalized Autoregressive Conditional Heteroscedasticity (GARCH), the Exponential Generalized Autoregressive Conditional Heteroscedasticity (EGARCH), and the Interactive Artificial Bee Colony (IABC), respectively. In the event study, the observation period is chosen to include 70 days for both pre/post-event. The Mean Absolutely Percentage Error (MAPE) value is used to examine the forecasting accuracy of the models. The experimental results indicate that the IABC constructed foreign exchange rate forecasting model is the most capable one to resist the impact caused by the specific events. In other words, the impact results in more significant forecasting error in the GARCH and the EGARCH models, but not in the IABC model.

Keywords: GARCH · EGARCH · IABC · Foreign exchange rate · Event study

© Springer International Publishing AG 2017
A.E. Hassanien et al. (eds.), *Proceedings of the International Conference on Advanced Intelligent Systems and Informatics 2016*, Advances in Intelligent Systems and Computing 533, DOI 10.1007/978-3-319-48308-5_73

1   Introduction

In economy fields, researchers use many tools for constructing models to simulate the economic conditions and generating the forecasting results. The most common used conventional tools is the time-series methods. The most popular time-series methods include the ARCH, the GARCH, and the EGARCH models. The time-series methods is capable to compose a model for the forecasting. The forecasting result is generated base on the historical data. In general situations, the outcome of the conventional forecasting model is easily affected by the incident or the event occurrence. In other words, when an event takes place, the forecasting result of the time-series model is disrupted and the forecasting result is with larger distortion than in the normal.

On the other hand, the fast development of the personal computer provides the very powerful computing power for people to do any calculation with computer in an easy way. The plentiful computing power provides the solid foundation for developing algorithms with automation. Thus, the computational intelligence is also raised in many industrial fields. In this paper, we use a newly invented Interactive Artificial Bee Colony (IABC) algorithm [1] to compose the forecasting model for the foreign exchange rate. The computational intelligence based foreign exchange rate forecasting model is compared with the conventional time-series models. We use the Mean Absolutely Percentage Error (MAPE) [2], which is proposed by Lewis in 1982, as the criterion to compare the accuracy of the forecasting results produced by different models. Three independent international trading related agreements are included as the studied event. The study subject is focus on the foreign exchange rate. The reason is that the foreign exchange rate plays an important role in the international trading. The more precise forecasting result brings in the more predictable trading surplus. The experimental results indicate that the computational intelligence based model is with higher resistance to the impact caused by the incident or the event occurrence.

The rest of the paper is composed as follows: the related literatures are briefly reviewed in Sect. 2, the experiment design is explained in Sect. 3, the experimental results and the discussion are given in Sect. 4, and the conclusion is made in the last section.

2   Literature Review

The basic assumption in time-series theory is that all current state of the data is somehow correlated with its historical data in different degrees. Based on this assumption, Engle (1982) proposed the Autoregressive Conditional Heteroskedasticity (ARCH) model [3]. He claims that the ARCH model is capable to be used in solving the problems with the volatility of time series. In the next few decades after ARCH is proposed, several scholars also propose the improved time-series models. For example, Bollerslev (1986) propose the Generalized Autoregressive Conditional Heteroscedasticity (GARCH) model [4] base on improving the ARCH model. The GARCH model is based on an assumption that the current conditional variance is a function of both the preceding period's conditional variance and the squares of the error terms, and the plus or minus sign of the error terms does not affect these conditional variances. Restrained by its own structure, the GARCH model is not able to depict the asymmetric feature shown by the volatility of the conditional variance of returns. Nevertheless, many distributions of returns are with the leptokurtic feature. In addition, residuals of stock market returns affect returns asymmetrically. It implies that the GARCH model is not able to explain the negative correlation between the stock returns and the changes in the volatility. In order to overcome the drawback in the GARCH model, Nelson proposes the Exponential Generalized Autoregressive Conditional Heteroscedasticity (EGARCH) model [5] in 1991. In short, the time-series models are widely used in the economy and finance analysis fields. Although they respectively have some drawbacks, they are the classical models for the economy related sequence in the forecasting.

To avoid a long lag in the conditional variance equation of the ARCH (p) model, Bollerslev (1986) [4] extends the ARCH model by leading the lagged conditional variances into the model construction. The extension made in the GARCH model makes the conditional heteroskedasticity of the current period affected by not only the conditional variances of last period, but also the average error of last period. The GARCH model can be described as follows:

$$\varepsilon_t = Y_t - x_t\alpha, \text{ where } Y_t|\Omega_t \sim N\left(x_t\alpha, \sigma^2\right) \tag{1}$$

$$\sigma_t^2 = \alpha_0 + \alpha_1\varepsilon_{t-1}^2 + \beta_1\sigma_{t-1}^2, \text{ where } \alpha_0 > 0, \ \alpha_1 \geq 0, \text{ and } \beta_1 \geq 0 \tag{2}$$

where Y_t is a function of exogenous variable x_t, α_0, α_1, and β_1 are constant coefficients, respectively, ε_t denotes the residual, σ_t^2 is the past squared residual of the past function, σ_{t-1}^2 is a condition of the variation in the past on a number j.

To further improve the performance of the GARCH model, Nelson (1991) proposes the EGARCH model, which can be described as follows:

$$Y_t = x_tB + \varepsilon_t, \text{ where } \varepsilon_t|\Omega_{t-1} \sim N(0, \sigma^2) \tag{3}$$

$$In\sigma_t^2 = \alpha_0 + \sum_{i=1}^{q}\left[\alpha_i\left(\left|\frac{\varepsilon_{i-i}}{\sigma_{t-i}}\right| - E\left|\frac{\varepsilon_{t-i}}{\sigma_{t-i}}\right| + \gamma\frac{\varepsilon_{t-i}}{\sigma_{t-i}}\right)\right] + \sum_{j=1}^{p}\beta_jin\sigma_{t-1}^2 \tag{4}$$

where σ_t^2 is a function of both the previous p period's conditional variance and the previous q period's squares of error terms. Moreover, σ_{t-j}^2 is the conditional variance of previous j periods, β_j is a fluctuation function of which reflects the past fluctuation, α_i denotes the fluctuation caused by the unexpected impacts, γ is the coefficient related to the asymmetric fluctuation, $\frac{\varepsilon_{t-i}}{\sigma_{t-i}}$ is the normalized residual, and p and q are the orders of the GARCH model, respectively.

Contrary to the conventional time-series model, the powerful computational power brought by the computers form a solid platform for developing the computational intelligence. The first Artificial Bee Colony (ABC) algorithm [6] is proposed by Karaboga in 2005. Four years later, Tsai et al. propose the IABC [1] algorithm by leading the universal gravitation concept into the algorithm. The IABC is originally proposed to solve numerical optimization problems. Nevertheless, by properly design the fitness function, the IABC algorithm is also capable to produce the forecasting result for the foreign exchange rate forecasting. In 2013, Chang et al. have utilized IABC to construct the foreign exchange rate forecasting model and obtained good results on the forecasting accuracy [7]. Our work is an extension on using the IABC model in the event study methodology. Assume we have i agents in a j dimensional space, and the solution matrix is denoted by X, the IABC algorithm can be depicted in 5 steps:

- Step 1. Initialization: Randomly spread the artificial agents into the solution space by Eq. (5):

$$X = min(X) + r[max(X) - min(X)] \tag{5}$$

where X is an i by j matrix, r is a random number of which $r \in [0, 1]$, and $min(\cdot)$ and $max(\cdot)$ denote the function extracting the minimum and the maximum value of X in all i rows, respectively.

- Step 2. Onlooker Bee Movement: There are two steps for moving the onlooker bees. Firstly, calculate the probability of food sources, which are provided by the employed bees, by Eq. (6).

$$P_i = \frac{F(\theta_i)}{\sum_{k=1}^{S} F(\theta_k)} \tag{6}$$

where θ_i is the position of the ith employed bee, S means the total number of the employed bees, and P_i stands for the probability for the onlooker bees choosing the ith employed bee to follows. Secondly, let N be a constant indicating the number of employed bees for an onlooker refers to in the movement process. The movement of the onlooker bee is calculated by Eq. (7):

$$\theta_q(t+1) = \theta_q(t) + \sum_{p=1}^{N} \tilde{F}_{qp} [\theta_q(t) - \theta_p(t)] \tag{7}$$

where θ_q denotes the position of the onlooker, t means the iteration number, $\tilde{F}_{qp}$ represents the normalized gravitation force, which is simply adopted from the probability of the food source, of the employed bee to the onlooker bee.

- Step 3. Scout Bee Movement: When an employed bee continuously has no follower onlooker beess in a predefined number of iteration, the employed bee becomes a scout and is moved by Eq. (8):

$$\theta_k(t+1) = \theta_{k_{min}}(t) + r[\theta_{k_{max}}(t) - \theta_{k_{min}}(t)] \tag{8}$$

where θ_k represents the position of the scout bee, r is a random number in the range of $[0, 1]$, and the *max* and the *min* denote the maximum and the minimum value occurs in all dimensions of θ_k, respectively.

- Step 4. Memory Update: If the newly found solution presents better fitness value than the original stored solution, replace the stored near best solution by the newly found one.
- Step 5. Termination Checking: Check whether the termination conditions are satisfied. If the answer is positive, output the latest near best solution and terminate the program; otherwise, go back to step 2 and repeat the process until the termination condition is satisfied.

3   Experiment Designs

3.1   Definition and Division of Events

This study aims to compare the exchange rate forecasting models adopted, after Taiwan signed FTAs such as ECFA, BIA, and ASTEP with other countries. Hence, the events are defined as the signing of these FTAs. The event periods, also called the event windows, need to be specified as well. An event period is a period of time during which an event may affect the dependent variable (foreign exchange rate). Naturally, the occurrence date of events should be included in the event window. In this study, such a date is each announcement date of the signing of the FTAs. Normally, besides the event date, an event period also includes a period of time before the event and that after the event. This phenomenon is due to the fact that post-event information reveals how the independent variable (foreign exchange market) has changed.

The reason to have an estimation period, also known as an estimation window, is that statistics during such a period can be used to estimate the value of the independent variable unaffected by any events. Such a value is referred to as the expected return. The abnormal return is the difference between the expected return and the actual return, the affected value of the independent variable. In this study, the pre-event window is defined as the 70 days prior to the events. On the contrary, the post-event window is the 70 days following the events. In addition, daily exchange rate data are adopted to build exchange rate forecasting models. The events included in this study are listed in Table 1 with brief introduction.

To find out which foreign exchange rate forecasting model presents the best predictive ability in the FTA events, we examine and compares the accuracy of the foreign exchange rate forecasting models adopted in this study under the effect of 3 FTA events. First, the time-series data is tested to see whether they are stationary by the two unit root tests, the augmented Dickey-Fuller (ADF) test and the Phillips-Perron (PP) test. Second, the Box-Jenkins approach is utilized to build an ARIMA model. The Akaike Information Criterion (AIC) value and the Schwarz Criterion (SC) value are the criteria for choosing lag periods. The Jarque-Bera (J-B) test is a goodness-of-fit test to see whether sample data have the skewness and the kurtosis that match a normal

Table 1. Events included in the analysis

No.	Event	Valid Date	Description
1	Economic Cooperation Framework Agreement (ECFA)	Sep. 12, 2010	The agreement is to enhance economic development and create open market
2	Taiwan-Japan Bilateral Investment Arrangement (BIA)	Jan. 01, 2012	The bilateral arrangement is to provide better investment environment for both
3	Agreement between Singapore and the Separate Customs Territory of Taiwan, Penghu, Kinmen and Matsu on Economic Partnership (ASTEP)	Apr. 19, 2014	The agreement represents the first signature around east-south countries and continues investing in the consumer welfare

distribution. Assume that the number of estimated parameters is n, and the total number of sample residuals is denoted by T. The J-B test result can be obtained by Eq. (9):

$$JB = \frac{T-n}{6}\left[S^2 + \frac{1}{4}(k-3)^2\right] \tag{9}$$

where S stands for the skewness and k denotes the kurtosis.

After creating the volatility models, the diagnostic test is conducted on the models to examine whether the standardized residuals already meet the requirement to be the white noise. Finally, the MAPE is employed as the criterion to analyze and compare the predictive ability of the foreign exchange rate forecasting models. Also, the Ljung-Box Q statistic is used to see whether these models are white noise:

$$Q(P) = n(n+2)\sum\nolimits_{k=1}^{P}\frac{1}{n-k_i^{p_i}} \sim \chi^2(P) \tag{10}$$

where n is the sample size and k denotes the lag periods. The statistic result is a chi-square distribution with p degrees of freedom (DOF).

Using the IABC algorithm to construct the foreign exchange rate forecasting model, a custom made fitness function for the training is necessary. The fitness function we use in the training process is revealed in Eq. (11):

$$\min f(W) = \sum\nolimits_{j=1}^{D}\left|\left(\sum\nolimits_{m=1}^{M} w_m \cdot v_{d,m}\right) - R_{actual,d}\right| \tag{11}$$

where $f(\cdot)$ indicates the fitness function, D denotes the total number of the reference days in the historical data, M is the total number of factors referenced in our forecasting model, w_m and $v_{d,m}$ are the weighting corresponding to the input data and the value of the referenced factors, respectively, and $R_{actual,d}$ stands for the actual foreign exchange rate on the day d.

The trained weighting is the basic component for producing the forecasting results by Eq. (12):

$$R_{f,d} = \sum_{m=1}^{M} w_m \cdot v_{d-1,m} \tag{12}$$

where $R_{f,d}$ denotes the forecasting outcome on day d.

Finally, the MAPE values of the foreign exchange rate forecasting outputs from different models are calculated by Eq. (13):

$$\text{MAPE} = \frac{1}{n} \sum_{t=1}^{n} \frac{\left| \widehat{S}_t - S_t \right|}{S_t} \times 100\% \tag{13}$$

where n denotes the sample size, $\widehat{S}_t$ indicates the expected exchange rate of period t, and S_t means the actual exchange rate of such a period.

4   Experiments and Experimental Results

For all target events, the event period is defined as 70 days before the event to 70 days after the event. Exchange rate forecasting models are built through the uses of time series-models (GARCH and EGARCH) and the IABC model. The input dataset is the foreign exchange rates collected from the historical data. The output of the GARCH and the EGARCH models are obtained by using the commercial models in EViews 8; and the IABC model is built with Matlab R2014b (8.4.0.150421). For training the IABC model, 32 artificial agents are used in the experiments with 600 iterations and 30 runs with different random seeds. The sliding window contains previous 7 trading-day data as the training input. The number of considered employed bee is set to 4, and the trained model, which produces the minimum forecasting error in the training phase is preserved as the final model for producing the MAPE in the testing phase with the test data. The historical TWD-USD foreign exchange rate used in the experiment is given

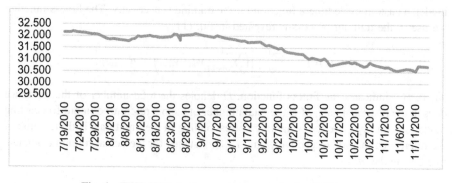

Fig. 1. TWD-USD foreign exchange rate in the ECFA event.

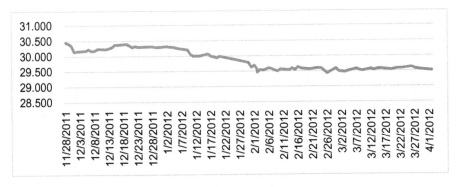

Fig. 2. TWD-USD foreign exchange rate in the BIA event.

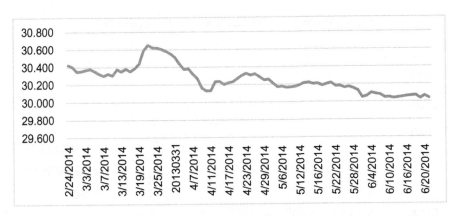

Fig. 3. TWD-USD foreign exchange rate in the ASTEP event.

in Figs. 1, 2 and 3. The exchange rate and the date are shown in the vertical and horizontal axis, respectively.

The MAPE value is calculated as the criterion for evaluating the predictive ability of the foreign exchange rate forecasting models. Based on the MAPE, the following figures present the predictive ability of these models. The results are presented in Figs. 4, 5 and 6. The vertical axis represents percentage changes in errors, and the horizontal axis denotes monthly averages.

The MAPE presents the distortion between two signals. Based on its definition, the smaller MAPE value implies the better forecasting quality. In the ideal case, the MAPE value should be 0. In Fig. 4, all of the GARCH, EGARCH and IABC models obviously fall within the standard range, suggesting that all these three models have a pretty good predictive ability. The IABC model has smaller MAPE values in all time intervals. This phenomenon means that it has an even better predictive ability that the time series models. The EGARCH model has smaller MAPE values than the GARCH model in August, September, October and November indicating that in these three

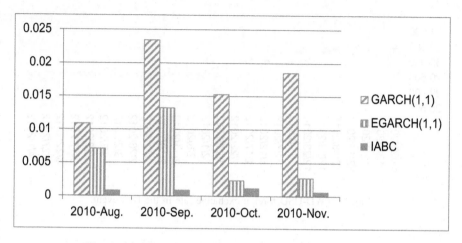

Fig. 4. MAPE values obtained in the event of signing ECFA.

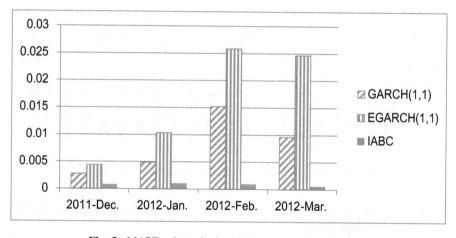

Fig. 5. MAPE values obtained in the event of signing BIA

months, the EGARCH model has better forecasting performance than the GARCH model.

In Fig. 5, the MAPE values obtained by the IABC model are stably with a very low value; the MAPE values obtained by the GARCH and the EGARCH models are also small because the maximum value is only a little exceed 0.025.

In Fig. 6, all of the GARCH and EGARCH models are within the standard range, indicating that these three models have a pretty good predictive ability. The IABC model has the smallest MAPE values in all time intervals. This phenomenon suggests that IABC model has the best predictive ability than the time series models. The GARCH model has smaller MAPE values than the EGARCH model in October, September, and December, indicating that in these three months, the GARCH model

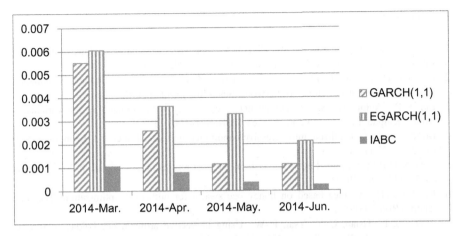

Fig. 6. MAPE values obtained in the event of signing ASTEP.

has better forecasting performance than the EGARCH model. On the contrary, the EGARCH model has smaller MAPE values in January, suggesting that in this month, the EGARCH model has better forecasting performance than the GARCH model.

5   Conclusions

In this paper, the event study method is used to examine 3 international trading related events including the ECFA, the BIA, and the ASTEP signed between Taiwan and other participants. In addition, the foreign exchange rate forecasting models are built by the time-series methods and the computational intelligence method, namely, the GARCH, the EGARCH, and the IABC, respectively. In the event study method, the observation period is chosen to include the 70-day pre-event and the 70-day post-event. The MAPE value is employed to examine the forecasting accuracy of the models. The experimental results indicate that the IABC constructed foreign exchange rate forecasting model is the most capable one to resist the impact caused by the specific events. In other words, the impact results in more significant forecasting error in the GARCH and the EGARCH models, but not in the IABC model. In the future work, we plan to use a wider diversity of foreign exchange rate forecasting models to study which model has the best predictive ability and the higher resistance to the impact caused by the incidents.

Acknowledgement. This work is partially supported by the Key Project of Fujian Education Department Funds (JA15323), Fujian Provincial Science and Technology Project (2014J01218), Fujian Provincial Science and Technology Key Project (2013H0002), and the Key Project of Fujian Education Department Funds (JA13211). The authors also gratefully acknowledge the helpful comments and suggestions from the reviewers, which have improved the presentation.

References

1. Tsai, P.-W., Pan, J.-S., Liao, B.-Y., Chu, S.-C.: Enhanced artificial bee colony optimization. Int. J. Innovative Comput. Inf. Control 5(12), 5081–5092 (2009)
2. Lewis, C.D.: Industrial and Business Forecasting Methods: A practical Guide to Exponential Smoothing and Curve Fitting. Butterworth Scientific, London (1982)
3. Engle, R.: Autoregressive conditional heteroscedasticity with estimates of the variance of United Kingdom inflation. Econometrica 50(4), 987–1007 (1982)
4. Bollerslev, T.: Generalized autoregressive conditional heteroscedasticity. J. Econometrics 31 (3), 307–327 (1986)
5. Nelson, D.B.: Conditional heteroscedasticity in asset returns: a new approach. Econometrica 59(5), 347–370 (1991)
6. Karaboga, D.: An idea based on honey bee swarm for numerical optimization, Technical report TR06 (2005)
7. Chang, J.-F., Hsiao, C.-T., Tsai, P.-W.: Using interactive artificial bee colony to forecast exchange rate. In: Proceedings of 2013 the Second International Conference on Robot, Vision and Signal Processing, pp. 133–136 (2013)

Distributed Influential Node Detection Protocol for Mobile Social Networks

Doaa AbdelMohsen$^{(\boxtimes)}$ and Mohammed Hamdy

Information Systems Department, Ain Shams University, Cairo, Egypt
{doaamohsen, m.hamdy}@cis.asu.edu.eg

Abstract. Mobile social networks enable users to share their common favorites as they are connected together using their mobile devices. These social networks are deployed on set of different network technologies like MANETs. Mobile social network success depends on the number of its users and their activities. However, this network may contain a large number of connected users, only a few numbers of them can be detected as influential users. The influential user is a user that has a significant effect or influence on other users. In this paper, a proposed protocol that runs fully distributed and can efficiently detect the influential users in mobile social networks. Mostly, the literature provides protocols that use centralized data processing to detect these users based on either friend relations or social-network graphs of all users. But, these approaches are resource exhausting and not scalable. To the best of our knowledge and according to the literature, the protocol can precisely detect influential users or nodes in a mobile social network. Results of the protocol performance analysis has been investigated and presented with a set of significant and promising contributions.

Keywords: Mobile social network · Influential · MANET

1 Introduction

A mobile Ad-hoc Network (MANET) is connected mobile devices are characterized by being autonomous. These devices are connected together by wireless links. Social networks are a kind of MANETs that connects users with others by common interests. Mobile users can socially communicate through their wireless devices; this is called mobile social network. An influential user is a person that has an ability to affect other users in the mobile social network. Dissemination of information among nodes within the network is one of the most important problems in the mobile Social network. To improve this feature, some protocols look for the influential user in the network.

There are many applications use the influential detection such as, (e.g., [1–3]). Mobile users may not similar in level of mobility. Some of the protocols use active users (high mobility level) as an initial user to improve information dissemination in mobile social networks, as they will be able to transmit messages to a larger number of users, rather than select it randomly. The problem is not only identifying the influential users but also the problem is to identify them centrally. Centralized detection of influential users causes a bottleneck in the network because of the high traffic and depends on the network connectivity. Also, most of all it requires complete social-contact graphs.

A.E. Hassanien et al. (eds.), *Proceedings of the International Conference on Advanced Intelligent Systems and Informatics 2016*, Advances in Intelligent Systems and Computing 533, DOI 10.1007/978-3-319-48308-5_74

There are two challenges when identifying these influential mobile users. First, mobile social networks are large-scale, and developing techniques require distributed way. Second, the power consumption in smart phones is limited.

Information dissemination is one of the most important challenges in the mobile social network. Identifying the influential nodes is a proposed solution for efficient dissemination. User behaviours help in determining the way of information dissemination in mobile social networks.

In this paper, we detect influential users in mobile social networks through a fully distributed technique. The influential users are defined as the users who have a high centrality in their social contact graphs. The proposed protocol aims to detect influential users in mobile social networks by running it in a background of each Smartphone. The proposed protocol periodically performs random walks by a few number of smart phones, and calculate the centrality of users by their random walk counters. The performance of the proposed protocol is evaluated through the simulation scenarios.

The remainder of this paper is organized as follows. The work related to the problem is presented in Sect. 2. Section 3 explains and discusses the proposed work. The simulation environment and results analysis are presented in Sect. 4. Finally, Sect. 5 has the paper conclusion.

2   Related Work

Identifying influential members is studied for information dissemination in social networks [4, 5, 8]. In [4, 8], the authors propose a protocol to solve problem of information dissemination by maximizing the dissemination of information in a social network using initial set of k users. They simulate their data as social network and present it as Markov random field. The set of k users used are obtained using collaborative filtering database.

The term random walk is introduced for the first time by [9]. On graphs, random walks are described as a user's walks from a source node to a destination node and in every step in his route; the user selects the next node to visit randomly from the neighbours of the current node. To calculate centrality in social networks random walks have been used. The random-walk centrality is proposed in [10], authors calculate how often a node has been visited in a graph by using all possible pair of nodes compared with all nodes in a network. The random-walk closeness centrality measurement is presented in [11], which measures the time that the random walk message is received in the same network by a node from other nodes.

In [6], the authors investigate how to maximize the opportunity of communication among the mobile users by using k number of users as a target set from all network users. A framework is proposed to search for these users through mobile social networks. They present a framework to update the community structure by history information. They also show that the performance of advanced protocols in delay tolerant networks can be improved by their framework. [7], a study of evaluating the several centrality measures performance in the network for identifying high-risk of infection with different degrees in an isolated network, random-walk and shortest-path [12]. Their results illustrate that in the risk prediction of infection the degree implements highly related to other network measures.

In [13], authors propose Sybil Guard protocol that uses a certain type of random walk to restrict the negative impact of Sybil attacks which every node selects randomly the next hop based on a pre-calculated permutation. Information dissemination is one of the important applications of mobile networks. Authors in [14], propose information dissemination in a peer-to-peer system, for mobile users to improve availability of data. This approach can help mobile devices to query on data from their neighbours when they can't access the internet with their connection. A protocol, called (SWIM) the Shared Wireless Infostation Model is presented in [15], that aims to increase the dissemination of information within a network is among mobile users. A system of radio-tagged whales is used to evaluate the performance of SWIM. Authors in [16], propose a technique to choose the best sources for content sharing among the mobile users. In [17], authors study the human interactions that may affect the performance of information dissemination in mobile social networks. They find the information dissemination depends on the devices' social status and the number and density of devices in the network.

The proposed researches in mobile social network detect the influential users using centralized protocols or using a complete social graph. In this paper, a protocol proposed to identify influential mobile users for enhance information dissemination through a fully distributed design.

3 Network Model

The paper objective is to detect the influential users on one partition of the mobile social network. This partition is a simulated partition. To maintain a real partition of MANET is too hard. So, a set of features which can form this "artificial partition is assumed". The assumed network model at a certain time is an undirected, unweight graph G (N, E) where N represents the set of nodes and E is the set of edges. G_x (N_x, E_x) represents one of the network partitions, where: G (N, E) = G_1(N_1, E_1)...∪ G_x(N_x, E_x)...∪G_k(N_k, E_k), N = N_1...∪N_x...∪N_k, and E = E_1...∪E_x...∪E_k. The network topology is varying according to the node movements. Each node covers a fixed radius of radio transmission range R, where R is constant over the network operation time and the participating mobile nodes should be uniquely identified.

4 Distributed Influential Node Detection Protocol

In this section, details on design and implementation of the proposed protocol are introduced and discussed.

4.1   Random Walks

Random walks are described as a user's walks from a source node to a destination node and in every step in his route; the user selects the next node to visit randomly from the neighbours of the current node [9]. Random walks are used to design a fully distributed protocol, for detecting influential users in mobile social networks. If periodically, random walks are initialized for a few number of smart phones, Influential users probably are visited by these random walks extremely than any normal user. The proposed protocol works as following. In Each t minute, a small test message on each user. This message contains Time-To-Live (TTL) field that pre-configured in the program. When Smartphone contacts its neighbour, if test message is found in its own queue, it will send it to its neighbour. When a test message is received, the Smartphone decreases its TTL by 1, and then push it in its queue, then waits for a contact to send the message to another neighbour. If a test message with t = 0 is found it will be discarded. The technique counts the random-walk in each Smartphone to record the times it has been visited by the others. The probability that influential users are visited is high based on the paradox of friendship. Random-walk counters are updated and reset in the upper layer applications. The proposed protocol performance depends on two parameters: L random walks length (i.e., how many users are visited by one random walk), and t the frequency of new test messages generation. The number of test messages traveling through the network and the accuracy of detected influential users are two important factors in reducing power consumption. So that short length of random walks is preferred.

4.2   Clustering with DB-Scan

Each node receives a data item from another node; this data item may be a disease so this node will be infected Fig. 1 showed the node status. This infected node sends data items to its neighbours within the same cluster and this may cause infection for the whole cluster.

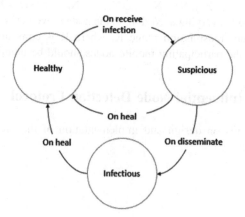

Fig. 1. Node status

Clusters are formed as a result of mobility. We propose to identify the infected clusters and the isolated one. DB-Scan (Density-based spatial clustering of applications with noise) [18] is not invited in ad-hoc networks clustering because of its centrality but it is invited now. DB-Scan is used to find the physical partitions which have a proper intense of number of nodes. If there are a group of nodes in some area, DB-Scan groups nodes together which are density-reachable, the nodes lie alone in low-density areas are marked as outliers' nodes. DB-Scan indicates easily the physical clusters and efficiently ignores the noise. Using DB-Scan, Clusters can easily be built with fully distributed way.

4.3 Election Mechanism

Once, the physical network proper partitions (clusters) have been discovered using DB-Scan, the random-walks of all users in these clusters is recorded. After time interval (T), these counters are shared within neighbourhoods. Then, based on a common sorted counters' list, the highest k counters should be elected by the neighbours.

4.4 Information Dissemination

In this work, since, detection of the influential users is proposed based on using the random walk counters without any global network structure, a fully distributed protocol for the Information dissemination problem is introduced. Mobile users run the proposed protocol periodically to share their random walk counters with other users. Based on the election method we can detect the k- influential users in each cluster. In this scenario, these influential mobile users can support propagating information after the moment they receive it. Social relations play a key role in data dissemination. The information dissemination is a challenging process. There are several factors can affect this process, such as the user privacy concerns and behaviour (status) like being busy with some work. All of those users may not help in information dissemination operation.

We evaluate the performance with different mobility levels and generate random walk test messages with different frequencies as well. Figure 2 shows protocol steps described as following:

Algorithm 1. Protocol Steps

```
Build clusters using DB-Scan
  ForEach Cluster do
    Elect cluster head using random walks
  END Foreach
  Disseminate head as influential
```

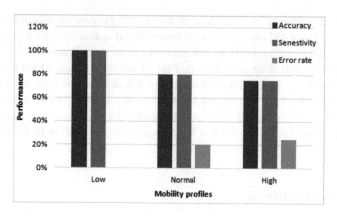

Fig. 2. Impact of changing mobility on protocol performance.

5 Performance Evaluation

An extensive simulation has been established to evaluate the proposed protocol. The following section introduces simulation setup, simulation results and analysis sections.

5.1 Simulation Setup

To evaluate the proposed protocol, the simulator is built using C# to simulate the social mobile network. A network of mobile devices clustered using DB-Scan in a university campus of maximum area 1000 m^2 is assumed. Nodes (users) move in the area according to random walk mobility model. Messages are generated at random source nodes to random destination nodes. The simulator runs for 10 h, each of results are average of 10 runs.

The protocol independent parameters are set to be L = 10, the random walk length, and t = 10 s where t is the frequency of sending test messages. Tables 1, 2 and 3 mentioned the values of each independent parameter in each scenario. Three experiment scenarios are conducted to show the performance of the proposed protocol. The following dependent performance metrics are introduced to reflect the performance of the proposed protocol:

Table 1. Network parameters for scenario 1.

Parameter	Value
Mobility of nodes (M)	Low- Normal - high
Number of nodes (N)	300
Network area (A)	1000 m2

Table 2. Network parameters for scenario 2.

Parameter	Value
Walk Length (L)	Short-Normal- Long
Number of nodes (N)	300
Network area (A)	1000 m2

Table 3. Network parameters for scenario 3.

Parameter	Value
Walk Length (L)	0–10 h
Number of nodes (N)	300
Network area (A)	1000 m2

$$\text{Sensitivity } (S) \text{ which, } S = \frac{TP}{P} \tag{1}$$

$$\text{Error Rate } (E) \text{ which, } E = \frac{FP + FN}{P + N} \tag{2}$$

$$\text{Accuracy } (A) \text{ which, } A = \frac{TP + TN}{P + N} \tag{3}$$

The proposed protocol can be presented as a detector. Since, detection is a classification process, the proposed Distributed Influential Node Detection Protocol can be considered as a classifier. To measure the performance of this classifier, Eqs. 1, 2 and 3 are used to measure the classifier sensitivity, error rate, and accuracy [19]. Because of the number of influential nodes is a very small proportion to the number of clusters the sensitivity measurement is used besides the accuracy of the classifier. True positives (TP) refer to the positive influential nodes that were correctly detected by the protocol. True negatives (TN) are the negative influential nodes that were correctly detected by the protocol. False positives (FP) are the negative influential nodes that were incorrectly detected as positive. False negatives (FN) are the positive influential nodes that were mislabeled as negative. P is all influential nodes that correctly detected whether positive or negative so, P = TP + FN and N is all influential nodes that incorrectly detected whether positive or negative so, N = FP + TN.

In the next section, the simulation scenarios are displayed in details, together with the results obtained in each case, and the discussion of the results.

5.2   Simulation Scenarios and Results

Three scenarios are used to measure the proposed protocol performance. The first scenario is used for ensuring that the protocol works accurately with all mobility profiles. Walk length is an important parameter in the proposed protocol so the second scenario shows that the greater walk length the greater the accuracy of the detector. The last one is used to prove that the protocol works accurately in an earlier time of the simulation as the proposed protocol works as an epidemical model.

Scenario 1: Impact of nodes mobility (M):. In this scenario, the mobility of nodes is varied from low to high by changing their velocity. As shown in Fig. 2, the accuracy and sensitivity of the proposed protocol in influential detection are found to be increasing but the error rate is decreasing with decreasing the mobility of nodes on the same network area. Increasing the mobility of nodes means that nodes become moving faster. Therefore, nodes can move from cluster to another repeatedly, and test messages become in large. So, the proposed protocol can't scan and detect accurately. This means that the error rate will increase.

Scenario 2: Impact of Walk Length (L):. In this scenario, the walk length of test messages is changed. As shown in Fig. 3, the accuracy and sensitivity of the proposed protocol in influential detection are found to be increasing and the error rate is decreasing with increasing the walk length of test messages. Increasing the walk length, means that message reach the largest number of nodes. Therefore, nodes can learn more from this messages who is the influential one in its cluster. So, the protocol is able to scan and detect with high accuracy and sensitivity. This means that the error rate will decrease.

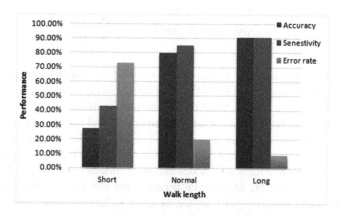

Fig. 3. Impact of changing walk length on protocol performance.

Scenario 3: Accuracy, Sensitivity, and Error rate through time (T):. In this scenario, the parameters accuracy, sensitivity and error rate are measured through time. The proposed protocol works as an epidemical model so we have to be sure that the proposed protocol works accurately through time. Figure 4 shows the parameters value through the whole simulation time (long term operation) after 45 s the parameters value didn't change until the end of the simulation. Figure 5 shows the short term operation from time 0 to 45 s, accuracy and sensitivity increased from 0 to 98 % through time so error rate in the beginning of simulation found to be 100 % and decreased through time that happened because nodes learn who are the influential nodes in their cluster along the simulation time using test messages disseminated in the network.

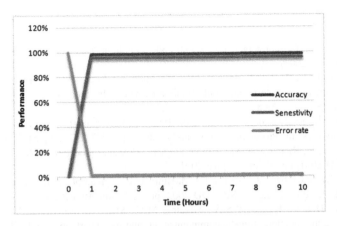

Fig. 4. Protocol performance through time (long term operation).

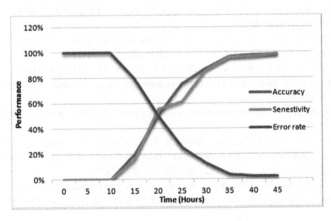

Fig. 5. Protocol performance through time (short term operation).

6 Conclusion

In this paper, we propose a fully distributed protocol to detect influential mobile users. The proposed protocol uses random walks to calculate the influential level of users based on the frequency of walks that visit their smart phones. We evaluate the performance of the proposed protocol using simulation scenarios for three standard measurements error rate, sensitivity, and accuracy. The simulation results show that the proposed protocol performs very close to the ideal measurements. For the information dissemination application, the proposed protocol performs better than random selection for users with low centrality to propagate information through the network.

References

1. Kleinberg, J.: The convergence of social and technological networks. Commun. ACM **51** (11), 66–72 (2008)
2. Gaonkar, J., Li, R.R., Choudhury, L.C., Schmidt, A.: Micro-blog: sharing and querying content through mobile phones and social participation. In: Proceedings of MobiSys 2008, pp. 174–186, June 2008
3. Motani, M., Srinivasan, V., Nuggehalli, P.S.: PeopleNet: engineering a wireless virtual social network. In: Proceedings of MobiCom 2005, pp. 243–257, August-September 2005
4. Domingos, P., Richardson, M.: Mining the network value of customers. In: Proceedings of SIGKDD 2001, pp. 57–66, August 2001
5. Kempe, D., Kleinberg, J., Tardos, E.: Maximizing the spread of influence through a social network. In: Proceedings of SIGKDD 2003, pp. 137–146, August 2003
6. Han, B., Hui, P., Kumar, V.S.A., Marathe, M.V., Shao, J., Srinivasan, A.: Mobile data offloading through opportunistic communications and social participation. IEEE Trans. Mob. Comput. **11**(5), 821–834 (2012)
7. Nguyen, N.P., Dinh, T.N., Tokala, S., Thai, M.T.: Overlapping communities in dynamic networks: their detection and mobile applications. In: Proceedings of MobiCom 2011, pp. 85–95, September 2011
8. Richardson, M., Domingos, P.: Mining knowledge-sharing sites for viral marketing. In: Proceedings of SIGKDD 2002, pp. 61–70, July 2002
9. Pearson, K.: The problem of the random walk. Nature **72**(1865), 294 (1905)
10. Newman, M.E.: A measure of betweenness centrality based on random walks. Social Netw. **27**(1), 39–54 (2005)
11. Noh, J.D., Rieger, H.: Random walks on complex networks. Phys. Rev. Lett. **92**(11), 118701 (2004)
12. Christley, R.M., Pinchbeck, G.L., Bowers, R.G., Clancy, D., French, N.P., Bennett, R., Turner, J.: Infection in social networks: using network analysis to identify high-risk individuals. Am. J. Epidemiol. **162**(10), 1024–1031 (2005)
13. Yu, H., Kaminsky, M., Gibbons, P.B., Flaxman, A.: SybilGuard: defending against sybil attacks via social networks. In: Proceedings of SIGCOMM 2006, pp. 267–278, September 2006
14. Papadopouli, M., Schulzrinne, H.: Effects of power conservation, wireless coverage, and cooperation on data dissemination among mobile devices. In: Proceedings of MobiHoc 2001, pp. 117–127, October 2001
15. Small, T., Haas, Z.J.: The shared wireless infostation model - a new ad hoc networking paradigm (or where there is a whale, there is a way). In: Proceedings of MobiHoc 2003, pp. 233–244, June 2003
16. McNamara, L., Mascolo, C., Capra, L.: Media sharing based on colocation prediction in urban transport. In: Proceedings of MobiCom 2008, pp. 58–69, September 2008
17. Zyba, G., Voelker, G.M., Ioannidis, S., Diot, C.: Dissemination in opportunistic mobile ad-hoc networks: the power of the crowd. In: Proceedings of INFOCOM 2011, pp. 1179–1187, April 2011
18. Ester, M., et al.: A density-based algorithm for discovering clusters in large spatial databases with noise. In: Kdd, vol. 96, no. 34 (1996)
19. Han, J., Pei, J., Kamber, M.: Data Mining: Concepts and Techniques. Elsevier, Amsterdam (2011)

Enhanced Algorithms for Fuzzy Formal Concepts Analysis

Ebtesam E. Shemis$^{(\boxtimes)}$ and Ahmed M. Gadallah$^{(\boxtimes)}$

Institute of Statistical Studies and Research, Cairo University, Giza, Egypt
Eptesam.elhossiny@hotmail.com, ahmgad10@yahoo.com

Abstract. Fuzzy formal concept analysis (FFCA) is a generalized form of traditional formal concept analysis (FCA) that exploits fuzzy set theory to process uncertain data efficiently. Generally, most real world applications incorporate uncertain data at least for some extent. Consequently, they need reliable approaches to discover potentially useful nontrivial knowledge. Commonly, FFCA aims mainly to reach such knowledge in form of fuzzy formal concepts. It is used widely in data analysis tasks, association rule discovery and extraction of essential ontology components. This paper proposes two enhanced algorithms for extracting fuzzy formal concepts based on fuzzy sets of objects and crisp sets of attributes. Such kind of FFCA best suits Ontology construction and association rule mining tasks. Commonly, extracting fuzzy concepts is considered the most time consuming process in FCA and FFCA. So, the proposed enhanced algorithms aim mainly to reduce the complexity and extraction time of fuzzy formal concepts' extraction process. The first enhanced algorithm best fits in case of the existence of symmetric correlated attributes. On the other hand, the second enhanced algorithm generally reduces the complexity as a result of reducing total number of generated fuzzy concepts. It works extremely better when the number of distinct intents of objects is relatively smaller. The results of testing the proposed enhanced algorithms show their added value.

Keywords: Fixpoint · Formal concept · Fuzzy concept · Formal concept analysis · Fuzzy formal concept analysis · Conceptual scaling and knowledge extraction

1 Introduction

Generally, formal concept analysis (FCA) is a branch of lattice theory that has been developed since 1980-ies [10]. It is a field of applied mathematics based on concepts and concept hierarchies [5,6]. In consequence, FCA has been exploited by various kinds of applications such as linguistics, information retrieval [1], economics and much more [7]. Commonly, classical FCA analyzes data in form of

© Springer International Publishing AG 2017
A.E. Hassanien et al. (eds.), *Proceedings of the International Conference on Advanced Intelligent Systems and Informatics 2016*, Advances in Intelligent Systems and Computing 533, DOI 10.1007/978-3-319-48308-5_75

binary formal context that describes binary relationship among a set of objects and a set of attributes. Such context is based on traditional crisp set theory where an element either belongs fully to a set (take a membership value 1) or does not belong at all (take a membership value 0). FCA is used to generate set of formal concepts, also called fixpoints [10,13]. Afterward, it builds the corresponding conceptual hierarchy known as formal concept lattice. Concept lattice represents the sub-concept/super-concept relationship (order relations) among the set of generated formal concepts. For association rule mining, FCA can efficiently produce attribute implications that help in extracting hidden association rules. Commonly, classical FCA provides only crisp scaling method to deal with multi-valued context that is not appropriate for human reasoning. The problem of crisp scaling is that it mainly depends on dividing attribute domain into several intervals. And hence, suffers from the problem of crisp boundaries. So it is hard to decide the boundaries between scaled attributes' intervals [6]. This problem can be perfectly tackled using the fuzzification process for scaling multi-valued attributes to fuzzy sets. Consequently, FFCA faces the limitations of classical FCA. Fuzzy set theory and fuzzy logic are able to handle imprecise data such that an element can belong to a set to some extent with a membership value in the range of [0, 1] [8,9]. FFCA can be widely utilized in fuzzy association rules' generation as it can be used to generate fuzzy rules base [27].

This paper contributes to the family of FFCA algorithms. It proposes two efficient enhanced algorithms for extracting fuzzy formal concepts from a fuzzy context. It is well known that extracting set of all formal concepts from large contexts is a time consuming problem whose counting number problem is $\#P$-complete [14]. But fortunately, if the relation (I) between object set(X) and attribute set(Y) is considerably small, one can get set of all formal concepts in a reasonable time even if $|X|$ and $|Y|$ are large [15].

The rest of this paper is organized as follows: Sect. 2 presents the foundations of formal concept analysis. Fuzzy formal concept analysis is addressed in Sect. 3. Section 4 introduces some related works. The proposed algorithms to extract fuzzy formal concepts are presented in Sect. 5. Consequently, Sect. 6 addresses some experiments for testing and evaluating the proposed algorithms. Finally, the conclusion is presented in Sect. 7.

2   Formal Concept Analysis

This section introduces FCA basic notions, more details are found in [2,5,6, 10]. Commonly, formal concept analysis aims to discover underlying clusters of objects and attributes within a dataset [16]. It accepts data inputs in the form of object-attribute values known as formal context. A formal context is a representation of a binary relation I between objects G and their attributes M. It can be represented as cross table of object rows, and column attributes (or vice versa). The intersected cell between each object g and attribute m has cross symbol only if $(g, m) \in I$ (donated as gIm). Given $A \subseteq G$ and $B \subseteq M$, the derivation of A and B is given by (1) and (2):

$$A \uparrow := \{m \in M \mid (gIm) \ \forall g \in A\}. \tag{1}$$

where $A \uparrow$ represents common attributes in M shared by all objects of A.

$$B \downarrow := \{g \in G \mid (gIm) \ \forall m \in B\}. \tag{2}$$

where $B \downarrow$ represents all objects that share all attributes of B.
Alternatively, $A \uparrow$ and $B \downarrow$ can be donated as A' and B' respectively.

Definition 1. *(A, B) is a formal concept of the context (G, M, I) iff $A \subseteq G$, $B \subseteq M$, $A' = B$ and $B' = A$. Where A and B are concept extent and intent respectively.*

Let (A_1, B_1) and (A_2, B_2) be formal concepts, then (A_1, B_1) is $\leq$ (A_2, B_2) if $A_1 \subseteq A_2$ and $B_2 \subseteq B_1$. The set of all concepts partially ordered by this relation is called concept lattice.

Traditional FCA can only deal with binary context but usually the values of a dataset attribute are within a specified domain of values which may be a range of values. In such case, a many-valued context should be handled. Rationally, to use FCA in such situation, conceptual scaling can be used to transform such context into binary one which unfortunately leads to the crisp boundaries problem.

Definition 2. *A many-valued context (G, M, V, I) is composed of G objects, M attributes, V attribute values and a ternary-relation $I \subseteq G \times M \times V$. An element of I, $(g, m, w) \in I$ indicates that the attribute m has the value w for the object g.*

3 Fuzzy Formal Concept Analysis

Generally, there are multiple points of view for FFCA and hence multiple basic definitions' sets. This section introduces some basic notions and definitions of FFCA [12,17]. In a fuzzy formal context $\Bbbk(G, M, I = \varphi(G \times M))$, I is a fuzzy set on domain of $G \times M$ such that each relation $(g, m) \in I$ has a membership value $\mu(g, m)$ in [0, 1]. A confidence threshold is an interval of $[\alpha_1, \alpha_2]$ where $0 \leq \alpha_1 \leq \alpha_2 \leq 1$ and is applied to fuzzy context $\Bbbk(G, M, I)$ to eliminate relation (g, m) if its membership value $\mu(g, m)$ is out of confidence threshold interval.

Definition 3. *A fuzzy formal concept of a fuzzy context $\Bbbk(G, M, I)$ is a pair $(A_f = \varphi(A), B)$ where $A \subseteq G$ is a fuzzy concept extent, $B \subseteq M$ is a fuzzy concept intent and $A \uparrow = B$ and $B \downarrow = A$. There are multiple definitions for $A \uparrow$ and $B \downarrow$. The first definition is given by (3) and (4) [11, 18, 19].*

$$A \uparrow := \{b \in B \mid \forall a \in A : \mu_I(a, b) \geq \mu_A(a)\}. \tag{3}$$

where $A \uparrow$ is the derivation of object set A and represents the concept intent.

$$B \downarrow := \{a / \mu_A(a) \mid \mu_A(a) = min_{b \in B}(\mu_I(a, b))\}. \tag{4}$$

where $B \downarrow$ is the derivation of attribute set B and represents the concept extent.

Another definition that uses an α threshold is presented in [2,17]. This definition is given by (5) and (6).

$$A \uparrow := \{m \in M \mid \forall g \in A : \alpha_1 \leq \mu_I \leq \alpha_2\}. \tag{5}$$

$$B \downarrow := \{g \in G \mid \forall m \in B : \alpha_1 \leq \mu_I(g,m) \leq \alpha_2\}. \tag{6}$$

where $A \uparrow = B$ represents the fuzzy concept intent and $B \downarrow = A$ represents the fuzzy concept extent.

4   Related Works

Generally, most algorithms related to fuzzy concepts' extraction fall under one of the following categories: crisply generated fuzzy concepts, fuzzy concepts with fuzzy extents/crisp intents, fuzzy concepts with crisp extents/fuzzy intents and fuzzy concepts with fuzzy extents/fuzzy intents. Most algorithms that generate fuzzy concepts take the benefit of utilizing current existing FCA algorithms to generate fuzzy concepts. These algorithms mainly transform the fuzzy context into an isomorphic crisp one [23]. Such transformation process varies from one approach to another. Some examples of the algorithms that follow this category can be found in [6,12,20].

Commonly, it is noted that the count of crisply generated fuzzy concepts is considerably less than the count of all fuzzy concepts (with fuzzy extents/fuzzy intents). Although transforming fuzzy context into binary one allows to use existing crisp algorithms, there is still an extra cost involved due to: (1) extra processing for transforming the context to crisp one and the subsequent concepts' conversion to fuzzy ones as well as (2) expanding the context results in increasing objects' number and hence significant increase in computation [12]. The algorithm presented in [6] transforms fuzzy context to corresponding crisp one by setting maximum membership value per linguistic variable to 1 and others to 0. Then it uses any traditional algorithm to get whole crisp intents set. Finally it fetches the corresponding fuzzy objects by getting intent↓ from the original fuzzy context. Approach presented in [12] transforms the fuzzy context into corresponding binary one by pair each object in fuzzy context with each non-zero membership value $\mu_I(O_i, b_j)$ and creates a new crisp object $O_i/\mu_I(O_i, b_j)$. Then it uses any traditional concept generation algorithm to generate all possible formal concepts. Finally, it converts the crisp objects back to fuzzy ones and removes redundant objects by selecting the largest membership value.

Additionally, some other algorithms that generate fuzzy concepts with fuzzy intents/fuzzy extents can be found in $B\check{e}loh\grave{a}vek$'s methods presented in [21, 22,24,25]. $B\check{e}loh\grave{a}vek$'s methods are based on the fact that the relation between fuzzy subsets is strongly linked to the implication notion. In consequence, one can formulate FFCA in terms of fuzzy algebras. In short, this approach describes the complete lattice L as following: L $= \langle L, \vee, \wedge, \otimes, \rightarrow, 0, 1 \rangle$ where $\langle L, \vee, \wedge, 0, 1 \rangle$ is a complete lattice, $\langle L, \otimes, 1 \rangle$ is an abelian monoid, $\rightarrow, \otimes$ are operations that form an adjoint pair (meaning, $a \otimes b \leq c \Leftrightarrow a \leq b \wedge c$). The set of all fuzzy sets in universe X is denoted by L^X where $A : X \rightarrow L$ is a mapping that assigns every $x \in X$ a truth value A(x) $\in$ L. In this approach, the number of generated fuzzy concepts depends mainly on the number of distinct membership values produced by the used implication function (Lukasiewicz/Gödel implications). Such kind of algorithms perfectly generates all possible fuzzy concepts but it generates inordinately large number of concepts so it consumes larger amount of time and hence is of limited practical use. Consequently, recent algorithms are proposed to reduce number of generated fuzzy concepts such as [3,4].

It can be noted that the algorithms designed to extract fuzzy concepts with crisp intent/fuzzy extent or fuzzy intent/crisp extent, usually produce relatively smaller number of fuzzy concepts in a reasonable time compared with full fuzzy concepts. These algorithms mainly reduce the complexity. Yet, they suffer from ignoring one fuzzy side of the concept (intent or extent). Some examples of such algorithms are presented in [2,6,12].

5 The Proposed Algorithms to Extract Fuzzy Formal Concepts

In this section, two algorithms are proposed to extract fuzzy formal concepts in terms of fuzzy extents and crisp intents. Both algorithms take a raw input context and a threshold interval as inputs and filter the input context with conditions on the membership values while processing. Consequently, they don't waste time to convert the entire context to the filtered one as a separate process that has complexity of O (N×M) where N and M are the counts of rows and columns respectively. In both proposed algorithms, the set operations performed on the extents are fuzzy set operations introduced by Zadeh [8] while the set operations performed on intents are traditional set operations.

The first proposed Algorithm 1 computes the extent membership value of each concept using the minimum function as described previously in (4). In consequence, it extracts intents using (3). It is based mainly on the attribute set and works more efficiently in case of existence of redundant attributes (more symmetric correlated attributes), otherwise it becomes much closer to the recent fuzzy CbO algorithm [12].

Algorithm 1. Attribute set based fuzzy concept extraction algorithm

---

Input: A fuzzy formal context $\Bbbk := (G, M, I = \varphi(G \times M))$ and a confidence threshold $T=[t_1, t_2]$.
Result: Set of fuzzy formal concepts C
initialize $C \longleftarrow \phi$;
for *each attribute* $i \in I$ **do**
    initialize $Cnew \longleftarrow \phi$;
    Get i' // each element in i' is a fuzzy extent $(O_k, Min_\mu(O_k, i))$;
    Set $New \longleftarrow 0$;
    if $i' \notin C.extent$ **then**
        $C \longleftarrow C \cup ((i', \mu_{i'}), i'')$;
        for $j=0$ to $(C.size - New)$ **do**
            $Inters \longleftarrow c_j.extent \cap i'$;
            if $Inters \notin C.extent$ **then**
                $C \longleftarrow C \cup (inters, inters')$;
                New++;
            end
        end
    end
end
if $M \notin C.intents$ **then**
    $C \longleftarrow C \cup \{M', M\}$;
end
if $G \notin C.extents$ **then**
    $C \longleftarrow C \cup \{G, G'\}$
end

On the other hand, the second proposed Algorithm 2 depends mainly on distinct intents per objects. So, it works more efficiently in case of the counts of distinct intents per objects are relatively small. Generally, the performed experiments show that Algorithm 2 reaches a great reduction in complexity when compared with some existing algorithms specially with large data sets. This backs to its dependency on the unique set of extents with maximal set of intents and ignorance of the subsets of intents with same extent set. Accordingly, despite of ignoring some concepts, it extremely enhances the overall performance (a trade off between precision and complexity). Consequently, the proposed Algorithm 2 reduces the total number of generated fuzzy concepts as a result of using (5) and (6) with lower cost than that obtained in [2,17].

Algorithm 2. Fuzzy concept extraction algorithm respecting the distinct intents of the object set

---

Input: A fuzzy formal context $\mathbb{k}:= (G, M, I = \varphi(G \times M))$ and a confidence threshold $T=[t_1, t_2]$.
Result: Set of fuzzy formal concepts C
initialize $C \longleftarrow \{M', M\}$;
for *each object* $g \in G$ **do**
    Set NewPerObject $\longleftarrow 0$;
    Get g';
    if $g' \notin C.intent$ **then**
        $C \longleftarrow C \cup ((g'', \mu_I), g')$;
        $NewPerObject + +$;
        for $j=0$ to $(C.size - NewPerObject)$ **do**
            Inters$\longleftarrow c_j.Intent \cap g'$;
            if $Inters \notin C.intent$ **then**
                $C \longleftarrow C \cup ((Inters', \mu_I), Inters)$;
                $NewPerObject + +$;
            end
        end
    end
end
if $G \notin C.extents$ **then**
    $C \longleftarrow C \cup \{G, G'\}$
end

---

6   Experiments and Evaluation

In this section, some experiments are conducted over a data set titled "countries investment confidence marks" available in [26]. A transformed subset of this dataset is also used in [6,20]. As a preparation phase, the original data set is fuzziffied respecting a set of predefined linguistic terms. The original dataset has 43 countries (objects) and 15 confidence criteria (attributes). All 15 confidence criteria take a value in the range [0, 4] such that 0 means less confidence mark (maximum risk) and 4 means minimum risk. The fuzzification process of the values of such attributes are done by applying three linguistic terms low, medium and high using the membership functions illustrated in Fig. 1. These membership functions are also used in [6]. Accordingly, the total count of attributes became 45 attributes as a result of the fuzzification process.

A snapshot of a subset of the original dataset (10 objects, 3 attributes) is presented in Table 1. As noted, attribute symbols are used for simplification such that: A, B and C denote political stability, general attitude towards the investors, and nationalization respectively. The corresponding fuzziffied context of the original dataset is presented in Table 2 using the linguistic terms defined in Fig. 1.

The proposed algorithms have been compared with fuzzy CbO algorithm [12]. Table 3 shows the result of comparison between proposed Algorithm 2 and fuzzy

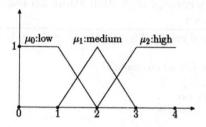

Fig. 1. Membership functions of linguistic terms {*low, medium and high*} defined over the linguistic variable marks

Table 1. Countries investment confidence marks

Country	A	B	C
Canada	3.7	2.7	3.5
USA	3.7	3.8	4
Mexico	2.9	2.2	2
Argentinia	1.1	1.7	1.7
Brazil	2.8	2.8	2.6
Chile	0.3	0	0
Colombia	2	2	1.7
Peru	1.6	1.2	1.1
Venezuela	2.5	2.1	2.1

Table 2. Fuzzified countries investment confidence marks

Country	A_{low}	A_{mod}	A_{high}	B_{low}	B_{mod}	B_{high}	C_{low}	C_{mod}	C_{high}
Canada	0	0	1	0	0.3	0.71	0	0	1
USA	0	0	1	0	0	1	0	0	1
Mexico	0	0.1	0.91	0	0.8	0.21	0	1	0
Argentinia	0.9	0.11	0	0.3	0.71	0	0.3	0.71	0
Brazil	0	0.21	0.8	0	0.21	0.8	0	0.41	0.6
Chile	1	0	0	1	0	0	1	0	0
Colombia	0	1	0	0	1	0	0.3	0.71	0
Peru	0.4	0.61	0	0.8	0.21	0	0.9	0.11	0
Venezuela	0	0.5	0.5	0	0.91	0.1	0	0.91	0.1
Australia	0	0	1	0	0.3	0.71	0	0	1

Table 3. Comparison between proposed Algorithm 2 and fuzzy CbO respecting threshold interval vs number of iterations

Threshold	Fuzzy CbO	Proposed Algorithm 2
[0, 1]	1,593,142	296,435
[0.3, 1]	227,114	69,913
[0.4, 1]	125,928	38,370
[0.5, 1]	71,573	26,056
[0.6, 1]	34,668	14,531
[0.7, 1]	17,357	9294
[0.8, 1]	10,161	6650

Table 4. Comparison between proposed Algorithm 1 and fuzzy CbO with regard to attribute redundancy rate vs number of iterations

Attribute redundancy rate %	Proposed Algorithm 1	Fuzzy CbO
0 %	964	967
20 %	967	1930
40 %	970	2893
60 %	973	3856
80 %	976	4819
100 %	979	5782

Table 5. Comparison between proposed Algorithm 2 and fuzzy CbO respecting threshold interval vs number of iterations

Objects' No.	Fuzzy CbO	Proposed Algorithm 1	Proposed Algorithm 2
10	10,241	8,446	239
20	57,796	52,739	2,876
30	132,382	125,954	16,204
43	227,072	227,069	69,913

CbO in terms of the number of performed iterations for different threshold intervals. On the other hand, the proposed Algorithm 1 doesn't show a big enhancement over fuzzy CbO with respect to different threshold intervals. However, it shows a big enhancement when tested over different percentages of attributes redundancy (symmetric correlation). The results are illustrated in Table 4. Also a test is performed with threshold interval [0.3, 1] on different object counts (10, 20, 30 and 43) and the test results are illustrated in Table 5 and shown in Figs. 2 and 3.

The experiments of the first proposed Algorithm 1 vs fuzzy CbO show that the first algorithm works extremely better in case of existence of symmetric

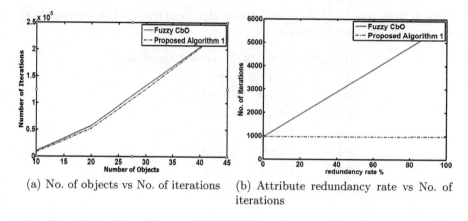

(a) No. of objects vs No. of iterations   (b) Attribute redundancy rate vs No. of
                                                iterations

Fig. 2. Experiments of fuzzy CbO vs proposed Algorithm 1

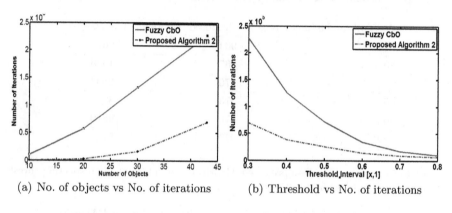

(a) No. of objects vs No. of iterations   (b) Threshold vs No. of iterations

Fig. 3. Experiments of fuzzy CbO vs proposed Algorithm 2

correlated attributes. In contrary, it becomes more closer to fuzzy CbO when no symmetric correlations exist between the dataset attributes. On the other hand, the second proposed algorithm is more efficient than both of first proposed one and fuzzy CbO. This is due to reducing the number of generated concepts such that it only persists the largest unique intents and omits other subsets of this largest intent with same extents (with tiny difference in membership values).

7   Conclusion

Generally, FFCA represents one important and challenging field of the data mining research area. Many researches have been introduced to enhance such field. Yet, there is still a need for more efficient approaches in order to enhance the whole process output and accelerate related processes. Commonly, the process of extracting fuzzy formal concepts is the most complex process in FFCA. In this paper, two proposed algorithms have been introduced to reduce the complexity associated with this process. The first proposed algorithm generates the same concepts as fuzzy CbO but with a reduction in complexity and hence the execution time. It works better in case of the existence of symmetric correlated attributes. On the other hand, the second proposed algorithm generates less number of fuzzy concepts due to elimination of the very closed concepts extents. It works more efficiently when the number of distinct intents of objects is relatively smaller. It is not affected by any redundant object intent as it works mainly on the distinct set of intents. The experiments show the added value of the proposed algorithms.

References

1. Kumar, C.A., Mouliswaran, S.C., Amriteya, P., Arun, S.R.: Fuzzy formal concept analysis approach for information retrieval. In: Proceedings of the Fifth International Conference on Fuzzy and Neuro Computing (FANCCO - 2015), pp. 255–271 (2015)
2. Yang, K.M., Kim, E.H., Hwang, S.H., Choi, S.H.: Fuzzy concept mining based on formal concept analysis. Int. J. Comput. **2**(3), 279–290 (2008)
3. Singh, P.K., Aswani Kumar, C.: A method for reduction of fuzzy relation in fuzzy formal context. In: Balasubramaniam, P., Uthayakumar, R. (eds.) ICMMSC 2012. CCIS, vol. 283, pp. 343–350. Springer, Heidelberg (2012)
4. Singh, P.K., Cherukuri, A.K., Li, J.: Concepts reduction in formal concept analysis with fuzzy setting using Shannon entropy. Int. J. Mach. Learn. Cyber. 1–11 (2014). doi:10.1007/s13042-014-0313-6
5. Wille, R.: Formal concept analysis as mathematical theory of concepts and concept hierarchies. In: Ganter, B., Stumme, G., Wille, R. (eds.) Formal Concept Analysis. LNCS (LNAI), vol. 3626, pp. 1–33. Springer, Heidelberg (2005)
6. Zheng, S., Zhou, Y., Martin, T.: A new method for fuzzy formal concept analysis. In: IEEE/WIC/ACM International Joint Conferences on Web Intelligence and Intelligent Agent Technologies, pp. 405–408 (2009)
7. Ganter, B., Stumme, G., Wille, R.: Formal Concept Analysis. LNCS (LNAI), vol. 3626. Springer, Heidelberg (2005). Carbonell, J.G., Siekmann, J. (eds.)
8. Zadeh, L.A.: Fuzzy sets. In: Information and Control, pp. 338–353 (1965)
9. Doubois, D., Ostasiewicz, W., Prade, H.: Fuzzy Sets: History and Basic Notions. Kluwer Academic, Boston (1999)
10. Ganter, B., Wille, R.: Formal Concept Analysis: Mathematical Foundation. Springer, Berlin (1999)
11. Martin, T., Shen, Y., Majidian, A.: Soft concept hierarchies to summarise data streams and hightlight anomalous changes. In: Hüllermeier, E., Kruse, R., Hoffmann, F. (eds.) Information Processing and Management of Uncertainty in Knowledge Based Systems. Communications in Computer and Information Science, vol. 81, pp. 44–54. Springer, Heidelberg (2010)
12. Majidian, A., Martin, T., Cintra, M.E.: Fuzzy formal concept analysis and algorithm. In: Proceedings of the 11th UK Workshop on Computational Intelligence (UKCI 2011), pp. 61–67, September 2011
13. Wille, R.: Restructuring lattice theory: an approach based on hierarchies of concepts. In: Rival, I. (ed.) Ordered Sets. NATO Advanced Study Institutes Series, vol. 83, pp. 445–470. Springer, Netherlands (1982)

14. Kuznetsov, S.: Interpretation on graphs and complexity characteristics of a search for specific patterns. Automomatic Documentation Math. Linguist. **24**(1), 37–45 (1989)

15. Krajca, P., Outrata, J., Vychodil, V.: Parallel algorithm for computeing fixpoints of Galois connections. Ann. Math. Artif. Intell. **59**, 257–272 (2010)

16. Belohlavek, R.: Algorithms for fuzzy concept lattices. In: Proceeding RASC 2002, Nottingham, UK, pp. 200–205, December 2002

17. Quan, T.T., Hui, S.C., Cao, T.H.: A fuzzy FCA based approach to conceptual clustering for automatic generation of concept hierarchy on uncertainty data. In: Proceeding CLA, pp. 1–12 (2004)

18. Martin, T., Majidian, A.: Beyond the known unknowns- finding fuzzy concepts for creative knowledge discovery. In: World Conference on Soft Computing, San Francisco (2011)

19. Martin, T., Siyao, Z., Majidian, A.: Fuzzy taxonomies for creative knowledge discovery. In: 8th International Semantic Web Conference (ISWC) (2009)

20. Bělohlávek, R., Sklenář, V., Zacpal, J.: Crisply generated fuzzy concepts. In: Ganter, B., Godin, R. (eds.) ICFCA 2005. LNCS (LNAI), vol. 3403, pp. 269–284. Springer, Heidelberg (2005). doi:10.1007/978-3-540-32262-7_19

21. Bělohlávek, R., Baets, B.D., Outrata, J., Vychodil, V.: Computing the lattice of all fixpoints of a fuzzy closure operator. IEEE Trans. Fuzzy Syst. **18**, 546–557 (2010)

22. Bělohlávek, R., De Baets, B., Outrata, J., Vychodil, V.: Lindig's algorithm for concept lattices over graded attributes. In: Torra, V., Narukawa, Y., Yoshida, Y. (eds.) MDAI 2007. LNCS (LNAI), vol. 4617, pp. 156–167. Springer, Heidelberg (2007)

23. Bělohlávek, R.: Reduction and a simple proof of characterization of fuzzy concept lattices. Fundamenta Informaticae **46**, 277–285 (2001)

24. Bělohlávek, R., Vychodil, V.: What is fuzzy concept lattice? In: CLA, pp. 34–45 (2005)

25. Bělohlávek, R., Vychodil, V.: Formal concept analysis and linguistic hedges. Int. J. Gen. Syst. **41**, 1–29 (2012)

26. Jambu, M.: Exploratory and Multivariate Data Analysis. Academic Press Inc., Orlando (1991)

27. Barea, V.L., Medina, J., Bulo, I.M.: Towards generating fuzzy rules via fuzzy formal concept analysis. In: 7th European Symposium on Computational Intelligence and Mathematics (ESCIM 2015), pp. 60–65 (2015)

A Comparative Study of Feature Selection and Classification Techniques for High-Throughput DNA Methylation Data

Alhasan Alkuhlani[✉], Mohammad Nassef, and Ibrahim Farag

Faculty of Computers and Information, Department of Computer Science,
Cairo University, Giza, Egypt
alhasan.alkuhlani@gmail.com, {m.nassef,i.farag}@fci-cu.edu.eg

Abstract. The high dimensionality of data is a common problem in classification. In this work, a small number of significant features is investigated to classify data of two sample groups. Various feature selection and classification techniques are applied in a collection of four high-throughput DNA methylation microarray data sets. Using accuracy as a performance metric, the repeated 10-fold cross-validation strategy is implemented to evaluate the different proposed techniques. Combining the Signal to Noise Ratio (SNR) and Wilcoxon rank-sum test filter methods with Support Vector Machine-Recursive Feature Elimination (SVM-RFE) as an embedded method has resulted in a perfect performance. In addition, the linear classifiers showed excellent results compared to others classifiers when applied to such data sets.

Keywords: Microarray · DNA Methylation · Feature selection · Classification · Cross-validation

1 Introduction

High-throughput DNA methylation (DNAm) microarrays are recently used to study the relationships between DNAm and diseases, such as cancer, in order to develop new drugs and to identify DNAm biomarkers related to particular cancer for diagnosis and prognosis purposes [1]. DNA methylation is a biochemical process which usually occurs in areas, namely CpG sites, of DNA sequence where a cytosine C base occurs beside a guanine G base [2]. It plays an important role in the development of diseases [3]. The classification of DNAm microarray data allows predicting tissue types as well as detecting significant methylation markers for disease diagnosis and drug discovery. Generally, the data sets from high-throughput methylation arrays contain an enormous number of CpG sites (referred in this study as features) and small sample size. The high dimensionality of data is a common problem in classification. To overcome this problem, the need to reduce dimensionality using feature selection has become necessary [4,5].

© Springer International Publishing AG 2017
A.E. Hassanien et al. (eds.), *Proceedings of the International Conference on Advanced Intelligent Systems and Informatics 2016*, Advances in Intelligent Systems and Computing 533, DOI 10.1007/978-3-319-48308-5_76

Feature selection (FS) techniques are used to select small informative and relevant features that are related to the problem of interest. Feature selection techniques can be categorized into three types: filter, wrapper and embedded methods. Filter methods usually rank features based on statistics or information theory measures. They are fast and independent from the classifier, but sometimes get redundant features that lack achieving the best classification performance. Wrapper methods depend on a predictor to select subsets based on classification performance. The limitation of this model is its intensive computational cost. Embedded methods try to combine the benefits of filter and wrapper methods and manage a trade-off between computational cost and accuracy [5].

Although FS and classification techniques have been designed for gene expression data, relatively few methods have been performed on DNAm data. Cai et al. [6] proposed ensemble-based FS techniques by hybrid Multi-category Receiver Operating Characteristic (Multi-ROC), Random Forests (RFs) and Maximum Relevance and Minimum Redundancy (mRMR) methods followed by machine learning methods to predict the lung adenocarcinoma (LADC), squamous cell lung cancer (SQCLC) and small cell lung cancer (SCLC) lung cancers subtypes. Ma and Teschendorff [7] offered a variational Bayesian beta-mixture model (VBBMM) to FS problem in the context of DNAm data. They demonstrated the benefit of VBBMM as a prioritization step in the context of detecting diagnosis markers in breast cancer. Using a high-throughput DNAm data, Meng et al. [8] explored a two-stage FS approach to select a small subset of CpG sites to classify two sample groups which are lung cancer and normal tissue samples. Firstly, they used two filtering methods, Principal Component Analysis (PCA) and Wilcoxon rank-sum test, to remove insignificant features, and then SVM-RFE was applied to detect the final methylation markers.

Amin et al. [9] proposed a method to analyse DNAm data of breast cancer subtypes. They implemented some non-specific-filtering and filtering FS methods to select significant hyper-methylated CPG sites for every breast cancer subtype. Formal Concept Analysis (FCA) was used as mathematical theory modeling to identify the relationship between the breast cancer subtypes and the involved hyper-methylated genes. They visualized that by the formal concept lattice. As a result, the common hyper-methylated genes between the breast cancer subtypes is detected from the formal concept lattice. Valavanis et al. [10] performed FS and classification framework on 450k human methylation data of breast cancer and B-cell lymphoma using evolutionary algorithms and Gene Ontology (GO) tree. Regarding gene expression microarray datasets, there are many research efforts that applied FS and classification [4,5,11–13].

In this paper, we investigate whether a small number of significant CpG sites is sufficient to classify data of two sample groups. We compared various FS and classification methods in a collection of four high-throughput DNAm data sets. We evaluated a couple of hybrid filter and embedded feature selection methods. Section 2 describes the used DNAm data sets and their preprocessing. In Sect. 3, the different FS and classification techniques are presented. The experimental results are discussed in Sect. 4. Section 5 concludes the paper.

2   DNAm Data Sets and Preprocessing

Illumina has developed three common HumanMethylation BeadChip platforms for DNAm studies which are GoldenGate, Infinium HumanMethylation27 (IHM-27k) and Infinium HumanMethylation450 (IHM-450k)BeadChip. The IHM-27k interrogates more than 27,000 CpG sites related to more than 14,000 genes [14]. The more recent IHM-450k array interrogates more than 480,000 CpG sites, that span 99 % of the RefSeq genes [15]. At each CpG site, the methylation fraction is summarized by Beta-value varied between 0 and 1 and calculated as:

$$\beta = \frac{M}{M + U + 100} \tag{1}$$

where M methylated intensity and U unmethylated intensity of each site.

2.1   Data Sets

We used four binary-class DNAm data sets that are based on the IHM-27k and IHM-450k platforms. Two of these data sets are from The Cancer Genome Atlas (TCGA) (https://tcga-data.nci.nih.gov/tcga/), whereas the others are from the Gene Expression Omnibus (GEO) (http://www.ncbi.nlm.nih.gov/geo/). The four data sets are described in Table 1.

Table 1. Data sets Information.

Data set	Platform	Source	# of features after preprocessing	# of samples
RCC Subtypes	IHM-450k	TCGA	299,211	150 KIRC vs. 150 KIRP
KIRP	IHM-450k	TCGA	305,056	276 cancer vs. 45 normal
Rheumatoid Arthritis	IHM-450k	GEO (GSE42861)	361,525	138 patient vs. 51 normal
Ovarian Cancer	IHM-27k	GEO (GSE19711)	25,652	77 cancer vs. 86 control

The first data set represents the Renal cell carcinoma (RCC), which is the most common type of kidney cancer. Approximately 85 % of all kidney cancers are RCC. The most common subtypes of RCC are kidney renal clear cell carcinoma (KIRC) and kidney renal papillary cell carcinoma (KIRP)[16]. We constructed this dataset by randomly selecting 150 tumor samples from each of KIRC and KIRP under the TCGA's IHM-450k platform. The KIRP data set is the second data set that contains 276 tumor and 45 normal samples from TCGA based on the IHM-450k platform.

Third, the rheumatoid arthritis data set is from GEO and was generated using the IHM-450k platform. The GEO accession number for this data set is GSE42861 [17]. We randomly selected 189 (138 patient and 51 normal) samples from the 691 samples. At last, the ovarian cancer data set is also from GEO with accession number GSE19711 [18]. It was generated using the IHM-27k platform. We selected a group of women with age 50 to 60 years to avoid heterogeneity between the age groups. The selected samples included 163 samples with 77 cancer and 86 healthy control.

2.2   Preprocessing

One of multiple steps of preprocessing the DNAm microarray data sets is removing every defective CpG site [19]. We filter out the sites with the missing values. Moreover, to guarantee that our selection is not Single Nucleotide Polymorphisms (SNPs) dependent; we eliminated the probes overlapped by SNPs. In addition, the Illumina HumanMethylation arrays contains cross-reactive probes. Thus, we excluded these probes to avoid confusion in the data obtained from such arrays [20]. The sites in the sex chromosomes are also removed. The number of CpG sites after preprocessing for each data set is displayed in Table 1.

3   Feature Selection and Classification Methods

In this study, We have used three filter FS methods, Signal to Noise Ratio (SNR), Wilcoxon Rank-Sum Test (WilcoxonT) and Chi-squared test (Chi), and two embedded methods, Random Forest Importance (RF) and Support Vector Machine-Recursive Feature Elimination (SVM-RFE). In addition, we have considered four classification techniques, Support Vector Machine (SVM), K-Nearest Neighbor (KNN), Naive Bayes (NB) and Linear Discriminant Analysis (LDA).

3.1   Feature Selection Methods

The Signal to Noise Ratio (SNR) has good performance on biomedical data [11]. SNR calculate each feature's significance independently where the more important the feature, the higher its SNR. The SNR for feature f:

$$SNR(f) = \frac{\mu_{f1} - \mu_{f2}}{\sigma_{f1} + \sigma_{f2}} \tag{2}$$

where μ_{f1}, μ_{f2} are means of the first and second class for feature f and σ_{f1}, σ_{f2} are standard deviations of the first and second class for feature f.

Wilcoxon Rank-Sum test (WilcoxonT) is a non-parametric statistical test, which can be used with data from any distribution [8]. We applied WilcoxonT using R on each feature. The features ranked by p-value in ascending order where the less the feature's p-value, the more its importance.

Chi-squared test (Chi) is a statistical test applied to data to identify features that its distribution differs between groups [21]. We performed Chi-squared using the *FSelector* R package [22].

Random forest (RF) is an ensemble machine learning method that consists of a number of individual classification trees and is widely used in bioinformatics and computational biology [23]. It performs FS by identifying the weight of importance of each feature. The importance of a feature was measured by computing the increased amount in misclassification when the data for this feature is permuted while all others are left unmodified [24]. We performed Random Forest Importance using the *FSelector* R package [22].

SVM-RFE [12] is a well-known embedded method that has shown perfect prediction ability in analyzing high-dimensional microarray data [5,8,13]. It is a recursive backward elimination algorithm based on linear SVM. At each iteration, the feature which lead to the lowest margin of class separation is eliminated. The output of the algorithm is the ranking for features based on their classification effects, where the feature with the biggest effect is the most significant.

3.2   Classification Methods

Support Vector Machine (SVM) [25] is the state-of-the-art supervised classification algorithm that have shown an excellent performance in high-dimensional microarray data classifications [8,13]. The basic concept of linear SVM is building up a hyperplane to maximize the margin of separation between two classes.

K-Nearest Neighbor (KNN) is a supervised machine learning algorithm which classifies a sample to a specific class based on the majority of its k closest neighbors [26]. The most common methods to calculate the distances for KNN are Euclidean Distance and Pearson Correlation.

Naive Bayes (NB) is probabilistic classifier that has been successfully applied in a number of application domains [27]. Its major features are independent of the variables given the class and perform well with the small training data.

Linear Discriminant Analysis (LDA) is a common classification and dimension reduction technique. It transforms the data into low dimension space. So, its model achieves a good prediction performance and well separated classes [28].

4   Experiments and Results

4.1   Experimental Settings

After preprocessing all the attempted data sets (Table 1), we applied seven FS techniques. The first two techniques, namely SNR and WilcoxonT, are implemented separately as filter methods. For each method, we selected the top 10 % and 40 % features from the IHM-450k and IHM-27k platforms respectively. The third FS technique, namely S&W, combined the first two methods and is calculated by the sum of the feature's corresponding ranks proposed in [29]. The calculation of the sum of ranks begins by assigning a rank for each feature in

subsets S_s and S_w resulting from SNR and WilcoxonT respectively, where the
top feature is assigned rank 1. For each overlapped feature i in S_s and S_w, the
new rank equals the sum of i's rank in both S_s and S_w. Next, we sorted the
overlapped features by the new rank in an ascending order to be the ranked
features resulted from the third FS method. After that, the Chi-squared (Chi),
random forest importance (RF) and SVM-RFE FS techniques are performed on
data constructed by the S&W's top 3,000 and 1,500 features of the IHM-450k
and IHM-27k platforms respectively. The remaining of this article refers to the
last three combined methods by S&W+Chi, S&W+RF and S&W+SVM-RFE
respectively. In addition, a random FS, namely Random, has been considered to
contrast the efficiency of the mentioned techniques.

The goal of FS is to select few significant features. So, for each data set, the
first X features produced from each of the seven mentioned FS methods are used
to construct 13 configured sub-data sets, where X equals 5, 10, 15, 20, 25, 30, 40,
50, 60, 70, 80, 90 or 100. For classification, the linear SVM ($gama=2$ and $cost=1$),
KNN ($k=5$), NB and LDA classification techniques were adopted. Briefly, for
each of the 4 data sets, we tried 364 combinations (4 classifiers × 13 configurations × 7 FS techniques). The repeated 10-fold Cross-Validation (CV) technique
is applied to evaluate each combination, using accuracy as the performance measurement. Figure 1 illustrates the general framework of the experiments.

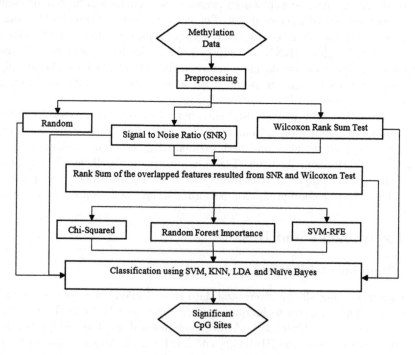

Fig. 1. The general framework of the experiments.

4.2   Results and Discussion

We applied the experimental settings for feature selection and classification on each data set. Figures 2, 3, 4 and 5 show the accuracy against the number of features for the proposed FS and classification approaches on RCC subtypes, KIRP, Rheumatoid Arthritis and Ovarian Cancer data sets, respectively. Each figure consists of four graphs, which represent results from the four applied classifiers. All the graphs have the same legend scheme for marking the seven used feature selection methods to allow an easy comparison.

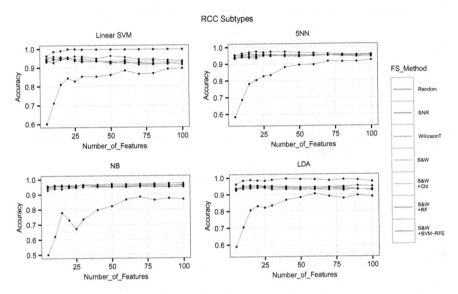

Fig. 2. Accuracy against the number of features for different FS and classification approaches (RCC dataset).

For the RCC subtypes data set (Fig. 2) and the KIRP data set (Fig. 3), the experimental results show that the S&W+SVM-RFE FS method slightly outperforms all the other methods. Almost all classifiers perform relatively equally on all the FS methods except the random method. Regarding the Rheumatoid Arthritis data set (Fig. 4) and the Ovarian Cancer data set (Fig. 5), it is clear that the (S&W+SVM-RFE) FS method outperforms all the other methods with the Linear SVM and LDA linear classifiers. That is because the linear separability of the selected features gives an advantage to the linear classifiers to outperform the others. As stated, most of the used FS methods are hybrid. The run time increases as we move from single to hybrid methods, however, the difference in run time is negligible between the hybrid methods.

The reader can notice that the lower bound for the classification accuracy of each data set is proportional to the ratio of its first class (50%, 86%, 73% and 47% respectively). That is clear from applying the random FS over a small

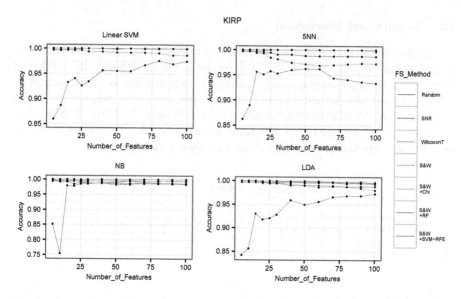

Fig. 3. Accuracy against the number of features for different FS and classification approaches (KIRP dataset).

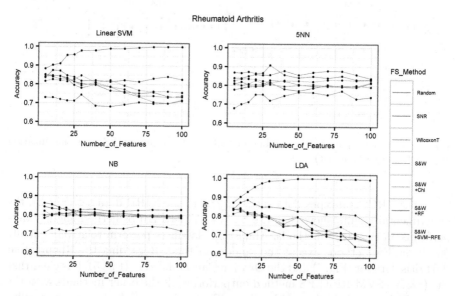

Fig. 4. Accuracy against the number of features for different FS and classification approaches (Rheumatoid Arthritis dataset).

number of features. Generally, the hybrid (S&W+SVM-RFE) FS with the linear classifiers showed excellent results. Table 2 lists the maximum classification accuracies with all the used FS techniques over the four data sets.

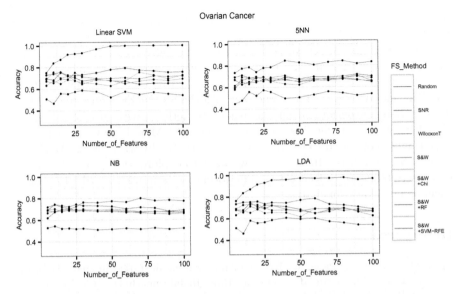

Fig. 5. Accuracy against the number of features for different FS and classification approaches (Ovarian Cancer data set).

Table 2. Maximum classification accuracies for the experiments performed over all the data sets.

Data set	Classifier	Random		SNR		WilcoxonT		S&W		S&W +Chi		S&W +RF		S&W +SVM- RFE	
		MA	#F	MA	#F	MA	#F	MA	#F	MA	#F	MA	#F	MA	#F
RCC Subtypes	SVM	89.47 %	100	94.73 %	20	96.07 %	5	95.93 %	10	96.00 %	10	96.20 %	50	99.87 %	20
	5NN	92.20 %	100	95.33 %	30	95.93 %	15	95.80 %	20	95.87 %	15	95.60 %	25	97.53 %	20
	NB	88.33 %	70	95.20 %	70	96.33 %	80	95.60 %	70	96.13 %	10	96.40 %	100	97.27 %	90
	LDA	90.40 %	60	94.27 %	20	94.80 %	10	95.67 %	15	95.47 %	10	95.40 %	20	99.53 %	40
KIRP	SVM	97.57 %	80	100.00 %	5	100.00 %	5	100.00 %	5	99.69 %	5	100.00 %	25	100.00 %	5
	5NN	96.26 %	50	100.00 %	5	100.00 %	5	100.00 %	5	99.88 %	5	100.00 %	25	100.00 %	5
	NB	98.94 %	70	99.57 %	70	99.50 %	20	99.69 %	15	99.69 %	25	99.94 %	5	100.00 %	5
	LDA	97.28 %	100	100.00 %	5	100.00 %	5	100.00 %	5	100.00 %	5	99.69 %	15	100.00 %	5
Rheumatoid Arthritis	SVM	74.43 %	30	84.61 %	10	85.18 %	5	85.17 %	5	84.63 %	30	87.40 %	10	100.00 %	90
	5NN	77.13 %	80	81.59 %	30	82.79 %	30	82.70 %	20	83.52 %	15	87.84 %	70	90.81 %	30
	NB	74.17 %	70	81.07 %	25	81.35 %	25	81.02 %	15	82.02 %	5	84.33 %	5	86.22 %	5
	LDA	73.78 %	25	83.53 %	5	84.13 %	5	83.88 %	5	82.44 %	10	88.55 %	15	99.89 %	60
Ovarian Cancer	SVM	58.72 %	30	75.22 %	15	75.11 %	25	75.54 %	10	75.12 %	70	79.19 %	60	100.00 %	70
	5NN	57.54 %	25	68.04 %	70	70.84 %	40	69.91 %	20	70.03 %	20	71.63 %	90	84.69 %	40
	NB	54.58 %	10	74.48 %	10	69.97 %	25	70.47 %	10	68.90 %	20	74.59 %	10	79.87 %	70
	LDA	60.62 %	40	74.72 %	15	72.53 %	25	73.10 %	15	68.87 %	70	78.11 %	60	97.52 %	80

MA= Maximum Accuracy, #F= Number of Features.

5   Conclusions

In this paper, we have tried various feature selection and classification techniques for four different high-dimensional DNAm data sets. The different sources and structures of data sets allowed us to evaluate the different FS and classification techniques considered in this work. We evaluated a couple of hybrid feature selection methods. (S&W+SVM-RFE) is one of these methods that combines the SNR and WilcoxonT filter methods with the SVM-RFE embedded method. This method outperformed all the other FS methods with all the attempted classifiers. Moreover, the linear classifiers, linear SVM and LDA, showed better performance compared to the other classifiers especially with the (S&W+SVM-RFE) FS method. For future work, we want to explore the biological meaning of the selected features that get high performance accuracies in order to propose them as biomedical biomarkers.

References

1. Li, D., Xie, Z., Le Pape, M., Dye, T.: An evaluation of statistical methods for dna methylation microarray data analysis. BMC Bioinform. **16**(1), 1 (2015)
2. Das, P.M., Singal, R.: DNA methylation and cancer. J. Clin. Oncol. **22**(22), 4632–4642 (2004)
3. Zhuang, J., Widschwendter, M., Teschendorff, A.E.: A comparison of feature selection and classification methods in dna methylation studies using the illumina infinium platform. BMC Bioinform. **13**(1), 59 (2012)
4. Lee, C.P., Leu, Y.: A novel hybrid feature selection method for microarray data analysis. Appl. Soft Comput. **11**(1), 208–213 (2011)
5. Saeys, Y., Inza, I., Larranaga, P.: A review of feature selection techniques in bioinformatics. Bioinformatics **23**(19), 2507–2517 (2007)
6. Cai, Z., Xu, D., Zhang, Q., Zhang, J., Ngai, S.M., Shao, J.: Classification of lung cancer using ensemble-based feature selection and machine learning methods. Mol. BioSyst. **11**(3), 791–800 (2015)
7. Ma, Z., Teschendorff, A.E.: A variational bayes beta mixture model for feature selection in dna methylation studies. J. Bioinform. Computat. Biol. **11**(04), 1350005 (2013)
8. Meng, H., Murrelle, E.L., Li, G.: Identification of a small optimal subset of CpG sites as bio-markers from high-throughput DNA methylation profiles. BMC Bioinform. **9**(1), 457 (2008)
9. Amin, I.I., Hassanien, A.E., Kassim, S.K., Hefny, H.A.: Big DNA methylation data analysis and visualizing in a common form of breast cancer. In: Hassanien, A.E., Azar, A.T., Snasael, V., Kacprzyk, J., Abawajy, J.H. (eds.) Big Data in Complex Systems. SBD, vol. 9, pp. 375–392. Springer, Heidelberg (2015)
10. Valavanis, I., Pilalis, E., Georgiadis, P., Kyrtopoulos, S., Chatziioannou, A.: Cancer biomarkers from genome-scale DNA methylation: Comparison of evolutionary and semantic analysis methods. Microarrays **4**(4), 647–670 (2015)
11. Gunavathi, C., Premalatha, K.: Cuckoo search optimisation for feature selection in cancer classification: a new approach. Int. J. Data Min. Bioinform. **13**(3), 248–265 (2015)
12. Guyon, I., Weston, J., Barnhill, S., Vapnik, V.: Gene selection for cancer classification using support vector machines. Mach. Learn. **46**(1–3), 389–422 (2002)

13. Zhou, X., Tuck, D.P.: MSVM-RFE: extensions of SVM-RFE for multiclass gene selection on DNA microarray data. Bioinformatics **23**(9), 1106–1114 (2007)
14. Bibikova, M., Le, J., Barnes, B., Saedinia-Melnyk, S., Zhou, L., Shen, R., Gunderson, K.L.: Genome-wide dna methylation profiling using infinium® assay. Epigenomics **1**(1), 177–200 (2009)
15. Bibikova, M., Barnes, B., Tsan, C., Ho, V., Klotzle, B., Le, J.M., Delano, D., Zhang, L., Schroth, G.P., Gunderson, K.L., et al.: High density dna methylation array with single CpG site resolution. Genomics **98**(4), 288–295 (2011)
16. Lipworth, L., Morgans, A.K., Edwards, T.L., Barocas, D.A., Chang, S.S., Herrell, S.D., Penson, D.F., Resnick, M.J., Smith, J.A., Clark, P.E.: Renal cell cancer histological subtype distribution differs by race and sex. BJU Int. **117**(2), 260–265 (2016)
17. Liu, Y., Aryee, M.J., Padyukov, L., Fallin, M.D., Hesselberg, E., Runarsson, A., Reinius, L., Acevedo, N., Taub, M., Ronninger, M., et al.: Epigenome-wide association data implicate DNA methylation as an intermediary of genetic risk in rheumatoid arthritis. Nat. Biotechnol. **31**(2), 142–147 (2013)
18. Teschendorff, A.E., Menon, U., Gentry-Maharaj, A., Ramus, S.J., Weisenberger, D.J., Shen, H., Campan, M., Noushmehr, H., Bell, C.G., Maxwell, A.P., et al.: Age-dependent dna methylation of genes that are suppressed in stem cells is a hallmark of cancer. Genome Res. **20**(4), 440–446 (2010)
19. Dedeurwaerder, S., Defrance, M., Bizet, M., Calonne, E., Bontempi, G., Fuks, F.: A comprehensive overview of infinium humanmethylation450 data processing. Briefings Bioinform. **15**(6), 929–941 (2013)
20. Chen, Y.A., Lemire, M., Choufani, S., Butcher, D.T., Grafodatskaya, D., Zanke, B.W., Gallinger, S., Hudson, T.J., Weksberg, R.: Discovery of cross-reactive probes and polymorphic CpGs in the illumina infinium humanmethylation450 microarray. Epigenetics **8**(2), 203–209 (2013)
21. Zhang, Q., Wu, H., Zheng, H.: Aberrantly methylated CpG island detection in colon cancer. J. Proteomics Bioinform. 2015 (2015)
22. Romanski, P., Kotthoff, L.: Fselector: Selecting attributes (2013). https://cran. r-project.org/web/packages/FSelector/. R package version 0.19
23. Strobl, C., Boulesteix, A.L., Zeileis, A., Hothorn, T.: Bias in random forest variable importance measures: Illustrations, sources and a solution. BMC Bioinform. **8**(1), 1 (2007)
24. Liang, J.D., Ping, X.O., Tseng, Y.J., Huang, G.T., Lai, F., Yang, P.M.: Recurrence predictive models for patients with hepatocellular carcinoma after radiofrequency ablation using support vector machines with feature selection methods. Comput. Methods Programs Biomed. **117**(3), 425–434 (2014)
25. Boser, B.E., Guyon, I.M., Vapnik, V.N.: A training algorithm for optimal margin classifiers. In: Proceedings of the Fifth Annual Workshop on Computational Learning Theory, pp. 144–152. ACM (1992)
26. Cover, T.M., Hart, P.E.: Nearest neighbor pattern classification. IEEE Trans. Inf. Theory **13**(1), 21–27 (1967)
27. Keller, A.D., Schummer, M., Hood, L., Ruzzo, W.L.: Bayesian classification of DNA array expression data. Technical Report UW-CSE-2000-08-01 (2000)
28. Huerta, E.B., Duval, B., Hao, J.K.: A hybrid LDA and genetic algorithm for gene selection and classification of microarray data. Neurocomputing **73**(13), 2375–2383 (2010)
29. Kuncheva, L.I.: A stability index for feature selection. In: Artificial Intelligence and Applications, pp. 421–427 (2007)

Exploiting Existed Medical Ontologies to Build Domain Ontology for Hepatobiliary System Diseases

Galal AL-Marzoqi[✉], Marco Alfonse, Ibrahim F. Moawad,
and Mohamed Roushdy

Faculty of Computer and Information Science,
Ain Shams University, Abbasia, Cairo, Egypt
galalalmarzoqi@gmail.com, marco@fcis.asu.edu.eg,
{ibrahim_moawad,mroushdy}@cis.asu.edu.eg

Abstract. The Hepatobiliary System is one of the important systems in the human body. This system has the ability to regulate the other systems. It may be affected by many pathologic conditions, which affect other organs negatively. Additionally, it plays an important role in many body functions like protein production. In this paper, we exploited the existed Ontologies in the medical domain to build a new Hepatobiliary System Diseases (HSD) Ontology. The Ontology has been built by using the BioPortal system to find the correct candidate Ontologies. This Ontology was developed in pathology domain and represented in the Web Ontology Language (OWL) that has recently become the standard language for the semantic web. The HSD Ontology development methodology includes five phases: HSD Query Extraction phase, Ontology Selection phase, Ontology Mapping and Partitioning phase, Adding Knowledge phase, and Ontology Merging and Validation phase. By developing the HSD Ontology in pathology domain, both intelligent systems and physicians can share, reason, and exploit this knowledge in different ways.

Keywords: Ontology selection · OWL ontology · Hepatobiliary System Diseases · Medical Ontology

1 Introduction

Ontology, in Philosophy, is the branch of metaphysics that deals with the nature of being, while it, in Logic, is the set of entities presupposed by a theory [1]. Medical Ontology is interested in resolving important issues such as the reusing and sharing of medical information [2]. Conceptually, it is a kind of controlled concepts of well-defined terms with limited relationships between those terms, capable of interpretation by both humans and computers [3]. Medical Ontology is now widely acknowledged that Ontologies can make a significant contribution to the design and implementation of information systems in the medical area. There are some Medical Ontologies developed to facilitate this purpose, such as (Open Biomedical Ontologies (OBO) [4], National Center for Biomedical Ontology (NCBO) BioPortal [5]. BioPortal is an open repository of biomedical Ontologies that provides access via web browsers

© Springer International Publishing AG 2017
A.E. Hassanien et al. (eds.), *Proceedings of the International Conference on Advanced Intelligent Systems and Informatics 2016*, Advances in Intelligent Systems and Computing 533, DOI 10.1007/978-3-319-48308-5_77

and Web services to Ontologies. The BioPortal contains Ontologies that range in subject matter such as anatomy, phenotype, experimental conditions, imaging, chemistry and health. It has a repository which includes 479 Ontologies (6,638,279 total classes) [6]. There are many tools for Ontology management in different formats (Protégé, OilEd, Apollo, RDFedt, etc.) [7].

In addition, human body systems consist of specific tissues, cells and organs that work together to perform specific functions. These systems are interconnected and dependent, so they can't work separately, such as Nervous System, Respiratory System, Immune System, Digestive System and Hepatobiliary System [8]. The Hepatobiliary system includes four organs: Liver, Gallbladder, Bile Duct and Pancreas. It is responsible for many processes inside the body. These processes are important to keep body regulated and healthy. Conceptually, it plays an important role in many body functions like protein production [9]. In this paper, we exploited the existed Ontologies in the medical domain to build a new Hepatobiliary System Diseases (HSD) Ontology. The BioPortal system was used to find the correct candidate Ontologies to build the HSD Ontology. This Ontology was developed in pathology domain and is represented in the Web Ontology Language (OWL) that has recently become the standard language for the semantic web.

This paper is organized as follows. Section 2 presents the related work. Section 3 presents the methodology followed in this research to build the Hepatobiliary System Diseases (HSD) Ontology. The methodology development of the HSD Ontology phases is described in Sect. 4. To show the benefits of the proposed HSD Ontology, Sect. 5 presents a beneficial scenario of liver disease causal relations. Finally, Sect. 6 concludes the most important points in this paper.

2   Related Work

Many research works have been achieved to build specific domain Ontologies for different diseases [10]. For example, Abdel-Badeeh M. Salem et al. [11] developed Ontology based knowledge representation for Liver Cancer that was built using the Protégé-OWL editing environment. Also, Riichiro Mizoguchi, et al. [12] developed a disease Ontology based on River Flow Model and a browsing tool for causal chains defined in it. Because the Ontology is based on Ontological consideration of causal chains, it could capture characteristics of diseases appropriately. The definition of disease as causal could be also very friendly to clinicians since it is similar to their understanding of disease in practice. Ibrahim F. Moawad et al. [13] developed Viral Hepatitis Ontology using the Ontology of Biomedical Reality (OBR) framework for the A, B, C and D viruses, which are the most widely spread among males and females. This Ontology is represented in the Web Ontology Language (OWL). Martinez-Romero et al. [14] presented BiOSS, a novel system for the selection of Biomedical Ontologies. BiOSS evaluates the adequacy of Ontology for a given domain according to three different criteria: (1) the extent to which the Ontology covers the domain; (2) the semantic richness of the Ontology in the domain; (3) the popularity of the Ontology in the Biomedical Community.

The main contribution of this paper is exploiting the existed Ontologies in the medical domain to build a new Hepatobiliary System Diseases (HSD) Ontology in pathology domain. The BioPortal system was used to find the correct candidate Ontologies to build HSD Ontology. By developing the HSD Ontology in pathology domain, both intelligent systems and physicians can share, reason, and exploit this knowledge in different ways.

3 HSD Ontology Methodology

In this section, we represent the Hepatobiliary System Diseases (HSD) Ontology building methodology that includes five phases: (1) HSD Query Extraction phase, where we queried Hepatobiliary system diseases from various sources such as domain experts, medical book, and trusted medical sites. After that, we have studied Hepato-biliary system diseases with their organs, symptoms\signs, treatments and disease causal relations. (2) Ontology Selection phase, we used BioPortal system to find the best candidate Ontologies that are related to our keywords. The output of this phase is the candidate Ontologies that can be used together to provide the best coverage to domain and presents candidate Ontologies in higher ranking. (3) Ontology Mapping and Partitioning phase, we mapped and partitioned, each candidate Ontologies (MESH, MEDDRA and SNOMED CT) by organs and their diseases. (4) Adding Knowledge phase, we have added knowledge symptoms\signs, treatments and disease causal relations into HSD Ontology. (5) Ontology Merging and Validation phase, the domain experts have been consulted to review results of Hepatobiliary system diseases, symptoms\signs, treatments and disease causal relations as shown in Fig. 1.

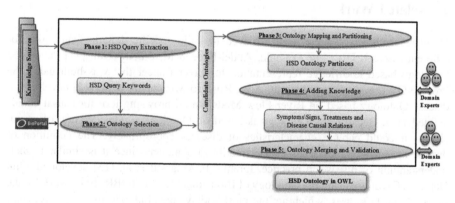

Fig. 1. The HSD ontology methodology

4 HSD Ontology Development Phases

In this section, we present the HSD Ontology methodology that includes five phases: HSD Query Extraction phase, Ontology Selection phase, Ontology Mapping and Partitioning phase, Adding Knowledge phase, Ontology Merging and Validation phase. In the following sub-sections, these phases are explained in details.

4.1 HSD Query Extraction Phase

In this phase, we have extracted the needed knowledge from several medical sources such as domain experts, medical book [15], and trusted medical sites [16–19]. After extracting the needed knowledge, we have studied the four organs Liver, Gallbladder, Bile Duct and Pancreas with their diseases to determine the symptoms/signs caused by these diseases, treatments and disease causal relations into the Hepatobiliary system diseases. The Liver organ includes 23 diseases and the Gallbladder organ includes 10 diseases, while the Bile Duct organ includes 5 diseases and Pancreas organ includes 6 diseases. Table 1 shows all organs and diseases in the Hepatobiliary system, which are represented as keywords to use in the next phase (Ontology Selection) to find the candidate Ontologies.

Table 1. Organs and their diseases in the Hepatobiliary System

HSD	
Organ	**Disease**
Liver	Alagille Syndrome, Autoimmune Hepatitis, Focal Nodular Hyperplasia, Liver Cirrhosis, Galactosemia, Gilbert Disease, Hemochromatosis, Hepatitis A, Hepatitis B, Hepatitis C, Hepatitis D, Hepatic Encephalopathy, Liver Neoplasms, Non-Alcoholic Fatty Liver Disease, Reye Syndrome, Glycogen Storage Disease Type I, Hepatolenticular Degeneration, Liver Failure, Echinococcosis, Liver Tumor, Hepatic Adenoma, Liver Cyst and Storage Disease.
Gallbladder	Bile Reflux, Biliary Dyskinesia, Cholangitis, Primary Sclerosing Cholangitis, Cholecystitis, Gallstones, Cholestasis, Gallbladder Neoplasms, Gallbladder Polyp and Porcelain Gallbladder.
Bile Duct	Bile Duct Neoplasms, Primary Biliary Cirrhosis, Biliary Atresia, Choledocholithiasis and Bile Duct Stricture.
Pancreas	Chronic Pancreatitis, Hereditary Pancreatitis, Pancreatic Neoplasms, Cystic Fibrosis, Acute Pancreatitis and Diabetes.

4.2 Ontology Selection Phase

In this phase, the Ontology selection includes three processes: Query Expansion, Ontology Evaluation and Ontology Combination and Ranking. The phase starts from the input data that may be supplied by the researcher, an agent, or automatically extracted from information resource. The primary goal of Ontology selection is retrieving a single Ontology or multiple Ontologies that represent the researcher's domain keywords [14, 20].

4.2.1 Query Expansion

In this section, we manually applied the expansion of each keyword with the same meaning or synonyms. Query Expansion includes three steps: Normalization and

Correction, Preferred Name and Semantic Expansion. The keywords are selected according to 4 Organs and 44 Diseases of the Hepatobiliary system in pathology domain as shown in Table 1.

- Normalization and Correction

In this step, we applied the keywords that are normalized to standard format, whereas all punctuations, double spaces and extra symbols are removed. Also, we used a spell checker that is applied to correct possible typographic and repair any spelling errors. Furthermore, we asked medical experts to correct any keyword that could not be corrected by the spell checker and selected the most preferred name to that keyword. There are two terminological resources (Medical Experts and UMLS which are specialized resource for the biomedical data field) are used to remove possible keywords with the same meaning and then each remaining keyword expanded with synonyms to increase the possibility of finding it into the candidate Ontologies.

- Preferred Name and Semantic Expansion

This phase includes two steps: Preferred Name means a keyword name that found in candidate Ontology, while, in Semantic Expansion step, the UMLS is used to expand each keyword with synonyms. These synonyms will be useful in Ontology Evaluation to increase chances of finding each keyword in the candidate Ontologies. For example, the result of the semantic expansion for keyword "Autoimmune Hepatitis" will be its synonyms (Autoimmune Chronic Hepatitides), (Hepatitides, Autoimmune Chronic), (Autoimmune Chronic Hepatitis), (Autoimmune Hepatitides), (Autoimmune Hepatitis), (Chronic Hepatitis, Autoimmune), (Chronic Hepatitides, Autoimmune), (Hepatitis, Autoimmune Chronic) and (Hepatitides, Autoimmune). Table 2 shows a sample of keywords that have been entered through the steps of Query Expansion.

Table 2. Sample of keywords in Query Expansion

Query Expansion			
Input (Keywords)	Normalization & Correction	Preferred Name	Semantic Expansion (Synonyms)
Liver	Liver	Liver	(Liver), (Livers)
Autoimune Hepatits	Autoimmune Hepatitis	Hepatitis, Autoimmune	(Autoimmune Chronic Hepatitides), (Hepatitides, Autoimmune Chronic), (Autoimmune Chronic Hepatitis), (Autoimmune Hepatitides), (Autoimmune Hepatitis), (Chronic Hepatitis, Autoimmune), (Chronic Hepatitides, Autoimmune), (Hepatitis, Autoimmune Chronic), (Hepatitides, Autoimmune).
Focal-Nodular Hyperplasa	Focal Nodular Hyperplasia	Focal Nodular Hyperplasia	(Hyperplasia, Focal Nodular), (Focal Nodular Hyperplasias), (Hyperplasia, Focal Nodular).
Acute Pancretits	Acute Pancreatitis	Acute Pancreatitis	(Acute Pancreatitis), (AP - Acute pancreatitis), (Acute pancreatitis (disorder)).

- Ontology Retrieval (Candidate Ontologies)

The Ontology retrieval step involves accessing the Ontology repository and obtaining the candidate Ontologies. For example, if we have searched about this

Table 3. Sample of keywords with their synonyms

Ontology Retrieval	
Semantic Expansion (Synonyms)	Candidate Ontologies
(Liver), (Livers)	MESH
(Autoimmune Chronic Hepatitides), (Hepatitides, Autoimmune Chronic), (Autoimmune Chronic Hepatitis), (Autoimmune Hepatitides), (Autoimmune Hepatitis), (Chronic Hepatitis, Autoimmune), (Chronic Hepatitides, Autoimmune), (Hepatitis, Autoimmune Chronic), (Hepatitides, Autoimmune), (Hepatitis, Autoimmune)	MESH
(Hyperplasia, Focal Nodular), (Focal Nodular Hyperplasias), (Hyperplasias, Focal Nodular), (Focal Nodular Hyperplasia)	MESH
(Liver Failure), (Hepatic Failure)	MEDDRA
(Acute Pancreatitis), (AP - Acute pancreatitis), (Acute pancreatitis (disorder)).	SNOMED CT

keyword (Focal Nodular Hyperplasia) with synonyms (Hyperplasia, Focal Nodular), (Focal Nodular Hyperplasias) and (Hyperplasias, Focal Nodular), to get the candidate Ontology (MESH). Table 3 shows a sample of keywords with their synonyms to get candidate Ontologies.

4.2.2 Ontology Evaluation, Combining and Ranking
Ontology Evaluation plays an important role in the Ontology selection process. It evaluates the candidate Ontologies by different ways: (a) Domain coverage (candidate Ontology covers the input keywords), (b) Semantic richness of Ontology in the medical domain and (c) Popularity of Ontology in the biomedical community. In BioPortal, candidate Ontologies are retreived in two different outputs: "Single Ontology" or "Combined Ontology". The Combined Ontology output represents Ontologies that can be used together to provide the best coverage to domain and presents candidate Ontologies in higher ranking. The system shows different groups of candidate Ontologies that cover some or all keywords. Table 4 shows best group that covers the input keywords (Coverage 100.0) and locats ranking (1). We have selected the best ranking (1) and coverage (100.0) [20].

Table 4. Combination ontology and ranking

Combination Ontology Ranking					
Ranking	Ontology	Final Score	Coverage	Semantic Richness	Popularity
1	MESH, MEDDRA, SNOMD CT	84.5	100.0	73.2	91.4
2	MESH, MEDDRA, RCD	83.2	97.8	73.1	91.1
3	MESH, MEDDRA, NCTI	82.9	97.1	73.8	91.2

4.3 Ontology Mapping and Partitioning Phase

In this phase, we mapped and partitioned each candidate Ontologies (MESH, MED-DRA and SNOMED CT) by organs and their diseases in Hepatobiliary system diseases. Table 5 shows candidate Ontologies partitioning to five classes: "Liver", "Gallbladder", "Bile Duct", "Pancreas" and "Other Class". After that, we took all organs and diseases related in Hepatobiliary system diseases of candidate Ontologies and deleted all diseases are not related. Additionally, we unified the common concept (organ name) to have single organ: Liver, Gallbladder, Bile Duct and Pancreas in candidate Ontologies.

Table 5. Candidate ontologies partitioning

MESH Ontology				
Liver	**Gallbladder**	**Bile Duct**	**Pancreas**	**Other Class**
Alagille Syndrome, Autoimmune Hepatitis, Focal Nodular Hyperplasia, Liver Cirrhosis, Galactosemia, Gilbert Disease, Hemochromatosis, Hepatitis A, Hepatitis B, Hepatitis C, Hepatitis D, Hepatic Encephalopathy, Liver Neoplasms, Non-Alcoholic Fatty Liver Disease, Reye Syndrome, Glycogen Storage Disease Type I, Hepatolenticular Degeneration, Liver Failure, Echinococcosis.	Bile Reflux, Biliary Dyskinesia, Cholangitis, Primary Sclerosing Cholangitis, Cholecystitis, Gallstones, Cholestasis, Gallbladder Neoplasms.	Bile Duct Neoplasms, Primary Biliary Cirrhosis, Biliary Atresia, Choledocholithiasis.	Chronic Pancreatitis, Hereditary Pancreatitis, Pancreatic Neoplasms, Cystic Fibrosis.	"Amino Acids, Peptides, and Proteins", "Anesthesia and Analgesia", "Archaea", "Bacteria", Bacterial "Infections and Mycoses", "Behavior and Behavior Mechanisms", "Biological Factors", "Cells", "Complex Mixtures", "Chemical Actions and Uses", "Chemically-Induced Disorders", "Dentistry", "Eye Diseases", "Enzymes and Coenzymes".
MEDDRA Ontology				
Liver	**Gallbladder**	**Bile Duct**	**Pancreas**	**Other Class**
Liver Tumor, Hepatic Adenoma.	Gallbladder Polyp, Porcelain Gallbladder.	Bile Duct Stricture	Acute Pancreatitis, Diabetes	"Cardiac disorders", "Ear and labyrinth disorders", "Endocrine disorders", "Eye disorders", "Gastrointestinal disorders".
SNOMED CT Ontology				
Liver	**Gallbladder**	**Bile Duct**	**Pancreas**	**Other Class**
Liver Cyst, Storage Disease.	-	-	-	"Clinical finding", "Event", "Observable entity", "Record artifact", "Specimen".

On the other hand, we mapped all diseases related to Hepatobiliary system diseases into candidate Ontologies to specific four organ classes: "Liver", "Gallbladder", "Bile Duct" and "Pancreas", and mapped the four organs classes into general "Organ" class to merge them into HSD Ontology under "Hepatobiliary System" class as shown in Fig. 2.

4.4 Adding Knowledge Phase

In this phase, we added knowledge from several medical sources such as domain experts, medical book and trusted medical sites. Then we studied all diseases that have been selected in the previous phase, with their symptoms/signs, treatments and disease causal relations into the Hepatobiliary system diseases.

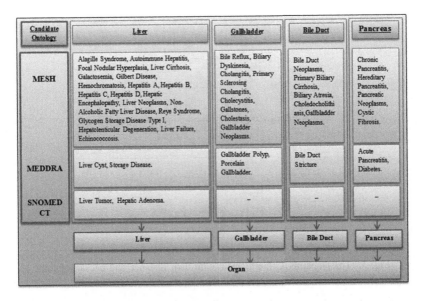

Fig. 2. Disease/Organ classes mapping

4.5 Ontology Merging and Validation Phase

In this phase, the domain experts have been consulted to review results of Hepato-biliary system diseases, symptoms\signs, treatments and disease causal relations. The domain experts have validated the mapping and partitioning, editing some of termi-nologies, and rephrasing some of the classes' names. Also, they have added other classes to Human Body tree such as: "Organ", "Liver Disease", "Gallbladder Disease", "Bile Duct Disease", "Pancreas Disease", "Treatment", etc. In addition, the domain experts have merged the "Organ" class under the "Hepatobiliary System" class. Also, they merged the "Hepatobiliary System" class under the "System" class.

- Developing HSD Ontology in OWL

We used the combination process as top down and bottom up processes [21]. We built HSD Ontology in pathology domain as complete manual process by protégé editor. This Ontology includes 230 classes, 258 instances and many relations. The Hepato-biliary_System class contains Organ subclass which includes 4 subclasses: "Liver", "Gallbladder", "Bile_Duct" and "Pancreas". The Liver class has Liver_Disease subclass that includes 23 subclasses such as Hepatitis_C class contains three subclasses: "Hepatitis_C_Symptom_and_Sign" subclass includes 31 instances, "Hepatitis_C_Treatment" subclass includes 2 instances and "Hepatitis_C_Causal_Relation" class includes 16 subclasses as shown in Fig. 3.

The main benefit of the developed HSD Ontology is containing of long lists of diseases, symptoms\signs, treatments and the relationships between diseases that give both physicians and students of medicine an easy way to diagnose and treat any disease accurately. In addition, HSD Ontology puts the physicians in a good point to know

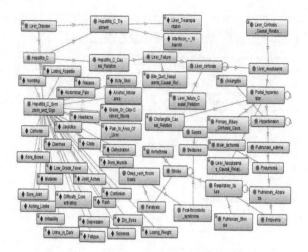

Fig. 3. The Hepatitis_C hierarchical with their instance and causal relations

many of serial complications of specific disease. It helps to simplify prognosis of the disease. This HSD Ontology provides an important source to understand the nature of Hepatobiliary system disorders.

5   Beneficial Scenario: Liver Disease Causal Relations

This scenario provides great benefit to the physicians and medical students to have information about prognosis and complications of Hepatobiliary system diseases. This scenario helps physicians to be aware about the future dangers of diseases. Hepatitis C is an infectious disease that may lead to very serious complications, if physicians did not deal with such cases seriously. For example, Hepatitis C could lead to Liver failure then Liver failure may develop Liver Cirrhosis then Liver Cirrhosis could lead to Liver Neoplasms etc. Beneficially, this scenario informs physicians about the next forms of specific diseases to help physicians to be completely enthusiastic to treat patients well.

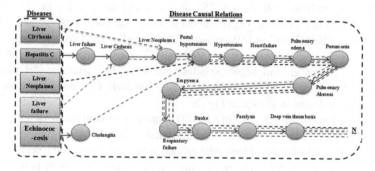

Fig. 4.   Liver disease causal relations

The scenario shown in Fig. 4 aims to help physicians and medical students to know the prognosis of the diseases that have causal relation.

6   Conclusion

The Hepatobiliary System is very important for human vital processes. It has the ability to regulate the other systems. In this paper, we exploited the existed Ontologies in the medical domain to build a new Hepatobiliary System Diseases (HSD) Ontology. The Ontology has been built by using the BioPortal system to find the correct candidate Ontologies. This Ontology was developed in pathology domain and is represented in the Web Ontology Language (OWL) that has recently become the standard language for the semantic web. The HSD Ontology development methodology includes five phases: HSD Query Extraction phase, Ontology Selection phase, Ontology Mapping and Partitioning phase, Adding Knowledge phase, and Ontology Merging and Validation phase. By developing the HSD Ontology in pathology domain, both intelligent systems and physicians can share, reason, and exploit this knowledge in different ways.

References

1. Bright, T.J., Furuya, E.Y., Kuperman, G.J., Cimino, J.J., Bakken, S.: Development and evaluation of an ontology for guiding appropriate antibiotic prescribing. J. Biomed. Inf. (2011)
2. Sánchez, D., Moreno, A.: Learning medical ontologies from the web. In: Riaño, D. (ed.) K4CARE 2007. LNCS (LNAI), vol. 4924, pp. 32–45. Springer, Heidelberg (2008)
3. Cicortas, A., Iordan, V.S., Fortis, A.E.: Considerations on construction ontologies. J. Ann. Comput. Sci. Ser. 1, 79–88 (2009)
4. Smith, B., Ceusters, W., Klagges, B., Köhler, J., Kumar, A., Lomax, J., Mungall, C., Neuhaus, F., Rector, A.L., Rosse, C.: Relations in biomedical ontologies. Genome Biol. 6 (2005)
5. Bail, S., Horridge, M., Parsia, B., Sattler, U.: The justificatory structure of the NCBO BioPortal ontologies. In: Aroyo, L., Welty, C., Alani, H., Taylor, J., Bernstein, A., Kagal, L., Noy, N., Blomqvist, E. (eds.) ISWC 2011. LNCS, vol. 7031, pp. 67–82. Springer, Heidelberg (2011). doi:10.1007/978-3-642-25073-6_5
6. Noy, N.F. Shah, N.H., Whetzel, P.L. et al.: BioPortal: ontologies and integrated data resources at the click of a mouse. In: International Conference on Biomedical Ontology (ICBO), 24 July 2009
7. Seongwook, Y., Anchit, A., Preetham, C., Paavany, J., Ashish, M., Shikha, S.: Survey about ontology development tools for ontology-based knowledge management. University of Southern California (2009)
8. Thirugnanam, M., Ramaiah, M., Pattabiraman, V., Sivakumar, R.: Ontology based disease information system. In: International Conference on Modelling Optimization and Computing, pp. 3235–3241 (2012)
9. Official Partner of the Liver Strong Foundation (2014). http://www.livestrong.com/article/119869-list-body-systems. Accessed 10 Dec

10. Galal, A.L., Marzoqi, M.A., Ibrahim, F., Moawad, M.R.: A survey on applying ontological engineering approach for hepatobiliary system diseases. In: International Conference on Information Technology (ICIT), pp. 370–375, Egypt (2015)
11. Alfonse, M., Aref, M.M., Salem, A.-B.M.: Ontology-based knowledge representation for liver cancer. In: Proceedings of the International eHealth, Telemedicine and Health ICT Forum for Educational, Networking and Business. Luxembourg, G. D. Luxembourg, ISSN 1818 -9334, pp. 821–825, 18–20 April 2012
12. Kozaki, K., Kou, H., Yamagata, Y., Imai, T., Ohe, K., Mizoguchi, R.: Browsing casual chain in a disease ontology. In: International Semantic Web Conference (Posters & Demos)'12 (2012)
13. Moawad, I.F., Marzoqi, G.A.L., Salem, A.-B.M.: Building OBR-based OWL ontology for viral hepatitis. In: Proceeding of Med-e-Tel, pp. 821–825. Luxembourg (2012)
14. Martinez-Romero, M., Vazquez-Naya, J.M., Pereira, J., Pazos, A.: BiOSS: A system for biomedical ontology selection. Comput. Methods Programs Biomed. **114**, 125–140 (2014)
15. Boon, N.A., Colledge, N.R., Walker, B.R.: Davidson's Principles & Practice of Medicine, 20th edn., pp. 935–994 (2006)
16. Liver Disease Information Center. http://www.liverfoundation.org/abouttheliver/info/
17. Gallbladder Attack. http://www.gallbladderattack.com/gallbladderdisease.shtml. Accessed Mar 2015
18. Medline Plus. https://www.nlm.nih.gov/medlineplus/bileductdiseases.html. Accessed April 2015
19. The National Pancreas Foundation. https://www.pancreasfoundation.org/patient-information/about-the-pancreas/common-disorders-of-the-pancreas/. Accessed April 2015
20. BioPortal. https://bioportal.bioontology.org. Accessed Mar 2016
21. Uschold, M., Gruninger, M.: Ontologies: principles, methods and applications. Knowl. Eng. Rev. **11** (1996)

An Approach
for Opinion-Demographic-Topology Based
Microblog Friend Recommendation

Sherin Moussa$^{(\boxtimes)}$

Department of Information Systems, Faculty of Computer
and Information Sciences, Ain Shams University, Cairo 11566, Egypt
sherinmoussa@cis.asu.edu.eg

Abstract. Through the tremendous increase of users on the microblogging social networks with their associated streams of content, the scarcity of one user's attention arises. The process of filtering such massive content and discovering who other users could be aligned with his own interests would consume much time. Thus, various mechanisms have been investigated to recommend friends by analyzing the posted content, social graph, or user profiles. In this paper, we propose a new approach for microblog friend recommendation based on the opinion, or sentiment, towards the topics in the microblogs combined with the social graph, in addition to the demographic data available in the user profiles, including age, gender, and location. We have deployed a cloud-based recommender service using R language for big data analytics, which applies our proposed approach to gather feedback from real Twitter users. Results show 0.77 average precision value, with 21 % increase rate considering opinion mining.

Keywords: Big data analytics · Opinion mining · Sentiment analysis · Recommender system · Microblogs · Topic models · Social graph · Twitter · Social network

1 Introduction

Microblogging social networks have encountered an impressive growth during the last decade, which allow short sentences exchange of either text, uploaded images, or video links. They don't only preserve social interconnections among users, but they also serve as a main informative channel of interesting content. Nowadays, Twitter has been recognized to be the most popular social network for microblogging services [1, 2]. Studies have been conducted to infer the possible opinion, or sentiment, of Twitter users that resides underneath their posted tweets [3–10]. By May 2015, the number of Twitter users has exceeded 500 million users, having 332 million active users that post around 1 billion tweets a day [11]. With this continuous surge of new users and content, it becomes a real challenge for a user to filter and decide the right users to follow and the right content to share with others that match his interests and preferences. Thus, many personalized recommender systems have been emerged simultaneously to guide Twitter users towards what could be of interest to them among this

© Springer International Publishing AG 2017
A.E. Hassanien et al. (eds.), *Proceedings of the International Conference on Advanced Intelligent Systems and Informatics 2016*, Advances in Intelligent Systems and Computing 533, DOI 10.1007/978-3-319-48308-5_78

exponentially-generated content and users [12]. Some recommendation systems have been dedicated to recommend possibly interesting content or hashtags to read, or to re-tweet. While others have focused to recommend candidate users to follow, either based on the content they post (content-based), or the social user-connections graph they have in common (topology-based). Both approaches were proven to perform similarly at different positions of ranking [1].

At this point, a user may wonder: "If a user tweets about the same topic I'm interested in, or has common friends between us, does this mean that her/his opinion about that topic is the same as mine? Definitely, I won't be willing to let a recommender system suggests new friends contradicting my sentiments! If the suggested friends indeed comply with me, are we of a similar age range? Same gender? Do we live nearby?" In such a situation, both opinion mining and content/topology-based recommendation systems should be integrated to satisfy this newly emergent need. In this study, we propose a new friend recommendation approach for microblogging networks, taking into consideration the opinion, or sentiment, towards the extracted topics from the posted microblogs, as well as the user's social graph. Recommended friends are then classified based on the demographic data in the user profiles. We have developed our proposed approach as a cloud-based recommender service using Twitter APIs and R language for big data analytics in order to gather feedback from real Twitter users.

The rest of the paper is organized as follows: Sect. 2 presents the main researches related to our study that were directed towards friend recommender systems on Twitter. Section 3 introduces our proposed approach, with a detailed explanation of its main components. Section 4 discusses our evaluation methodology and the associated outcomes for the proposed approach. Section 5 then summarizes our study and presents our future work.

2   Related Work

Different approaches have been considered for friend recommendations on Twitter, the widely-accessible microblog social network. In [13], an adaptive social network model driven by social recommendation was proposed, leading to scale-free leadership networks in online societies. Artificial agent-based simulations were used, assuming that users with broad interests and good judgments were considered to likely become popular leaders for the others. However, this paradigm was challenged by the findings that social influence often plays a more important role than similarity of past activities.

In [14, 15], Twittomender was developed using Lucene platform to identify and promote interesting users to be friends (followees) on Twitter, where both tweets and followee/follower user ids were mined by following the links between users on the social graph. Content-based recommendation categorization was applied on the tweets of the users, followees, and followers, whereas the collaborative-based considered the ids of followees, and followers. In [16], a followee recommender system was proposed using an algorithm that explored the topology of followers/followees network of Twitter. The users followed by the followers of a certain user's followees were assumed to be possibly recommended. Candidates were then sorted according to different weighting features. Whereas in [17], a user recommender system on Twitter was introduced based on the bag-of-signal model, which relied on the wavelet transform signal processing technique to define an efficient pattern-based similarity function among users, adding the time dimension to the number of occurrences of the weighted informative topics. An online matrix factorization approach named RMFO was introduced in [18], where Weighted Regularized Matrix Factorization (WRMF) collaborative filtering technique was investigated to create user specific rankings and to recommend topics to users at real-time for tweets data based on individual preferences inferred from the user's past interactions. In [19], Social Regularization concept was presented to denote all social constraints on recommender systems, where a low-rank matrix factorization approach was applied to design a matrix factorization objective function with social regularization. Vector Space Similarity (VSS) and Pearson Correlation Coefficient (PCC) were used to define the similarity between users based on the items they rated in common.

A user recommendation technique was proposed in [20] based on the sentiment-volume-objectivity (SVO) weighting function for the user interests. Naïve Bayes classifier was considered for sentiment analysis at sentence level, assuming that a sentence matches the whole tweet, with only one opinion related to an entity. Another method was described in [21] to infer sentiment-based communities by proposing a community-based approach for user recommendation, while considering users' attitudes towards their own interests without getting the whole social structure. SVO user profiling approach and the weighted Tanimoto similarity metric were used to evaluate user similarity for each topic. Whereas a clustering algorithm based on modularity optimization and the Adamic-Adar tie strength technique were then applied to discover heavily connected users and to finally suggest the most relevant users respectively.

On the other hand, RecLand was developed in [22] as an extension for Katz and TwitterRank based on the topological score for link prediction recommendation approach. Semantic information on users, their relationships and user authority were integrated to recommend users that match their topical interests. A novel two-step method of recommending celebrities was introduced in [23] based on genetic algorithm and color psychology. A cost function of a user was trained and optimized to achieve the ideal weights, and then the celebrities were rated based on the follower and following counts, background color and description parameters using Max Luscher color theory to find psychological similarity of different users. Whilst in [24, 25], another friend recommender system was presented in the social bookmarking domain by analyzing the user behavior represented in the bookmarks tagged by a user and the frequency of each used tag. Tag-based and Resource-based user profiling were applied, where Tag-based similarity was calculated among a target user and the other users using PCC. In [26], a holistic hybrid algorithm using a proposed logistic regression model was introduced to recommend users on Twitter according to content-based, collaborative-based and user-based information (i.e., number of tweets, number of followers and number of followees). The main drawback of this approach is that popular users become more popular, where other less-popular users may have more interesting content to the user but never recommended.

Authors in [27] investigated another friend recommender system using a concept-sensitive hash function. User profiles were modelled as a set of pseudo-cliques, whose vertices were the terms extracted from the user's tweets based on the co-occurrence matrix. A distance metric was then applied between two sets of pseudo cliques to find relevant pseudo cliques in the graph. In [28], natural language processing, fuzzy set concept, and google translate APIs were combined to introduce a recommendation system for Twitter users moving to a new society or culture. The system translated history tweets in the user's mother language to extract interests and then recommended followees, followers and hashtags based on those interests in his new neighborhood. Another approach recommending new followees to Twitter users was proposed in [29] through learning their historical friends-adding patterns. Scores were computed based on a user's past social graph and his interactions with other connected users, and then passed into a learning machine along with the lately added list of followees of the user. A learning to rank approach was then applied to detect the best combination of recommendation strategies the user considered to add new followees in the past. Whereas in [30], reciprocal relationships were used to develop a friend recommender system. A user-based collaborative filtering algorithm was modified to generate name recommendations based on the geographical location.

While most of the current hybrid algorithms combine content and/or collaborative features plainly - by computing their average or product in order to make recommendations using LDA or TF-IDF weighting schemes, our approach is set apart from previous researches in four ways: (1) We propose a novel combination between the user's topics of interest, his opinion and sentiment analyzed about those interests, in addition to his social graph in order to recommend friends. (2) We apply a different weighting scheme proven to be effective. (3) We consider demographic data for classifying and ranking recommendations. (4) A user-defined-parameter recommendation service.

3 Opinion-Demographic-Topology-Based Friend Recommender

Opinion-Demographic-Topology (ODT)-based microblog friend recommendation approach consists of two main components as explained below. Both the processing engine and the data layers are cloud-based, where JSON web services are used to connect between the web interface and the server-side as presented in Fig. 1 for the system architecture. R language packages are used to perform the big data analytic techniques required [31, 32].

3.1 ODT-Based Web Interface

This is the main web application interface through which users interact with the proposed friend recommender approach. Users are requested to create their ODT-based profiles, where their Twitter accounts are associated. Users can build their own queries to search for certain topics and get friend recommendations based on the entered criteria according to our ODT-based friend recommendation engine. The resultant analytics with any related visual representation are then displayed.

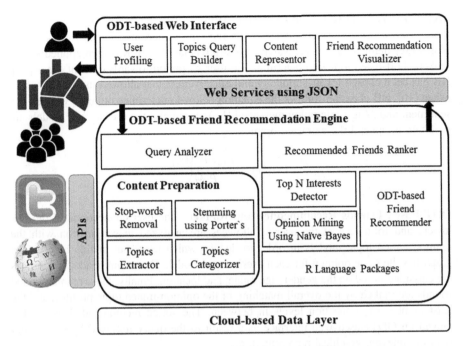

Fig. 1. ODT-based microblog friend recommender system architecture

3.2   ODT-Based Friend Recommendation Engine

This core component is triggered upon receiving a query. The entities specified in the query are analyzed to retrieve the relevant topic profiles previously prepared as clarified in the content preparation module. In the following, we discuss its main modules.

Content Preparation. Upon the user's ODT-based profile association with his Twitter account through Twitter API, all the user's and his followees' tweets are retrieved for further processing, without those of his followers. This is due to two facts: (1) each user automatically receives the tweets posted by his followees, who share content related to their interests. (2) Following another user is a free decision made by the user. Hence, it seems rational to consider that the followees' tweets could offer some insights about a user's interests. However, it is inaccurate to associate this user with his followers' content, as the user cannot control what they post or who follows them. Therefore, Stop words, negligible words, punctuation characters, and numbers are removed, emotions are replaced, and then Porter's algorithm is applied for stemming [33]. Latent Dirichlet Allocation including followees (LDA+f) scores are used to construct the weighting scheme, which was proved to outperform LDA model [26]. Topicmodels R package is considered to extract the topics mentioned in both the user's and his followees' tweets, which applies LDA model [34]. Wikipedia is integrated to serve as a knowledge base to resolve disambiguated entities. Topic profiles are then generated, in which topics are categorized to build ranked sub-trees for each entity. Thus, this user's profile is now associated with his topic profiles for later analysis.

Top N Interests Detector and Opinion Mining. This module ranks the generated topic profiles per user profile, determining the top N interesting topics for each user. N is a user-defined parameter. Naïve Bayes (NB) classifier in R is then applied to this ranked short-listed topics to determine the user's opinion towards each of them [32]. NB has been chosen for its simplicity and effectiveness that allowed its widely-spread usage [10]. NB classifier follows the equation below to determine opinion, where s is a sentiment and M is a microblog message [4]. Accordingly, the user's topic models are modified to append the extracted opinion.

$$P(s|M) = \frac{P(s).P(M|s)}{P(M)}$$

ODT-Based Friends Recommender and Ranker. The user connections of the followers and followees of the user's followees are analyzed in conjunction with the opinion-mined topic models in order to deduce which users out of this huge social graph has same opinion for each user's topic model. The recommended candidates are classified based on the gender, age range, and location. Finally, those recommended friends are ranked by the overall matching of the opinion-mined topic profiles, and the least number of connection levels in between. The sorted recommended list is displayed on ODT-based web interface as a result to the query retrieved by the system with the relevant graphical representation.

4 Experiments and Results

We have carried out an offline evaluation using a Twitter dataset for 78,637 users (nodes) and 1,710,106 interconnections (edges). Each user has a minimum of 50 tweets over a period of 1 year. We have created their associated ODT-based profiles and constructed the required topic profiles with opinion appended. We have selected a diverse set of users representing active users, passive users, moderate users, with high, moderate, and few number of followers, and high, moderate, and few number of followees. The selected list of users represented 40 % of the whole dataset, which we considered as the test users. Hence, the pool of users to recommend from for each of the test users represented 60 % of the whole dataset, where the original interconnections between the test users and the recommendation pool users are made hidden.

Experiments have investigated the precision value of our ODT-based friend recommendation approach, where it represents the successful rate of suggesting a friend from the recommendation pool that is already in the original social graph of a test user. We have studied the results based on the top 4, 6, 8, and 10 interesting opinion-matched topic profiles, taking into consideration the depth of the social graph to the second, third, and fourth connection levels in order to generate a list of 5, 10, and 15 recommended users. Results are ranked according to the previously stated parameters. As shown in Fig. 2, the precision values range from 0.65 till 0.73 for the different 5, 10, and 15 recommendation list size of opinion-matched topic profiles, with an average value of 0.7. This indicates that the number of topic profiles doesn't effectively change

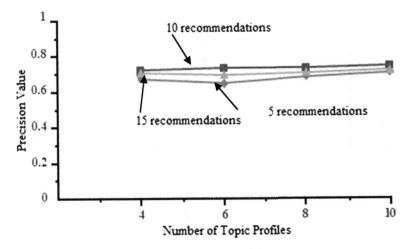

Fig. 2. Precision value for the different number of topic profiles

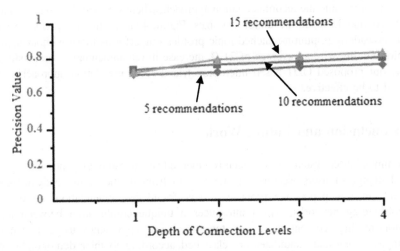

Fig. 3. Precision value for the different connection levels

the successful rate of suggesting friends rather than the matching between opinions among their topic profiles using our proposed approach.

Figure 3 shows the precision values for the different depth levels of social graph connections (i.e. levels of followers and followees) versus the 5, 10, and 15 recommendation list sizes, where the average number of opinion-matched topic profiles has been considered. Experiments show that the deeper the connection levels engaged, the higher the precision value. The average precision value is 0.77 and the highest precision value is 0.88, which is obtained when the fourth connection level is included for a recommendation list size of 15. We have also investigated the effect of applying

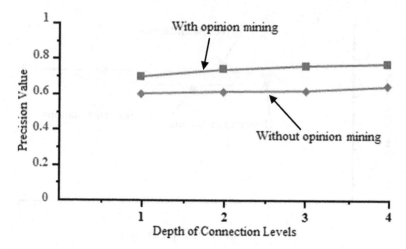

Fig. 4. Precision value for opinion-matched versus non topic profiles

opinion mining into the recommendation approach, where we used the average value of the 5, 10, and 15 recommendation list sizes. Figure 4 shows that the precision values while considering opinion-matched topic profiles outperforms those without applying opinion mining with an average of 21 % increase in recommendation successful rate. Thus, our proposed ODT-based microblog friend recommendation approach is considered to be effective.

5   Conclusion and Future Work

Recommendation systems have recently emerged in microblogging networks to face the challenge of users' scarcity of attention in front of the dramatically increasing content. In this paper, we propose a new friend recommendation approach for microblogging networks, which introduces a unique combination between topics' opinion mining, content-based and topology-based approaches using R language packages. Suggested candidates are classified according to their demographic data. A cloud-based recommendation service connected to Twitter and Wikipedia APIs is developed to evaluate the proposed approach. Experiments show that the number of topic profiles doesn't effectively change the successful rate of suggesting friends rather than the opinions matched between the topic profiles. However, the more social graph levels engaged for followees and followers, the higher the successful rate of suggesting friends, where the highest precision value is 0.88. In addition, considering opinion mining in our proposed friend recommendation approach has achieved an average of 21 % increase in recommendation successful rate.

As our future work, we will investigate other personalized techniques to enhance the quality of results. In addition, we will study our approach on a different type of social networks, i.e. Instagram, where the nature of content is remarkably different.

References

1. Armentano, M.G., Godoy, D., Amandi. A.: Towards a followee recommender system for information seeking users in twitter. In: Workshop on Semantic Adaptive Social Web (SASWeb 2011). CEUR Workshop Proceedings, vol. 730, pp. 27–38 (2011)
2. Chen, K., Chen, T., Zheng, G., Jin, O., Yao, E., Yu, Y.: Collaborative personalized tweet recommendation. In: 35th International ACM SIGIR Conference on Research and Development in Information Retrieval, pp. 661–670 (2012)
3. Gokulakrishnan, B., Priyanthan, P., Ragavan, T., Prasath, N., Perera, A.: Opinion mining and sentiment analysis on a twitter data stream. In: 2012 IEEE International Conference on Advances in ICT for Emerging Regions (ICTer), pp. 182–188 (2012)
4. Pak, A., Paroubek, P.: Twitter as a corpus for sentiment analysis and opinion mining. In: LREc, vol. 10, pp. 1320–1326 (2010)
5. Bifet, A., Frank, E.: Sentiment knowledge discovery in twitter streaming data. In: Pfahringer, B., Holmes, G., Hoffmann, A. (eds.) DS 2010. LNCS, vol. 6332, pp. 1–15. Springer, Heidelberg (2010)
6. Grosse, K., Chesnevar, C.I., Maguitman, A.G.: An argument-based approach to mining opinions from Twitter. In: AT 918, pp. 408–422 (2012)
7. Agarwal, A., Xie, B., Vovsha, I., Rambow, O., Passonneau, R.: Sentiment analysis of twitter data. In: Workshop on Languages in Social Media, pp. 30–38. Association for Computational Linguistics (2011)
8. Thelwall, M., Buckley, K., Paltoglou, G.: Sentiment in Twitter events. J. Am. Soc. Inf. Sci. Technol. 62(2), 406–418 (2011)
9. Wakade, S., Shekar, C., Liszka, K.J., Chan, C.: Text mining for sentiment analysis of Twitter data. In: International Conference on Information and Knowledge Engineering (IKE), The Steering Committee of the World Congress in Computer Science, p. 1. Computer Engineering and Applied Computing (WorldComp) (2012)
10. Vinodhini, G., Chandrasekaran, R.M.: Sentiment analysis and opinion mining: a survey. Int. J. 2(6), 282–292 (2012)
11. Wikipedia: https://en.wikipedia.org/wiki/Twitter
12. Kywe, S.M., Lim, E.-P., Zhu, F.: A survey of recommender systems in Twitter. In: Aberer, K., Flache, A., Jager, W., Liu, L., Tang, J., Guéret, C. (eds.) SocInfo 2012. LNCS, vol. 7710, pp. 420–433. Springer, Heidelberg (2012)
13. Zhou, T., Medo, M., Cimini, G., Zhang, Z., Zhang, Y.: Emergence of scale-free leadership structure in social recommender systems. PLoS One 6(7), e20648 (2011)
14. Hannon, J., McCarthy, K., Smyth, B.: Finding useful users on twitter: twittomender the followee recommender. In: Clough, P., Foley, C., Gurrin, C., Jones, G.J., Kraaij, W., Lee, H., Mudoch, V. (eds.) ECIR 2011. LNCS, vol. 6611, pp. 784–787. Springer, Heidelberg (2011)
15. Hannon, J., Bennett, M., Smyth, B.: Recommending twitter users to follow using content and collaborative filtering approaches. In: Fourth ACM Conference on Recommender Systems, pp. 199–206. ACM (2010)
16. Armentano, M.G., Godoy, D.L., Amandi, A.A.: A topology-based approach for followees recommendation in Twitter. In: Workshop Chairs, p. 22 (2011)
17. Arru, G., Gurini, D. F., Gasparetti, F., Micarelli, A., Sansonetti, G.: Signal-based user recommendation on twitter. In: 22nd International Conference on World Wide Web Companion, pp. 941–944. International World Wide Web Conferences Steering Committee (2013)

18. Diaz-Aviles, E., Drumond, L., Gantner, Z., Schmidt-Thieme, L., Nejdl, W.: What is happening right now... that interests me? online topic discovery and recommendation in twitter. In: 21st ACM International Conference on Information and Knowledge Management, pp. 1592–1596. ACM (2012)

19. Ma, H., Zhou, D., Liu, C., Lyu, M.R., King, I.: Recommender systems with social regularization. In: Fourth ACM International Conference on Web Search and Data Mining, pp. 287–296. ACM (2011)

20. Gurini, D.F., Gasparetti, F., Micarelli, A., Sansonetti, G.: A sentiment-based approach to twitter user recommendation. In: RSWeb@ RecSys (2013)

21. Gurini, D.F., Gasparetti, F., Micarelli, A., Sansonetti, G.: iSCUR: interest and sentiment-based community detection for user recommendation on Twitter. In: Dimitrova, V., Kuflik, T., Chin, D., Ricci, F., Dolog, P., Houben, G.-J. (eds.) UMAP 2014. LNCS, vol. 8538, pp. 314–319. Springer, Heidelberg (2014)

22. Dahimene, R., Constantin, C., Mouza, C.D.: Recland: a recommender system for social networks. In: 23rd ACM International Conference on Conference on Information and Knowledge Management, pp. 2063–2065. ACM (2014)

23. Tajbakhsh, M.S., Aghababa, M.P., Solouk, V., Akbari-Moghanjoughi, A.: Friend recommendation based on the Luscher color theory: Twitter use case. In: 2013 IEEE Malaysia International Conference on Communications (MICC), pp. 218–221. IEEE (2013)

24. Manca, M., Boratto, L., Carta, S.: Design and architecture of a friend recommender system in the social bookmarking domain. In: Science and Information Conference (SAI), pp. 838–842. IEEE (2014)

25. Manca, M., Boratto, L., Carta, S.: Mining user behavior in a social bookmarking system-a delicious friend recommender system. In: DATA, pp. 331–338 (2014)

26. Guimarães, S., Ribeiro, M.T., Assunção, R., Meira Jr., W.: A holistic hybrid algorithm for user recommendation on twitter. J. Inf. Data Manage. 4(3), 341 (2013)

27. Bamba, P., Subercaze, J., Gravier, C., Benmira, N., Fontaine, J.: The twitaholic next door: scalable friend recommender system using a concept-sensitive hash function. In: 21st ACM International Conference on Information and Knowledge Management, pp. 2275–2278. ACM (2012)

28. AlMeshary, M., Abhari, A.: A recommendation system for Twitter users in the same neighborhood. In: 16th Communications & Networking Symposium, p. 1. Society for Computer Simulation International (2013)

29. Islam, M., Ding, C., Chi, C.: Personalized recommender system on whom to follow in Twitter. In: 2014 IEEE Fourth International Conference on Big Data and Cloud Computing (BdCloud), pp. 326–333. IEEE (2014)

30. Jamil, N., Alhadi, A.C., Noah, S.A.: A collaborative names recommendation in the Twitter environment based on location. In: 2011 International Conference on Semantic Technology and Information Retrieval (STAIR), pp. 119–124. IEEE (2011)

31. R Documentation: https://stat.ethz.ch/R-manual/R-devel/library/stats/html/00Index.html

32. Meyer, D., Dimitriadou, E., Hornik, K., Weingessel, A., Leisch: Package 'e1071'. http://cran.rproject.org/web/packages/e1071/index.html

33. Porter, M.F.: An algorithm for suffix stripping. Program 14(3), 130–137 (1980)

34. Hornik, K., Grün, B.: topicmodels: an R package for fitting topic models. J. Stat. Softw. 40(13), 1–30 (2011)

Plant Recommender System Based on Multi-label Classification

Alaa Tharwat[1,2,4(✉)], Hani Mahdi[1], and Aboul Ella Hassanien[3,4]

[1] Faculty of Engineering, Ain Shams University, Cairo, Egypt
engalaatharwat@hotmail.com
[2] Faculty of Engineering, Suez Canal University, Ismailia, Egypt
[3] Faculty of Computers and Information, Cairo University, Giza, Egypt
[4] Scientific Research Group in Egypt (SRGE), Cairo, Egypt
http://www.egyptscience.net

Abstract. In this paper, a plant recommender system using 2D digital images of leaves is proposed. This system made use of feature fusion technique and the multi-label classification method. Feature fusion technique is used to combine the color, shape, and texture features. Invariant moments, color moments, and Scale Invariant Feature Transform (SIFT) are used to extract the shape, color, and texture features, respectively. The multi-label classification method is capable of classifying samples in more than one class. In multi-label classification method, the nearest neighbor classifier with different metrics is used to match the unknown image with the training images and assigns five different class labels (i.e. recommendations) for each unknown image. The proposed approach was tested using Flavia dataset which consists of 1907 colored images of leaves. The experimental results proved that the accuracy of feature fusion method was much better than all other single features. Moreover, the experiments demonstrated their robustness to provide reliable recommendations.

Keywords: Recommender systems · Multi-label classification · Plant identification · Feature fusion

1 Introduction

Plants play an important role in the life for different purposes such as air, climate, water, and food security. Plants have different species/classes, which are subject to the danger of extinction. Therefore, classify plant species is needed to protect plant resources. Hence, plant identification techniques have become a hot topic in the recent years [1]. Manual plant identification is expensive, time-consuming, and requires more efforts from labors and experts. On the contrary, automatic plant identification based on machine learning is fast, cheap, and accurate; hence, it is important for different parties: agriculture, chemistry, and pharmacological.

© Springer International Publishing AG 2017
A.E. Hassanien et al. (eds.), *Proceedings of the International Conference
on Advanced Intelligent Systems and Informatics 2016*, Advances in Intelligent
Systems and Computing 533, DOI 10.1007/978-3-319-48308-5_79

There are many studies have been done for the plant identification based on computational models. Gaber *et al.* used two variants, i.e. one-dimensional and two-dimensional, of Linear Discriminant Analysis (LDA) and Principal Component Analysis (PCA) to extract the features from the 2D images of Flavia dataset. The results of the two-dimensional approach (i.e. 2D-PCA and 2D-LDA) achieved accuracy rate up to 90 % while the one-dimensional approach (i.e. 1D-PCA and 1D-LDA) achieved accuracy ranged from 69 % to 75 % [1]. Caglayan *et al.* proposed plant identification model based on color and shape features of the leaf images to classify 32 different kinds of plants. They used Support vector Machines (SVM), k-Nearest Neighbor (k-NN), Random Forest, and Naive Bayes classifiers and the best accuracy achieved reached to 96 % [2]. In [3], Arun *et al.* used the PCA as a feature extraction and dimensionality reduction method and k-NN classifier. They used Flavia dataset and achieved accuracy ranged from 78 % to 81.3 %. Alaa *et al.* combined the features of color (color moments), shape (invariant moments), and texture (Scale Invariant Feature Transform (SIFT)) features to build a classification model [4]. They used Falvia dataset and their proposed model based on the combined features (i.e. feature fusion) achieved results better than all other individual features.

Recommendation systems are used to give different suggestions about items or services to attract more clients than traditional ones. Recommendation system has two main categories, namely, content-based and collaborative filtering. In the content-based method, the items or products are recommended to a user based on the items he has interested previously, while collaborative filtering methods use the preferences of the user who target the recommendation and the opinions of the other users (i.e. neighbors) who has similar preferences. There are many data mining methods have been investigated in the application domain of recommender systems. Multi-label classification method has been used in different applications and it is used in the recommender system domain [5].

The multi-label classification method is used to assign a set of labels to an unknown sample. This method can be used to classify a user in one or more profiles and also to predict the ratings of the items. In this paper, a plant identification model is proposed. This model has three phases. In the first phase, the color, shape, and texture features of the 2D images of leaves are extracted, normalized, and then fused. The features are then reduced in the second phase using LDA, which also discriminate between different classes. In the third phase, the Nearest Neighbor (NN) classifier is used to classify an unknown sample. However, we have modified NN classifier to generate different ranks instead of on class label. These ranks are then combined, i.e. rank level aggregation, to determine the ranks of all classes regarding the unknown sample.

The rest of the paper is organized as follows. Section 2, introduces the preliminaries of this research. Section 3, presents the details of proposed model. Experimental results and discussion are presented in Sect. 4. Finally, conclusions are summarized in Sect. 5.

2   Preliminaries

2.1   Multi-label Classification

The aim of multi-label classification algorithms is to assign more than one class label for each unknown sample. In other words, each unknown sample is classified to more than one class. These algorithms are needed in many applications such as semantic classification of scenes and educational content in E-learning environments. For example, an image can represent more than one concept such as beach and late afternoon at the same time. Therefore, several labels can be assigned to a learning pattern. Multiple similar cases can be found in the literature [6].

Mathematically, in a multi-label classification algorithm, the finite set of labels is denoted by, $L = \{\lambda_j : j = 1, \ldots, q\}$ and the training samples are denoted by $D = \{(x_i; Y_i), i = 1, \ldots, m\}$, where x_i is i^{th} training sample and $Y_i \subset L$ is the subset of labels of the sample i. The subset of labels of the sample i is denoted by a binary vector, $Y_i = \{y_1, y_2, \ldots, y_q\}$, where $y_j = 1$ indicates the presence of the label λ_i in the set of relevant labels for x_i. Finally, the goal of the classification multi-labels it is to assign the correct set of labels for a new sample x_{test}, that is, to predict if the label λ_j must be assigned or not to the example x_{test} [5].

In multi-label classification, there are two main steps: multi-label classification (MLC) and label ranking (LR). MLC refers to the learning model which is used to divide the space of labels into relevant or irrelevant labels, while LR step is used to sort the labels according to their relevance for a given sample [6].

2.2   Recommender Systems

Recommender systems (RS) are widely used in E-commerce. As mentioned in Sect. 1, RS is classified into collaborative filtering and content-based approach. In collaborative filtering method, nearest neighbor algorithms are used to predict the items preferences for a user based on the opinions of other users. These opinions are the rating scores which obtained from different users [7]. Collaborative filtering has two algorithms as follows:

- Memory-based (or user-based) which is also known as the nearest neighbor algorithm. In this algorithm, the users' items are treated by means of statistical techniques to find the users with similar preferences to predict the preferences or recommendations for the current user. This method is easy to implement, but it is slow in the large datasets [7].
- Model-based (or item-based) algorithms which use the data mining techniques to predict the user preferences. However, this method can only be applied in a homogeneous domain. Moreover, this algorithm suffers from sparsity due to the number of ratings that need predict is much greater than the number of the obtained ratings [5,7]. Another limitation is the complexity of the algorithm which increased linearly with both the number of users and items.

2.3   Feature Extraction Method

The aim of feature extraction methods is to measure some properties of the object and transform these properties to numeric values. There are different types of feature extraction method. In this research, we will use color, shape, and texture features.

Color Features: Color features algorithms are widely used in different applications. In this method, the image's colors are transformed into numeric values. Color features are robust against rotation and scaling. In addition, it is easy to implement, fast and it needs low storage requirements compared with texture features [2]. In this paper, color moments will be used as color features.

Color Moments: Color moments features consist of the first order (i.e. mean), second order (i.e. variance), and third order (i.e. skewness). Hence, color moments are represented by only nine numbers that represent three moments for each color components [2]. The first color moment of the kth color component ($k = 1, 2, 3$) is denoted by, $M_k^1 = \frac{1}{XY} \sum_{x=1}^{X} \sum_{y=1}^{Y} f_k(x, y)$, where XY represents the total number of pixels of the image and $f_k(x, y)$ is the color value of the kth color component of the image pixel (x, y). The hth moment of kth color component can be calculated as follows, $M_k^h = \frac{h}{XY} \sum_{x=1}^{X} \sum_{y=1}^{Y} (f_k(x, y) - M_k^1)^{\frac{1}{h}}$, $h = 2, 3, \ldots$.

Shape Features: This type of features is used to measure the similarity between shapes. These similarity measurements can be classified into two main categories: region-based and contour-based methods. In the region-Based methods, the whole area of an image or object is used to extract the features, while in contour-based methods, the extracted features represent the contour of the shape of the object such as shape descriptors (e.g. length irregularity, circularity, and complexity) [2]. Shape features methods are easy to implement and relatively robust against transformation and rotation. In this paper, invariant moments method will be used as a shape feature extraction method.

Invariant Moments: Invariant moments was first introduced in [8]. Invariant moments method is robust against rotation and translation, but it is sensitive to noise, which limits the use of moments in many applications. The set of seven moments (ϕ_n) can be calculated as reported in [8,9].

Texture Features: In texture features method, the texture of each region in the image is described and then transformed in numeric values. It has two types of texture features, namely, sparse descriptor and dense descriptor. In the sparse descriptor methods, the keypoints (i.e. interest points) in a given image are first located and then a local patch around these keypoints are constructed and then the features are extracted. *Scale Invariant Feature Transformation* (SIFT) [10] and Speed up Robust Feature (SURF) [11] are the most well-known algorithms in this type. In the dense descriptor-based method, local features are extracted from every point over the input image. *Local Binary Pattern* (LBP) [12] and *Weber's Local Descriptor* (WLD) [13] are two well-known algorithms in the dense descriptor-based methods [14]. In this paper, SIFT algorithm is used as a texture feature extraction method.

Scale Invariant Feature Transform (SIFT): SIFT algorithm has five main steps. Firstly, the scale-space function from the input image is constructed. This is achieved by convolution the input image with Gaussian filters of varying scales. The *Difference of Gaussian* (DoG) is represented as the difference between two nearby scales separated by a constant multiplicative factor k as follows, $D(x, y, \sigma) = L(x, y, k\sigma) - L(x, y, \sigma)$, where $I(x, y)$ is the input image, $L(x, y, \sigma) = G(x, y, \sigma) * I(x, y)$ and $L(x, y, k\sigma) = G(x, y, k\sigma) * I(x, y)$ are two images that produced from the convolution of Gaussian filters with σ and $k\sigma$, respectively, and $G(x, y, \sigma) = \frac{1}{2\pi\sigma^2} exp[-\frac{x^2+y^2}{\sigma^2}]$ is the Gaussian function. Secondly, the interest points in DoG pyramids which are called keypoints are located. These keypoints represent the local maxima or minima of $D(x, y, \sigma)$ by comparing each point with the pixels of all its neighbors. Thirdly, a low contrast poorly localized keypoints are eliminated. Fourthly, one or more orientation are assigned to the keypoints based on local image properties. For each keypoint, an orientation histogram is calculated from the gradient orientations of the sample points within a region around that keypoint. In the orientation histogram, the peaks represent dominant directions of the local gradients and highest peak with any other local peaks are used to create the orientation of the keypoint. Fifthly, the descriptor for each patch is created [10, 15].

2.4   Linear Discriminant Analysis (LDA)

LDA is one of the supervised dimensionality reduction methods. The main goal of LDA is to transform the data from a high dimensional space to a lower dimensional space by preserving the most discriminative data. The LDA searches for space, which maximizes the ratio of between-class variance (S_b) to the within-class variance (S_w). The within-class variance represents the variance between the samples of the same class while the between-class variance represents the variance or distance between different classes [1, 16].

2.5   Feature Fusion

The aim of the feature fusion technique is to combine different feature vectors to give more representative features for the patterns. The features are combined by concatenating it into one feature vector. Because the features are collected from different sources; hence, the features need to be normalized using Z_{Score} method to make the features compatible. Z_{score} normalization method is used to map the input features into a distribution with mean of *zero* and standard deviation of *one* as follows, $\acute{f_i} = \frac{f_i - \mu_i}{\sigma_i}$, where f_i represents the ith feature vector, μ_i and σ_i are the mean and standard deviation of the ith feature vector, respectively, $\acute{f_i}$ is

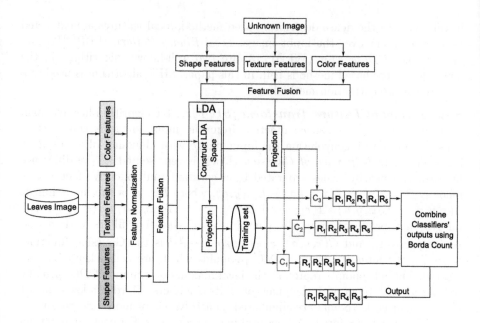

Fig. 1. A block diagram of the proposed model.

the ith normalized feature vector [17–19]. Moreover, due to high dimensionality of the combined features, dimensionality reduction technique such as LDA or PCA are used to reduce the dimension of features [18].

3   Proposed Approach

The proposed approach consists of three phases. In the first phase, the color moments as color features, invariant moments to extract shape features, and SIFT to extract texture features for the 2D images of leaves are extracted (see Fig. 1). Due to different ranges of the features, in this phase, the features are normalized. Next, the normalized features are fused into one feature vector. Due to high dimensionality, in the second phase, LDA is used to reduce the dimension of features and remove irrelevant features. In the last and third phase, Nearest Neighbor (NN) classifier is used to classify an unknown sample. Instead of assign one class label to each unknown sample, the NN classifier is modified to determine the nearest five classes to the unknown sample. Moreover, in this phase, three different metrics (Euclidean, cosine, and city-block) are used with NN classifier and the outputs (i.e. ranks) of the three classifiers are combined using Borda count method. More details about Borda count method is reported in [20].

4   Experimental Results and Discussion

In this section, two experimental scenarios are performed. The aim of the first experiment is to identify a leaf plant image using single feature extraction method (i.e. shape, color, or texture), while the aim of the second experiment is to identify the leaf image using all features (i.e. feature fusion approach).

4.1   Experimental Setup

In this research, Flavia leaves images dataset was used. Flavia dataset consists of 20 different plants species/classes and each class has 20 images with size 1600×1200. The images are in different orientation, illumination and quality as shown in Fig. 2.

Fig. 2. Sample of different leaves' images (one sample from each class or plant).

In all experiments, invariant moments, color moments, and SIFT feature extraction methods are used to extract the shape, color, and texture features, respectively. Moreover, the number of training images was five, while the remaining images (20 images) were used to test the proposed model. The accuracy rate (i.e. the ratio of the total number of correctly classified sample to the total number of test samples) is used to test the performance of our proposed method.

4.2   Experimental Scenarios

Single Feature Extraction Method: This experiment is designed to test all single feature extraction method, i.e. color moments, invariant moments, and SIFT features. Tables 1, 2, and 3 summarize the results obtained from this scenario. As shown from tables, many notices can be seen. First, the shape features achieved the lowest accuracy, while the texture features achieved the best results. Further analysis of the shape results shows that the accuracy of the first rank (i.e. single label) was 30 %, while total accuracy that was calculated from all ranks (i.e. multi-label) was $(30 + 30 + 15 + 5 + 5 = 85\%)$. In other words, the accuracy of the proposed model increased by assigning more class labels, i.e. recommendations, to the unknown samples. This results proved that how the multi-label classification is more suitable because it assigns more class labels with different ranks or weights to the unknown sample instead of one class label and hence it increased the accuracy of the proposed model and by determining the most relevant class labels instead of determining only one class label. Second, the proposed model using color feature achieved results better than shape features and total accuracy was 90 %. Third, the results were improved when the texture features were used and the total accuracy was 100 %.

Table 1. Results of the proposed model using color features.

Unknown Samples	True class label	Rank1	Rank2	Rank3	Rank4	Rank5
I_1	C_1	C_4 ★★★★	C_9 ★★★	C_{14} ★★	C_{20} ★★	C_6 ★
I_2	C_2	C_7 ★★★★★	C_2 ★★★★	C_{11} ★★★	C_3 ★	C_{13} ★
I_3	C_3	C_{17} ★★★★★	C_3 ★★★★	C_{12} ★★	C_2 ★★	C_{19} ★
I_4	C_4	C_4 ★★★★★	C_{14} ★★★★	C_1 ★★★	C_{15} ★★	C_9 ★
I_5	C_5	C_5 ★★★★	C_6 ★★★	C_9 ★★★	C_8 ★★	C_{14} ★
I_6	C_6	C_{14} ★★★★	C_6 ★★★	C_9 ★★	C_{15} ★★	C_1 ★
I_7	C_7	C_{13} ★★★★★	C_{11} ★★★★	C_7 ★★★	C_5 ★★	C_2 ★
I_8	C_8	C_5 ★★★★	C_8 ★★★★	C_6 ★★	C_7 ★★	C_9 ★
I_9	C_9	C_8 ★★★★	C_9 ★★★	C_{14} ★★	C_{20} ★★	C_1 ★
I_{10}	C_{10}	C_{10} ★★★★★	C_{15} ★★★★	C_6 ★★	C_1 ★★	C_9 ★
I_{11}	C_{11}	C_{11} ★★★★★	C_7 ★★★★	C_2 ★★★	C_3 ★	C_{17} ★
I_{12}	C_{12}	C_{12} ★★★★	C_{12} ★★★★	C_{19} ★★★	C_{17} ★★	C_3 ★
I_{13}	C_{13}	C_{13} ★★★★★	C_{11} ★★★	C_7 ★★	C_2 ★★	C_8 ★
I_{14}	C_{14}	C_{14} ★★★★★	C_1 ★★★	C_4 ★★	C_{15} ★★	C_6 ★
I_{15}	C_{15}	C_{15} ★★★★	C_1 ★★★	C_{14} ★★	C_6 ★★	C_{10} ★
I_{16}	C_{16}	C_8 ★★★★	C_7 ★★★	C_{16} ★★	C_9 ★★	C_6 ★
I_{17}	C_{17}	C_{17} ★★★★★	C_3 ★★★★	C_{12} ★★	C_{11} ★★	C_{18} ★
I_{18}	C_{18}	C_{18} ★★★★	C_{12} ★★★★	C_{19} ★★★	C_{17} ★★	C_2 ★
I_{19}	C_{19}	C_{19} ★★★★	C_3 ★★★	C_{12} ★★	C_{12} ★★	C_{18} ★
I_{20}	C_{20}	C_{16} ★★★★★	C_8 ★★★★	C_7 ★★	C_5 ★★	C_9 ★
Accuracy (%)		$\frac{10}{20}$ (50%)	$\frac{6}{20}$ (30%)	$\frac{2}{20}$ (10%)	0	0

Table 2. Results of the proposed model using shape features.

Unknown Samples	True class label	Rank1	Rank2	Rank3	Rank4	Rank5
I_1	C_1	C_1 ★★★★★	C_{19} ★★★	C_4 ★★	C_{14} ★	C_{13} ★
I_2	C_2	C_{18} ★★★★★	C_9 ★★★	C_8 ★★	C_{12} ★★	C_2 ★
I_3	C_3	C_7 ★★★★	C_{17} ★★★	C_5 ★★	C_{20} ★★	C_2 ★
I_4	C_4	C_4 ★★★★★	C_{13} ★★★	C_{19} ★★★	C_{11} ★★	C_6 ★
I_5	C_5	C_{17} ★★★★	C_7 ★★★	C_5 ★★★	C_9 ★★	C_2 ★
I_6	C_6	C_6 ★★★★★	C_{13} ★★★	C_{15} ★★	C_{15} ★	C_{14} ★
I_7	C_7	C_{17} ★★★	C_5 ★★★	C_9 ★★	C_7 ★★	C_2 ★
I_8	C_8	C_8 ★★★★★	C_{12} ★★★★	C_{18} ★★★	C_9 ★★	C_{17} ★
I_9	C_9	C_9 ★★★★	C_{17} ★★★	C_5 ★★	C_2 ★★	C_7 ★
I_{10}	C_{10}	C_{14} ★★★★★	C_{10} ★★★	C_{15} ★★	C_6 ★	C_{16} ★
I_{11}	C_{11}	C_7 ★★★★	C_{20} ★★★	C_{13} ★★	C_5 ★★	C_3 ★
I_{12}	C_{12}	C_8 ★★★★★	C_{12} ★★★★	C_{18} ★★★	C_5 ★★	C_{17} ★
I_{13}	C_{13}	C_7 ★★★★	C_{20} ★★★	C_{13} ★★	C_5 ★★	C_3 ★
I_{14}	C_{14}	C_6 ★★★★	C_{14} ★★★	C_{15} ★★	C_{16} ★	C_{13} ★
I_{15}	C_{15}	C_{14} ★★★★★	C_{10} ★★★	C_{15} ★★	C_{16} ★	C_6 ★
I_{16}	C_{16}	C_8 ★★★★★	C_{12} ★★★★	C_{18} ★★★	C_9 ★★	C_{17} ★
I_{17}	C_{17}	C_{17} ★★★★★	C_5 ★★★	C_7 ★★	C_9 ★★	C_2 ★
I_{18}	C_{18}	C_{18} ★★★★	C_9 ★★★	C_2 ★★	C_{17} ★★	C_8 ★
I_{19}	C_{19}	C_{14} ★★★★	C_{19} ★★★	C_4 ★★	C_6 ★	C_{13} ★
I_{20}	C_{20}	C_7 ★★★★	C_{20} ★★★	C_{17} ★★	C_5 ★★	C_9 ★
Accuracy (%)		$\frac{6}{20}$ (30%)	$\frac{6}{20}$ (30%)	$\frac{3}{20}$ (15%)	$\frac{1}{20}$ (5%)	$\frac{1}{20}$ (5%)

Table 3. Results of the proposed model using texture features.

Unknown Samples	True class label	Rank1	Rank2	Rank3	Rank4	Rank5
I_1	C_1	C_1 ★★★★★	C_{14} ★★★★	C_{15} ★★★	C_{16} ★★	C_9 ★
I_2	C_2	C_2 ★★★★★	C_8 ★★★	C_{19} ★★★	C_{12} ★★	C_{14} ★
I_3	C_3	C_3 ★★★★★	C_{17} ★★★★	C_{20} ★★★	C_8 ★	C_9 ★
I_4	C_4	C_4 ★★★★★	C_{14} ★★★	C_{15} ★★	C_6 ★★	C_1 ★
I_5	C_5	C_5 ★★★★★	C_7 ★★★★	C_{20} ★★	C_{16} ★★	C_{14} ★
I_6	C_6	C_6 ★★★★★	C_4 ★★★★	C_{14} ★★★	C_{16} ★★	C_1 ★
I_7	C_7	C_7 ★★★★★	C_5 ★★★★	C_{16} ★★	C_4 ★★	C_{20} ★
I_8	C_8	C_8 ★★★★★	C_6 ★★★★	C_{12} ★★	C_{18} ★★	C_{19} ★
I_9	C_9	C_9 ★★★★★	C_4 ★★★	C_{16} ★★	C_6 ★★	C_2 ★
I_{10}	C_{10}	C_{10} ★★★★★	C_{15} ★★★★	C_{14} ★★★	C_1 ★★	C_4 ★
I_{11}	C_{11}	C_{11} ★★★★★	C_8 ★★★★	C_{12} ★★	C_6 ★★	C_9 ★
I_{12}	C_{12}	C_{12} ★★★★★	C_{19} ★★★	C_8 ★★	C_2 ★★	C_{16} ★
I_{13}	C_{13}	C_{13} ★★★★★	C_2 ★★★	C_8 ★★	C_{19} ★★	C_{20} ★
I_{14}	C_{14}	C_2 ★★★★★	C_{14} ★★★★	C_{15} ★★★	C_1 ★★	C_4 ★
I_{15}	C_{15}	C_{15} ★★★★★	C_{14} ★★★	C_{10} ★★	C_1 ★★	C_9 ★
I_{16}	C_{16}	C_{16} ★★★★★	C_{14} ★★★★	C_2 ★★	C_{15} ★★	C_8 ★
I_{17}	C_{17}	C_{17} ★★★★★	C_{20} ★★★★	C_{19} ★★	C_9 ★★	C_8 ★
I_{18}	C_{12}	C_{18} ★★★★★	C_{18} ★★★	C_8 ★★	C_{16} ★★	C_{19} ★
I_{19}	C_{19}	C_{19} ★★★★★	C_{16} ★★★	C_{12} ★★	C_2 ★★	C_{18} ★
I_{20}	C_{20}	C_{20} ★★★★★	C_{17} ★★★	C_{16} ★★	C_{14} ★★	C_7 ★
Accuracy (%)		$\frac{18}{20}$ = (90%)	$\frac{2}{20}$ = (10%)	0	0	0

Table 4. Results of the proposed model using all features.

Unknown Samples	True class label	Rank1	Rank2	Rank3	Rank4	Rank5
I_1	C_1	C_1 ★★★★★	C_{14} ★★★★	C_4 ★★★	C_{15} ★★	C_6 ★
I_2	C_2	C_2 ★★★★★	C_8 ★★★★	C_{19} ★★★	C_{12} ★★	C_{18} ★
I_3	C_3	C_3 ★★★★★	C_{17} ★★★★	C_{20} ★★★	C_9 ★★	C_{19} ★
I_4	C_4	C_4 ★★★★★	C_1 ★★★★	C_6 ★★	C_{14} ★★	C_9 ★
I_5	C_5	C_5 ★★★★★	C_7 ★★★★	C_{20} ★★★	C_{16} ★★	C_9 ★
I_6	C_6	C_6 ★★★★★	C_8 ★★★	C_4 ★★	C_1 ★★	C_{16} ★
I_7	C_7	C_7 ★★★★★	C_5 ★★★★	C_{19} ★★★	C_9 ★	C_{16} ★
I_8	C_8	C_8 ★★★★★	C_6 ★★★★	C_2 ★★	C_{12} ★★	C_{18} ★
I_9	C_9	C_9 ★★★★	C_2 ★★★	C_{16} ★★	C_4 ★★	C_5 ★
I_{10}	C_{10}	C_{10} ★★★★★	C_{15} ★★★★	C_{14} ★★★	C_1 ★★	C_{16} ★
I_{11}	C_{11}	C_{11} ★★★★★	C_8 ★★★★	C_9 ★★	C_{12} ★★	C_8 ★
I_{12}	C_{12}	C_{12} ★★★★★	C_{19} ★★★★	C_2 ★★★	C_{18} ★★	C_8 ★
I_{13}	C_{13}	C_{13} ★★★★★	C_2 ★★★★	C_8 ★★	C_{19} ★	C_{11} ★
I_{14}	C_{14}	C_{14} ★★★★★	C_{15} ★★★★	C_{16} ★★	C_1 ★	C_4 ★
I_{15}	C_{15}	C_{15} ★★★★★	C_{10} ★★★★	C_{14} ★★★	C_1 ★★	C_{16} ★
I_{16}	C_{16}	C_{16} ★★★★★	C_{14} ★★★★	C_{15} ★★★	C_8 ★★	C_2 ★
I_{17}	C_{17}	C_{17} ★★★★★	C_{20} ★★★★	C_9 ★★	C_{18} ★★	C_{19} ★
I_{18}	C_{18}	C_{18} ★★★★★	C_{19} ★★★	C_{20} ★★★	C_{12} ★★	C_9 ★
I_{19}	C_{19}	C_{19} ★★★★★	C_{12} ★★★	C_2 ★★★	C_{18} ★★	C_8 ★
I_{20}	C_{20}	C_{20} ★★★★★	C_{17} ★★★★	C_9 ★★	C_7 ★	C_1 ★
Accuracy (%)		$\frac{20}{20}$ = (100%)	0	0	0	0

Feature Fusion: The aim of this experiment is to evaluate the proposed model using feature fusion approach. In this experiment, the color, shape, and texture features are combined into one feature vector and the accuracy is tested to show how it improved the accuracy through using many different features. Table 4 summarizes the results obtained from this scenario. As shown, the total accuracy was 100 % and 100 % of the accuracy was found in the first rank class label, which reflects that how the robustness of the proposed model. Moreover, the proposed model using all features achieved results better than all other individual features.

To further evaluate our proposed model, we compared it by some state-of-the-art approaches reported in [1,4] which used Flavia dataset. In terms of accuracy, the proposed model achieved the best accuracy (about 100 %), while the proposed models [1,4] achieved 95 % and 70 %, respectively. Moreover, the proposed model built a recommender system, which offers or suggests five class labels with different ratings.

To conclude, the feature fusion method achieved results better than all other single feature extraction methods and the multi-label classification is suitable to build a recommendation system for plant identification system, which offers more than one decision for the unknown sample.

5   Conclusions

In this paper, we have proposed a plant recommender system to identify plants using 2D leaves' images. In this model, color, shape, and texture features were used for feature extraction. Moreover, all individual features were combined into one feature vector. Nearest neighbor classifiers were used with different metrics to calculate the nearest five class labels to the unknown sample (i.e. multi-label classification) and then Borda count method was used to aggregate all ranks to determine the final ranks (i.e. label ranking). The proposed model works as a multi-label classification method to build the plant recommender system.

The experimental results have proved that texture feature extraction method achieved the best results compared with color and shape features. Moreover, the feature fusion method achieved accuracy (100 %) better than all single feature extraction methods. Moreover, the recommender system offers different class labels with different ranks or ratings. In the future work, we are going to test our approach against a large database of leaves' images. This would allow us to evaluate whether our approach will give the same good results and improve it in the future in order to develop an expert system for plant identification.

References

1. Gaber, T., Tharwat, A., Snasel, V., Hassanien, A.E.: Plant identification: Two dimensional-based vs. one dimensional-based feature extraction methods. In: Herrero, Á., Sedano, J., Baruque, B., Quintián, H., Corchado, E. (eds.) 10th International Conference on Soft Computing Models in Industrial and Environmental Applications. AISC, vol. 368, pp. 375–385. Springer, Heidelberg (2015). doi:10.1007/978-3-319-19719-7_33
2. Caglayan, A., Guclu, O., Can, A.B.: A plant recognition approach using shape and color features in leaf images. In: Petrosino, A. (ed.) ICIAP 2013, Part II. LNCS, vol. 8157, pp. 161–170. Springer, Heidelberg (2013)
3. Priya, C.A., Balasaravanan, T., Thanamani, A.S.: An efficient leaf recognition algorithm for plant classification using support vector machine. In: International Conference on Pattern Recognition, Informatics and Medical Engineering (PRIME), pp. 428–432. IEEE (2012)

4. Tharwat, A., Gaber, T., Awad, Y.M., Dey, N., Hassanien, A.E.: Plants identification using feature fusion technique and bagging classifier. In: Gaber, T., Hassanien, A.E., El-Bendary, N., Dey, N. (eds.) The 1st International Conference on Advanced Intelligent System and Informatics (AISI2015), November 28-30, 2015, Beni Suef, Egypt. AISC, vol. 407, pp. 461–471. Springer, Heidelberg (2016). doi:10.1007/978-3-319-26690-9_41

5. Carrillo, D., López, V.F., Moreno, M.N.: Multi-label classification for recommender systems. In: Pérez, J.B., et al. (eds.) Trends in Practical Applications of Agents and Multiagent Systems, pp. 181–188. Springer, Heidelberg (2013)

6. Tsoumakas, G., Katakis, I., Vlahavas, I.: Random k-labelsets for multilabel classification. IEEE Trans. Knowl. Data Eng. **23**(7), 1079–1089 (2011)

7. Schafer, J.B., Konstan, J.A., Riedl, J.: E-commerce recommendation applications. In: Kohavi, R., Provost, F. (eds.) Applications of Data Mining to Electronic Commerce, pp. 115–153. Springer, Heidelberg (2001)

8. Hu, M.K.: Visual pattern recognition by moment invariants. IRE Trans. Inf. Theory **8**(2), 179–187 (1962)

9. Elfattah, M.A., ELsoud, M.A.A., Hassanien, A.E., Kim, T.: Automated classification of galaxies using invariant moments. In: Kim, T., Lee, Y., Fang, W. (eds.) FGIT 2012. LNCS, vol. 7709, pp. 103–111. Springer, Heidelberg (2012)

10. Tharwat, A., Gaber, T., Hassanien, A.E., Shahin, M.K., Refaat, B.: SIFT-based arabic sign language recognition system. In: Abraham, A., Krömer, P., Snasel, V. (eds.) Afro-European Conf. for Ind. Advancement. AISC, vol. 334, pp. 359–370. Springer, Heidelberg (2015)

11. Ahmed, S., Gaber, T., Tharwat, A., Hassanien, A.E., Snáel, V.: Muzzle-based cattle identification using speed up robust feature approach. In: 2015 Proceedings of the International Conference on Intelligent Networking and Collaborative Systems (INCOS), pp. 99–104. IEEE (2015)

12. Tharwat, A., Gaber, T., Hassanien, A.E.: Two biometric approaches for cattle identification based on features and classifiers fusion. Int. J. Image Min. **1**(4), 342–365 (2015)

13. Gaber, T., Tharwat, A., Hassanien, A.E., Snasel, V.: Biometric cattle identification approach based on weber's local descriptor and adaboost classifier. Comput. Electron. Agric. **122**, 55–66 (2016)

14. Chen, J., Shan, S., He, C., Zhao, G., Pietikainen, M., Chen, X., Gao, W.: WLD: A robust local image descriptor. IEEE Trans. Pattern Anal. Mach. Intell. **32**(9), 1705–1720 (2010)

15. Tharwat, A., Mahdi, H., Hennawy, A., Hassanien, A.E.: Face sketch recognition using local invariant features. In: 2015 7th International Conference of Soft Computing and Pattern Recognition (SoCPaR), pp. 117–122. IEEE (2015)

16. Tharwat, A., Mahdi, H., Hennawy, A.E., Hassanien, A.E.: Face sketch synthesis and recognition based on linear regression transformation and multi-classifier technique. In: Gaber, T., Hassanien, A.E., El-Bendary, N., Dey, N. (eds.) The 1st International Conference on Advanced Intelligent System and Informatics (AISI2015), November 28-30, 2015, Beni Suef, Egypt. AISC, vol. 407, pp. 183–193. Springer, Heidelberg (2016). doi:10.1007/978-3-319-26690-9_17

17. Ibrahim, A., Tharwat, A.: Biometric authenticationmethods based on ear and finger knuckle images. Int. J. Comput. Sci. Issues (IJCSI) **11**(3), 134–138 (2014)

18. Semary, N.A., Tharwat, A., Elhariri, E., Hassanien, A.E.: Fruit-based tomato grading system using features fusion and support vector machine. In: Filev, D., Jabłkowski, J., Kacprzyk, J., Krawczak, M., Popchev, I., Rutkowski, L., Sgurev, V., Sotirova, E., Szynkarczyk, P., Zadrozny, S. (eds.) Intelligent Systems'2014. AISC, vol. 323, pp. 401–410. Springer, Heidelberg (2015). doi:10.1007/978-3-319-11310-4_35

19. Tharwat, A., Ibrahim, A., Hassanien, A.E., Schaefer, G.: Ear recognition using block-based principal component analysis and decision fusion. In: Kryszkiewicz, M., Bandyopadhyay, S., Rybinski, H., Pal, S.K. (eds.) PReMI 2015. LNCS, vol. 9124, pp. 246–254. Springer, Heidelberg (2015). doi:10.1007/978-3-319-19941-2_24

20. Sharif, M.M., Tharwat, A., Hassanien, A.E., Hefny, H.A., Schaefer, G.: Enzyme function classification based on borda count ranking aggregation method. In: Ryżko, D., Gawrysiak, P., Kryszkiewicz, M., Rybiński, H. (eds.) Machine Intelligence and Big Data in Industry. SBD, vol. 19, pp. 75–85. Springer, Heidelberg (2016). doi:10.1007/978-3-319-30315-4_7

An Ontology-Based Framework for Linking Heterogeneous Medical Data

Rashed Salem, Basma Elsharkawy[(✉)], and Hatem Abdel Kader

Faculty of Computers and Information, Information Systems Department,
Menoufia University, Shibin al Kawm, al Minufiyah, Egypt
basma_mohamed11@hotmail.com

Abstract. Clinical records contain a massive heterogeneity number of data, generally written in free-note without a linguistic standard. Other forms of medical data include medical images with/without metadata (e.g., CT, MRI, radiology, etc.), audios (e.g., transcriptions, ultrasound), videos (e.g., surgery recording), and structured data (e.g., laboratory test results, age, year, weight, billing, etc.). Consequently, to retrieve the knowledge from these data is not trivial task. Handling the heterogeneity besides largeness and complexity of these data is a challenge. The main purpose of this paper is proposing a framework with two-fold. Firstly, it achieves a semantic-based integration approach, which resolves the heterogeneity issue during the integration process of healthcare data from various data sources. Secondly, it achieves a semantic-based medical retrieval approach with enhanced precision. Our experimental study on medical datasets demonstrates the significant accuracy and speedup of the proposed framework over existing approaches.

Keywords: Data integration · Heterogeneity · Image retrieval · CBIR · Semantic ontology

1 Introduction

Big data is a collection of data sets in a large variety of domains [15], healthcare is one of such domains. There are different types of data including structured, semi-structured, and unstructured. Statistically, 80 % of medical data are unstructured, which further complicates the management of these data [11, 21]. The major source for healthcare applications is patient records. Data integration is the task of combining different data sources, and providing a unified view of the data. Such integrated data are needed to be standardized and kept in a repository, i.e., data warehouses, for ease of retrieval and analytics later [16].

However, integrating data from a variety of sources is not a trivial task, due to the large volumes of heterogeneous data during mapping, ranking, and key matching [9]. Moreover, structural and semantic heterogeneity is another problem that faces data integration [10, 13]. In this paper, we address the problem of resloving structural and semantic heterogeneity for healthcare applications. While structural heterogeneity addresses schema conflicts, semantic heterogeneity addresses, meaning conflicts,

© Springer International Publishing AG 2017
A.E. Hassanien et al. (eds.), *Proceedings of the International Conference on Advanced Intelligent Systems and Informatics 2016*, Advances in Intelligent Systems and Computing 533, DOI 10.1007/978-3-319-48308-5_80

e.g., synonyms and homonyms conflicts. Fortunately, semantic Web can be exploited to resolve semantic heterogeneity issue.

Using semantic Web, the same concepts which given by several words, i.e., synonyms, as well as the different concepts given by the same word, i.e., homonyms, can be defined.

Semantic technologies (e.g., Ontology) provide major solutions to semantic interoperability in healthcare systems. Moreover, ontologies can deliver solutions for image retrieval. They seek to map the low-level image features with high- level ontology concepts. Compared with content-based and keyword-based image retrieval, ontology-based retrieval concentrates on capturing semantic content. Furthermore, ontology plays an important role to represent the knowledge as a set of concepts within a domain and the relationships between pairs of concepts. Ontologies can be used to support a variety of tasks in different domains including knowledge representation, natural language processing, information retrieval, database integration, digital libraries, etc.

This paper proposes a framework in which different medical data types are merged into a unified format. The proposed framework tackles heterogeneity issue during the data integration process. By the proposed framework physicians can build a patient's history record, and thus helps physicians in decision-making. The proposed framework keeps all patient history without losing any data. Furthermore, this paper proposes a semantic-based framework for medical data retrieval.

This paper is organized as follows; Sect. 2 presents a literature review of clinical data management. The proposed semantic-based framework for resolving heterogeneity and retrieval of healthcare data is discussed in Sect. 3. Section 4 presents handling several cases of medical data. Implementation and discussion of the proposed framework are provided in Sect. 5. Section 6 concludes the work and highlights the future work.

2 Related Work

There are many approaches proposed in the literature for managing clinical records, including, chunking, data-driven, free-text assignment codes, content- based image retrieval, and semantic-based image retrieval. The chunking approach is proposed to identify non-recursive words and base noun phrases in the text, i.e., a key issue in symptom and disease identification as a term. Thus, it extracts structured data from clinical records easily. Chunking handles data annotated by medical domain using a chunk annotation scheme with extra credibility, which involves symbols as noun phrases (NPs), main verb (MVs), and a common annotation for adjectival and adverbial phrases (APs) [19]. However, there is a very limited amount of annotated text of this kind available for healthcare systems [22].

In data-driven approaches, a driving data element that is an independent variable should be selected. The independent variable is used to determine the other linked patient information. Syndrome/sub-syndrome classification and 3-digit ICD-9 final diagnosis code are used to determine the driving data element. The data element that is used to realize the patient's record would have a clear and easily recognized relationship. The mapping from data element is well defined if the patient has been grouped by a single value of this data element. The challenge is not only in data storage and access, but also in scalability of healthcare sources [4].

Medical image retrieval can help physicians in finding information that assist them in decision making. Medical image retrieval systems extract features as color, texture, shape and spatial relationships. Image features are extracted from the full image and then are indexed. The variety of medical image types make the process of retrieval is a non-trivial task. For instance, radiology images face many difficulties [5]. Particularly, such radiology images contain rich information and specific features that need to be carefully recognized for medical image analysis.

Image retrieval systems are generally classified into two major approaches. The first approach searches local or global image features such as color or texture. The other approaches add keywords to images as an annotation. Content-based Image Retrieval (CBIR) approach is considered a rapidly advancing research area. It depends on searching similarity of image features from a database based on the color, shape and texture [5]. Images are presented as a query against image database. The similarity between image features in the database are retrieved with the help of indexing images [7, 20]. The indexing of images provides a rapid path for searching image databases [3]. However, there is still a "Semantic gap" between what users need and what CBIR systems can achieve. In particular, there are no sensible means by which queries can be presented to CBIR systems [14, 18].

The Semantic-based Image Retrieval (SBIR) systems include several components of information extraction, such as a textual description and visual feature, and a semantic image retrieval. The extraction process of SBIR is based on low-level features of images to identify objects. Open issues are the nature of digital images, as well as descriptions of images, i.e., high-level concepts such as rat and dogs. However, the main problem is the semantic gap discrepancy between low-level features and high-level concepts [12]. Moreover, different users at a different time may give different interpretations for the same image [1, 17]. Table 1 provides a summarized comparison among medical retrieval approaches and our proposed framework.

3   Proposed Semantic-Based Framework for Integrating Medical Data

Clinical data are represented in structured and unstructured form. Surgical producers, treatment and drugs are examples of structured data. Structured data can be computerized and allow performing analysis of data, queries and aggregation for patient records. Generally, structured data are represented in a restricted field through a record or file is called structured data. This involves data which contained in relational databases and tables. They are organized in a mightily mechanized and manageable structure. Structured data are prepared for seamless integration into a database or well structured file format. Structured data needs to stay comparatively simplistic and uncomplicated [11]. Moreover, structured data depend on a data model. The data model specifies how data will be generated, stored, processed and accessed [6]. Structured data are generated through constrained choices in the form of data entry, which overall

Table 1. Medical data type retrieval approaches.

	Data-driven	Chunking	SBIR	CBIR	Proposed Framework
Data type	Free text	Free text	Image	Image	Image, Free text, audio, and video
Data missing	Lossy	Lossy	Lossy	Lossy	Loss-less
Challenges	(1) Data scalability (2) Data access	(1) Identification of medical concepts (2) Clarification of medical concepts relations	(1) The different forms of images (2) Lack of relation between objects and the meaning	(1) Semantic understanding of media is visual (2) Integrating, Searching, Selecting	(1) Grew up of global ontology
Precision	Inaccurate due to data missing	Inaccurate in medical concepts	Inaccurate with medical images	Inaccurate with medical images	Accurate
Performance	Degrading with large database	Degrading with medical concepts	Degrading with general concepts	Degrading with large database	High performance
Scalability	Low scale	Low scale	Low scale	Low scale	High scale

drop-down menus, check boxes, and pre-filled templates. This type of data is easily searchable and aggregated, can be analyzed and reported, and is linked to other information resources. The high cost and performance limitations of storage, memory and processing allows relational databases and spreadsheets using structured data are the only way to effectively manage data [6]. Unlike, unstructured clinical data may contain free clinical notes and multimedia content such as medical images and voice. These files may have an interior structure, nevertheless they are still considered as an unstructured form because the data which they contain don't care appropriate sorted into a database. The concept of "big data" is widely associated with unstructured data. Big data denote to extremely large datasets that are difficult to analyze with traditional tools [2, 6].

There are a variety of challenges for handling clinical notes, including ungrammatical, short phrases and abbreviations. The proposed framework helps in solving these challenges of clinical notes, and the heterogeneity of clinical record. In addition to structured medical data, the proposed framework merges medical images, clinical notes and audio data type into a unified framework. Then, physicians can perform a "DL language" or "SPARQL" query to access the different data types. The proposed medical image retrieval framework handles medical images as url and gives a label for each medical image. In the case of medical reports, which consist of both medical images and text, the process of text extraction is performed firstly before performing medical image processing.

4 Multiform Medical Data Handling

The proposed framework focuses on integrating data from heterogeneous medical, data sources. Herein, we introduce how the proposed framework handles unstructured medical data such as clinical notes, medical images, physician's reports and audio data type.

In Fig. 1, patient record has been handled either as clinical note, medical image, audio type or patient report. Then the local ontology will be merged with the global one. The physicians or administrator allow doing any query on the global ontology.

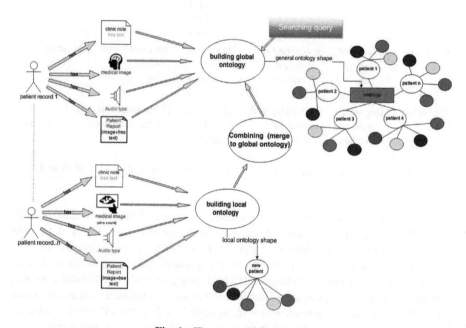

Fig. 1. The proposed framework

4.1 Clinical Text - Free Form

Clinical text has a wealthy detailed information of great possibility usage to scientists and health service researchers. Text schema describes the structure of a text file and how a text document is read or written in a raw format. The structure of text stream defines either fixed column widths or columns which are separated by delimiters. To convert a text schema to XML Schema, a specify separator, field separator and the field names of the text file should be determined. In other side, XML Schemas contain annotations for providing additional information, such as medical information. The conversion process between XML and CSV has been done automatically by detecting all repeated elements in XML that are used for splitting data to rows.

The proposed framework can handle free text as a block. Clinical note is extracted from clinical notes of physician's prescriptions. While the input is free text, the expected output merges all patient data and performing query to get all data merged for the patient. There are three processes in managing clinical notes.

Firstly, extracting clinical notes process has done from physician's prescriptions and the output of this process is in unstructured form. Secondly, transforming process for unstructured form has done to get semi-structured form as to facility the dealing when building an ontology. Finally, transforming XML form to get a structured form, as in "CSV file" which allow to build RDF "ontology" file where queries can be carried out to access the medical data.

4.2   Medical Image Data Type

The second form is medical images in healthcare application. The major challenge in this case is how to handle these large number of images with their various formats and then merging them with other medical data types. The proposed framework helps in solving this challenge by building an ontology for these images and access any of the images of its label through the global ontology. Physicians can get medical image in an ontology by its label or URL, see Fig. 2(a). DICOM images as an example, consist of two parts, i.e., text header and binary image. Medical image metadata as well as field names or image's URL is indexed and transformed into XML elements.

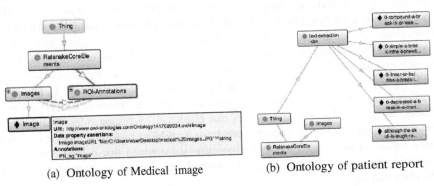

(a) Ontology of Medical image   (b) Ontology of patient report

Fig. 2. Both ontology of medical images and patient report

4.3   Audio Media Data Type

Audio media type is represented in health care as medical diagnosis from physicians abroad or encounters between physicians and patients. There are five processes to complete this stage. The first process is performing speech to text process. The second process is acting as the first process in free text data type. Output from the first process should be converted to semi-structured form as XML form. Then a structured form represented in CSV file has been created and global ontologies have been built, which we can perform query simply to access all media types.

4.4   The Medical Patient Report

Output report may contain free text and medical images. Thus, images are extracted firstly from reports and then ontology is created for both image and extracted free text. This case involves handling both free text and medical images as discussed previously. The proposal framework deals with patient report depending on the extracted free text of the report and create an ontology for free text see Fig. 2(b). Medical images have been handled as a normal image, but also we can get pure image and build its ontology.

5   Implementation and Discussion

The proposed framework is implemented with freely and open-source tools, i.e., semantic Web technologies. The proposed ontology is configured as follows: number of classes, number of individuals and number of properties are 21,58 and 68, respectively. Moreover, maximum depth, maximum number of children and average number of children are 5, 75 and 4, respectively. Indeed, a local ontology for each media type of medical information is created, and then such local ontologies are merged into global ones. The global ontology being built is efficiently scalable; new patient records can be added and merged easily. Thus, all patients medical information is integrated into their history without loss of any data. The global ontology merges other healthcare domain ontology such as accounts, hospital budget, geographical places analogies, etc. By merging all this information, a background for patient history is complete, which can help physicians in decision making. Furthermore, the technician can apply queries against the global ontology using either URI, labels, DL language or SPARQLE. Searching inside the ontology is tested several times, the searching process is simple, and the results are returned quickly and accurately as shown in Fig. 3. The main challenge in retrieval system was the correlation between medical concepts. Fortunately, the proposal framework tackles this challenge by adding rules, and relations among concepts. In Fig. 3, three patients have a problem with long, but two of them have cancer in the lung. Therefore, physicians get alarmed that patient 3 may have cancer in the lung. Accordingly, the proposal framework can help physicians in decision making.

To implement CBIR approach for comparison with the proposed framework, dataset of 68 medical images is used. Note that, the main problem facing retrieval systems is semantic gap and the relationships among medical concepts [8]. The proposed framework tackles this issue, including semantic gap and related concepts. CBIR systems retrieve medical images according to the distance similarity between the query image and the dataset as shown in Fig. 4. By the experiments, CBIR query is an image showed in Fig. 4(a). The results of a CBIR system depend on the distance similarity between the query image and the dataset. Figure 4(b) shows the result from CBIR.

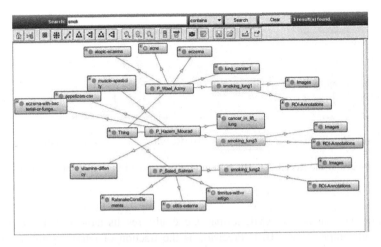

Fig. 3. Relationships of medical concepts

However, it can be noticed easily that query is a chest medical images has cancer and the results show different images, including broken hand, thus it leads to low accuracy.

However, the proposed framework retrieves all medical images according to their labels. For example, if applying searched against the ontology using labels or annotator "smoking lung", physicians retrieve medical images of three patients have a smokers lung. Moreover, physicians can observe that patient 1 and 2 has a ray cancer from the retrieved results. Thus, they conclude that patient 3 may be infected by cancer as shown in Fig. 3. Therefore, the proposed framework exploits related concepts and helps physicians in decision making. To calculate the accuracy of the proposed framework against CBIR, Fig. 5(a) and (b) shows the precision and recall measures, receptively. The precision of the proposed framework is better than CBIR, although the proposal framework takes into account the human errors. Moreover, the recall of the proposed framework outperforms the CBIR due to the challenges of CBIR which tackled by the proposed framework. Furthermore, Fig. 5(c) shows the consumed time for

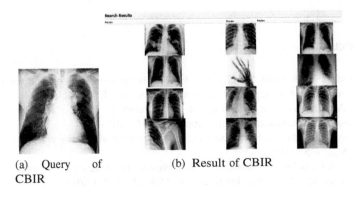

(a) Query of CBIR   (b) Result of CBIR

Fig. 4. CBIR-based retrieval

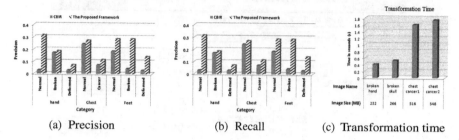

| (a) Precision | (b) Recall | (c) Transformation time |

Fig. 5. Precision and Recall of CBIR system vs. the proposed framework and it's transformation time

transforming images into XML format. Generally, results indicate that the proposed framework outperforms CBIR. Precision is the fraction of retrieved images that are relevant to physician's information need.

6 Conclusion

Managing unstructured clinical data is one of the major problems in healthcare systems. The heterogeneity of clinical data is considered as a critical roadblock to achieving integration and interoperability between systems. In this paper, we proposed a semantic-based framework for managing heterogeneous medical data, including free clinical notes, audio data, and medical images. It tackles the heterogeneity by unifying different medical data types into a unified form and building ontology. The ontology keeps all patient history and enables queries for patient and disease history. The manipulation of the proposed framework is demonstrated by the different medical cases.

The future step is to handle medical videos, which can be transformed into medical images, and integrate them with different medical data types. Moreover, we plan to enrich the framework with spatial and temporal information of patients to discover new insights from analytics.

References

1. Alkhawlani, M., Elmogy, M., El Bakry, H.: Text-based, content-based, and semantic-based image retrievals: A survey. Int. J. Comput. Inf. Technol. **4**(1), 8–66 (2015)
2. Belle, A., Thiagarajan, R., Soroushmehr, S.M., Navidi, F., Beard, D.A., Najarian, K.: Big data analytics in healthcare. Biomed. Res. Int. **2015** (2015)
3. Bhamare, D.P., Abhang, S.A.: Content based image retrieval: A review. Int. J. Comput. Sci. Appl. **8**(2), 1–5 (2015)
4. Buczak, A.L., Babin, S., Moniz, L.: Data-driven approach for creating synthetic electronic medical records. BMC Med. Inform. Decis. Mak. **10**(1), 59 (2010)

5. Chaudhari, R., Patil, A.: Content based image retrieval using color and shape features. Int. J. Adv. Res. Electr. Electron. Instrum. Eng. **1**(5), 386–392 (2012)
6. Grover, N.: 'Big Data'-architecture, issues, opportunities and challenges. IJCER **3**(1), 26–31 (2014)
7. Haldurai, L., Vinodhini, V.: A study on content based image retrieval systems. Int. J. Innovative Res. Comput. Commun. Eng. **3**(3) (2015)
8. Jobay, R., Sleit, A.: Quantum inspired shape representation for content based image retrieval. J. Sign. Inf. Process. **5**(02), 54 (2014)
9. Kadadi, A., Agrawal, R., Nyamful, C., Atiq, R.: Challenges of data integration and interoperability in big data. In: 2014 IEEE International Conference on Big Data (Big Data), pp. 38–40. IEEE (2014)
10. Kang, L., Yi, L., Dong, L.: Research on construction methods of big data semantic model. In: Proceedings of the World Congress on Engineering, vol. 1 (2014)
11. Katal, A., Wazid, M., Goudar, R.: Big data: issues, challenges, tools and good practices. In: 2013 Sixth International Conference on Contemporary Computing (IC3), pp. 404–409. IEEE (2013)
12. Kaur, H., Jyoti, K.: Survey of techniques of high level semantic based image retrieval. Int. J. Res. Comput. Commun. Technol. IJRCCT **2**(1), 15–19 (2013). ISSN: 2278-5841
13. Kienast, R., Baumgartner, C.: Semantic Data Integration on Biomedical Data Using Semantic Web Technologies. INTECH Open Access Publisher (2011)
14. Kulkarni, P., Kulkarni, S., Stranieri, A.: A novel architecture and analysis of challenges for combining text and image for medical image retrieval. Int. J. Infonomics (IJI) (2014)
15. Pooja, S.J., Gupta, R.: Big data: advancement in data analytics. Int. J. Comput. Technol. Appl. **5**(4), 1466–1469 (2014)
16. Priyanka, K., Kulennavar, N.: A survey on big data analytics in health care. Int. J. Comput. Sci. Inform. Technol. **5**(4), 5685–5688 (2014)
17. Rahimzadeh, R., Farzan, A., Fathabad, Y.F.: A survey on semantic content based image retrieval and CBIR systems. Int. J. Tech. Phys. Probl. Eng. (IJTPE) (2014). Published by International Organization of IOTPE
18. Sasikala, S., Gandhi, R.S.: Efficient content based image retrieval system with metadata processing. Int. J. Innovative Res. Sci. Technol. **1**(10), 72–77 (2015)
19. Savkov, A., Carroll, J., Cassell, J.: Chunking clinical text containing noncanonical language. In: ACL 2014, p. 77 (2014)
20. Jadhav Seema, H., Sunita, S., Hari, S.: Content based image retrieval system with semantic indexing and recently retrieved image library. Int. J. Adv. Comput. Technol. (IJACT) (2012)
21. Sun, J., Reddy, C.K.: Big data analytics for healthcare. In: Proceedings of the 19th ACM SIGKDD International Conference on Knowledge Discovery and Data Mining, pp. 1525–1525. ACM (2013)
22. Uzuner, O., Yetisgen, M., Stubbs, A.: Biomedical/Clinical NLP. In: COLING 2014, pp. 1–2 (2014)

A Hybrid Grey Wolf Based Segmentation with Statistical Image for CT Liver Images

Abdalla Mostafa[1,6(✉)], Ahmed Fouad[2,6], Mohamed Houseni[3], Naglaa Allam[3], Aboul Ella Hassanien[4,6], Hesham Hefny[1], and Irma Aslanishvili[5]

[1] Institute of Statistical Studies and Research, Cairo University, Cairo, Egypt
abdalla_mostafa75@yahoo.com
[2] Faculty of Computers and Information, Suez Canal University, Suez, Egypt
[3] National Liver Institute, Menofia University, Monufia, Egypt
[4] Faculty of Computers and Information, Cairo University, Cairo, Egypt
[5] Computer Science Department, Tbilisi State University, Tbilisi, Georgia
[6] Scientific Research Group in Egypt (SRGE), Giza, Egypt

Abstract. Liver segmentation is a main step in all automated liver diagnosis systems. This paper aims to propose an approach for liver segmentation. It combines the usage of grey wolf optimization, statistical image of liver and simple region growing to segment the whole liver. Starting with Grey Wolf optimization algorithm, it calculates the centroid values of different clusters in CT images. A statistical image of liver is used to extract the potential area that liver might exist in. Then the segmented liver is enhanced using simple region growing technique (RG). A set of 38 images, taken in pre-contrast phase, was used to segment the liver and test the proposed approach. Similarity index is used to validate the success of the approach. The experimental results showed that the overall accuracy offered by the proposed approach, results in 94.08 % accuracy.

Keywords: Grey wolf optimization · Region growing · Segmentation

1 Introduction

The observation of nature and swarms have inspired the creation of different meta-heuristic algorithms. The behaviour of the swarms individuals is simple, but the amazing co-ordination between them presents a wonderful structured social organization. Grey wolf is one of the swarm inspired algorithms that focuses on the astonishing collective behaviour of swarm's members. Researchers have been trying to trace the steps of swarms behaviour to find sophisticated methods in order to solve complex optimization problems.

L. Sankari combined the Glowworm swarm optimization (GSO) algorithm with Expectation-Maximization (EM) algorithm to handle the problem of the maxima in EM. GSO clusters the image to find the initial seed points and pass it to EM algorithm for segmentation [7]. S. Jindal used Bacterial foraging optimization algorithm (BFOA), inspired by a special type of bacteria called

© Springer International Publishing AG 2017
A.E. Hassanien et al. (eds.), *Proceedings of the International Conference on Advanced Intelligent Systems and Informatics 2016,* Advances in Intelligent Systems and Computing 533, DOI 10.1007/978-3-319-48308-5_81

Escherichia coli. BFOA has an advantage of needing no thresholding in the process of image segmentation. It also reduces the computational complexity of segmentation process [2]. Y. Linag et al., combined the Ant Colony Optimization (ACO) algorithm and Otsu with expectation and maximization algorithm to find out multilevel threshold for segmenting the objects which have complex structure. They combined the parametric EM with the non-parametric ACO [3]. A. Mostafa et al., combined simple region growing (RG) technique with Artificial Bee Colony optimization technique (ABC). The efficiency of RG is improved when using ABC as a clustering technique [5].

The remainder of this paper is ordered as follows. Sect. 2 gives an overview about the nature inspired optimization technique of Grey Wolf Optimization algorithm(GW). Section 3 describes in detail the steps of the proposed approach for liver segmenting the liver and getting the ROIs. Sect. 4 shows the experimental results of the proposed approach. Finally, conclusions and future work are discussed in Sect. 5.

2   Grey wolf optimization

Grey wolf is a population-based algorithm, proposed by Mirjalili et al., mimics the leadership hierarchy and hunting mechanism followed by the grey wolves group [4]. Grey wolves are social animals that tend to live in a group called pack. They follow a social hierarchy to transfer commands and perform hunting. The social hierarchy consists of four levels. The first level **Alpha** (α) takes the leadership of the pack. The second level called **Beta** (β) is the coordinators between commander and other members. The third level called **Delta** (δ) is subordinate wolves. They are not alpha nor beta wolves. They submit to the alpha and beta and follow their commands. The fourth (lowest) level is called **Omega** (ω). They do not seem important as the other dominant members.

The following subsections present the mathematical models of the social hierarchy, tracking, encircling and attacking prey as follows.

2.1   Social Hierarchy

In the grey wolf optimizer algorithm (GWO), the fittest solution is considered the alpha α, while the second and the third fittest solutions are respectively called beta β and delta δ . The rest of the solutions are considered omega ω. In GWO algorithm, the hunting operation is controlled by α, β and δ, and the ω solutions follow these three wolves.

2.2   Encircling Prey

During the hunting, the prey is encircled by the grey wolves. The following equations present the mathematical model of the behaviour of encircling as follows.

$$\vec{D} = |\vec{C} \cdot \vec{X_p}(t) - \vec{X}(t)| \tag{1}$$

$$\vec{X}(t+1) = \vec{X_p}(t) - \vec{A} \cdot \vec{D} \tag{2}$$

Where t is the current iteration, $\vec{A}$ and $\vec{C}$ are coefficient vectors, $\vec{X_p}$ is the position vector of the prey, and $\vec{X}$ indicates the position vector of a grey wolf. The vectors $\vec{A}$ and $\vec{C}$ are calculated as follows:

$$\vec{A} = 2\vec{a} \cdot \vec{r_1} - \vec{a} \tag{3}$$

$$\vec{C} = 2 \cdot \vec{r_2} \tag{4}$$

Where components of $\vec{a}$ are linearly decreased from 2 to 0 over the course of iterations and $\vec{r_1}$, $\vec{r_2}$ are random vectors in $[0, 1]$

2.3 Hunting

Usually, the operation of hunting is guided by the alpha α. But, the beta β and delta δ might participate in hunting occasionally. It is observed that often, fewer than half of wolves are actually involved with physically bringing down the prey. In the mathematical model of hunting behaviour, it is assumed that the alpha α, beta β and delta δ have better knowledge than the other members about the potential location of the prey. The first three best solutions are saved, while the other agents are forced to update their positions according to the best solution. The positions update depends on the position of the best search agents. This is shown in the following equations [4].

$$\vec{D_\alpha} = |\vec{C_1}.\vec{X_\alpha} - \vec{X}|,$$
$$\vec{D_\beta} = |\vec{C_2}.\vec{X_\beta} - \vec{X}|, \tag{5}$$
$$\vec{D_\delta} = |\vec{C_3}.\vec{X_\delta} - \vec{X}|$$

$$\vec{X_1} = \vec{X_\alpha} - \vec{A_1} \cdot (\vec{D_\alpha}),$$
$$\vec{X_2} = \vec{X_\beta} - \vec{A_2} \cdot (\vec{D_\beta}), \tag{6}$$
$$\vec{X_3} = \vec{X_\delta} - \vec{A_3} \cdot (\vec{D_\delta}),$$

$$\vec{X}(t+1) = \frac{\vec{X_1} + \vec{X_2} + \vec{X_3}}{3}. \tag{7}$$

2.4 Search for Prey (Exploration)

Exploration is the ability to look for new solution away from the current solution in search space [6]. The position of α, β and δ controls the process of exploration in GWO algorithm. These wolves diverge from each other when they search for prey and converge when they attack the prey. The exploration process is modelled by initializing $\vec{A}$ with random values greater than 1 or less than − 1. This forces the searching agent to diverge from the prey and search globally. When $|A| > 1$, the wolves are obliged to diverge from the prey to find a fitter prey.

2.5   Attacking Prey (Exploitation)

Exploitation is the process of local search. It searches the area which is adjacent to the current solution [6]. When the surrounded prey stops moving, the grey wolf attacks. The vector $\overrightarrow{A}$ is a random value in interval $[-2a, 2a]$. The value of a decreases from 2 to 0 during the algorithm iterations. The wolves move and attack the prey when $|A| < 1$. This represents the exploitation process. The following subsection presents the GWO algorithm as follows.

2.6   Grey Wolf Optimization Algorithm

Grey wolf optimization algorithm is described step by step as follows.

- **Step 1.** Parameters initialization is the first step in standard grey wolf optimizer algorithm. It sets the initial values of the population size n, the parameter a, coefficients $\overrightarrow{A}$ and $\overrightarrow{C}$ and the maximum number of iterations max_{itr}.
- **Step 2.** Initialize the counter of iteration t.
- **Step 3.** Generate the initial population n randomly and evaluate each search agent $\overrightarrow{X_i}$ in the population by calculating its fitness function $f(\overrightarrow{X_i})$.
- **Step 4.** Assign the values of the first three best solutions to $\overrightarrow{X_\alpha}$, $\overrightarrow{X_\beta}$ and $\overrightarrow{\delta}$, respectively.
- **Step 5.** Repeat the following steps until the termination criterion is satisfied.

    - **Step 5.1.** Update each solution (search agent) in the population as shown in Eq. 7.
    - **Step 5.2.** Decrease the parameter a from 2 to 0.
    - **Step 5.3.** The coefficients $\overrightarrow{A}$ and $\overrightarrow{C}$ are updated as shown in Eqs. (3) and (4), respectively .
    - **Step 5.4.** Evaluate each search agent in the population by calculating its fitness function $f(\overrightarrow{X_i})$.

- **Step 6.** Update the first three best solutions $\overrightarrow{X_\alpha}$, $\overrightarrow{X_\beta}$ and $\overrightarrow{X_\delta}$, respectively.
- **Step 7.** Increase the iteration counter $t = t + 1$.
- **Step 8.** Repeat the overall process until the termination criteria is satisfied.
- **Step 9.** Produce the best found search agent (solution) $\overrightarrow{X_\alpha}$.

3   The Proposed CT Liver Segmentation Approach

The proposed CT liver segmentation approach consists of three main phases, including the preprocessing phase, clustering using Grey wolf algorithm, statistical image and region growing phases. Figure 1 shows the sequent steps of the proposed approach for liver segmentation. It is described in detail in the following section, including all involved steps and the characteristics of each phase.

Algorithm 1 shows the steps of the proposed approach.

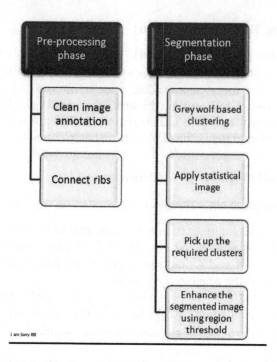

Fig. 1. The proposed liver segmentation approach: phases

Algorithm 1. The proposed liver segmentation approach

1: Clean the image from the patient's information and machine bed.
2: Connect ribs to separate the liver tissues from the adjacent flesh and muscles tissues.
3: Apply GWO algorithm to cluster the intensity values of the image.
4: Multiply the resulting clustered image by the prepared statistical image.
5: Pick up some pixels from the resulting image which represent the required clusters of liver.
6: Separate the required cluster in a separate image, convert it into binary and multiply the binary image by the original image to get the initial segmented image.
7: Enhance the segmented image using region growing to extract the whole liver.
8: Use similarity index (SI) to calculate the accuracy of segmentation process.

The following subsections will describe the proposed algorithm steps in details as follows.

3.1 Preprocessing Phase

The first step in preprocessing phase is to prepare the statistical image which will be used later to get the potential area that liver might be in. The next subsection describes how to create the statistical image and cleaning the image.

Statistical Image Preparation. The statistical image is the summation of all possible occurrence of liver in CT images. Simply, the dataset of liver images is segmented manually from the abdomen for each image. Each image is converted into a binary image with the values of zeros and ones. All of the binary images of the dataset are summed in one matrix. For each element in the matrix, the value is divided by the number of the dataset elements. Now, the resulting matrix represents the probability of liver cells occurrence (existence) in any image in the dataset. It has a range of values between zero and one. Finally, two types of images can be extracted. The first one is a binary image of all pixels that have a value more than zero, which represents the all possible occurrence of the liver in a CT image. The second binary image is the one that has all pixels that have the value of one, which represents the initial seed points that can be passed to the region growing method.

Image Cleaning Phase depends on using filters for smoothing and stressing certain features of the manipulated image. In this paper, mean filter is used for smoothing the image in cleaning step, and contrast stretching is used in the step of connecting ribs.

Image cleaning is the first preprocessing where the image is cleaned from its annotations. Patient's information and machine's bed are removed to ease the next step of connecting ribs. This is done using mean filter ans some morphological operations. In brief, the morphological operations of opening, closing and erosion can be used to clean the image in different aspects. It removes the characters of annotation in the image, removes the machine's bed, and also erodes the body skin and a part of flesh and muscles from the abdomen. See the algorithm of connecting ribs in detail in [9].

3.2   Grey Wolf Phase

Grey wolf optimization can be used as a clustering technique, helping to segment the liver in CT images. The major key in GWO algorithm is the parameter setting. There are three parameters including the number of defined clusters, the population size, and the number of iterations. First of all, it is obvious that the number of clusters has an effect on the resulting image using different optimization techniques. The intensity of the image has two extreme values, represented by white bones and black background. Also, there are different values of liver boundaries, lesion and other organs. The experiments showed that seven clusters proved the best efficiency in segmentation as explained later. Also, there is a need to investigate the number of population size of artificial wolfs in the colony. The last parameter is the number of iterations which definitely affects the speed of application. A fitness function is used to find out the new solutions in every iteration. The best solutions will be the centroids of the resulting clusters which are called global parameters. Algorithm 2 shows the steps of using GWO to segment the liver.

Algorithm 2. Grey Wolf phase

1: Set the value of different parameters of clusters, colony search agents and iterations.
2: Apply the algorithm of grey wolf optimization on the image.
3: Get the vector of global parameters, that represent the intensity values of clusters' centroids.
4: Replace every centroid value with the related cluster label.
5: Display the coloured clustered image to make it easier for the next step.
6: Pick up some points inside the liver representing the main clusters.
7: Extract the required clusters in a binary image.
8: Fill the holes using morphological operations.
9: Multiply the resulting binary image by the original image. This results in the segmented image, which may include other organs that fall in the same liver clusters.

The advantage of using GWO as a clustering technique is that there is no noise effect on segmentation. Because noise is always represented by small pixels spreading in the image, and it is filled by morphological operations step.

3.3   Statistical Image and Region Growing Phase

As the intensity values of flesh, muscles and some other organs, may fell in the range of the cluster or clusters of liver. The resulting image of Grey wolf phase may include tissues of flesh, muscles, stomach, kidney, spleen ... etc. So, the usage of the statistical image can remove a great part of these tissues. The first binary statistical image, representing the statistical occurrence, is multiplied by the resulting GWO phase image. Then, It is multiplied by the original image and passed to region growing method.

The techniques of region-based applies the concept of similarity difference to extract regions. Level set and fast marching use an initial contour, evolved outwards, using a smoothing function [1]. Watershed segmentation method applies the idea of water drops which falls in a basin. It uses the gradient of the image to find out whether every pixel belongs to one or more local minimum [8]. Region growing can be used for one region extraction as level set and fast marching. But, it has some advantages, compared to other methods, including simplicity, speed and low computational cost. In this paper, region growing has been chosen to enhance the GWO segmented image.

4   Experimental Results and Discussion

The experiments of the proposed approach of segmentation, using GWO with statistical image will be covered in this section. A set of 38 CT images were used to apply and test the different phases of the proposed approach. The images of the abdomen of the patients were taken in the early arterial phase of CT scan.

Figure 2 shows the resulting image from picking up the required clusters that represent the liver. The chosen clusters are multiplied by the original image. The user does not have to choose the cluster of the small regions of lesions inside the

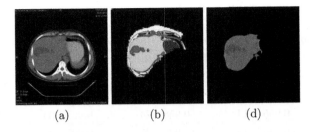

(a)   (b)   (d)

Fig. 2. Picking up the required clusters from the image resulting from applying the statistical image: (a) Original image. (b) Statistical occurrence image. (c) The picked up clusters

liver. The small fragments representing the un-chosen clusters of lesion might be holes inside the liver. When the liver is extracted, these holes can be filled easily.

In the last process of segmentation, the image segmented by GWO and statistical image, is enhanced using simple region growing technique. Figure 3 shows the difference between the segmented image and annotated one. It shows a very thin boundary of the liver as a difference between the resulting image of the proposed approach and the manual segmented image.

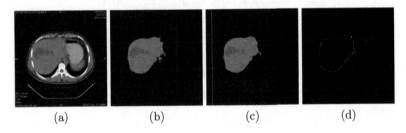

(a)   (b)   (c)   (d)

Fig. 3. Enhanced RG segmented image: (a) Original image. (b) Picked up clusters image. (c) Enhanced image by region growing, (d) Difference image

The experiment of liver segmentation using the proposed approach of grey wolf, evaluated by similarity index, showed a good efficiency. It showed a better performance than other compared segmentation methods. The implementation of GWO clustering algorithm combined with the statistical image has an output of good initial segmented liver. The resulting image is enhanced using simple region growing technique.

Table 1 shows the results of the implementation of the proposed approach. We can mention here that the usage of GWO is not affected by any ration of noise, because noise is always small regions or pixels which may represent different cluster, but included within the area of the main liver cluster. This results in small holes, which is filled by the morphological operations when the liver clusters are picked up.

Table 1. Results of proposed approach

Image no.	GWO	Image no.	GWO	Image no.	GWO	Image no.	GWO
1	0.985	11	0.951	21	0.91	31	0.925
2	0.941	12	0.933	22	0.962	32	0.941
3	0.878	13	0.954	23	0.943	33	0.923
4	0.955	14	0.958	24	0.921	34	0.946
5	0.946	15	0.939	25	0.907	35	0.959
6	0.973	16	0.966	26	0.943	36	0.956
7	0.951	17	0.954	27	0.958	37	0.965
8	0.896	18	0.887	28	0.936	38	0.96
9	0.952	19	0.919	29	0.954		
10	0.951	20	0.908	30	0.946		
Result	0.9408						

Table 2. Comparison between the proposed approach and other approaches

Ser	Approach	Result
1	Region growing (RG)	84.82
2	Wolf local thresholding + RG	91.17
3	Morphological operations +RG	91.20
4	K-means +RG	92.38
5	Level set	92.10
6	Artificial Bee Colony (ABC)	93.73
7	Proposed approach (GW)	94.08

Table 2 compares the results of the proposed approach to other approaches applied on the same dataset. The compared approaches are region growing, wolf local thresholding, morphological operations, K-means, and Artificial Bee Colony. It shows that the proposed approach achieved the best result.

5   Conclusion and Future Work

This paper is interested in the extraction of the whole liver which includes the phases of preprocessing and segmentation. Preprocessing is involved in cleaning image annotations and connecting ribs. Segmentation phase relies on the grey wolf algorithm as a clustering technique dividing the image into a number of clusters. Using the statistical image reduces the area that liver might exist in, removing some other organs that have the same intensity values. The user picks up some seed points inside the liver that determines the extracted clusters of liver. The extracted clusters are multiplied with the original to get the initial segmented region of liver. This region is enhanced using simple region growing segmentation technique. The segmentation using the proposed approach has a considerable average accuracy rate 94.08 % using SI measure.

References

1. Caselles, V., Kimmel, R., Sapiro, G.: Geodesic active contours. Int. J. Comput. Vis. **22**, 61–79 (1997)
2. Jindal, S.: A systematic way for image segmentation based on bacteria foraging optimization technique (Its implementation and analysis for image segmentation). Int. J. Comput. Sci. Inf. Technol. **5**(1), 130–133 (2014)
3. Liang, Y., Yin, Y.: A new multilevel thresholding approach based on the ant colony system and the EM algorithm. Int. J. Innov. Comput. Inf. Control **9**(1) (2013)
4. Mirjalili, S., Mirjalili, S.M., Lewis, A.: Grey wolf optimizer. Adv. Eng. Softw. **69**, 46–61 (2014)
5. Mostafa, A., Fouad, A., Abd Elfattah, M., Ella Hassanien, A., Hefny, H., Zhue, S.Y., Schaefer, G.: CT liver segmentation using artificial bee colony optimisation. In: 19th International Conference on Knowledge Based and Intelligent Information and Engineering Systems, Procedia Computer Science, Singapore, vol. 60, pp. 1622–1630 (2015)
6. Palupi Rini, D., Mariyam Shamsuddin, S., Sophiyati Yuhaniz, S.: Particle swarm optimization: technique, system and challenges. Int. J. Comput. Appl. (0975 8887) **14**(1), 19–27 (2011)
7. Sankari, L.: Image segmentation using glowworm swarm optimization for finding initial seed. Int. J. Sci. Res. (IJSR) **3**(8), 1611–1615 (2014)
8. Vanhamel, I., Pratikakis, I., Sahli, H.: Multiscale gradient watersheds of color images. IEEE Trans. Image Process. **12**(6), 617–626 (2003)
9. Zidan, A., Ghali, N.I., Hassanien, A., Hefny, H.: Level set-based CT liver computer aided diagnosis system. Int. J. Imaging Robot. **9**, 26–36 (2013)

Implementation of Elliptic Curve Crypto-System to Secure Digital Images over Ultra-Wideband Systems Using FPGA

Mohammed F. Albrawy[1(✉)], Ali E. Taki El-Deen[2],
Rasheed M. El-Awady[3], and Mohy E. Abo-Elsoaud[3]

[1] Communications and Electronics Department,
Mansoura University, Mansoura, Egypt
Mohammed.albrawy@yahoo.com

[2] IEEE Senior Member, Comms. and Electronics Department,
Alexandria University, Alexandria, Egypt
a_takieldeen@yahoo.com

[3] IEEE Senior Member, Comms. and Electronics Department,
Mansoura University, Mansoura, Egypt

Abstract. In this paper, a simple implementation for elliptic curve equation on Field Programmable Gate Array (FPGA) will be proposed. As Elliptic Curve Cryptography (ECC) offers a smaller key size without sacrificing security level. A brief survey on applying the main equation of Elliptic Curve (EC) with different values of the coefficients a and b. A comparison between results depended on the correlation coefficient of each value. Value of a and b implemented on FPGA according to correlation results between plaintext image and ciphertext image on MATLAB. This EC equation will be applied to an ultra-wide band (UWB) system to secure transmission of data in a wireless channel. Here, a brief survey on UWB technology has been implemented with software simulation for a secured system based on ECC algorithm.

Keywords: FPGA · ECC · UWB · Image encryption · Information security

1 Introduction

Over many years, mathematicians studied elliptic curves to solve a diverse range of problems. Now, there is a rich and beautiful history related to elliptic curves. Later, elliptic curve systems commercially acceptance when its protocols included in the security products of companies [1].

In order of that Elliptic curve systems could be used in communication systems security. As telecommunication system like Ultra-wideband (UWB) become a main part of our daily life. The key issue for the cryptographic community is the method of broadcasting the encrypted data to a large set of users, but only the subset of privileged users can decrypt the data[10]. Federal Communications Commission(FCC) regulates that the frequency for the UWB technique is from 3.1 GHz to 10.6 GHz [2].

This paper is organized as follows: Sect. 2 presents UWB system with a proposed block diagram. Section 3 shows a detailed explanation for image encryption over ECC.

© Springer International Publishing AG 2017
A.E. Hassanien et al. (eds.), *Proceedings of the International Conference
on Advanced Intelligent Systems and Informatics 2016*, Advances in Intelligent
Systems and Computing 533, DOI 10.1007/978-3-319-48308-5_82

Section 4 demonstrates resultant software implementation while Sect. 5 is the hardware implementation using FPGA. Finally a conclusion of the whole work represented in Sect. 6.

2   Ultra Wide-Band Transmission System

Ultra wide-band is a system based essentially on the principle of co-existence with other narrowband systems [11] which shown in Fig. 1. UWB basically is a technology for wireless communication, localization, telemedicine, and wireless sensors [3].

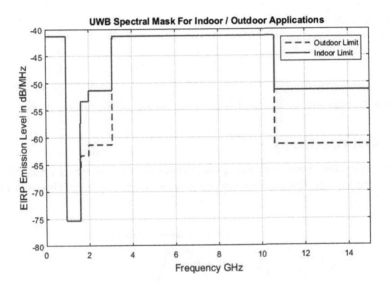

Fig. 1. UWB spectral mask by FCC

UWB taken on the character of "pure impulse signal" with signal energy spread over large amount of spectrum. Generation of a suitable signal is the initial step in any radio communication link, after that it is modulated with demanded data.

UWB signal design concern with bandwidth 10 dB to ensure compliance with the requirements of bandwidth minimum [9], the 20-dB bandwidth to make certain that the signal remains below the 20-dB band edge corners on the UWB communications power spectral density (PSD) mask, and the frequency of the highest radiation emission, which must be below the maximum allowed PSD [4].

UWB advantages summarized in sharing spectrum, gigantic bandwidth, low S/N ratios, immunity to detection and intercept and simple transceiver architecture [5]. Beside benefits that UWB brings to wireless communications, additional level of security can be achieved by supplying strong crypto-system like elliptic curve cryptography. Figure 2 shows the proposed block diagram of the secured system.

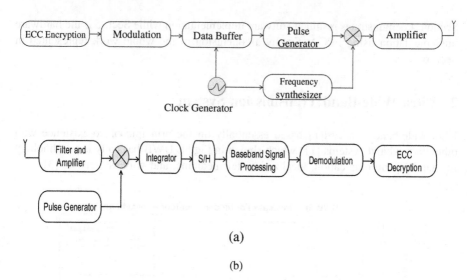

(a)

(b)

Fig. 2. Ultra wide-band secured system block diagram (a) Transmitter. (b) Receiver.

2.1 Short Pulse Generation

Before generation of any ultra-short pulse, the desired wave shape must be determined first for the system. The convenience and ease of generation which shown in Fig. 3.

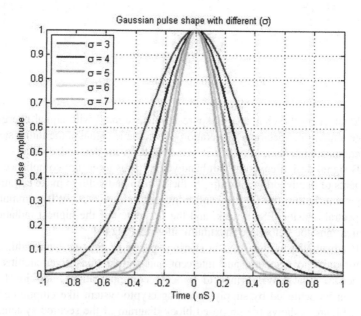

Fig. 3. Gaussian Pulse with different σ value

$$f(t) = Ae^{-\frac{(t-\mu)^2}{2\sigma^2}} \tag{1}$$

$$\tau = 2\pi.\sigma \tag{2}$$

where:

A: Pulse amplitude
t: Time
σ: Standard deviation
μ: Mean

Equation 1 describes the generation of a Gaussian pulse but Pulse width can be defined in seconds by Eq. 2.

The Gaussian pulse width is on the order of one or less than a nanosecond which shown in Fig. 3 with different width., and its energy is spread over a broad range of frequencies with a very low power spectral density.

3   Image Security Over Elliptic Curve Cryptography

An elliptic curve takes the general form as:

$$E: y^2 = x^3 + ax + b\ [7], \tag{3}$$

where x, y are co-ordinates of Galois Field GF (P), and a, b are integer modulo P, satisfying

$$4a^3 + 27b^2 \neq 0 (\text{mod P})\ [8], \tag{4}$$

Where P is a modular prime integer that makes the EC of a finite field. An elliptic curve E over GF (P) consist of the points (x, y) defined by Eqs. (3) and (4), along with an additional point called O (point at infinity) in EC. Equation (3) used for image encryption with the main condition as in Eq. (4).

Let's create elliptic group with prime number P = 4397 and different values of coefficients a and b to encrypt plaintext image.

3.1   Appling Elliptic Curve Cryptography

While applying elliptic curve as a crypto-system, it's necessary to pay attention to some considerations such as:

- Galois Field creation with a prime number P,
- Specifying the base point from the Galois Field GF (P = 4397) therefore private key depends on, and (3) Encoding plaintext data according to the generated GF.
- Encoding process converts each pixel value to its corresponding value in the GF. Algorithm 1 shows the manner of generating Galois field.

Algorithm 1. Galois field creation

Input: Prime number P , integers $x, y \in [1 : P - 1]$, integers a, b
Output: $Z \leftarrow (x, y)$
1. For i from 1 to P do $RHS = (x^3 + a x + b) \bmod P$
2. For j from 1 to P do $LHS = (y^2) \bmod P$
3. If RHS= LHS then $Z \leftarrow (x_i, y_j)$
5. Return(Z)

Next step is showing the base point. Algorithm 2 demonstrates the best and shortest way to allocate the suitable points from the GF that could be selected to be the base point. Base point is a unified value that is generally accepted by both transmitter and receiver.

Algorithm 2. Picking out base point

Input: Prime number P, Galois Field $Z = (x, y)$, Multiplier i
Output: BP = Z
1. For i from 1 to $P - 1$ do R = $i * Z$
2. For j from 1 to $length(Z)$ do
2.1 If $R = Z_j$ then $K_i \leftarrow i$
3. If length (K) > 3 then BP $\leftarrow Z_i$
4. Return (BP)

After that a suitable private key using Algorithm 3 could be selected from P−1 integers where not all these integers appropriate to be a private key because multiplication equation that is stated as a condition must be verified. Also, we can consider Algorithm 3 is a test for every integer n from 1 to P−1 at equation Q = n Z to see the result of multiplying that integer in any point at GF (P).

Algorithm 3. Picking out available private keys

Input: Prime number P, Galois Field $Z = (x, y)$
Output: K
1. For i from 1 to $P - 1$ do R = $i * Z_i$
2. For j from 1 to $length(Z)$ do
3. If $R = Z_j$ then $K_i = i$
4. Return (K)

Now it's time to prepare data for ciphering by applying image encoding. Encoding image presented in Algorithm 4. Encoding includes a point multiplication process.

Algorithm 4. Plaintext Encoding

Input: Galois Field $Z = (x, y)$, Image X, Auxiliary base parameter r, Integer k=[1:20]

Output: Y= (x_pnt, y_pnt)

1. For i from 1 to $length(x)$ do $x\_pnt = X \times r + k$
2. If $x\_pnt = x$ then $y\_pnt = y$ else $k = k + 1$
3. Go to 1
4. $Y \leftarrow (x\_pnt, y\_pnt)$
5. Return (Y)

Multiplication basically converted to doubling and addition process. Case-1 illustrates how point doubling performed. For example, let's consider $Q = 2Z$ then multiplication converted to addition or point doubling where $Q = Z + Z$. But what about $n = 3$? Here, case-2 starts where $Q = 3Z = Z+2Z$. Let us rename it to $Z = Z_1$ and $2Z = Z_2$ which makes the multiplication process converts to addition $Q = Z_1 + Z_2$.

Case-1: $Z_1 = Z_2 = Z$

$$\lambda = \frac{3x_1^2 - a}{2y_1} \, mod\, P \tag{5}$$

$$x_3 = \lambda^2 - 2x_1 \, mod\, P \tag{6}$$

$$y_3 = \lambda(x_1 - x_3) - y_1 \, mod\, P \tag{7}$$

Case-2: $Z_1 \neq Z_2$

$$\lambda = \frac{y_2 - y_1}{x_2 - x_1} \, mod\, P \tag{8}$$

$$x_3 = \lambda^2 - x_1 - x_2 \, mod\, P \tag{9}$$

$$y_3 = \lambda(x_1 - x_3) - y_1 \, mod\, P \tag{10}$$

Algorithm 5. Elliptic Curve Crypto-system

Input: Prime number (P), Encoded image(Y), Random number (R), Private_Key (PK), Base point (BP)

Output: Ciphertext C

1. Public_key $\leftarrow$ PK $\times$ BP
2. A $\leftarrow$ R $\times$ Public_key
3. C $\leftarrow$ A + Y
4. Return (C)

Public key is now determined by multiplying chosen base point by a special key fundamental point. Algorithm 5 presents steps to perform encryption process using Elliptic curve cryptography.

4   Software Implementation with Results

The results of applying EC equation illustrated in Table 1 for a digital image with different values of a in rows and b in columns. The correlation coefficient between plaintext and ciphertext was obtained for all results. When a = 0, the best correlation coefficient was 0.00036875 when b = −2 while it was 0.000025933 with a = 1 and b = −1. Therefore, we can say that the best ever through this results is a = 1 and b = −1 with the lowest correlation coefficient.

Now cipher image with the lowest correlation where a = 1 and b = -1 will be transmitted over UWB system, but first it has to be converted to binary form.

Table 1. Appling Elliptic Curve Equation with different values for a and b.

	b = -2	b = -1	b = 0	b = 1	b = 2
a = 1					
	0.00036875	0.0011	0.0036	-0.0021	0.0163
a = 2					
	-0.0039	0.000025933	-0.0061	0.007	0.0015
a = 3					
	0.00089036	0.0045	-0.005	0.0041	0.0078

4.1   Transmitting Image

Transmission will be on one of these pixels that equals to $(10)_{10}$. From system diagram in Fig. 2, the data will multiply by the Gaussian pulse as a base band signal. Figure 4 illustrates the resulting base band pulse multiplied by Cos $(2\pi f)$.

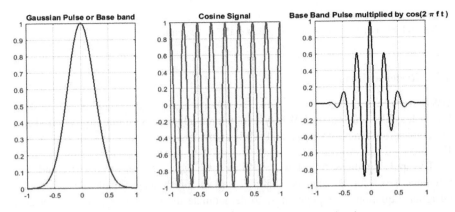

Fig. 4. Multiplication of Gaussian pulse by Cos $(2\pi f)$.

Cosine wave generates from frequency synthesizer in the block diagram. The cosine signal frequency centered at the desired frequency of f_c. Next step is to multiply the base band signal from the binary data. According to pixel value, the corresponding binary form is $(00001010)2$ which will modulate using any modulation scheme OOK, PPM, PAM and bi-phase modulation.

Direct Sequence systems can be classified into Pulse Position Modulation (PPM) and Pulse Amplitude Modulation (PAM). Mathematical expressions have been provided in the published literature which include both types of systems in [6].

If bi-phase modulation is utilized, two data lines are required, one to trigger a positive pulse to transmit a "1" and another to trigger a negative pulse to transmit a "0". Figure 5(a) shows modulation of data using the bi-phase modulation scheme.

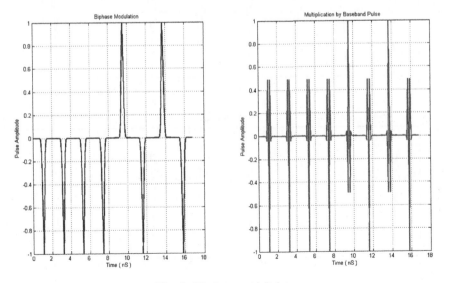

Fig. 5. Bi-phase modulation

In Fig. 5(b) baseband pulse multiplied by the modulated signal in Fig. 5(a) generating pulses which is ready now to be radiated throw antenna.

4.2   Recieved Image

UWB signal has an inherent noise-like behavior which makes robust security systems highly feasible. That makes UWB system theoretically a noise immunity system. But if we considered a random noise add while transmission to the data signal as seen in Fig. 6.

By applying the simplest way to filter our signal by performing 0.3 and –0.3 reference level for all signals. Figure 7 shows thresholded signal and demodulatedsignal. Equation 11 explains how to demodulate the resultant signal to its original Gaussian pulse signal.

$$g\prime(t) = cos(2\pi f) \times r(t) \tag{11}$$

Where

$g\prime(t)$: expected demodulated Gaussian pulse signal.
r(t): received signal.

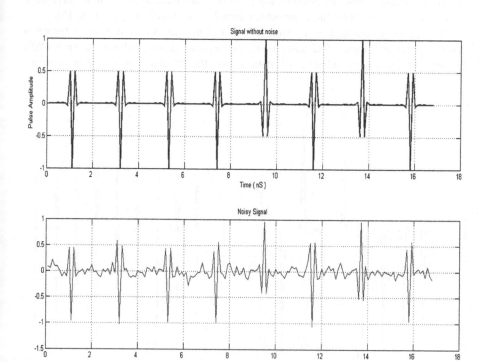

Fig. 6. Ideal signal and noised signal

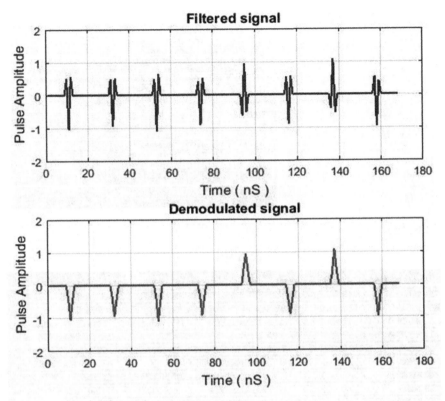

Fig. 7. Threshold noised signal and demodulation result

It's easy now to perform DAC to signal. If signal higher than 0.5 so it's one else its zero. That worked successfully on MATLAB and output was $(00001010)_2 = (10)_{10}$ which is the pixels' value.

5   EC Implementation on FPGA

We applied EC equation using a = 1 and b = −2. Figure 8(a) shows the simulation results for VHDL implementation based on Mentor Graphics Tools. We split the main equation into R.H.S with (y^2 mod P) and L.H.S with (x^3 + x−2 mod P) where P is a large prime number [12]. The chosen prime number is P = 509 so finite field results depend on P value. To verify the equation, we have to specify a point from the finite field. Simulation results in Fig. 7 depended on two points from the finite field (319,193) and (93,338) that make right side equal to left side.

In Fig. 8(b), coefficients are changed to a = 1 and b = −1 that will change all values for the finite field even though P is still 509. Two points are used (507, 206) and (507, 45) where first point is out of finite field and the second lies on it.

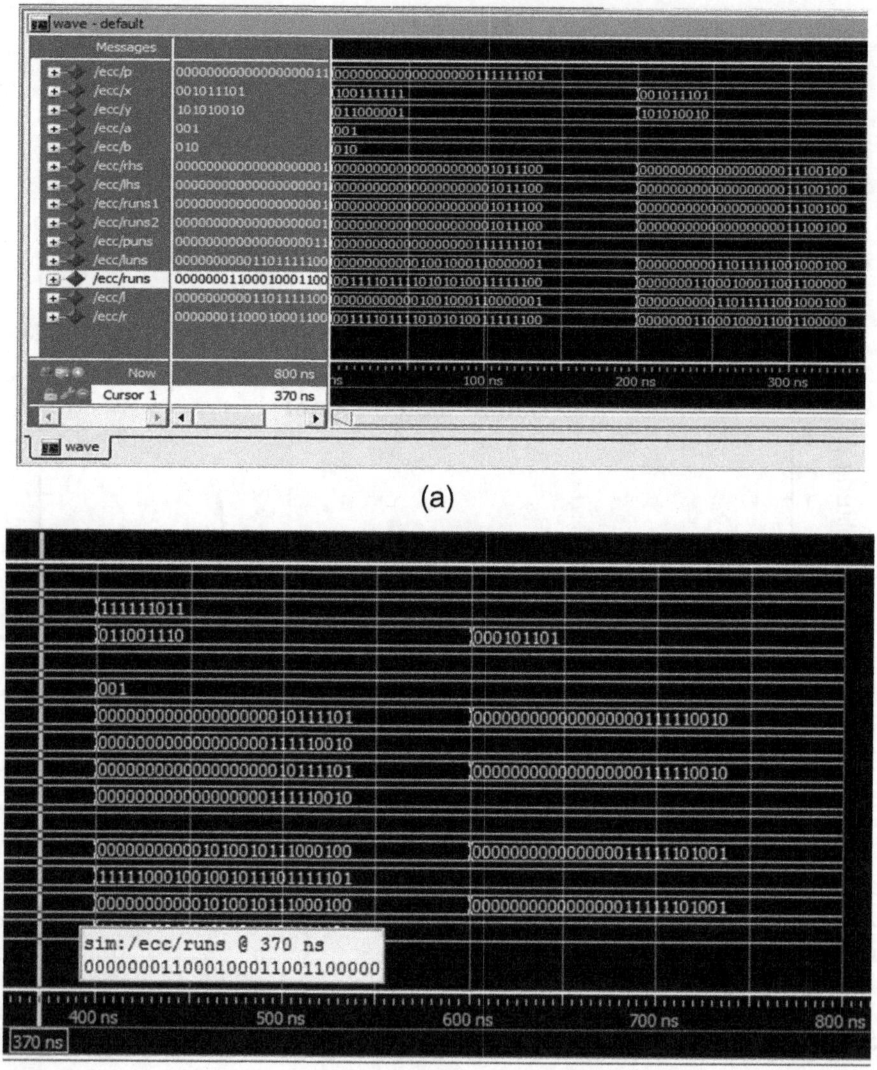

(a)

(b)

Fig. 8. Modelsim results for (a) a = 1 and b = −2, (b) a = 1 and b = −1

From simulation results, it's obvious that point (507,206) make both sides unequal and EC equation does not verified. That's why we have to observe all chosen points such as choosing base point mentioned in Algorithm 2.

Figure 9 presents the resultant RTL schematic design and architecture circuit. The device used here was: Xilinx - Spartan3E:3S500EFG320. Table 2 clarify the actual usage resources for our device.

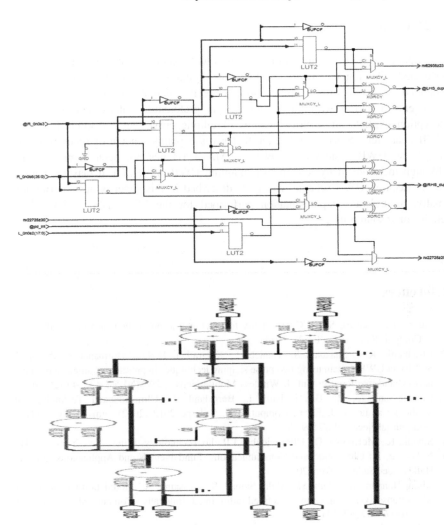

Fig. 9. Hardware Desig for EC Equation (Above) Architecture for EC Schematic Design. (Bottom) RTL Schematic Design.

Table 2. Resource usage on Xilinx - Spartan3E

Resource	Used	Available	Utilization
IOS	105	232	45.26 %
Global Buffers	0	24	0.00 %
Function Generators	68	9312	0.73 %
CLB Slices	34	4656	0.73 %
DFFs or Latches	0	9776	0.00 %
Block RAMs	0	20	0.00 %
Block Multipliers	4	20	20.00 %

6 Conclusion

In this paper, a brief description of securing a wide band system using ECC has been performed. The implementation of ECC on digital image in MATLAB has been explained in detail with a comparison between different values of EC equation coefficients a and b with the results of the preferential values. Our scheme of encryption exploits all security features of elliptic curves and is applicable to all ASCII characters. Since no reference that gives explains scheme of securing digital data throw UWB system using ECC in MATLAB has not been coming across, (although there are many references explaining UWB and ECC), It may be claimed that the secured UWB scheme with ECC described in this paper is our contribution. Finally, a hardware implementation performed for the EC equation with a brief details for resource usage.

References

1. Gupta, K., Silakari, S.: ECC over RSA for asymmetric encryption: a review. IJCSI Int. J. Comput. Sci. Issues, 8(3-2) (2011)
2. El-Khamy, R.S., Shaaban, S., Ghaleb, I., Kheirallah, H.N.: Performance of the IEEE 802.15.4a UWB system using two pulse shaping techniques in presence of single and double narrowband interferences. Int. J. Wireless Mobile Netw. (IJWMN) 5(3), 35–44 (2013)
3. Chávez-Santiago, R., Balasingham, I., Bergsland, J.: Ultrawideband Technology in Medicine: a survey. J. Electr. Comput. Eng. Volume 2012 (2012). https://www.hindawi.com/journals/jece/2012/716973/
4. Siwiak, K., McKeown, D.: Ultra-wideband Radio Technology. Wiley, New York (2004)
5. Nekoogar, F.: Ultra-Wideband Communications: Fundamentals and Applications. Prentice Hall, Upper Saddle River (2005)
6. Abdul Hannan, M., Islam, M., Abdul Samad, S., Hussainn, A.: Modulation techniques for RFID transceiver using software defined radio. Int. J. Innovative Comput. Inf. Control 8(10 (A)), 6667–6692 (2012)
7. Deligiannidis, L.: Implementing elliptic curve cryptography. International Conference Frontiers in Education: CS and CE, FECS 2015 (2015)
8. Bos, J.W., Halderman, J.A., Heninger, N., Moore, J., Naehrig, M., Wustrow, E.: Elliptic curve cryptography in practice. In: Christin, N., Safavi-Naini, R. (eds.) FC 2014. LNCS, vol. 8437, pp. 157–175. Springer, Heidelberg (2014). doi:10.1007/978-3-662-45472-5_11
9. AL-Tamimi, H.M., Hamza, H.A.: Generation of UWB waveforms with adaptive mitigation of multiple narrowband interference in IR UWB Systems. Int. J. Sci. Eng. Res. 6(3), 34–44 (2015). ISSN 2229-5518 34
10. Nagalakshmi, N., Rajalakshmi, S.: Enabled security based on elliptic curve cryptography with optimal resource allocation schema in cloud computing environment. In: 2015 IEEE Seventh National Conference on Computing, Communication and Information Systems (NCCCIS) (2015). ISBN 978-1-4799-8990-4

11. Bahrami, H., Abdollah Mirbozorgi, S., Ameli, R., Rusch, L.A., Gosselin, B.: Flexible, polarization-diverse UWB antennas for implantable neural recording systems. IEEE Trans. Biomed. Circuits Syst. **10**(1), 38–48 (2015)
12. Thiranant, N., Lee, Y.S., Lee, H.: Performance comparison between RSA and elliptic curve cryptography-based QR code authentication. In: 29th International Conference on Advanced Information Networking and Applications Workshops. IEEE (2015) doi:10.1109/WAINA. 2015.62

Combining Public-Key Encryption with Digital Signature Scheme

Mohammad Ahmad Alia[(⊠)]

Computer Information Systems Department, Faculty of Sciences and IT,
Al-Zaytoonah University of Jordan, Amman, Jordan
dr.m.alia@zuj.edu.jo

Abstract. This paper presents the possibilities of combing public-key encryption and digital signature algorithms which are actually based on different mathematical hard problems. Since the output of the combination produces an Encrypted signed message. In general, most of the currently used public-key algorithms are computationally expensive with relatively lengthy key requirement due to the dependency on the number theory. Therefore, it's important to show a combinational protocols which are based on different mathematical hard problem. In some sense, difficult to solve. In the combined scheme, we present the powerful and practical encryption digital signature scheme and its security level and execution time.

Keywords: Cryptography · NP-Hard problem · Digital signature · Public key encryption

1 Introduction

Cryptography algorithms can be classified into two board categories, secret key (one key, single key, symmetric) algorithms and public key (two key, asymmetric) algorithms (refer to Fig. 1). In general, Cryptography protocol employs public key cryptosystem to exchange the secret key and then uses faster secret key algorithms to ensure confidentiality, integrity, non-repudiation, authentication, and accessibility of the data stream [4, 5]. In 1976, Diffie-Hellman [2, 6] developed the concept of the public-key cryptosystem (refer to Figs. 1 and 2), and the first asymmetric encryption protocol is the RSA [7] algorithm which was published in 1978. As well as, there are many others asymmetric encryption algorithms issued since the RSA. Among them are Rabin [8], ElGamal [9], and Elliptic Curve [10].

In public-key Cryptography, Every public-key algorithm is actually based on a mathematical problem. These problems are called "mathematical hard problems" and are classified into two major groups according to the Cryptography standards. These groups are P (Polynomial) and NP (Non-deterministic polynomial). The P hard problem is defined when the problem is solved in polynomial time. Whereby, if the validity of a proposed solution can be checked only in polynomial time then the problem is considered as an NP hard mathematical problem [1–3]. Basically, public-key algorithms are classified into many major types depending on the mathematical hard problem. These problems are the discrete logarithm problem (DLP), the

© Springer International Publishing AG 2017
A.E. Hassanien et al. (eds.), *Proceedings of the International Conference on Advanced Intelligent Systems and Informatics 2016*, Advances in Intelligent Systems and Computing 533, DOI 10.1007/978-3-319-48308-5_83

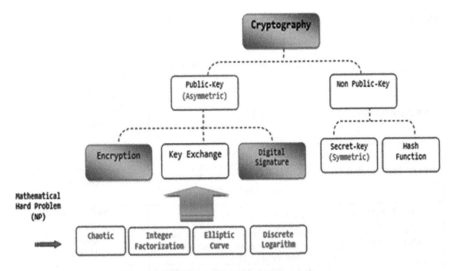

Fig. 1. Main branches of public-key scheme

integer factorization problem (IFP), the Elliptic Curve discrete logarithm problem (ECDLP), the chaotic hard problem, etc. This following survey study will help us to identify the strength of the used public-key algorithms according to their mathematical hard problem (refer to Fig. 1).

In asymmetric encryption protocols, there is a pair of keys, one of which is known to encrypt the plaintext and called as the public key, while the other key is known as the private key and is used to decrypt the encrypted plaintext.

As discussed earlier, every public-key encryption algorithm is based on a NP mathematical problem that is in some sense difficult to be solved, therefore the public-key algorithms are classified according to their hard problems in this subsection. Table 1 show the key size for prime based algorithms (RSA, DSA, etc.) and integer based algorithm (ECC, Chaotic) algorithms, regarding to the resistance to brute force attacks. The keys space for RSA and DSA were calculated based on the number of primes existed for particular key sizes [14].

This paper developed a public-key encryption with digital signature scheme which are based on mathematical hard problem. The paper discusses the possibility of creating a combinational public key encryption and digital signature protocol.

2 The Proposed Encryption and Digital Signature Protocol

The combination between encryption algorithm and digital signature algorithm provides confidentiality, Integrity, authentication, and non-reputation services for messages. In this article, both algorithms are defined as any of the different mathematical problems; since it's possible to combine encryption based IFP with digital signature based DLP in some cases. However, in the proposed solution the sender and the receiver must generate their own private and public keys. The sender must compute

Table 1. Public-Key Encryption and Digital Signature protocols (Efficiency and Key Size) [14]

Encryption and Digital Signature Algorithms		
NP-Hard Problem	Efficiency	Typical Key Size for High Performance
Integer Factorization	The speed in RSA is considered much slower than other symmetric cryptosystems	Large Prime Number (1024-bit)
	Rabin operations are more efficient than RSA	
Discrete Logarithm	ElGamal and DSA is probabilistic.	Large Prime Number (1024-bit)
Elliptic Curve	The discrete logarithm problem on elliptic curve cryptosystem is more difficult than the other mathematical problem	Short Key (128-bit)
Chaos- Fractal	The fractal based public-key cryptosystem provides high level of security at a much low cost, in term of key size and execution time.	Short Key (128-bit)

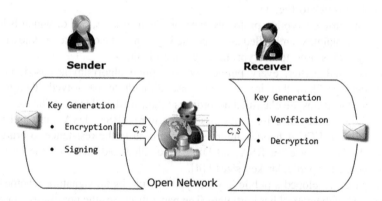

Fig. 2. Encryption and signature combinational protocol

his/her keys for digital signature purpose while the receiver will compute the encryption and decryption keys, in parallel with the sender.

As shown by Fig. 2, the message is selected by the sender and then encrypted and signed to be sent over the insecure network to the receiver. After receiving the encrypted and signed message, the receiver should verify the signature authenticity before decrypting the ciphertext.

2.1   Combination of RSA and DSA Protocol

Figure 3 shows the combinational protocol between RSA and DSA algorithms, since RSA is based on integer factorization problem while DSA is based on discrete logarithm problem.

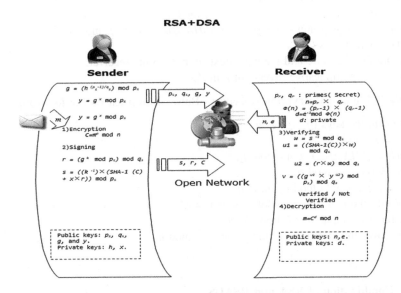

Fig. 3. RSA + DSA protocol

Key generation algorithm (generated by sender, Alice) - Alice must do:

1. Choose a prime number (p_s), where $2^{L-1} < p_s < 2^L$ for $512 \leq L \leq 1024$ and L a multiple of 64.
2. Choose a prime numbers (q_s), where q_s divisor of $(p-1)$, and $2^{159} < q_s < 2^{160}$.
3. Compute g as follows: $g = (h^{(p-1)/q}_s) \bmod p_s$, where $1 < h < (p_s - 1)$, and $g > 1$.
4. Choose a random integer x, with $0 < x < q_s$.
5. Compute y as follows: $y = g^x \bmod p_s$.
   Send $(p_s, q_s, g,$ and $y)$ to Bob (verifier).

Key generation algorithm (generated by receiver, Bob) - Bob must do:

6. Choose two prime numbers (p_r, q_r) randomly, secretly, and roughly of the same size.
7. Compute the modulus n as follows: $n = p_r \times q_r$.
8. Compute the $\Phi(n)$, as follows: $\Phi(n) = (p_r-1) \times (q_r-1)$.
9. Choose the public key e, such that $1 < e < \Phi(n)$, and GCD $(e, \Phi(n)) = 1$.
10. Compute the decryption key d, where $d = e^{-1} \bmod \Phi(n)$.
    Determine the public keys (e, n) and determine the private keys $(\Phi(n), d)$.

Encryption and Signing (sender - Alice) - Alice must do the following:

11. Obtain the public keys (e, n).
12. Determine the message m to be encrypt such that $0 < m < n$.
13. Encrypt the message as follows, $c = m^e \bmod n$.
14. Choose a random integer k, with $0 < k < q_s$.
15. Compute r as follows $r = (g^k \bmod p_s) \bmod q_s$.
16. Compute s as follows: $s = ((k^{-1}) \times (SHA\text{-}1(C) + x \times r)) \bmod q_s$.
    The signature is (r, s).
    Send the signature and the ciphertext (c, r, s) to the receiver.
    k^{-1} is a multiplicative inverse of k in Z_q.

Verifying and Decryption (receiver - Bob) - Bob must do the following:

17. Obtain the keys $(p_s, q_s, g, \text{and } y)$.
18. $w = s^{-1} \bmod q_s$.
19. $u1 = ((SHA\text{-}1(C)) \times w) \bmod q_s$.
20. $u2 = (r \times w) \bmod q_s$.
21. $v = ((g^{u1} \times y^{u2}) \bmod p_s) \bmod q_s$.
    Verify the message m as follows: is $v = r$?.
22. Obtain the ciphertext c from Alice.
23. Recover the message as follows, $m = c^d \bmod n$.

2.2   Combination of RSA and RSADS

Figure 4 shows the combinational protocol between RSA and RSADS algorithms, since both RSA algorithms are based on integer factorization problem.

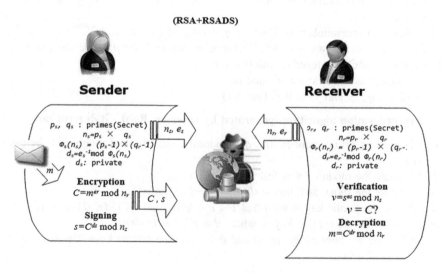

Fig. 4. RSA + RSADS protocol

Key generation algorithm (generated by sender, Alice) - Alice must do:

1. Choose two prime numbers (p_s, q_s) randomly, secretly, and roughly of the same size.
2. Compute the modulus n as follows: $n_s = p_s \times q_r$.
3. Compute the $\Phi(n)$, as follows: $\Phi_s(n_s) = (p_s-1) \times (q_s-1)$.
4. Choose the public key e, such that $1 < e_s < \Phi_s(n_s)$, and GCD $(e_s, \Phi_s(n_s)) = 1$.
5. Compute the decryption key d_s, where $d_s = e_s^{-1} \bmod \Phi_s(n_s)$.
   Determine the public keys (e_s, n_s) and determine the private keys $(\Phi_s(n_s), d_s)$.

Table 2. Performance evaluation between integers and primes based public key algorithms

Description	Integer based Algorithm		Primes based Algorithm	
	Key Size	Time (Millisecond)	Key Size	Time (Millisecond)
Key generation		26		580
E/D and DS		12		17
	56-bit		512 - bit	
Key generation		32		1032
E/D and DS		18		281
	80-bit		1024-bit	
Key generation		108		3395
E/D and DS		43		660
	112-bit		2048-bit	
Key generation		144		6980
E/D and DS		90		9658
	128-bit		3072-bit	
Key generation		8763		10465
E/D and DS		7050		15462
	192-bit		7680-bit	
Key generation		60187		36442
E/D and DS		76440		108386
	256-bit		15360-bit	

*E/D: Encryption and Decryption
*DS: Signing and Verification

Key generation algorithm (generated by receiver, Bob)- Bob must do:

6. Choose two prime numbers (p_r, q_r) randomly, secretly, and roughly of the same size.
7. Compute the modulus n as follows: $n_r = p_r \times q_r$.
8. Compute the $\Phi(n)$, as follows: $\Phi_r(n_r) = (p_r-1) \times (q_r-1)$.
9. Choose the public key e, such that $1 < e_r < \Phi_r(n_r)$, and GCD $(er, \Phi_r(n_r)) = 1$.
10. Compute the decryption key d_r, where $d_r = er^{-1} \bmod \Phi_r(n_r)$.
    Determine the public keys (e_r, n_r) and determine the private keys $(\Phi_r(n_r), d_r)$.

Encryption and Signing (sender - Alice) - Alice must do the following:

11. Determine the message m to be encrypt such that $0 < m < n_r$.
12. Encrypt the message as follows, $c = m^{er} \bmod n_r$.
13. Sign the ciphertext C as: $c = C^{ds} \bmod n_s$
    Send the signature and the ciphertext (C, s) to the receiver.

Verifying and Decryption (receiver - Bob) - Bob must do the following:

14. $v = s^{es} \bmod n_s$
    Verify the message m as follows: is $v = C$?
15. Obtain the ciphertext C from Alice.
16. Recover the message as follows, $m = C^{dr} \bmod n_r$.

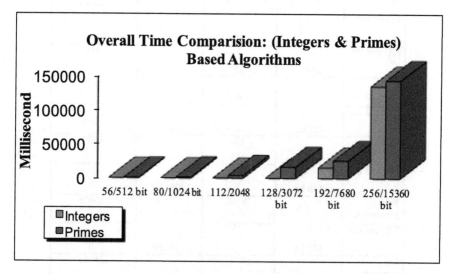

Fig. 5. Overall time comparison between integers and primes based public key algorithms' time

3 Performance Evaluation Based on Equivalent Key Sizes for Fractal and Public-Key Encryption Protocol

We compare the performance of the integer based public-key algorithm against the well-known prime's public-key algorithms (such as RSA and DSA) for the combined encryption and digital signature cryptosystems. Table 2 shows the performance for both approaches. Both protocols were coded in C++ with GMP library. Also, Miller-Rabin algorithm [18] is implemented for primarily test which was coded using C ++ and GMP as well. However, the comparison between integers and primes based public-key cryptosystems shows that integers based public key encryption and digital signature algorithms performs better than primes based public key algorithms in general. As Fig. 5 indicate, integers based public-key algorithm (ECC), provides higher level of security at a much lower cost, both in term of key size and execution time. Moreover,

4 Conclusions

This paper presents and implemented a scheme of combining public-key encryption and digital signature algorithms and proposed an encryption digital signature protocol. However, an overall running time that compare between integers and primes based public key algorithms' time were presented.

Acknowledgments. The authors would like to thank Al Zaytoonah University of Jordan for supporting this study.

References

1. Alia, M.A.: Survey on mathematical hard problems based public-key cryptosystems. World Acad. Sci. Eng. Technol. **68**, 395–402 (2010)
2. Alia, M.A.: Cryptosystems based on chaos theory. In: International Symposium on Chaos, Complexity and Leadership, 17–19 December 2013 (2013)
3. RSA Laboratories, What is a Hard Problem. RSA the Security Division of EMC (2007)
4. Branovic, I., Giorgi, R., Martinelli, E.: Memory performance of public-key cryptography methods in mobile environments. In: ACM SIGARCH Workshop on Memory Performance: Dealing with Applications, Systems and Architecture (MEDEA 2003), New Orleans, LA, USA, pp. 24–31 (2003)
5. Menezes, A., Van Oorschot, P., Vanstone, S.: Handbook of Applied Cryptography, pp. 4–15, 516. CRC Press (1996)
6. Diffie, W., Hellman, M. E.: New directions in cryptography. IEEE Trans. Inf. Theory **IT-22**, 644–654 (1976)
7. Rivest, R.A., Shamir, A., Adleman, L.: A method for obtaining digital signatures and public-key cryptosystems. Commun. ACM **21**(2), 120–126 (1978)

8. Rabin, M.O.: Digitalized signatures and public-key functions as intractable as factorization: the ACM digital library. Technical report. UMI Order Number: TR-212, Massachusetts Institute of Technology (1979)
9. ElGamal, T.: A public-key cryptosystem and a signature scheme based on discrete logarithms. IEEE Trans. Inf. Theory **IT-31**(4), 469–472 (1985)
10. Koblitz, N.: Elliptic curve cryptosystems. Math. Comput. **48**, 203–209 (1987)
11. Stallings, W.: Cryptography and Network Security Principles and Practices. Pearson Education, 3 edn. (2003)
12. Al-Tuwaijry, F.A., Barton, S.K.: A high speed RSA processor. In: IEEE Conference, 2–6 September, Loughborough, UK, pp. 210–214 (1991)
13. Burnett, S., Paine, S.: RSA Security's Official Guide to Cryptography. Osborne/ McGraw-Hill, Berkeley (2001)
14. Elaine, B., Barker, W., Burr, W., Polk, W., Smid, M.: Recommendation for Key Management–Part 1: General NIST Special Publication 800-57 (2006)
15. Chaum, D., van Antwerpen, H.: Undeniable signatures. In: Brassard, G. (ed.) CRYPTO 1989. LNCS, vol. 435, pp. 212–216. Springer, Heidelberg (1990)
16. Burrows, J.H.: Digital Signature Standard (DSS). In: Federal Information Processing Standards Publication 186, Computer Systems Laboratory, National Institute of Standards and Technology, Fips Pub, vol. 186, pp. 1–5 (1994)
17. Public Law, Electronic Signatures in Global and National Commerce Act. Weekly Compilation of Presidential Documents, Public Law, vol. 36, pp. 106–229 (2000)
18. MediaWiki: Literate Programs, Miller-Rabin (2006). http://en.literateprograms.org/Miller-Rabin_primality_test_(C,_GMP)

Intelligent Robotic and Control Track

The Challenges of Reinforcement Learning in Robotics and Optimal Control

Mohammed E. El-Telbany$^{(\boxtimes)}$

Computers and Systems Department, Electronics Research Institute, Giza, Egypt
telbany@eri.sci.eg

Abstract. Reinforcement Learning (RL) is an emerging technology for designing control systems that find optimal policy, through simulated or actual experience, according to a performance measure given by the designer. This paper discusses a widely used RL algorithm called Q-learning. This paper discuss how to apply these algorithms to robotics and optimal control systems, where several key challenges must be addressed for it to be useful. We discuss how Q-learning algorithm can adapted to work in continuous states and action spaces, the methods for computing rewards which generates an adaptive optimal controller and accelerate learning process and finally the safe exploration approaches.

Keywords: Reinforcement learning · Robotics · Optimal control · Neural networks

1 Introduction

In RL, the learning agent interacts with an initially unknown environment and modifies its action policies to maximize its cumulative rewards. Thus, RL provides an efficient framework to solve learning control problems which are difficult or even impossible for supervised learning and traditional dynamic programming (DP) methods. The aim of DP is to compute optimal policies given a perfect model of the environment as an Markov Decision Processes(MDP). Traditional DP algorithms have some limitations due to their assumption of a perfect model and for great computational costs. Actually, RL algorithms are viewed as an adaptive DP or approximate DP, with less computation and without assuming a perfect model of the environment [1]. From the perspective of automatic control, the DP/RL framework comprises a nonlinear and stochastic optimal control problem [2]. Moreover, RL is an important way for adaptive optimal control [3,4]. The RL has been widely studied not only in the machine learning and neural network community but also in robotics and control theory [1,5–10]. Both of robotics and optimal control systems fields have benefited from concepts of RL [11,12]. Several robotic systems in [13–15] have implemented this technique to develop stabilizing controllers, showing that performance equalled and in some cases exceeded that of hand-tuned controllers. Others have used RL to

© Springer International Publishing AG 2017
A.E. Hassanien et al. (eds.), *Proceedings of the International Conference on Advanced Intelligent Systems and Informatics 2016*, Advances in Intelligent Systems and Computing 533, DOI 10.1007/978-3-319-48308-5_84

develop optimal strategies for teams of multiple robots [16,17]. Both RL and optimal control address the problem of finding an optimal policy. However, optimal control assumes perfect knowledge of the systems description in the form of a model. The advantage of RL is that it is robust to changes in a system, such as malfunctioning actuators, changes in models parameters, and external disturbances. However, the RL is generally a hard problem and many of its challenges are particularly apparent in the robotics and control systems applications [12]. These challenges can be summarized as follows:

- **Continuous State and Action Space**: Robots have sensors that output data in a continuous domain.
- **Environment Uncertainty**: The system dynamics or parameters may be unknown and subject to noise, so the controller must be robust and able to deal with uncertainty.
- **Reward Function Design**: Crafting reward functions (i.e. *reward shaping*) may be difficult since rewards can rely upon successful completion of a task. This may lead to a sparsity of instances actually influencing the learning of a robotic system [18].
- **Safe Exploration Strategies**: Robotic hardware is typically not suited for RL's trial-and-error nature. Current RL approaches in robotic tasks typically need prior knowledge on the task.

The outline of this paper is as the following. Section 2 discuss the optimal control problem. Section 3 presents the notation and theory of RL learning. Section 4, the popular Q-learning algorithm will be presented. The description of challenges faces RL algorithms for robotics and optimal control systems described in Sect. 5. Finally, Sect. 6, closing out the paper by discussion and conclusions.

2 Introduction to Optimal Control

Consider a discrete-time system whose dynamics over T stages is described by a time-invariant equation

$$s_{t+1} = f(s_t, a_t), \qquad t = 0, 1, 2, ..., T - 1 \tag{1}$$

where for all t, the state s_t is an element of the state space $\mathcal{S}$ and the action a_t is an element of the action space $\mathcal{A}$. $T \in \mathbb{N}_0$ is referred to as the *optimization horizon*. The transition from t to $t + 1$ is associated with an instantaneous cost signal $c_t = c(s_t, a_t) \in \mathfrak{R}$ which is assumed to be bounded by a constant B_c, and for every initial state s_0 and for every sequence of actions, the discounted cost over T stages is defined as

$$C_T(s_0) = \sum_{t=0}^{T} \gamma^t c(s_t, a_t) \tag{2}$$

where $\gamma \in [0, 1]$ is the discount factor. In this context, an optimal control sequence or policy $u_0^*, u_1^*, u_2^*, ..., u_{T-1}^*$, is a sequence of actions that minimizes the cost over T stages. There are three class of policies, *open-loop* policies which select at time t the action u_t based only on the initial state s_0 of the controller and the current time $(a_t = \pi_0(t, x_0))$, *closed-loop* policies which select the action a_t based on the current time and the current state $(a_t = \pi_c(t, x_t)))$, and *closed-loop stationary* policies for which the action is selected only based on the knowledge of the current state $(a_t = \pi_s(x_t))$.

3 Reinforcement Learning Theory

RL has been defined as the study of planning and learning in a scenario where a learner actively interacts with the environment to achieve a certain goal [19]. The set of all possible actions is defined as $a \in A$ and the set of all possible states is $s \in S$. The environment is modelled as a Markov Decision Process (MDP). That is, for each action that the agent takes, the agent perceives a reward $r(s, a)$ for executing a while in s, and a next state in the environment $s' = \delta(s, a)$. For each state-action pair (s, a), there is a transition probability δ which describes the distribution of states from which s' is drawn. The objective of the agent is to choose actions which maximize its reward and hence help it achieve its goal. Since the agent only receives rewards related to the action just taken, some other mechanism must be used to account for long-term reward feedback. The agent must deem whether it is more appropriate to explore further unknown states $s \in S$ and actions $a \in A$ or utilize the information already collected in order to optimize (maximize) the current reward $r(s, a)$. One of the fundamental obstacles in RL is the *exploration-exploitation* dilemma. The particular action that an agent will choose given a state is given by a mapping $\pi : S \rightarrow A$ known as the policy. RL algorithms attempt to find the optimal policy π^* which maximizes rewards. This policy can be quantified in terms of what is known to be a *policy value* V_π, which is given as

$$V_\pi = E[\sum_{\tau=0}^{T-t} r(s_{t+\tau}, \pi(s_{t+\tau})) | s_t = s]. \tag{3}$$

This policy value is the expected reward $r(s, a)$ parametrized by a state s and action a that is yielded when starting at s using policy π. It is summed over a finite time horizon. The optimal policy is in terms of the policy value, moreover $V^* = V_{\pi^*}(s) = \max_\pi V_\pi(s)$. Therefore V^* is the optimal cumulative reward that the learner may expect to receive while starting at s and executing π^*. The next relevant quantity is the optimal state-action value function $Q^*(s, a)$, which is defined for all $(s, a) \in S \times A$ as the expected return for taking action a at state s and then following the optimal policy. It is given in terms of the optimal policy value as

$$Q^*(s, a) = E[r(s, a)] + \gamma \sum_{s' \in S} Pr[s'|s, a] V^*(s'). \tag{4}$$

When system dynamics model and reward function are known a priori, the automatic learning process of a controller is often called *model-based* RL. In classical control techniques restrictive assumptions on the system model is needed such as linearity and determinism. In model-based RL methods, the models are generally non-linear and stochastic and do not require an analytical model but learning the state transition function $\delta(s, a)$ and reward function $r(s; a)$ instead. Such models are often easier to construct than deriving an analytical model, especially when the behaviour is stochastic. When the system dynamics and reward function are often unknown in practice, the *model-free* RL techniques are used to directly learn the optimal policy from interacting with the environment. The most popular method are Temporal Difference (TD) learning [1], Q-learning [20], SARSA [1] which is *critic-only* algorithms. The evolutionary algorithms have been applied to search for policies directly [22] which is *actor-only* algorithms. Finally, *actor-critic* methods [1,21], which combine both of the policy and value function and each one is improved separately. In the next section, the Q-learning algorithm is described that seeks out this optimal state-action value function. Thus, some methods seek to approximate Q^* by exploring the state-action space and thereby obtain π^*.

4   Q-Learning

The Q-learning is a model-free algorithm for solving optimal control problems on-line, subject to partial or imperfect knowledge of the system state and models, where the objective is to estimate the optimal state-action value function Q^*. The Q-learning can be utilized to learn an approximate state-action value function $Q(s_t, a_t)$ that is iteratively updated by the rule,

$$Q(s, a) \leftarrow \alpha Q(s, a) + (1 - \alpha)[r(s, a) + \gamma \max_{a' \in A} Q(s', a')], \tag{5}$$

where the parameter α is a learning rate, $0 < \alpha \leq 1$ and $\gamma \in [0, 1)$ is a factor used to discount future rewards. A more detailed description of the procedure is shown in Algorithm 1, where SelectAction(π, s) must ensure that each state can be visited infinitely many times, like in an ϵ-greedy approach. That is, a random action is chosen with probability ϵ, and the action is chosen optimally as $\arg\max_{a \in A} Q(s, a)$ with probability $1 - \epsilon$.

5   Q-Learning in Robotics and Optimal Control

5.1   Scaling up

The successful application of RL techniques to robotics and optimal control is limited by the fact that, in most robotic and control tasks, the state and action spaces are continuous, multidimensional, and in essence, too large for conventional RL algorithms to work. The well known *curse of dimensionality* makes infeasible using a tabular representation of the Q-learning function, which is the classical approach that provides convergence guarantees. One could attempt to approximate a continuous space by finely discretizing it, but this increases

---

Algorithm 1. Q-Learning for Reinforcement Learning

---

   procedure Q-LEARNING(π) **returns** an action
      $Q \leftarrow Q_0$
      for t=0; t<T; t++ **do**
         $s \leftarrow$ SelectState(π, s)
         for each step of epoch t **do**
            $a \leftarrow$ SelectAction(π, s)
            $r' \leftarrow$ Reward(s, a)
            $s' \leftarrow$ NextState(s, a)
            $Q(s, a) \leftarrow Q(s, a) + \alpha[r' + \gamma \max_{a'} Q(s', a') - Q(s, a)]$
            $s \leftarrow s'$
         end for
      end for
      return $\max_{a'} f(Q[s', a'])$
   end procedure

---

the amount of memory required to store Q-values and the learning time for Q-learning becomes intractable. In addition, in a grid-based representation there is no notion of similarity between neighbouring states, so discrete representations do not generalize well [24]. Generalization can be introduced by mapping states to action-values via a *function approximator*. When a function approximation technique is used to generalize among similar states, the convergence of the algorithm is compromised, since updates unavoidably affect an extended region of the domain, that is, some situations are modified in a way that has not been really experienced, and the update may degrade the approximation. Actually, there are several existing systems use a neural network as the function approximator [24–26] with success. In [25], each possible action has an associated neural network which approximates the reward of executing that action from the state input to the neural net. The states are forwarded to the inputs of the neural network and outputs are the estimates of the Q-values. During each iteration of the learning algorithm, the current state of the system is forwarded to the inputs of each neural network, but the weights are only updated for the network whose action was selected. The networks are trained using other networks resulting in dependencies preventing the specification of general rules to make the choices. Moreover, multiple networks require more training data because of the increased number of free parameters. The weight correction error for the single step Q-learning is:

$$e_t = r_t + \gamma \cdot \max_{a \in A} Q(x_{t+1}, a_{t+1}) - Q(x_t, a_t) \tag{6}$$

In [26], inspired from *deep learning architecture* as preprocessing step for reducing the dimension and a way of extracting the information out of the image automatically, a neural network is trained with a variant of the Q-learning algorithm, with stochastic gradient descent to update the weights based on experience replay. The output layer has neurons equal to the number of actions for robot control algorithm. Using continuous actions gives more accurate and predictable movement, reduces wear, and improves energy consumption in practice [24]. There exist a number of approaches for incorporating both continuous state and action spaces into Q-learning. In [27], propose a continuous state and action space approach that uses only one neural networks as a function approximator. In [28], states and actions are coarsely discretized, and linear interpolation is used to represent intermediate states. In [24], a single neural network takes the current state as input, and the outputs correspond to control points of an action-reward interpolator. The Q-value of any given action can be smoothly, approximately computed using the interpolator, which is known as a *Wire-Fitted Neural Network*, to waypoint navigation of a boat in simulation with success. Actor-critic methods [21,23] require two function approximators, which must both be trained. One estimates the long-term utility for each state, and the other learns to choose the optimal action in each state. In [13] a system for RL-based optimal control of uncertain non-linear systems is described, where the actor and critic each consist of a neural network. In [29], a locally weighted learning (LWL) used directly to represent the Q-Function of the learning problem with a LWL system using a *kd*-tree for a faster look up of the neighboured inputs. Since the Q-Values are changing with the policy, special algorithms has to be used to update the already existing examples in memory. Likewise, Carden [31] uses Nadaraya-Watson kernel regression [32] to estimate the Q-value of a state-action pair based on past experiences and shows convergence to the true Q-values within some tolerance with probability. This type of regression estimates the value at a particular point as a weighted average of nearby observations using non-negative kernel function to assign weights to observations.

5.2   Environment Uncertainty

Robots must be able to optimally adapt to uncertainty and unforeseen changes in order to tackle stochastic and dynamic environments. Overcoming the uncertainty of world hidden state is handled using Bayesian RL [30] which is based on partially observable Markov decision process (POMDP). Also, handling the uncertainty about possible outcomes of the action is handled by *experience replay* [25], where the robot' previous experiences is used multiple times to update the action. Another approach is to use *eligibility traces* which taking into account increasing uncertainty about future rewards [1].

5.3   Reward Function Design

Another challenge faced RL in robotics an control is the designing of suitable reward functions. Most of efficient rewards that direct the systems quickly to success which consider the cost of real-world experience are realized by *designer*. In optimal control the objective is to determine the policies that will cause a process to satisfy the physical constraints and at the same time optimize some performance criterion which expressed as an integral over time of a cost function of states and actions of physical processes [33]. This also implies finding the trajectory of states corresponding to the optimal policy. This is similar to goal of Q-learning which aims to maximize the reward received by a robot over some time-horizon. Sutton [4] claims that RL is equivalent to direct adaptive optimal control. When an accurate model of the system to be controlled is not available, adaptive optimal control methods can be applied to trajectory tracking and optimal control cases. In optimal control, a cost is typically given as:

$$J = \phi(x(t_f), t_f) + \int_{t_0}^{t_f} L(x(t), u(t), t)dt \tag{7}$$

where $x(t)$ is the state trajectory, $u(t)$ is the control sequence, and ϕ is a terminal cost. An equivalent accumulated RL reward over some trajectory can be given as $-L$, and $-\phi$ is a reward given at an absorbing state. A natural choice for L then is $L = \frac{1}{2}x(t)^T Q x(t) + u(t)^T R u(t)$, where Q and R are positive definite matrices. This quadratic cost is often associated with the Linear Quadratic Regulator (LQR) [34] and is employed by Ng [35] for RL-based stabilization of an autonomous helicopter. This choice of reward encourages the system to be stable while also encouraging small control outputs and smooth transitions. Another issue is the concept of *reward shaping* has been considered in robotics and control [7, 36–39] which is used to speed up the learning process, and to tackle the temporal credit assignment problem in a more efficient way. The idea of reward shaping is to give additional feedback to the robot' controller to improve its convergence rate. To overcome the misleading problems of reward shaping, a potential-based reward function $F(s, a, s')$ was proposed [18] as the difference of some potential function Φ when moving from state s to s'

$$F(s, a, s') = \gamma \Phi(s) - \Phi(s') \tag{8}$$

where γ is a discount factor. The definition of F and Φ was extended by Wiewiora *et al.* [45] by consider *Look-ahead advice* which shapes robots reward and robot policy must choose the action with the maximum sum of both Q-value and potential as follows:

$$Q(s, a) \leftarrow \alpha Q(s, a) + (1 - \alpha)[r(s, a) + F(s, a, s', a') + \gamma \max_{a' \in A} Q(s', a')], \tag{9}$$

5.4   Safe Exploration Strategies

The learner success is based on the exploration strategy to discover rewarding states and trajectories. Balancing the exploration/exploitation trade-off is one of the most difficult problems in RL for robotics and control [40]. Specially, in real-world applications, exhaustive exploration is impossible and dangerous which is often problematic. Balancing the exploitation-exploration trade-off method is ϵ-greedy the agent performs a random action for exploration with probability ϵ and exploits otherwise by executing a greedy policy. Another method is the R-max algorithm [41] which learns a maximum-likelihood tabular model of the environment. The algorithm classifies each state-action as known or unknown according to the number of times it was visited. It is important for exploration techniques in RL to ensure acceptable system performance and consider the safety of the robot or controlled system in RL dangerous tasks which is not guaranteed [42]. In a robotics context, arbitrary exploration is not desired if not discouraged since the robot can easily be damaged. Therefore, the classical RL paradigm in a robotics context is not directly applicable since exploration needs to take hardware constraints into account. Two ways of implementing cautious exploration are to either avoid significant changes in the policy [43] or to explicitly discourage entering undesired regions in the state space [44]. In [42], the safe exploration techniques are: (1) *Optimal control approaches* which utilizing the expected return or its variance safety criterion.(2) *Labelling-based approaches* utilizes some kind of state/action labelling and use *safety function* and *backup policy*. Finally, (3) *Prior knowledge-based approaches* which initialize the search using the prior knowledge by using *Learning from Demonstration* (LfD) methods such as *Apprenticeship Learning* where the reward function is learned using LfD [8]. However, safe exploration without prior knowledge is practically impossible and prior knowledge almost always helps to achieve faster convergence.

6   Conclusions and Discussions

Implementing the RL learning algorithms for robotics and optimal control applications faces several challenges that need to be addressed in order for RL to be realized. The most significant theoretical difficulty is the inability of current RL algorithms to solve problems with large state-action spaces in acceptable time. In this respect, the state space of as task for a robot with more than three degrees of freedom can generally be considered large for RL. Though many adaptations of Q-learning have been developed to deal with continuous state and action spaces, most have not been shown to provably converge to the optimal Q-values. The most important practical difficulty is related to the trial-and-error nature of RL. Performing occasional random actions are essential for RL to improve upon the task solution in a largely unsupervised way. Such explorative actions can lead to control signals and system states that quickly wear down or directly damage the robot. This lead to the emergence of safe-exploration approaches which without prior knowledge is practically impossible. In practice, this limits the time that learning can be performed on a real robot. The usage of prior knowledge makes it possible to speed-up and guided the robot to feasible states. In the slowly changing environment, Q-learning and SARSA learning algorithms that have non-greedy policy and a fixed learning rate can handle the changes. In uncertain or fast changing environment, a big discount factor assist to forget the past environment and therefore help to learn the current environment.

References

1. Sutton, R., Barto, B.: Reinforcement Learning: An Introduction. MIT Press, Cambridge (1998)
2. Cao, X.: Stochastic Learning and Optimization. Springer, Heidelberg (2009)

3. Bertsekas, D.: Neuro-Dynamic Programming. Athena Scientific, Belmon (1996)
4. Sutton, R., Barto, A., Williams, R.: Reinforcement learning is direct adaptive optimal control. IEEE Control Syst. **12**, 19–22 (1992)
5. Tesauro, G.: TD-Gammon, a self-teaching backgammon program achieves master-level Play. Neural Comput. **6**(2), 215–219 (1994)
6. Szepesvri, C.: Algorithms for Reinforcement Learning. Morgan and Claypool, San Rafael (2010)
7. Randlv, P., Alstrm, P.: Learning to drive a bicycle using reinforcement learning and shaping. In: Proceedings of the Fifteenth International Conference on Machine Learning, pp. 463–471 (1998)
8. Abbeel, P., Coates, A., Quigley, M., Ng, A.: An application of reinforcement learning to aerobatic helicopter flight. In: Advances in Neural Information Processing Systems, vol. 19. MIT press (2007)
9. Wang, F., Zhang, H., Liu, D.: Adaptive dynamic programming: an introduction. IEEE Comput. Intell. Mag. **4**(2), 39–47 (2009)
10. Busoniu, L., Babuska, R., De Schutter, B., Ernst, D.: Reinforcement Learning and Dynamic Programming Using Function Approximators. CRC Press, New York (2010)
11. Kaelbling, L., Littman, M., Moore, A.: Reinforcement learning: a survey. J. Artif. Intell. Res. **4**, 237–285 (1996)
12. Kober, J., Bagnell, J., Peters, J.: Reinforcement learning in robotics: a survey. Int. J. Robot. Res. **32**, 1238–1274 (2013)
13. Bhasin, S.: Reinforcement learning and optimal control methods for uncertainnonlinear systems. Ph.D., University of Florida (2011)
14. Hester, T., Quinlan, M., Stone, P.: RTMBA: a real-time model-based reinforcement learning architecture for robot control. In: International Conference on Robotics and Automation, ICRA 2010, pp. 85–90 (2012)
15. Kim, H., Jordan, M., Sastry, S., Ng, A.: Autonomous helicopter flight via reinforcement learning. In: Advances in Neural Information Processing Systems (2003)
16. El-Telbany, M.: Reinforcement Learning Algorithms For Multi-Robot Organization Ph.D. thesis, Faculty of Engineering, Cairo Univrsity (2003)
17. Erfu, Y., Dongbing, G.: Multiagent reinforcement learning for multi-robot systems: a survey. Technical report (2004)
18. Ng, A.: Shaping and policy search in Reinforcement Learning. Ph.D., Universityof California, Berkeley (2003)
19. Mohri, M., Rostamizadeh, A., Talwalkar, A.: Foundations of Machine Learning. The MIT Press, Cambridge (2012)
20. Watkins, C., Dayan, P.: Q-learning. Mach. Learn. **8**, 279–292 (1992)
21. Grondman, I., Busoniu, L., Lopes, G., Babuka, R.: A survey of actor-critic reinforcement learning: standard and natural policy gradients. IEEE Trans. Syst. Man Cybern Part C: Appl. Rev. **4**(2), 39–47 (2012)
22. Heidrich-Meisner, V., Lauer, M., Igel, C., Riedmiller, M.: Reinforcement learning in a nutshell. In: 15th European Symposium on Artificial Neural Networks (ESANN2007), pp. 277–288 (2007)
23. van Hasselt, H.: Reinforcement learning in continuous state and action spaces. In: Wiering, M., van Otterlo, M. (eds.) Reinforcement Learning. ALO, vol. 12, pp. 205–248. Springer, Heidelberg (2012)
24. Gaskett, C.: Q-Learning for Robot Control. Ph.D., Australian National University (2002)
25. Lin, L.: Self-improving reactive agents based on reinforcement learning, planning and teaching. Mach. Learn. Austr. Natl. Univ. **8**, 293–321 (1992)

26. Mnih, V., Kavukcuoglu, K., Silver, D., Graves, A., Antonoglou, I., Wierstra, D., Riedmiller, M.: Playing atari with deep reinforcement learning. CoRR (2013)
27. Hagen, S., Krose, B.: Neural Q-learning. Neural Comput. Appl. **21**(2), 81–88 (2003)
28. Takahashi, Y., Takeda, M., Asada, M.: Continuous valued Q-learning for vision-guided behavior. In: International Conference on Multisensor Fusion and Integration for Intelligent Systems (1999)
29. Smart, W.: Making Reinforcement Learning Work on Real Robots. Ph.D., BrownUniversity (2002)
30. Duff, M.: Optimal learning: computational procedures for bayes adaptive Markov decision processes. Ph.D. dissertation, University of Massachusetts (2002)
31. Carden, S.: Convergence of a Q-learning variant for continuous states and actions. J. Artif. Intell. Res. **49**, 705–731 (2014)
32. Nadaraya, E.: On estimating regression. Theory Prob. Appl. **9**(1), 141–142 (1964)
33. Kirk, D.: Optimal Control Theory. An Introduction. Prentice Hall, Englewood Cliffs (1970)
34. Anderson, B., Moore, J.: Optimal Control: Linear Quadratic Methods. Prentice Hall, Upper Saddle River (1989)
35. Ng, A.Y., Coates, A., Diel, M., Ganapathi, V., Schulte, J., Tse, B., Berger, E., Liang, E.: Autonomous inverted helicopter flight via reinforcement learning. In: Ang, M.H., Khatib, O. (eds.) Experimental Robotics IX. STAR, vol. 21, pp. 363–372. Springer, Heidelberg (2006). doi:10.1007/11552246_35
36. Ng, A., Harada, D., Russell, S.: Potential-based shaping in model-based reinforcement learning. In: Proceedings of the 16th International Conference on Machine Learning (1999)
37. Matric, M.: Reward functions for accelerated learning. In: Proceedings of the 11th International Conference on Machine Learning, pp. 181–189 (1994)
38. Konidaris, G., Barto, A.: Autonomous shaping: Knowledge transfer in reinforcement learning. In: Proceedings of the 23th International Conference on Machine Learning (2006)
39. Asmuth, J., Littman, M., Zinkov, R.: Potential-based shaping in model-based reinforcement learning. In: Proceedings of AAAI Conference on Artificial Intelligence (2008)
40. Thrun, S.: The role of exploration in learning control. In: White, D., Sofg, D. (eds.) Handbook of Intelligent Control: Neural, Fuzzy and Adaptive Approaches (1992)
41. Brafman, R., Tennenholtz, M.: R-max a general polynomial time algorithm for near optimal reinforcement learning. J. Mach. Learn. Res. **3**, 213–231 (2002)
42. Garcia, J., Fernandez, F.: Safe exploration of state and action spaces in reinforcement learning. J. Artif. Intell. Res. **45**, 515–564 (2012)
43. Peters, J., Mulling, K., Altun, Y.: Relative entropy policy search. In: Proceedings of the National Conference on Artificial Intelligence, pp. 1607–1612 (2010)
44. Deisenroth, M., Rasmussen, C., Fox, D.: Learning to control a low-cost manipulator using data-efficient reinforcement learning. In: Proceedings of the International Conference on Robotics: Science and Systems (2011)
45. Wiewiora, E., Cottrell, G., Elkan, C.: Principled methods for advising reinforcement learning agents. In: ICML, pp. 792–799 (2003)

Robust Control for Asynchronous Switched Nonlinear Systems with Time Varying Delays

Ahmad Taher Azar[1,2]([⊠]) and Fernando E. Serrano[3]

[1] Faculty of Computers and Information, Benha University, Benha, Egypt
[2] Nanoelectronics Integrated Systems Center (NISC), Nile University, Cairo, Egypt
ahmad_t_azar@ieee.org, ahmad.azar@fci.bu.edu.eg
[3] Central American Technical University, Zona Jacaleapa, Tegucigalpa, Honduras
serranofer@eclipso.eu

Abstract. In this article a novel robust controller for the control of switched nonlinear systems with asynchronous switching is proposed considering state delays. The proposed approach improves the actual methodologies found in literature in which the disturbance rejection properties of these two methodologies consider a disturbance equal to zero but the proposed robust controller considers any kind of disturbances that makes this strategy to surpass other similar methodologies. The main objective is that the robust controller stabilizes the studied system in matched and unmatched modes considering the dwell time in order to obtain an exponentially stable closed loop system. Another characteristic of the proposed control strategy is that a conmutative control law in both matched and unmatched cases is designed with a linear part, where the gain matrices for the linear part are obtained by linear matrix inequalities LMI's along with a nonlinear controller part.
abstract environment.

Keywords: Asyncronous control · Switched systems · Robust control · Nonlinear control · Time delayed systems

1 Introduction

Asynchronous switched control of linear and nonlinear systems have been studied in recent years in which there are several interesting results, considering that an asynchronous switching occurs when the modes of the controller and the system are differents. Some of these studies found in literature are related to linear systems such as [3] where the control of the studied system with average dwell time is proposed for the synchronous and asynchronous cases. Another interesting study can be found in [5] where a H_∞ controller [15] of switched delayed system with average dwell time is evinced. In [1] there is another example for the stabilization of time delayed linear systems [9] under asynchronous switching, a robust H_∞ controller is implemented following a similar procedure as the previous study. In [2] an asynchronous finite time H_∞ controller with mode dependent

© Springer International Publishing AG 2017
A.E. Hassanien et al. (eds.), *Proceedings of the International Conference on Advanced Intelligent Systems and Informatics 2016*, Advances in Intelligent Systems and Computing 533, DOI 10.1007/978-3-319-48308-5_85

dynamic state feedback is shown so the controller is designed implementing its disturbance rejection properties. Lyapunov-Krasovskii functionals for the design of asynchronous switched controller for nonlinear systems can be found in [4] in order to test and ensure the closed loop exponential stability, implementing an average dwell time. In [6] the average dwell time is implemented again for the exponential stability proof and in the case of the nonlinear controller, this includes delays in the states. In [7] a L_∞ robust asynchronous controller like the proposed system in this article is shown, where as similar to the previous study the switched system consists of a linear and nonlinear part and the delays are considered only in the linear part. In the case of intelligent controllers to solve this kind of problems there are several studies such as [8] where an asynchronous fuzzy controller for switched nonlinear systems via switching Lyapunov functions implementing a Takagi-Sugeno model is used to estimate each subsystem, so the average dwell time is used obtaining the gain matrices solving the LMI's. There are other approaches such as sliding mode control [10,12,14] and intelligent control techniques [13] that can be implemented for the stabilization of this kind of systems in future studies. In this article a novel asynchronous robust controller for switched nonlinear time delayed systems is proposed. The objective of this study is to design a conmutative controller in the presence of disturbances, contrary to other control approaches for nonlinear systems in which the disturbance is considered as zero $w = 0$ [7]. Another contribution of this article is that the time delays are considered not only in the linear states these are also considered in the nonlinear states something that is not found in the literature as far as the author knowledge so with these results the outcomes obtained in other studies are significantly improved. The design procedure consists in selecting appropriated Lyapunov-Krasovskii functionals in order to consider the delayed states in the linear and nonlinear parts, dividing the problem in two parts the matched case and unmatched case so the Lyapunov functions derivatives are obtained to corroborate the closed loop exponential stability. The robust conmutative control laws are obtained for both switching cases, the matched and unmatched cases, so the controller consists of two parts one linear state feedback part, obtaining the gain matrices solving the required LMI's [16] and a nonlinear feedback control law obtained for each switching case. The article is divided in three sections; Sect. 2 shows the problem formulation, in the following section, the robust controller design is shown, in Sect. 4 a simulation is evinced, and in the last section the conclusions of this study are presented.

2   Problem Formulation

Consider [7]

$$\dot{x}(t) = A_{\sigma(t)}x(t) + A_{d\sigma(t)}x(t - d(t)) + f_{\sigma(t)}(x(t), x(t - d(t))) + B_{\sigma(t)}u(t) \\ + G_{\sigma(t)}w(t). \quad (1)$$

Where

$x(t) \in \mathbb{R}^n$, $u(t) \in \mathbb{R}^m$, $w(t) \in \mathbb{R}^p$, $x(t_0 + \theta) = \varphi(\theta)$ and $\theta \in [-\tau, 0]$ and $0 \le d(t) \le \tau$, $\dot{d}(t) \le \eta < 1$ and $max_w(w(t)) = \gamma$

In order to design the proposed controller is important to define the following switching instances of the system as found in [3,7]

$$\sigma : (t_0, \sigma(t_0)), (t_1, \sigma(t_1))...(t_k, \sigma(t_k)) \tag{2}$$

and the controller switching instance are defined as [3,7]

$$\sigma : (t_0, \sigma(t_0)), (t_1 + \Delta_1, \sigma(t_1))...(t_k + \Delta_k, \sigma(t_k)) \tag{3}$$

where $0 < \Delta_k < inf_{k \ge 1}(t_{k+1} - t_k)$. Apart from these equations it is important to define the average dwell time and chattering bound [4,6,7].

Definition 1. *For any* $0 < t \le T$ *define* $N_{\sigma(t)}(T, t)$ *be the number of switching numbers over* (t, T). *So* $N_{\sigma(t)}(T, t) \le N_0 + \frac{T-t}{\tau_a}$ *is met for* $\tau_a > 0$ *and* $N_0 \ge 0$ *then* τ_a *is the average dwell time and* N_0 *is the chattering bound.*

With these definitions the proposed approach could be designed following the required design procedure.

3   Robust Asynchronous Controller Design for Nonlinear Switched Systems with Time Varying Delays

The results of this article are explained in this section, so to derive the proposed approach the problem must be divided in the matched and unmatched cases. The following theorem depicts the asynchronous robust controller design for the studied system. It is important to remark that the asynchronous robust control law is divided into a linear part (state feedback) and a nonlinear part, obtaining the gains solving the required LMI's.

Theorem 1. *An asynchronous robust controller is obtained if the following LMI's are solved for* $(K_{i\sigma(t)}, K_{j\sigma(t)})$.

$$\begin{bmatrix} 2P_i A_{\sigma(t)} + \alpha P_i + 2P_i B_{\sigma(t)} K_{\sigma(t)} + Q_i & 2P_i A_{d\sigma(t)} \\ 0 & -Q_i e^{-\alpha\tau} \end{bmatrix} < 0$$

$$\begin{bmatrix} 2P_j A_{\sigma(t)} - \beta P_j + 2P_j B_{\sigma(t)} K_{j\sigma(t)} + Q_j & 2P_j A_{d\sigma(t)} \\ 0 & -e^{\beta\tau} Q_j \end{bmatrix} < 0. \tag{4}$$

with τ_a

$$\tau_a > \tau_a^* = \frac{(\alpha + \beta)T_{max} + ln(\mu)}{\alpha}. \tag{5}$$

Proof. Consider the case when $t \in [t_{k-1} + \Delta_{k-1}, t_k]$ by selecting the following Lyapunov functional [7] where $P_i > 0$ and $Q_i > 0$ are diagonal matrices:

$$V_{i1} = x^T(t)P_i x(t) + \int_{t-d(t)}^{t} e^{-\alpha(t-s)} x^T(s) Q_i x(s) ds. \tag{6}$$

Taking the time derivative of (6) and substituting system (1)

$$
\begin{aligned}
\dot{V}_{i1} \leq\ & 2x^T(t)P_iA_{\sigma(t)}x(t) + 2x^T(t)P_iA_{d\sigma(t)}x(t-d(t)) + 2x^T(t)P_if_{\sigma(t)}(x(t), x(t-d(t))) \\
& + 2x^T(t)P_iB_{\sigma(t)}u(t) + 2x^T(t)P_iG_{\sigma(t)}w(t) + x^T(t)Q_ix(t) \\
& - x^T(t-\tau)Q_ix(t-\tau)(1-\eta)e^{-\alpha\tau} - \alpha\int_{t-d(t)}^{t} e^{-\alpha(t-s)}x^T(s)Q_ix(s)ds.
\end{aligned}
\tag{7}
$$

Making $\overline{x} = [x^T(t), x^T(t-d(t))]$ and defining a control law with a nonlinear part $u_{nl\sigma(t)}$

$$U(t) = K_{i\sigma(t)}x(t) + u_{nl\sigma(t)}. \tag{8}$$

Then

$$
\begin{aligned}
\dot{V}_{i1} \leq\ & \overline{x}^T\Phi\overline{x} + 2x^T(t)P_if_{\sigma(t)}(x(t), x(t-d(t))) \\
& + 2x^T(t)P_iB_{\sigma(t)}u_{nl\sigma(t)}(t) + 2x^T(t)P_iG_{\sigma(t)}w(t) \\
& \qquad\qquad\qquad\qquad\qquad - \alpha V_{i1}.
\end{aligned}
\tag{9}
$$

where

$$\Phi = \begin{bmatrix} 2P_iA_{\sigma(t)} + \alpha P_i + 2P_iB_{\sigma(t)}K_{\sigma(t)} + Q_i & 2P_iA_{d\sigma(t)} \\ 0 & -Q_ie^{-\alpha\tau} \end{bmatrix} < 0. \tag{10}$$

The following property is important to obtain the asynchronous robust control law

Property 1. *The nonlinear part of system (1) has the following property*

$$f_{\sigma(t)}(x(t), x(t-d(t))) - f_{\sigma(t)}(x(t), x(t-\tau)) \cong 0. \tag{11}$$

To obtain the robust control law is necessary to implement [11]

$$inf_{u\in u(y,t)}sup_{x\in Q(y,t)}sup_{w\in W}[L_fV_{i1}(x, u, w, t) + \alpha_v(x,t)] < 0. \tag{12}$$

Defining $\alpha_v(x,t) = x^T(t)x(t)$, substituting (9) in (12) and implementing Property 1 yields

$$
\begin{aligned}
=\ & 2x^T(t)P_if_{\sigma(t)}(x(t), x(t-d(t))) \\
& + 2x^T(t)P_iB_{\sigma(t)}u_{nl\sigma(t)}(t) + 2x^T(t)P_iG_{\sigma(t)}w(t) \\
& \qquad\qquad\qquad\qquad - \alpha V_{i1} + x^T(t)x(t).
\end{aligned}
\tag{13}
$$

Implementing Property 1 the following nonlinear control law is obtained

$$u_{nl\sigma(t)} = -B_{\sigma(t)}^{-1}f_{\sigma(t)}(x(t), x(t-\tau)) - B_{\sigma(t)}^{-1}G_{\sigma(t)}\gamma_v - \frac{1}{2}B_{\sigma(t)}^{-1}P_i^{-1}x(t) \tag{14}$$

where $\gamma_v \in \mathbb{R}^p$ is a constant vector that is used to satisfy the following condition in order to meets the robust stability requirements

$$sup_\gamma sup_w[2x^T(t)P_iG_{\sigma(t)}w(t) - 2x^T(t)P_iG_{\sigma(t)}\gamma_v] \cong 0. \tag{15}$$

so substituting the nonlinear robust control law (14) in (13) the Lyapunov function derivative obtained is:

$$\dot{V}_{i\sigma(t)} = \begin{cases} 0 \ \|x\| = 0 \\ -\alpha V_{i\sigma(t)} \ \|x\| > 0 \end{cases} \tag{16}$$

with the following conmutative controller

$$U(t) =$$
$$\begin{cases} K_{i\sigma(t)}x(t) \ \|x\| = 0 \\ -B_{\sigma(t)}^{-1} f_{\sigma(t)}(x(t), x(t-\tau)) - B_{\sigma(t)}^{-1} G_{\sigma(t)}\gamma_v - \frac{1}{2} B_{\sigma(t)}^{-1} P_i^{-1} x(t) \ \|x\| > 0 \end{cases} \tag{17}$$

When $t \in [t_k, t_k + \Delta_k]$ the following Lyapunov-Krasovskii functional is implemented [7] $P_j > 0$ and $Q_j > 0$ are diagonal matrices

$$V_{j1} = x^T(t) P_j x(t) + \int_{t-d(t)}^{t} e^{\beta(t-s)} x^T(s) Q_j x(s) ds. \tag{18}$$

Taking the time derivative of (18) and substituting (1)

$$\dot{V}_{j1} \le 2x^T(t) P_j A_{\sigma(t)} x(t) + 2x^T(t) P_j A_{d\sigma(t)} x(t-d(t)) + 2x^T(t) P_j f_{\sigma(t)}(x(t), x(t-d(t)))$$
$$+ 2x^T(t) P_j B_{\sigma(t)} u(t) + 2x^T(t) P_j G_{\sigma(t)} w(t) + x^T(t) Q_j x(t)$$
$$- x^T(t-\tau) Q_j x(t-\tau)(1-\eta) e^{\beta\tau} + \beta \int_{t-d(t)}^{t} e^{\beta(t-s)} x^T(s) Q_j x(s) ds. \tag{19}$$

Making $\overline{x} = [x^T(t), x^T(t-d(t))]$ and defining a control law with a nonlinear part $u_{nl\sigma(t)}$

$$U(t) = K_{j\sigma(t)} x(t) + u_{nl\sigma(t)} \tag{20}$$

Then

$$\dot{V}_{j1} \le \overline{x}^T \Phi \overline{x} + 2x^T(t) P_j f_{\sigma(t)}(x(t), x(t-d(t)))$$
$$+ 2x^T(t) P_j B_{\sigma(t)} u_{nl\sigma(t)}(t) + 2x^T(t) P_j G_{\sigma(t)} w(t)$$
$$+ \beta V_{j1}. \tag{21}$$

where

$$\Phi = \begin{bmatrix} 2P_j A_{\sigma(t)} - \beta P_j + 2P_j B_{\sigma_t} K_{j\sigma(t)} + Q_j & 2P_j A_{d\sigma(t)} \\ 0 & -e^{\beta\tau} Q_j \end{bmatrix} < 0 \tag{22}$$

with (12) the asynchronous robust control law is obtained, with $\alpha_v(x,t) = x^T(t) x(t)$, yielding

$$= 2x^T(t) P_j f_{\sigma(t)}(x(t), x(t-d(t)))$$
$$+ 2x^T(t) P_j B_{\sigma(t)} u_{nl\sigma(t)}(t) + 2x^T(t) P_j G_{\sigma(t)} w(t)$$
$$+ \beta V_{j1} + x^T(t) x(t). \tag{23}$$

Implementing Property 1 the following nonlinear control law is obtained

$$u_{nl\sigma(t)} = -B_{\sigma(t)}^{-1}f_{\sigma(t)}(x(t), x(t-\tau)) - B_{\sigma(t)}^{-1}G_{\sigma(t)}\gamma_v - \frac{1}{2}B_{\sigma(t)}^{-1}P_j^{-1}x(t) \quad (24)$$

where $\gamma_v \in \mathbb{R}^p$ is a constant vector that is used to satisfy the following condition in order to meets the robust stability requirements

$$sup_\gamma sup_w [2x^T(t)P_jG_{\sigma(t)}w(t) - 2x^T(t)P_jG_{\sigma(t)}\gamma_v] \cong 0 \quad (25)$$

so substituting the nonlinear robust control law (24) in (23) the Lyapunov function derivative obtained is:

$$\dot{V}_{j\sigma(t)} = \begin{cases} 0 \ \|x\| = 0 \\ \beta V_{j\sigma(t)} \ \|x\| > 0 \end{cases} \quad (26)$$

with the following conmutative controller

$$U(t) =$$
$$\begin{cases} K_{j\sigma(t)}x(t) \ \|x\| = 0 \\ -B_{\sigma(t)}^{-1}f_{\sigma(t)}(x(t), x(t-\tau)) - B_{\sigma(t)}^{-1}G_{\sigma(t)}\gamma_v - \frac{1}{2}B_{\sigma(t)}^{-1}P_j^{-1}x(t) \ \|x\| > 0 \end{cases} \quad (27)$$

Now to test the exponential stability the following steps must be implemented [6]: considering that

$$[\tau_l, \tau_{l+1}] = T \uparrow (\tau_l, \tau_{l+1}) \bigcup T \downarrow (\tau_l, \tau_{l+1}) \quad (28)$$

where $T \uparrow (\tau_l, \tau_{l+1})$ and $T \downarrow (\tau_l, \tau_{l+1})$ represents the increasing and decreasing intervals of the Lyapunov functionals. Considering (16) and (26) over all the interval $[\tau_l, \tau_{l+1}]$ and using the following property

Property 2. *The Lyapunov functional for different switching instants has the following property*

$$V_k(x_{\tau_l}) < \mu V_l(x_{\tau_l}) \quad (29)$$

where $\sigma(\tau_l) = l$ and $\sigma(\tau_l^-) = m$

Therefore

$$V_{\sigma(t)}(x_t) \le e^{-\alpha T \downarrow (\tau_l, \tau_{l+1}) + \beta T \uparrow (\tau_l, \tau_{l+1})} V_{\sigma(\tau_l)}(\tau_l) \quad (30)$$

$$\le e^{-\alpha[T \uparrow (t-\tau_l) + T \downarrow (t-\tau_l)]} \cdot e^{\alpha T \uparrow (t-\tau_l)} \cdot e^{\beta T \uparrow (t-\tau_l)} V_{\sigma(\tau_l)}(\tau_l) \quad (31)$$

$$\le e^{-\alpha(t-\tau_l)} e^{(\alpha+\beta)T \uparrow (t-\tau_l)} V_{\sigma(\tau_l)}(\tau_l) \quad (32)$$

Considering Definition 1 and Property 2

$$\le e^{-\alpha(t-\tau_0)} e^{N_\sigma(\alpha+\beta)T_{max}} \mu^{N_\sigma} V_{\sigma(t_0)}(t_0) \quad (33)$$

where $T_{max} = max_l T \uparrow (\tau_{l+1} - \tau_l)$ for any switching instant $\tau_1, \tau_2, ..., \tau_l$. Rearranging (33) yields

$$\le e^{(N_\sigma(\alpha+\beta)T_{max} + N_\sigma ln\mu)} e^{-(t-t_0)[\alpha - (\alpha+\beta)\frac{T_{max}}{\tau_a} - \frac{ln\mu}{\tau_a}]} V_{\sigma(t_0)}(t_0) \quad (34)$$

$$\leq e^{(N_\sigma(\alpha+\beta)T_{max}+N_\sigma ln\mu)}e^{-(t-t_0)\lambda}V_{\sigma(t_0)}(t_0) \tag{35}$$

where

$$\lambda = \alpha - (\alpha+\beta)\frac{T_{max}}{\tau_a} - \frac{ln\mu}{\tau_a} > 0 \tag{36}$$

So the system is exponentially stable with the following condition

$$\tau_a > \tau_a^* = \frac{(\alpha+\beta)T_{max} + ln\mu}{\alpha} \tag{37}$$

This completes the proof.

4   Simulation Example

The following problem is solved to test the theoretical results obtained in this article. The following matrices and nonlinear functions for system (1) are used for these purposes:

$$A_2 = \begin{bmatrix} -0.32 & 0.0 \\ 0.0 & -0.43 \end{bmatrix}, A_{d1} = \begin{bmatrix} -0.4 & 0.0 \\ 0.0 & -0.5 \end{bmatrix}, A_{d2} = \begin{bmatrix} -0.43 & 0.0 \\ 0.0 & -0.52 \end{bmatrix}$$

$$B_1 = \begin{bmatrix} 0.8 & 0.0 \\ 0.0 & 0.9 \end{bmatrix}, B_2 = \begin{bmatrix} 0.89 & 0.0 \\ 0.0 & 0.91 \end{bmatrix}, G_1 = \begin{bmatrix} -0.3 & 0.0 \\ 0.0 & -0.2 \end{bmatrix}, G_2 = \begin{bmatrix} -0.31 & 0.0 \\ 0.0 & -0.21 \end{bmatrix}$$

$$f_1 = \begin{bmatrix} x_1(t)e^{-2x_2(t)} + x_1(t-d(t))^2 \\ x_2(t) + exp^{-3x_2(t-d(t))} \end{bmatrix}, f_2 = \begin{bmatrix} 1.4x_1e^{-2.7x_2(t)} + x_1(t-d(t))^2 \\ 0.6x_2 + 0.9e^{-3x_2(t-d(t))} \end{bmatrix} \tag{38}$$

And the matrices $P_i = P_j = Q_i = Q_j = I$ where I is the identity matrix for $i = 1, 2$ and a step disturbance $w(t) = [step(0.00001), step(0.00001)]$ The stabilized variables are shown in Fig. 2 with the switching modes shown in Fig. 1 showing that the system is clearly stabilized.

$$K_{1matched} = K_{1unmatched} = \begin{bmatrix} 1.6 & 0.0 \\ 0.0 & 1.8; \end{bmatrix} \tag{39}$$

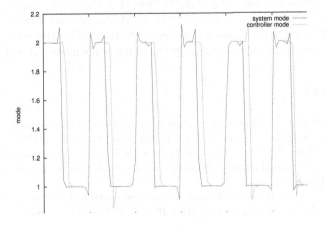

Fig. 1. Switching modes

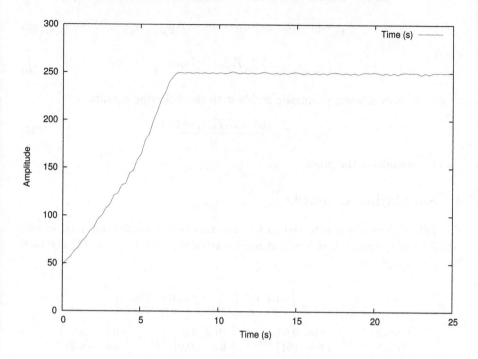

Fig. 2. variable x_2

$$K_{2matched} = K_{2unmatched} = \begin{bmatrix} 1.78 & 0.0 \\ 0.0 & 1.82; \end{bmatrix} \tag{40}$$

5   Conclusion

In this article an asynchronous robust controller for switched nonlinear system with state delays is shown. The results yielded in this study improves the outcomes obtained in studies found in literature. A conmmutative controller for the matched and unmatched cases are implemented to stabilize the system and these theoretical results are corroborated in the numerical simulation section.

References

1. Lian, J., Ge, Y.: Robust output tracking control for switched systems under asynchronous switching. Nonlinear Anal. Hybrid Syst. **8**, 57–68 (2013)
2. Liu, H., Shen, Y., Zhao, X.: Asynchronous finite-time control for switched linear systems via mode-dependent dynamic state-feedback. Nonlinear Anal. Hybrid Syst. **8**, 109–120 (2013)
3. Zhang, L., Gao, H.: Asynchronously switched control of switched linear systems with average dwell time. Automatica **46**, 953–958 (2010)

4. Wang, Y.E., Sun, X.M., Wu, B.: Lyapunov Krasovskii functionals for switched nonlinear input delay systems under asynchronous switching. Automatica **61**, 126–133 (2015)
5. Wang, Y.E., Sun, X.M., Zhao, J.: Asynchronous H ∞ control of switched delay systems with average dwell time. J. Franklin Inst. **349**, 3159–3169 (2012)
6. Zhai, S., Yang, X.S.: Exponential stability of time-delay feedback switched systems in the presence of asynchronous switching. J. Franklin Inst. **350**, 0016–0032 (2013)
7. Zhengrong, X., Chen, Q., Huang, S., Xiang, Z.: Robust L ∞ reliable control for uncertain switched nonlinear systems with time delay under asynchronous switching. Appl. Math. Comput. **222**, 658–670 (2013)
8. Zheng, Q., Zhang, H., Huang, S., Xiang, Z.: Asynchronous H ∞ fuzzy control for a class of switched nonlinear systems via switching fuzzy Lyapunov function approach. Neurocomputing **182**, 0925–2312 (2016)
9. Azar, A.T., Serrano, F.E.: Deadbeat control for multivariable discrete time systems with time varying delays. In: Azar, A.T., Vaidyanathan, S. (eds.) Chaos Modeling and Control Systems Design. SCI, vol. 581, pp. 97–132. Springer, Heidelberg (2015). doi:10.1007/978-3-319-13132-0_6
10. Azar, A.T., Serrano, F.E.: Adaptive sliding mode control of the furuta pendulum. In: Azar, A.T., Zhu, Q. (eds.) Advances and Applications in Sliding Mode Control systems. SCI, vol. 576, pp. 1–42. Springer, Heidelberg (2015). doi:10.1007/978-3-319-11173-5_1
11. Freeman, R.A., Kokotovic, P.V.: Robust Nonlinear Control Design State Space and Lyapunov Techniques. Birkhäuser Basel, Boston (1999)
12. Mekki, H., Boukhetala, D., Azar, A.T.: Sliding modes for fault tolerant control. In: Azar, A.T., Zhu, Q. (eds.) Advances and Applications in Sliding Mode Control systems. SCI, vol. 576, pp. 407–433. Springer, Heidelberg (2015). doi:10.1007/978-3-319-11173-5_15
13. Azar, A.T., Vashist, R., Vashishtha, A.: A rough set based total quality management approach in higher education. In: Zhu, Q., Azar, A.T. (eds.) Complex System Modelling and Control Through Intelligent Soft Computations. Studies in Fuzziness and Soft Computing, vol. 319, pp. 389–406. Springer, Heidelberg (2015)
14. Zhu, Q., Azar, A.T.: Anti-synchronization of identical chaotic systems using sliding mode control and an application to Vaidyanathan Madhavan chaotic systems. In: Azar, A.T., Zhu, Q. (eds.) Advances and Applications in Sliding Mode Control Systems. SCI, vol. 576, pp. 527–547. Springer, Heidelberg (2015)
15. Azar, A.T., Serrano, F.E.: Stabilization and control of mechanical systems with backlash. In: Handbook of Research on Advanced Intelligent Control Engineering and Automation. Advances in Computational Intelligence and Robotics (ACIR), vol. 575. IGI Global, USA (2015)
16. Azar, A.T., Serrano, F.E.: Design and modeling of anti wind up PID controllers. In: Zhu, Q., Azar, A.T. (eds.) Complex System Modelling and Control through Intelligent Soft Computations. Studies in Fuzziness and Soft Computing, vol. 319, pp. 1–44. Springer, Heidelberg (2015)

Path Planning for Line Marking Robots Using 2D Dubins' Path

Ibrahim A. Hameed[(✉)]

Faculty of Engineering and Natural Sciences,
Department of Automation Engineering (AIR),
Norwegian University of Science and Technology,
NTNU in Ålesund, Postboks 1517, 6025 Ålesund, Norway
ibib@ntnu.no

Abstract. This paper proposes a path planning algorithm based on 2D *Dubins' path* for the construction of a curvature continuous trajectory for the autonomous guidance of line marking robots in football stadiums. The algorithm starts with four corner points representing the playable football field and generates a set of waypoints representing various parts of the field layout such as touch and goal lines, goal and penalty area, center line and mark, corner and penalty arcs, center mark and center circle, and penalty marks. A complete, continuous and smooth path is then generated by connecting these waypoints using 2D Dubins' path in a way to ensure that the generated path takes into account the dynamic constraints of the vehicle (such as maximum curvature and velocity), keep the vehicle at a safe distance from obstacles, and not harm the field grass. The efficiency of the algorithm is tested using simulation and in reality. Results showed that the algorithm is able to reliably plan a safe path in real time able to command the line marking robot with high accuracy and without the need for human guidance. The path planning algorithm developed in this paper is implemented in both Matlab and Python.

Keywords: Guidance and navigation · Robotics in sports · FIFA · Line marking and illustration · Dubins' curve · Football

1   Introduction

The central idea of most robotics applications is to eliminate the need for a human operator. The most obvious reason is; (1) to save labour and reduce cost where robots can conduct certain tasks far more rapidly and with greater precision than can human, (2) the human is bad for the production process such as in semiconductor and food handling, and pharmaceuticals where the use of robots reduce contamination, and (3) the product is bad for the human such as handling hazardous materials in chemical and radioactive plants where the use

© Springer International Publishing AG 2017
A.E. Hassanien et al. (eds.), *Proceedings of the International Conference on Advanced Intelligent Systems and Informatics 2016*, Advances in Intelligent Systems and Computing 533, DOI 10.1007/978-3-319-48308-5_86

of robots minimise risks to human operators. Another reason is that in aging societies that many countries are facing, the need arises to allocate work to smart machines, as in the case of industrial robots [1].

To date, robots have been very successful at manipulation in controlled environments such as in factories and in laboratories. Historical attempts of producing robotic arms went back to mid 50s when the first programmable robot arm for high speed handling was designed by George Devol in 1954 [2]. The use of robotics in laboratory, pharmaceutical industry and environmental monitoring went back to late 80s [3]. The use of robotic has been extended, since then, to include domains such as industrial maintenance and repair [4], outdoor robots for domains such as agriculture [5–7], animal-farming [8], mining [9], construction [10,11], power plants [12], oil and gas industry [13], space exploration [14], security and defense [15,16], etc.

Path-planning systems are of great significance when it comes to the performance and mission accomplishment of practically every type of outdoor robotics. Determination of a collision free path that a robot can follow between start and goal positions through obstacles cluttered in a workspace is central to the design of each autonomous robot to accomplish tasks without or with minimal human guidance [17]. Coverage Path Planning, on the other hand, is the task of determining a path that passes over all points of an area or volume of interest while avoiding obstacles. This task is integral to many robotic applications, such as vacuum cleaning robots, painter robots, autonomous underwater vehicles, demining robots, lawn mowers, automated harvesters, window cleaners, etc. [18,19]. An optimised path for a robot to navigate in an environment is the one which takes into account both the physical constraints of the vehicle, and the workspace constraints, such as obstacles or environmental forces [20,21].

Hameed et al. (2011) presented a promising application of field robotics for automating the periodical operations frequently carried out in football stadiums such as lawn cutting, lawn stripping and line marking in football stadiums [22]. The manual operation of these tasks is very expensive, very frequent, and requires very skilled and well trained personnel. In an attempt to reduce the cost, line marking or field panting is done less frequently through the use of long lasting paints. These types of paints contains high level of metal acetate salts which can last longer, however, it is more expensive and can kill the turf. A proposed solution is to use low cost paints which cannot harm the grass and painting more frequent. This solution can be feasible if this task can be performed autonomously using low weight vehicle able to work around the clock with high precision and with minimum human intervention. The line marking robot is a small three or four wheeled vehicle equipped with a line marking spray nozzle able to provide homogeneously painted lines and curves of constant width (i.e., 12 cm width). The proposed system is expected to reduce the amount of paint needed per each football field to an absolute minimum and hence reduce cost and possible environmental impact.

In this paper, a complete curvature and continuous path is constructed by connecting the field layout components such as straight lines, arcs and circles using Dubins' paths. The generated trajectory takes into account both the physical constraints of the vehicle such as the minimum turning radius of the vehicle and the workspace constraints such as obstacles. The remainder of the paper is organized as follows: standard FIFA football field layout and dimension are presented in Sect. 2. In Sect. 3, the algorithm for generating waypoints representing field layout is presented. Dubins' path and its use in connecting layout components to construct a continuous, smooth, and safe trajectory for robot navigation are presented in Sect. 4. In Sect. 5, the ability of the developed algorithm in generating a complete and smooth trajectory for a simulated field is presented. Finally, concluding remarks and future work are presented in Sect. 6.

2   Standard FIFA's Football Field Dimensions and Layout

The Fédération Internationale de Football Association (FIFA) (i.e., International Football Association in English) is an association governed by Swiss law founded in 1904 and based in Zurich. It has 209 member associations and its goal is the constant improvement of football game around the world. The rule of FIFA is to define and amend the Laws of the game on behalf of the global football community for the global game to grow and thrive. The official dimensions of football fields for international matches are specified in Law 1 of the official FIFA Laws of the Game, known as the field of play law. According to the FIFA rules for international matches, dimensions and parts of the field of play are defined as follows [23]:

2.1   Field Markings

The field of play must be rectangular and marked with lines. These lines belong to the areas of which they are boundaries. The two longer boundary lines are called touch lines. The two shorter lines are called goal lines. The field of play is divided into two halves by a halfway line, which joins the midpoints of the two touch lines. The centre mark is indicated at the midpoint of the halfway line. A circle with a radius of 9.15 m is marked around it, as it is shown in Fig. 1a.

2.2   Field Dimensions

The length of the touch line must be greater than the length of the goal line. The two touch lines must be of the same length, and be between 90 and 120 m in length (or between 100 and 110 m for international matches). The two goal lines must be of the same length, and be between 45 and 90 m (or between 64 and 75 m for international matches). All lines must be equally wide, not to exceed 12 cm, as it is shown in Fig. 1b.

2.3   Goal Area

Two lines are drawn at right angles to the goal line, 5.5 m from the inside of each goalpost. These lines extend into the field of play for a distance of 5.5 m and are joined by a line drawn parallel with the goal line. The area bounded by these lines and the goal line is the goal area.

2.4 Penalty Area

Two lines are drawn at right angles to the goal line, 16.5 m from the inside of each goalpost. These lines extend into the field of play for a distance of 16.5 m and are joined by a line drawn parallel with the goal line. The area bounded by these lines and the goal line is the penalty area. Within each penalty area, a penalty mark is made 11 m from the midpoint between the goalposts and equidistant to them. An arc of a circle with a radius of 9.15 m from the centre of each penalty mark is drawn outside the penalty area.

2.5 Corner Arcs

A flagpost is placed at each corner of the field. A quarter circle with a radius of 1 m (from each corner flagpost is drawn inside the field of play).

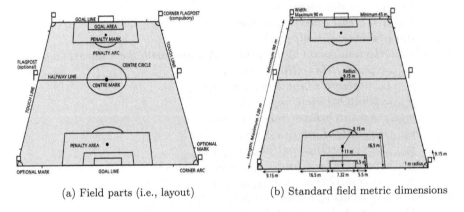

(a) Field parts (i.e., layout)   (b) Standard field metric dimensions

Fig. 1. Standard FIFA field of play dimensions and layout

3 Waypoint Generation of the Field Layout

In this section, waypoints representing the field of play parts are generated. The input to the algorithm are four corner points of the field of play and the outputs are the coordinates of all field parts.

3.1 Field of Play Coordinates

The algorithm receives four coordinates of the green rectangle (i.e., playable field), $P_i = (x_i, y_i)$ where $i = 1, 2, 3$, and 4. The two lower and upper goal lines are of length, w, and are represented by line segments $\overline{P_1 P_2}$ and $\overline{P_3 P_4}$ respectively. The two right and left touch lines are of length, l, and are represented by $\overline{P_2 P_3}$ and $\overline{P_4 P_1}$ respectively. To generate a rectangle field, width and length offset value is calculated over each goal and touch lines using the following two equations, respectively:

$$w_{ij} = |\sqrt{(x_i - x_j)^2 + (y_i - y_j)^2} - w|/2 \tag{1}$$

$$l_{ij} = |\sqrt{(x_i - x_j)^2 + (y_i - y_j)^2} - l|/2 \tag{2}$$

Two function, based on simple linear algebra, for line representation, $line_{ij} = createLine(p_i, p_j)$, for creating a line segment using its staring and ending points, p_i and p_j, and a function for finding a point p on a line segment at a distance d from its staring point, $p = pointOnLine(line, d)$, are used to find the four corner points of the field of play [24]. The corner points, C_i for $i = 1, 2, 3$, and 4, shown in Fig. 2a, can be generated in the following order:

1. *Represent the playable field (i.e., main rectangle):*
    $line_{12} = createLine(p_1, p_2)$,
    $line_{23} = createLine(p_2, p_3)$,
    $line_{34} = createLine(p_3, p_4)$, and
    $line_{41} = createLine(p_4, p_1)$.
2. *Find intersection points of the field of play lines with the main rectangle:*
    $p_{w_{12}} = pointOnLine(line_{12}, w_{12})$,
    $p_{w_{12}+w} = pointOnLine(line_{12}, w_{12} + w)$,
    $p_{l_{23}} = pointOnLine(line_{23}, l_{23})$,
    $p_{l_{23}+l} = pointOnLine(line_{23}, l_{23} + l)$,
    $p_{w_{34}} = pointOnLine(line_{34}, w_{34})$,
    $p_{w_{34}+w} = pointOnLine(line_{34}, w_{34} + w)$,
    $p_{l_{41}} = pointOnLine(line_{41}, l_{41})$, and
    $p_{l_{41}+l} = pointOnLine(line_{41}, l_{41} + l)$.
3. *Find corner points of the field of play:*
    $C_1 = pointOnLine(createLine(p_{l_{41}+l}, p_{l_{23}}), w_{12})$,
    $C_2 = pointOnLine(createLine(p_{l_{41}+l}, p_{l_{23}}), w_{12} + w)$,
    $C_3 = pointOnLine(createLine(p_{l_{23}+l}, p_{l_4}), w_{12})$, and
    $C_4 = pointOnLine(createLine(p_{l_{23}+l}, p_{l_4}), w_{12} + w)$.

3.2   Centre Line and Centre Mark Coordinates

The centre line divides the field into two halves by a halfway line, which joins the midpoints of the two touch lines. The two midpoints can be obtained as follows:

$MP_{23} = pointOnLine(createLine(C_2, C_3), l/2)$, and
$MP_{41} = pointOnLine(createLine(C_4, C_1), l/2)$.

The centre mark is indicated at the midpoint of the halfway line and is obtained as follows:

$CM = pointOnLine(createLine(MP_{23}, MP_{41}), w/2)$.

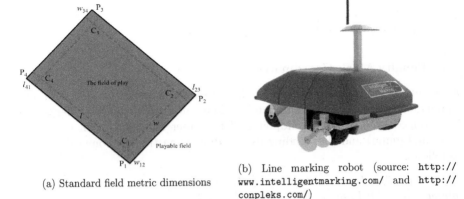

(a) Standard field metric dimensions

(b) Line marking robot (source: http://www.intelligentmarking.com/ and http://conpleks.com/)

Fig. 2. Coordinate generation of the field of play and line marking robot

3.3   Goal Area Coordinates

It is the area bounded by two lines drawn at right angles to the goal line, 5.5 m from the inside of each goalpost (i.e., 9.16 m from the midpoint of the goal line) and extended into the field of play for a distance of 5.5 m and are joined by a line drawn parallel to the goal line. The coordinates of the lower and upper goal areas are calculated as follows:

$$g_{l1} = pointOnLine(createLine(C_1, C_2), w/2 + 9.16),$$
$$g_{l4} = pointOnLine(createLine(C_1, C_2), w/2 - 9.16),$$
$$g_{u1} = pointOnLine(createLine(C_2, C_3), w/2 + 9.16),$$
$$g_{u4} = pointOnLine(createLine(C_2, C_3), w/2 - 9.16),$$
$$g_{l2} = pointOnLine(createLine(g_{l1}, g_{u4}), 5.5),$$
$$g_{u3} = pointOnLine(createLine(g_{l1}, g_{u4}), l - 5.5),$$
$$g_{l3} = pointOnLine(createLine(g_{l4}, g_{u1}), 5.5), \text{ and}$$
$$g_{u2} = pointOnLine(createLine(g_{l4}, g_{u1}), l - 5.5).$$

3.4   Penalty Area Coordinates

It is the area bounded by two lines drawn at right angles to the goal line, 16.5 m from the inside of each goalpost (i.e., 20.16 m from the midpoint of the goal line) and extended into the field of play for a distance of 16.5 m and are joined by a line drawn parallel with the goal line. The coordinates of the lower and upper penalty areas are calculated as follows:

$$p_{l1} = pointOnLine(createLine(C_1, C_2), w/2 + 20.16),$$
$$p_{l4} = pointOnLine(createLine(C_1, C_2), w/2 - 20.16),$$
$$p_{u1} = pointOnLine(createLine(C_2, C_3), w/2 + 20.16),$$
$$p_{u4} = pointOnLine(createLine(C_2, C_3), w/2 - 20.16),$$
$$p_{l2} = pointOnLine(createLine(p_{l1}, p_{u4}), 16.5),$$

$$p_{u3} = pointOnLine(createLine(p_{l1}, p_{u4}), l - 16.5),$$
$$p_{l3} = pointOnLine(createLine(p_{l4}, p_{u1}), 16.5), \text{ and}$$
$$p_{u2} = pointOnLine(createLine(p_{l4}, p_{u1}), l - 16.5).$$

3.5   Penalty Mark Coordinates

A penalty mark is made 11 m from the midpoint between the goalposts and equidistant to them. The coordinates of lower and upper penalty marks can be obtained by finding the midpoints of the lower and upper goal lines (i.e., gmpl and gmpu) and then finding its coordinates on the line connecting them as follows:

$$gmpl = pointOnLine(createLine(C_1, C_2), w/2),$$
$$gmpu = pointOnLine(createLine(C_3, C_4), w/2),$$
$$lpm = pointOnLine(createLine(gmpl, gmpu), 11.0), \text{ and}$$
$$upm = pointOnLine(createLine(gmpl, gmpu), l - 11.0).$$

3.6   Centre Circle Coordinates

It is a circle with a radius of 9.15 m marked around the centre mark (CM). The coordinates of the centre circle can be obtained for a suitable precision using the parametric equation of a circle which is given by:

$$(x, y) = (x_0, y_0) + r(\cos(\theta), \sin(\theta)) \tag{3}$$

where x and y are the circle coordinates, x_0 and y_0 are the centre mark coordinates, the circle radius $r = 9.15$ m, and the circle angle θ in radians where $0 \leq \theta \leq 2\pi$. In this paper, an angle of a step size of 0.01π is used, bigger values will draw the circle faster but you might notice imperfections, and vice versa.

3.7   Corner and Goal Arcs Coordinates

An arc is a part of the circle with an angle in radians $\theta_{arc} \in [\theta_{start}, \theta_{end}]$ and therefore it can be generated for a suitable angle step size using Eq. 3. θ_{start} and θ_{end} can be easily obtained in terms of the starting and ending point of the arc using the equation:

$$\theta_{start} = \tan^{-1} \frac{y_1 - y_0}{x_1 - x_0} \tag{4}$$

$$\theta_{end} = \tan^{-1} \frac{y_2 - y_0}{x_2 - x_0} \tag{5}$$

where (x_0, y_0), (x_1, y_1), and (x_2, y_2) are the arc centre, arc staring, and arc ending points, respectively. Corner arcs, for example, has a radius of 1 m, the starting and ending points of the arc at C_1 can be found as follows:

$$ARC_{start}^{C_1} = pointOnLine(createLine(C_1, C_2), 1), \text{ and}$$
$$ARC_{end}^{C_1} = pointOnLine(createLine(C_4, C_1), l - 1).$$

4   Dubins' Path

Dubins' path in this paper are used to connect the components of the generated field layout such as line segments, arcs and circles in order to generate a continuous, smooth, and complete path which can be used by the robot to navigate throughout the line marking (or field painting) process. Given two points in the plane, the initial and final points, P_i and P_f, respectively. Each point is associated with its own orientation angle, α and β, respectively, which defines the prescribed direction of motion. The combination of (P_i, α) and (P_f, β), are known as the initial and final configurations. Given (P_i, α) and (P_f, β), the task is to find the shortest smooth path from P_i to P_f such that it starts and ends with the directions of motions α and β, respectively, and the path curvature is limited by $1/\rho$ where ρ is the minimum turning radius of the vehicle or the robot under consideration.

The first complete solution to this problem was first reported by Dubins in 1957 [25] where he showed that the shortest feasible path consists of exactly three path segments of a sequence CCC or CSC, where C for *circle* is an arc of radius ρ, and S for *straight* is a line segment which can then be decomposed into a set of six candidate paths. The optimum path was then found by explicitly computing all paths on the list and then comparing them which may become a problem in applications where computation time is critical, such as in real-time robot motion planning. Instead, Shkel and Lumelsky presented a logical classification scheme that allows one to extract the shortest path from the Dubins set directly, without explicitly calculating the candidate paths [26]. In this paper, a function for finding the optimal path connecting any initial and final configurations, $path_{P_i, P_f} = Dubins((P_i, \alpha), (P_f, \beta), \rho)$, is developed.

5   Simulation Results

Assume that a field has the coordinates, starting from lower left corner and in counterclockwise direction, $(-30.7, 45.9)$, $(54.1, 130.7)$, $(110.7, 74.1)$, and $(25.9, -10.7)$ and a robot with a minimum turning radius of $2\,\text{m}$ with some random heading angle is located at $(-20.4, 54.6)$. The algorithm started by generating the waypoints of the playable field layout and then used Dubins' path to connect the layout components. The generated trajectory for an accuracy of $0.01\,\text{m}$ is shown in Fig. 3 and a simulation can be seen on Youtube[1]. The complete trajectory can then be used to autonomously guide the line marking robot shown in Fig. 2b throughout the line marking process.

---

[1] https://www.youtube.com/watch?v=iKLNoAniZGw.

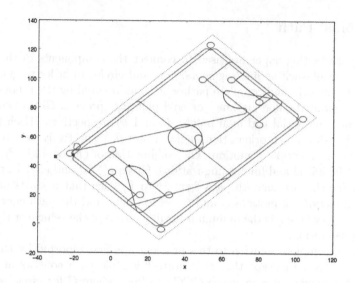

Fig. 3. Coordinate generation of the field of play.

6   Conclusions

In this paper, a path planning algorithm is used to generate the waypoints representing the layout of the field of play according to FIFA's standard dimensions. Dubins' path is used to connect the layout components in a way to generate a complete, continuous, short, and smooth trajectory which can be used safely to navigate the robot throughout the line marking process. Trajectory generation based on clothoid arcs is left for future work.

References

1. Kimitoshi, Y., Ryohei, U., Shunichi, N., Mitsuharu, K., Kei, O., Kiyoshi, M., Masaru, I., Isao, S., Masayuki, I.: Home-assistant robot for an aging society. Proc. IEEE **100**(8), 2429–2441 (2012)
2. Michael, E.M.: Evolution of robotic arms. J. Robot. Surg. **1**(2), 103–111 (2007)
3. Robert, B.: Robots in the laboratory: a review of applications. Ind. Robot Int. J. **39**(2), 113–119 (2012)
4. Parker, L., Draper, J.: Robotics applications in maintenance and repair. In: Handbook of Industrial Robotics, 2nd edn. Wiley (1999)
5. van Henten, E.J., Hemming, J., Tuijl, B.V., Kornet, J., Meuleman, J., Bontseman, J., Os, E.V.: An autonomous robot for harvesting cucumbers in greenhouses. Auton. Robots **13**(3), 241–258 (2002)
6. van Henten, E.J., Bac, C.W., Hemming, J., Edan, Y.: Robotics in protected cultivation. In: Proceedings of the 4th IFAC Conference on Modelling and Control in Agriculture, Horticulture and Post Harvest Industry, IFAC Proceedings Volumes, vol. 46, no. 18, pp. 170–177 (2013)

7. Astrand, B., Baerveldt, A.: An agricultural mobile robot with vision-based perception for mechanical weed control. Auton. Robots **13**(1), 21–35 (2002)
8. Andersen, N., Braithwaite, I., Blanke, M., Sorensen, T.: Combining a novel computer vision sensor with a cleaning robot to achieve autonomous pig house cleaning. In: Proceedings of the 44th IEEE Conference on Decision and Control (CDC), Seville, Spain, pp. 8831–8836 (2005)
9. Gary, S., Stentz, A.: A robotic system for underground coal mining. In: Proceedings of the IEEE International Conference on Robotics and Automation (ICRA), pp. 633–638 (1992)
10. Werfel, J., Bar-Yam, Y., Rus, D., Nagpal, R.: Distributed construction by mobile robots with enhanced building blocks. In: Proceedings of the 2006 IEEE International Conference on Robotics and Automation (ICRA), Orlando, FL, pp. 2787–2794 (2006)
11. Zavadskas, E.K.: Automation and robotics in construction: international research and achievements. Autom. Constr. **19**(3), 286–290 (2010)
12. Abidi, M., Eason, R., Gonzalez, R.: Autonomous robotic inspection and manipulation using multisensor feedback. Comput. J. **24**(4), 17–31 (1991)
13. Anisi, D.A., Skourup, C.: A step-wise approach to oil and gas robotics. In: Proceedings of the 1st IFAC Workshop on Automatic Control in Offshore Oil and Gas Production, IFAC Proceedings Volumes, vol. 45, no. 8, pp. 47–52 (2012)
14. Yim, M., Roufas, K., Duff, D., Zhang, Y., Eldershaw, C., Homans, S.: Modular reconfigurable robots in space applications. Auton. Robots **14**(2), 225–237 (2003)
15. Masłowski, A.: Intelligent mobile robots supporting security and defense. In: Proceedings of the International Conference IMEKO TC-17 Measurement and Control in Robotics, NASA Space Center, Huston, Texas, USA (2004)
16. Carroll, D.M., Nguyen, C., Everett, H.R., Frederick, B.: Development and testing for physical security robots. In: Proceedings of SPIE 5804, Unmanned Ground Vehicle Technology VII, 550 (2005). doi:10.1117/12.606235
17. Parker, L.E.: Path planning and motion coordination in multiple mobile robot teams. In: Encyclopedia of Complexity and System Science, pp. 5783–5800 (2009)
18. Galceran, E., Carreras, M.: A survey on coverage path planning for robotics. Rob. Auton. Syst. **61**(12), 1258–1276 (2013)
19. Hameed, I.A.: Intelligent coverage path planning for agricultural robots and autonomous machines on three-dimensional terrain. J. Intell. Rob. Syst. **74**(3–4), 965–983 (2014)
20. Hameed, I.A., Bochtis, D., Sørensen, C.A.: An optimized field coverage planning approach for navigation of agricultural robots in fields involving obstacle areas. Int. J. Adv. Rob. Syst. **10**, 231–2013 (2013). InTech
21. Lekkas, A.M., Dahl, A.R., Breivik, M., Fossen, T.I.: Continuous-curvature path generation using fermat's spiral. Model. Ident. Control **34**(4), 183–198 (2013). ISSN 1890–1328
22. Hameed, I.A., Sorrenson, C.G., Bochtis, D., Green, O.: Field robotics in sports: automatic generation of guidance lines for automatic grass cutting, striping and pitch marking of football playing fields. Int. J. Adv. Rob. Syst. **8**(1), 113–121 (2011). InTech, ISSN 1729–8806
23. FIFA: Laws of the game 2015/2016, pp. 1–140 (2016). http://www.fifa.com
24. Hameed, I.A., Bochtis, D.D., Sørensen, C.G., Nøremark, M.: Automated generation of guidance lines for operational field planning. Biosyst. Eng. **107**(4), 294–306 (2010)

25. Dubins, L.E.: On curves of minimal length with a constraint on average curvature and with prescribed initial and terminal positions and tangents. Am. J. Math. **79**, 497–516 (1957)
26. Shkel, A.M., Lumelsky, V.: Classification of the dubins set. Rob. Auton. Syst. **34**(4), 179–202 (2001)

Author Index

© Springer International Publishing AG 2017
A.E. Hassanien et al. (eds.), *Proceedings of the International Conference
on Advanced Intelligent Systems and Informatics 2016*, Advances in Intelligent
Systems and Computing 533, DOI 10.1007/978-3-319-48308-5

Printed in the United States
By Bookmasters